AF560738

ENCYCLOPAEDIA OF CHEMICAL THERMODYNAMICS

ENCYCLOPAEDIA OF CHEMICAL THERMODYNAMICS

Vol. II

By

K.K. Jain

DISCOVERY PUBLISHING HOUSE PVT. LTD.
NEW DELHI-110 002

Published by:
Namit Wasan
DISCOVERY PUBLISHING HOUSE PVT. LTD.
4383/4B, Ansari Road, Darya Ganj
New Delhi-110 002 (India)
Phone : +91-11-23279245; 23253475; 43596065
E-mail : discoverybooksindia@gmail.com
discoverypublishinghouse@gmail.com
namitwasan9@gmail.com
web : www.discoverypublishinggroup.com

Edition: **2020**

ISBN: 978-81-8356-351-2 (Set)

Encyclopaedia of Chemical Thermodynamics

Printed at:
Infinity Imaging Systems
Delhi

Preface

There are very few books available on chemistry which present the subject matter in a concise manner so that it is interesting and within grasp of B.Sc. (Hons.) and postgraduate students.

Keeping the above approach in mind this book has been planned. It is hoped that it would be useful to chemistry students at all levels. A logical and rigorous approach has been adopted but too much philosophical discussion has been avoided in order to make the material always within the grasp of students.

Each chapter of the book contains a discussion of the subject matter and important mathematical relations followed by suitable solved examples to illustrate the basic principles. At the end of each chapter a set of problems has been given to help the students become familiar with the concepts and principle presented in the text.

Although it has not been possible to give specific reference to contributions which inspired us in writing this book, we do knowledge the help of many who contributed directly or through their writings.

The comments, criticism and suggestions from the readers will be very much appreciated.

Author

CONTENTS

1

Kinetic Theory of Gases

INTRODUCTION

A simple gas theory that *a gas consists of a very large number of molecules in rapid motion in a vacant space* can be very readily employed to derive and justify these laws mathematically. This theory is termed as the *Kinetic Molecular Theory of Gases.*

THE PRINCIPLE OF CORRESPONDING STATES

In 1881, van der Waals demonstrated that if the pressure volume and temperature of a gas are expressed in terms of its critical pressure critical volume and critical temperature, then an important generalization called the principle of corresponding states would be obtained.

According to van der Waals equation, we have

$$\left(P + \frac{a}{V^2}\right)(V - b) = RT \qquad ...(1)$$

This equation has constants 'a' and 'b' which cannot be measured directly. Therefore, this equation is not applicable to all gases and in order to make it so, these constants have to be removed.

It has been found experimentally that near the critical point all gases give the same type of isotherms as carbon dioxide, but the actual position of the critical isotherms gets varied with the critical constant of the gas studied. The similarity is great enough to postulate that if for each different gas a suitable scale of pressure, volume and temperature is selected, it will be possible to superimpose all the critical isotherms on one another. If this were done of the critical isotherms it becomes probable that all other isotherms would coincide.

Under such circumstances, there should be one isothermal diagram and one fundamental equation for all gases. Such a scale will be one in which the pressure, volume, and temperature will be expressed as fractions of the critical pressure, volume and temperature.

In order to obtain such an equation, the values of P, V and T must be expressed as corresponding fractions of P_c, V_c and T_c, the critical pressure, volume and temperature,

$$\pi = P/P_c \quad \text{or} \quad P = \pi P_c$$

$$\phi = V/V_c \quad \text{or} \quad V = \phi V_c$$

$$\theta = T/T_c \quad \text{or} \quad T = \theta T_c$$

where π, ϕ and θ are referred to as *reduced pressure, reduced volume and reduced temperature respectively.*

On substituting the values of P, V and T in the van der Waals equation for one mole of an ideal gas, we get

$$\left(\pi P_c + \frac{a}{\phi^2 V_c^2}\right)(\phi V_c - b) = R_q T_c \qquad ...(2)$$

$$P_c = \frac{a}{27\,b^2}, \quad V_c = 3b$$

$$\text{and} \quad T_c = \frac{8a}{27\,Rb}$$

On substituting these values in equation (2), we get,

$$\left(\pi \frac{a}{27b^2} + \frac{a}{9\phi^2 b^2}\right)(3\phi\, b - b) = R\theta \frac{8a}{27Rb}$$

$$\text{or} \quad \frac{a}{27b^2}\left(\pi + \frac{3}{\phi^2}\right) \times b(3\phi - 1) = \frac{a}{27b} 8\theta$$

$$\text{or} \quad \frac{a}{27b}\left(\pi + \frac{3}{\phi^2}\right)(3\phi - 1) = \frac{a}{27b} 8\theta$$

$$\text{or} \quad \left(\pi + \frac{3}{\phi^2}\right)(3\phi - 1) = 8\theta$$

This equation involving neither of the characteristic constants a, b, T_c, P_c and V_c is known as van der Waals reduced equation of state. This equation is applicable to all substances (liquid or gaseous) irrespective of their nature, because it is not involving either of the characteristic constants.

It follows from the above equation that *fi two substances are at the same reduced temperature* (θ) *ad pressure (π) they must have the same reduced volume (φ).* This statement is called the law of corresponding states or principle of corresponding states.

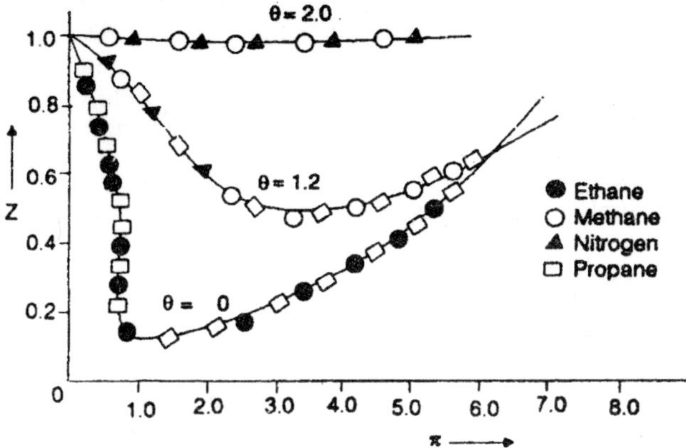

Fig. 1.1 : The compressibility factor as a function of reduced pressure and reduced temperature.

Significance of Law of Corresponding States

It has a great significance in the study of the relationship between physical properties ad chemical constitution of various liquids. As it is known that the pressure has practically no effect on liquids, so while studying the relation between physical properties and chemical constitution the determination of these properties should be performed at the same reduced temperature.

It has been found that the *boiling point of a liquid at atmospheric pressure is almost 2/3rd of its critical temperature* (T_c) *when expressed in absolute scale.* Consequently, boiling point is considered as the *corresponding temperature.* So, various physical properties of liquids such as molecular volumes of various liquids should be compared at their respective boiling points.

In actual practice one cannot use the above reduced equation of state directly. One makes use of the graphs between compressibility factor 'Z' and the reduced pressure at different reduced temperatures. The same graph is applicable to all gases. This can be seen from the following considerations, since $Z = \frac{PV}{RT}$.

Writing this in terms of reduced terms,

We have $$Z = \frac{(\pi P_c)(\phi V_c)}{R(\theta T_c)} = \left(\frac{P_c V_c}{RT_c}\right)\frac{\pi\phi}{\theta} = \frac{3}{8}\frac{\pi\phi}{\theta}$$

Thus, if two gases are having the same reduced temperature and reduced pressure, then their reduced volume will be the same, thus R.H.S. is independent of the as, thus all gases will have the same value of 'Z'. If isotherms are made to plot in terms of the reduced variables, the same curve should be obtained for all gases. The principle of corresponding states thereby enables the properties of real gases to be plotted o a single diagram (Fig. 1.1).

LIQUEFACTION OF GASES

For every known gas there is a temperature above which it is impossible to liquefy a gas, o matter how great the pressure may be. This temperature is known as the critical temperature. The minimum pressure necessary to liquefy a gas at its critical temperature is its critical pressure.

There are many gases such as ammonia, chlorine, sulphur dioxide, carbon dioxide, etc. which can be liquefied by the application of a suitable pressure alone because they are having high critical temperatures. On the other hand, there are gases such as hydrogen, oxygen, nitrogen and helium which cannot be liquefied by the simple application of a suitable pressure alone because they are having very low critical temperatures. These gases were regarded as *permanent* because the real significance of critical temperature was misunderstood for a longtime. But now it is known that these gases could be liquefied if they are first cooled below their respective critical temperatures and then made to compress.

The various methods used for the liquefaction of gases are as follows:

(1) *By the use of freezing mixture:* Faraday (1823) was able to liquefy a number of SO_2, Cl_2, CO_2, etc. by subjecting them to high pressures at low temperatures. He used a V-shaped tube, in one arm of which the gas was prepared and i the other end it liquefied under its own pressure and with the help of external cooling (Fig. 1.2). But Faraday was not able to liquefy the gases like hydrogen, oxygen and nitrogen, even when the pressure

was as high as 2790 atmospheres. These gases were, therefore, known as *permanent gases.*

In 1869, Andrews studied the effect of pressure on CO_2 at different temperatures and observed that for each gas there exists a temperature, called *critical temperature*, above which it is not possible to liquefy the gas by compression no matter how great may be the pressure. The ignorance of this fact was the main reason why the early attempts to liquefy the so-called 'permanent' gases got failed.

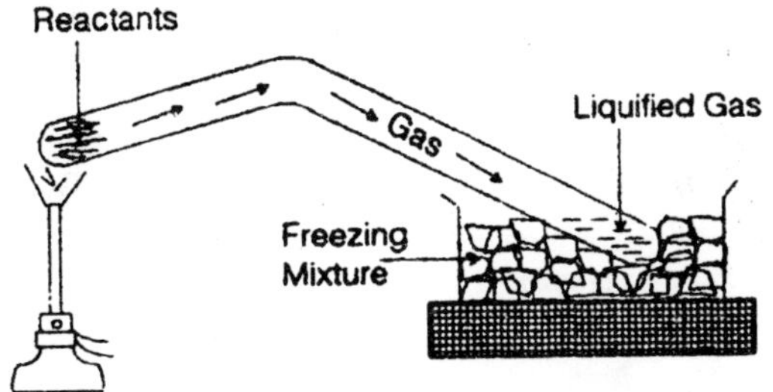

Fig. 1.2 : Faraday's method of liquefaction of gases.

(2) *By the use of volatile refrigerants:* This method is based on the fact that when a volatile compound evaporates rapidly, it absorbs its latest heat of vaporisation from the surroundings and consequently temperature of the surroundings becomes low.

Pictet was successful in liquefying CO_2 by the cooling caused by the evaporation of liquid SO_2 around it. Thilosier (1834) succeeded in solidifying liquid CO_2 by allowing it to evaporate freely under atmospheric pressure.

Thilosier, obtained temperature – 100 to – 110°C by allowing a mixture of a solid CO_2 (dry ice) and ether (called Thilosier's mixture) to evaporate under reduced pressure.

Faraday was able to liquefy most of the common gases by using Thilosier's mixture. He however failed to liquefy O_2, N_2, H_2, etc. and the called them permanent gases.

(3) *By the use of Joule-Thomson effect* (By expansion of cold compressed gases).

Joule-Thomson Effect

The English physicists, James Joule and William Thomson (later Lord Kelvin), observed that if a *gas under high pressure is allowed to expand into a region of low pressure it suffers a fall in temperature.* This phenomenon is termed as Joule Thomson effect.

Joule-Thomson effect *provides further support to the view that attractive forces do exist between gas molecules.* When the gas expands, the molecules fall apart from one another. Therefore, work has to be done so as to overcome the cohesive or attractive forces which tend to hold the molecules together. Hence, work is done by the system at the expense of the kinetic energy of the gaseous molecules. Consequently, the kinetic energy would decrease and since this is proportional to temperature, cooling results. It may be noted that in this case no external work is being done by the as in expansion.

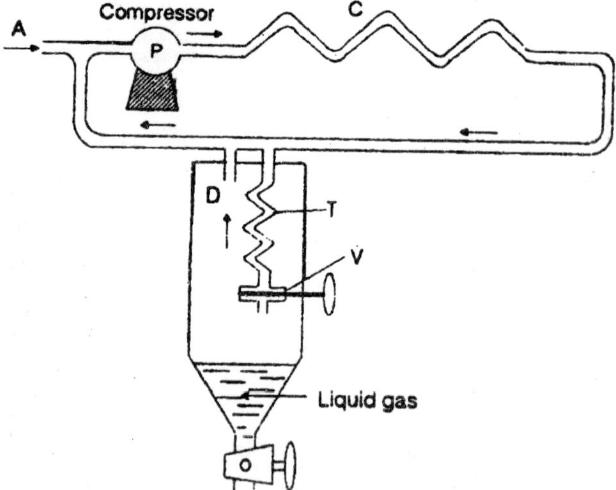

Fig. 1.3 : Linde's apparatus for liquefaction of gases.

Inversion Temperature

Experiment has revealed that gases become cooler during the Joule-Thomson's expansion *only if they are below a certain temperature called the* inversion temperature, T_i. The inversion temperature is characteristic of each gas. It is related to van der Waals constant a and b of the gas connected by the following relation:

$$T_i = \frac{2a}{Rb}$$

At the inversion temperature, there occurs no Joule-Thomson effect.

Thus, if a gas under pressure is allowed to pass through a porous plug and expands adiabatically into a region of very low pressure *at the inversion temperature,* there occurs neither fall no rise in temperature. If, however, the expansion occurs *above* the inversion temperature, there occurs a small *rise* of temperature and if it occurs *below* the inversion temperature there occurs a small *fall* of temperature.

In most gases, the inversion temperature lies within the range of ordinary temperature. Hence, they are cooled in Joule-Thomson's expansion. Hydrogen and helium, however, have *very low inversion temperatures.* Thus, at ordinary temperatures, these gases get *warmed* up instead of getting cooled in the Joule-Thomson's expansion. But if hydrogen is first cooled to – 80°C which is its inversion temperature and helium is fist cooled to – 240°C which is its inversion temperature, then these gases also get cooled on expansion in accordance to the Joule-Thomson effect. Linde process of the liquefaction of gases, like air, is based on Joule-Thomson effect.

The actual steps of the process will be clear from the Fig. 1.4. Purified dry gas is first of all compressed to a pressure of 200 atmospheres. The compressed gas is then allowed to pass through coils C, cooled by water or ammonia. The compressed and cooled gas is allowed to pass through a long spiral type of tube T. After travelling through it the gas gets diffused through a valve into a chamber (D) where it expands suddenly. The pressure inside the chamber is 50 atmospheres The control valve V is adjusted from outside. The as is cooled ad is made to pass back to the compression over the incoming gas. Thus the incoming gas gets further cooled before expansion takes place inside the chamber. On repeating the cycle several times, the temperature of the expanding gas would go sufficiently down ad a part of it gets condensed. The liquid collects at the bottom of the chamber from where it is drawn off. The whole arrangement is insulated by wool or left to avoid heating of the tubes.

By Adiabatic Expansion

If a gas is allowed to expand against an external pressure (say against a piston of an engine), it does some work. If a system is closed and insulated so that neither heat could enter the system nor go out of it, then the expansion would be almost adiabatic. This work done against the external pressure would be at the expanse of kinetic energy of the gas molecule. Consequently, a fall in temperature takes place. This principal combined with Joule-Thomson effect finds use in Claude's process for the liquefaction of air (Fig. 1.4).

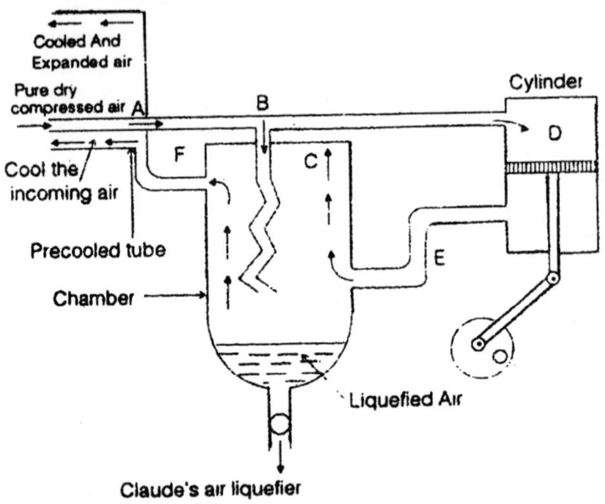

Fig. 1.4

Air free from CO_2 and moisture is made to compress to 40 - 50 atmospheres and then cooled by passing through a precooler. The gas then splits into two steams at B. A pat of the air enters the condenser C directly, where the other part enters the cylinder D where it undergoes expansion adiabatically, whereby the temperature of the gas falls. The cooled gas then enters the liquefying chamber via the side tube E.

As this cooled gas goes up in the chamber, it cools the air passing down the Jet J whereby further cooling takes place because of Joule-Thomson effect. The cooled gas then goes out through the precooler tube T, where it cools the incoming air. This cycle continues till sufficient low temperature gets reached and air would begin to liquefy. The liquefied air coolects at the bottom of the condenser.

The Claude's process has been observed to be much more economical than Linde's because a part of the work in compression can be recovered while in Linde's process work is carried out i free expansion and is lost. The work gained can be used to work the compressors.

Production of Low Temperatures by Adiabatic Deagnetization of a Paramagnetic Substance

It is possible to attain very low temperatures in the velocity of absolute zero by :

(i) Joule-Thomson expansion of hydrogen

(ii) Using liquid helium, and

(iii) By *making use of the magnetic properties of paramagnetic substances* such as the salts of the rare earths.

The para-magnetism of these substances is attributed to the presence of odd (unpaired) electrons.

The method of adiabatic demonetization of a paramagnetic substance was first of all suggested i 1926 by Debye and independently i 1927 by Giauque. This method is based on the fact that *the temperature of thermally isolated paramagnetic substance drop shwne jit gets demonetized.* This can be best understood by considering the following equation for a magnetic system:

$$dS = \frac{C_b dT}{T} + \varepsilon_0 V\left(\frac{\partial M}{\partial T}\right)_B dB \qquad ...(1)$$

where M refers to the magnetization in the presence of the external magnetic field B; V the volume, ε_0 the vacuum permeability and C_B the heat capacity of the system at constant magnetic field.

The magnetization of a paramagnetic substance at constant magnetic field B always decreases with increase in temperature T so that the partial derivative $(\partial M/\partial T)_B$ is always negative. Hence, when the substance is demagnetized reversibly ad adiabatically (so that dB is negative and dS = 0), we get

$$\frac{dT}{T} = -\frac{\varepsilon_0 V}{C_B}\left(\frac{\partial M}{\partial T}\right)_B dB \qquad ...(2)$$

or

$$\left(\frac{\partial T}{\partial B}\right)_S = \frac{\varepsilon_0 TV}{C_B}\left(\frac{\partial M}{\partial T}\right)_B \qquad ...(3)$$

Eq. (3) is known to form the basis of the principle of adiabatic demagnetization which gives the rate of change of temperature with magnetic field upon reversible adiabatic demagnetization. It can be seen from Eq. (2) that dT/T is negative so that dT is also negative. Hence a paramagnetic substance would cool upon adiabatic demagnetization.

The temperature-entropy (T-S) curve for a paramagnetic salt, such as gadolinium sulphate octahydrate, is depicted in Fig. 1.5.

From the curve it can be seen that in the unmagnetized state (when B = 0), the spins of the unpaired electrons of the paramagnetic material get oriented in a random manner in all directions. Thus, there occurs a disorder.

If, however, the substance is made to magnetize, the spins, and hence the magnetic moments associated with them, tend to align themselves with the magnetic field, thus reducing disorder. In other words, it means that the unmagnetized substance is having greater entropy than the magnetized substance at a given temperature. This can be seen from Fig. 1.5.

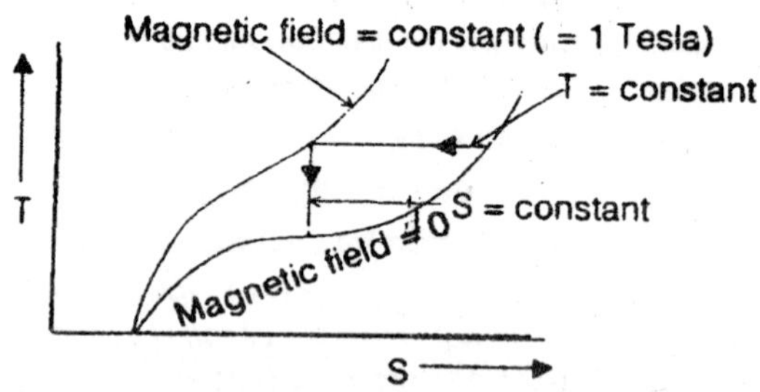

Fig. 1.5 : The T-S curve for a paramagnetic salt.

In the first stage of the process, the paramagnetic substance, in contact with a liquid helium bath at 1 K, would be *isothermally magnetized.* Under isothermal conditions (T = constant, dT = 0), Eq. (1) gets reduced to

$$TdS = \mu_0 TV(\partial M/\partial T)_2 dB \qquad ...(4)$$

As now dB is positive and $(\partial M/\partial T)_B$ is always negative, an increase in B causes an overflow of heat and hence a *decrease* in entropy. This is exhibited by the T = constant line on the T-S curve of Fig. 17.

In the second stage, the substance is removed and *adiabatically demagnetized.* Now it cools. This is depicted by the S = constant line on the T-S curve. The extent of cooling has been found to depend upon the strength of the applied magnetic field, the intensity of magnetization and the heat capacity of the paramagnetic salt.

The experimental set-up used for carrying out adiabatic demagnetization is shown in Fig. 1.6. The paramagnetic substance A is made to suspend inside a cryostat which gets cooled by liquid helium boiling under reduced pressure. During the first isothermal stage, the space D is having helium gas; this space will permit heat exchange between the substance and liquid helium. During the second stage, the space D gets evacuate to establish adiabatic conditions. The cycle of isothermal magetization, followed by adiabatic demagnetization, gets repeated to get millidegree temperatures.

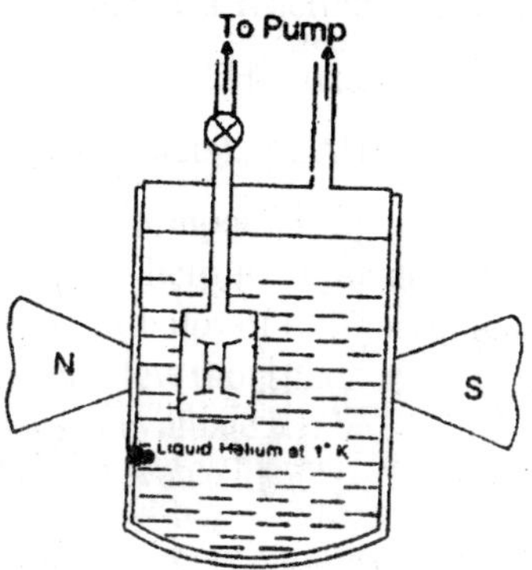

Fig. 1.6 : Experimental set-up for adiabatic demagnetization of a paramagnetic salt.

Even in the absence of external field B, which is usually of the order of 1-4 tesla, there exists a local magnetic field (or residual field) B_0 which is present in the paramagnetic substance which is arising mainly from magnetic interactions between the dipoles. B_0 is very small compared with B, its magnitude being of the order of 0.01 tesla. If T_1 refers to the initial temperature, then the final temperature T_2 would be given by the formula.

$$T_2/T_1 = B_0/B$$

If $$T_1 = 1\,K,$$

$$B_0 = 10^{-2} \text{ tesla, } B = 1 \text{ tesla, then } T^2 \approx 10^{-2}\,K$$

We can remark here that the process of adiabatic demagnetization is also applicable to the magnetic moments of *nuclei*. The application of adiabatic demagnetization to nuclear magnetic moments has allowed a temperature of 10^{-6} K to be reached. However, one should remember that *it would not be possible to attain absolute zero of temperature by using a succession of adiabatic cooling processes*. This can be understood from the T-S curve of Fig. 1.6.

In 1949 W.F. Giauque won nobel prize in Chemistry for his contributions to thermodynamics and the production of low temperatures by adiabatic demagnetisation of a paramagnetic salt.

INTERMOLECULAR FORCES OR VAN DER WAALS FORCES

It is now well established that forces of attraction exist between *neutral molecules* as well. These are termed as intermolecular forces or cohesive forces or van der Waals forces. These forces arise from three in types of interactions, viz.,

(1) *Dipole-Dipole Interaction*

(2) Dipole-Induced Dipole Interaction, and

(3) Induced Dipole-Induced Dipole Interaction.

Dipole—Dipole Interaction

The *polar molecules* which, though neutral, are having permanent dipoles. In such molecules the van der Waals forces are mainly attributed to electrical interaction between the dipoles known as *dipole-dipole interaction*. For instance, gases like ammonia, sulphur dioxide, hydrogen fluoride, hydrogen chloride, etc., are having permanent dipoles as a result of which appreciable dipole-dipole interaction exists between the molecules of these cases.

The magnitude of this type of interacting has been found to depend upon the dipole moment of the molecule concerned. The greater the dipole moment the greater would be the dipole-dipole interaction. Due to the attractive iteraction, these gases can be readily liquefied.

The average interaction energy of the two molecules with permanent dipole moments μ_1 and μ_2 is given by the expression

Interaction energy,

$$\phi(r) = -2\left(\frac{\mu_1\mu_2}{4\pi\varepsilon_0}\right)^2\left(\frac{1}{6r}\right)\left(\frac{1}{3kT}\right) \quad ...(1)$$

where r refers to their separation, k Boltzmann constant and $4\pi\varepsilon_0$ the permittivity factor of the vacuum.

Dipole—Induced Dipole Interaction

A polar molecule may some times be able to polarise a neutral molecule which lies in its vicinity and thus brings about dipolarity i that molecule just as a magnet is able to induce magnetic dipolarity i a neutral piece of iron lying close by. The induced dipole then undergoes interaction with the dipole moment of the first molecule and thereby the two molecules get attracted together. The magnitude of this interaction would, evidently, depend upon the magnitude of the dipole moment of the polar molecule and the polarizability of the neutral molecule. The average interaction energy for the two molecules is given by the following expression:

$$\text{Interaction energy, } \phi(r) = -\frac{\mu_1^2\alpha_2}{(4\pi\varepsilon_0)^2 r^6} \quad ...(2)$$

where μ_1 refers to the permanent dipole moment of molecule 1, α_2 refers the polarizability of molecule 2 and r, as usual, refers to the separation of the molecules. $4\pi\varepsilon_0$ as mentioned above, refers to the permittivity factor.

Induced Dipole—Induced Dipole Interaction. London Forces or Dispersive Forces

The van der Waals forces are also known to exist even in non-polar molecules like O_2 ad N_2 and also in non-polar monoatomic molecules like He, Ne, Ar, tc. This attraction has been evident from the condensation of these gases into liquids at sufficiently high pressures and low temperatures. This could not be understood for may years.

In 1930, F. London gave a satisfactory explanation for the existence of forces of attraction between *no-polar* molecules. This explanation is based on quantum mechanics. According to this view, electrons of a neutral molecule keep on oscillating with respect to the nuclei of the atoms.

Because of this, at a given instant, positive charge will be concentrated in one region and negative charge in another region of the same molecule. Hence, a non-polar molecule would become momentarily self polarised. This polarised molecule maybe able to induce a dipole moment in a neighbouring molecule with antiparallel orientation, as depicted in Fig. 1.7.

The electrostatic forces of attraction between induced dipoles and the original dipoles (due to electron oscillation) are termed as London forces. These forces are also termed as dispersive forces because the well known phenomenon of dispersion of light is also related to these dipoles.

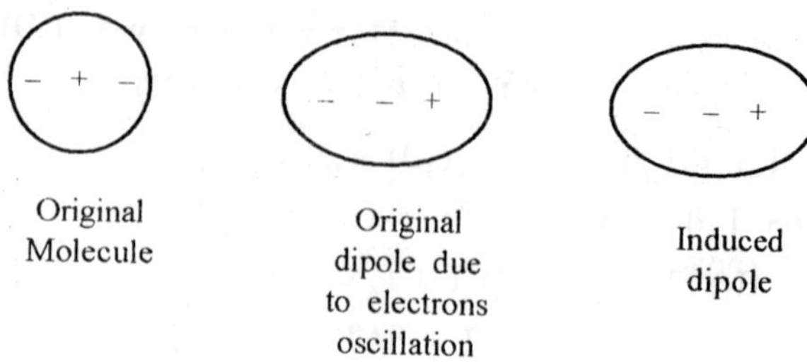

Fig. 1.7

For a pair of adjacent molecules, London forces have been found to vary inversely as the *seventh power* of the distance between them, *i.e.*, $F \propto 1/r^7$. The magnitude of these forces, evidently, would be very appreciably small.

The van der Waals attraction is *non-polar molecules is therefore exclusively ascribed* to London forces. In polar molecules having permanent dipoles, however, other electrostatic forces because of dipole-dipole interaction also contribute towards the van der Waals attraction.

This is named kinetic molecular theory because any model for explaining the behaviour of gases must involve both the molecular concept and motion of the particles. This theory was developed by Bernoulli, Clausius, Kronig, Maxwell, Kelvin, Boltzman and others. For the sake of generality, the theory assumes the properties of an ideal gas and is more or less based upon the following basic postulates:

POSTULATES OF THE KINETIC MOLECULAR THEORY OF GASES

Kinetic theory of gases is based upon the following assumptions:

(*i*) *All gases are made up of very large umber of minute particles called molecules.*

(*ii*) The molecules are small and separated from one another by large distances so that the actual volume occupied by the molecules is almost negligible as compared to the total volume of the gas.

(*iii*) The molecules are always moving in all directions in straight lines with very high velocities. They keep on colliding against each other (molecular collisions) and against the walls of the vessel at very small intervals of time.

(*iv*) The molecules are spherical and perfectly elastic, and there occurs o loss of kinetic energy due to mutual collisions or collisions with the walls of the container. The kinetic energy may be transferred from one molecule to another but it is not converted into any other form of energy such as heat.

(*v*) The pressure exerted by a gas on the walls of the containing vessel is due to the bombardment of the molecules of the gas on the walls of the containing vessel.

(*vi*) The average kinetic energy of the molecules of gas is directly proportional to the absolute temperature.

(*vii*) There are no attractive forces between molecules or between molecules and the walls of the vessel in which the gas is contained. The molecules move completely independent of one another.

(*viii*) The motion imparted to the molecules by gravity is negligible in comparison to the effect of the continued collisions between them.

(*ix*) The laws of classical mechanics (in particular the Newton's second law of option) are applicable to gaseous molecules.

Justification for the Postulates

The above postulates are sufficiently justified by the following experiences.

1. It is known that gases can be compressed by the application of pressure and can also expand indefinitely filling the whole vessel no matter how large it may be. This reveals that gases are made of very large umber of minute particles separated from each other by large distances. Further these particles occupy very small volumes as compared to the total volume of the gas. In a gas like oxygen, hydrogen or nitrogen, at N.T.P. the volume occupied by the molecules themselves is only 0.014 percent of the total volume. The remaining 99.986 per cent of the volume is merely an empty space.

   [Postulates (*i*) and (*ii*)]

2. Dust particles i air are found to be in constant irregular zig-zag motion termed *Brownian movement*. This is an evidence for the very fast straight line motion of the molecules.[Postulate (*iii*)]

3. Rapid motion of a large number of molecules must involve frequent collisions between then and these collisions must be elastic as otherwise there will occur constant loss of kinetic energy and the molecules must come to stand still which does not happen. [Postulates (*iii*) and (*iv*)]

4. If a gas is contained in a closed vessel, "it is expected that moving molecules will keep on hitting the walls of the vessel continuously thus exerting pressure." [Postulate (*v*)]

5. Molecules are very close in a compressed gas and intermolecular attraction should be high but this attraction must be negligible because even a highly compressed gas expands freely and fills the whole available space.

[Postulate (*vii*)]

6. Gases intermix freely with other gases in all proportions. Even a heavier gas rises up to intermix with a lighter gas. This indicates that the effect of gravity is negligible i comparison to the effect of the continued collisions between them. [Postulate (*viii*)]

7. On heating, the temperature of gas rises. Heat energy supplied increases the speed of the molecules and their average kinetic energy increases thereby causing the rise of temperature. [Postulate (*vi*)]

Derivation of the Kinetic Equation for Gases

BY taking the postulates of kinetic theory into consideration, it is possible by applying the laws of classical mechanics to derive a mathematical expression which can explain the various gas laws. This expression is known as the *Kinetic Gas Equation*.

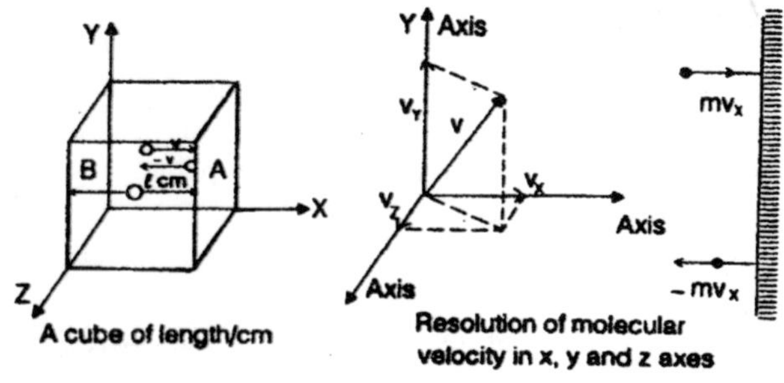

Fig. 1.8

Suppose a definite mass of a gas is contained in a cobical vessel (Fig. 1.8) and suppose that

the length of each side of the cube = l cm

the total number of molecules = N

the mass of one molecule = m

The selection of the cubical shape of the container is purely a matter of convenience and involves no loss of generality because the pressure of a gas is independent of the shape of the container.

It is regarded that the container are smooth and get oriented perpendicular to the x, y and z-axes of a Cartesian coordinate system. The motion of molecules in the container, at any instant, is totally random.

Although, the molecules are moving in every possible direction the average velocity v of a molecule at any moment can be resolved into three directions X, Y, Z at right angles to one another (Fig. 1.8).

Suppose a single molecule is moving with a velocity v.

Suppose v_x, v_y and v_z be the components of velocity v in the x, y ad z directions parallel to the three edges of the cube.

These three components get related to the resultant velocity v by the expression give below.

$$v = v_x^2 + v_y^2 + v_z^2 \quad ...(1)$$

Suppose the motion of this molecule occurs between the two opposite faces A and B parallel to the x-axis. It strikes the face A at regular intervals.

The velocity of the molecule before it strikes the face A is v_x. But the molecule is perfectly elastic. Therefore it will rebound with the same velocity with the sign changed, *i.e.*, $-v_x$.

$$= 2mv_x \text{ gm. cm/sec} \quad ...(2)$$

In order to strike the same face (A) against, the molecule will have to move to opposite face 'B' and come back. It means that the particle has to travel a distance 2l cm for each successive collision on the same face.

Thus, by traversing 2l cm., one impact is produced. By traversing 1 cm., 1/2l impacts get produced.

$\therefore$ By traversing v_x cm, $\frac{v_x}{2l}$ impacts get produced per second.

(v_x is velocity of the molecule that is the distance travelled i one second.)

Hence the number of impacts made by one molecule in one second

$$= \frac{v_x}{2l}$$

$\therefore$ Total change ion momentum per second (due to $\frac{v_x}{2l}$ impact) on face A

$$= 2mv_x \times \frac{v_x}{2l}$$

$$= \frac{mv_x^2}{l} \text{ gm cm/sec}^2$$

Total change in momentum per second because of the impacts of the molecule on two opposite faces A and B, along x-axis is therefore, double, *i.e.*, equal to

$$\frac{mv_x^2}{l} + \frac{mv_x^2}{l} = \frac{2mv_x^2}{l} \quad ...(3)$$

Also the total change of momentum per second due to impact of a single molecule on the two opposite faces along y-axis and two opposite faces along z-axis will be give as follows:

$$\frac{2mv_y^2}{l}, \frac{2mv_z^2}{l} \text{ respectively.}$$

Therefore, the total change in momentum per second o all the six faces of the cube, is equal to

$$= \frac{2mv_x^2}{l} + \frac{2mv_y^2}{l} + \frac{2mv_z^2}{l}$$

or $$= \frac{2m}{l}(v_x^2 + v_y^2 + v_z^2)$$

or $$= \frac{2mv^2}{l} \text{ gm cm/sec}^2 \quad ...(4)$$

$$(\because \; v_x^2 + v_y^2 + v_z^2 = v^2)$$

Thus total change of momentum, per second due to one molecule

$$= \frac{2mv^2}{l} \text{ gm cm/sec}^2$$

According to Newton's second law of motion, the rate of momentum is force *i.e.*,

$$\therefore \qquad \text{Force} = \frac{2mv^2}{l} \text{ dynes}$$

The pressure 'P' (force per unit area) is the total force divided by area of six faces of the cube ($6l^2$), *i.e.*,

$$P = \frac{\text{Force}}{6l^2} = \frac{2mv^2}{l} \times \frac{1}{6l^2} = \frac{1}{3}\frac{mv^2}{l^3}$$

or

$$P = \frac{1}{3}\frac{mv^2}{l^3} \qquad ...(5)$$

But l^3 is the volume of the cube 'V'

$$P = \frac{1}{3}\frac{mv^2}{V}$$

Hence Pressure exerted by one molecule is

$$P = \frac{1}{3V} mv^2 \qquad ...(6)$$

The total pressure exerted by N different molecules of same mass ad having different velocities $v_1, v_2, v_3 \ldots v_N$ will be as follows:

$$P = P_1 + P_2 + P_3 + \ldots + P_N$$

$$= \frac{1}{3V} mv_1^2 + \frac{1}{3V} mv_2^2 + \frac{1}{3V} mv_3^2 + \ldots + \frac{1}{3V} mv_N^2$$

[from Equation ... (6)]

$$P = \frac{1}{3V} m(v_1^2 + v_2^2 + v_3^2 + \ldots + v_N^2) \qquad ...(7)$$

If we multiply and divide the right had side of equation (7) by , we get

$$P = \frac{mN}{3V} \frac{(v_1^2 + v_2^2 + v_3^2 + \ldots v_N^2)}{N}$$

$$= \frac{mN}{3V} \times (\text{mean square velocity})$$

or $$P = \frac{mN}{3V} \times (\text{root mean square velocity})^2$$

$$= \frac{mNc^2}{3V}$$

where N refers to the total number of molecules and $\sqrt{\bar{c}^2}$, the root mean square velocity give by

$$\sqrt{\bar{c}^2} = \sqrt{\frac{v_1^2 + v_2^2 + v_3^2 + \ldots v_N^2}{N}}$$

$$PV = \frac{1}{3} mN\bar{c}^2 \qquad \ldots(8)$$

This is known as the Kinetic Gas Equation

It may be expressed as $PV = \frac{1}{3} M\bar{c}^2$ where M is the total mass of the gas.

or $$P = \frac{1}{3}\frac{m}{V}\bar{c}^2 = \frac{1}{3} D\bar{c}^2$$

where D is density of the gas.

Kinetic Energy and Temperature

Translational kinetic energy of a body of mass m with velocity $\sqrt{\bar{c}^2}$ is give by as follows:

$$K.E. = \frac{1}{2} m\bar{c}^2$$

As the velocity of a molecule increases with the rise in temperature, it becomes possible to define mean kinetic energy in terms of temperature of the gas.

Suppose one mole of an ideal gas is under consideration. Then the total number of molecules is N (Avogadro's number).

$\therefore$ Kinetic energy per mol of the gas for N particles is given by

$$K.E. = \frac{1}{2} mN\bar{c}^2 \qquad \ldots(9)$$

The kinetic gas equation (8) for N molecules of a gas is

$$PV = \frac{1}{2} mN\bar{c}^2$$

or $$PV = \frac{2}{3} \times \frac{1}{2} mN\bar{c}^2$$

or $$= \frac{2}{3} K.E. \qquad (\because \frac{1}{2} mNc^2 \text{ is the kinetic energy})$$

On comparing this equation with ideal gas equation PV = RT for 1 mole of the gas, we obtain

$$RT = \frac{2}{3} K.E.$$

or $$K.E. = \frac{3}{2} RT$$

or $$K.E. \propto T \qquad ...(10)$$

Hence the kinetic energy of one mole of any gas is directly proportional to its absolute temperature.

This is also the Maxwell's generalisation, which states that "*Kinetic energy of translation of an ideal gas is independent of the nature of the gas, its pressure and depends only upon the temperature.*"

But $$K.E. \propto \bar{c}^2$$

$$\therefore \quad \bar{c}^2 \propto T$$

or $$\bar{c}^2 \propto \sqrt{T} \qquad ...(11)$$

Hence molecular velocity of any gas is proportional to the square root of the absolute temperature. The molecular motion is, therefore, often known as *thermal motion of the molecules.*

At absolute zero, the thermal motion ceases completely and kinetic energy of the gas would become also zero.

Deviation of Gas Laws from the Kinetic Equation

(a) Boyle's Law: *This law states, "Temperature remaining constant, the volume of a give mass of a gas is inversely proportional to the pressure."* mathematically it can be expressed as

$$V \propto 1/P$$

$$PV = \text{constant}$$

This law can be derived from the kinetic equation

$$PV = \frac{1}{3} mN\bar{c}^2$$

This can be written as

$$PV = \frac{2}{3} \times \frac{1}{2} mN\bar{c}^2 \quad ...(i)$$

But $\frac{1}{2} mN\bar{c}^2$ = average kinetic energy of gas which is proportional to the absolute temperature.

i.e. $$\frac{1}{2} mN\bar{c}^2 \propto T$$

or $$\frac{1}{2} mN\bar{c}^2 = KT$$

Substituting in (i) above

$$PV = \frac{2}{3} . KT$$

In this equation $\frac{2}{3}$ KT would be a constant for a give mass of gas at a constant temperature.

$\therefore$ $$PV = \text{constant}$$

or $$V \propto \frac{1}{P}.$$ This is Boyle's law.

(b) Charle's Law: *According to this law, "The volume of a give mass of a gas is directly proportional to the absolute temperature at a constant pressure."*

We have already derived

$$PV = \frac{2}{3} kT$$

or $$V = \frac{2}{3} \times \frac{k}{P}. T$$

or $$\frac{V}{T} = \frac{2k}{3P}$$ Thus for a give mass of a gas

if pressure P, remains constant, the expression $\frac{2k}{3P}$ is constant.

$\therefore$ At constant pressure

$$\frac{V}{T} = K$$

or $$V = K \,.\, T.$$

Hence $V \propto T$. This is Charle's law.

(c) Avogadro's Hypothesis: *This law states, "Equal volumes of all gases under similar conditions of temperature ad pressure contain equal number of molecules."*

For any two gases, the kinetic equation can be put as follows:

(i) $$P_1V_1 = \frac{1}{3} m_1 N_1 \bar{c}_1^2$$

(ii) $$P_2V_2 = \frac{1}{3} m_1 N_1 \bar{c}_1^2$$

On rewriting:

(i) $$P_1V_1 = \frac{2}{3} \times \frac{1}{2} m_1 N_1 \bar{c}_1^2$$

(ii) $$P_2V_2 = \frac{2}{3} \times \frac{1}{2} m_1 N_1 \bar{c}_2^2$$

Since $P_1 = P_2$ and $V_1 = V_2$

$$\therefore \quad \frac{1}{2} m_1 N_1 \bar{c}_1^2 = \frac{1}{2} m_1 N_1 \bar{c}_2^2 \qquad ...(a)$$

Now if the temperature is also constant, the average kinetic energy per molecule must be the same.

or $$\frac{1}{2} m_1 \bar{c}_1^2 = \frac{1}{2} m_2 \bar{c}_2^2 \qquad ...(b)$$

On dividing (a) by (b), we have

$$N_1 = N_2$$

This is Avogadro's Law

(d) The Ideal Gas Equation: On Combining Boyle's law, Charles law ad Avogadro's law we find that the volume of a gas depends on the pressure, temperature and number of moles, as follows:

$V \propto 1/P$ (at constant T and n) (Boyle's law)

$V \propto T$(at constant P and n) (Charles' law)

$V \propto n$(at constant T and P) (Avogadro's law)

Thus, V should be proportional to the product of these three terms, *i.e.*,

$$V \propto nT/P$$

or $$V = R(nT/P)$$

or $$PV = RT \qquad ...(i)$$

where R, the proportionality constant, is called the gas constant. Eq. (i) is called the ideal gas equation.

The Gas Constant: As one mole of a ideal gas occupies a volume of 22.414 dm^3 at N.T.P. (0°C ad 1 atmp pressure), we find from Eq. (i) the gas constant

$$R = \frac{PV}{nT} = \frac{(1\ \text{atm})(22.414\ \text{dm}^3)}{(1\ \text{mol})(273.15\ \text{K})}$$

$$= 0.08206\ \text{dm}^3\ \text{atm}\ \text{K}^{-1}\ \text{mol}^{-1}$$

In SI units, 1 atom = $1.01325 \times 10^5\ \text{N m}^{-2}$, so that

$$R = \frac{PV}{nT}$$

$$= \frac{(1.03125 \times 10^5\ \text{N m}^{-2})(22.414 \times 10^{-3}\ \text{m}^3)}{(1\ \text{mol})(273.15\ \text{K})}$$

$$= 8.314\ \text{N m K}^{-1}\ \text{mol}^{-1}$$

$$= 8.314\ \text{J K}^{-1}\ \text{mol}^{-1} \qquad (\because\ J = Nm)$$

Note: 1 dm^3 = 1×10^{-3} m^3 = 1000 cm^3 = 1 litre.

(e) Dalton's Law of partial pressure: This law states, "*Total pressure P of a number of gases i a enclosed space is equal to the sum of the individual (partial) pressures of each gas, provided the gases do not react with one another.*" Thus

$$P = p_1 = p_2 + p_3 + p_4 + ... + p_N$$

Gas pressure is the force per unit area exerted by the molecules of the gas against the walls of the container.

Suppose N_1 molecules, each of mass m_1 of a gas A, are contained in a vessel of volume V.

The according to the kinetic theory equation the pressure is given by

$$p_a = \frac{m_1 N_1 \bar{c}_1^2}{3V} \qquad ...(i)$$

where $\bar{c}_1$ is the root mean square velocity of the molecules of the gas A.

Now, suppose N_2 molecules, each of mass m_2 of another gas B, are contained i the same vessel at the same temperature and there is no other gas present at that time. The pressure of the gas will b given by

$$p_b = \frac{m_2 N_2 \bar{c}_2^2}{3V} \quad \text{...(ii)}$$

where $\bar{c}_2$ is root mean square velocity of the molecules of the gas B.

If both the gases are present in the same vessel, the total pressure P will be give by

$$P = \frac{m_1 N_1 \bar{c}_1^2}{3V} + \frac{m_2 N_2 \bar{c}_2^2}{3V} = p_a + p_b$$

Similarly if three, four or ore gases were present the total pressure P will be given by

$$P = p_a + p_b + p_c + p_d \cdots$$

This is Dalton's law of partial pressures.

(f) Gaham's Law of Gaseous Diffusion: The law may be stated *"Under similar conditions of temperature and pressure, the rates of diffusion of gases are inversely proportional to the square roots of their densities."*

Mathematically,

$$\frac{r_1}{r_2} = \sqrt{\frac{D_2}{D_1}}$$

According to the kinetic equation

$$PV = \frac{1}{3} m N \bar{c}^2$$

or

$$\bar{c} = \sqrt{\frac{3PV}{mN}}$$

$$= \sqrt{\frac{3PV}{M}} \quad \text{[mN = molecular weight of the gas.]}$$

$$= \sqrt{\frac{3P}{D}} \quad \left[\because \; D = \frac{M}{V}\right]$$

$$\bar{c} = \sqrt{\frac{1}{D}} \text{ if P is kept constant.}$$

$\bar{c}$ is the mean velocity of the molecules and rate of diffusion of a gas will be dependent upon it at a particular temperature.

$$r \propto \bar{c}$$

$$r \propto \sqrt{\frac{1}{D}} \text{ if P is kept constant.}$$

$\bar{c}$ is the mean velocity of the molecules and rate of diffusion of a gas will be dependent upon it at a particular temperature.

$$r \propto \bar{c}$$

$$r \propto \sqrt{\frac{1}{D}}$$

This is Graham's Law.

Calculation of Molecular Velocity

In calculating various properties of the gases, one makes use of the following three velocities.

(i) Root mean Square velocity (RMSV): *It may be defined as the square root of the mean of the velocities of the molecules.* If v_1 v_2 ... etc. are the velocities possessed by different molecules, then

$$\sqrt{\bar{c}^2} = \sqrt{\frac{v_1^2 + v_2^2 + \ldots + v_N^2}{N}}$$

The expression to calculate this velocity can be obtained from the kinetic gas equation, according to which

$$PV = \frac{1}{3} mN\bar{c}^2$$

For 1 mole of the gas

$$PV = RT \text{ and } N = NA \text{ thus}$$

$$\sqrt{\bar{c}^2} = \sqrt{\frac{3PV}{mNa}} = \sqrt{\frac{3RT}{mNa}} = \sqrt{\frac{3RT}{M}} = \sqrt{\frac{3kT}{m}} = \sqrt{\frac{3P}{d}}$$

where K is Boltzmann constant k = R/Na

(2) Average Velocity (c): *It is defined as the average of the various velocities possessed by the molecules, i.e., as*

$$\bar{c} = \frac{v_1 + v_2 + ... + v_N}{N}$$

It can be calculated by using the expression

$$\bar{c} = \sqrt{\frac{8RT}{\pi M}}$$

$$\bar{c} = \sqrt{\frac{8RT}{\pi M}} = \sqrt{\frac{8kT}{\pi m}}$$

(3) Most Probably Velocity (MPV): It may be defined *as the velocity corresponding to the maximum fraction of the molecules.* It is given by the following expression

$$MPV = \sqrt{\frac{2RT}{M}} = \sqrt{\frac{2PV}{M}} = \sqrt{\frac{2P}{d}}$$

The relations between these velocities can be easily worked out

$$\frac{MPV}{\bar{c}^2} = \frac{\sqrt{2RT/M}}{\sqrt{8RT/\pi M}} = \sqrt{\frac{22}{28}} = \frac{1}{1.128}$$

$$\frac{MPV}{RMSV} = \frac{\sqrt{2RT/M}}{\sqrt{3RT/M}} = \sqrt{\frac{2}{3}} = \frac{1}{1.224}$$

Thus MPV : $\bar{c}$: RMSV : : 1 : 1 . 128 : 1.224.

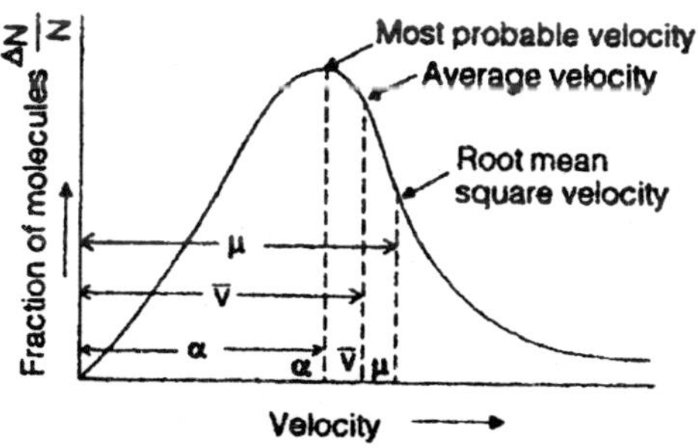

Fig. 1.9 : Relative values of most probable velocity, Average ($\bar{c}$) velocity $\sqrt{\bar{c}^2}$ and Root mean square velocity.

i.e. RMSV > c > MPV. This is also shown in Fig. (1.9). The relation between average and RMSV will be

$$\text{Average} = \left(\frac{1.128}{1.224}\right) MPV = 0.9213\ MPV.$$

COMPRESSIBILITY FACTOR

It is known that the equation of state, PV = nRT, derived from the postulates of the kinetic theory, *is valid for an ideal gas only.* Real gases tend to obey this equation only approximately ad that too under conditions of low pressure and high temperature.

The higher the pressure and the lower the temperature, the greater would be the deviations from the ideal behaviour. In general, the most easily liquefiable and highly soluble gases exhibit larger deviations. Hence gases like carbon dioxide, sulphur dioxide and ammonia exhibit much larger deviations than hydrogen, oxygen, nitrogen, etc.

It is best to represent the deviations from ideal behaviour i terms of the compressibility factor, Z, which is defined as

$$Z = \frac{PV}{(PV)_{ideal}} = \frac{PV}{nRT} \qquad ...(1)$$

For an ideal gas, Z = 1 under all conditions of temperature and pressure. The deviation of Z from unity is, therefore a measure of the imperfection of the gas under consideration.

The graphs plotted for the compressibility factors determined for a number of gases over a range of pressures at a constant temperature (0°C) have been depicted. At extremely low pressures, the value of Z for all the gases are close to unity which implies that the gases behave almost ideally. At very high pressures, all the gases have Z more than unity indicating that the gases are less compressible than an ideal gas. This is attributed to the fact that at high pressures. the molecular repulsive forces are dominant.

It can be seen from Fig. 1.10 that at moderately low pressures, carbon monoxide, methane and ammonia are more compressible than an ideal gas, *i.e.* $Z < 1$. This is attributed to the fact that at low pressures, the long range attractive forces are dominate and favour compression. The compressibility factor Z goes on decreasing with increase in pressure, passes through a minimum at a certain stage and then starts increasing with increase in pressure.

The gases would now becomes less compressible than ideal gas, *i.e.*, $Z > 1$. It can be seen from the figure that although carbon monoxide and methane exhibit marked devications from ideal behaviour only at high pressures, yet ammonia exhibits large deviation even at low pressure.

It can be seen that hydrogen and helium at 0°C are less compressible *at all pressures* than the ideal gas, *i.e.*, $Z > 1$. However, if the temperature is sufficiently low (*e.g.*, below – 165°C for hydrogen and below – 240°C for helium) these gases would also give the same type of Z – P plots as are shown by ammonia, carbon monoxide and methane at 0°C.

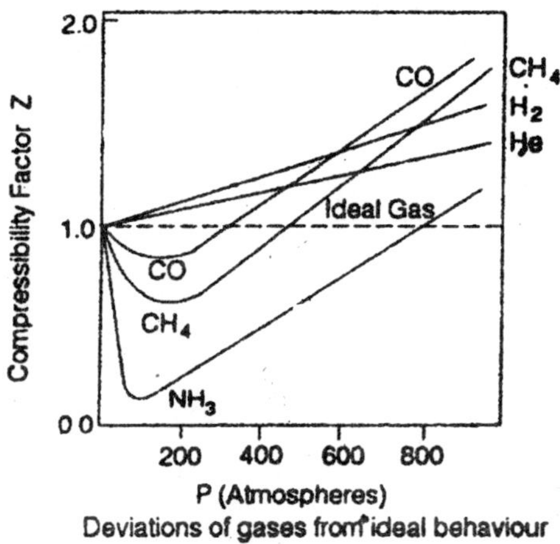

Fig. 1.10 : Deviations of gases from ideal behaviour.

On the other hand, if the temperature becomes sufficiently high, the Z-P plots of ammonia, carbon monoxide and methane will be almost similar to those of hydrogen and helium at 0°C *i.e.*, the value of Z will increase continuously with increasing pressure.

EXPANSIVITY AND COMPRESSIBILITY

When fluids (gases and liquids) are heated, they expand but the expansion of gases is much more than that of liquids. Similarly, they can be compressed. The variation of volume V with temperature T, keeping pressure P constant is termed as coefficient of thermal expansion or the coefficient of isobaric expansion or simple, expansivity, α of the fluid. Thus,

$$\alpha = \frac{1}{V}\left(\frac{\partial V}{\partial T}\right)_P \quad ...(1)$$

Similarly, the variation of V with P, keeping T constant, is termed *as the coefficient of isothermal compressibility* or, simply, compressibility, β of the fluid, thus,

$$\alpha = -\frac{1}{V}\left(\frac{\partial V}{\partial T}\right)_T \qquad ...(2)$$

Generally, as the volume of a gas gets decreased as pressure increases it means that the quantity $(\partial V/\partial P)_T$ is negative. Hence, a minus sign is included in the definition so as to make β positive. It is to be noted that α has the dimensions of T^{-1} and β those of P^{-1}.

Let us derive the value of α and β for an ideal gas.

For n moles of an ideal gas, we have

$$PV = nRT \qquad ...(i)$$

or

$$V = nRT/P \qquad ...(ii)$$

On differentiating Eq. (ii) with respect to T at constant P, we get,

$$\left(\frac{\partial V}{\partial T}\right)_P = \frac{nR}{P} = \frac{PV}{TP} = \frac{V}{T}$$

$$\therefore \qquad \alpha = \frac{1}{V}\left(\frac{\partial V}{\partial T}\right)_P = \left(\frac{1}{V}\right)\left(\frac{V}{T}\right) = \frac{1}{T}$$

Again, differentiating Eq. (ii) with respect to P at constant T. We get,

$$\left(\frac{\partial V}{\partial P}\right)_T = \frac{nRT}{P^2} = \left(\frac{nRT}{P}\right)\left(-\frac{1}{P}\right) = -\frac{V}{P}$$

$$\therefore \qquad \beta = -\frac{1}{V}\left(\frac{\partial V}{\partial P}\right)_T = \left(-\frac{1}{V}\right)\left(-\frac{V}{P}\right) = \frac{1}{P}.$$

DEVIATION FROM IDEAL GAS BEHAVIOUR

If we study the actual behaviour of the gases, it can be see that no gas *is ideal or perfect* in this sense although some gases tend to approach perfection as the temperature rises above their boiling points.

Since we know that no gas is ideal or perfect, the equation PV = nRT is obeyed as an approximation by real gas at low pressure while at high pressure and low temperature there occurs much deviation from this ideal behaviour.

Regnault and Amagat attempted to study the effect of change of P on PV of real gases like H_2, He, N_2O_2 and CO_2 under different conditions of temperatures and demonstrated that they did not obey

Boyle's law, more particularly at high pressure and low temperatures. Their results have been depicted in Fig. 1.11.

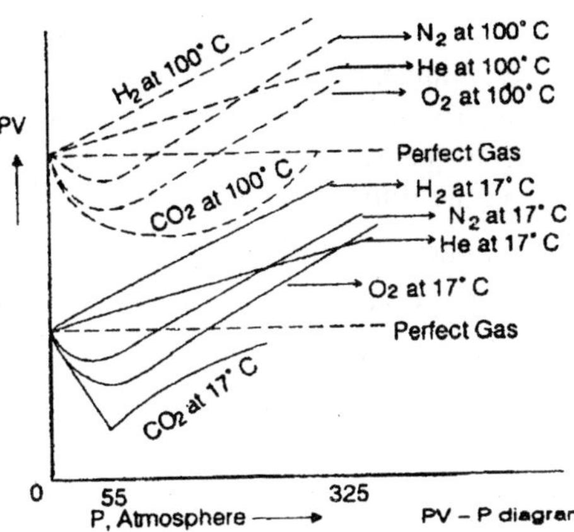

Fig. 1.11 : PV-P diagram.

From the diagram the following conclusions can be drawn:

(i) *At low Pressure:* Left hand portions of the curves show the devications at low pressures. With the rise of pressure, PV increases for hydrogen and helium whereas it decreases for 23 and CO2. Same is the case with other gases. *Thus at low pressure PV for all gases excepting H_2 and He, is having a lower value than expected of an ideal gas.*

(ii) *At high pressure:* Right-had portions of the curves show the deviations at high pressures. For H_2 and He, PV gets increased regularly with the rise of pressure but for N_2 and CO_2 it first decreases, passes through a minimum where it remains constant for a while and the rises. At very high pressure PV for all gases is having a higher value than that expected of an ideal gas.

Thus at high pressures PV for all gases including H_2 and He is having higher value than expected for an ideal gas.

The above results can be interpreted as follows:

(a) At moderate pressures, there occur a negative or under perfect deviation. Hydrogen and helium do not appear to show under perfect behaviour at ordinary temperature. However, at a much lower temperature hydrogen and helium also give a curve of the same type as the other gases.

(b) All the curves approach the ideal value as the pressure approaches zero, *i.e.*, at low pressure all gases behave as ideal gases.

(c) At very high pressures, there occurs a positive or over perfect devication.

(d) At very high pressures gases do not eve approximately obey gas laws.

(iii) *Effect of temperature:* In there are two sets of curves, one at 170186 and the other at 100°C. It is evident from these curves that the deviations are greater at 17°C while in case of CO_2 a sharp break occurs as the liquefaction point gets reached.

In general if the temperature gets changed, the shape of the curve would remain the same but the *deviations would become more pronounced as the temperature falls.*

The extent of under perfect behaviour is more or less dependent on how near the gas is to the temperature at which it will liquefy. This explains why CO_2 exhibits much greater deviation than N_2 while H_2 (which liquefies at a much lower temperature) exhibits over perfect behaviour. It means that deviations decrease with increase in temperature. It may be emphasized that Boyle's law is obeyed only at very low pressure and moderately high temperature.

For every gas there occurs a characteristic temperature when PV-P curve runs considerably parallel to the pressure axis from zero to moderate values of pressure. It implies that *at this temperature the value of PV remains constant for an appreciable range of pressure* signifying that at this temperature Boyle's law gets obeyed for a considerable range of pressure. This temperature is termed as the Boyle's temperature.

Explanation of Deviation

So as to explain deviations from ideal behaviour, it becomes essential to modify the kinetic theory of gases. The following two postulates of the kinetic theory do not hold good under all conditions. We will make an attempt to examine them more critically.

Postulate No. 1: The volume occupied by the gas molecules themselves is negligibly small as compared to the total volume occupied by the gas. One can justify this postulate only under ordinary conditions of temperature and pressure.

By calculations it can be see that in some of the common gases, the volume occupied by the molecules themselves, under ordinary conditions, is early 0.014 percent of the total volume of the gas. This is a negligible fraction. But, if the pressure becomes too high (say, 100 atmosphere or more), the total volume of the gas will decrease appreciably while the volume of the molecules will remain almost the same because the molecules are incompressible. *Therefore under conditions of high pressure, the volume occupied by the gas molecules cannot be neglected in comparison with the total volume of the gas.*

The same thing is expected to happen when the temperature is decreased to a large extent. The total volume of the gas gets decreased considerably, no doubt, but the volume occupied by the molecules themselves remains practically the same.

In this case, too, *the volume occupied by the molecules would be no longer be negligible. Thus the postulate No. 1. is not valid at high pressures ad low temperatures.*

Postulate No. 2 : The forces of attraction between gas molecules are negligible. This assumption is expected to be valid at low pressures or at high temperatures because under these conditions the molecules li far apart from once another.

But at high pressures or at low temperatures, the volume becomes small and molecules lie closer to one another. The intermolecular forces of attraction, therefore, become appreciable and cannot be ignored.

Hence, *the postulate No. 2, is not valid under conditions of high pressure and low temperature.*

It becomes necessary, therefore, to apply suitable corrections to the ideal gas equation so as *to make it applicable to real gases.*

EQUATIONS OF STATE FOR REAL GASES

A number of equations of state have been given to describe the P-V-T relationship in real gases. The earliest ad the best known equation is the one which is given by van der Waals.

The van der Waals Equation of State

In 1873, van der Waals gave his famous equation of state for a non-ideal (*i.e.*, imperfect gas). He modified the ideal gas equation by postulating that the gas molecules were not mass points but behaved like rigid spheres having a certain diameter and that there existed *intermolecular forces of attraction between them.* The two correction terms introduced by van der Waals have been described below.

1. *Correction due to volume of gas molecules:* It is known that the derivation of the ideal gas equation PV = nRT is based on the assumption that the gas molecules are mass points, *i.e.*, they do not have any finite volume. van der Waals discarded this assumption and suggested that a correction term *nb* should be subtracted from the total volume V so as to get the *ideal volume which is compressible.*

In order to understand the meaning of the correction term b, consider two gas molecules as unpentrable and incompressible spheres, each of which is having a diameter σ as shown in Fig. 1.12. It can be seen that the centres of the two spheres cannot approach each other more closely than the distance σ.

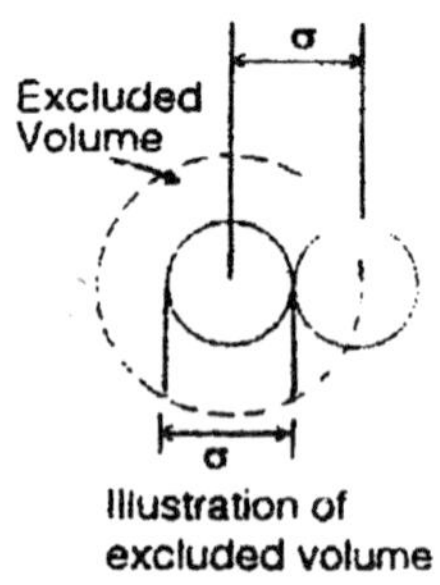

Fig. 1.12

For this pair of molecules, therefore, a sphere of radius σ and therefore of volume $\frac{4}{3}\pi\sigma^3$ constitutes what is termed as the excluded volume. The excluded volume per molecule would be therefore half the above, volume, viz., $\frac{2}{3}\pi\sigma^3$. The actual volume of one gas molecule of radius r is $\frac{4}{3}\pi\sigma^3 = \frac{4}{3}\pi(\sigma/2)^3 = \frac{1}{6}\pi\sigma^3$. ...(1)

∴ Excluded volume per molecule = $\frac{2}{3}\pi\sigma^3 = 4 \times \frac{1}{6}\pi\sigma^3$ = 4 times the actual volume oft he gas molecule.

The excluded volume would be *thus 4 times the actual volume of the molecule.* Evident, the excluded volume per mol of the gas would be $N_A \times 4 \times \frac{4}{3}\pi r^3 = b$ where N_A refers to the Avogadro's number. The compressible volume per mole of the gas would thus be V – b. If volume V of the gas is having n moles, the excluded volume would be nb. Thus, the ideal volume which is compressible is V-nb, the volume b per mole is also termed as co-volume.

2. *Correction due to intermolecular forces of attraction:* In the derivation of the ideal gas equation, an assumption has been made that there are no intermolecular forces of attraction, we will consider a molecule which is lying somewhere in the midst of the vessel, as shown at the point A (Fig. 1.13).

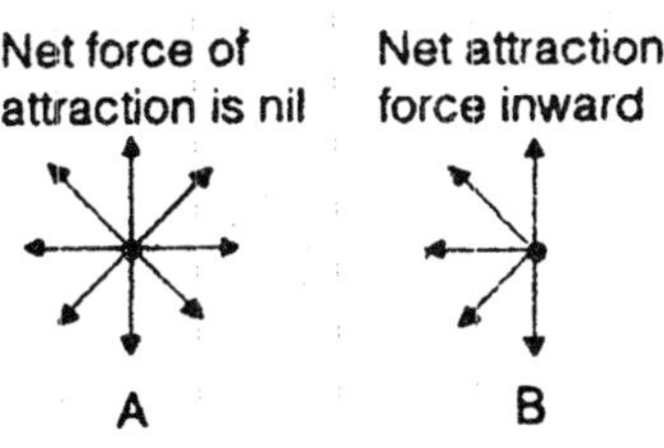

Fig. 1.13

Thus molecule is being attracted uniformly on all sides by the neighbouring molecules. These forces are able to neutralize one another ad there occurs no resultant attractive force on the molecule. However as the molecule approaches the wall of the vessel, as shown at B, it would experience attractive forces form the bulk of the molecules behind it.

Hence, it means that it will strike the wall with a lower velocity and will exert a lower pressure than it would have done if there was no force of attraction. It is, therefore, necessary to add a certain quantity to the pressure of the gas i order to obtain the *ideal pressure*. The correct pressure, therefore, world be P + p.

Calculation of Correction Factor p

The force of attraction exerted on a single molecule which is about to strike the wall evidently is dependent upon the *number of molecules per unit volume in the bulk of the gas, i.e.*, directly upon the *density* of the gas. Further, the number of molecules striking the wall at any given instant is also dependent directly upon the *density* of the as. Thus, the total inward attractive pull on the molecules which gives a measure of the correction factor, p, is proportional to the square of the density (ρ) of the gas, *i.e.*,

$$p \propto \rho^2 \qquad ...(2)$$

But density is inversely proportional to the volume and if V refers to the volume occupied by one mole of a as, the value of p for one mole of a gas will be inversely proportional to the square of volume.

Hence $p \propto 1/V^2$

or $p = a/V^2$ ...(3)

where a refers to a *constant* depending upon the nature of the gas.

The kinetic gas equation for *one mole* of a real gas, therefore, takes the following form

$$(P + a / V_m^2)(V_m - b) = RT \qquad ...(4)$$

where V_m refers to the *molar volume* of the gas.

This is termed as van der Waals equation. The constant a and b are termed as van der Waals constants. These are characteristic of each gas.

Eq. (4) has been found to be valid for one mole of a gas only. If there are moles of a gas occupying volume V, then, the excluded volume will be given by b and the compressible volume, therefore, will be V – nb. The pressure correction factor p for n moles, in the light of Eq. 2, will be proportional to $n^2\rho^2$, *i.e.*

$$p \propto n^2\rho^2 \text{ ç } n^2 \times 1/V^2$$

Thus, $p = an^2/V^2$ ...(5)

Thus, the van der Waals equation for n moles of a as may be put as follows:

$$\left(P + \frac{n^2 a}{V^2}\right)(V - nb) = mRT \qquad ...(6)$$

Eq. 6 has been found to be more accurate than the ideal as equation PV = nRT for expressing the PVT behaviour of real gases. Thus, if one mol of carbondioxide is taken at 47°C and compress it to different pressures, the volume, as observed by experiment, is found to be closer to that calculated from van der Waals equation that to that calculated from the ideal gas equation. The departure from ideal gas equation would become more and more wide as the pressure continues to rise (Table 1).

Table 1 : Volume of Carbon Dioxide at 47°C under Different Pressures.

Pressure (atm)	*Volume (dm^3) of the gas*		
	Observed experimentally	*Calculated from van der Waals equation*	*Calculated from ideal gas equation*
1	27.20	26.20	26.30
10	2.52	2.53	2.63
40	0.54	0.55	0.66
100	0.098	0.10	0.20

Units for van der Waals Constants

The unit for the van der Waals constants a and b would depend upon the units in which *P* and *V* are expressed. It follows from Eq. 5 that the constant a is expressed by pV^2/n^2, *i.e.* pressure × $(\text{volume})^2/\text{mol}^2$.

Table 2 : Van der Walls constants for some common gas.

Gas	*a (dm^6 atm. mol^{-2})*	*b (dm^3 mol^{-1})*
Ammonia	4.17	0.0371
Argon	1.35	0.0322
Carbon dioxide	3.59	0.0427
Carbon monoxide	1.49	0.0399
Chlorine	6.49	0.0562
Ethane	5.49	0.0638
Ethylene	4.47	0.0571
Helium	0.034	0.0237
Hydrogen	0.024	0.0266
Hydrogen chloride	3.67	0.0408
Hydrogen bromide	4.45	0.0433
Methane	2.25	0.0428
Neon	0.21	0.0171
Nitric oxide	1.34	0.0279
Nitrogen	1.39	0.0319
Oxygen	1.36	0.0318
Sulphur dioxide	6.71	0.0564
Water	5.46	0.0305

If pressure is expressed in atmospheres and volume in dm^3 the value of a will be in atm dm^6 mol^{-2}. As regards b, it is incompressible volume per mole, of a gas. Hence it must have the same units as volume per mole, e.g, dm^3 mol^{-1}.

The van der Waals constants for some gases have been included in Table 2.

The constant a may be regarded as *measure of the van der Waals forces of cohesion existing* between the molecules of a given gas. The greater the value of a, the greater is the strength of the van der Waals forces. The values of a for hydrogen and helium are very small, being 0.024 and 0.034 respectively, indicating that the cohesive forces (van der Waals forces) in these gases are very weak.

The values of a for ammonia, carbon dioxide, chlorine and sulphur dioxide are very high being 4.17, 3.59, 6.49 and 6.71, respectively. This reveals that van der Waals forces in these gases are very strong. The greater the value of a, the greater would be the ease with which a gas can be liquefied.

EXPLANATION OF BEHAVIOUR OF REAL GASES OF THE BASIS OF VAN DER WAALS EQUATION

$$\left(P+\frac{a}{V^2}\right)(V-b) = RT \qquad \text{...van der Waals equation of state}$$

or $$PV+\frac{a}{V}-Pb+\frac{ab}{V^2} = RT$$

On Neglecting $\frac{ab}{V^2}$ a product of two very small quantities, we obtain

$$PV - + \frac{a}{V} = RT = P'V'$$

where P′ and V′ refer to the ideal pressure and volume of the gas respectively.

(i) At low pressure: Small addition of pressure due to mutual attraction plays an important part while b is negligible as compared to large total volume V. The factor a/V, therefore overweights Pb *i.e.*,

$$\frac{a}{V} > Pb$$

$$\frac{a}{V} - Pb = Z, \text{ a positive quantity.}$$

From eq. (1), we have

$$PV + Z = RT = P'V'$$

$$PV = RT - Z = P'V' - Z$$

i.e., the observed PV is less than P'V'.

The observed product PV at low pressure would be less than P'V'. the product of pressure and volume if the gases were ideal.

At low pressures, therefore PV decreases with increase in pressure in case of N_2, O_2 and CO_2, etc.

(ii) At high pressure: If pressure is high, the volume is small and small volume occupied by molecules themselves cannot be neglected. Whereas in comparison with the high pressure, small addition of pressure due to mutual attraction is negligible. Hence Pb overweights a/V, *i.e.*,

$$Pb > \frac{a}{V}.$$

i.e., $\frac{a}{V} - Pb = -y$, a negative quantity

$$\therefore \qquad PV + (-y) = P'V'$$

$$PV - y = P'V'$$

$$PV = P'V' + y$$

i.e., the observed PV is greater than the value expected from an ideal gas

This explains the fact that PVl. after reaching a minimum, increases with further increase in pressure.

(iii) At ordinary (medium) pressure: It is seen that at ordinary temperature the factor $\frac{a}{V^2}$ becomes predominant at low pressures whereas the factor Pb predominates at high pressures. Therefore, these two factors will balance each other at some intermediate range of pressures in which the gas will show the ideal behaviour.

Hence at medium pressure, $Pb = \frac{a}{V}$

The equation (1) is reduced to

$$PV = RT = P'V'$$

and the gas behaves like an ideal gas.

(iv) At high temperature: At high temperatures, V is very large and hence both Pb and $\frac{a}{V}$ will be negligibly small, so that we have

$$PV = RT = P'V'$$

i.e., at high temperatures the behaviour of a real gas nearly agrees with the expected for ideal gas.

This explains why the deviations become less at high temperature.

(v) At low temperature: Both P and V are small, hence both pressure (attraction) and volume corrections are appreciable, with the result that the deviations are more pronounced.

(vi) Exceptional behaviour of hydrogen and helium: The molecules of these gases are of very small mass and consequently the force of attraction between the molecules is always negligible.

Hence neglecting the term $\frac{a}{V}$ equation (1) gets reduced to

$$PV - Pb = RT = P'V'$$

$$PV = P'V' + Pb$$

i.e., the product PV would become greater than P'V' event at low pressures.

EXPERIMENTAL DETERMINATION OF CRITICAL CONSTANTS

(A) Determination of Critical Temperature (T_c) and Critical Pressure (P_c)

Andrews determine critical temperature and critical pressure of a gas from its isotherms at different temperatures. This is a simple method which is based on the principle that *at critical temperature the surface of separation between liquid and vapour phases disappears.* This method is used when the substance is in liquid state at ordinary temperature.

The apparatus consists of a U-shaped hard glass tube having one limb smaller than the other. The tube is filled partially with mercury enclosing a small quantity of dry air in the longer limb. The upper portion of the longer tube is graduated. A small quantity of liquid and its vapour is kept over the mercury in the shorter limb and the mouth is sealed. The portion of to he shorter tube having liquid vapour, is then enclosed i a jacket, the temperature of which may be changed slowly and gradually when desired by circulating a suitable liquid from a thermostat.

The temperature of jacket is gradually increased. As the liquid is being heated, its surface tension is slowly decreasing and the boundary (or meniscus) would become progressively flatter and flatter. When the critical temperature is reached, the boundary or meniscus between the liquid and the vapour flickers and suddenly would disappear. The temperature at which this takes place is noted. The tube is then cooled a bit slowly and temperature is again noted when the gas becomes turbid for a moment and a small amount of liquid quickly settles at the bottom with the subsequent reappearance of the meniscus. The mean of the two temperature radians is regarded to be the critical temperature (T_c). At the critical temperature the reading of the corresponding pressure from the manometer is also noted. The mean of the two pressure reading corresponding to the two temperatures would five the critical pressure (P_c).

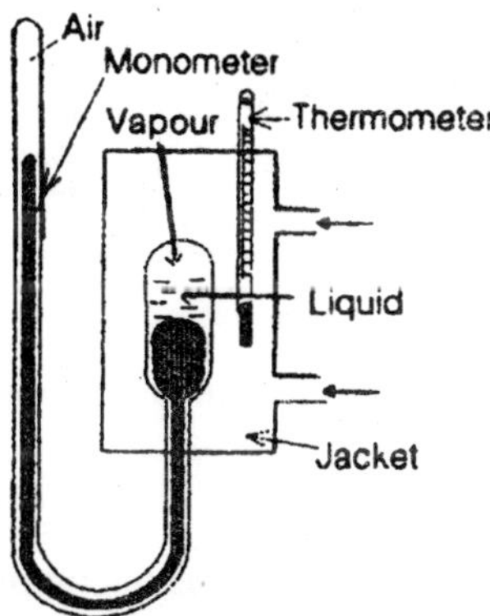

Fig. 1.14 : Cagniard de la Tour's apparatus.

B. Determination of Critical Volume (V_c)

It is not possible to determine the critical volume (V_c) of the gas experimentally because slight changes in pressure and temperature of the gas give rise to large changes in volume. So it becomes difficult to determine the V_c accurately by experiment.

However, the critical volume (V_c) can be determined indirectly by using the well known rule of rectilinear diameter due to Cailletet Mathias known as Cailletet and mathias rule which states that *the mean value of densities of any substance in the liquid state and its saturated vapour at the same temperature is a linear function of temperature i.e., the mean of the densities of liquid and its saturated vapour would be lying on a straight line called the rectilinear diameter.*

Thus if the density of a liquid is plotted against temperature, a curve of the type BC (Fig. 1.15) is obtained. If the density of the saturated vapour of the liquid is plotted against temperature, curve AC is obtained. The two curves meet at a point C which must correspond to the density at critical temperature. At point C, the liquid and its vapour become co-existent and identical and this point must give the critical density. In practice some difficulty arises in getting the upper portions of the two curves as the conditions existing there are uncertain. Hence, the upper portions of the curves would be almost approximate.

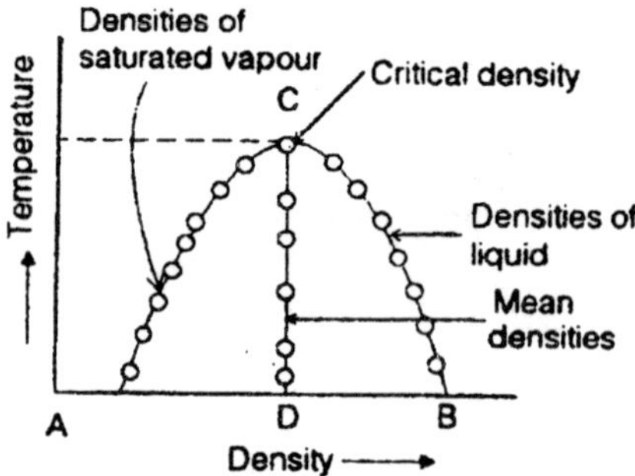

Fig. 1.15 : Determination of critical density (Dc).

Consequently the densities of the liquid and its saturated vapours would be determined at a various temperatures and plotted against T up to a few degrees below T_c (critical temperature).

The curve is then made to complete. The mean densities are then plotted against the various temperatures (T) when a rectilinear (straight) line DC is obtained. The rectilinear diameter is then made to extend to meet the curve at C. The point C, therefore, corresponds to T_c (critical temperature) and critical density (D_c) of the substance. The value V_c is then obtained as follows:

$$\text{Critical volume } (V_c) = \frac{\text{Mol. wt. of sustance}}{\text{Critical density}} = \frac{M}{D_c}$$

Critical constants of some common gases are give in table 3.

Table 3 : Critical constants of gases.

Gas	*Critical Temperature, °K*	*Critical Pressure, atm*	*Critical Volume, ml.*
Helium, He	5.2	2.26	60.68
Hydrogen, H_2	33.2	12.28	64.51
Nitrogen, N_2	126.0	33.5	90.03
Carbon monoxide, CO	134.4	34.6	90.03
Argon, Ar	150.7	48.2	78.43
Oxygen, O_2	154.3	49.7	74.42
Methane, CH_4	190.2	45.6	98.76
Carbon dioxide, CO_2	304.3	72.2	95.65
Nitrous oxide, N_2O	306.9	71.9	96.91
Hydrochloric acid, HCl	324.5	81.6	89.90
Hydrogen sulphide, H_2S	373.5	89.0	126.86
Ammonia, NH_3	405.5	111.5	72.03
Chlorine, Cl_2	417.1	76.1	123.91
Sulphur dioxide, SO_2	430.3	77.6	124.75

The P-V Isotherms of Carbon Dioxide—Critical Phenomenon

The curves representing the variation of volume and pressure at constant temperature are termed as Isotherms. (Greek : isos = 'equal'). In the case of an ideal gas the isotherms are almost rectangular hyperbolas (Fig. 41.16).

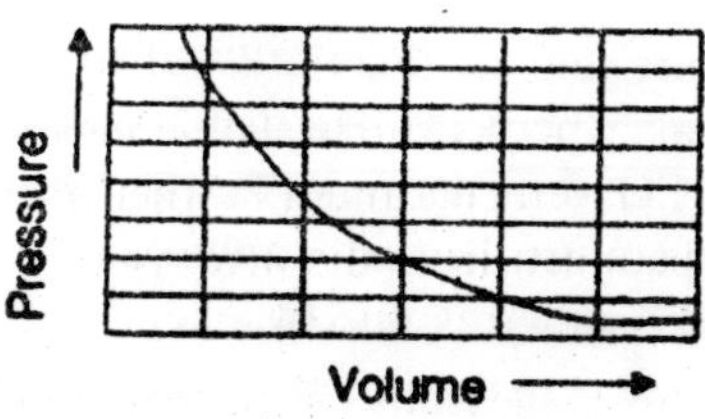

Fig. 1.16 : Rectangular hyperbola—isotherm in case of an ideal gas.

In 1869 T. Andrews attempted to study the effect of pressure on volume of CO_2 at different temperatures. In this experiment, some CO_2 gas was enclosed in a capillary tube and the volume occupied by the gas was measured at different pressures upto about 400 atmospheres.

His apparatus consisted of a sort of hydraulic press as shown in Fig. 1.17. Two similar capillaries were taken. One was filled partially with carbon dioxide and a pellet of mercury was introduced in the tube to act as a seal. In the other capillary, air was filled and sealed by mercury. These capillaries were then fixed into metal compression chambers by liquid tight joint. The chambers were interconnected ad carried screw plugers at the bottom to apply pressure. The upper portions of the tubes were surrounded by water bath to keep the temperature constant.

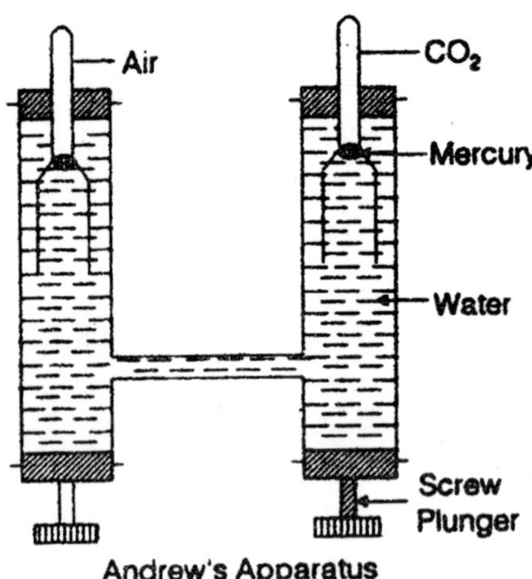

Fig. 1.17 : Andrew's apparatus.

If the volume readings were plotted against the pressure a set of curves of the type were obtained. Each curve would correspond to one particular temperature and is, therefore, called isotherm. The curves should be rectangular hyperbolas like that expected for an ideal gas. It is quite evident that CO_2 does not yield rectangular hyperbolas although with increasing temperature the curves approximate to one. We will now consider the various isotherms one by one.

(a) *At 13.1°C:* From the isotherm it can be seen that carbon dioxide is entirely gaseous at low pressures (*e.g.*, along AB) ad the volume decreases with increasing pressure in the normal fashion (*i.e.*, AB is roughly hyperboloid). The point A would correspond to gaseous CO_2 occupying a certain volume under a certain pressure. At low pressures, if the pressure gets increased, the volume would decrease as shown by the curve AB.

At B liquefaction starts; vapour gets converted into liquid of very much greater density and the volume gets decreased suddenly as the volume occupied by a liquid would be much less than that of gas so that BC would be almost parallel to the volume axis-indicating the sudden decrease in volume occurring at the same pressure (49.8 atmospheres) as more and more of the gas changes into liquid.

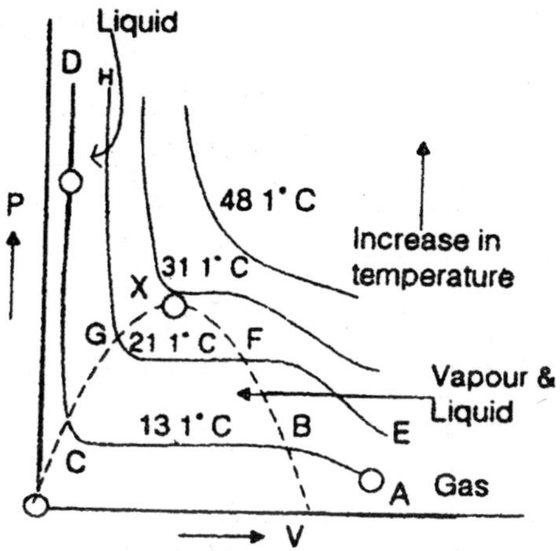

Fig. 1.18 : Isotherms of CO_2 near the critical temperature.

Liquefaction gets completed at C and CD reveals the small effect of pressure on a liquid. The steep line CD which is almost vertical represents the compressibility of a liquid and reveals that the liquids are much less compressible, *i.e.*, increase in pressure will produce a very small decrease in volume. In order to summaries, along curve AB there is only vapour, along CD only liquid and along BC liquid and vapour co-exist. As BC is parallel to the volume axis, there can be only one pressure at which liquid and vapour co-exist. This pressure is termed as the *vapour pressure* of the gas.

(b) The next isotherm EFGH at 21.1°C is also similar with the difference that the horizontal portion where liquefaction occurs is slightly short (*i.e.*, the range of volumes over which gas and liquid can co-exist is smaller) and the vapour pressure higher (*i.e.*, the pressure at which liquefaction commences is higher).

(c) As the temperature gets raised, this horizontal portion would become less and less till at 31.1°C it reduces almost to a point X. Above this temperature there occurs no trace of horizontal portion, *i.e.*, beyond 31.1°C there occurs no indication of liquefaction at all.

Anderws applied a pressure of 300 to 400 atmospheres but could to liquefy CO_2 if the temperature was beyond 31.1°C. Below 31.1°C a pressure of 75 atmospheres was quite sufficient to liquefy carbon dioxide.

Point X refers to the critical point for CO_2. The distinction between liquid and vapour state disappears at this point and CO_2 exists in a state called the critical state.

The isotherm passing through critical point is termed as the critical isotherm. From the experiment Andrews was able to conclude that if the temperature of CO_2 is above 31.1°C it cannot be liquefied no matter how high the pressure may be. This maximum temperature (31.1°C) at which liquefaction can be brought about is termed as critical temperature.

The pressure required to liquefy a gas at the critical temperature is termed as the critical pressure. If the temperature is below the critical temperature, the gas will liquefy at a somewhat lower pressure. All other gases have been found to behave similarly. The critical phenomenon observed b Andrews for CO_2 can be observed for any gas.

CRITICAL CONSTANTS

(i) ***Critical temperature of a gas (T_c)*** *may be defined as the minimum temperature which must be reached before it can be liquefied by the application of pressure.* A gas cannot be liquefied by pressure above its critical temperature, no matter how high the applied pressure may be. Each gas is having definite critical temperature, *e.g.*, critical temperature of CO_2 is 31.1MC.

(ii) ***Critical pressure of a gas (P_c)*** *may be defined as the minimum pressure necessary for the liquefaction of a gas at critical temperature.*

(iii) ***Critical volume of a gas (V_c)*** *may be defined as the volume occupied b 1 mole of the gas or liquid under critical pressure and under critical temperature.*

CONDITIONS FOR LIQUEFACTION OF GASES

From the Andrew's experiment very useful conditions for the liquefaction of a gas may be stated as :

(i) The temperature must be at or below the critical temperature.

(ii) A suitable pressure must be applied equal to the critical pressure if the temperature is at the critical temperature but becoming less if the temperature is reduced below this value.

van der Waals Equation as an Improvement Over Simple Laws

Van der Waals equation is an improvement over simple laws due to following reasons:

(i) *Van der Waals equation explains the devication from Boyle's law:* Regnault and Amagat studied the behaviour of cases and observed that they do not obey simple gas laws under different conditions of temperature and pressure, more particularly at low temperature and high pressure. They also plotted graphs between the product of pressure and volume ad pressure and found that PV product was not constant as desired by Bole's law. These deviations from ideal behaviour were explained by van der Waals equations.

(ii) *It also explains critical phenomenon.* This equation satisfactorily explains critical phenomenon. observed in the case of gases. The theoretical isotherm (*i.e.*, the curves between V and P at constant T) obtained from the values of V and the experimental values of 'a' and 'b' from van der Waals equation were found to be quite similar. The theoretical isotherms obtained from simple gas laws are quite different from the experimental isotherms.

PRINCIPLE OF CONTINUITY OF STATES

Although, the transition from liquid to vapour and *vice versa* generally takes place suddenly and the two states can be sharply distinguished from each other as can be see from AB and CD parts of 13.1°C isotherm for carbon dioxide, yet where occurs a possibility of passing from one state of aggregation to the other without involving sudden change.

Hence for further theoretical considerations it becomes essential to show that the liquid stated does not refer to a sharp and discontinuous transition from gaseous state but rather a continuation of the gaseous phase into the region of very strong intermolecular attractions where the volume becomes very small. The idea of a contiguity from liquid to gaseous state will become evident from the following consideration:

The dome-shaped curve depicted b the dashed line connects the end points of the horizontal portions of the isotherms. As a horizontal portion represents two phases in equilibrium, all points inside the dome correspond to equilibrium between liquid and vapour. For such points it always becomes possible to draw any sharp distinction between liquid and vapour, because there is a definite boundary surface between the two phases.

For points lying outside the dome only there is only one phase (liquid or vapour) ad here it is not possible to draw any sharp distinction between liquid and vapour. This fact is termed as the *principle of contiguity of states* and can be demonstrated with reference to Fig. 1.19. Suppose, that carbon dioxide gas under atmospheric pressure is made to heat a temperature above the critical temperature, as to 48°C and that is then subjected to a pressure greater than critical pressure, as, to 100 atmospheres, the system will change volume as shown by the isotherm 48°C.

When the temperature is lowered to 13.1°C while maintaining the pressure at 100 atmospheres, a point on the curve CD will be reached a curve which applies to liquid carbon dioxide. At no point in the process there is any abrupt change of state; the change from gas to liquid will be contiguous. Similarly, it is possible to start with liquid carbon dioxide under the conditions represented, say by the point C.

If we now raise the pressure o the liquid to 10 atmospheres and the temperature to 48°C the condition of the system will be represented by a point on that isotherm; and on reducing the pressure to that of the atmosphere ad then allowing the temperature to fall to 13.1°C, we shall be having carbon dioxide as a gas; and at no time during the whole process any discontinuity of properties will be observed, or an separation into liquid and vapour. The passage from the liquid to the gaseous state would be continuous.

An important conclusion derived from the principle of continuity of states is that one may think of a liquid near the critical point as just a highly compressed gas. If a equation of state can be found which accurately describes the gas close to its critical temperature and pressure that equation should also be able to describe the liquid.

VAN DER WAALS EQUATIONS AND CRITICAL STATE—RELATIONS BETWEEN VAN DER WAALS CONSTANTS AND CRITICAL CONSTANTS

A very good confirmation of the general character of van der Waal's equation may be provided b its application to critical phenomenon.

The van der Waals equation may be put as follows:

$$\left(P + \frac{a}{V^2}\right)(V - b) = RT \text{ (putting } = 1) \quad ...(1)$$

$$PV + \frac{a}{v} - Pb - \frac{ab}{V^2} = RT$$

On multiplying throughout by V^2, we obtain

$$PV^3 + aV - PbV^2 - ab = RTV^2$$

or $$PV^3 + aV - PbV^2 - ab - RTV^2 = 0$$

On arranging in descending powers of V, we get

$$PV^3 - PBV^2 - RTV^2 + aV - ab = 0$$

On dividing b P, we get

$$V^3 - bV^2 - \frac{RTV^2}{P} + \frac{aV}{P} - \frac{ab}{P} = 0$$

$$V^3 - V^2 - \left(b + \frac{RT}{P}\right) + \frac{aV}{P} - \frac{ab}{P} = 0 \quad ...(2)$$

This is a third degree equation in terms of V (cubic equation in V) and below a certain value of temperature T the equation can be having one of three real roots. With the rise of temperature these roots would approach one another and become identical ultimately.

This can be readily understood if a theoretical P – V graph is plotted from the van der Waals equation. The curves obtained are of the type ABEFGCD as depicted in Fig. 1.19. The three real unequal roots correspond to point B, F and C. The points B and C represent saturated vapour and liquid respectively, but F is having no physical significance.

As the temperature gets raised the S-shaped dotted portion of the curve gets reduced to a point. The values for the volume of liquid and the gas approach each other until at this volume of liquid and the gas approach each other until at this point, the three values of V become identical. This point would correspond to the critical temperature T_c and the volume V_c.

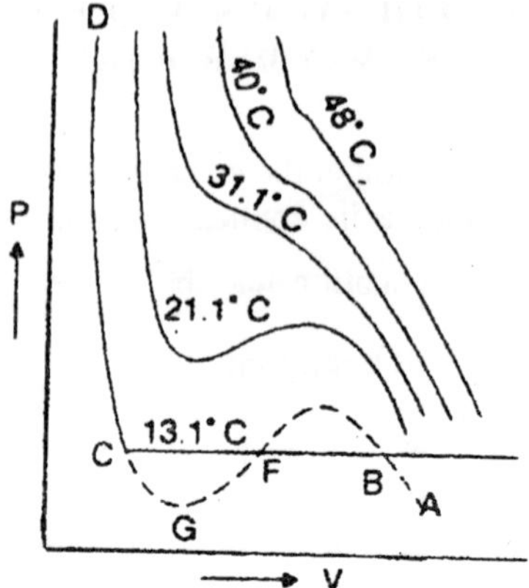

Fig. 1.19 : Isothermals of CO_2 from Van der Waals equation.

At the critical point, the three roots of the van der Waals equation would become identical and equal to V_c. But this refers to the value of the three roots of the van der Waals equation at the critical point.

$$(V - V_c)^3 = 0 \qquad ..(3)$$

or

$$V^3 - 3V_cV^2 + 3V_c^2V - V_c^3 = 0 \qquad ...(4)$$

On substituting these values in eq. (2), we get,

$$V^3 - V^2\left(b + \frac{RT_c}{P_c}\right) + \frac{aV}{P_c} - \frac{ab}{P_c} = 0 \qquad ...(5)$$

As both eqs. (4) and (5) are identical then the coefficients of the equal powers of V i the two equations may be equated:

$$3V_c = b + \frac{RT_c}{P_c} \qquad ...(6)$$

$$3V_c^2 = \frac{a}{P_c} \qquad ...(7)$$

$$V_c^3 = \frac{ab}{P_c} \qquad ...(8)$$

On dividing Eq. (8) by Eq. (7), we get,

$$\frac{V_c^3}{3V_c^2} = \frac{ab}{P_c} \times \frac{P_c}{a}$$

or

$$\frac{V_c}{3} = b$$

$$V_c = 3b \qquad ...(9)$$

Substituting the value of V_c from Eq. (9) into Eq. (7), we obtain

$$\frac{a}{P_c} = 3 \times (3b)^2$$

or
$$\frac{a}{P_c} = 27b^2$$

or
$$P_c = \frac{a}{27b^2} \qquad ...(10)$$

Substituting the values of V_c and P_c in equation (6), we get

$$3 \times 3b = b + \frac{RT_c}{a/27b^2}$$

$$9b = b + \frac{RT_c}{a/27b^2}$$

or
$$8b = RT_c \times \frac{27b^2}{a}$$

or
$$T_c = 8b \times \frac{a}{27b^2} \times \frac{1}{R}$$

$$T_c = \frac{8a}{27bR} \qquad ...(11)$$

Derivation of $P_cV_c = \frac{3}{8} RT_c$ from van der Waals Equation and calculation of van der Waals constants 'a' and 'b' in terms of T_c and P_c.

From van der Waals equation, we get

$$V^3 - V^2\left(b + \frac{RT_c}{P_c}\right) + V\frac{a}{P_c} - \frac{ab}{P_c} = 0 \qquad ...(12)$$

and
$$V^3 - 3V_cV^2 + 3V_c^2V - V_c^3 = 0 \qquad ...(13)$$

On equating the coefficients of (12), and (13), we get

$$3V_c = \frac{RT_c}{P_c} + b \qquad ...(14)$$

$$3V_c^2 = \frac{a}{P_c} \qquad ...(15)$$

$$V_c^3 = \frac{ab}{P_c} \qquad ...(16)$$

On dividing Eq. (16) by (15), we get

$$V_c = 3b \quad \text{or} \quad b = \frac{1}{3}V_c \qquad ...(17)$$

On substituting the value of b in (14), we obtain

$$3V_c = \frac{RT_c}{P_c} + \frac{V_c}{3}$$

or $$3V_c - \frac{V_c}{3} = \frac{RT_c}{P_c}$$

or $$\frac{8}{3}V_c = \frac{RT_c}{P_c}$$

or $$P_cV_c = \frac{3}{8}RT_c \qquad ...(18)$$

On rearranging Eq. (18), we get

$$V_c = \frac{3}{8}\frac{RT_c}{P_c}$$

On squaring, we obtain

$$V_c^2 = \frac{9}{64}\frac{R^2T_c^2}{P_c^2} \qquad ...(19)$$

On substituting the value of $V_c^2 = \frac{a}{3P_c}$ from Eq. (15) into Eq. (19), we get,

$$\frac{a}{3P_c} = \frac{9}{64}\frac{R^2T_c^2}{P_c^2}$$

$$a = \frac{27R^2T_c^2}{64P_c} \qquad ...(20)$$

On substituting the value of $V_c = 3b$ in Eq. (14), we obtain,

$$9b = \frac{RT_c}{P_c} + b$$

or $$8b = \frac{RT_c}{P_c}$$

or $$b = \frac{RT_c}{8P_c} \qquad ...(21)$$

SOLVED PROBLEMS

Example 1:

For a fluid show that (¶P/¶T)$_V$ = a/b where a is expansivity and b is compressibility of the gas. Hence, show that for an ideal gas.

$$(\partial P/\partial T)_V = R/V$$

Solution:

As a gas can be defined by any two of the three parameters P, V and T, we can write

$$V = f(T \, . \, P)$$

As V is a state function, hence its total differential can be written as follows:

$$dV = \left(\frac{\partial V}{\partial T}\right)_P dT + \left(\frac{\partial V}{\partial P}\right)_T dP \qquad ...(1)$$

By definition, $\alpha = \frac{1}{V}\left(\frac{\partial V}{\partial T}\right)_P$ so that $\left(\frac{\partial V}{\partial T}\right)_P = \alpha V$

and $\beta = -\frac{1}{V}\left(\frac{\partial V}{\partial P}\right)_T$ so that $\left(\frac{\partial V}{\partial P}\right) = -\beta V$

On substituting for α and β in Eq. (i) we get

$$dV - \alpha \, VdT - \beta \, VdP$$

When V is constant, dV = 0. Hence,

$$\alpha \, VdT - \beta \, VdP = 0$$

or $$V(\alpha \, dT - \beta \, dP) = 0$$

But $V \neq 0$, hence

$$\alpha \, dT - \beta \, dP = 0$$

or $$\alpha \, dT = \beta \, dP$$

or $$\left(\frac{\partial P}{\partial T}\right)_V = \frac{\alpha}{\beta}$$

For an ideal gas we know, $\alpha = 1/T$ and $\beta = 1/P$. Hence,

$$\left(\frac{\partial P}{\partial T}\right)_V = \frac{\alpha}{\beta} = \frac{1/T}{1/P} = \frac{P}{T} = \frac{R}{V}$$

This is to be proved.

Example 2:

Oxygen has a density of 1.429 gm. per litre at N.T.P. Calculate the R.M.S. and A.V. of oxygen molecules.

Solution:

Now $\qquad$ RMSV $= \sqrt{\dfrac{3P}{D}}$

In this case, $\qquad P = 1$ atmosphere

$= 76 \times 13.6 \times 981$ dynes/sq. cm.

$D = 1.422$ gm/litre

$= \dfrac{1.429}{1000}$ gm/c.c $= 0.001429$ gm/c.c.

Substituting the values in the above equation, we have

$$\text{RMSV} = \sqrt{\frac{3 \times 76 \times 13.6 \times 981}{0.001429}} = 46100 \text{ cm/sec}$$

This is the R.M.S. (root mean square mean velocity) of oxygen molecules.

Now A.V. (Average velocity)

$= 0.9213 \times$ R.M.S. velocity

$= 0.9213 \times 46{,}100 = 42458$ cm/sec.

Example 3(a):

Calculate the reduced pressure, volume and temperature of CH_4 gas when one mole of gas is contained in a 5 litre flask under 5 atmospheric pressure. What will be the temperature of the as when

Critical temperature $\quad T_c = 190.2°K$

Critical volume $\quad V_c = 0.0988$ *litre/mole and*

Critical pressure $\quad P_c = 45.6$ *atm.*

Solution:

We know,

Reduced pressure $\qquad \pi = P/Pc$

Here P = 5 atm $\qquad P_c = 45.6$ atm.

Substituting the values, $\pi = 5/45.6 = 0.1096$

Reduced volume $\phi = V/V_c$

Here V = 5 litres and $V_c = 0.0988$ litre/mole

substituting the values $\phi = \dfrac{5}{0.0988} = 50.607$

According to the law of corresponding states.

$$\left(\pi + \frac{3}{\phi^2}\right) \times (3\phi - 1) = 8\theta$$

where θ is the reduced temperature

i.e.
$$\theta = \frac{\left(\pi + \dfrac{3}{\phi^2}\right) \times (3\phi - 1)}{8}$$

Substituting the values of π and ϕ.

$$\theta = \frac{\left[0.1096 + \dfrac{3}{(50.607)^2}\right](3 \times 50.607 - 1)}{8}$$

$$= 2.0888$$

We know that reduced temperature

$$\theta = \frac{T}{T_c}$$

$$T = T_c \times \theta$$

Here $T_c = 190.2°K$

Substituting the values.

Temperature $= 190.2 \times 2.0888 = 397.3°K$.

Some Examples are given in Four Types

Type I: *Calculation of velocity when temperature alone is given.*

By kinetic gas equation:

$$PV = \frac{1}{3} mn\bar{c}^2$$

$$(R.M.S.V.) = \sqrt{\frac{3PV}{mn}} = \sqrt{\frac{3PV}{M}} = \sqrt{\frac{3RT}{M}}$$

M denotes the molecular weight of the gas and R = 8.3×10^7 ergs

$$(\text{RMSV}) = \sqrt{\frac{3 \times 8.3 \times 10^7 \times T}{M}}$$

$$= 1.58 \times \sqrt{\frac{T}{M}} \times 10^4 \text{ cm/sec.}$$

Example 3(b):

Calculate the root mean square and average velocity of carbon dioxide at 1000°C.

Solution:

$$(\text{RMSV}) = 1.58 \times \sqrt{\frac{T}{M}} \times 10^4 \text{ cm/sec.}$$

Now T = 273 + 1000 = 1273 K; M = 44

$$\therefore \quad (\text{RMSV}) = 1.58 \times \sqrt{\frac{1273}{44}} \times 10^4 \text{ cm/sec.}$$

$$= 84870 \text{ cm/sec.}$$

This gives the root mean square velocity.

Now average velocity = 0.9213 × R.M.S.V. of CO_2

Average velocity = 0.9213 × 84870 = 78190 cm/sec.

Type II : *Calculation of molecular velocity at N.T.P.*

From the kinetic gas equation we have derived that

$$\text{RMSV} = \sqrt{\frac{3PV}{M}}$$

Knowing the pressure and volume at N.T.P. we can calculate the value of RMSV as exemplified below.

Example 4:

Calculate the root mean square velocity of oxygen at N.T.P.

Solution:

For oxygen at N.T.P.

P = 76 cms = 76 × 13.6 × 981 dynes/sq-cm

V = 22.4 litres = 22400 cc.

Substituting the values in the equation:

$$(\text{RMSV}) = \sqrt{\frac{3PV}{M}}$$

we have,

$$\text{RMSV} = \sqrt{\frac{3 \times 76 \times 13.6 \times 981 \times 22400}{32}}$$

$$= 46140 \text{ cm/sec.}$$

Type III : *Calculation of molecular velocity when pressure and temperature. (i.e., not at N.T.P.) are given.*

In such calculations, the first step involves the conversion of gram molecular volume (22.4 litres at N.T.P.) to the volume under the different pressure and temperature. Having found the new volume, this value is substituted in the equation:

$$\text{RMSV} = \sqrt{\frac{3PV}{M}}$$

This calculation will become clear from the following example:

Example 5:

A vessel of 50 litre capacity contains 10 moles of steam under 5 atmospheres of pressure. Calculate the temperature of the steam using van der Waals equation if for water a = 5.46 atm (litre/mole)^2 and b = 0.031 litre/mole.

Solution:

We know $\left(P + \frac{n^2a}{V^2}\right)(V - nb) = nRT$

i.e.

$$T = \left(P + \frac{n^2a}{V^2}\right)\left(\frac{V - nb}{nR}\right)$$

Here $P = 5$ atm, $n = 10$ moles

$V = 10$ litres and $a = 5.46 \text{ atm} \left(\frac{\text{litre}}{\text{mole}}\right)^2$

and $b = 0.03$/lite/mole.

Substituting the values,

$$T = \left(5 + \frac{(10)^2 \times 5.46}{(50)^2}\right)\left(\frac{50 - 10 \times 0.031}{10 \times 0.08}\right)$$

$$= 311.9°K$$

Example 6(a):

The critical pressure and critical temperature of a gas obeying van der Waals equation have values 73 atms and 37°C. Calculate constants 'a' and 'b'. The value of the gas constant, R is 0.082 atm degree^{-1}.

Solution:

We have $$T_c = \frac{8a}{27Rb} \quad ...(i)$$

$$P_c = \frac{a}{27b^2} \quad ...(ii)$$

Dividing (i) by (ii) we have,

$$\frac{T_c}{P_c} = \frac{8a}{27Rb} \times \frac{27b^2}{a} = \frac{8b}{R}$$

Now $$T_c = 31° + 273 = 304°K$$

$$P_c = 0.082 \text{ atm degree}^{-1}$$

Hence $$\frac{304}{73} = \frac{8b}{0.082}$$

$\therefore$ $$b = 0.4269$$

Substituting the value of b in (i) *i.e.*,

$$T_c = \frac{8a}{27Rb}. \text{ we get}$$

$$304 = \frac{8a}{27 \times 0.082 \times 0.04269}$$

$$a = 3.591.$$

Example 6(b):

Calculate the root mean square velocity of nitrogen at 27°C and 700 mm pressure.

Solution:

(i) Conversion of G.M.V. of new volume

At N.T.P.	At New Conditions
$P_1 = 760$ mm	$P_2 = 700$ mm

$V_1 = 22400$ c.c. $V_2 = ?$

$T_1 = 273$ °K $T_2 = 273 + 27 = 300$ °K

By gas equation

$$\frac{P_1V_1}{T_1} = \frac{P_2V_2}{T_2}$$

$$V_2 = \frac{P_1V_1}{T_1} \times \frac{T_2}{P_2}$$

$$= \frac{22400 \times 760}{273} \times \frac{300}{700} = 26720 \text{ c.c.}$$

(ii) Calculating the value of RMSV.

Now $P = 700$ mm $= 70$ cm

$= 70 \times 13.6 \times 981$ dynes/cm

$V = 26720$ c.c.; $M = 28$ gm

Substituting the values in the equation:

$$\text{RMSV} = \sqrt{\frac{3PV}{M}}$$

$$\text{RMSV} = \sqrt{\frac{3 \times 70 \times 13.6 \times 981 \times 26720}{28}}$$

$= 51710$ c/sec.

Type IV : *Calculation of molecular velocity when pressure and density are given.*

$$PV = \frac{1}{3} mnc^2 = \frac{1}{3} Mc^2$$

$$\text{RMSV} = \sqrt{\frac{3PV}{M}} = \sqrt{\frac{3P}{D}}$$

where $\frac{V}{M}$ denotes density (D) of the gas in gm/c.c.

Example 7:

Calculate the pressure exerted by one mole of carbon dioxide gas in a 1.32 dm^3 vessel at 48°C using :

(a) the ideal gas equation and

(b) the van der Waals equation. The van der Waals constants are;

$$a = 3.59\ dm^6\ atm\ mol^{-2}$$

and $$b = 0.0427\ dm^3\ mol^{-1}.$$

Solution:

(a) For an ideal gas, PV = nRT, so that

$$P = \frac{nRT}{V}$$

$$= \frac{(1\ mol)\ (0.08206\ dm^3\ atm\ K^{-1}\ mol^{-1})(321\ K)}{1.32\ dm^3}$$

$$= 19.9\ atm.$$

(b) For a van der Waals as,

$$P = \frac{nRT}{V - nb} - \frac{n^2a}{V^2}$$

$$= \frac{(1\ mol)\ (0.08206\ dm^3\ atm\ K^{-1}\ mol^{-1})(321\ K)}{(1.32\ dm^3) - (1\ mol)(0.0427\ dm^3\ mol^{-1})}$$

$$= \frac{(1\ mol)^2 (3.59\ dm^6\ atm\ mol^{-2})}{(1.32\ dm^3)^2}$$

$$= 2062\ atm - 2.06\ atm = 18.56\ atm$$

The experimental value of the pressure is 16.40 atm. Thus, van der Waals equation gives a better agreement with experiment than the ideal gas equation.

Example 8:

Two moles of ammonia are enclosed in a five litre flask at 27°C. Calculate the pressure of ammonia using van der Waals equation. For ammonia, a = 4.17 atm. litre^{-2} mole^{-2} and b = 0.0371 litre mole^{-1}.

Solution:

V is the volume of n (= 2) moles of the gas = 5 litres.

a = 4.17 atms litre^{-2} mole^{-2} (correction due to inter-molecular attraction).

b = 0.0371 litre mole^{-1} (volume correction)

R = 0.082 litre atms. absolute degree^{-1} mole^{-1}.

$$T = 27 + 273 = 300MK$$

number of moles (n) = 2

According to van der Waals equation,

$$\left(P + \frac{a}{V^2}\right)(V - b) = \text{Rt, valid for one mole.}$$

$$\left(P + a\frac{n^2}{V^2}\right)(V - nb) = \text{nRT valid for n moles.}$$

i.e.

$$P = \frac{nRT}{V - nb} - \frac{an^2}{V^2}$$

Substituting the values of a, b, R, V and T in equation.

$$P = \frac{2 \times 0.082 \times 300}{5 - 2 \times 0.0371} - \frac{4.17 \times 2 \times 2}{5 \times 5}$$

$$= 9.98 - 0.67 = 9.31 \text{ atm.}$$

2

Applications of Electrochemistry

INTRODUCTION

These can be applied to all titrations in which there occurs a sharp change in the conductivity at the end point. *The determination of the end point of a titration by means of conductivity measurement is termed as conductometric titration.*

In a conductometric titration the titrant is added from the burette and the conductivity readings corresponding to various increments of titrant are plotted against the volume of titrant. Two curves will be obtained which will intersect each other at a point, called "*end point*" or "*equivalence point*".

When the reaction is not quantitative, there is some curvature in the curve near the end point which may be due to hydrolysis, dissociation of the product or appreciable solubility in case of precipitation reactions.

It should be borne in mind that the resistance, and thus the conductance of an electrolytic solution is profoundly affected by changes in temperature. The resistance decreases by about 1% to 2% for each degree increase in temperature and for this reason it is desirable to carry out a conductometric titration at an approximately constant temperature. If absolute measurements are to be made a constant temperature bath is, of course, necessary. The concentration of the titrant must be 10 times as the solution being titrated. This is done to keep the volume charge small. If it cannot be done, a correction to the reading must be applied, *i.e.*,

Actual conductivity = $\left(\frac{v + V}{V}\right)$ × observed conductivity where v is the volume of titrant or reagent added and V is the original volume.

Advantages of Conductometric Titrations

Among the many advantages that conductometric titrations possess over the ordinary volumetric methods, we may mention the following:

(i) They are particularly useful in the case of coloured or turbid liquids, where ordinary indicators cannot function.

(ii) They can be employed for the analysis of dilute solutions and also for very weak acids.

(iii) No special precautions are necessary as the titration approaches the equivalence point, since the latter is found graphically.

(iv) It is not necessary to measure the actual conductance, as we may use any quantity that is proportional to it, *e.g.*, the reading on a Wheatstone bridge or resistance box. The same can be directly plotted against the volume of the alkali solution used up.

Apparatus for Conductometric Titrations

(a) *Measuring Circuits:* In order to prevent concentration changes due to reaction at the electrodes, the resistance of solution is generally measured with alternating currents and frequencies from 60 to 10000 c/sec have been used for ordinary conductance measurements, although, work at high frequencies finds may applications. Most of the circuits used are of the Wheatstone bridge type. Stock has published an excellent critical review of the apparatus used for conductometric titrations, specially for microtitrations.

(b) *Commercial Apparatus :* Fairly simple commercially available equipment is accurate enough for most conductometric titrations. Its main advantages are that it is convenient and can be set up and used without any knowledge of electricity or electronics.

(c) *Electrodes and Cells :* A number of titration cells have been recommended of which those of Britton and Kano are the most practical. The simplest titration cell consists of a dipping electrode, a beaker of the appropriate size and a mechanical stirring device (Fig. 2.1). For precipitation titrations, the electrodes are mounted vertically to prevent excessive amounts of precipitate from collecting n the electrode surfaces. Because a large effective surface area minimises polarisation, most conductivity electrodes are plated with a layer of platinum black. The platinisation of electrodes also tends to decreases the cell capacitance and allows the establishment of a sharp balance point. On the other hand, platinised electrodes have a greater tendency to adsorb substances from solution. Difficulties are encountered with dilute solution or solution containing surface active agents; bright platinum of lightly platinised electrodes are used in such cases.

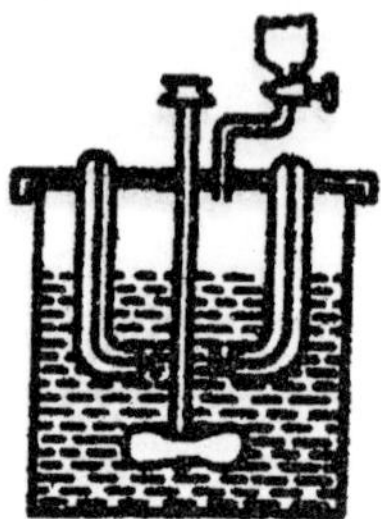

Fig. 2.1

(d) *Special Methods :* Although conductance is generally measured by means of a.c. apparatus, it has been shown by a number of workers that d.c. can be used provided that the cell has two unpolarisable electrodes, such as Ag/AgCl electrode. Eastmen has showed that direct current method is capable of yielding results that agree to 10% with those obtained with the usual a.c. method.

Differential Conductance Titration

Such titration have been investigated by Daval and Duval. They used two cells connected in two branches of a Wheatstone bridge, both cells contained identical sample solution and both were titrated with the same reagent, but one cell always had 0.05 to 0.1 ml more titranti in it than the other.

The titration curves for each cell were plotted on graph and their point of intersection gave one end point.

Types of Conductometric Titrations

Acid-base Titrations

(a) *Conductometric titrations of strong acids and strong bases: A* typical strong acid/strong base titration may be exemplified by:

$$H^+ + Cl^- + Na^+ + OH^- \rightarrow Na^+ + Cl^- + H_2O.$$

The net result being the replacement of hydrogen ions by sodium ions. As we know that equivalent ionic conductance of hydrogenise is much greater than that of the sodium ion, so that the conductance of the solution at the beginning of titration is large but decreases up to the equivalence point; this is shown in Fig. 2.2. At the equivalence point, the conductance is equal to that of a pure solution of sodium chloride and is at a minimum. Beyond the equivalence point, the conductance increases in direct proportion to the excess of sodium hydroxide added.

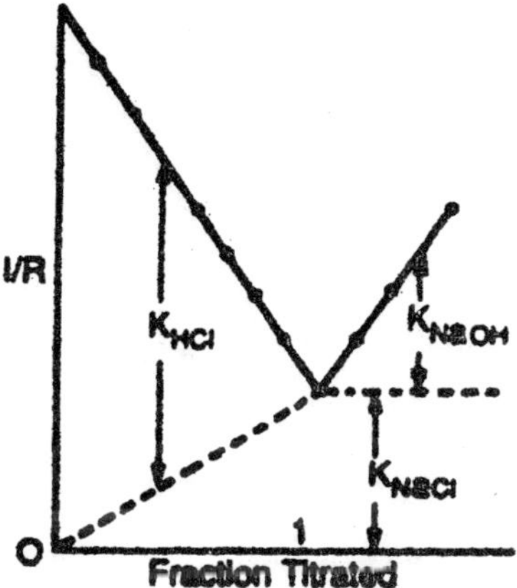

Fig. 2.2 : Conductometric titration curve of a strong acid with a strong base.

The slope of the first branch of the titration curve, and thus sharpness with which the end point may be established, depends on differences between the equivalent ionic conductance of the hydrogen ion and the cation of the base used, in this case the sodium ion. As the equivalent ionic conductance of the hydrogen ion is much greater than that of any other cation, strong acids can almost always be successfully titrated to a conductometric end point.

On *special important advantage of* conductometric titration is that the *relative change in conductance is almost independent of the concentration of strong acid titrated.* Thus, dilute solution can be titrated with almost the same accuracy as concentrated solution. The titration curve is displaced along the 1/R axis, but the sharpness is not generally affected. In this respect, conductometric acid-base titrations differ from potentiometric titrations.

Although the titration involving strong acids with strong bases was among the first conductometric titrations attempted, the conductometric method is not much used for such determinations, because a large number of indicator methods are available, even coloured solutions are now mostly titrated potentiometrically with a glass electrode.

(b) *Strong acid with a weak base:* Let us consider the titration of HCl (strong acid) with NH_4OH (weak base). When ammonium hydroxide is added to HCl, the conductivity decreases first due to the replacement of the fast moving H^+ ions by slow moving NH_4^+ ions.

$$H^+Cl^- + NH_4OH \rightarrow NH_4^+ + Cl- + H_2O.$$

After the end point is passed, further addition of NH_4OH does not change the conductance because NH_4OH, a weakly ionised electrolyte, has a very small conductivity compared with that of the acid or salt (Fig. 2.3 and 2.4).

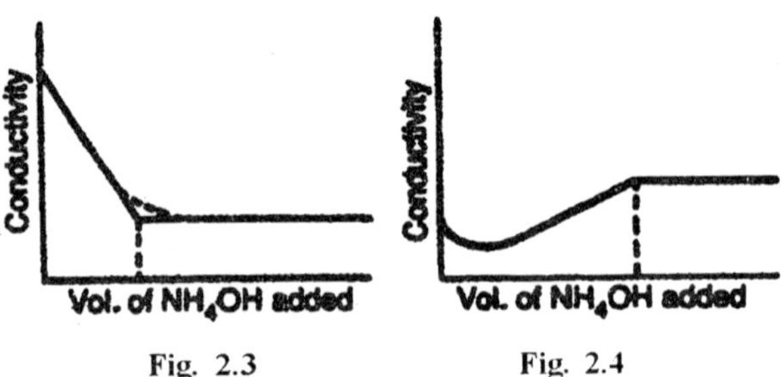

Fig. 2.3 Fig. 2.4

(c) *Weak acid with a strong base:* Let us consider the titration of the weak acid like CH_3COOH with strong base NaOH. When a small amount of NaOH is added to CH_3COOH, the conductivity decreases initially and then increases with the further addition of NaOH.

$$CH_3COO^- + H^+ + Na^+ + OH^- \rightarrow CH3COO^- + Na^+ + H_2O$$

When the neutralisation of acid is complete, further addition of alkali introduces excess of OH^- ions. The conductance of the solution, therefore, begins to increase more sharply.

(d) *Conductometric titration of weak acid and weak bases:* When weak acids are titrated with strong bases, curves such as 1 and 2 in (Fig. 2.5) are obtained. Before the equivalence point the conductivity is governed by the extent of ionisation (and by the concentration) the conductivity first decreases as the anion formed suppresses the ionisation, passes through a minimum, and then increases upon the equivalence point because of the conversion of non-conductivity weak acid into its conducting salt. Extensive information about the effect of the dissociation constant on the titration curves of both weak acids and bases has been compiled by Britton. Bruni and Eastman have carried out mathematical analysis of the titration curves of weak electrolytes.

Near the equivalence point, the graphs of such titrations are often curved because the incompleteness of reaction allows extra hydroxide ions (or hydrogen ions if a weak base is being titrated) to be present, thus increasing the conductance; the more incomplete the reaction, the more is curvature.

Conductometric titrations utilise measurements far enough removed from the end point for a fair amount of curvature near the end point to be tolerated. In fact many titration reaction that are too incomplete for potentiometric end point detection can be used conductometrically. For instance, boric acid can be titrated conductometrically with sodium hydroxide whereas the pH values change at end point is not large enough for potentiometric or indicator methods to be used. Weak acids with dissociation constants as small as 10^{-10} can readily be determined, especially in fairly concentrated solution. (Fig. 2.6).

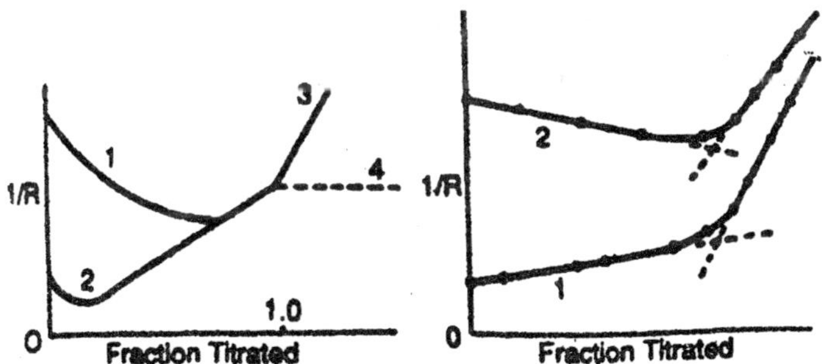

Fig. 2.5 : Conductometric titrations of weak acids and weak bases.

Fig. 2.6 : Replacement conductometric titrations.

End-point Calculation Method: After the conductometric titration curve has been plotted in usual way, estimate the end point by drawing two straight lines (the usual graphical method). Arbitrarily choose values for f_1, f_2, and f_3 such that f_1 and f_2 are before the end point and f_3 is after the end point and calculate f_4 by using the appropriate formula.

For a weak acid/weak base titration:

$$\frac{f_2 + f_1 - 2f_1f_2}{(1 - f_2)(1 - f_1)} = \frac{f_4 + f_3 - 2}{(f_2 - 1)(f_4 - 1)}$$

For a strong acid/weak base titration,

$$\frac{1 - f_2f_1}{(1 - f_2)(1 - f_1)} = \frac{f_4 + f_3 - 2}{(f_2 - 1)(f_4 - 1)}$$

Draw straight lines through f_1 and f_2 and through f_3 and f_4. The intersection of these lines is the new end point. Repeat the process using the new estimated end point. The new end points found in this way will converge rapidly so long as the first estimated end point was not very greatly in error.

Replacement Titration

When a strong acid reacts with the sodium or potassium salts of week acids, the weakest acid will be replaced first. Such a replacement reaction is exemplified by

$$HCl + CH_3COONa \rightarrow NaCl + C_3H_3COOH$$

The investigation of the replacement of weak acid from its salt was first investigated by Dutoit.

Analogous reaction for the replacement of the weak bases from their salts by strong bases can, of course, be cited.

Kolthoff has studied this type of reaction extensively, and curves for two of Kolthoff's system are shown in (Fig. 2.6). Curve I shows the increase in conductivity during the replacement of acetic acid by hydrochloric acid to form sodium chloride; the increase is due to the fact that the mobility of chloride ion is greater than that of acetate ion.

After the equivalence point, the conductively increases more rapidly, owing to the additions of excess of hydrochloric acid.

Kolthoff has calculated that accurate titrations with 0.01 M solutions are possible if the dissociation constant of the liberated acid is less than 5×10^{-5}; Lapshin has, however, titrated the salt of weak acids and bases whose dissociation constants are greater than 10^{-4}.

The main application of replacement titrations is in the determination of alkaloids. As alkaloids are weak bases that are often isolable in water, indicator methods do not work well.

Precipitation Titrations

Precipitation titrations can be effectively followed by conductometric methods, though not so well as acid-base titrations which are characterised by sharp breaks because both the hydrogen ion and hydroxyl ion have very high equivalent ionic conductances. In precipitation titrations, one pair of ions is substituted for another and as one experimenter has a choice of reagents, good results can usually be obtained. If a cation is to be precipitated, titrant whose cation has the smallest possible mobility is selected and, if an anion is to be precipitated, a titrant whose anion has as small a mobility as possible, in this way the maximum possible change in conductance during a titration is assured.

Let us consider the titration between $AgNO_3$ and KCl. The reaction involved may be represented as

$$K^+ + Cl^- + Ag^+ NO_3^- \rightarrow K^+ + NO_3^- + AgCl.$$

In the early stages of the titration , addition of silver nitrate, the conductance does not change very much because the Cl^- ions are replaced by NO_3^- ions: both have almost same ionic conductances. After the end point is passed, the excess of the added salt causes a sharp increase in conductance.

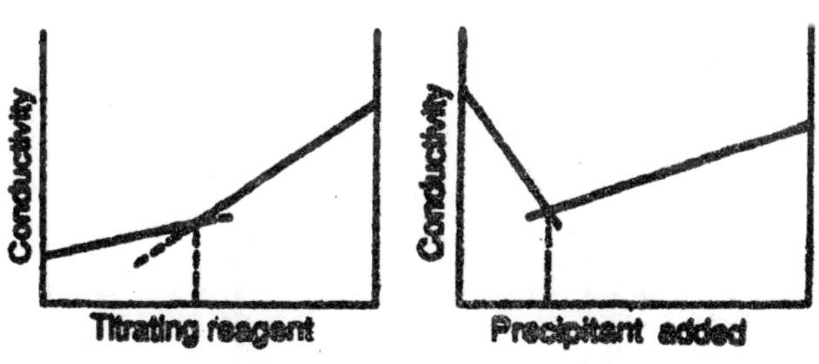

Fig. 2.7 Fig. 2.8

If both the products of the reaction are sparingly soluble, the curve obtained takes up the form as shown in Fig. 2.8. An example of this is the reaction between $MgSO_4$ and $Ba(OH)_2$. The reaction is as follows:

$$Mg^{2+}SO_4^{2-} + Ba_2^+ + 2OH^- \rightarrow Mg(OH)_2 + BaSO_4$$

In this type of titrations, conductivity decreases in the beginning but increases after the end point has passed.

Precipitate titrations cannot be performed very accurately. The main reasons for this are as follows:

(i) Due to slow separation of the precipitate.

(ii) Due to the removal of titrated solute by adsorption on precipitate.

Some efforts have been made to eliminate the errors that are due to adsorption, slowness of precipitation and electrode fouling, and are common to a number of conductometric precipitation titrations carrying out the titrations at elevated temperature, *e.g.*, in boiling solution, this apparently gives accurate results but there is insufficient evidence to judge these methods critically.

Complexometric Titrations

Conductivity has long been used in the study of complex formation. Werner, for example, supported many of the postulates of his coordination theory with conductivity measurements. Under controlled conditions conductivity measurements can be used to show the type of electrolyte, *e.g.* univalent, a complex salt gives in solution.

Conductometric titrations have also been applied to the study of complex salts; for instance, Job titrated roseo-cobaltic sulphate $[Co(NH_3)_5H_2O]_2(SO_4)_3$ with barium hydroxide, finding two breaks in the titrations curve.

Let us illustrate the complexometric titration by titrating KCl with $Hg(ClO_4)_2$. This is shown in Fig. 2.9. This curve shows two breaks one due to the formation of $HgCl_4^{2-}$ and the other one at the end of the reaction due to the formation of K_2HgCl_4.

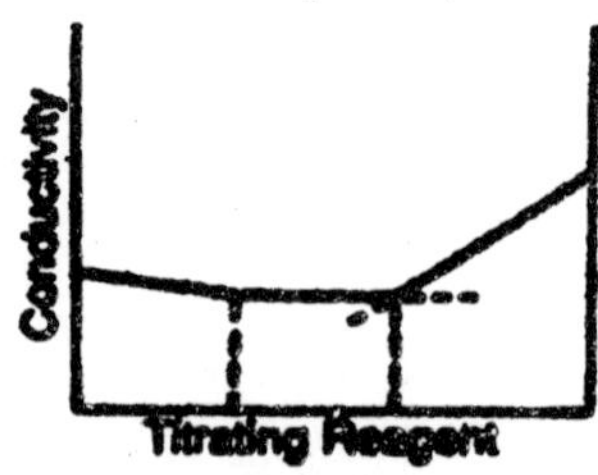

Fig. 2.9

$$Hg(ClO_4)_2 + 4KCl \rightarrow HgCl_4^{2-} + 2K^+ + 2KClO_4$$

$$HgCl_4^{2-} + 2K^+ \rightarrow K2HgCl_4$$

OSTWALD'S DILUTION LAW

As dissociation of weak electrolytes being a reversible process, the rate of formation of ions and their removal by recombination will be governed by the law of mass action. The applicability of the law of mass action to equilibria involving weak electrolytes was first shown by Wilhelm Ostwald and a form of it developed below in terms of dilution instead of concentration is known as Ostwald's dilution law which we shall discuss now. Consider the ion formation by an incomplete dissociation of an electrolyte AB.

$$\begin{array}{ccc} AB & \rightleftharpoons & A^+ + B^- \\ C(1-\alpha) & & C\alpha \quad C\alpha \end{array}$$

If C moles of the electrolyte (AB) are dissolved per litre of the solution and α is the degree of dissociation, then at equilibrium, the concentration of

$$A^+ \text{ ions} = C\alpha$$

$$B^- \text{ ions} = C\alpha$$

and $\quad AB \text{ molecules} = C(1-\alpha).$

Applying the law of mass action to the above equilibrium, we get

$$\frac{C\alpha \times C\alpha}{C(1-\alpha)} = K \text{ or } K = \frac{C\alpha^2}{(1-\alpha)} \qquad ...(1)$$

where K is called the dissociation constant. The choice of units for the value of K will depend on the number of ions formed from one dissociating molecule. The volume, V, of a solution containing one mole or one gram formula weight of an electrolyte is called the *dilution for the solution*. When a solution has C moles of an electrolyte per litre, dilution V of the solution is given by

$$V = 1/C \text{ litre per mole}$$

Equation (1) may now be written in terms of V instead of C values:

$$K = \frac{\alpha^2}{(1-\alpha)V} \text{ moles per litre.} \qquad (2)$$

Equation (2) is a *mathematical expression of Ostwald's dilution law* for equilibrium state. This law holds good for weak electrolytes since the value of 'K' is constant for them. Thus, Ostwald's dilution law is nothing but the law of mass action applied to ionic equilibrium existing in the solution of weak electrolyte. For weak electrolytes such as acetic acid or ammonium hydroxide, the value of the degree of dissociation is very small.

Therefore, $1 - \alpha \cong 1.$

The equation (1) then simplifies to $K = C\alpha^2$

$$\therefore \qquad \alpha^2 = \frac{K}{C}$$

or $$\alpha = \sqrt{(K/C)} \qquad ...(3)$$

From equation (3), it follows that the degree of dissociation of a weak electrolyte is inversely proportional to the square root of its concentration. For equation (2) when α is small,

$$\alpha = \sqrt{(KV)} \qquad ...(4)$$

From the above expression, it follows that the degree of dissociation of weak electrolyte is directly proportional to the square root of its dilution, (*i.e.*, the volume containing 1 gm mole of the electrolyte).

Experimental verification of the dilution law: The dilution formula may be experimentally verified by determining the value of K and substituting the value of α at different dilution V in the expression $\alpha^2/(1 - \alpha)V$. The value comes out to be constant in case of weak electrolytes within reasonable limits.

Hence α, the degree of dissociation, is determined by substituting the value of equivalent conductivity at λ_r different dilutions V and the value of equivalent conductivity at infinite dilution λ_∞ in the formula:

$$\alpha = \frac{\lambda_v}{\lambda_\infty}$$

Determination of dissociation constant of acetic acid: Acetic acid is a weak electrolyte and when dissolved in water dissociates as

$$CH_3COOH \rightleftharpoons CH_3COO^- + H^+$$

If C gm moles of acetic acid are dissolved/litre of the solution and α is the degree of dissociation, then the dissociation constant

$$K = \frac{C\alpha^2}{(1-\alpha)}$$

If λ_V is the equivalent conductance of the CH_3COOH solution at the given dilution and λ_∞ that at infinite dilution, then

$$\alpha = \lambda_V/\lambda_\infty$$

The value of λ_V is determined by conductivity measurement at dilution V. The value of λ_∞ for acetic acid can be calculated with the help of Kohlrausch's law, *i.e.*,

$$\lambda_\infty \text{ for } CH_3COOH = \lambda \text{ for } CH_3COO^- + \lambda \text{ for } H^+$$

On substituting the values of λ_V, λ_∞ and V, the value of K can be calculated. At 20°C the value of K for acetic acid is found to be 1.8×10^{-5}.

Limitations of Ostwald's Dilution Law

(1) Ostwald's dilution law is obeyed by weak electrolytes only. Strong electrolytes like HCl, NaOH, salts etc. do not obey this law. In such cases the values of dissociation constant K are never constant but rapidly fall with dilution.

(2) The values of degree of dissociation of an electrolyte as determined from colligative properties and from conductance measurements often disagree.

(3) The law is a direct corollary from the Arrhenius theory. But the Arrhenius theory itself fails when applied to the strong electrolytes. The behaviour of the strong electrolytes, however, is explained better by the modern theory known as Debye—Huckely theory.

Basically, the failures of Ostwald's dilution law are due to inapplicability of Arrhenius theory of dissociation for some solutions of electrolytes.

DEGREE OF DISSOCIATION

The degree of dissociation *is the fraction fo the total number of molecules dissociated into ions, i.e.,*

$$\alpha = \frac{\text{No. of molecules dissociated into ions}}{\text{Total number of molecules taken}}$$

We know that the conductivity of a solution is due to the presence of ions in solution. Greater the number of ions, greater is the conductivity.

$\therefore$ λ_V, the equivalent conductivity at a particular dilution

$\propto$ No. of the ions present at this dilution

As ions are produced due to the dissociation of molecules of the electrolyte, $\lambda_v \propto$ No. of molecules dissociated into ions at a particular dilution.

Similarly, $\lambda_\infty \propto$ No. of ions at infinite dilution. ...(i)

As at infinite dilution all molecules are in ionic form

$\lambda_\infty \propto$ Total number of molecules taken. ...(ii)

Dividing (i) by (ii), we get

$$\frac{\lambda_v}{\lambda_\infty} = \frac{\text{No. of molecules dissociated into ions at a particular dilution}}{\text{Total number of molecules taken}}$$

$$= \alpha \text{ (the degree of dissociation)}$$

Thus, *degree of dissociation, α, at any dilution, is the ratio of the equivalent conductivity at that dilution to the equivalent conductivity at infinite dilution.*

The value of λ_v can be obtained by direct measurement of conductivity; while λ_∞ can be calculated with the help of Kohlrausch's law, *i.e.*,

$$\lambda_\infty = \lambda^+ + \lambda^-$$

Hence, the value of α, the degree of dissociation, can be computed.

DISSOCIATION CONSTANT

According to Ostwald's dilution law for a weak binary electrolyte

$$\frac{\alpha^2}{(1-\alpha)V} = K$$

where, α is the degree of dissociation when 1 mole of electrolyte is dissolved in V litres of water. The constant K is called the dissociation constant or ionization constant of the electrolyte.

For the dissociation of acetic acid, a weak binary electrolyte, dissociation is represented as

$$\underset{(1-\alpha)\text{ moles}}{CH_3COOH} \rightleftharpoons \underset{\alpha\text{ g-ion}}{CH_3COO^-} + \underset{\alpha\text{ g-ion}}{H^+}$$

Suppose 1 g mole of acetic acid is dissolved in V litres of water. If α is the degree of dissociation at this dilution, then at equilibrium

$$[CH_3COOH] = (1 - \alpha)V; [CH_3COOH^-] = [H^+] = \alpha/V.$$

Therefore, on applying law of mass action we get Oswald's dilution formula as

$$\frac{[CH_3COO^-][H^+]}{[CH_3COOH]} = \frac{\alpha^2}{(1-\alpha)\,V} = K_a$$

where K_a is the *dissociation constant* of acetic acid.

We know $\alpha = \lambda_V/\lambda_\infty$, therefore the above equation takes the form

$$\frac{\lambda_V^2}{\lambda_\infty(\lambda_\infty - \lambda_V)V} = K_a$$

The value of λ_v is determined by conductivity measurement at dilution V.

The value of λ_∞ for acetic acid can be calculated with the help of Kohlrausch's law, *i.e.*,

$$\lambda_\infty\, CH_3COOH = \lambda CH_3COO^- + \lambda H^+$$

One substituting the values of λ_y, λ_∞ and V, the value of K_a can be calculated. At 20°C the value of K_a for acetic acid is found to be 1.8×10^{-5}.

ABSOLUTE IONIC VELOCITIES : IONIC MOBILITIES

Ionic mobility possesses a fixed value for each ion, no matter of which electrolyte it forms a part. It may be emphasised that ionic conductance is not the same thing as ionic mobility. These are, however, proportional to each other. If u_a^o and u_c^o are the actual velocities of positive and negative ions of a given electrolyte at infinite dilution, then

$$\lambda_a^o = ku_a^o \quad ...(1)$$

$$\lambda_c^o = ku_c^o \quad ...(2)$$

where k is a constant.

Absolute ionic velocity of an ion represents its *velocity in cms^{-1} under a potential gradient of 1 volt cm^{-1}*. It has shown that under a potential gradient of 1 volt cm^{-1}, the value of constant k is the change in one g equivalent of the ion *i.e.* 96500 coulombs (or 1 Faraday). Mathematically,

$$\text{Absolute ionic mobility} = \frac{\text{Ionic conductance}}{96500} \quad ...(3)$$

Mobilities of H^+ and OH^-

These ions have exceptionally high ionic mobilities in hydroxylic solvents like water, alcohols etc. In water, H^+ is strongly hydrated forming H_3O^+, which is able to transfer H^+ to the neighbouring hydrogen bonded water molecules by rearrangement of hydrogen bonds. Thus, H^+ is transferred from one end of the series of the other end without actually migrating through the solution. Similarly, the high mobility of OH^- in water is due to H^+ transfer between OH^- and water.

Experimental Determination of Ionic Velocities

Ionic mobility can be measured directly by a technique based on the movement of a visible boundary between two electrolytes (Fig. 2.10). Suppose it is required to determine the ionic velocity of K^+. The KCl may be one of the principal electrolyte and the indicator electrolyte may be taken as $CdCl_2$ having one (*i.e.*, Cl^-) ion common.

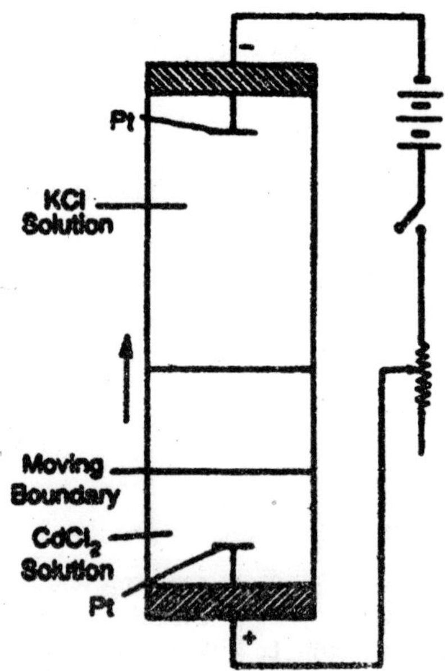

Fig. 2.10 : Determination of ionic mobility.

A decinormal solution of KCl is held over a solution of similar concentration of $CdCl_2$, in a tube of uniform bore. A current is passed through the solution. K^+ ions move towards the cathode. These are followed by the slow moving Cd^{2+} ions and the boundary moves. Let the boundary move through a distance, x cm in t second when the fall of potential in electric fields is E volts cm^{-1}. The mobility of H^+ is given by

$$u = \frac{x}{t \times E} \text{ cm s}^{-1}\text{/volt cm}^{-1} \quad ...(4)$$

ELECTROCHEMISTRY

A substance through which an electric current can pass is known as conductor. On the contrary, a substance through which no electric current can pass is known as an insulator or dielectric. An electric current in a conductor can consist of (a) a flow of electron only (b) a flow of ions only and (c) a flow of both electrons and ions. The first type of conductor is known as metallic conductor or electronic conductor. The second and third types of conductor are known as electrolytic (or ionic) and mixed conductors, respectively.

[I] Metallic or Electronic Conductor

When electricity passes through such type of conductor, it produces it produce no observable effect on it. It only raise the temperature of the conductor. Metals are typical example of such conductors. Certain sulphides, *e.g.*, PbS, CdS also belong to this category. In metallic conductors, there is no change in the chemical properties of the conductor.

[II] Electrolytic or Ionic Conductor

When electricity passes through such a conductor, it decomposes the conductor. Substance of this type whose solutions conduct the electricity and which are decomposed by the passage of electricity are called electrolytes. A substance whose solution does not conduct electricity is called a non-electrolyte, such as solutions of canesugar, alcohol etc.

[III] Mixed Conductor

In such a conductor, the electric current is partly electronic and partly ionic. β-Ag_2S, CuCl, γ-CuBr, metallic sodium in ammonia are examples of mixed conductors. In such a case, the ratio of electronic to ionic conductors varies with the concentration.

[IV] Electrolysis

The process of decomposition of an electrolyte by the passes of electric current through it is known as electrolysis. For bringing about electrolysis, we need a vessel known as 'electrolysis vessel'. In this vessel, an electrolytic solution is taken and two metallic rods are dipped in it. These two rods are then connected to the two terminals, (*i.e.*, + and –) of the battery. These metallic rods which lead the electric current to and form the electrolyte are called electrodes. The electrode connected with the positive end of the battery is called the positive electrode or anode, whereas the rod connected to the negative end of the battery is known as negative electrode or cathode. This complete arrangement is known as electrolytic cell. The current enters the electrolyte through the anode and leaves through the cathode.

The electrically charged atoms or radical that move to the electrodes are called ions. Ion travelling towards the cathode is known as cation and ion travelling towards anode is known as anion. The liquid surrounding the cathode is known as catholyte and that round the anode as anolyte.

[V] Laws of Electrolysis

Faraday (1834) first stated two laws on the basis of his studies on electrolysis.

1. *First Law :* According to this law, "the amount of substance (W) liberated on the electrode is directly proportional to the quantity of electricity (Q) passed."

Mathematically, $W \times Q$

If a current of C amperes is passed for t secs, then the quantity of electricity is given by current strength X time *i.e.*, c.t.

$$\therefore \quad w \propto, C \cdot t = Z_C t$$

where, Z = proportionality constant, known as 'electrochemical equivalent.'

$$C = 1 \text{ ampere}$$

$$\text{and } t = 1 \text{ sec,}$$

$$\text{then } Z = W \text{ gms}$$

So electrochemical equivalent may be defined as, "the amount of substance (in g) liberated on the electrode on passing a current of 1 ampere for 1 second or an passing 1 coulomb of electricity."

Electrochemical equivalents of hydrogen, silver and copper are 0.00001036, 0.001118 and 0.0003292, respectively.

2. *Second Law :* According to this law, "if some quantity of electricity is passed through different electrolytes, then the amount of substance liberated on the respective electrodes will be in the ratio of their equivalent weights."

If same quantity of electricity is passed through cells containing HCl, $AgNO_3$ and $CuSO_4$, then the amounts of hydrogen, silver and copper deposited on the respective cathode will be in the ratio of this equivalent weights. Hence,

$$\frac{\text{Amount of substance liberated}}{\text{Equivalent wt. of the substance}} = \frac{\text{wt. of hydrogen}}{\text{Eq. wt. of hydrogen}}$$

$$= \frac{\text{wt. of silver}}{\text{Eq. wt. of silver}} = \frac{\text{wt. of copper}}{\text{Eq. wt. of copper}}$$

or

$$\frac{\text{Wt. of hydrogen}}{1.008} = \frac{\text{Wt. of silver}}{107.88} = \frac{\text{Wt. of copper}}{31.78}$$

[IV] Faraday

It is unit of electricity and is defined as, 'the quantity of electricity (in coulombs) required to deposit or liberate 1 gm. equivalent of any substance on passing electric current.' One faraday is equivalent to 96,500 coulombs.

One the basis of faraday unit, we can find the magnitude of the elementary charge on an electron, provided Avogardro's number is known. We know that 1 gm atom of silver requires 96,500 coulombs of electricity and the number of atoms in 1 g atom of silver is 6.023×10^{23}. So,

$$\text{Elementary change } (e) = \frac{96.500}{6.023 \times 10^{23}} \text{ Coulombs}$$

$$= \frac{96.500}{6.023 \times 10^{23}} \times 3 \times 10^{9} \text{ e.s.u.}$$

$$= 4.18 \times 10^{-10} \text{ e.s.u.}$$

[VII] Practical Electrical Units

The student who is acquainted with the elementary principles of electricity and magnetism is aware that electrical units are measured in various systems of units, such as electrostatic, electromagnetic and particle units. For our present purpose, it is convenient to employ practical units, a few of which we proceed to describe.

The practical unit of current is ampere which is defined as 'unvarying current which when passed through a solution of nitrate of silver in water produces silver at the rate of 0.001118 g per second.' The absolute ampere and international ampere are nearly identical.

The practical unit of quantity of electricity is coulomb, which is defined as, 'the amount of electricity conveyed in one second by a current of one international ampere.'

$$1 \text{ Coulomb} = 1/10 \text{ e.m.u.}$$

The practical unit of resistance is ohm and is defined as, 'the resistance offered to a steady current by a uniform column of mercury of length 106.300 cm at a temperature 0°C and the mass being 14.4521 g.'

Fig. 2.11 represented in the practical unit of electromotive force and of potential is volt and is defined as "the difference of potential steadily applied to a conductor, which has a resistance of 1 ohm and produces a current of 1 international ampere."

$$1 \text{ volt} = 1/300 \text{ e.m.u.}$$

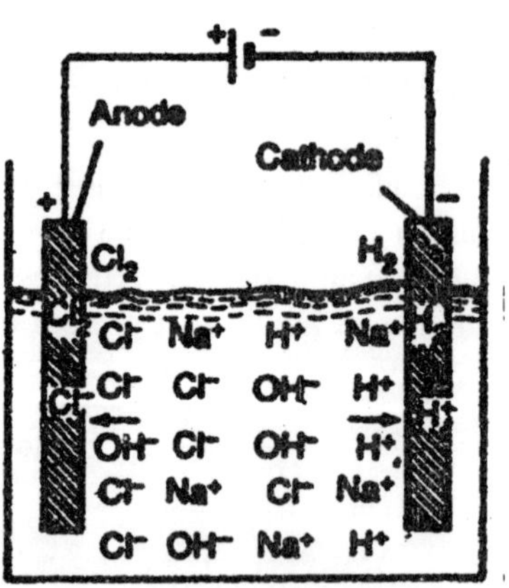

Fig. 2.11 : Mechanism of electrolytic conduction.

For practical purpose, the international values are taken to be identical with the absolute value.

For steadily, direct currents, we have Ohm's law given by

$$E = RC$$

i.e. the emf., E of the circuit measured in volts, is equal to the product of the resistance, R in ohms and current, C in amperes. Ohm's law is not only applicable to a complete electric circuit but also to any part of the circuit.

Movement of Ions during Electrolysis

Who have already said that the flow of electricity in an aqueous solution results due to the movement of ions.

When an electric field is applied to electricity in an aqueous solution results due to the movement of ions.

When an electric field is applied to electrolytic solution, the movement of ions occurs towards respective electrodes. This can be demonstrated by the following experiment shown in Fig. 2.12.

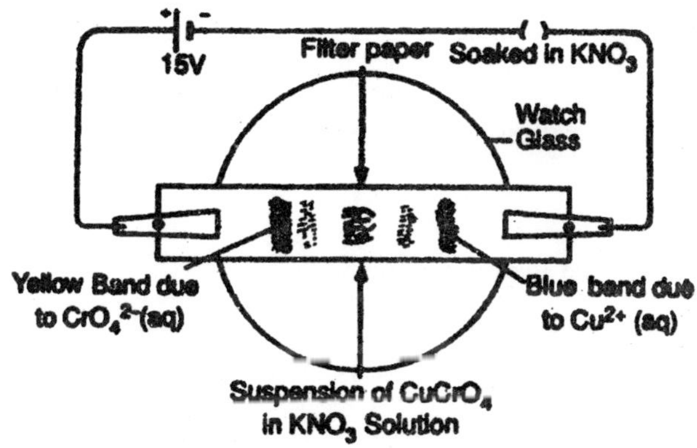

Fig. 2.12 : Movement of ions during electrolysis.

A filter paper strip (about 3 × 8 cm) is cut out. It is moistened with a little dilute potassium nitrate solution. Then, it is kept on a clean watch glass. A battery of 15 volts is connected across the filter paper by using a pair of crocodile clips. At the centre of filter paper strip, a suspension of yellow copper chromate in potassium nitrate is placed. When the electric current is applied across the filter paper strip, two zones are observed on the filter paper. A blue zone is seen moving towards the cathode which is due to the migration of Cu^{2+} ions. A yellow zone due to chromate ion, CrO_4^{-2}(aq) is seen moving towards the anode.

The above experiment can also be done with coloured, water-soluble ionic compounds such as potassium permanganate, cobalt chloride, nickel chloride, etc.

RESISTANCE OR CONDUCTANCE

Metallic conductors as well as electrolytes obey Ohm's law, which states that "*The strength of current (I) flowing through a conductor is directly proportional to the potential difference (E) applied across the conductor and inversely proportional to the resistance (R) of the conductor.*"

Thus, the mathematical form of Ohm's law becomes as

$$I = E/R.$$

In relation to electrolytes, the term conductance (C) is used more frequently than resistance. Conductance implies the ease with which electric current can flow through a conductor. It is therefore, defined as the reciprocal of resistance,

Conductance = 1/R.

Units: It is expressed in units of reciprocal or inverse ohms, ohms^{-1} or mhos.

Resistance

The resistance R of a conductor is

(i) directly proportional to its length (l cm) and

(ii) inversely proportional to its area of cross of cross section.

Hence $$R \propto \frac{l}{a} \quad \text{or} \quad R = \rho\frac{l}{a} \qquad ...(1)$$

where ρ is a constant depending on the nature of the material of the conductor, and is called the "*specific resistance*" or "*resistivity.*" If l = 1 cm and a = l sq. cm, then ρ = R ohms.

Hence, specific resistance may be defined as:

"*The resistance of a uniform column of the material of the conductor having a length of 1 cm and a cross section of 1 sq. cm.*"

Units: As we know that

$$\rho = R\frac{a}{l} = \frac{\text{Ohm (cm)}^2}{\text{cm}} = \text{ohm. cm} \qquad ...(2)$$

Specific Conductance

The specific conductance of a conductor is the reciprocal of specific resistance, and is generally denoted by K (Greek, Small Kappa). Equation (2) can be written as

$$R = \frac{1}{K}\frac{1}{a} \text{ ohms} \qquad \left[\therefore \rho = \frac{1}{K}\right] \quad ...(3)$$

$$K = \frac{1}{R} \times \frac{1}{a} \text{ ohms}^{-1} \text{ com}^{-1}$$

If l = 1, and a = 1 sq. cm. equation (3) becomes as

$$K = \frac{1}{R} = \rho$$

In the case of solution, the definition will be the same as of a conductor. In simple words, it is defined as:

"The conductivity offered by a solution of length 1 cm area of unit cross section."

Units: We know that $K = 1/\rho$

Hence the unit of K will be

$$\frac{1}{\text{ohm. cm}} = \text{ohm}^{-1} \text{ cm}^{-1} = \text{mho/cm.}$$

Equivalent Conductance

It may be defined as,

"The conductance of a solution containing 1 gm. equivalent of an electrolyte when placed between two sufficiently large electrodes which are 1 cm apart." It is denoted by λ_r, where V is the volume in c.c. containing one gram equivalent of electrolyte dissolved in it and is measured in reciprocal ohm, or mho.

Molecular Conductance

It is defined as,

"The conductance of a solution containing 1 gm mole of the electrolyte when placed between two sufficiently large electrodes placed one cm apart."

It is denoted by symbol μ, and is measured in mhos,

Relation Between Specific Conductance and Equivalent Conductance

Consider a rectangular metallic vessel with opposite sides exactly 1 cm apart (Fig. 2.13). If now 1 c.c. of solution is placed in this vessel, the area of the opposite faces of the cube converted by the solution will be 1 sq. cm. The measured conductance of the solution will be its specific conductance. If 1 c.c. of the solution is placed in the above vessel containing 1 gm. equivalent of the electrolyte, then by definition, the measured conductance will also be equal to the equivalent conductance,

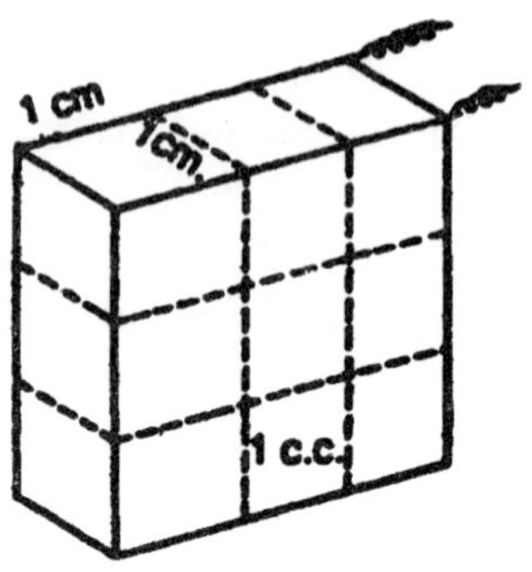

Fig. 2.13

$$\lambda = K$$

Now if 9 c.c. of pure solvent are added so that the total volume now occupied by the solution is 10 c.c.; the (because 10 c.c. of the diluted solution still contains 1 gm. equivalent of the electrolyte). But by definition of the specific conductance, the measured conductance of the diluted solution will be ten times the specific conductance,

$$\therefore \qquad \lambda = K \times 10$$

Similarly if the solution is diluted 100 times its volume, the measured conductance will still be equivalent conductance, but will be 100 times the specific conductance.

Thus, $$\lambda_v = K \times 100$$

In general, if the solution contains 1 gm equivalent of the electrolyte dissolved in V c.c. of the solution, then

$$\lambda_r = K \times V$$

$\because$ Equivalent conductance = specific conductance × volume of solution in c.c. containing 1 gm. equivalent of the electrolyte.

Relation between Molecular conductance and Specific Conductance: As discussed above, a similar type of the relation also exists between molecular conductance and specific conductance *i.e.*,

$$\mu_r = K \times V$$

or Molecular conductance = [Specific conductance]

× [Volume of solution in c.c. containing

1 gm. mole of electrolyte]

Effect of Dilution

(i) *Conductance:* The conductance of s solution is due to the presence of ions in solution. Since on dilution the degree of dissociation of electrolyte in increased and more ions are produced in solution, it is thus expected that the conductance of a solution increase on dilution.

(ii) *Specific Conductance:* Specific conductance depends upon the number of ions present per c.c. of the solution. Since on dilution the degree of dissociation increases but the number of ions per c.c. decreases, it is therefore expected that the specific conductance of a solution should decrease on dilution.

(iii) *Equivalent Conductance:* The value of equivalent conductance increases on dilution. This increase is due to the fact that equivalent conductance is the product of specific conductance and the volume V of the solution containing 1 gm equivalent of the electrolyte. Since the decreasing value of specific conductance is more than compensated by the increasing value of V, the value of λ_y will increase with dilution.

(iv) *Molecular Conductance: As* discussed in equivalent conductance, the molar conductance of an electrolyte increases with dilution.

(v) The variation of equivalent conductance, with dilution is shown in Fig. (2.14) where the equivalent conductance of different dilutions is plotted against $\sqrt{(c)}$, where c is the concentration of the solution in moles per litre.

From Fig. (1.4), it is evident that the equivalent conductance tends to become maximum as the concentration decreases or dilution increases.

However the maximum conductivity in case of strong electrolytes like KCl is reached at relatively higher concentrations (or lower dilutions) that the case of weak electrolytes like CH_3COOH.

Equivalent Conductance at Infinite Dilution

The maximum value of the equivalent conductance of an electrolyte obtained at high dilutions is known as the equivalent conductance at infinite dilution. It is denoted by λ_∞. By infinite dilution we mean that the solution is so dilute that any further dilution does not produce any charge in the equivalent conductance of the solution.

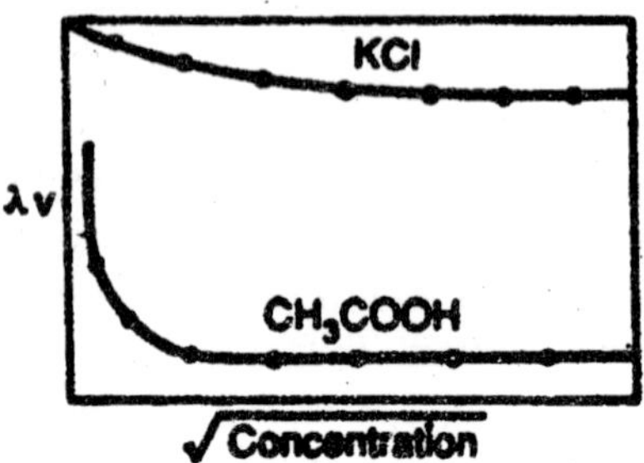

Fig. 2.14

Measurement of Specific Conductance

We know $$R = \rho \times \frac{l}{a}$$

$$\frac{1}{\text{obs. conductance}} = \frac{1}{\text{sp. conductance}} \times \frac{l}{a}$$

$$\text{Sp. conductance} = \text{Obs. conductance} \times \frac{l}{a}$$

$$= \text{Obs. conductance} \times \text{cell constant.}$$

The ratio l/a is known as *cell constant* and is generally denoted by x. It depends upon the area of cross section of electrodes and the distance between them. Thus, the determination involves the measurement of conductance of the solution and of the cell constant.

Measurement of Conductance: The conductance of a solution can be expressed by

$$\frac{1}{R} = x(c_1\lambda_1 + c_2\lambda_2 + c_3\lambda_3 \ldots + c_1\lambda_1)$$

Here c_i terms are the concentrations of various ions in solution, λ_i terms are numerical constants characteristic of ions and x is a proportionality factor that takes account of the geometry of the cell. Because of interionic interactions and solvent-ion interactions, the values of λ_i depend slightly on concentration, and extrapolation to infinite dilution is required to obtain values that are physical constants of ions. Thus with proper calibration, a single measurement can give the concentration of any electrolyte.

The conductance of a solution can be measured by finding out the resistance by means of a Whetstone bridge arrangement. However when D.C. current is used in the solutions, the following difficulties may arise:

(i) During the conduction of D.C. current, there starts electrolysis which results in a change in the concentration of the solution.

(ii) Products of electrolysis get accumulated at the electrodes and set up a back E.M.F. Above mentioned difficulties are overcome by using A.C. current in the following manner.

(i) *The polarisation produced due to A.C. current in one direction is neutralised by the effect of current in other direction.*

(ii) *No electrolysis takes place.*

When an A.C. current is used, the following modifications are nade:

(i) *An A.C. current of frequency fo 1000-4000 cycles per second is generally used, an oscillator has to be used.*

(ii) *An alternating current cannot be employed to detect the null-point and is therefore replaced by a Head phone.*

(ii) *In order to minimise the polarisation still further, the platinum electrodes in the cell are coated with finely divided platinum. This gives rise to a larger surface to the electrodes and consequently the polarisation is retarded.*

Method : The apparatus used for the purpose is a modified form of the Wheatstone bridge and is shown in Fig. 2.15.

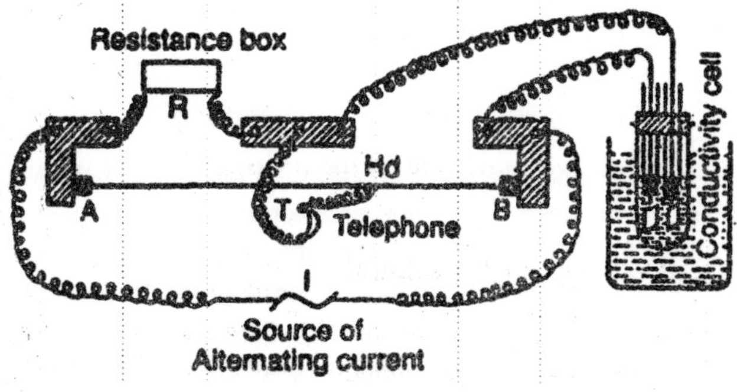

Fig. 2.15 : Measurement of Eq. conductance.

(i) The solution whose conductance is to be measured is taken in the conductivity cell. The conductivity cell consists of electrodes which are platinum coated with finely divided platinum. These are welded to platinum wires which are sealed through two glass tubes. The tubes are firmly fixed in an ebonite cover so that the distance between the electrodes is not altered during the experiment. The copper wires of the circuit dip in mercury placed in glass tubes to complete the circuit. Some other types of conductivity cells are shown in Fig. 2.16.

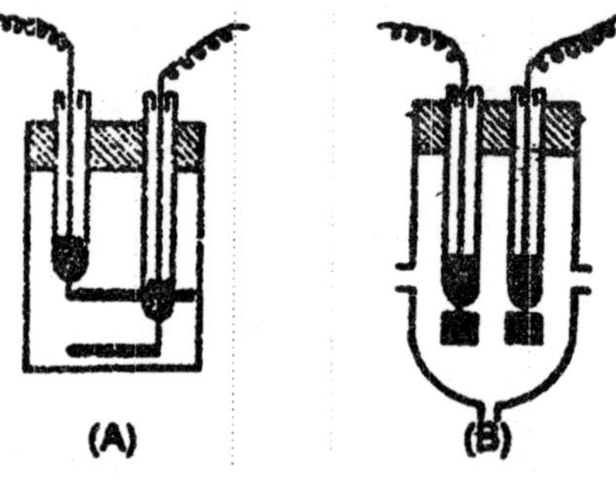

Fig. 2.16

(ii) The cell is connected to a resistance box R on one side and to a long thin wire AB stretched along a scale on the other side.

(iii) Some known resistance 'R' is taken out of the resistance box. An alternating current of about 1000 - 4000 cycles per second is passed through the solution. The sliding contact d is moved on the wire AB so that minimum or no sound is heard in the head phone. Then,

$$\frac{\text{Resistance of the solution}}{\text{Resistance R}} = \frac{\text{Length Bd}}{\text{Length Ad}}$$

Thus, knowing R, Ad and Bd the resistance of the solution can be calculated.

$$\therefore \text{ Conductivity of the solution} = \frac{1}{\text{Resistance of the solution}}$$

Determination of Cell Constant: The cell constant is determined by substituting the value of sp. conductance of N/50 KCl solution at 25°C as accurately determined by Kohlrausch (0.002765) and the value of conductance is observed with the given cell in the relationship:

$$\text{Cell constant} = \frac{0.002765}{\text{obs. conductance}}$$

Measurement of Equivalent Conductance: Knowing specific conductance as above, the equivalent conductance of the solution can be measured by measuring the specific conductance K with volume V in c.c. containing 1 gm. equivalent of the electrolyte

$$\lambda_r = K \times V.$$

KOHLRAUSCH'S LAW

The values of equivalent conductances at infinite dilution for certain pairs of electrolytes at 18°C are given in Table 2.1.

Table 2.1

Pairs of electrolytes with the same anion	λ_0 *Ohm*$^{-1}$	*Difference Ohm*$^{-1}$	*Pairs of electrolytes with the same cation*	λ_0 *Ohm*$^{-1}$	*Difference Ohm*$^{-1}$
KF	111.2		KCl	130.0	
		21.1			18.8
NaF	90.1		KF	111.2	
KCl	130.0		NaCl	108.9	
		21.1			18.8
NaCl	108.9		NaF	90.1	

From the above table it follows that

(i) When potassium is replaced by sodium in any of the electrolytes, there is always a difference of 21.1 ohm^{-1} in equivalent conductances at infinite dilution irrespective of the nature of the anion with which they are associated.

(ii) When chlorine is replaced by fluorine, there is always a difference of 18.8 ohm^{-1} in equivalent conductances at infinite dilution irrespective of the nature of the cation with which they are associated. The above observations led Kohlrausch (1875) to formulate a law known after his name as Kolhrausch's law. This law can be stated as follows:

"The equivalent conductance at infinite dilution for any electrolyte is the sum of equivalent conductances for the cation and anion at infinite dilution."

Mathematically, Kohlrausch's law can be put as

$$\lambda_{\infty} = \lambda_{\infty}^{+} + \lambda_{\infty}^{-}$$

where λ_{∞}^{+} and λ_{∞}^{-} are the equivalent conductances of the cation and anion respectively at infinite dilution.

Applications of Kohrausch's Law

(i) *Calculation of Equivalent Conductance at Infinite Dilution for Weak Electrolytes:* Equivalent conductance at infinite dilution can be calculated by a graphical method. But the graphical method is not applicable to weak electrolytes as they are very slightly ionized. But Kohlrausch's law can be used to determine the value of λ_{∞} for weak electrolytes. For example, equivalent conductance of acetic acid at infinite dilution can be calculated from the equivalent conductances at infinite dilution of hydrochloric acid, sodium acetate and sodium chloride as follows:

$$\lambda_{\infty}HCl = \lambda_{\infty}H^{+} + \lambda_{\infty}Cl^{-} \quad \text{(say x)} \quad ...(1)$$

$$\lambda_{\infty}CH_3COONa = \lambda_{\infty}Na+ + \lambda_{\infty}CH_3COO^{-} \quad \text{(say y)} \quad ...(2)$$

$$\lambda_{\infty}NaCl = \lambda_{\infty}Na^{+} + \lambda_{\infty}Cl^{-} \quad \text{(say z)} \quad ...(3)$$

Add equations (1) and (2) and subtract (3), we get

$$\lambda_{\infty}HCl + \lambda_{\infty}CH_3COONa - \lambda_{\infty}NaCl$$

$$= \lambda_{\infty}H^{+} + \lambda_{\infty}Cl^{-} + \lambda_{\infty}Na^{+} + \lambda_{\infty}CH_3COO^{-} - \lambda_{\infty}Na^{+} - \lambda_{\infty}Cl^{-}$$

$$x + y - z = \lambda_{\infty}H^{+} + \lambda_{\infty}CH_3COO^{-}$$

$$= \lambda_{\infty}CH_3COOH$$

or $\lambda_{\infty}CH_3COOH = x + y - z.$

Since the values of x, y and z are experimentally determined, we can easily calculate the value of λ_{∞} for acetic acid.

(ii) *Calculation of Degree of Dissociation of Weak Electrolytes :* The value of λ_0 helps to calculate the degree of dissociation and the dissociation constants of weak acids and weak bases.

For example, the degree of dissociation α_c of a weak electrolyte at the concentration c equivalents per litre may be given by the following relation:

$$\alpha_{-} = \frac{\lambda_c}{}$$

where λ_c is the equivalent conductance of electrolyte a^+ concentration c and λ_0 is the equivalent conductance of the same electrolyte at infinite dilution. Hence, measurement of λ_c permits evaluation of α_c if λ_0 known.

(iii) *Determination of Solubility of Springly Soluble Salts* : For ordinary purposes ionic salts like AgCl, $BaSO_4$, $CaCO_3$, Ag_2CrO_4, $PbSO_4$, PbS, $Fe(OH)_3$, etc. are regarded as insoluble, but they do have a *very small* but *definite* solubility in water. The solubility of such sparingly soluble salts is obtained by determining the specific conductivity (K) of a saturated salt solution. Since only a very small amount of salt is present in solution. The equivalent conductivity at such high dilution can practically be taken as λ_∞, *i.e.*,

For very sparingly soluble salts; $\lambda_r = \lambda_\infty = K.V$

where V is the volume in cm3 containing 1 g equivalent of salt.

The value of λ_∞ can be calculated with the help of Kohlrausch's law *i.e.*,

$$\lambda_\infty = \lambda_+ + \lambda_-$$

Thus, knowing the values of K and λ_∞, the value of V can be calculated:

$$V = \frac{\lambda_r}{K} \simeq \frac{\lambda_\infty}{K} = \frac{\lambda_+ + \lambda_-}{K}$$

But V cm^3 of saturated solution contains = 1 g eqt. of salt

$\therefore$ 1000 cm^3 (or litre) = solution contains = 1000/V g eqt. of salt

Hence, solubility of salt,

$$S = \frac{1000}{V} \text{ g equivalent/litre}$$

$$= \frac{1000\,E}{V} \text{ g/litre} = \frac{1000 \times E \times K}{(\lambda_+ + \lambda_-)} \text{ g/litre}$$

where E is the equivalent weight of salt.

APPLICATION OF CONDUCTIVITY MEASUREMENTS

(a) Basicity of Organic Acids : From an examination of the conductances of the sodium salts of a number of organic acids, Ostwald (1887) discovered the empirical relation

$$\lambda_{1024} - \lambda_{32} = 10.8b \quad ...(1)$$

where λ_{1024} and λ_{32} are equivalent conductances of the salt at 25°C at dilutions of 1024 and 32 litres per equivalent, respectively and b is the basicity of the acid.

Procedure : To prepare N/32 solution of the sodium salt of an acid, a N/16 NaOH solution is accurately prepared. A known volume of it, say 50 ml is neutralised by adding dropwise a concentrated solution of the acid, using an indicator paper. When the neutralisation is complete, the volume is made upto 100 ml. This gives N/32 solution of the salt and the N/1024 solution is prepared by a careful dilution of the solution.

Measure the resistance of N/32 and N/1024 solution of the taking all the usual precautions. Using cell constant, calculate the equivalent conductances at both the dilutions. Then, calculate the basicity of the acid using the formula given above.

Limitations

(i) This rule applies to some organic acids but does not succeed with inorganic acids,

(ii) This method fails when applied to very weak acids whose salts are considerably hydrolysed in solution.

(b) Degree of Hydrolysis : When the salt of a weak acid or base is dissolved in water, hydrolysis occurs, so that the conductance of the solution is now partly due to the ions of the salt and partly due to the ions of the salt and partly due the ions (H^+ or OH^-) of the acid or base formed by hydrolysis. An excess of the weak acid or base in the presence of its salts can be regarded as completely unionised. Therefore, addition of weak acid or base to the salt solution suppresses the hydrolysis but the ionisation will be unaffected.

Aniline hydrochloride ($C6H_5NH_3^+Cl^-$), the salt of a weak base and a strong acid, is hydrolysed in water giving $(1 - \alpha)$ equivalent of unhydrolysed salt and α equivalents of free acid and free base, where α is the degree of hydrolysis.

$$\underset{1-\alpha}{C_6H_5NH_3^+} \rightleftharpoons \underset{\alpha}{C_6H_5NH_2} + \underset{\alpha}{H^+}$$

The weak undissociated base aniline does not contribute to the total equivalent conductance λ of the solution, which may, therefore, be written as

$$\lambda = (1 - \alpha)\lambda_c + \alpha\lambda_{HCl}$$

or $$\alpha = \frac{\lambda - \lambda_c}{\lambda_{HCl} = \lambda_c} \qquad ...(2)$$

which λ_c and λ_{HCl} are the equivalent conductances of the unhydrolyzed salt and free acid at the same concentration. λ_c as has been pointed out, can be determined by measuring the equivalent conductance in the presence of excess aniline.

The value of λ_{HCl} at concentrations M/32, M/64, M/128 are known to be respectively 393, 399 and 401. Thus α being determined from eq. (2) the hydrolysis constant K_h is easily obtained from the following relation:

$$K_h = \frac{\alpha^2 C}{1 - \alpha}$$

(c) Solubility of a Sparingly Soluble Salt : This is based on the simple fact that the saturated solution of a sparingly soluble salt is so dilute that the electrolyte present there in may be regarded as completely ionised. The relationship between concentration C (in gm. equivalents per litre), the equivalent conductance λ, and the specific conductance K, is given by the equation

$$\lambda = K \times \frac{1000}{C} \qquad ...(3)$$

The solution being saturated, C represents the solubility (in gram equivalents per litre). Since the solution (though saturated with respect to the almost insoluble salt) is so dilute, it may be regarded as being practically ionised at infinite dilution. Hence λ may be put equal to λ_∞. This as we know, is equal to $\lambda_{(cation)} + \lambda_{(anion)}$, the figures for which are available from published tables. Hence the solubility of the sparingly soluble salt, C, is

$$C = \frac{100\,K}{\lambda_{(cation)} + \lambda_{(anion)}} \text{ gm equivalent/1000 ml} \qquad ...(4)$$

So, the only experimental work required is to determine the conductance of a saturated solution.

(d) Ionic Product of Water, K_w : This can be calculated from the specific conductance of pure water, as determined experimentally, and its equivalent conductance, as calculated from the λ_0 values of its two ions. The equivalent conductance at infinite dilution λ_∞ is defined by the equation

$$\lambda_\infty = 100 \frac{K_\infty}{C} \text{ mhos cm}^2 \qquad ...(5)$$

where K_∞ is the specific conductance at infinite dilution and C the concentration in gm. equivalent per litre. The most accurate experimental value of specific conductance of purest water so far available is 5.54×10^{-8} mho cm^{-1} at 25°C. This may be put equal to K as pure water may be regarded as a solution of ionised water at infinite dilution.

The values of equivalent conductances of the two ions of water at infinite dilution are:

$$\lambda_\infty(H^+) = 349.8 \text{ mhos cm}^2$$

$$\text{and } \lambda_\infty(OH^-) = 198.6 \text{ mhos cm}^2$$

Hence the equivalent conductance at infinite dilution, λ_∞, of water may be equal to 349.8 + 198.6 *i.e.* 548.4 mhos cm^2. Substituting the values of K_∞ and λ_∞ in equation (5), we get value of C for each of the ions of water:

$$[H^+] = \frac{1000 \times 5.54 \times 10^{-4}}{548.4} = 1.01 \times 10^{-7}$$

$$[OH^-] = \frac{1000 \times 5.54 \times 10^{-4}}{548.4} = 1.01 \times 10^{-7}$$

Hence the ionic product of water.

$$K_w = [H^+][OH^-]$$

$$= 1.01 \times 10^{-7} \times 1.01 \times 10^{-7} = 1.02 \times 10^{-14}$$

(e) Degree of Dissociation of Weak Electrolyte : The degree of dissociation of ionisation α, of a weak electrolyte is given by

$$\alpha = \frac{\lambda_v}{\lambda_\infty} = \frac{\text{Equivalent conductance at the given dilution}}{\text{Equivalent conductance at infinite dilution}}$$

λ_r for a weak electrolyte is determined experimentally, while the λ_∞ is calculated with the help of Kohlrausch's law.

(f) Transport Number : See transport number.

CONDUCTANCE IN NON-AQUEOUS SOLVENTS

Various substances (generally in liquid form) other than water which are used for dissolving the other compounds are known as *non-*

aqueous solvents. In earlier studies of measurement of conductance, water was generally utilised due to:

(a) its wide liquid range

(b) ease of handling

(c) its plentiful occurrence in nature

(d) its unusual solvating characteristics and

(e) its high dielectric constant

The work on the measurements of conductances or substances in non-aqueous solvents first appeared in 1892.

Later on a large number of papers appeared in 1897 when Cady started his studies on liquid ammonia as a solvent. The conductance of non-aqueous solutions depend upon the following two factors:

(a) Effect of Dielectric Constant : The dielectric constant of a solvent can affect the dissociating power. This effect was realised by Thomson and Nernst and is known as Thomson-Nernst rule. According to this rule,

"*The force exerted between the charged particles say e_1 and e_2 is inversely proportional to the dielectric constant (D) of the medium and to the square of the distance (d) between the particular.*"

$$F = \frac{1}{D}\frac{e_1e_2}{d^2}$$

From the above relation it follows that E is the amount of energy required to separate two ions of charges e_1 and e_1 which are d cm apart. If the dielectric of the solvent is large the energy required to separate two ions will be smaller and therefore, the ionisation can take place easily.

Thus, formic acid, hydrogen cyanide, nitromethane, ethylene glycol, methyl alcohol, etc., with a high value of dielectric constant should behave as good ionising solvents whereas the solvents with a low dielectric constant such as hydrocarbons and halogenated hydrocarbons should be poorest ionising solvents.

In actual practice, we observe that the above expectations are in agreement with experimental observations which may be seen from the following Table 1.2:

Table 2.2

Solvent	*Dielectric Constant*	*Degree of Ionisation in 2000 litres dilution* $(C_2H_5)_3N$ *Picrate*	$(C_3H_5)_1NH$ *Picrate*	$(C_2H_5).NH_2.HCl$
H_2O	81.1	0.99	0.97	0.97
CH_3OH	35.4	0.94	...	0.92
C_2H_5OH	25.4	0.89	...	0.88
CH_3CN	36.4	0.95	0.90	0.31
$(CH_3)_2CO$	20.7	0.88	0.96	0.10
C_2H_5N	12.4	0.74	0.53	0.04
$CHCl_3$	10.9	0.47	0.005	0.00

In media of low dielectric constant, one observes, that for dilute solutions, the value of λ_r diminishes (in the usual manner) with increasing concentration of the solute. However, a minimum value of λ_r is attained and again there occurs an increase in λ_r as concentration increases. The dielectric constant of a solvent depends upon its dipole moment. Fajan has proposed that the ionisation is large in those solvents which contain a large cation and a small anion.

(*b*) **Effect of Viscosity:** In 1906, Walden reported the quarternary ammonium salts like $(C_2H_5)_6NI$ can be dissolved in a number of solvents other than water. He also studied the conductance of solutions of those compounds and gave a relation connecting equivalent conductivity of the solution, viscosity and dielectric constant of the medium.

$$\lambda_\infty \eta = \text{constant} \quad ...(1)$$

where λ_∞ is the equivalent conductance of a solution in non-aqueous solvents and η is the viscosity of this solvent. Equation (1) is known as Walden's rule which may be stated in words as:

"*The product of limiting equivalent conductance of a given solute and viscosity of the pure solvent is constant.*"

The constant in equation (1) is independent of temperature and nature of the solvents and the value of $(C_2H_5)_4NI$ in more than 30 solvents at 0°C is 0.70, *i.e.*,

$$\lambda_\infty \eta = 0.70 \quad ...(2)$$

The above rule of Walden does not apply to water and glycol as solvents. For **water**, $\lambda_\infty \eta = 1$ and the value of constant was observed to vary with the nature of the solute. From equation (1) it follows that one can calculate the equivalent conductance of a solute at infinite dilution from the viscosity of the solvent and the value of constant which can be determined.

Limitations of Walden's Rule

(1) In general, Walden's rule is not universally valid. It is not applicable to highly associated solvents and also not to the solvents of high viscosity.

(2) The value of constant in Walden's rule equation (1) is found to vary with variation of temperature.

Another Walden's Relation

Walden also showed that there is a relation among $(\lambda_\infty - \lambda_r)$, viscosity and dielectric constant. His mathematical relation is

$$\lambda_\infty - \lambda_r = \frac{K_e}{D\eta} \qquad ...(3)$$

where K_e is constant which depends only on the nature of the solute and its concentration but not on the nature of the solvent. He reported that the value of K_c is approximately given by

$$K_e \sim 51.5\ C^{0.45} \qquad ...(4)$$

where C is the concentration of the solute in gm. equivalent per litre.

Walden also reported that equation (1) may be modified for strong electrolytes in various solvents such as

$$= \frac{41.5}{D} C^{1/2} \qquad ...(5)$$

Equation (5) can be used successfully to interpret the conductivity of dilute solution of strong electrolytes in non-aqeuous solvents of relatively high dielectric constants.

For example, CH_3OH(D = 30.3): C_2H_5OH (D = 24.1); $C_6H_5NO_2$ (D = 35.5); NH_3 liquid at – 40°C (D = 23).

Onsagar's Equation

This may be written in the form

$$\lambda_r = \lambda_\infty - (A + B\lambda_\infty \sqrt{C}) \qquad ...(6)$$

where A and B are constants dependent only on the nature of the solvent and the temperature. When the above equation was applied to non-aqueous solutions, it was observed to be in full agreement with theoretical requirements for the chlorides, and thiocynates of the alkali metals in methyl alcohol solution. Other electrolytes, such as nitrates, tetralkyl ammonium salts and salts of higher valency, however, exhibit appreciable deviations. These discrepancies become more pronounced for the solvents of law dielectric constants. In order to explain these deviations, equation (6) was modified to:

$$\lambda_r = \alpha[\lambda_\infty - (A + B\lambda_\infty)\sqrt{(\alpha C)}\,] \quad ...(7)$$

Equation (7) may be written as

$$\lambda_r = \alpha\lambda'_r \quad ...(8)$$

where $$= \lambda_\infty - (A + B\lambda_\infty\sqrt{(\alpha C)}\,) \quad ...(9)$$

From equation (9), it follows that is smaller than λ_r for solvents of low dielectric constants (non-aqueous solvents). From equation (8), it follows that for medium of low dielectric constant the degree of dissociation α is numerically equal to $\lambda_r/$ instead of λ_r/λ_∞ as proposed by Arrhenius.

SOLVED EXAMPLES

Example 1:

A cell, whose resistance when filled with 0.1 M-KCl solution is 192.3 ohms was found to be 6.306 ohms when filled with 3.186 × 10^{-3} M-NaCl solution at 25°C. Calculate :

(i) the cell constant,

(ii) the specific and equivalent conductivities of NaCl solution, and

(iii) the degree of dissociation of NaCl solution at this dilution. [Given: Sp. conductivity of 0.1 M-KCl is 0.01289 ohm^{-1} cm^{-1} at 25°C. Equivalent conductivity of solution and chloride ions at 25°C are respective 50.11 and 76.34 ohms^{-1} cm^2.

Solution:

Cell constant, (x) = Sp. Conductivity × Resistance

$$= 0.01289 \text{ ohm}^{-1}\text{ cm}^{-1} \times 192.3 \text{ ohm}$$

$$= 2.479\ cm^{-1}.$$

Sp. conductivity of 3.186×10^{-3} M-NaCl solution (K)

$$= \text{Cell Constant/Resistance}$$

$$K = \frac{2.479\ cm^{-1}}{6.306\ ohm} = 3.931 \times 10^{-4} ohm^{-1} cm^{-1}.$$

But for 3.186×10^{-3} M-NaCl solution,

$$V = \frac{1000}{3.186 \times 10^{-3}}\ cm^2$$

$$\therefore \quad \lambda_r = K \times V$$

$$= \frac{3.931 \times 10^{-4} \times 1000}{3.186 \times 10^{-3}}$$

$$= 123.4\ ohm^{-1}\ cm^2\ (g\ eqt)^{-1},$$

Now $\quad \lambda_\infty NaCl = \lambda_{Na+} + \lambda_{Cl-} = 50.11 + 76.34$

$$= 126.45\ ohm^{-1}\ cm^2\ (g\ eqt)^{-1}$$

$\therefore$ Degree of dissociation,

$$\alpha = \frac{\lambda_r}{\lambda_\infty} = \frac{123.4}{126.45} = 0.976 \text{ or } 97.6\%.$$

Example 2:

At 25°C, the specific conductivity of pure water is 5.5×10^{-8} mho cm^{-1}. The ionic conductance of H^+ and OH^- ions at this temperature are 349.8 and 198.5 mhos cm^2 respectively. Calculate the value of the ionisation constant of water.

Solution:

$$\lambda_{\infty\ H2O} = \lambda_{H+} + \lambda_{OH-}$$

$$= 349.8 + 198.5$$

$$= 548.3\ mho\ cm^2$$

But conc. of water = 18 g/18 cm^3 = 1 mole/18 cm^3

or $\quad V = 18\ cm^3$

$$\therefore \quad \lambda_r = K \times V$$

$$= 5.5 \times 10^{-8} \times 18$$

$$= 9.9 \times 10^{-7}\ ohm^{-1}\ cm^2\ eqt^{-1}$$

$$\therefore \qquad \alpha = \frac{\lambda_r}{\lambda_\infty}$$

$$= \frac{9.9 \times 10^{-7}}{548.3} \simeq 1.8 \times 10^{-9}$$

It dissociation of water were complete, then concentration of H^+ or OH^- ions would have been 1000/18 g-ion per litre.

But actual degree of dissociation is 1.8×10^{-9}, then

$$[H^+] = [OH^-] = \frac{1000}{18} \times 1.8 \times 10^{-9}$$

$$= 10^{-7} \text{ g-ion/litre}$$

Also $\quad [H_2O] = 1000/18$ g-mole/litre,

$$\therefore \qquad K = \frac{[H^+][OH^-]}{[H_2O]}$$

$$= \frac{1 \times 10^{-7} \times 1 \times 10^{-7}}{1000/18}$$

$$= 1.8 \times 10^{-16}.$$

Example 3:

The molar conductivity at infinite dilution of $AgNO_3$, NaCl and $NaNO_3$ is 116.5, 110.3 and 105.2 ohm^{-1} cm^2 mol^{-1} respectively. The conductivity at AgCl in water is 2.4×10^{-6} ohm^{-1} cm^{-1} and water used in experiment is 1.16×10^{-6} ohm^{-1} cm^{-1}. Find out the solubility of AgCl in g dm^{-3}.

Solution:

$$\lambda_\infty = \frac{1000\, K_{salt}}{C}$$

Here we have

$$(\lambda_\infty)_{AgCl} = (\lambda_\infty)_{AgNO3} + (\lambda_\infty)_{NaCl} - (\lambda_\infty)_{NaNO3}$$

$$= 116.5 + 110.3 - 105.2 = 121.6$$

$$K_{salt} = K_{solution} - KH_2O$$

$$= (2.4 \times 10^{-6} - 1.16 \times 10^{-6})\ ohm^{-1}\ cm^{-1}$$

$$= 1.24 \times 10^{-6}\ ohm^{-1}\ cm^{-1}$$

Substituting these values in the above equation

$$121.6 = \frac{1000 \times 1.24 \times 10^{-6}}{C}$$

or
$$C = \frac{1000 \times 1.24 \times 10^{-6}}{121.6}$$
$$= 1.019 \times 10^{-5} \text{ equivalent per dm}^3$$
$$= (1.019 \times 10^{-3} \times 143.5) \text{ g per dm}^3$$
$$= 1.4622 \times 10^{-3} \text{ g dm}^{-3}.$$

Example 4:

The ionic conductance of Na^+ is 50.1 ohm^{-1} cm^2 mol^{-1}. Calculate the ionic mobility of Na^+.

Solution:

Since,

$$\text{Ionic mobility} = \frac{\text{Ionic conductance}}{96500}$$
$$= \frac{(50.1 \text{ ohm}^{-1}\text{cm}^2\text{mol}^{-1})}{(96500 \text{ coulombs mol}^{-1})}$$
$$= 5.1917 \times 10^{-4} \frac{\text{cm}^2}{\text{ohm coulomb}}$$
$$= 5.1917 \times 10^{-4} \frac{\text{cm}^2}{\text{ohm } ampere}$$
$$= 5.1917 \times 10^{-4} \text{ cm}^2 \text{ V}^{-1} \text{ s}^{-1}$$
$$= 5.1917 \times 10^{-84} \text{ m}^{-2} \text{ V}^{-1} \text{ s}^{-1}$$

Example 5:

At 291 K the ionic velocities of Ag^+ is 0.00057 cm s^{-1} and that of NO_3^- is 0.00063 cm s^{-1}. What is the value of λ_∞ for $AgNO_3$ at 291 K? If the specific conductivity of 0.1 N $AgNO_3$ solution is 0.00947 ohm^{-1} cm^{-1}, find the degree of dissociation at this dilution.

Solution:

$$\text{Absolute ionic mobility} = \frac{\text{Ionic conductance}}{96500}$$

$\therefore$
$$\lambda_{Ag+} = 0.00057 \times 97500$$
$$\lambda_{NO3-} = 0.00063 \times 96500$$

$$\therefore \quad \lambda_{Ag+} + \lambda_{NO3-} = 96500(0.00057 + 0.00063)$$

$$= 96500 \times 0.0012 = 115.8$$

$$\lambda_c = \frac{100 \times K}{C} = \frac{1000 \times 0.00947}{0.1} = 94.7$$

$$\therefore \quad \alpha = \frac{\lambda_c}{\lambda_\infty} = \frac{94.7}{115.8} = 0.817.$$

Example 6:

When a conductance cell was filled with a 0.02 mol dm^{-2} aqueous solution of KCl, which had a specific conductance of 0.2767 S m^{-1}, it had a resistance of 82.39 ohm at room temperature. When the same cell was filled with a 2.5 × 10^{-3} mol dm^{-3} solution of K_2SO_4, its resistance was 325 ohm. Calculate the cell constant and the specific conductance of K_2SO_4 solution.

Solution:

For KCl, $c = 0.02 \text{ mol dm}^{-3} = 0.02 \times 10^3 \text{ mol m}^{-3}$

$R = 82.39$ ohm;

$K = 0.2767 \text{ S m}^{-1}$

$= 0.2767 \text{ ohm}^{-1} \text{ m}^{-1}$

Since K = cell constant/R, we have

Cell constant $= KR$

$= (0.2767 \text{ ohm}^{-1} \text{ m}^{-1})(82.39 \text{ ohm})$

$= 22.80 \text{ m}^{-1} = 0.2280 \text{ cm}^{-1}$

For K_2SO_4 solution,

K = Cell constant/R

$$= \frac{22.80 \text{ m}^{-1}}{325 \text{ ohm}}$$

$= 7.01 \times 10^{-2} \text{ ohm}^{-1} \text{ m}^{-1}$

$= 7.01 \times 10^{-2} \text{ S m}^{-1}$

Example 7:

A solution of copper (II) sulphate is electrolysed between copper electrodes by a current of 10.0 ampere for 60 minutes. Wheat changes take place at the electrodes and in the solution?

Solution:

The electrode reactions are as follows:

At anode $Cu \rightarrow Cu^{2+} + 2e$

At cathode $Cu^{2+} + 2e \rightarrow Cu$

$$\text{Number of moles of electrons} = \frac{(10.0\ \text{A})(3600\ \text{s})}{(96{,}500\ \text{coulomb / mole})} = 0.373\ \text{F}$$

As it take 2 moles of electrons to react with each mole of copper (I) or copper metal the number of moles of copper dissolved or deposited is 0.865, *i.e.*,

$$= (0.373\ \text{mole e})\ \frac{(1\ \text{mole Cu})}{(2\ \text{moles e})} = 0.1865\ \text{mole Cu}$$

Thus, 0.1865 mole of copper is dissolved from anode, 0.1865 mole of copper is deposited at the cathode, and the original copper (II) ion concentration of the solution remains unchanged.

Example 8:

When a constant current was passed through a solution of $AuCl_4^-$ *ions between gold electrodes for a period of 1.0 minute, the cathode increased in weight by 0.1314 g. How much charge was passed? What was the current, I?*

Solution:

At the cathode reduction of gold (III) to gold metal takes place.

$$AuCl_4^- + 3e \longrightarrow Au + 4Cl^-$$

No. of moles of Au

$$= \frac{0.1314\ \text{g}}{197\ \text{g mol}^{-1}} = 0.667 \times 10^{-3}\ \text{mol}$$

$$Q = (0.667 \times 10^{-3}\ \text{mol Au})\ \frac{(3\ \text{mole e})}{(\text{Au mol})}$$

$$= 2.001 \times 10^{-3}\ \text{F}$$

$$I = \frac{Q}{t} = \frac{(2.001 \times 10^{-3}\ \text{faraday})(96500\ \text{coulomb / faraday})}{60\ \text{seconds}}$$

$$= 3.22\ \text{A}.$$

3

Thermodynamics Relations Based on Second Law

THERMODYNAMIC EQUATION OF STATE

Any relationship between the thermodynamic function (E or H) and the PVT data is called the thermodynamic equation of state.

(i) **Energy as a function of T and V.** For any reversible process involving only P.V work, the combined form of the first and second laws can be written as

$$TdS = dq_{rev} = dE + PdV \qquad ...(1)$$

E is a state function and dE is exact differential, given as

$$dE = \left(\frac{\partial E}{\partial T}\right)_V dT + \left(\frac{\partial E}{\partial V}\right)_T dV = C_V dT + \left(\frac{\partial E}{\partial T}\right)_V dT + \left(\frac{\partial E}{\partial V}\right)_T dV$$

$C_V = (\partial E/\partial T)_V$ = heat capacity at constant volume

$$\therefore\ dS = \frac{C_V}{T} dT + \frac{1}{T}\left[P + \left(\frac{\partial E}{\partial V}\right)_T\right] dV \qquad ...(2)$$

Entropy is also a state function S = S (T, V) and dS is exact differential given as,

$$dS = \left(\frac{\partial S}{\partial V}\right)_V dT + \left(\frac{\partial S}{\partial V}\right)_T dV \qquad ...(3)$$

Comparing the coefficient of dT and dV separately of equations (2) and (3), we get

$$\left(\frac{\partial S}{\partial T}\right)_V = \frac{C_V}{T} = \frac{1}{T}\left(\frac{\partial E}{\partial T}\right)_V \qquad ...(4)$$

$$\left(\frac{\partial S}{\partial V}\right)_V = \frac{1}{T}\left[P + \left(\frac{\partial E}{\partial V}\right)_T\right] \qquad ...(5)$$

Differentiating equation (4) with respect to V at constant T, we get

$$\left(\frac{\partial^2 S}{\partial V \partial T}\right) = \frac{\partial}{\partial V}\left(\frac{C_V}{T}\right)_T = \frac{1}{T}\frac{\partial}{\partial V}\left[\left(\frac{\partial E}{\partial T}\right)_V\right]_T \qquad ...(6)$$

On differentiating equation (5) with respect to T at constant volume, we get

$$\frac{\partial^2 S}{\partial T \partial V} = \frac{\partial}{\partial T}\left[\frac{1}{T}\left\{P + \left(\frac{\partial E}{\partial V}\right)_T\right\}\right]_V$$

$$= -\frac{1}{T^2}\left(\frac{\partial E}{\partial V}\right)_T + \frac{1}{T}\left(\frac{\partial P}{\partial T}\right)_V + \frac{1}{T}\frac{\partial^2 E}{\partial T \partial V} \qquad ...(7)$$

Since S is a state function and dS is exact differential, therefore, Euler's theorem of exactness must be valid i.e.,

$$\frac{\partial^2 S}{\partial V \partial T} = \frac{\partial^2 S}{\partial T \partial V}$$

On applying this condition, equations (7) and (6) give

$$\frac{1}{T}\frac{\partial^2 E}{\partial V \partial T} = -\frac{P}{T^2} - \frac{1}{T^2}\left(\frac{\partial E}{\partial V}\right)_T + \frac{1}{T}\left(\frac{\partial P}{\partial T}\right) + \frac{1}{T}\frac{\partial^2 E}{\partial T \partial V} \qquad ...(8)$$

Since dE is exact differential

$$\therefore \quad \frac{\partial^2 E}{\partial V \partial T} = \frac{\partial^2 E}{\partial T \partial V}$$

Applying this equality in equation (8), we get

$$\left(\frac{\partial E}{\partial V}\right)_T = T\left(\frac{\partial P}{\partial T}\right)_V - P \qquad ...(9)$$

This expression is known as the thermodynamic equation of state and is true for all substances under all conditions and follows directly from the acceptance of the validity of the first and second laws. Equation (9) can be written in another form if we realise that $(\partial P/\partial T)_V = \alpha/\beta$, where $\alpha = V^{-1}(\partial V/\partial T)_P$, and $\beta = -V^{-1}(\partial V/\partial P)_T$.

$$\left(\frac{\partial E}{\partial V}\right)_T = T\frac{\alpha}{\beta} - P$$

Let us apply it to an ideal gas and then to a van der Waals gas.

Thermodynamic equation of state applied to an ideal gas. We have argued that the energy of an ideal gas is independent of volume and have defined an ideal gas as one obeying the two relationships PV = nRT and $(\partial E/\partial V)_T = 0$. With the assistance of the thermodynamic equation of state we can now prove the latter condition.

For an ideal gas PV = nRT

$$\therefore \left(\frac{\partial P}{\partial T}\right)_V = \frac{nR}{V} = \frac{P}{T}$$

On substituting the expression $(\partial P/\partial T)V = PT$ into equation (10) we get,

$$\left(\frac{\partial E}{\partial V}\right)_T = T\left(\frac{\partial P}{\partial T}\right)_V - P = T\frac{P}{T} - P = 0 \qquad ...(10)$$

Van der Waals equation. It is given as

$$P = \frac{nRT}{V-nb} - \frac{an^2}{V^2}$$

where a is the attraction coefficient and b is the excluded volume per mole; a measure of bulk of the gas molecules. Using the thermodynamic equation of state we can deduce an expression for the internal pressure $(\partial E/\partial V)_T$ for the gas.

For van der Waals equation

$$\left(\frac{\partial P}{\partial T}\right)_V = \frac{nR}{V-nb}$$

On substituting this expression into thermodynamic equation of state we obtain

$$\left(\frac{\partial E}{\partial V}\right)_T = \frac{nRT}{V-nb} - P = \frac{nRT}{V-nb}\left(\frac{nRT}{V-nb} - \frac{an^2}{V^2}\right)$$

$$\left(\frac{\partial E}{\partial V}\right)_T = \frac{an^2}{V^2} \qquad ...(11)$$

For one mole of a gas, equation (11) reduces to

$$\left(\frac{\partial E}{\partial V}\right)_T = \frac{a}{V^2}$$

or $\quad dE = \frac{a}{V^2} dV$

Integration gives $\Delta E = -a \left(\frac{1}{V^2} - \frac{1}{V_1}\right)$

The energy of a van der Waals gas is related to the attraction coefficient a. The energy increase with an isothermal expansion of the gas. When the molecules are close to each other, the volume is smaller, the intermolecular forces reduce the energy of the gas.

(ii) **Enthalpy as function of temperature and pressure.**

The enthalpy is defined as

$H = E + PV$

$\therefore dH = dE + PdV + VdP$

The combined first and second laws give $dE + PdV = TdS$, therefore

$$TdS = dH - VdP \qquad ...(12)$$

H being a state function H = H (T, P) its total differential is given by

$$dH = \left(\frac{\partial H}{\partial T}\right)_P dT + \left(\frac{\partial H}{\partial P}\right)_T dP = C_p dT + \left(\frac{\partial H}{\partial T}\right)_P dP \qquad ...(13)$$

Here $C_p = (\partial H/\partial T)_P$

Putting this expression into equation (12), we get

$$dS = \frac{C_P}{T} dT + \frac{1}{T}\left[\left(\frac{\partial H}{\partial P}\right)_T - V\right] dP \qquad ...(14)$$

Since S = f (T, P)

$$\therefore dS = \left(\frac{\partial S}{\partial T}\right)_P dT + \left(\frac{\partial S}{\partial P}\right)_T dP \qquad ...(15)$$

Comparing the coefficients of dT and dP separately in equations (28) and (15), we get

$$\left(\frac{\partial S}{\partial T}\right)_P = \frac{C_P}{T} = \frac{1}{T}\left(\frac{\partial H}{\partial T}\right)_P \qquad ...(16)$$

$$\left(\frac{\partial S}{\partial P}\right)_T = z\frac{1}{T}\left[\left(\frac{\partial H}{\partial P}\right)_T - V\right] \qquad ...(17)$$

On differentiating equation (16) with respect to P at constant T, we get

$$\frac{\partial^2 S}{\partial P \partial T} = \frac{1}{T}\frac{\partial^2 H}{\partial P \partial T} \qquad ...(18)$$

On differentiation w.r. to T at constant P equation (17) yields

$$\frac{\partial^2 S}{\partial T \partial P} = -\frac{1}{T^2}\left[\left(\frac{\partial H}{\partial P}\right)_T - V\right] + \frac{1}{T}\left[\frac{\partial^2 H}{\partial T \partial P} - \left(\frac{\partial V}{\partial T}\right)_P\right] \qquad ...(19)$$

Since S and H and state functions, therefore,

$$\frac{\partial^2 S}{\partial T \partial P} = -\frac{\partial^2 S}{\partial T \partial P}, \text{and } \frac{\partial^2 H}{\partial T \partial P} = \frac{\partial^2 H}{\partial P \partial T}$$

Applying this condition, equations (18) and (19) yield

$$\left(\frac{\partial H}{\partial P}\right)_T = V - T\left(\frac{\partial V}{\partial T}\right)_P \qquad ...(20)$$

The expression (20) is thermodynamic equation of state.

For ideal gas PV = nRT

$$\therefore \left(\frac{\partial V}{\partial T}\right)_P = \frac{nR}{P} = \frac{V}{T}$$

From equation (20), we have

$$\left(\frac{\partial H}{\partial P}\right)_T = V - T\left(\frac{\partial V}{\partial T}\right)_P = V - T\frac{V}{T} = 0$$

Now we shall consider a few more applications of the results of the results of this section to some simple cases:

(a) **Relation between C_P and C_V**

We developed expressions for the difference between the heat capacities at constant pressure and at constant volume, as

$$C_P - C_V = \left[\left(\frac{\partial E}{\partial V}\right)_T + P\right]\left(\frac{\partial V}{\partial T}\right)_P \qquad ...(21)$$

But $\left(\frac{\partial E}{\partial V}\right)_T = T\left(\frac{\partial P}{\partial T}\right)_V - P$ ...(22)

$$\therefore \qquad C_P = C_V = T\left(\frac{\partial P}{\partial T}\right)_V\left(\frac{\partial V}{\partial T}\right)_P \qquad ...(23)$$

But $(\partial V/\partial T)_P = \alpha V$, where α is the coefficient of thermal expansion. Therefore, equation (23) may be written as,

$$C_P - C_V = TV\,\alpha\left(\frac{\partial P}{\partial T}\right)_V \qquad ...(24)$$

V is a function of T and P,

$V = f(T, P)$

$$\therefore\ dV = \left(\frac{\partial V}{\partial T}\right)_P dT + \left(\frac{\partial V}{\partial P}\right)_T dP$$

At constant volume, this equation reduces to

$$\left(\frac{\partial V}{\partial T}\right)_P (\partial T)_V + \left(\frac{\partial V}{\partial P}\right)_T (\partial P)_V = 0$$

$$\therefore\ \left(\frac{\partial P}{\partial T}\right)_V = \frac{(\partial V/\partial T)_P}{(\partial V/\partial P)_T} = \frac{\alpha}{\beta} \qquad ...(25)$$

where β is the compressibility coefficient defined as $b = -V^{-1}(\partial V/\partial P)_T$, and $a = V^{-1}(\partial V/\partial T)_P$

Using equations (24) and (25), We get

$$C_P - C_V = VT\frac{\alpha^2}{\beta}$$

This equation is equally true for solids, liquids and gases.

For water the density is maximum at 3.98° C, and $C_P = C_V$. This can be shown with the help of equation (23) and the definition of density (ρ). The density is defined as mass (m) per unit volume:

$\rho = m/V$,

$$\text{hence} \quad \frac{d\rho}{dT} = \frac{d}{dT}(m/V) = -\frac{m}{V^2}(dV/dT)$$

For maximum, the derivative $d\rho/dT$ must be zero, so should be its right side. Since m and V^2 are not zero, therefore dV/dT should be zero. Equation (23) gives $C_P - C_V = 0$, and $C_P = C_V$.

For an ideal gas $\alpha = T^{-1}$, $b = P^{-1}$, therefore $C_P - PV/T = nR$.

(b) **Joule-Thomson coefficient for a van der Waals gas**

Joule-Thomson coefficient is the pressure dependence of temperature under isoenthalpic conditions, and is given as

$$\mu\,\pi = \left(\frac{\partial T}{\partial P}\right)_H = -\frac{1}{C_P}\left(\frac{\partial H}{\partial P}\right)_T \qquad ...(26)$$

From equations (22) and (26), we get,

$$\mu\pi = \frac{1}{C_P}\left[T\left(\frac{\partial V}{\partial T}\right)_P - V\right] = \frac{V}{C_P}(\alpha T - 1)$$

where $\alpha V = (\partial V/\partial T)_P$

The temperature, at which $\mu\pi$ is zero, is called the inversion temperature given by

$T\pi = V(\partial V/\partial T)_P = 1/\alpha$

For one mole, van der Waal's equation of state is written as

$$PV = RT + Pb - \frac{a(V-b)}{V^2}$$

If $V - b = V$ and $V \cong \frac{RT}{P}$ then

$$PV = RT + Pb - \frac{Pa}{RT}$$

and $$V = \frac{RT}{P} + b - \frac{a}{RT}$$

$$\therefore \left(\frac{\partial V}{\partial T}\right)_P = \frac{R}{P} + \frac{a}{RT^2}$$

On substituting the expression for $(\partial V/\partial T)_P$ into (23) we get,

$$\left(\frac{\partial H}{\partial P}\right)_T = V - T\left(\frac{\partial V}{\partial T}\right)_P = V - \frac{RT}{P} - \frac{a}{RT}$$

$$= \left(\frac{RT}{P} + b - \frac{a}{RT}\right) - \frac{RT}{P} - \frac{a}{RT} = b - \frac{2a}{RT}$$

But $(\partial H/\partial P)_T = -C_P\mu\pi$

$$\therefore \mu\pi = \frac{1}{C_P}\left(\frac{2a}{RT} - b\right)$$

For very low temperatures the first term will dominate and the coefficient will be positive, while for higher temperatures the coefficien will be negative. The *inversion temperature* will be $T\pi = \frac{2a}{Rb}$ when mp is zero.

(c) **Variation of C_V with Volume at Constant T**

$$C_V = \left(\frac{\partial E}{\partial T}\right)_V$$

Differentiating with respect to V at constant T, we get

$$\left(\frac{\partial C_V}{\partial V}\right)_T = \frac{\partial}{\partial V}\left[\left(\frac{\partial E}{\partial T}\right)_V\right] = \frac{\partial}{\partial T}\left[\left(\frac{\partial E}{\partial V}\right)_T\right]_V$$

But $\left(\frac{\partial E}{\partial V}\right)_T = T\left(\frac{\partial P}{\partial T}\right)_V - P$

$$\therefore \quad \left(\frac{\partial C_V}{\partial V}\right)_T = \frac{\partial}{\partial T}\left[T\left(\frac{\partial P}{\partial T}\right)_V - P\right]_V$$

$$= T\frac{\partial^2 P}{\partial T^2} + \left(\frac{\partial P}{\partial T}\right)_V - \left(\frac{\partial P}{\partial T}\right)_V = T\left(\frac{\partial^2 P}{\partial T^2}\right)_V$$

For an ideal gas PV = nRT and $(\partial P/\partial T)_V$ = nR/V, therefore

$\left(\frac{\partial^2 P}{\partial T^2}\right)_V = 0$, and hence

$\left(\frac{\partial C_V}{\partial V}\right)_T = 0$ for an ideal gas.

For a van der Waals gas $P = \frac{nRT}{V-nb} - \frac{an^2}{V^2}$, and $(\partial P/\partial T)_V = \frac{nR}{V-nb}$

therefore

$\left(\frac{\partial^2 P}{\partial T^2}\right)_V = 0$, and hence

$\left(\frac{\partial C_V}{\partial V}\right)_T = T\left(\frac{\partial^2 P}{\partial T^2}\right)_V = 0$ for a van der Waal's gas.

(d) For **free expansion** of a gas the Joule coefficient μ_J is defined by $\mu_J = (\partial T/\partial V)_E$. With the help of thermodynamic equation of state we can correlate μ_J with some measurable physical properties of a gas. E is a state function and hence

$$dE = \left(\frac{\partial E}{\partial T}\right)_V dT + \left(\frac{\partial E}{\partial V}\right)_T dV = C_V dT + \left(\frac{\partial E}{\partial V}\right)_T dV$$

For constant energy process dE = 0, and

$$C_V\,(\partial T)_E = -\left(\frac{\partial E}{\partial V}\right)_T (\partial V)_E$$

$$\text{or}\quad \left(\frac{\partial T}{\partial V}\right)_E = -\frac{1}{C_V}\left(\frac{\partial E}{\partial V}\right)_T$$

But $(\partial T/\partial V)_E = \mu_J$ and $(\partial E/\partial V)_T = T\,(\partial P/\partial T)_V - P$

$$\therefore\ \mu_J = -\frac{1}{C_V}\left[T\left(\frac{\partial P}{\partial T}\right)_V - P\right]$$

Since $(\partial P/\partial T)_V = \alpha/\beta$

$$= \mu_J = \frac{P - \alpha T/\beta}{C_V}$$

For an ideal gas $\alpha/\beta = P/T$, therefore $\mu_J = 0$. That is, the free expansion of an ideal gas is isothermal.

THE LAWS OF THERMODYNAMICS

The success of thermodynamics in various fields of science is based on the wide application of the laws of thermodynamics which are as well established, as widely accepted and as free from experimental anomalies as any known scientific laws.

First Law dE = dq + dw

The change in internal energy (dE) in any system is the sum of the heat entering the system (dq) and the work done on the system (dw). The internal energy of an isolated system is constant and dq = 0, dw = 0, dE = 0.

Second Law dS = dq_{rev}/T

The change in entropy (dS) during any change in a system is equal to the heat entering the system (when the change is performed reversibly) divided by the absolute temperature T. In all actual processes the entropy of an isolated system always increases ($dS > 0$).

Third Law

The entropy at absolute zero (0K) is zero for any pure crystalline substance *i.e.*, S^0 (0K) = 0.

Zeroth Law

This law states that if two systems A and B are thermal equilibrium with a third system C then A and B will also be in thermal equilibrium when placed together. This law like other laws is based on human experiences. We can explain this law further as follows. Suppose we have three systems A, B and C placed together. A and C are in contact so that the properties of both change and finally the thermal equilibrium is attained and two come at the same temperature; C is in contact with B also then again thermal equilibrium is attained. If this is so then A and B should also be in thermal equilibrium. That is A, B and C should be at the same temperature.

If A (T_A) *** C (T_C) B (T_B),

then A (T_A) *** B (T_B), or if $T_A = T_C = T_B$, then $T_A = T_B$.

Thus the law provides a rational definition of temperature.

In thermodynamic studies the temperature's unit is taken as kelvin (K). The kelvin is strictly defined as the fraction 1/273.16 of the temperature interval between the absolute zero (– 273.16°C) and the triple point of water. In practice the temperature in kelvin is obtained by adding 273.15 to the temperature in degree celcius, e.g.,

25°C = 25 + 273. 15 = 298.15 K.

The use of the symbol °K is not recommended. The symbol of temperature in kelvin scale is T.

$= 1.56 \times 10^{-3}\ dm^3\ mol^{-1}$

CALCULATION OF ENTROPY CHANGES

In calculation of entropy changes we must take into consideration the entropy of system and the surroundings. The combination of system and surroundings correspond to an isolated system. The surroundings are usually considered to be the (i) heat reservoir that can add to, or subtract heat from the system, or (ii) mechanical device which can do work on, or accept work from the system. Thus the total entropy change of the isolated system ($\Delta S_{universe}$) is taken as

$$\Delta S_{universe} = \Delta S_{system} + \Delta S_{surroundings}$$

Entropy Change of System (ΔS_{system})

Entropy change of the system in a reversible isothermal change can be calculated by using the relation $dS = dq_{rev}/T$. But this equation cannot be used as such to calculate the entropy of irreversible changes. For this we have to devise the ways to carry the change in reversible ways. This is valid because the entropy is a state function and the changes in entropy of the system are the same irrespective the way the change is brought about.

Entropy Change of the Surroundings ($\Delta S_{surroundings}$)

The entropy changes for the surroundings can be calculated if we know the gain or loss in the heat of the surroundings. For this case, the equation $dS = dq/T$ can be used to calculate $\Delta S_{surroundings}$ because the surroundings are so large that the removal or addition of heat can be taken as reversible processes, as it will not alter the properties of the surroundings to any greater extent. [For analogy we take the case that removal or addition of a few drops of water will not change the water content of ocean, (Surroundings)].

$$\Delta S_{surroundings} = -\frac{q_{system}}{T} = \frac{q_{surroundings}}{T}$$

Now we shall consider the calculation of entropy change for some processes.

Phase changes. As an example of phase changes we may consider the melting of a solid. At melting point, the liquid and solid exist together in equilibrium. The addition of a little heat would melt some of the solid, the removal of a little heat would solidify some of the liquid, but the solid liquid equilibrium will be maintained. This process can be considered as reversible and entropy change can be given by

$$\Delta S_f = \int_1^2 \frac{dq_{rev}}{T_f}$$

Since the temperature is constant during the process, it may be taken outside the integral, giving

$$\Delta S_f = \frac{1}{T_f}\int dq_{rev}$$

The integral is the total heat absorbed which for this constant pressure process is the enthalpy of fusion, so that

$$\Delta S_f = \frac{\Delta H_f}{T_f} \quad ...(1)$$

In general, for any phase transformation taking place at constant temperature and pressure, e.g. [solid $\rightleftharpoons$ liquid (melting); liquid $\rightleftharpoons$ vapour (vaporization); solid $\rightleftharpoons$ vapour (sublimation); solid (α) $\rightleftharpoons$ solid (β) (transition) it must be true that

$$\Delta S_t = \frac{\Delta H_t}{T_t} \quad ...(2)$$

where ΔH_t is the enthalpy change of the transformation and T_t is the constant temperature at which the transformation takes place.

Entropy Change for Reversible Isothermal Expansion of Ideal Gases

Let the volume of n moles of an ideal gas be changed from V_1 to V_2 at constant temperature. Since for this isothermal expansion, the internal energy does not change, the heat absorbed can be calculated from the first law expression as

$$dq_{rev} = -dw_{rev} = PdV = \frac{nRT}{V}dV$$

Therefore, for this process the entropy change of the system can be evaluated as

$$\Delta S = \int_{V_1}^{V_2} \frac{dq_{rev}}{T} = \int_{V_1}^{V_2} nR\left(\frac{dV}{V}\right) = nR\ln\left(\frac{V_2}{V_1}\right) \quad ...(3)$$

Since at constant temperature $P_1/P_2 = V_2/V_1$, we can also write for an isothermal expansion from pressure P_1 to P_2

$$\Delta S = -nR\ln\left(\frac{P_2}{P_1}\right)$$

$$= -nR \times 2.303\log\left(\frac{P_2}{P_1}\right) \quad ...(4)$$

Entropy Change on Heating or Cooling of a Substance

Let us consider n moles of a substance at a temperature T_1. Let its temperature be change to T_2 keeping the volume constant. If this change is brought about reversibly then we can use the defining equation of entropy. Heat change for this process is

$$dq_{rev} = C_V dT$$

where C_V is the heat capacity at constant volume.

Entropy change is

$$\Delta S = \int_{T_1}^{T_2} \frac{dq_{rev}}{T} = \int_{T_1}^{T_2} \frac{C_V dT}{T}$$

If C_V does not vary with temperature, then

$$\Delta S = C_V \ In \left(\frac{T_2}{T_1}\right) = nC_{V,m} \ In \left(\frac{T_2}{T_1}\right) \qquad ...(5)$$

For temperature changes from T_1 to T_2 at constant P,

$$\Delta S = \int_{T_1}^{T_2} C_P \left(\frac{dT}{T}\right) = nC_{P,m} \ In \left(\frac{T_2}{T_1}\right) \qquad ...(6)$$

if the heat capacity at constant pressure (C_P) is taken as independent of temperature. Here $nC_{P,m} = C_P$.

Reversible Adiabatic Changes

If any system undergoes a reversible adiabatic change, then by definition, $dq_{rev} = 0$ at all stages of the process, so that

$$\Delta S = q_{rev}/T = 0$$

Therefore, reversible adiabatic processes are often called *isentropic* processes. However, in the case of irreversible adiabatic expansion though qirr is zero but ΔS is not zero, because $\Delta S \neq q_{irr}/T$.

The entropy change in a system of an ideal gas being heated or cooled from T_1 to T_2 and expanding from V_1 to V_2, is given by

$$\Delta S = C_V \ In \frac{T_2}{T_1} + nR \ In \frac{V_2}{V_1}$$

This equation is valid for all types of changes which an ideal gas is permitted to undergo. For adiabatic reversible change.

$$\left(\frac{T_2}{T_1}\right) = \left(\frac{V_1}{V_2}\right)^{nR/CV} \quad \text{for an ideal gas}$$

$$\therefore \ \ln\frac{T_2}{T_1} = \frac{nR}{C_V}\ln\left(\frac{V_1}{V_2}\right) = \frac{nR}{C_V}\ln\left(\frac{V_2}{V_1}\right)$$

$$\text{Hence } \Delta S = -\frac{nR}{C_V}.C_V\ln\frac{V_2}{V_1} + nR.\ln\frac{V_2}{V_1} = 0$$

That is, any increase in entropy due to expansion is exactly balanced by an equal decrease in entropy due to lowered temperature in an adiabatic reversible expansion of an ideal gas.

For adiabatic irreversible change.

$$\frac{T_2}{T_1} = \frac{C_V + (P_2/P_1)nR}{C_P}$$

Therefore, ΔS for irreversible adiabatic change is not zero, this is so because the final temperature (T_2) in adiabatic irreversible change and adiabatic reversible change are different.

ENTROPY CHANGE FOR SIMULTANEOUS HEATING AND EXPANSION OF AN IDEAL GAS

(a) **Temperature and volume are changed**. Let us consider n moles of an ideal gas initially at temperature T_1 and volume V_1. Let its temperature and volume be changed to a final value of T_2 and V_2 respectively. The entropy change can be calculated by devising the following reversible paths for the process.

Path A

Step 1. Expand the gas isothermally at temperature T_1 from volume V_1 to V_2. In this case internal energy change would be zero and the first law will give

$$dq_{rev} = -dw = PdV = \frac{nRT_1}{V}dV$$

$$\therefore \ \Delta S_1 = \int_{V_1}^{V_2}\frac{dq_{rev}}{T} = \int_{V_1}^{V_2}\frac{nRT_1}{T_1} = nR\ \ln\left(\frac{V_2}{V_1}\right)$$

Step 2. Heat the gas from T_1 to T_2 at constant volume V_2 then

$$dq_{rev} = C_V dT$$

$$\therefore \ \Delta S_2 = \int_{T_1}^{T_2}\frac{dq_{rev}}{T} = C_V\int_{T_1}^{T_2}\frac{dT}{T}$$

$$= C_V \text{ In} \left(\frac{T_2}{T_1}\right) = nC_{V,m} \left(\frac{T_2}{T_1}\right)$$

C_V is assumed to be independent of T

The entropy change for the system is

$$\Delta S_{system} = \Delta S_1 + \Delta S_2$$

$$\Delta S_{system} = nR \text{ In} \left(\frac{V_2}{V_1}\right) + nC_{V,m} \text{ In}\left(\frac{T_2}{T_1}\right) \quad ...(1)$$

Path B

Step 1. Heat the gas at constant Volume V_1 from T_1 to T_2

$$\Delta S_1 = nC_{V,m} \text{ In} \left(\frac{T_2}{T_1}\right)$$

Step 2. Expand the gas isothermally at temperature T_2 from volume V_1 to V_2

$$\Delta S_2 = nR \text{ In} \left(\frac{V_2}{V_1}\right)$$

$$\therefore \Delta S_{system} = \Delta S_1 + \Delta S_2 = nC_{V,m} \text{ In} \left(\frac{T_2}{T_1}\right) + nR \text{ In} \left(\frac{V_2}{V_1}\right) \quad ...(2)$$

Thus we see that the final expression for the entropy change comes out to be the same irrespective of the fact whether we first change the volume and then temperature or the reverse. This suggests that entropy is a state function.

Path C

Step 1. Heat the gas at constant volume from T_1 to an intermediate temperature T.

Step 2. Expand isothermally at T from V_1 to an intermediate volume V′.

Step 3. Heat at constant volume from T′ to T_2.

Step 4. Expand isothermally at T_2 from V′ to V_2.

For step 1: $dq_{rev} = nC_{V,m}\, dT$

$\Delta S_1 = nC_{V,m} \text{ In } (T'/T_1)$

For step 2 : $\Delta S_2 = nR \ln (V'/V_1)$

For step 3 : $\Delta S_3 = nC_{V,m} \ln (T_2/T')$

For step 4 : $\Delta S_4 = nR \ln (V2/V')$

The entropy change for the system is

$$\Delta S_{system} = \Delta S_1 + \Delta S_2 + \Delta S_3 + \Delta S_4$$

$$= nC_{V,m} \ln\left(\frac{T'}{T_1}\right) + nR \ln\left(\frac{V'}{V_1}\right) + nC_{V,m} \ln\left(\frac{T_2}{T'}\right) + nR \ln\left(\frac{V_2}{V'}\right)$$

$$= nC_{V,m} \ln\left(\frac{T_2}{T_1}\right) + nR \ln\left(\frac{V_2}{V_1}\right) \qquad ...(3)$$

(b) **Temperature and pressure are changed.** Let n moles of an ideal gas, initially at temperature T_1 and pressure P_1 be brought to a final temperature T_2 and pressure P_2. We can use the following reversible path between two states.

Step 1. Heat the gas at constant pressure P_1 from T1 to T_2, then

$$dq_{rev} = nC_{P,m}\, dT$$

Step 2. Expand the gas isothermally at temperature T_2 from P_1 to P_2, then

$$dq_{rev} = -\,dw = PdV$$

For step 1: $\Delta S_1 = \int_{T_1}^{T_2} nC_{P,m} \frac{dT}{T}$

$$= nC_{P,m} \ln\left(\frac{T_2}{T_1}\right)$$

(If C_P is independent of T).

For step 2: $\Delta S_2 = \int_{P_1}^{P_2} \frac{PdV}{T_2} = -\int_{P_1}^{P_2} \frac{VdP}{T_2} = -\int_{P_1}^{P_2} nRd \ln P$

$$= -\,nR \ln\left(\frac{P_2}{P_1}\right)$$

$$\Delta S = \Delta S_1 + \Delta S_2 = nC_{P,m} \ln\left(\frac{T_2}{T_1}\right) - nR \ln\left(\frac{P_2}{P_1}\right) \qquad ...(4)$$

This result can be transformed into the form of equation (3) by using the following relations:

$$C_{P,m} = C_{V,m} + R$$

$$\frac{P_2}{P_1} = \frac{T_2V_1}{T_1V_2}$$

$$\therefore \Delta S = n\,(C_{V,m} + R)\ \text{In}\left(\frac{T_2}{T_1}\right) - nR\ \text{In}\left(\frac{T_2}{T_1}\right) - nR\ \text{In}\left(\frac{V_1}{V_2}\right)$$

$$\Delta S = nC_{V,m}\ \text{In}\left(\frac{T_2}{T_1}\right) + nR\ \text{In}\left(\frac{V_2}{V_1}\right)$$

Entropy change when a van der Waals gas is heated and expanded simultaneously. Let n moles van der Waals gas initially at temperature T_1 and volume V_1 be brought to temperature T_2 and volume V_2. Combined first and second laws give,

$$TdS = dq_{rev} = dE + PdV \qquad ...(5)$$

But $E = f\,(T, V)$

$$\therefore dE = \left(\frac{\partial E}{\partial T}\right)_V dT + \left(\frac{\partial E}{\partial V}\right)_T dV$$

$$= C_V\ dT + \left(\frac{\partial E}{\partial T}\right)_V dV \qquad ...(6)$$

From equation (5) and (6), we get

$$dS = \frac{C_V}{T} dT + \frac{1}{T}\left[\left(\frac{\partial E}{\partial V}\right)_T + P\right] dV \qquad ...(7)$$

For a van der Waals gas

$$P = \frac{nRT}{V - nb} - \frac{an^2}{V^2}$$

and $$\left(\frac{\partial E}{\partial V}\right)_T = \frac{an^2}{V^2}$$

$$\therefore \qquad dS = \frac{nC_{V,m}}{T} dT + \frac{nR}{V - nb} dV$$

Taking $C_{V,m}$ as independent of T and integrating between suitable limits, we have

$$S_2 - S_1 = \Delta S = nC_{V,m}\ \text{In}\ \frac{T_2}{T_1} + nR\ \text{In}\left[\frac{V_2 - nb}{V_1 - nb}\right] \qquad ...(8)$$

If the gas expands isothermally then dT = 0 and the entropy change is given by

$$\Delta S = nR \ In \left[\frac{V_2 - nb}{V_1 - nb}\right]$$

ΔS can be calculated if b is known. Here we see that the entropy change depends upon the size of the molecules (b).

If $C_{V,m}$ depends upon temperature as given by

$C_{V,m} = \alpha + \beta T + \gamma T_2$, then

$C_{V,m}/T = \alpha/T + \beta + \gamma T$

and $\int_{T_1}^{T_2} \frac{C_{V,m}}{T} dT = \alpha \int_{T_1}^{T_2} \frac{dT}{T} + \beta \int_{T_1}^{T_2} dT + \gamma \int_{T_1}^{T_2} TdT$

$$= \alpha \ In \ (T_2/T_1) + \beta \ (T_2 - T_1) + \frac{1}{2}\gamma\left(T_2^2 - T_1^2\right)$$

Therefore, for the n moles of a van der Waals gas the entropy change is given by

$$\Delta S = n\alpha \ In \ (T_2/T_1) + n\beta \ (T_2/T_1) + \frac{1}{2}\gamma\left(T_2^2 - T_1^2\right) + nR \ In \left[\frac{V_2 - nb}{V_1 - nb}\right]$$

Similarly, for a gas obeying the equation of state. $\frac{PV}{RT} = 1 + \frac{B(T)}{V}$, we can calculate ΔS.

We have derived

$$dS = \frac{C_V}{T} dT + \frac{1}{T}\left[\left(\frac{\partial E}{\partial V}\right)_T + P\right] dV$$

and $\left[\frac{\partial E}{\partial V}\right]_T = T\left[\frac{\partial P}{\partial T}\right]_V - P$

$$\therefore \quad dS = \frac{C_V dT}{T} + \left[\frac{\partial P}{\partial T}\right]_V dV \qquad ...(9)$$

For the given gas

$$\left(\frac{\partial P}{\partial T}\right)_V = \frac{R}{V} + \frac{R}{V^2} B(T) + \frac{RT}{V^2}\left(\frac{dB}{dT}\right)_V \qquad ...(10)$$

Equations (10) and (9) give,

$$dS = \frac{C_V}{T}dT + \left[\frac{R}{V} + \frac{R}{V^2}B(T) + \frac{RT}{V^2}\left(\frac{dB}{dT}\right)_V\right]dV \qquad ...(11)$$

Integrating equation (11) we get,

$$\Delta S = C_V \text{ In } \frac{T_2}{T_1} + R \text{ In } \frac{V_2}{V_1} - RB\left[\frac{1}{V_2} - \frac{1}{V_1}\right] - RT\frac{dB}{dT}\left[\frac{1}{V_2} - \frac{1}{V_1}\right]$$

$$\Delta S = C_V \text{ In } \frac{T_2}{T_1} + R \text{ In } \left[\frac{V_2}{V_1}\right] + R\left[\left(B + T\frac{dB}{dT}\right)\right]\left[\frac{1}{V_1} - \frac{1}{V_2}\right]$$

Thus, it is possible to calculate the entropy change for a gas obeying virial equation of state, provided we know the experimental conditions and the data for B (T).

THERMODYNAMIC FUNCTIONS

The thermodynamic quantities given in Table 1. are quite sufficient to describe the properties of a system. However, new functions of simpler form can be defined, as given in Table 1. These are found to be more useful.

Table 1 : Various Thermodynamic Functions

Definition	*Function*	*Name*
$dE = dq + dw$	E	Internal energy
$dS = \frac{dq_{rev}}{T}$	S	Entropy
$H = E + PV$	H	Enthalpy
$G = H - TS$	G	Gibbs free energy
$A = E - TS$	A	Helmholtz free energy (work function)
$\mu_i = (\partial G/\partial n_i)_{T, P, nj} \quad j \neq i$	μ	Chemical potential
$C_P = (\partial H/\partial T)_P$	C_P	Heat capacity at constant pressure
$C_V = (\partial E/\partial T)_V$	C_V	Heat capacity at constant volume

HOMOGENEOUS FUNCTION

Any function f (x, y, z) is said to be homogeneous of degree m of the variables x, y, z, if

$$f(\lambda x, \lambda y, \lambda z) = \lambda^m f(x, y, z) \quad ...(1)$$

for all positive values of l. This identity implies that an increase in all variables by a factor λ produces an increase in the function by a factor of λ^m. If the degree of homogeneity is unity (m = 1), the operation of multiplying each of the independent variables by a parameter λ produces the same effect as multiplying the overall function by the same factor. All the extensive properties of a system are homogeneous of first degree of moles.

Volume is an extensive property and also it is a homogeneous function of first degree of mole, because when we double the moles of the substance its volume is also doubled.

That is for

$$V = f(T, P, n)$$

keeping T and P constant if n is doubled then the new volume V′ is given by,

$$V' = f(T, P, 2n) = 2V$$

Thermodynamic functions namely, volume, energy, enthalpy, entropy, free energy and the work function are extensive properties and hence are homogeneous function of the first degree in moles. The intensive properties are homogeneous function of zeroth degree of moles. For example, pressure is an intensive property and hence a homogeneous function of zeroth degree *i.e.*

$$P(\lambda n_1, \lambda n_2 ...) = \lambda^0 P(n_1, n_2 ...)$$

Euler's Theorem on Homogeneous Functions

This theorem states that if the function f (x, y, z) is a homogeneous function of degree m then

$$x\frac{\partial f}{\partial x} + y\frac{\partial f}{\partial y} + z\frac{\partial f}{\partial z} = mf(x, y, z) \quad ...(2)$$

and for a first degree function (m = 1), we have

$$x\frac{\partial f}{\partial x} + y\frac{\partial f}{\partial y} + z\frac{\partial f}{\partial z} = f(x, y, z) \quad ...(3)$$

These equations can be generalised for any number of independent variables. This theorem is very useful to deduce the relations involving partial molar quantities.

STATE FUNCTIONS

Thermodynamic functions are state functions. Magnitude of changes in such functions are independent of how the change is accomplished. These depend only upon initial and final states of the system. An analogy of state function can be found in mountain climbing. Suppose that we want to climb a mountain peak 1 km above our lodge. This decision defines our initial (lodge) and final (mountain peak) states. There may be various paths up the mountain but the vertical distance of the peak from the lodge is all the time 1 km. The height of the mountain peak cannot be altered by choosing the different paths though the actual distance travelled and the amount of work and fatigue will be different depending upon the path followed. Vertical distance in this example corresponds to thermodynamic functions, E, H, S, G, μ, T , P, V. These all are state functions. We cannot alter the energy, enthalpy change etc. of the system by choosing different sequence of processes provided the initial and final states for different paths are not altered.

Properties of State Functions –Exact Differentials

State functions give exact differentials. These exact differentials can be integrated between limits, without regard for the actual change that occur as the system moves from one limiting condition to the other. The properties of a state function can be understood if we take a concrete example of volume of definite amount of the substance. It depends upon temperature (T) and pressure (P). Mathematically we can speak that volume is a function of temperature and pressure *i.e.*,

$V = f(T, P)$ (for a definite amount)

If T and P are given a particular value of V is thereupon fixed. If T and P are changed the value of V will accordingly be changed. For this reason T and P are called *independent variables* and V is called the *dependent variable*.

The change in V can be estimated provided the derivatives of the function V with respect to T and P are known. The derivative is the rate of change of dependent variable with the independent variable. For the present case the two derivatives are :

$(\partial V/\partial T)_P$ = rate of change of volume with temperature at constant pressure.

$(\partial V/\partial P)_T$ = rate of change of volume with, pressure at constant temperature.

If the derivative is positive the function V increases with the increase of the independent variable; if the derivative is negative the function decreases with the increase of the independent variable. If the derivative is zero the function is independent of the variables, and the value of the function may be maximum or minimum under that condition.

The change of volume due to infinitesimal change of temperature (dT) at constant pressure is equal to the rate of change of volume with temperature $(\partial V/\partial T)_P$ multiplied by dT, and is given by

$$dV = (\partial V/\partial T)_P\, dT$$

The change in volume due to infinitesimal change of pressure (dP) at constant temperature is given by

$$dV = (\partial V/\partial P)_T\, dP$$

If both the temperature and pressure are changed simultaneously then the total change in volume (dV) is given by

$$dV = \left(\frac{\partial V}{\partial T}\right)_P dT + \left(\frac{\partial V}{\partial P}\right)_T dP$$

This is called the total differential of the function (V). This equation was obtained by assuming that *whether we first change temperature then pressure or, first change the pressure then temperature, the final total change in volume is the same*. This is true because V is a *state function* and dV is *an exact differential*. In the language of mathematics the condition of exactness is

$$\frac{\partial}{\partial P}\left[\left(\frac{\partial V}{\partial T}\right)_P\right]_T = \frac{\partial}{\partial T}\left[\left(\frac{\partial V}{\partial P}\right)_T\right]_P$$

or $$\frac{\partial^2 V}{\partial P \partial T} = \frac{\partial^2 V}{\partial T \partial P}$$

This relation is applicable for any state function ϕ.

Let $\phi = f(x, y)$, then total differential $d\phi$ is given by

$$d\phi = \left(\frac{\partial \phi}{\partial x}\right)_y dx + \left(\frac{\partial \phi}{\partial y}\right)_x dy = M(x,y)\,dx + N(x,y)\,dy$$

where $M(x, y) = \left(\frac{\partial \phi}{\partial x}\right)_y$ and $N(x, y) = \left(\frac{\partial \phi}{\partial y}\right)_x$ ...(1)

If dϕ is an exact differential then

$$\left(\frac{\partial M}{\partial y}\right)_x = \left(\frac{\partial N}{\partial x}\right)_y \quad ...(2)$$

$$\therefore \frac{\partial^2 \phi}{\partial y dx} = \frac{\partial^2 \phi}{\partial x \partial y} \quad ...(3)$$

Here ϕ can be considered as any thermodynamic function (E, H, G etc). and (x, y) may be thermodynamic T, P, V etc. The relations given by equations (1, 2) and (1.3) are called the Euler's theorem of exactness. Any function and dY is inexact differential.

Some other properties of derivatives :

(i) They can be turned upside down, the subscript remaining the same.

$$1 \Big/ \left(\frac{\partial \phi}{\partial x}\right)_y = \left(\frac{\partial x}{\partial \phi}\right)_y$$

However, $1 \Big/ \left(\frac{\partial^2 \phi}{\partial x^2}\right) \neq \frac{\partial x^2}{\partial^2 \phi}$

(ii) They can be multiplied together cancelling numerator with denominator

provided the subscripts are the same.

$$\left(\frac{\partial \phi}{\partial x}\right)_y \left(\frac{\partial x}{\partial z}\right)_y = \left(\frac{\partial \phi}{\partial z}\right)_y$$

But $\left(\frac{\partial \phi}{\partial x}\right)_y \left(\frac{\partial x}{\partial z}\right)_t \neq \left(\frac{\partial \phi}{\partial z}\right)$

THIRD LAW OF THERMODYNAMICS AND EVALUATION OF ABSOLUTE ENTROPY

So far we have been concerned with the evaluation of entropy changes in different processes under different conditions but can we find the absolute value of entropy of an element or a compound? The thermodynamic relation at hand to evaluate entropy change is

$$dS = \frac{C_P}{T} dT$$

$$\int_{T_1}^{T_2} dS = S(T_2) - S(T_1) = \int_{T_2}^{T_1} \frac{C_P}{T} dT$$

$$\text{or } S(T_2) = S(T_1) + \int \frac{C_P}{T} dT \qquad ...(1)$$

S (T_2) can be evaluated provided the quantities on R. H. S. of this equation are known. C_P can be measured experimentally over a range of temperature, but the problem remains to find a value of S (T_1). This problem is solved by assigning a zero value of S (T_1) at absolute zero as stated by the **third law of thermodynamics** in the following form:

"Every substance has a finite positive entropy, but at 0K the entropy may become zero, and does so become in case of a perfectly crystalline substance." The statement is confined to pure crystalline solids because theoretical arguments and experimental evidence have shown that the entropy of solutions and supercooled liquids is not zero at 0K. For pure crystalline solids the law has been verified repeatedly, and at present little doubt remains about the general validity of the above statement of the law. Thus as $T \rightarrow 0$, $S(T) \rightarrow 0$. With this convention equation (1) may written as:

$$S^0(T) = \int_0^T \frac{C_P}{T} dT \qquad ...(2)$$

S^0 (T) is absolute entropy of the substance at a temperature TK. It is a positive quantity. Its value can be calculated if we integrated R H S of equation (2) from T = 0 to any desired temperature T. To be able to integrate we must know the values of C_P right from absolute zero to the temperature concerned. The measurement of C_P upto the absolute zero is inconvenient. For this purpose the method of extrapolation is employed. The values of C_P/T are plotted against T and the curve is extended upto T = 0. The area under the curve from T = 0 to T = T gives the absolute value of entropy of the substance.

In many cases the value of heat capacity down to 0K can be obtained by the use of Debye equation

$$C_V = 3R\left[\frac{4}{5}\pi^4 (T/\theta)^3\right] = a(T/\theta)^3$$

If a substance (X) undergoes phase transformation between T = 0 to T = T, then all the contributions of the absolute entropy must be considered. For example, the absolute entropy of a gaseous substance (X) at 298 K and 1 atm may be calculated by considering the entropy contributions due to the :

1. heating the substance from $T = 0$ to $T = T_t$, the transition point,
2. transition from one form (α) to another from (β) at $T = T_t$,
3. heating the substance from $T = T_t$ to $T = T_m$, the melting point,
4. melting the substance at a constant temperature $T = T_m$,
5. heating the substance from $T = T_m$ to $T = T_b$, the boiling point,
6. evaporation of the substance at a constant temperature $T = T_b$,
7. heating the substance from $T = T_b$ to $T = 298$ K, the temperature in question.

Various Changes

$$X\ (T = 0, \alpha) \rightarrow X\ (Tt, \alpha) \rightleftharpoons X\ (T_t, \beta) \rightarrow X\ (T_m, S) \rightleftharpoons X\ (T_m, l)$$

$$X\ (298, g) \leftarrow X\ (T_b, g) \rightleftharpoons X\ (T_b, l)$$

Entropy Contributions

$$S^0\ (T) = \int_0^{T_t} C_P(\alpha)\, d\ In\ T + \frac{\Delta H_t}{T_t} + \int_{T_t}^{T_m} C_P(\beta)\, d\ In\ T + \frac{\Delta H_{fus}}{T_m}$$

$$+ \int_{T_m}^{T_b} C_P(l)\, d\ In\ T + \frac{\Delta H_{vap}}{T_b} + \int_{T_b}^{298} C_P(g)\, d\ In\ T$$

The steps involved in the evaluation of entropy for nitrogen are summarized in Table 1.

Table 2 : Evaluation of Entropy of Nitrogen from Heat Capacity

	Contribution	$JK^{-1}mol^{-1}$
1.	Extrapolation from 0 – 10 K and $\int C_p d$ In T for solid α from 10 to 35.61 K	27.17
2.	Transition, solid $\alpha \rightarrow$ solid β 228.9/35.61	6.43
3.	$\int C_p d$ In T for solid β from 35.61 to 63.14 K	23.38
4.	Fusion, 720.9/63.14	11.42
5.	$\int C_p d$ In T for liquid from 63.14 to b.pt.	11.41
6.	Vaporization, 5535.0/77.32	72.13
7.	$\int C_p d$ In T for liquid from 77.32 to 298 K	39.20

	Entropy of Ideal gas at 298 K and 1atm	191.14

ENTROPY OF REACTION

With the convention that the entropy of each substance is zero at absolute zero, every substance is assigned a positive entropy at any other temperature. These values are tabulated for elements and compounds. For any chemical reaction the change of entropy ΔS can be evaluated by using the relation

$$\Delta S = \Sigma S^0 \text{ (products)} - \Sigma S^0 \text{ (reactants)} \quad ...(1)$$

For a general reaction of the type $aA + bB \rightarrow mM + nN$, the entropy change is given by

$$\Delta S = [\ mS^0 (M) + nS^0 (N)] - [aS^0 (A) + bS^0 (B)]$$

where S^0, s are the entropies of the given pure substances at a given temperature.

Variation of ΔS^0 with Temperature

If the value of entropy of reaction at one temperature $\Delta S(T_1)$ is known its value at some temperature $\Delta S(T_2)$ can be calculated from the heat capacities of the reaction species.

The calculation is done by using the relation $(\partial S/\partial T)_P = C_P/T$ or $dS = (C_P/T)\, dT = C_P d \ln T$. For the general case

$$\Delta S^0 = mS^0 (M) + nS^0 (N) - aS^0 (A) - bS^0 (B)$$

Differentiating w.r. to T at constant P, we get

$$\left(\frac{\partial \Delta S^0}{\partial T}\right)_P = m\left(\frac{\partial S^0(M)}{\partial T}\right)_P + n\left(\frac{\partial S^0(N)}{\partial T}\right)_P - a\left(\frac{\partial S^0(A)}{\partial T}\right)_P - b\left(\frac{\partial S^0(B)}{\partial T}\right)_P$$

$$= m\, C_P (M)/T + n\, C_P(N)/T - a\, C_P (A)/T - b\, C_P (B)/T$$

$$= \Delta\, C_P/T \quad ...(2)$$

where $\Delta C_P = [m\, C_P(M) + n\, C_P (N)] - [a\, C_P (A) + bC_P (B)]$

$$= \Sigma C_P \text{ (products)} - \Sigma C_P \text{ (reactants)}$$

$$\therefore d(\Delta S^0) = \Delta C_P \frac{dT}{T} = \Delta C_P d \ln T \quad ...(3)$$

Integration between proper limits gives

$$\Delta S^0 (T_2) - \Delta S^0 (T_1) = \int_{T_1}^{T_2} \Delta C_P d \ln T \quad ...(4)$$

If the variation of ΔC_P with temperature is known, equation (4) can be integrated. For approximate purposes ΔC_P may be assumed to be independent of T, and integration gives

$$\Delta S^0 (T_2) - \Delta S^0 (T_1) = \Delta C_P \text{ In } (T_2/T_1)$$

or $$\Delta S^0 (T_2) = \Delta S^0 (T_1) + \Delta C \; 2.303 \log (T_2/T_1) \quad ...(5)$$

THE CONCEPT OF RESIDUAL ENTROPY

Certain chemical reactions between crystals do not give $\Delta S = 0$ at 0K, which seems to indicate that exceptions to this law exist. Nearly all these exceptional reactions involve either ice, carbon monoxide, nitrous oxide or hydrogen. In the crystalline state, there four substances appear to retain a finite entropy at 0K. Thus any reaction in which one which 1mole of ice is a product gives $\Delta S = 3.3 \text{ JK}^{-1}$, at 0K and any reaction in which 1mole of ice is a reactant gives $\Delta S = -3.3 \text{ JK}^{-1}$ at 0K.

This shows that if entropies of zero are assigned to the other substances involved, then ice at 0K appears to have a residual entropy of 3.3 JK^{-1}. Similarly, at 0K, CO has a residual entropy of 5.8 JK^{-1} N_2O has a residual entropy of 5.8 JK^{-1} and hydrogen has a residual entropy of 6.2 JK^{-1}. This suggests that the molecules in the crystals of the substances are not in true thermodynamic equilibrium at 0K. The molecular origin of residual entropy is well understood with the help of statistical thermodynamics.

A number of liquids are readily supercooled below their thermodynamic freezing points and can exist in this state even at 0K. For such substance the entropy change of the reaction, Crystal $\rightarrow$ Liquid, at 0K can be determined and it has been found that liquid has a greater entropy than the crystal. Thus the third law is also not applicable to supercooled liquids e.g. glass.

ENTROPY AT EQUILIBRIUM

Natural processes are accompanied with an increase in the total entropy of the system and its surroundings. This gives us the equilibrium condition when the system cannot undergo any spontaneous change. At this state the entropy of the isolated system is maximum. Mathematically it can be explained with the help of the combined equation of the first and second laws of thermodynamics:

(a) $dS = \dfrac{dE + PdV}{T}$ for a reversible process

But $dE = C_V\, dT + (\partial E/\partial V)_T\, dV$

$$\therefore\ dS = C_V\ \frac{dT}{T} + \frac{1}{T}\left[(\partial E/\partial V)_T + P\right]dV$$

In an isolated system dE = 0, dT = 0 any dV = 0, therefore

$(dS)_{E,V} = 0$

For an irreversible change, since $(PdV)_{rev} > (PdV)_{irr}$,

$$\therefore\ dS > \frac{dE + PdV_{irr}}{T}$$

In an isolated system, dE = 0, dV = 0 and hence

$(dS)_{E,V} > 0$ ...(1)

The combined result may be written as

$(dS)_{E,V} \geq 0$

The state of equilibrium correspond of maximum entropy of the universe.

(b) If the entropy and volume are constant, then dS = 0 dV. and we should have

$(dE)_{S,V} = 0$ for a reversible change.

For an irreversible change $TdS > dE + PdV_{irr}$ or

$dE < TdS - PdV_{irr}$

$\therefore\ (dE)_{S,V} < 0$ ...(2)

The system maintained at constant entropy and constant volume will attain the equilibrium at a state of minimum energy.

(c) At constant entropy and pressure another condition of equilibrium in terms of enthalpy may be given. We have the relation dH = TdS – VdP, at constant S and P, dS = 0, dP = 0

$\therefore\ (dH)_{S,P} = 0$

For irreversible change $(dH)_{S,P} < 0$.

Thus the state of equilibrium is attained when the energy is minimum and the entropy is maximum. These are the necessary conditions for equilibrium. The sufficient condition of equilibrium is that if any virtual displacement (δ) is made in the variables of the system, then entropy in the new state will be less than the old equilibrium state and δS < 0, and the energy in the new state will be larger than the old equilibrium state and δE > 0. This can be understood by considering the Fig. 3.1(a) and 2 (b) for a process A $(T_1, V_2, E_1, S_1) \rightarrow (T_2, V_2, E_2, S_2)$.

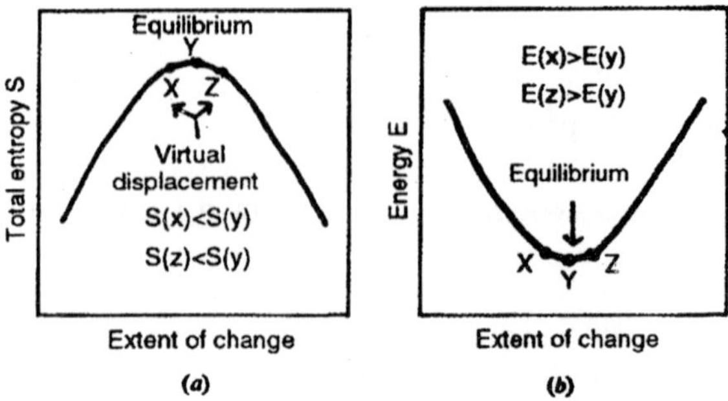

Fig. 3.1 : The condition for equilibrium (a) The entropy is maximum and $\delta S < S < 0$; (b) The energy is minimum and dE > 0.

When equilibrium is attained, the entropy is maximum and energy is minimum. Now suppose that by some means, the equilibrium is displaced to new states X or Z. From the figure it is quite clear that the entropy in the state X is less than the entropy in the state Y i.e. S (X) < S(Y) and δS in negative.

Similarly S (Z) < S (Y) and $\delta S < 0$. The energy in either of the displaced state is greater than the energy at equilibrium i.e. E (X) > E (Y) and E (Z) > E (Y), therefore $\delta E > 0$ is the sufficient condition for equilibrium.

Necessary condition . $(dS)_{E, V} > 0$; $(dE)_{S,V} < 0$.

Sufficient condition. $\delta S < 0$; $\delta E > 0$.

RELEVANCE OF THE SECOND LAW

The original statement of the second law of thermodynamics is much concerned with the development of the theoretical aspects of heat engines specially the Carnot's cycle. This has got a very little implication in chemical processes. But, the further developments and various derivations involving the concept of this law are of immense use for chemists. For example, the direction of a particular spontaneous transformation can be ascertained with the help of second law of thermodynamics which is not predicted by first law of thermodynamics.

SPONTANEOUS PROCESS

Most of the natural processes are spontaneous. In fact all human activities and all that happens in the universe involve a significant degree of this concept. For example, the production of wheat from seeds, the formation of clouds in the sky on a sunny day, the fall of rain from clouds, the growth and multiplication of micro-organism when food is digested, are all spontaneous processes.

All those processes which take place of their own accord are called spontaneous process. As such all the spontaneous processes are irreversible in nature. Once having undergone a spontaneous transformation in a particular direction, the process will not take place in the reverse direction spontaneously. The spontaneity however never means that the process is instantaneous. It simply stands that the change is favourable in a particular direction. A few other examples of spontaneous processes are as follows. These will explain as to why do we need a *second law*? A plate dropped on to the floor breaks and the pieces scatter; when we take ice-cream our lips get cold; petroleum vapour and oxygen when sparked react in a car engine producing exhaust gases; water runs downhill; a drop of perfume in a room diffuses quickly to effect the entire atmosphere; aqueous solutions of NaOH and HCl react to produce NaCl (aq) and heat. All these processes are known to proceed in one direction. The question is that why the reverse of the considered processes do not occur? Can we believe on our eyes to see the pieces of broken plate spontaneously reorganise themselves to reproduce the original plate? Has any one experienced ever hot while taking ice cream? Is it possible exhaust gases CO, CO_2, H_2O can recombine to produce petroleum and oxygen? Is it possible for the dispersed vapours of the perfume to get converted into the droplets and return to the bottle under ordinary conditions? Can water flow of itself upwards? Can we get aqueous solutions of NaOH and HCl on dissolving NaCl in water?

None of these processes is known to be spontaneous as we know by our experiences. However, in the light of the first law of thermodynamics none of these processes seem to be impossible, because in these processes the energy in conserved, converted from one form into another. This implies that the first law of thermodynamics is incapable of predicting the direction of a process. Now the question arises that what law of nature compels the spontaneous processes to proceed in a particular direction? How and why the systems are aware of the directions in which to proceed? What is the driving force that leads the process in a given direction? The answers to all these questions can be provided by considering the second law of thermodynamics.

Statements of the Second Law of Thermodynamics

1. "Heat cannot spontaneously pass from a colder to a warmer body." (R.J.E. Clausius)

2. "It is impossible to transfer heat from a colder system to a warmer without other simultaneous changes occurring in the two systems, or in their environment." (P.S. Epstein)
3. "Every system which is left to itself will, on the average, change toward a condition of maximum probability." (G. N. Lewis)
4. "The state of maximum entropy is the most stable state for an isolated system." (Enrico Fermi).
5. "In any irreversible process the total entropy of all bodies concerned is increased." (G. N. Lewis)
6. "In an adiabatic process the entropy either increases or remains unchanged:

$$\Delta S \geq 0$$

Where the upper sign (>) refers to the irreversible, the lower (=) to the reversible case." (P. S. Epstein)
7. "The entropy function of a system of bodies tends of increase in all physical and chemical processes occurring in nature, if we include in the system all such bodies which are affected by the change." (Saha)
8. "The entropy increases towards a maximum and the energy of the universe in constant." (R.J.E. Clausius)
9. "Gain in information is loss in entropy." (G. N. Lewis)
10. "Entropy is times arrow." (A. Eddington)
11. "Every system left to itself changes rapidly or slowly in such a way to approach a definite final state of rest: No system of its own will change away from the state of equilibrium, except through the influence of external agencies" (G. N. Lewis)
12. " There is a general tendency in nature for energy to pass from more available to less available forms." (J. A. V. Butter)
13. "When an actual process occurs it is impossible to invent a means of restoring every system concerned to its original condition." (G. N. Lewis)
14. "There exists a characteristic thermodynamic function called entropy. The difference of a system in states (1) and (2) is given by the expression

$$S_2 - S_1 = \int_1^2 dq/T$$

over any reversible path,

and $S_2 - S_1 > \int_1^2 dq/T$

over any irreversible path.

The entropy is a property of the state only. Its value for an isolated system can never decrease." (R. E. Gibson).

THERMODYNAMIC TERMS

System

The thermodynamics system is the part of the physical universe to which we direct our attention for the purpose of experimental or theoretical investigation.

Surroundings

The surroundings are the remainder of the universe which can exchange matter of energy with the system. Usually, the surroundings are restricted to a region in the immediate vicinity of the system under investigation.

Boundary

The region or interface separating the system from the surroundings is called the boundary.

Let us consider a gas (or liquid) contained in a cylinder and let the cylinder be placed in a constant temperature bath maintained at 25°C, as shown in Fig. 3.2. In this case the substance (liquid or gas) constitutes the system, the walls of the cylinder are the boundaries and the constant temperature bath is the surrounding. Since we shall encounter various types of systems as well as walls or boundaries, it will be helpful at this point to define some of the more important type of walls.

(i) *A rigid wall.* A wall whose shape and position are fixed.

(ii) *An impermeable wall.* A wall that prevents passage of matter.

(iii) *A permeable wall.* A wall that allows that passage of matter and consequently, also of energy.

(iv) *An adiabatic wall.* It is wall which, when held rigid will prevent the passage of matter or energy provided no external force field are present (e. g. thermos bottle).

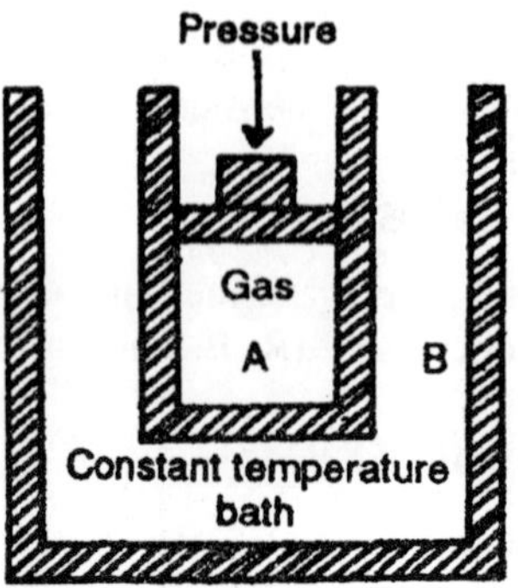

Fig. 3.2

(v) *A diathermic or diathermal wall*. It is a wall which, when held rigid will prevent the passage of matter but allow the passage of energy in the absence of any external force field. Thus it is a wall that is impermeable but not adiabatic. A plain metal container such as an aluminium can in an example of a container whose walls are diathermal.

Depending upon the type of the wall (boundary) of container the following types of systems can be considered.

(i) *An open system*. It is a system enclosed by a permeable wall that can exchange both the energy and the matter with the surroundings across the boundary. Thus $dE \neq 0$, $dm \neq 0$. (dm = change of mass; dE = change of energy).

(ii) *A closed system*. It is a system enclosed by impermeable walls. In system the mass remains constant but the energy may change. Thus $dm = 0$; $dE \neq 0$.

(iii) *An isolated system*. It is a system enclosed by rigid, adiabatic walls and not acted upon by any external force field. Thus in an isolated system neither the energy nor the mass can be exchanged with the surroundings across the boundary and $dE = 0$; $dm = 0$.

Consider a vessel containing a volatile liquid in contact with its vapours. This constitutes an open system as energy (heat) is taken by the system (liquid) for vaporization and the vapours can escape out of the vessel.

Thus system exchanges both the energy and the mass with the surroundings [Fig. 3.3(a)]. If the vapours are not allowed to leave the vessel then the system would be closed one [Fig. 3.3(b)]. However, if the walls of the vessel containing liquid are thermally insulated and the mouth of the vessel is closed then the system can exchange neither energy nor matter with the surroundings [Fig. 3.3(c)].

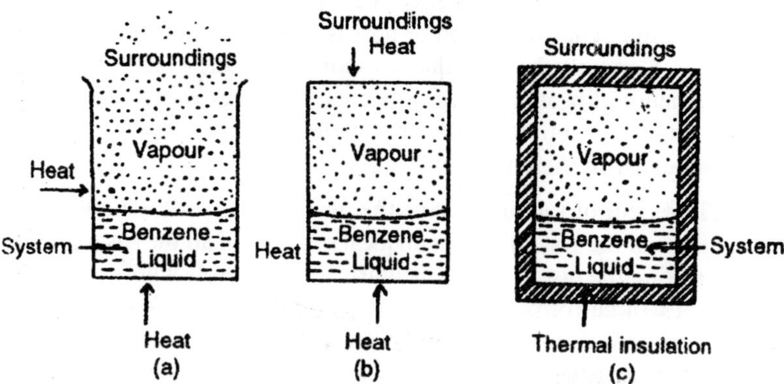

Fig. 3.3 : Types of systems : (a) open, (b) closed and (c) isolated.

THE CONCEPT OF ENTROPY

The concept of entropy is vital for thermodynamics. It is a Greek word with no equivalent in common language. It was invented by R.J.E. Clausius. He denoted it by the symbol S. **Entropy** stands for trope (a greek word meaning change), the suffix 'en' is written to identify it with *energy*. We have an intuitive feel for the concept of energy: a capacity to do work. However, there is no such intuitive grasp of entropy.

Nevertheless, we are quite familiar with the effects the entropy has been used for. Entropy is a measure of randomness or disorderness in a molecular system. More is randomness larger is the value of entropy. More regular is the arrangement lower is the entropy. Entropy is also a measure of unavailable energy. Sometimes entropy is also recognised as an index of exhaustion. Its significance can be appreciated by considering some general observations. In a cyclic process heat cannot be completely converted into work. The efficiency of working is never hundred per cent. The magnitude of work of irreversible expansion is less than the reversible expansion. Why is it so ? Why the entire heat absorbed cannot be converted into useful work? Wherelse the rest of the unused heat has gone? Why the system's capacity to do work is reduced in an irreversible process? The answers to these questions can partly be given as follows.

When heat is absorbed by a gas, the kinetic energy of the gas molecules will be increased and they will move faster, and collide with (i) each other, (ii) the container and (iii) the face of the piston, thereby the piston will be pushed. Only the third type of collision will bring about the expansions will not bring effective expansion of the gas. Thus some collisions and hence a fraction of heat is not utilized for the expansion of the gas. Certainly the work due to expansion is note exactly equivalent to the amount of heat absorbed. Since the heat goes waste only because of the random motion of the gas molecules, we can correlate the randomness with the unavailable work.

The extent of randomness or the unavailable work is indicated by the entropy of the system. Under an ideal situation, if all the gas molecules were striking the piston perpendicularly then certainly whole of the heat absorbed will be converted into work as shown in Fig. 3.4(a). But this is only possible if there is a regular and only one possible arrangement of the gas molecules. In other situations when the molecules are in random motion, a fraction of heat is converted into such a factor which is a measure of randomness of the molecular system. *So entropy is the measure of randomness of a molecular system.*

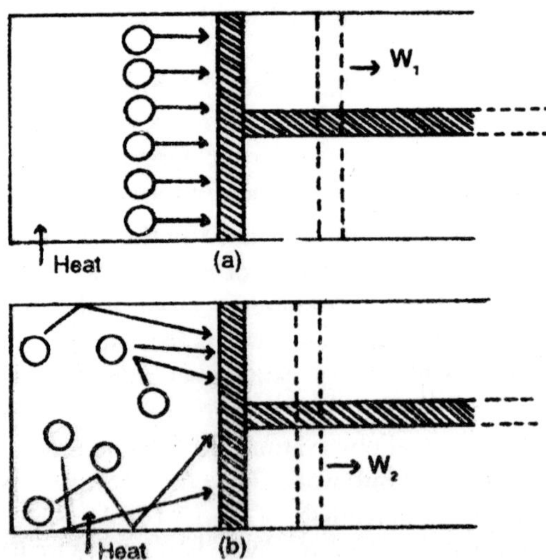

Fig. 3.4 : Conversion of heat work. (a) Ideal case of complete conversion. (b) Partial conversion due to randomness (dotted lines represent the position of the piston after expansion and $w_1 > w_2$).

This statement is also in line with the observations at the boiling point of a fluid. Under ordinary conditions when a liquid is heated its temperature increases, at the boiling point the temperature remains constant will whole of the liquid is transformed into vapour state with the continuous supply of heat. The question is that why the temperature at the boiling point remains constant? Where is the heat being utilised? The answer can be given if we consider the molecular arrangement in the liquid and vapour state at a given temperature.

The molecular arrangement in the liquid state is a bit more regular than the vapour state of the same substance. To disturb the regularity in the liquid state some work must be done. This work done on the system increases the disorderness but not the temperature at the boiling point. Since at the boiling point there is more disorderness in the vapour phase than the liquid, we can say the entropy of the vapour is greater than that of the liquid state of the same substance. Agreeing on the same lines we can say that the entropy of the liquid state is greater than that of the solid state of the same substance. In general, for any substance; S (vapour) > S (liquid) > S (solid) at a given temperature and pressure.

Mathematical Treatment of Entropy Concept

The logical concept of entropy can be given a mathematical foundation. This treatment is essential because we wish to recognise entropy (S) as a thermodynamic function. For this purpose we shall start with the first law of thermodynamics applied to the changes in states of an ideal gas. (Later on this can be extended for other substances also).

Let us consider an ideal gas that is undergoing a reversible change from an initial state T_1, V_1 to a final state T_2, V_2. For infinitesimal changes, the first law of thermodynamics can be written as,

$$dq_{rev} = dE - dw_{rev} \qquad ...(i)$$

If only PV work is considered then

$$-dw_{rev} = PdV$$

and $dq_{rev} = dE + PdV$

For an ideal gas $dE = C_V\ dT$, and $P = n\ RT/V$, hence

$$dq_{rev} = Cv\ dT + (nRT/V)\ dV \qquad ...(ii)$$

To know the total heat change we must integrate this equation. But we see that the integration is not possible because the second term on right side of the equation involves two variables T and V, whereas we have to carry the integration with respect to V only. However, if both sides of the equation (ii) are multiplied by 1/T, then integration can be carried. Thus

$$\frac{dq_{rev}}{T} = C_V \frac{dT}{T} + nR\frac{dV}{V} \qquad ...(iii)$$

Integrating equation (iii) between the limits

$V = V_1$, $V = V_2$, $T = T_1$, $T = T_2$, we have

$$\int_1^2 \frac{dq_{rev}}{T} = C_V \ln\left(\frac{T_2}{T_1}\right) + nR\ln\left(\frac{V_2}{V_1}\right) \qquad ...(iv)$$

For an ideal gas, $P_1V_1/T_1 = P_2 V_2/ T_2$, hence

$$\int_1^2 \frac{dq_{rev}}{T} = C_P \ln\left(\frac{T_2}{T_1}\right) + nR\ln\left(\frac{P_1}{P_2}\right) \qquad ...(v)$$

Here $C_P = (C_V + nR)$ = hear capacity at constant pressure. We find that

(i) dq_{rev} is not an exact differential because

$(\partial CV/\partial V)_T \neq (\partial n RT/\partial T)v$

(ii) (dq_{rev}/T) becomes an exact differential and can be integrated, hence 1/T is the integrating factor.

(iii) Since dq_{rev}/ T is an exact differential therefore the function whose change is given by dqrev/T must be a state function. Let this state function be S and hence

$$dq_{rev}/T = dS \qquad ...(vi)$$

Now the exactness of dS and state dependence of S can be shown as follows:

dS is exact differential. From equations (iii) and (vi) we have

$$dS = (C_V/T)\, dT + (nR/V)dV \qquad ...(viii)$$

At constant T, dT = 0 and

$(\partial S/\partial V)_T = nR/V$

Differentiating with respect to T at constant V we have

$$\frac{\partial^2 S}{\partial T\partial V} = \frac{\partial}{\partial T}(nR/V)_V = 0 \qquad ...(ix)$$

When V is constant dV = 0, and

$$\left(\frac{\partial S}{\partial T}\right)_V = (CV/T) \qquad ...(x)$$

On differentiation with respect to V at constant T, equation (x) gives

$$\frac{\partial^2 S}{\partial V \partial T} = \frac{\partial}{\partial V}(C_V / T)_T = 0 \qquad ...(xi)$$

From equation (ix) and (xi) we see that

$$\frac{\partial^2 S}{\partial T \partial V} = \frac{\partial^2 S}{\partial V \partial T}$$

Hence dS is an exact differential and S is a state function.

S *is a state function,* we have shown that

$$\int_1^2 \frac{dq_{rev}}{T} = C_V \text{ In } (T_2/T_1) + nR \text{ In } (V_2/V_1)$$

The right hand side of this equation depends only on the state variables T_1, V_1 and T_2, V_2, therefore the quantity on left side should also be a function only of state variables. This is

$$\int_1^2 \frac{dq_{rev}}{T} = \int_1^2 dS = S_2(T_2, V_2) - S_1(T_1, V_1) = \Delta S \qquad ...(xii)$$

$$\therefore \Delta S = C_V \text{ In } (T_2/T_1) + nR \text{ In } (V_2/V_1)$$

This state function S is called entropy function and like other thermodynamic functions the changes in entropy depend only on the state of the system and are independent of the path followed.

Equations (vi) and (xii) are the mathematical statements of the second law of thermodynamics and are used to define the entropy function. Here it must be regarded that this mathematical concept of entropy ($dS = dq_{rev}/T$) holds only for a REVERSIBLE change and change for an irreversible change $dS \neq dq_{irr}/T$.

COMBINED FROM OF THE FIRST AND SECOND LAWS OF THERMODYNAMICS

The mathematical form of the first law is

$$dq = dE - dw = dE + PdV \qquad ...(1)$$

(if only PV work is considered).

The mathematical expression for second law is

$dq_{rev} = TdS$ ...(2)

Equation (1) and (2) give

$TdS = dE + PdV$

or $dE = TdS - PdV$ ...(3)

This is the combined form of the first and the second laws of thermodynamics. This is very important and useful in the deduction of various other concepts of thermodynamic functions.

Enthalpy and Entropy

Enthalpy is expressed as

$H = E + PV$

$\therefore\ dH = dE + PdV + VdP$

Putting $dE + PdV = TdS$ we get

$dH = TdS + VdP$ ...(4)

Energy as a Function of S and V

Equation (3) clearly indicates that the change in energy depends on the change of entropy (dS) and the change of volume (dV). Thus mathematically E is a function of S and V

$E = f(S, V)$

The total differential of E is given as

$$dE = \left(\frac{\partial E}{\partial S}\right)_V dS + \left(\frac{\partial E}{\partial V}\right)_s dV \qquad ...(5)$$

Also $dE = TdS - PdV$ ...(6)

Comparing the coefficients of dS and dV separately in equations (5) and (6) we get,

$$\left(\frac{\partial E}{\partial S}\right)_V = T \qquad ...(7)$$

and $$\left(\frac{\partial E}{\partial V}\right)_s = -P \qquad ...(8)$$

Since dE is an exact differential therefore Euler's of exactness when applied with equation (6) gives

$$\left(\frac{\partial T}{\partial V}\right)_s = -\left(\frac{\partial P}{\partial S}\right)_V$$

From equation (3) we see that for constant energy dE = 0 and

$$\left(\frac{\partial S}{\partial V}\right)_E = \frac{P}{T}$$

Enthalpy as a Function of S and P

$$dH = TdS + VdP \qquad ...(9)$$

This equation clearly shows that the change in enthalpy (dH) depends on the changes in entropy (dS) and pressure (dP). Thus

$$H = f(S, P)$$

$$dH = \left(\frac{\partial H}{\partial S}\right)_P dS + \left(\frac{\partial H}{\partial P}\right)_s dP \qquad ...(10)$$

Comparing the coefficients of dS and dP respectively in equations (10) and (9), we obtain

$$\left(\frac{\partial H}{\partial S}\right)_P = T$$

$$\text{and} \left(\frac{\partial H}{\partial S}\right)_s = V \qquad ...(11)$$

ENTROPY CHANGE FOR ISOLATED SYSTEM

For reversible isolated processes the change of entropy is zero whereas for irreversible isolated processes the entropy change is greater than zero. This is shown as follows.

Let us suppose that small changes are made in the system reversibly and irreversibly. The work done by the system on the surrounding in the reversible process $(-dw_{rev})$ i.e.

$$(-dw_{rev}) > (-dw_{irr}) \qquad ...(1)$$

Since energy (E) of the system is a state function and dE is exact differential and must be the same for a given change whether it is performed reversibly or irreversibly. That is $dE_{rev} = dE_{irr} = dE$.

From the first law we have,

$$dE = dq_{rev} + dw_{rev}$$

$$\text{and } dE = d_{irr} + dw_{irr}$$

Since $(-dw_{rev}) > (-dw_{irr})$, R. H. S is positive

$\therefore\ dq_{rev} > dq_{irr}$

This is, in any small change a system will absorb more heat under reversible conditions than under irreversible conditions.

On dividing both sides of equation (1) by T, we get

$$\frac{dq_{rev}}{T} > \frac{dq_{irr}}{T}$$

But (dqrev/T) = dS (definition from 2nd law)

$$\therefore\ dS > \frac{dq_{irr}}{T}$$

$$\text{or } dS > \frac{dE - dw_{irr}}{T} \qquad ...(2)$$

$$\text{and } dS = \frac{dE - dw_{rev}}{T} \qquad ...(2b)$$

In an isolated system there is no exchange of energy, matter, heat etc., across the boundary, i.e., dq = 0, dw = 0, dE = 0.

Introducing these conditions into equations (2b) and (2a), we get

$(dS)_{rev} = 0$ and $(dS)_{irr} > 0$

or $(\Delta s)_{rev} = 0$ and $(\Delta s)_{irr} > 0$ ...(3)

The entropy changes expressed by equation (3) indicate as if it were not a state function? But it is. The why ΔS for reversible and irreversible isolated system are different? The explanation is like is like this. An isolated system includes the system under investigation and the surroundings in its immediate vicinity.

To calculate any property of an isolated system , therefore, we must account for the properties of both the system and the surroundings. The entropy change for the isolated system includes the changes in the system as well as surroundings, and the total entropy change is called the entropy of the universe.

ΔS (universe) = ΔS (system) + ΔS (surroundings)

For the system $D\,S^{syst}_{rev} = D\,S^{syst}_{irr}$. It is the DS (surrounding) which is different for reversible and irreversible changes.

SOLVED EXAMPLES

Example 1:

Calculate the standard entropy of formation of ethanol as represented by the following equation

$$2C \text{ (graphite)} + 3H_2 \text{ (g)} + 1/2\ O_2 \text{ (g)} \rightarrow C_2H_5OH \text{ (l)}$$

Solution:

The entropy change for the given reaction is

$\Delta S^0 = S^0 (C_2H_5OH) - [2S^0 (\text{C graphite}) + 3S^0 (H_2) + 1/2\ S^0 (O_2)]$

using the S0 values from the appendix we have

$\Delta S^0 = (160.7 \text{ JK}^{-1} \text{ mol}^{-1}) - (2 \times 5.7 + 3 \times 130.6 + 1/2 \times 205.1)$ $\text{JK}^{-1}\text{mol}^{-1}$

$= (160.7 - 505.7) \text{ JK}^{-1}\text{mol}^{-1} = -345.0 \text{ JK}^{-1}\text{mol}^{-1}$

Example 2:

(c) Calculate the entropy change when 1 mole of an ideal gas expands reversibly from an initial volume of 1 dm^3 to a find volume of 10 dm^3 at a constant temperature of 298 K.

Solution:

$$\Delta S = nR \ln\left(\frac{V_2}{V_1}\right) = 2.303 \text{ nR} \log\left(\frac{V_2}{V_1}\right)$$

Here $n = 1$; $V_1 = 1 \text{ dm}^3$, $V_2 = 10 \text{ dm}^3$

$\Delta S = R \ln 10 = 2.303 \text{ R} \log 10$

$= 2.303 \times (8.314 \text{ JK}^{-1} \text{ mol}^{-1})$

$19.15 \text{ JK}^{-1} \text{ mol}^{-1}$.

Example 3:

Prove that the volume of an ideal gas is a homogeneous function of zeroth degree in pressure and temperature.

Solution:

Volume of a fixed amount of a gas is a function of T and P.

$V = f(T, P)$ ...(i)

Applying the Euler's theorem of homogeneity we have

$$T\frac{\partial V}{\partial T} + P\frac{\partial V}{\partial P} = mV \qquad ...(ii)$$

For an ideal gas of known amount, PV = nRT and

$$\left(\frac{\partial V}{\partial T}\right)_P = \frac{nR}{P},$$

$$\left(\frac{\partial V}{\partial P}\right)_T = -\frac{V}{P}$$

Substituting the expressions for these derivatives into equation (ii) we obtain

$$\frac{nRT}{P} - V = mV$$

$$\text{or } \frac{nRT}{P} - \frac{nRT}{P} = mV = 0 \; (\because V = nRT/P)$$

since V is not zero, therefore m must be zero. Hence for an ideal gas volume is a homogeneous function of the zeroth degree in pressure and temperature *i.e.*,

$$V(\lambda P, \lambda T) = \lambda^o V(T, P)$$

This is obvious from the relation V = nRT/P also. Suppose we double both the temperature and pressure, the volume of the ideal gas will not be changed.

For other extensive properties we shall have

(i) $A = A(T, V, n_1, n_2)$ ...Helmholtz free energy

$\therefore A(T, \lambda V, \lambda n_1, \lambda n_2) = \lambda A$

since T is an intensive property, therefore, it is not multiplied by λ whereas V being an extensive property is multiplied by λ when the moles (n_i's) are so multiplied.

$S = S(E, V, n_1, n_2)$ ...Entropy

$S(\lambda E, \lambda V, \lambda n_1, \lambda n_2) = \lambda S$

$G = G(T, P, n_1, n_2)$

$G(T, P, \lambda n_1, \lambda n_2) = \lambda G$ ...Gibbs free energy

T and Pare intensive variables and hence are not multiplied by λ.

(iv) $H = H(S, P, n_1, n_2)$ ...Enthalpy

$\therefore$ H (λS, P, λn_1, λn_2) = λH

(v) E = E (S,V, n_1, n_2) ...Intrinsic energy

$\therefore$ E (λS, λV, λn_1, λn_2) = λE

(vi) V = V (T, P, n_1, n_2) ...Volume

$\therefore$ V(T, P, λn_1, λn_2) = λV

The treatment of homogeneity enables us to define the intensive and extensive properties in more precise form. Let us consider a consider a set of intensive variables x, y, z, and a set of extensive variables X, Y, Z, then

(i) The function F (x, y, z,...X, Y, Z...) is an extensive variable if and only if F (x, y, z, λX, λY, λZ) = λF (x, y, z, X, Y, Z) for arbitrary positive λ.

For example, volume is an extensive property because

V (T, P, λn_1, λn_2) = λV (T, P, n_1, n_2).

(ii) The function f (x, y, z, X, Y, Z) is an intensive property if and only if f (x, y, z, λX, λY, λZ) = f (x, y, z, X, Y, Z) for arbitrary positive λ.

Example 4(a):

10 g ice at 0° C are added to 20 g water at 90° C in a thermally insulated flask of negligible of negligible heat capacity. The heat of fusion of ice is 5980 J mol^{-1}. What is the final temperature, ΔS_{sytem} and $\Delta S_{surroundings}$? $C_{P,m}(H_2O, l)$ = 75.42 JK^{-1}mol^{-1}.

Solution:

When ice and water are mixed then ice melts and the temperature of the mixture is changed. The change may be considered step by step as follows:

(i) (10/08) mol H_2O (s, 273 K) $\xrightarrow{\text{Melt}}$ (10/18) mol H_2O (l,273K)

$$q\ (i) = n.\ \Delta H_{fus} = 5980 \times \frac{10}{18}\ J = 3322.2\ J$$

The water at 273 K is heated to a temperature T_2K.

(ii) (10/18) H_2O (l, 273 K) $\xrightarrow{\text{Heat}}$ (10/18) H_2O (l,T_2K)

q (ii) = $nC_{P,m}$ (H_2O) ($T_2 - T_1$)

$= (10/18)\ (75.42\ JK^{-1}\ mol^{-1})\ (T_2 - 273)$

$= 41.9\ (T_2 - 273)$

(iii) $(20/18)$ mol H_2O (l, 363 K) $\xrightarrow{\text{Cool}}$ $(20/18)$ mol H_2O (l, T_2 K)

$q\ (iii) = (20/18)\ mol \times (75.42\ JK^{-1}\ mol^{-1}) \times (T_2 - 363)$

$= 83.4\ (T_2 - 363)$

$\therefore$ 10g H_2O (s, 0°C) + 20g H_2O (l, 90°C) $\rightarrow$ 30g H_2O (l, T_2K)

Total heat change in the system q = q (i) + q (ii) + q (iii)

Since the change is adiabatic, therefore q = 0, and q (i) + q (ii) + q (iii) = 0

$$\frac{10}{18} \times 5980 + \frac{10}{18} \times 75.42 \times (T_2 - 273) + \frac{20}{18} \times 75.42\ (T_2 - 363) = 0$$

Solving for T_2 we get $T_2 = 306$ K.

Entropy changes:

$$\Delta S\ (i) = q_1/273 = \frac{5980 \times 10}{273 \times 18} = 12.169\ JK^{-1}$$

$$\Delta S\ (ii) = n_1 C_{P,m}\ In\ (T_2/T_1)$$

$$= \frac{10}{18} \times 75.42\ In\ \frac{306}{273} = -4.782\ JK^{-1}$$

$$\Delta S\ (iii) = n_2 C_{P,m}\ In\ (T_2/T_1)$$

$$= \frac{20}{18} \times 75.42\ In\ \frac{306}{363} = -14.317\ JK^{-1}$$

$$\Delta S = \Delta S\ (i) + \Delta S\ (ii) + \Delta S\ (iii) = 2.634\ JK^{-1}$$

$\Delta S_{Surrounding} = 0$, since the process is adiabatic, $q_{surr} = 0$.

Example 4(b):

Derive an expression for the coefficient of thermal expansion (α) for a gas that obeys van der Waals equation of state.

Solution:

For one mole of van der Waals gas

$$\left(P + \frac{a}{V^2}\right)(V - b) = RT$$

or $PV^3 - PbV^2 + aV - ab - RTV^2 = 0$

Differentiating with respect to T at constant P, we get,

$$3PV^2\left(\frac{\partial V}{\partial T}\right)_P - 2PbV\left(\frac{\partial V}{\partial P}\right)_P + a\left(\frac{\partial V}{\partial T}\right)_P - 2RTV\left(\frac{\partial V}{\partial T}\right)_P - RV^2 = 0$$

Solving for $\left(\frac{\partial V}{\partial T}\right)_P$, we have

$$\left(\frac{\partial V}{\partial T}\right)_P = \frac{RV^2}{3PV^2 - 2PbV + a - 2RTV} \qquad \text{...(i)}$$

Now α is defined as

$$\alpha = \frac{1}{V}\left(\frac{\partial V}{\partial T}\right)_P \qquad \text{...(ii)}$$

Putting the expression for $(\partial V/\partial T)_P$ from into (ii), we have

$$\alpha = \frac{1}{V}\left[\frac{RV^2}{3PV^2 - 2PbV + a - 2RTV}\right]$$

$$= \frac{R}{3PV - 2Pb + \frac{a}{V} - 2RT} \qquad \text{...(iii)}$$

van der Waals equation can also be written as

$$RT = PV - Pb + \frac{a}{V} - \frac{ab}{V^2} \qquad \text{...(iv)}$$

Putting the expression for RT into equation (iii), we get

$$\alpha = \frac{R}{3PV - 2Pb + \frac{a}{V} - 2\left(PV - Pb + \frac{a}{V} - \frac{ab}{V^2}\right)}$$

$$= R\left[PV - \frac{a}{V} + \frac{2ab}{V^2}\right]^{-1}$$

Example 124(c):

One mole of an ideal gas is taken through a reversible carnot cycle. Predict the sign of ΔT, ΔP, ΔV, ΔE, ΔH, ΔS in each step without preforming the actual calculations.

Solution:

A reversible Carnot cycle consists of (a) isothermal expansion, (b) adiabatic expansion, (c) isothermal compression, and (d) adiabatic compression.

(a) $\Delta T = 0$ (isothermal); $\Delta V > 0$ (expansion); $\Delta P < 0$ (since $V \propto 1/P$); $\Delta H = 0$, $\Delta E = 0$ since for ideal gas $E = E(T)$ and $H = H(T)$, $\Delta S > q > 0$ for isothermal expansion.

(b) $\Delta V > 0$, (expansion), $\Delta P < 0$, $\Delta E < 0$, $\Delta H < 0$, $\Delta T < 0$ (adiabatic expansion), $\Delta S = 0$ $(q = 0)$

(c) $\Delta T = 0$, $\Delta V < 0$, $\Delta P > 0$, $\Delta E = 0$, $\Delta H = 0$, $\Delta S < 0$.

(d) $\Delta T > 0$, $\Delta V < 0$, $\Delta P > 0$, $\Delta E > 0$, $\Delta H > 0$, $\Delta S < 0$.

Example 5:

(a) *Derive an expression for* β *for a van der Waals gas.*

Solution :

$$\beta = \frac{1}{V}\left(\frac{\partial V}{\partial P}\right)_T$$

van der Waals equation is written as,

$$PV^3 - PbV^2 + aV - ab - RTV^2 = 0$$

Differentiating with respect to P at constant T,

$$3PV^2\left(\frac{\partial V}{\partial P}\right)_T + V^3 - 2PbV\left(\frac{\partial V}{\partial P}\right)_T - bV^2 + a\left(\frac{\partial V}{\partial P}\right)_T - 2RTV\left(\frac{\partial V}{\partial P}\right)_T = 0$$

Solving for $\left(\frac{\partial V}{\partial P}\right)_T$, we get

$$\left(\frac{\partial V}{\partial P}\right)_T = \frac{V^2(b-V)}{3PV^2 - 2PbV + a - 2RTV}$$

Also, for van der Waals gas,

$$RTV = PV^2 + a - PbV - \frac{ab}{V}$$

$$\therefore \left(\frac{\partial V}{\partial P}\right)_T = \frac{V^2(b-V)}{3PV^2 - 2PbV + a - 2\left(PV^2 + a - PbV - \frac{ab}{V}\right)}$$

$$= \frac{b - V}{P - a/V^2 + 2ab/V^3}$$

$$\therefore \qquad \beta = -\frac{1}{V}\left(\frac{\partial V}{\partial P}\right)_T = -\frac{b - V}{PV - a/V + 2ab/V^2}$$

(b) *Show that volume is a state function for a gas obeying the equation:*

$P = RTV^{-1} - aV^{-2}$.

Solution:

Volume (V) is a state function if dV is exact differential; dV is an exact differential if $(\partial^2 V/\partial P\,\partial T) = (\partial^2 V/\partial T\,\partial P)$. Differentiating equation $P = RTV^{-1} - aV^{-2}$, we get

$$dp = RV^{-1}dT - RTV^{-2}\,dV + 2aV^{-3}dV$$

$$= RV^{-1}dT + (2aV^{-3} - RTV^{-2})dV$$

$$\therefore\ (\partial V/\partial P)_T = (2aV^{-3} - RTV^{-2})^{-1} = (aV^{-3} - PV^{-1})^{-1} \qquad \text{...(i)}$$

$$\text{and } (\partial V/\partial T)_P = -RV^{-1}/\ (2aV^{-3} - RTV^{-2}) = -\ R\ [aV^{-2} - P]^{-1} \qquad \text{...(ii)}$$

Differentiating equation (i) with respect to T at constant P, we get

$$\frac{\partial^2 V}{\partial T \partial P} = -\ (aV^{-3}\ PV^{-1})^{-2} \times \left[-3aV^{-4}\left(\frac{\partial V}{\partial T}\right)_P + PV^{-2}\left(\frac{\partial V}{\partial T}\right)_P\right]$$

$$= V^2\ (aV^{-2} - P)^{-2} \times V^{-2}\ (P - 3aV^{-2}) \times R\ (aV^{-2} - P)^{-1}$$

$$= R\ (aV^{-2} - P)^{-3}\ (P - 3aV^{-2}) \qquad \text{...(iii)}$$

Differentiating equation (ii) with respect to P at constant T, we have

$$\frac{\partial^2 V}{\partial P \partial T} = R\ (aV{-}2{-}P){-}2 \times \left[-2aV^{-3}(\partial V/\partial P)_T{}^{-1}\right]$$

$$= -\ R\ (aV^{-2} - P)^{-2} \times [2aV^{-3}(aV^{-3}PV^{-1})^{-1} + 1]$$

$$= R\ (P - 3aV^{-2})\ (aV^{-2} - P)^{-3} \qquad \text{...(iv)}$$

Thus $[\partial^2 V/\partial T\ \partial P] = [\partial^2 V/\partial P\ \partial T]$

Example 6:

For an ideal gas (PV = nRT) show that (1/T) is an integrating factor for dw = PdV.

Solution:

Given that dw = PdV ...(i)

PV = nRT for ideal gas.

$\therefore$ PdV + VdP = nRdT

and $dV = \frac{nR}{P}dT - \frac{V}{P}dP$ ...(ii)

Substituting the expression for dV in equation (i) we get

$$dw = nRdT - VdP = nRdT - \frac{nRT}{P}dP$$

Now $\frac{\partial}{\partial P}(nR)_T = 0$

and $\frac{\partial}{\partial T}\left(-\frac{nRT}{P}\right)_P = -\frac{nR}{P}$

Therefore, dw is inexact differential.

Let $f = \frac{1}{T}$ be an integrating factor, then

$$f.\ dw = \frac{1}{T}\ nR\ dT - \frac{nR}{P}dP$$

$$\therefore \frac{\partial}{\partial P}\left[\frac{nR}{T}\right]_T = 0$$

and $\frac{\partial}{\partial T}\left[\frac{nR}{P}\right]_P = 0$

Thus $f.\ dw = \frac{nR}{T}dT - \frac{nR}{P}dP$ is an exact differential and $\frac{1}{T}$ is and integrating factor.

Example 7(a):

Calculate the entropy change when 0.5 dm^3 of an ideal gas ($C_{V,m}$ = 12.6 $JK^{-1}\ mol^{-1}$) at 298 K and 1 atm is allowed to expand to double its volume and simultaneous heated to 373 K.

Solution:

ΔS for the system is the same whether the changes take place in one step or in stages. Thus ΔS system is given by

$$\Delta S = nC_{V,m} \ln \frac{T_2}{T_1} + nR \ln \frac{V_2}{V_1}$$

$C_{V,m} = 12.6 \text{ JK}^{-1} \text{ mol}^{-1}$, $T_1 = 298\text{K}$,

$T_2 = 373$ K, $V_1 = 0.5 \text{ dm}^3$,

$V_2 = 1.0 \text{ dm}^3$,

$P_1 = 1$ atm,

$$n = P_1V_1/RT_1 = \frac{[(1\text{atm}) \times 0.5 \text{ dm}^3]}{(0.082\text{dm}^3\text{atmK}^{-1}\text{mol}^{-1}) \times (298\text{K})}$$

$= 0.02$ mol

$$\therefore \Delta S = (0.02 \text{ mol}) \times (12.6 \text{ JK}^{-1}\text{mol}^{-1}) \times 2.303 \log \frac{373\text{K}}{298\text{K}}$$

$$+ (0.02 \text{ mol}) \times (8.314\text{JK}^{-1}\text{mol}^{-1}) \times 2.303 \log \frac{1.0\text{dm}^3}{0.5\text{dm}^3}$$

$= \mathbf{0.173 \text{ JK}^{-1}}$.

Example 7(b):

If $E = f(T, V)$ and dE is an exact differential then show that $\left(\frac{\partial E}{\partial V}\right)_T = T\left(\frac{\partial P}{\partial T}\right)_V - P$. *relationship can be derived by using the relation $dq = dE + PdV$ and taking $(1/T)$ as an integrating factor.*

Solution:

Since E = f (T, V)

$$\text{Therefore, } dE = \left(\frac{\partial E}{\partial T}\right)_V dT + \left(\frac{\partial E}{\partial V}\right)_T dV$$

Given that dq = dE + PdV

$$\text{Now } dE = \left(\frac{\partial E}{\partial T}\right)_V dT + \left(\frac{\partial E}{\partial V}\right)_T dV$$

$$\therefore dq = \left(\frac{\partial E}{\partial T}\right)_V dT + \left[\left(\frac{\partial E}{\partial V}\right)_T + P\right] dV$$

Let f = 1/T be the integrating factor, then f dq must be an exact differential.

$$fdq = \frac{1}{T}\left(\frac{\partial E}{\partial T}\right)_V dT + \frac{1}{T}\left[\left(\frac{\partial E}{\partial V}\right)_T + P\right] dV$$

$$= M\ dT + N\ dV$$

where, $M = \frac{1}{T}\left(\frac{\partial E}{\partial T}\right)_V$; $N = \frac{1}{T}\left[\left(\frac{\partial E}{\partial T}\right)_T + P\right]$

For the above differential to be exact we must have

$$\left(\frac{\partial M}{\partial V}\right)_T = \left(\frac{\partial N}{\partial T}\right)_V$$

or $$\frac{\partial}{\partial V}\left[\frac{1}{T}\left(\frac{\partial E}{\partial T}\right)_V\right]_T = \frac{\partial}{\partial T}\left[\frac{1}{T}\left(\frac{\partial E}{\partial T}\right)_T + \frac{P}{T}\right]_V$$

or $$\frac{1}{T}\frac{\partial^2 E}{\partial V \partial T} = \frac{1}{T}\frac{\partial^2 E}{\partial T \partial V} - \frac{1}{T^2}\left(\frac{\partial E}{\partial V}\right)_T - \frac{P}{T^2} + \frac{1}{T}\left(\frac{\partial P}{\partial T}\right)_V \quad ...(i)$$

Since dE is an exact differential, therefore,

$$\frac{\partial^2 E}{\partial T \partial V} = \frac{\partial^2 E}{\partial V \partial T} \quad ...(ii)$$

From equations (i) and (ii), we get

$$-\frac{1}{T^2}\left(\frac{\partial E}{\partial V}\right)_T - \frac{P}{T^2} + \frac{1}{T}\left(\frac{\partial P}{\partial T}\right)_V = 0$$

or $$\left(\frac{\partial E}{\partial V}\right)_T = T\left(\frac{\partial P}{\partial T}\right)_V - P$$

***Example** 7(c):*

One mole of supercooled water at – 10°C and 1 atm pressure turns into ice at –10°C. Calculate the entropy change in the surroundings and the system and the net entropy change. Heat capacity of water and ice at 1 atm may be taken as constant at 75.42 JK^{-1} mol^{-1} and 37.20 JK^{-1} mol^{-1} respectively.

ΔH_{fus} (273 K) = 6008 J mol^{-1}.

Solution:

The system is H_2O (l, – 10°C, 1 atm) → H_2O (s, –10°C, 1 atm). The given process is not a reversibly one so the entropy change for this cannot be calculated directly. But we can consider a series of reversible steps leading form the liquid water at – 10°C to solid ice at – 10°C.

The various steps are as follows:

Step 1. Heat reversibly the super cooled water at – 10°C to 0°C keeping the pressure constant.

Step 2. Allow the solidification of water at 0°C to ice at 0°C.

Step3. Cool the ice reversibly from 0°C to – 10°C. Thus

Step 1. $H_2O\ (l, -10°\ C, 1\ atm) \xrightarrow{\text{Heat}} H_2O\ (l, 0°\ C, 1\ atm)$

The heat required to raise the temperature of supercooled water by dT is

$$dq_{rev} = C_P\ dT = dH$$

The entropy change in raising the temperature from 263 K to 273 K

$$\Delta S_1 = \int_{263}^{273} \frac{dq_{rev}}{T} = C_P H_2O(l)\ \text{In} \frac{273}{263}$$

$$= 2.303 \times (75.42\ JK^{-1}mol^{-1}) \times \log \frac{273}{263}$$

$$= 2.81\ JK^{-1}\ mol^{-1}$$

Step 2: $H_2O\ (l, 0°C, 1\ atm) \rightarrow H_2O\ (s, 0°C, 1atm)$

$$\Delta S_2 = -\frac{\Delta H_{fus}}{T}$$

$$= -\frac{6008 Jmol^{-1}}{273K} = -22.0\ JK^{-1}mol^{-1}$$

Step 3: $H_2O\ (s, 0°C, 1atm) \xrightarrow{\text{COOl}} H_2O\ (s, -10°C, 1atm)$

Heat change: $dq_{rev} = C_P\ (H_2O, s)\ dT = dH$

Entropy change : $\Delta S_2 = C_P\ (H_2O, s)\ \text{In} \frac{263}{273}$

$$= 2.303 \times 37.20 \times \log \frac{263}{273}$$

$$= -1.38\ JK^{-1}\ mol^{-1}$$

On adding the equations of steps 1, 2 and 3, we get

$H_2O\ (l, -10°C, 1\ atm) \rightarrow H_2O\ (s; -10°C, 1\ atm)$

and $\Delta S = \Delta S_1 + \Delta S_2 + \Delta S_3$

$= 2.81 - 22.0 - 1.38$

$= -$ **20.57** $JK^{-1}\ mol^{-1}$

Here we see that there has been a decrease in entropy of water upon crystallization at $-10°C$, though the process is irreversible. This example emphasizes the fact that sign of the entropy change for the system plus surroundings, and not merely for any component, is related to irreversibility. Thus we must consider the entropy change for the system as well as that of surroundings. Now ΔS_{surr} can be calculated by considering the surroundings to be large enough so that the exchange of heat can be considered as a reversible process. Thus

$\Delta S_{surr} = \Delta H_{surr}/T = -\Delta H_{system}/T$

$\Delta H_{syst} = \Delta H_1 + \Delta H_2 + \Delta H_3$

ΔH_1 = Heat gained by the system in heating water from 263 K to 273 K

$= C_P\ (H_2O, l) \times \Delta T = (74.\ 42\ JK^{-1}\ mol^{-1}) \times (10\ K)$

$= 754.2 J\ mol^{-1}$

$\Delta H_2 = -\Delta H,\ fus = -6008\ mol^{-1}$

ΔH_3 = Heat lost by the system when it is cooled from 273 K to 263 K

$= C_P\ (H_2O, s) \times \Delta T = -(37.2\ JK^{-1}\ mol^{-1}) \times (10K)$

$= -372.0\ J\ mol^{-1}$

$\Delta H_{syst} = (754.2 - 6008 - 372)\ J\ mol^{-1}$

$= -5626\ J = -\Delta H_{surr}$

$$\Delta S_{surr} = \frac{5626}{263K} = 21.39\ JK{-}1$$

$\Delta S_{universe} = \Delta S_{syst} + \Delta S_{surr}$

$= -20.57\ JK^{-1} + 21.39.JK^{-1}$

$= +0.82\ JK^{-1}$

Example 8:

(a) *A function ϕ is defined as $\phi(x, y) = x^2 y^3 + x$. Write its partial derivatives and total differential $d\phi$. Test whether $d\phi$ is an exact differential or not.*

Solution:

ϕ (x, y) = $x^2y^3 + x$, Partial derivatives :

$$\left(\frac{\partial \phi}{\partial x}\right)_y = 2xy^3 + 1$$

$$\left(\frac{\partial \phi}{\partial y}\right)_x = 3x^2y^2$$

Total differential $d\phi = \left(\frac{\partial \phi}{\partial x}\right)_y dx + \left(\frac{\partial \phi}{\partial y}\right)_x dy$

$= (2xy^3 + 1)\, dx + 3x^2y^2\, dy$

In order that df bean exact differential, we have to prove

$$\frac{\partial^2 \phi}{\partial y \partial x} = \frac{\partial^2 \phi}{\partial x \partial y}$$

Now $\left(\frac{\partial \phi}{\partial x}\right)_y = 2xy^3 + 1$

and $\left(\frac{\partial \phi}{\partial y}\right)_x = 3x^2y^2$ give

$$\frac{\partial}{\partial y}\left[\left(\frac{\partial \phi}{\partial x}\right)_y\right]_x = \frac{\partial}{\partial y}(2xy^3 + 1)_x = 6\, xy^2$$

$$\frac{\partial}{\partial x}\left[\left(\frac{\partial \phi}{\partial y}\right)_x\right]_y = \frac{\partial}{\partial x}\left(3x^2y^2\right)_y$$

$$= 6xy^2 = \frac{\partial}{\partial y}\left[\left(\frac{\partial \phi}{\partial x}\right)_y\right]_x$$

Hence $d\phi$ is an exact differential.

(b) *Prove that in the equation* $dz = (52\, x^3y + 10\, y^5)\, dx + (13\, x^4 + 50\, xy^4)\, dy$, *dz is an exact differential.*

Example 9:

Two moles of an ideal gas are expanded isothermally at 298 K from a volume of V to 2.5V. Find the value of ΔS_{gas}, and ΔS_{total} the for following:

(a) Reversible expansion.

(b) Irreversible expansion when 400 J mol^{-1} are less absorbed than in (a).

(c) Free expansion.

Solution:

(a) $\Delta S_{gas} = nR \ In \ V_2/V_1$

$V_2 = 2.5 \ V, \ V_1 = V,$

n = 2 mol, hence

$\Delta S_{gas} = 2.303 \times (2 \ mol) \times (8.314 \ JK^{-1} \ mol^{-1}) \log 2.5$

$= 15.258 \ JK^{-1}$

$q_{rev} = nRT \ In \ V_2/V_1 = 4540.9 \ J$

$\Delta S_{surr} = q_{surr}/T = - q_{syst}/T$

$= - 4540.9 \ J/298 \ K$

$= - 15. \ 238 \ JK^{-1}$

$\Delta S_{total} = \Delta S_{gas} + \Delta S_{surr}$

$= 15.238 - 15.238 = 0$

(b) $\Delta S_{gas} = 2.303 \ nR \log V_2/V_1 = 15.238 \ JK^{-1}$

$q_{irr} = q_{rev} - 2 \times 400 \ J \ mol^{-1} = 4540.9 - 800$

$= 3740.9 \ J$

$\Delta S_{surr} = q_{surr}/T = - q_{syst}/T$

$= -3740.9 \ J/298 \ K$

$= - 12.533 \ JK^{-1}$

$\Delta S_{total} = \Delta S_{gas} + \Delta S_{surr}$

$= 15.298 - 12.553 = 2.685 \ JK^{-1}$

(c) $\Delta S_{gas} = 2.303 \ nR \log V_2/V_1 = 15.238 \ JK^{-1}$

$q = 0$ hence $\Delta S_{surr} = 0$, and

$\Delta S_{total} = \Delta S_{gas} = 15.238 \ JK^{-1}$.

Example 10(a):

Illustrate the method to evaluate the entropy change for the following process:

$H_2O(l, 298\ K, 1.01 \times 10^5\ Nm^{-2}) \rightarrow H_2O\ (v, 398\ K, 0.5 \times 10^5\ Nm^{-2})$.

Solution:

The given process is irreversible, therefore, we cannot use the equation $dS = dq_{rev}/T$. However, this system can be split into various reversible systems can be split into various reversible systems for the convenience. This is a valid procedure since S is a state function. We proceed as follows:

(a) Heat reversibly and isobarically the water from 298 K to 373 K, (b) allow the evaporation of water under reversible isothermal and isobaric condition, (c) heat water vapour reversibly and isobarically from 373 K to 398 K, (d) allow the isothermal and reversible expansion of the water vapour. The processes and the corresponding entropy changes are :

(a) $H_2O\ (l, 298\ K, 1.01 \times 105\ Nm^{-2}) \xrightarrow{\text{Heat}} H_2O\ (l, 373K, 1.01 \times 10^5\ N\ m^{-2})$

$\Delta S\ (a) = C_{P,m}\ (H_2O, l)\ In\ (373/298)$

(b) $H_2O\ (l, 373\ K, 1.01 \times 10^5\ Nm^{-2}) \rightleftharpoons H_2O\ (v, 373\ K, 1.01 \times 10^5\ N\ m^{-2})$

$\Delta S\ (b) = \Delta H_{vap}/373$

(c) $H_2O\ (v, 373\ K, 1.01 \times 10^5\ Nm^{-2}) \xrightarrow{\text{Heat}} H_2O\ (v, 398\ K, 1.01 \times 10^5\ N\ m^{-2})$

$\Delta S\ (c) = C_{P,m}\ (H_2O, v)\ In\ (398/373)$

(d) $H_2O\ (v, 398\ K, 1.01 \times 10^5\ Nm^{-3}) \xrightarrow{\text{Expand}} H_2O\ (v, 398\ K, 0.5 \times 10^5\ N\ m^{-2})$

$$\Delta S(d) = -\ R\ In\ \frac{0.5 \times 10^5\ Nm^{-2}}{1.01 \times 10^5\ Nm^{-2}}$$

Adding equation (a) to (d) we get,

$H_2O\ (l, 298\ K, 1.01 \times 10^5\ Nm^{-2}) \rightarrow H_2\ O\ (v, 398\ K\ 0.5 \times 10^5\ Nm^{-2})$

$\Delta S = \Delta S\ (a) + \Delta S\ (b) + \Delta S\ (c) + \Delta S\ (d)$

$$= C_{P,m}\ (H_2O, l)\ In\ \frac{373}{298} + \frac{\Delta H_{vap}}{373} + C_{P,m}\ (H_2O, v)$$

$$\text{In}\ \frac{373}{298} + R\ \text{In}\frac{1.01}{0.5}$$

On substituting the values of heat capacities of liquid and water vapour, ΔS can be calculated. In the step (d) we assumed the water vapour to behave as an ideal gas.

Example 10(b):

One mole van der Waals gas is allowed to expand isothermally and reversibly at 273 K from an initial volume of 1.0 dm³ to 50.0 dm³. Calculated the entropy change for the gas. The van der Waals constants are

$a = 6.5$ atm $dm^6\ mol^{-2}$

$b = 0.056\ dm^3\ mol^{-1}$.

Solution:

Given that the expansion is reversible and isothermal, therefore $\Delta S = q_{rev}/T$ can be used to calculate entropy change.

For 1 mole van der Waals gas

$q_{rev} = RT\ \text{In}\ (V_2 - b)/\ (V_1 - b)$

$\therefore\ \Delta S = R\ \text{In}\ (V_2 - b)/\ (V_1 - b)$

$$= (8.314\ JK^{-1}\ mol^{-1}) \times 2.303 \times \log\frac{(50 - 0.056)}{(1 - 0.056)}$$

$= 32.97\ JK^{-1}mol^{-1}$

Example 11:

A gas obeys the equation of state $\left(P + \frac{a}{V^2}\right)V = RT$. *Derive the cyclic rule for this gas.*

Solution:

The given equation of state can be written as

$$P = \frac{RT}{V} - \frac{a}{V^2} \qquad \text{...(i)}$$

Differentiation gives,

$$dP = \frac{R}{V}dT - \frac{RT}{V^2}dV + \frac{2a}{V^3}dV$$

$$= \frac{R}{V}dT - \left(\frac{RT}{V^2} - \frac{2a}{V^3}\right)dV \qquad \text{...(ii)}$$

As constant volume dV = 0 and hence equation (ii) yields.

$$\left(\frac{\partial P}{\partial T}\right)_V = \frac{R}{V} \qquad \text{...(iii)}$$

When pressure is constant, dP = 0 and equation (ii) reduces to

$$\left(\frac{\partial T}{\partial V}\right) = \left(\frac{RT}{V^2} - \frac{2a}{V^3}\right) \Big/ \frac{R}{V} \qquad \text{...(iv)}$$

For constant temperature process, dT = 0 and equation (ii) can be written as

$$\left(\frac{\partial V}{\partial P}\right)_T = -\frac{1}{\left(\frac{RT}{V^2} - \frac{2a}{V^3}\right)} \qquad \text{...(v)}$$

On multiplying equations (iii), (iv) and (v) we get

$$\left(\frac{\partial P}{\partial T}\right)_V \left(\frac{\partial T}{\partial V}\right)_P \left(\frac{\partial V}{\partial P}\right)_T$$

$$= -\left(\frac{R}{V}\right) \times \frac{\left(\frac{RT}{V^2} - \frac{2a}{V^3}\right)}{\left(\frac{R}{V}\right)} \times \frac{1}{\left(\frac{RT}{V^2} - \frac{2a}{V^3}\right)} = -1$$

or $$\left(\frac{\partial P}{\partial T}\right)_V \left(\frac{\partial T}{\partial V}\right)_P \left(\frac{\partial V}{\partial P}\right)_T + 1 = 0$$

This is cyclic rule.

Example 12(a):

Evaluate the change in entropy when 3 moles of a gas are heated from 27° C to 7227° C at a constant pressure of 1 atm. The molar heat capacity of the gas is 23.7 JK^{-1} mol^{-1}.

Solution:

n = 3 mol. T_1 = 300 K, T_2 = 100 K,

$C_{P,m}$ = 23.7 J_{K-1} mol_{-1}

$$\Delta S = nC_{P,m} \times 2.303 \log (T_2/T_1)$$
$$= (3 \text{ mol}) \times (23.7 \text{ JK}^{-1} \text{ mol}^{-1}) \times 2.303 \log (100 \text{ K}/300 \text{ K})$$
$$= \mathbf{85.5 \text{ JK}^{-1}}$$

Example 12(b):

Two blocks of copper metal are of the same size but are at different temperatures T_1 and T_2. These block are brought together and allowed to attain thermal equilibrium i.e. both are allowed to come to the same final temperature T. Show that the entropy change is given by

$$\Delta S = C_P \ \text{In} \left[\frac{(T_2 - T_1)^2}{4T_1T_2} + 1 \right]$$

Solution:

Let the metallic block A be at temperature T_1 and block B at temperature T_2 in the initial state. Let $T_2 > T_1$.

Heat lost by block B = $C_{PB} (T_2 - T)$

Heat gained by block A = $C_{PA} (T_2 - T_1)$

At equilibrium

$$C_{PB} (T_2 - T) = C_{PA} (T - T_1)$$

$$\therefore T = \frac{T_1 + T_2}{2} \qquad \because C_{PB} = C_{PA} =$$

Decrease of entropy of block B

$$= \int_{T_2}^{T} C_P d\text{In}T$$

$$\Delta S_B = C_P \ \text{In} \frac{T}{T_2}$$

Increase of entropy of block A

$$\Delta S_A = C_P \ \text{In} \ \frac{T}{T_1}$$

Total entropy change $\Delta S = \Delta S_A + \Delta S_B$

$$= C_P \left[\text{In} \frac{T}{T_1} + \text{In} \frac{T}{T_2} \right]$$

$$= C_P \left[\text{In} \frac{(T_1 + T_2)^2}{4} - \text{In} T_1 T_2 \right]$$

$$= C_P \text{ In } \frac{T_1^2 + T_2^2 + 2T_1T_2}{4T_1T_2}$$

Adding and subtracting $2T_1T_2$ to the In part of the above equation, we get

$$\Delta S = C_P \text{ In} \frac{T_1^2 + T_2^2 + 2T_1T_2 + 4T_1T_2}{4T_1T_2}$$

$$= C_P \text{ In} \left[\frac{(T_2 - T_1)^2}{4T_1T_2} + 1 \right]$$

Now $(T_1 - T_2)^2$ is always positive and hence the quantity within bracket on right side of the above equation is greater than one. Therefore right side is a positive should be spontaneous.

Example 12(c):

Calculate the entropy change for the transformation

12 (s, 1atm, 298 K) → I_2 (v, 1atm, 457 K)

Given that : ΔH_{fus}, m = 15.68 kJ mol^{-1} at the m.pt 113.6°C

$\Delta H_{vap,\,m}$ = 25.522 kJ mol^{-1} at the b.pt 184° C.

$C_{P,\,m}$ (12, s) = 54.684 + 13.431 × 10–4 (t –25) J deg–1 mol–1

$C_{P,\,m}$ (12, s) = 81.588 JK^{-1} mol^{-1}.

Solution:

Split the process at

(i) J_2 (s, 1 atm, 298 K) $\xrightarrow{\text{heat}}$ I_2 (s, 1 atm, 113.6°C)

$$\Delta S \text{ (i)} = \int_{298}^{386.6} \frac{C_{P,m(I_2,S)}}{T} dT$$

$C_{P,m}$ (I_2, S) = 54.684 + 13.431 × 10^{-4} (t + 273 – 273 –25)

$$\therefore \Delta S \text{ (i)} = 54.684 \int \frac{dT}{T} + 13.4.31 \times 10^{-4} \int \frac{TdT}{T}$$

$$- 13.431 \times 10^{-4} \times 298 \int \frac{dT}{T}$$

$= (54.684 - 13.431 \times 10^{-4} \times 298) \ln \frac{386.6}{298}$

$+ 13.431 \times 10^{-4} (386.6 - 298)$

$= 14.248 \text{ JK}^{-1}\text{mol}^{-1}$

(ii) I_2 (s, 1atm, 113.6°C) $\xrightarrow{\text{melt}}$ I_2 (l, 1atm, 113.6°C)

$\Delta S \text{ (ii)} = \Delta H_{fus}, m/T_m = 15648 \text{ J mol}^{-1}/386.6 \text{ K}$

= 40.474 JK–1 mol–1

(iii) I_2 (l, 1atm, 113.6°C) $\xrightarrow{\text{heat}}$ I_2 (l, 1atm, 184°C)

$\Delta S \text{ (iii)} = \int C_{P,m}(I_{2,} l)\frac{dT}{t}$

$= 81.588 \text{ JK}^{-1} \text{ mol}^{-1} \ln (457 \text{ K}/386.6 \text{ K})$

$= 13.649 \text{ JK}^{-1} \text{ mol}^{-1}$

(iv) I_2 (l, 1atm, 457 K) $\xrightarrow{\text{vaporise}}$ I_2 (v, 1atm, 457 K)

$\Delta S \text{ (iv)} = \Delta H_{vap,\, m} / T_b$

$= 25522 \text{ J mol}^{-1}/457 \text{ K}$

$= 55.846 \text{ JK}^{-1} \text{ mol}^{-1}$

Total entropy change for the transformation

$\Delta S = \Delta S \text{ (i)} + \Delta S \text{ (ii)} + \Delta S \text{ (iii)} + \Delta S \text{ (iv)}$

$= 124.218 \text{ JK}^{-1} \text{ mol}^{-1}$.

Example 13:

0.1kg nitrogen gas at 298 K are held by a piston under 30 atm pressure. The pressure is suddenly released to 10 atm and the gas expands adiabatically. If $C_{V,\, m} = 20.8\ J^{K-1}\ mol^{-1}$, what is the final temperature and volume? Calculate ΔS system for this expansion. What would be the value of ΔS (surroundings)?

Solution:

The change is adiabatic (q = 0) but irreversible as the expansion is sudden against a constant pressure of 10 atm. Applying the first law we shall get

$C_V (T_2 -/T_1) = w = - P_2 (V_2 - V_1)$

$P_1V_1/ T_1 = P_2V_2/T_2 = nR$ gives

$C_V (T_2 - T_1) = nR (T_1P_2/P_1 - T_2)$

$$\therefore T_2 = \frac{nC_{V,m} + nRP_2/P_1}{nC_{P,m}} \times T_1$$

$$= \left(\frac{20.8JK^{-1}mol^{-1} + 8.314JK^{-1} \times 10/30}{29.114JK^{-1}mol^{-1}}\right) \times 298\ K$$

= **241 K**

and $V_2 = P_1V_1T_2/T_1P_2$

$$= \frac{30 \times V_1 \times 241}{298 \times 10} = \mathbf{2.42\ V_1}$$

Entropy change for the system is given by

$$\Delta S\text{system} = nC_{V,m}\ \text{In}\frac{T_2}{T_1} + nR\text{In}\frac{V_2}{V_1}\ (\text{Eq. } 45)$$

$$= n\left(20.8JK^{-1}mol^{-1} \times 2.303\log\frac{241K}{298K}\right.$$

$$\left. + 8.314\ JK^{-1}\ mol^{-1}\ 2.303 \times \log\frac{2.42V_1}{V_1}\right)$$

$= n (-4.416\ JK^{-1}mol^{-1} + 7.348\ JK^{-1}\ mol^{-1})$

$= 3.571\ mol \times 2.932\ JK^{-1}\ mol^{-1}$

$= 10.47\ JK^{-1}$

$(\because n\ 0.1/0.028 = 3.571\ mol)$

$\Delta S_{Surroundings} = q_{surr}/T = 0$

Example 14:

The molar entropy of an ideal gas ($C_{P,m}$ = 20.9 $JK^{-1}mol^{-1}$) at 298K is 146.0 JK^{-1} mol^{-1}. Find its value at 500 K.

Solution:

We may assume it as a process of heating at constant pressure. Therefore, $dS = dq_{rev}/T = dH/T$ at constant P

$$dS = C_P\ \frac{dT}{T}$$

On integrating we get

$$\int_{S_1}^{S_2} dS = CP\int_{T_1}^{T_2} d\ In\ T$$

$$\therefore\ S\ (T_2) - S\ (T_1) = C_P\ In\ \frac{T_2}{T_1}$$

For the present case

$T_1 = 298$ K, $t_2 = 500$ K,

S (298 K) = 146.0JK^{-1} mol^{-1}

$\therefore$ S (500 K) = (146.0 JK^{-1} mol^{-1}) + (20.9 JK^{-1} mol^{-1})

$$\times\ 2.303 \times \log \frac{500K}{298K}$$

= 146.0 JK^{-1} mol^{-1} + 10.81 JK^{-1} mol^{-1}

= 156.81 JK^{-1} mol^{-1}

Example 15:

Calculate the entropy change when 10 m^3 of an ideal gas ($C_{P,m}$ = 25R) at 27° C and 1.01 × 10^5 N m^{-2} pressure are heated at constant pressure to 127°C.

Solution:

Given that

T_1 = 300 K, T_2 = 400 K, $C_{P,m}$ = 2.5 R,

P = 1.01 × 10^5 Nm^{-2},

Heat absorbed dq = dH = CP. dT = TdS

$$\therefore\ dS = C_P\ \frac{dT}{T} = C_P d\ In\ T$$

$$\int_1^2 dS = \Delta S = C_P\int d\ InT = C_P\ In\ \frac{T_2}{T_1} = nC_{P,m}In\frac{T_2}{T_1}$$

$$\text{Now, } n = \frac{PV}{RT} = \frac{(1.01\times 10^5\,Nm^{-2})\times 10\times 10^{-3}\,m^3}{(8.314JK^{-1}mol^{-1})\times(300K)}$$

$$\therefore\ \Delta S = (0.405\ mol) \times (2.5 \times 8.314\ mol^{-1}) \times 2.303\ \log\frac{400}{300}$$

= 2.422 JK^{-1}

Example 16(a):

1 mole of an ideal gas is allowed to expand isothermally at 27° C until its volume is tripled. Calculate ΔS_{syst} and ΔS_{univ} under the following conditions: (a) The expansion is carried out reversibly, (b) the expansion is a free expansion.

Solution:

(a) Isothermal reversible expansion: entropy change can be calculated by using the defining equation

$dS = dq_{rev}/T$ or $\Delta S = q_{rev}/T$.

In the present case

$q = -w = 2.303\ RT \log V_2/V_1$

$= 2.303 \times 8.314 \times 300 \log 3V_1/V_1$

$= 2740.6\ J\ mol^{-1}$

$\therefore \Delta S_{syst} = q_{rev}/T = 2740.6/300 = 9.135\ JK^{-1}\ mol^{-1}$

$\Delta S_{surr} = -\Delta S_{syst}$; here $\Delta S_{univ} = 0$

(b) In case of free expansion we can not use the equation $dS = dq_{rev}/T$. However, S being a state function, ΔS can be evaluated by

$\Delta S = 2.303\ R \log V_2/V_1 = 9.135\ JK^{-1}\ mol^{-1}$

For surrounding $q = 0$, and hence $\Delta S_{surr} = 0$

$\Delta S_{univ} = \Delta S_{syst} =$ **9.135 JK⁻¹ mol⁻¹**

Example 16(b):

(a) *Calculate the entropy change involved in the conversion of 1mole of ice at 0°C and 1 atm to liquid at 0° C and 1 atm; the enthalpy of fusion per mole of ice is* 6008 J mol^{-1}.

Solution:

The process is reversible.

H_2O (s, 0°C, 1 atm) → H_2O (l, 0° C, 1 atm)

$$\therefore \Delta S_{fus} = \frac{\Delta H_{fus}}{T_m}$$

$T_m = 273$ K; $\Delta H_{fus} = 6008\ J\ mol^{-1}$

$$\therefore \Delta S_{fus} = \frac{6008\text{J mol}^{-1}}{273\text{K}} = 22\ \mathbf{JK^{-1}}\ \text{mol}^{-1}$$

This is an increase in the entropy of the system. This means that the entropy of liquid state is greater than that of the solid state of H_2O. Since the randomness in liquid is greater than solid, hence entropy is a measure of randomness.

Example 16(c):

(b) *Calculate the change of entropy for the process*

H_2O (1, 373 K, 101325 NM^{-2}) $\rightarrow$ H^2O (v, 373 K, 101325 Nm^{-2})

ΔH_{vap} = 40850 J mol^{-1} at 373 K.

Solution:

The given process is isothermal and isobaric phase transformation.

$$\therefore \Delta S_{vap} = \frac{\Delta H_{vap}}{T_b} = \frac{40850\text{J mol}^{-1}}{373\text{K}} = \mathbf{109.5JK^{-1}mol^{-1}}$$

Since ΔS_{vap} = S(vapour) – S (liquid) = 109.5 JK^{-1} mol^{-1}

$\therefore$ S(vapour) = S(liquid) + 109.5 JK^{-1} mol^{-1}.

This observation is in tune with the fact that there is more randomness in the vapour state than the liquid state. Hence entropy is a measure of disorderness.

Example 16(d):

Evalvate the entropy charges for the following reversible processes:

(a) 1mole H_2O (l, 1atm, 100° C) $\rightleftharpoons$ 1 mole H_2O (v, 1 atm, 100°C)

ΔH_{vap} = 40850 J mol^{-1}

(b) 1 mole Sn (a, 1 atm, 13° C) $\rightleftharpoons$ 1 mole Sn (β, 1 atm, 13° C)

ΔH_{trans} = 2090 J mol^{-1}

Solution:

$$\text{(a)}\quad \Delta S_{vap} = \frac{\Delta H_{vap}}{T_b}$$

Here ΔH_{vap} = 40850 J mol^{-1}

$T_b = 100 + 273 = 373$ K

$$\therefore \Delta S_{vap} = \frac{40850 \text{Jmol}^{-1}}{373\text{K}} = 109.5 \text{ JK–1 mol}^{-1}$$

(b) $\Delta H_{trans} = 2090$ J mol^{-1} and $T_{trans} = 13 + 273 = 286$ K

$$\therefore \Delta S_{trans} = \frac{2090 \text{Jmol}^{-1}}{286\text{K}} = 7.31 \text{ JK}^{-1}\text{mol}^{-1}$$

Example 17:

A gas obeys the equation of state $P = nRT/V - n^2 RB/V^2$. Show that

(i) $\left(\frac{\partial S}{\partial V}\right)_T$ = nR/V; (ii) $(\partial E/\partial V)_T = n^2BR/V^2$, (iii) $(\partial H/\partial V)_T = 2n^2BR/V^2$ *where B is a constant.*

Solution:

(i) $(\partial S/\partial V)_T$ suggests that S = S (T, V), therefore,

$$dS = \left(\frac{\partial S}{\partial T}\right)_V dT + \left(\frac{\partial S}{\partial V}\right)_T dV$$

From the first and second laws we get

$$dS = dq_{rev}/T = \frac{1}{T}(dE + PdV)$$

But E = E (T,V) and hence

$$dE = \left(\frac{\partial E}{\partial T}\right)_V dT + \left(\frac{\partial E}{\partial V}\right)_T dV = C_V dT + \left(\frac{\partial E}{\partial V}\right)_T dV$$

$$\therefore dS = \frac{C_V}{T} dT + \frac{1}{T}\left[\left(\frac{\partial E}{\partial V}\right)_T + P\right] dV \qquad \text{...(iii)}$$

Comparing the coefficients of dV in equations (i) and (iii) it is seen that

$$\left(\frac{\partial S}{\partial V}\right)_T = \frac{1}{T}\left[\left(\frac{\partial E}{\partial V}\right)_T + P\right]$$

$$\text{But } \left(\frac{\partial E}{\partial V}\right)_T = T\left(\frac{\partial P}{\partial T}\right)_V - P \qquad \text{...(iii)}$$

$$\therefore \left(\frac{\partial S}{\partial V}\right)_T = \left(\frac{\partial P}{\partial T}\right)_V$$

For the given gas $(\partial P/\partial T)_V = nR/V$

$$\therefore \left(\frac{\partial S}{\partial V}\right)_T = nR/V \qquad \text{...(iv)}$$

(ii) Putting $(\partial P/\partial T)_V = nR/V$ in thermodynamic equation of state we get

$$(\partial E/\partial V)_T = T\frac{nR}{V} - P = n^2RB/V^2 \qquad \text{...(v)}$$

(iii) $H = E + PV$

$$\therefore \left(\frac{\partial H}{\partial V}\right)_T = \left(\frac{\partial E}{\partial V}\right)_T + P + V\left(\frac{\partial P}{\partial V}\right)_T \qquad \text{...(vi)}$$

For the given equation $P = nRT/V - n^2 RB/V^2$

$$\left(\frac{\partial P}{\partial V}\right)_T = - nRT/V^2 + 2n^2 RB/V^3$$

$$\therefore V\left(\frac{\partial P}{\partial V}\right)_T = - nRT/V + 2n^2RB/V^2 \qquad \text{...(vii)}$$

Also $(\partial E/\partial V)_T = n^2 RB/V^2$

On substituting the derivatives in equation (vi) we get

$$\left(\frac{\partial H}{\partial V}\right)_T = \frac{n^2RB}{V^2} + P - \frac{nRT}{V} + P - \frac{2n^2RB}{V^2}$$

$$\text{But } P - \frac{nRT}{V} = \frac{n^2RB}{V^2} \quad \therefore \quad \left(\frac{\partial H}{\partial V}\right)_T = \frac{2n^2RB}{V^2}$$

Example 18:

A gas obeying the equation of state $P(V - B) = RT$ undergoes a change from the initial state T_1, V_1 to a final state T_2, V_2. Derive an expression for the entropy change of this gas. The variation of heat capacity at constant volume is given by $C_V = a + bT + cT^2$.

Solution:

$$dS = \frac{dE + PdV}{T}, \; dE = C_V\, dT + (\partial E/\partial V)_T\, dV \text{ give}$$

$$dS = \frac{C_V}{T}dT + \frac{1}{T}[(\partial E/\partial V)_T + P]dV \qquad \text{...(i)}$$

Thermodynamic equation of state $(\partial E/\partial V)_T + P = T(\partial P/\partial T)_V$, when applied, equation (i) yields,

$$dS = \frac{C_V}{T} dT + (\partial P / \partial T)_V dV \quad \text{...(ii)}$$

For the given gas P = RT/(V – B), and (∂P/∂T)V = R/(V – B)

Substituting equation (iii) into (ii) we obtain

$$dS = \frac{C_V}{T} dT + R \frac{dV}{V - B} \quad \text{...(iv)}$$

$$\because \; C_V = a + bT + cT^2$$

$$\therefore \; \frac{C_V}{T} = \frac{a}{T} + b + cT \quad \text{...(vi)}$$

$$dS = \frac{a}{T} dT + bdT + cTdT + R \frac{dV}{V - B}$$

On integrating for proper limits, we get

$$\int_1^2 dS = a \int_{T_1}^{T_2} \frac{dT}{T} + b \int_{T_1}^{T_2} dT + c \int TdT + R \int_{V_1}^{V_2} \frac{dV}{V - B}$$

$$\therefore \; \Delta S = a \ln \frac{T_2}{T_1} + b (T_2 - T_1) + \frac{c}{2}(T_2^2 - T_1^2) + R \ln \frac{V_2 - B}{V_1 - B}$$

Example 19:

(a) *Starting with the fundamental relations prove that*

$$\left(\frac{\partial S}{\partial V}\right)_T = \left(\frac{\partial P}{\partial T}\right)_V = \frac{\alpha}{\beta},$$

where a = $V^{-1} (\partial V/\partial T)_P$, $\beta = V^{-1} (\partial V/\partial P)_T$

Solution:

(a) Combined first and second laws of thermodynamics give

TdS = dE + PdV

$$\text{But } dE = C_V \, dT + \left(\frac{\partial E}{\partial V}\right)_T dV$$

$$\therefore \; TdS = C_V \, dT + \left[\left(\frac{\partial E}{\partial V}\right)_T + P\right] dV \quad \text{...(i)}$$

At constant T, dT = 0 and

$$T\left(\frac{\partial S}{\partial V}\right)_T = \left(\frac{\partial E}{\partial V}\right)_T + P \quad \text{...(ii)}$$

Thermodynamic equation of state reads

$$(\partial E/\partial V)T + P = T\left(\frac{\partial P}{\partial T}\right)_V$$

$$\therefore \quad T\left(\frac{\partial S}{\partial V}\right)_T = T\left(\frac{\partial P}{\partial T}\right)_V$$

$$\text{or} \quad \left(\frac{\partial S}{\partial V}\right)_T = \left(\frac{\partial P}{\partial T}\right)_V \qquad \text{...(iv)}$$

Also V = f (T, P)

$$\therefore \; dV = \left(\frac{\partial V}{\partial T}\right)_P dT + \left(\frac{\partial V}{\partial P}\right)_T dP = \alpha V dT - \beta V dP$$

At constant volume dV = 0 and

$$\left(\frac{\partial P}{\partial T}\right)_V = \frac{\alpha}{\beta} \qquad \text{...(v)}$$

From equation (iv) and (v), we get

$$\left(\frac{\partial S}{\partial V}\right)_T = \left(\frac{\partial P}{\partial T}\right)_V = \alpha/\beta$$

(b) *For a given substance a = 1.24 × 10–3 K^{-1} and b = 9.3 × 10*–*5 atm–1 at 290 K and 1 atm. Assuming a and b to be constant, calculate the change in molar volume which should produce an entropy change of 2.1 JK^{-1} mol^{-1} at 290 K.*

Solution:

(b) We have derived $\left(\frac{\partial S}{\partial V}\right)_T = \left(\frac{\partial P}{\partial T}\right)_V = \alpha/\beta$

$$\text{or} \quad \frac{\Delta S}{\Delta V} = \alpha/\beta$$

$$\therefore \; \Delta V = \Delta S \; \beta/\alpha$$

$$= \frac{(2.1 JK^{-1} mol^{-1}) \times (9.3 \times 10^{-5} atm^{-1})}{(1.24 \times 10^{-3} K^{-1})}$$

$$= 0.1583 \; mol^{-1} \; atm^{-1} = \frac{0.1583 Jmol^{-1} atm^{-1}}{101.325 Jatm^{-1} dm^{-3}}$$

4

Free Energy Functions

PURPOSE OF NEW FUNCTIONS

The spontaneous processes have a tendency to achieve a position of minimum energy and maximum entropy. However, for a system it is frequently impossible to achieve both minimum energy and maximum entropy simultaneously. There are many endothermic process (gain of energy) which are spontaneous (evaporation of a liquid). Even there are a thermal processes known (e.g. $AgClO_2(s) \rightarrow Ag\ (s) + 1/2Cl_2\ (g) \pm O_2$ (g) $\Delta H = 0$] which are spontaneous. In such spontaneous process it is the entropy factor which over weighs the enthalpy (or energy) change of the system. But we also know the spontaneous processes in which the entropy of the system (ΔS_{syst}) is negative. For example, the dissolution of HCl gas in water. $HCl\ (g) + H_2O(l) \rightarrow OH_3 + (aq) + Cl^-(aq)$ is accompanied with a decrease of entropy $\Delta S_{syst} = -78\ JK^{-1}\ mol^{-1}$. In this case it is the enthalpy change ($\Delta H = -150\ kJ\ mol^{-1}$) which outweighs the decrease of entropy. From this discussion it is now quite clear that none of ΔH, ΔE, or ΔS_{syst} alone can decide the direction of a spontaneous process if we focus our attention on the system only. But ΔS_{univ} is a quantity which certainly decides the direction of a process: at constant E and V in an isolated system S is maximum. This is an odd situation in chemistry, because reactions are rarely studied under the condition energy. Experiments are conveniently performed either at constant T and P or at constant T and V. Hence it is desirable to introduce new functions which can decide the direction of spontaneous processes.

In all spontaneous processes it is a common observation that for endothermic change the entropy change is positive and high; if entropy change is negative then the process is highly exothermic. Thus there is always a compromise between the heat of reaction (ΔE or ΔH) and the entropy change of the system. Such a compromise between ΔE and ΔS, or ΔH and ΔS is represented by $\Delta A = \Delta E - TDS$ and $\Delta G = \Delta H - TDS$, so that for all spontaneous processes $\Delta E - TDS$ is negative (minimum energy and entropy, or by any other device) or $\Delta H - TDS$ is negative. Here A is called **Helmholtz free energy** (after German physiologist and physicist Hermann Ludwig Ferdinand von Helmholtz), and G is **Gibbs function** (after American physicist Josiah Williard Gibbs). These two free energy functions are defined by

$$A = E - TS \quad ...(1)$$

$$\text{and } G = H - TS \quad ...(2)$$

Here E is the energy, H enthalpy, S entropy, and T the temperature in kelvin. A and G are state functions and dA and dG are exact differentials.

PROPERTIES OF HELHOLTZ FREE ENERGY

(1) $A = E - TS$ (Definition)

and $dA = dE - TdS - SdT$ (differential change)

But $dE = dq + dw$ (1st law)

For a reversible isothermal process

$dq = TdS$

$\therefore \; dA \; TdS + dw_{rev} - TdS - SdT = dw_{rev} - SdT$

If T is constant $dT = 0$, and

$(dA)_T = dw_{rev}$

or $(\Delta A)_T = w_{rev}$ (Finite changes) ...(3)

or $-(\Delta A)_T = -w_{rev}$

The decrease in Helmholtz free energy is equal to the maximum amount of work which a system can do isothermally. Because of this relation between A and w, the Helmholtz free energy is called as *work function* (**Arbeit** is a German word which means work hence the symbol **A for work function**).

(2) $A = E - TS$

$dA = dE - TdS - SdT = dw - SdT$

$= -PdV - SdV - SdT$ (If PV work is considered)

At constant volume, dV, $= 0$ and

$(\partial A/\partial V)T = -P$

(3) $dA = -PdV\ SdT$

Since A is a state function and dA is exact differential applying Euler's theorem of exactness, we get

$(\partial S/\partial V)_T = (\partial P/\partial T)_V = \alpha/\beta$ ...(4)

Here $\alpha = V^{-1}(\partial V/\partial T)_P$ and $\beta = -V^{-1}(\partial V/\partial P)_T$

This is Maxwell's equation. This equation is very useful to study the variation of entropy with volume at constant temperature. Experimentally it is impossible to measure the change of entropy with change of volume as we do not have any apparatus which can measure entropy directly. But it is convenient to study the variation of pressure with temperature at constant volume. Hence, using Maxwell equation the experimentally difficult quantities can be calculated easily

(4) $(dA)_T = -PdV$

For ideal gas, $P = nRT/V$

$\therefore dA = -nRTd\ In\ V$

Integration gives

$\Delta A = -nRT\ In\ (V_2/V_1) = -nRT\ In\ (P_1/P_2)$

The change in A can be calculated for a given system if the initial and final states are given. This equation can also be used to calculate ΔA at constant temperature when an ideal solution is diluted from a known concentration ($C_1 = n/V_1$) to some other concentration $C_2 = n/V_2$. Since for an ideal system $(\Delta A)T = (\Delta G)_T$, hence

$\Delta G = nRT\ In\ C_1/C_2$, is also true.

(5) **Variation of A with T at constant V**

(i) $A = E - TS$

$dA = dE - TdS - SdT$

$dA = -SdT - PdV$

At constant volume $dV = 0$

$\therefore (\partial A/\partial T)_V = -S$

Hence, $A = E + T\left(\frac{\partial A}{\partial T}\right)_V$

or $A - T \left(\frac{\partial A}{\partial T}\right)_V = E$

Dividing both sides by $- T^2$, we get

$$\frac{-A + T(\partial A / \partial T)_V}{T^2} = - \frac{E}{T^2}$$

L.H.S. is the differential $\frac{\partial}{\partial T}\left(\frac{A}{T}\right)$

$$\frac{\partial (A/T)}{\partial T} = \frac{\partial (A/T^{-1})}{\partial T} = T^{-1}\left(\frac{\partial A}{\partial T}\right) - AT^{-2}$$

$$= \frac{T(\partial A / \partial T) - A}{T^2}$$

$$\therefore \frac{\partial}{\partial T}\left(\frac{A}{T}\right)_V = - \frac{E}{T^2} \qquad ...(5)$$

(ii) Consider a system in state 1 which is changed isothermally to state 2, so that

$$A_1 = E_1 + T\left(\frac{\partial A_1}{\partial T}\right)_V$$

and $A_2 = E_2 + T\left(\frac{\partial A_2}{\partial T}\right)_V$

The change in A in given by

$$A_2 - A_1 = \Delta A = (E_2 - E_1) + T \frac{\partial}{\partial T} (A_2 - A_1)_V$$

$$= \Delta E + T \left(\frac{\partial \Delta A}{\partial T}\right)_V$$

or $- \Delta A + T\left(\frac{\partial \Delta A}{\partial T}\right)_V = - \Delta E$

On dividing by $- T_2$ and replacing L.H.S. by differential form we get

$$\frac{\partial}{\partial T}\left[\left(\frac{\Delta A}{T}\right)\right]_V = - \frac{\Delta E}{T^2} \qquad ...(6)$$

PROPERTIES OF GIBBS FREE ENERGY

The Gibbs free energy G is defined by

$G = H - TS$

where H is the enthalpy defined by $H = E + PV$, therefore

$G = E + PV - TS$

Differential change in G is given by

$dG = dE + PdV + VdP - TdS - SdT$

From the first and second laws of thermodynamics, we know that

$dE = dq + dw = TdS + dw_{rev}$

On substituting the dE expression into dG expression we get

$dG = TdS + dw_{rev} + PdV + VdP - SdT - TdS$

At constant temperature and constant pressure $dT = 0$, $dP = 0$,

hence $(dG)_{T, P} = dw + PdV = dw - (-PdV)$

where dw includes all types of work, $-PdV$ = mechanical work and therefore,

$dw - (-PdV)$ = non-mechanical work = dwnon mech = dwnet

$\therefore (dG)_{T, P} = dw_{net}$

and $(\Delta G)_{T, P} = w_{net}$ ...(7)

or $-\Delta G = (-w_{net})$

The decrease in Gibbs free energy is equal to the amount of maximum work which a system can do isothermally over and above the mechanical work.

The logic for the nomenclature free energy can now be made clear. We know that in an adiabatic process the entire energy change (ΔE) in a system is converted into work: $\Delta E = w_{ad}$. However, under other conditions less than the entire energy change can be converted into work; the amount which is just unavailable to be converted into work is $T\Delta S$, because this amount is used to increase the randomness of the system at a constant T. The difference $\Delta E - T\Delta S$ and $\Delta H - T\Delta S$ are the free energies in the sense that if all the processes are reversible and isothermal then the complete conversion of free energies to work is possible. This is more clear from the relations: $(\Delta E - T\Delta S) = \Delta A = w_{rev}$ or $(\Delta H - T\Delta S) = \Delta G = w_{net}$. Therefore, w can be thought of as the amount *of work available from a reversible change of state in a system. It is (ΔG or ΔA) sometimes called the available work or availability.*

Another relation can be deduced as follows:

$G = H - TS$

$dG = dH - TdS - SdT$

But $H = E + PV$

and $dq = dE + PdV = TdS$

$\therefore\ dH = dE + PdV + VdP = TdS + VdP$

Hence $dG = -SdT + VdP$ ...(8)

At constant P, $dP = 0$ and from equation (8) we get,

$(\partial G/\partial T)_P = -S$

At constant T, $dT = 0$ and equation (8) gives,

$(\partial G/\partial P)_T = V$

Applying Euler's theorem of exactness for dG, we get from equation (8)

$(\partial S/\partial P)_T = -(\partial V/\partial T)_P = -\alpha V$

$\alpha = V^{-1}(\partial V/\partial T)_P$

Variation of free energy with pressure at constant T

$G = H - TS = E + PV - TS$

$dG = dE + PdV + VdP - TdS - SdT$

For a reversible process first and second laws give

$dE = TdS - PdV$ (only PV work)

Therefore $dG = -SdT + VdP$ ...(9)

At constant temperature, $dT = 0$, and equation (9) reduces to

$(dG)_T = VdP$

If the state of the system is changed from 1 to 2, then

$$\int_1^2 dG = \int_1^2 VdP$$

$$G_2 - G_1 = \Delta G = \int_1^2 VdP$$

Right hand side can be integrated if the variation of V with P is known. In case of an ideal gas V = nRT/P, and

$$\Delta G = nRT \int_1^2 \frac{dP}{P} = nRT \int_1^2 d \ln P$$

$$\therefore G_2 - G_1 = \Delta G = nRT \ln (P_2/P_1)$$

This equation can be used to calculate the free energy changes for expansion or compression. Also if the free energy of any substance is known under one pressure (P_1) then it is possible to calculate the free energy at some other pressure (P_2).

Relation between A and G. The Helmholtz and Gibbs free energy function are defined by

$$A = E - TS$$

$$G = H - TS$$

But H and E are related by the equation

$$H = E + PV$$

$$\therefore G = E + PV - TS = (E - TS) + PV$$

$$\text{Or } G = A + PV$$

Let the system be changed from state 1 to state 2, then the free energy functions, pressure and volume in the two states are

Initial state	Final State
G_1, A_1, P_1, V_1	G_2, A_2, P_2, V_2
$\therefore G_1 = A_1 + P_1V_1$ and	$G_2 = A_2 + P_2V_2$

Hence $G_2 - G_1 = A_2 - A_1 + (P_2V_2 - P_1V_1)$

or $\Delta G = \Delta A + (P_2V_2 - P_1V_1)$

(i) For an ideal gas PV = nRT

$$\therefore \Delta G = \Delta A + nR (T_2 - T_1)$$

For an isothermal change $T_2 = T_1 = T$ and $\Delta T = 0$, therefore

$(\Delta G) T = (\Delta A)T$ for an ideal gas.

Thus $\Delta G = \Delta A = - RT \ln (V_2/V_1) = - RT \ln (P_1/P_2)$

(ii) For a gas obeying the equation of state $P (V - b) = RT$, ΔG will not equal ΔA except at a temperature when b = 0

$$(\Delta A)_T = -\int_{V_1}^{V_2} PdV = -\int_{V_1}^{V_2} \frac{RT}{V-b} dV$$

$$= -RT \ln \frac{V_2 - b}{V_1 - b} = RT \ln (P_2/P_1)$$

$$(\Delta G)_T = \int_{P_1}^{P_2} VdP = \int_{P_1}^{P_2} \frac{RT}{P} dP + \int_{P_1}^{P_2} bdP$$

$$= RT \ln (P_2/P_1) + b(P_2 - P_1)$$

(iii) In the same way we can show that for a gas obeying the van der Waals equation of state:

$(P + an^2/V_2)(V - nb) = nRT$, the free energy changes ΔA and ΔG at constant T are given by

$$(\Delta A)_T = -nRT \ln \frac{V_2 - nb}{V_1 - nb} - an^2\left(\frac{1}{V_2} - \frac{1}{V_1}\right)$$

$$(\Delta G)T = -nRT \ln \frac{V_2 - nb}{V_1 - nb} - 2an^2\left(\frac{1}{V_2} - \frac{1}{V_1}\right)$$

$$+ n^2 bRT\left[\frac{1}{V_2 - nb} - \frac{1}{V_1 - nb}\right]$$

The two equations can be deduced by using the relations $(dA)_T = -P.dV$ and $(dG)_T = VdP$ as follows.

For a van der Waals gas $P = nRT/(V - nb) - an^2/V^2$, hence at constant T

$$(\Delta A)_T = -\int_{V_1}^{V_2} PdV$$

$$= -nRT \int_{V_1}^{V_2} \frac{1}{V - nb} dV + an^2 \int_{V_1}^{V_2} \frac{1}{V^2} dV$$

$$= -nRT \ln (V_2 - nb)/(V_1 - nb) - an_2 (1/V_2 - 1/V_1)$$

Now $dP = -\{nRT/(V - nb)^2\}dV + (2an^2/V^3)\, dV$

$$\therefore (\Delta G)_T = \int_{V_1}^{V_2} VdP$$

$$= -nRT\int_{V_1}^{V_2} \frac{V}{(V - nb)^2} dV + 2an^2 \int_{V_1}^{V_2} \frac{1}{V^2} dV$$

On integrating the first integral by parts and remembering that

$V_2/(V_2 - nb) - V_1/(V_1 - nb) = nb/(V_2 - nb) - nb/(V_1 - nb)$

We shall get the desired result.

Example 1:

4 moles of an ideal gas are compressed isothermally at 300 K from 2.02×10^5 N m^{-2} to 4.04×10^5 N m^{-2} pressure. Calculate the free energy change for the process.

Solution:

Given n = 4 mol, T = 300 K,

$P_1 = 2.02 \times 10^5$ Nm^{-2},

$P_2 = 4.04 \times 10^5$ N m^2

R = 8.5314 JK^{-1} mol^{-1}

$$\Delta G = nRT \text{ In } \frac{P_2}{P_1}$$

$$= (4 \text{ mol}) \times (8.314 \text{ JK}^{-1}\text{mol}^{-1}) \times 300\text{K} \times 2.303 \log \frac{4.04 \times 10^5 \text{ Nm}^{-2}}{2.02 \times 10^5 \text{ Nm}^{-2}}$$

= **6916.6 J**

TEMPERATURE DEPENDENCE OF FEE ENERGY

Gibbs-Helmholtz Equation

Gibbs-Helmholtz equation relates the free energy change to the enthalpy change and the rate of change of free energy with temperature. This equation may be applied to any change whether physical or chemical that takes place at constant pressure.

Consider the relations

$G = H - TS$

$H = E + PV$

$G = E + PV - TS$

For an infinitesimal change,

$dG = dE + PdV + VdP - TdS - SdT$

By the first law (if the external work is due to volume change only) we have

$$dE + PdV = dq$$

$$\therefore\ dG = dq + Vdq - TdS - SdT$$

But for a reversible change dq = TdS

so that $dG = -SdT + VdP$ ...(10)

In an isobaric change, dP = 0, and equation G = H – TS, yields

$$G = H + T\left(\frac{\partial G}{\partial T}\right)_P$$

$$\text{or } G - T\left(\frac{\partial G}{\partial T}\right)_P = H \qquad ...(11)$$

This is referred to as Gibbs-Helmholtz equation.

On dividing this equation by T^2, we obtain

$$\frac{G - T(\partial G/\partial T)_P}{T^2} = \frac{H}{T^2}$$

Now $-\dfrac{\partial(G/T)}{\partial T}$ gives the left hand side of the equation (by the usual formula for differentiating a fraction) therefore,

$$\left[\frac{\partial(G/T)}{\partial T}\right]_P = \frac{H}{T^2} \qquad ...(12)$$

$$\text{Hence } \left(\frac{G}{T}\right)_P = -\int\frac{H}{T^2}dT + I_1$$

This is another form of the Gibbs-Helmholtz equation, where I_1 is an integration constant.

Equation (12) can be reset into another form. By the method of differentiation we know that

$$d\,(T^{-1}) = -T^{-2}\,dT \text{ and } dT = -T^2\,d(1/T)$$

On substituting this into equation (12), we get

$$\left(\frac{\partial(G/T)}{\partial(1/T)}\right)_P = H$$

For a change from state 1 to state 2, we have

$$G_1 = H_1 + T\left(\frac{\partial(G_1)}{\partial T}\right)_P$$

$$G_2 = H_2 + T\left(\frac{\partial(G_2)}{\partial T}\right)_P$$

$$\text{and } \Delta G = \Delta H + T\left[\frac{\partial(\Delta G)}{\partial T}\right]_P$$

$$\therefore \left[\frac{\partial(\Delta G/T)}{\partial T}\right]_P = -\frac{\Delta H}{T^2} \quad ...(13)$$

Integrating equation (13), we get

$$\frac{\Delta G}{T} = -\int\frac{\Delta H}{T^2}dT + I_2$$

This is also Gibbs-Helmoholtz equation, I_2 is integration constant.

For approximate purposes if we assume that ΔH is independent of T then

$$\frac{\Delta G}{T} = \frac{\Delta H}{T} + I_2$$

If we consider the temperature dependence of ΔH as given by Kirchhoff's law, the integrated form of Gibbs Helmholtz equation may be written as

$$\int d(\Delta G/T) = -\int\frac{\Delta H}{T^2}dT$$

$$= -\int\left[\frac{\Delta\alpha}{T} + \frac{\Delta\beta}{2} + \frac{\Delta\gamma}{3}T + \frac{\Delta H_0}{T^2}\right]dT + I$$

$$\frac{\Delta G}{T} = -\left[\Delta\alpha \text{ d In } T + \frac{\Delta\beta}{2}T + \frac{\Delta\gamma}{6}T^2 - \frac{\Delta H_0}{T}\right] + I$$

$$\text{or } \Delta G = \Delta H_0 - \Delta\,\alpha\, T \text{ In } T + IT - \frac{1}{2}\Delta\beta T^2 - \frac{1}{6}\Delta\gamma T^3$$

Here I is the integration constant and may be evaluated if we know ΔG value at a given temperature. ΔH_0 is the value of ΔH at $T = T_1$ as used in the Kirchhoff's equation

$$\Delta h = \Delta\alpha T + \frac{1}{2}\Delta\beta T^2 + \frac{1}{3}\Delta\gamma T^3 + \Delta H_0$$

Similarly we can deduce various relations for A :

$$A - T\,(\partial A/\partial T)_V = E$$

$$\left[\frac{\partial(A/T)}{\partial(1/T)}\right]_V = E$$

$$\left[\frac{\partial(\Delta A/T)}{\partial T}\right]_V = -\frac{\Delta E}{T^2}$$

Application of the Gipps-Helmholtz Equation

Relation between electrical and chemical energy. Consider Daniell's cell

$Zn \mid ZnSO_4$ (soln | | $CuSO_4$ (soln) | Cu

At the zinc electrode, zinc ions pass into solution leaving the equivalent charge of electrons on the metal. The reverse occurs at the copper electrode where copper ions are reduced by electrons transferred through the external circuit from the zinc to the copper electrode. At various junctions (neglecting the potential difference at the liquid-liquid junction)

Zn electrode : $Zn \rightleftharpoons Zn^{2+}$ (aq) + $2e^-$ (oxidation)

Cu electrode : Cu^{2+} (aq) + $2e^- \rightleftharpoons$ Cu (reduction)

Metal junction : $2e^-$ (Zn) = $2e^-$ (Cu) (redox)

When the circuit is closed there is an e.m.f. in it equal to the sum of all the potential differences at the various junctions. Thus each potential difference corresponds to the net work of one part to the reaction, and the whole e.m.f. is, therefore, equivalent to the net work of the whole reaction which is equal to $-\Delta G$.

If the e.m.f. of the reversible cell is E volts and the quantity of electricity passing is n Faradays (or nF Coulomb) then the electrical work done by the system is equal to $n \times F \times E$ joules. Hence

$$-\Delta G = nFE = -w_{net}$$

This relationship may be substituted in the Gibbs-Helmholtz equation to get

$$-\,nFE = \Delta H - nFT\left[\frac{\partial E}{\partial T}\right]_P$$

Here $(\partial E/\partial T)_P$ is the rate of change of e.m.f. with temperature.

Example 2:

For Daniell cell calculate ΔG if

$E = 1.10$ volt at 25° C and $n = 2$;

F = 96500 *Coulomb.*

Solution:

For this case we have

$\Delta G = -\,nFE$

$= -2 \times (96500 \text{ Coulomb}) \times 1.10 \text{ volt}$

$= -\,212300$ volt Coulomb

$= -$ **212300 J**

Example 3:

For the Weston standard cell calculate ΔG, ΔH and ΔS using the following data:

E = 1.01463 V at 25°C; n = 2; F = 96500 C

$(\partial E/\partial T)_P = -\,5.0 \times 10\text{–}5 \text{ VK}^{-1}$.

Solution:

$\Delta G = -\,nFE$

$= -\,2 \times 96500 \times 1.01463 \text{ V}$

$= -\,195815 \text{ CV} = -$ **195815 J**

$$\Delta S = -\left(\frac{\partial \Delta G}{\partial T}\right)_P = nF\left(\frac{\partial E}{\partial T}\right)_P$$

$= 2 \times (96500 \text{ C}) \times (-5.0 \times 10^{-5} \text{ VK}^{-1})$

$= -\,9.65 \text{ CVK}^{-1}$

$= -$ **9.65 JK^{-1}**

$$\Delta H = -\,nFE + nFT\left(\frac{\partial E}{\partial T}\right)_P = \Delta G + T\Delta S$$

$= (-195815 \text{ J}) + (298 \text{ K}) \times (-9.65 \text{ JK}^{-1})$

$= -$ **198691 J**

FREE ENERGY OF THE SYSTEM AND ENTROPY OF UNIVERSE

For spontaneous process ΔS_{univ} is greater than zero, ΔS_{univ}, accounts for the entropy of system as well as surroundings. In most of the experiments it is very inconvenient to account for the changes in the surroundings. This difficulty is overcome by using another function of the system alone which can be evaluated by consulting standard tables on thermodynamic properties. Moreover, this function automatically takes care of entropy of system of system as well as surroundings. It is the free energy function A or G.

(ΔA) System and (ΔS) Universe

Let us consider an isolated system consisting of system and the surroundings (thermostant) as shown in Fig. 4.1(a). System is a vessel with a fixed volume and containing a substance which can undergo a chemical reaction or any other change. The surrounding which is a thermostant is considered to have a large capacity so that any heat given to it or taken from it by the system will not change the bath temperature appreciably.

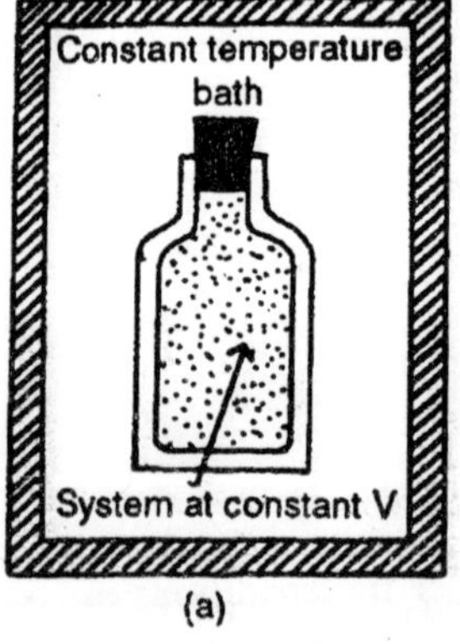

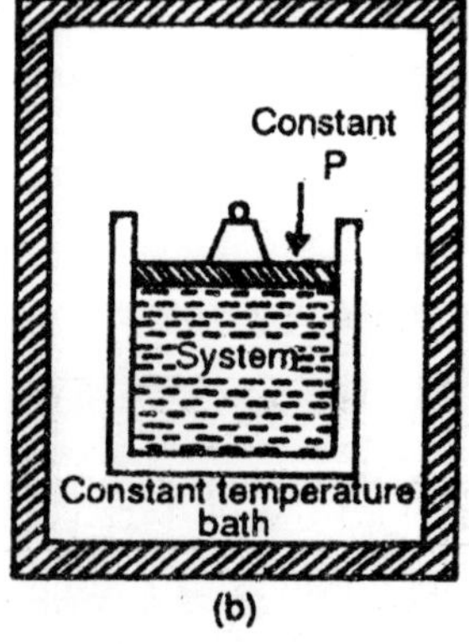

Fig. 4.1 : Spontenеous change (a) at constant temperature and volume $(\Delta A)_{T,V} < 0$, and (b) at constant temperature and pressure $(\Delta G)_{T,P} < 0$.

Let us suppose that the contents of the vessel (system) undergo some change and the energy of the system is changed by ΔE. Since the volume of the vessel is constant no work is done by the system and w = 0. First law for this process gives

$$\Delta E = qv$$

where q is the heat absorbed by the system from the surroundings.

$$q_{system} = q_{surroundings}$$

The entropy change of the surrounding (thermostant) resulting from the removal of heat q is given by

$$\Delta S_{surroundings} = q_{surroundings}/T = -(q_{system}/T)$$

$$= -(\Delta E_{system}/T)$$

where T is the temperature of the thermostant.

If ΔS is the entropy change of the system (within the vessel), then the entropy of the (universe) isolated system will be given by

$$\Delta S_{universe} = \Delta S_{system} + \Delta S_{surroundings}$$

$$= \Delta S_{system} - \Delta E system/T$$

$$= -\frac{1}{T}\left(\Delta E_{system} - T\Delta S_{system}\right)$$

$$= -\Delta A_{system}/T$$

since the Helmholtz free change for an isothermal process is given by $\Delta A = \Delta E - T\Delta S$.

This proves that the change in the Helmholtz free energy of the system at constant temperature and volume accounts automatically for the entropy changes of the system and surroundings. Since, for a spontaneous process the entropy change of the universe must be positive, we see that (ΔA) for the system must be negative. That is, in general for spontaneous process $\Delta A_{system} < 0$ (constant T. V).

(ΔG)System and (ΔS) Universe

If the experiment is carried out under the conditions of constant temperature and constant pressure then we can show that $(\Delta G)_{system}$ accounts for the entropy changes of the system and surroundings. If this is proved then for spontaneous process ΔG must be negative.

Consider an isolated system as shown in Fig. 4.1(b). The surrounding is a thermostat of large heat capacity and the system is a cylinder containing a substance and fitted with a movable piston which exerts a constant pressure on the contents of the cylinder. Let the substance undergoes a change accompanied by an enthalpy change (ΔH) and an entropy change ΔS. The heat absorbed by the system from the surroundings is equal to the enthalpy change ΔH.

$$q_{system} = \Delta H_{system} = -\, q_{surroundings}$$

The entropy of the surroundings would be,

$$\Delta S_{surrounding} = \frac{q_{surrounding}}{T} = -\frac{\Delta H_{system}}{T}$$

The entropy change of the entire system is

$$\Delta S_{total} = \Delta S_{universe} = \Delta S_{system} + \Delta S_{surroundings}$$

$$= \Delta S_{system} - \frac{\Delta H_{system}}{T}$$

$$= -\frac{\Delta H_{system} - T\Delta S_{system}}{T} = -\frac{\Delta G_{system}}{T}$$

since the Gibbs free energy change is given by

$$\Delta G = \Delta H - T\Delta S$$

Since for a spontaneous process the entropy change of the universe must be positive therefore ΔG_{system} must be negative. That is, for spontaneous change at constant T and P

$(\Delta G)system < 0$ (constant T, P)

Thus the use of ΔG is preferred to the use of ΔS to decide the nature of chemical reactions or any other change.

EFFICIENCY OF ENERGY CONVERSION

We can understand the significance of free energy function (G) in another way. Free energy change is defined by

$$\Delta G = \Delta H - T\Delta S$$

$$\text{or } \Delta H = \Delta G + T\Delta S \qquad ...(14)$$

Here ΔH is the total enthalpy change in the system, and ΔG is the net work obtainable and TΔS is a measure of unavailable energy. It is on account of TΔS term, we are not able to bring complete conversion of energy into useful work, as is asserted by the statement of the second law of thermodynamics. Equation (14) suggests a measure of efficiency of energy conversion as follows:

$$\text{Efficiency of energy conversion} = \frac{\text{Net work obtainble}}{\text{Total energy}} = \frac{\Delta G}{\Delta H}$$

$$= \frac{\Delta H - T\Delta S}{\Delta H} = -\frac{T\Delta S}{\Delta H} \quad ...(15)$$

Since ΔS is positive for spontaneous changes, equation (15) suggests that efficiency of energy conversion is always less than one except when ΔS is zero e.g. in fuel cell. The efficiency of various energy conversion devices are follows:

Device	Heat engine	Electricity to light	Dry cell	Fuel cell
Efficiency%	30	40	70	100

In biochemical process the free energy changes are used to calculate the efficiency. For example the efficiency of the degradation of glucose to carbon dioxide and water via glycolysis is calculated by using the following data:

$$C_6\,H_{12}O6\ (s) + 6O_2\ (g) \rightarrow 6CO_2\ (g) + 6H_2O(l),$$

$$\Delta G^o = -2879\ \text{kJ mol}^{-1}$$

$$ADP^{3-} + H^+ + HPO_4^{2-} \rightarrow ATP^{4-} + H_2O,$$

$$\Delta G^o = 31.4\ \text{kJ mol}^{-1}$$

But we know that 38 moles ATP the currency of energy are synthesized as a result of complete degradation of 1 mole of glucose:

$$C_6H_{12}O_6 + 38H^+\ 38ADP^{3-} + 38HPO_4^{2-} + 6O_2$$

$$\rightarrow 38ATP^{4-} + 6CO_2 + 44H_2O$$

$$\text{Hence, efficiency} = \frac{\text{energy stored in ATP molecules}}{\text{total energy released}}$$

$$= \frac{38 \times 31.4\text{kJ}}{2879\text{kJ}} = 0.42$$

$$= 42\%$$

Example 4:

One mole ideal gas at 27°C expands isothermally and reversibly from an initial volume of 2 dm³ to a final volume of 20 dm³ against a pressure that is gradually reduced. Calculate q, w, ΔE, ΔA, ΔG and ΔS.

Solution:

Process is carried out reversibly and isothermally. Also given that $n = 1$ mol, $V_1 = 2\ dm^3$, $V_2 = 20 dm^3$, $T = 300$ K

$$\text{(i) } w = -RT \ln \frac{V_2}{V_1} = 2.303\ (-8.314\ JK^{-1}\ mol^{-1}) \times (300K) \times \log \frac{20dm^3}{2.0dm^3}$$

$= 5744\ Jmol^{-1}$

(ii) Since gas is ideal and expansion is isothermal ($dT = 0$), $\Delta E = 0$

(iii) First law gives, $q = \Delta E - w = 0 + 5744\ J\ mol^{-1}$

$= 5745\ Jmol^{-1}$

(iv) $\Delta H = \Delta E + \Delta(PV) = \Delta E + \Delta(nRT) = 0 + 0 = 0$

$$\text{(v) } \Delta A = -nRT \ln \frac{V_2}{V_1} = w - 5744\ mol^{-1}$$

$$\text{(vi) } \Delta G = nRT \ln \frac{P_2}{P_1} = -nRT \ln \left(\frac{V_2}{V_1}\right) = -5744\ J\ mol^{-1}$$

$$\text{(vii) } \Delta S = \frac{q_{rev}}{T} = \frac{5744 Jmol^{-1}}{300K} = 19.15 JK^{-1} mol^{-1}$$

Entropy change can also be calculated as

$$\Delta S = R \ln \frac{V_2}{V_1}$$

$$= (8.314 JK^{-1} mol^{-1}) \times 2.303 \times \log \frac{20}{2}$$

$= 19.15\ JK^{-1}\ mol^{-1}$

and also as

$$\Delta S = \frac{\Delta H - \Delta G}{T} = \frac{0 + 5744 Jmol^{-1}}{300K}$$

$= 19.51 JK^{-1} mol^{-1}$

If the process is investigated for isolated system then we must consider the changes in the properties of the surroundings too, which are equal and opposite of those of the system in a reversible process. However, this is not true in the case of irreversible processes.

Example 5:

One mole ideal gas expands isothermally at 27°C into an evacuated vessel so that the pressure drops from 1.01×10^6 Nm^{-2} to 1.01×10^5 Nm^{-2}. Calculate the change in thermodynamic quantities. ΔE, ΔH, ΔA, ΔG and ΔS. What ΔSuniverse?

Solution:

The process is isothermal but not reversible.

w = 0, because the gas expands against zero pressure (evacuated vessel) and no external work can be done.

ΔE = 0, because the gas is ideal and the expansion is isothermal.

First law gives $q = \Delta E - w = 0 - 0 = 0$

$\Delta H = \Delta E + \Delta(PV) = \Delta E + \Delta(nRT) = 0 + 0 = 0$

$$\Delta G = RT\ \text{In}\left(\frac{P_2}{P_1}\right)$$

$$= (8.314\ JK^{-1}mol^{-1}) \times (300\ K) \times 2.303\ \log\frac{1.01 \times 10^5\ Nm^{-2}}{1.01 \times 10^6\ Nm^{-2}}$$

$= -\ 5744\ Jmol^{-1}$

$$\Delta A = -\ RT\ \text{In}\ = nRT\ \text{In}\ \frac{V_2}{V_1} = -5744\ Jmol^{-1}$$

$$(\Delta S)_{system} = R\text{In}\frac{V_2}{V_1} = -\ R\text{In}\ \frac{P_2}{P_1} = 19.15\ JK^{-1}mol$$

$$(\Delta S)_{surr} = \frac{-q_{system}}{T} = 0$$

$(\Delta S)_{univ} = (\Delta S)_{system+}\ (\Delta S)_{surr} = 19.15 JK^{-1}mol^{-1} + 0$

$= \mathbf{19.15\ JK^{-1}mol^{-1}}$

Thus the entropy of an isolated system in an irreversible process is positive.

Example 6:

One mole toluene is vaporized at its boiling point 111°C. The heat of vaporization at this temperature is 363.3 $J g^{-1}$. Calculate the maximum work done against 1 atm q, ΔH, ΔE, ΔG, and ΔS for the vaporization of 1 mole toluene. (Assume the vapour to behave as ideal gas).

$C_6H_5CH_3$ (liquid) → $C_6H_5CH_3$ (vapour)

Solution:

The process is isothermal, isobaric and reversible,

$T = 111 + 273 = 384$ K

$-w = P\Delta V = P(V^v ... V^l)\ PV^v = RT$ $[\because V^v >> V^l]$

$= (8.314\ JK^{-1}\ mol^{-1}) \times (384\ K) = 3193\ mol^{-1}$

$q_P = \Delta H_{vap} = (363.3\ Jg^{-1}) \times (92g\ mol^{-1})$

$= 33423.6\ J\ mol^{-1}$

$\Delta E = \Delta H - P\Delta V = (33424J\ mol^{-1} - 3193\ Jmol^{-1})$

$= 30231\ Jmol^{-1}$

$$\Delta G = \int VdP = n\ RT\ In\ \frac{P_2}{P_1} = nRT\ In\ \frac{1}{1} = 0$$

$$DS = \frac{q_{rev}}{T} = \frac{\Delta H_{vap}}{T_b} = \frac{33424Jmol^{-1}}{384K}$$

$= 87.04\ JK^{-1}\ mol^{-1}$

Example 7:

Calculate the change in Gibbs free energy for the process

$H_2O(l, -10°C) \rightarrow H_2O\ (S, -10°C)$

Given information:

The vapour pressures of water at –10°C is 2.15 mm Hg, and the vapour pressure of ice at –10°C is 1.95 mm.

Solution:

The given process is irreversible. To calculate the free energy change the process may be carried out by the following reversible steps:

(i) A mole of water is transferred at –10°C from liquid to saturated vapour (v. p. = 2.15 mm Hg). Since the two phases are in equilibrium. $\Delta G_1 = 0$.

(ii) The water vapour is allowed to expand from 2.15 to 1.95 mm at –10°C

$$\Delta G_2 = RT\ In\ \frac{P_2}{P_1}$$

$$= 2.303 \times (8.314 \text{ JK}^{-1} \text{ mol}^{-1}) \times (263 \text{ K}) \times \log \frac{1.95}{2.15}$$

$= -\ 214 \text{ J mol}^{-1}$

(iii) A mole of water is transferred at –10°C from vapour at P = 1.95 mm to ice at – 10°C. Since the two phases are in equilibrium, $\Delta G_3 = 0$.

Total free energy change = $\Delta G_1 + \Delta G_2 + \Delta G_3$

$= 0 + (-214 \text{ J mol}^{-1}) + 0$

$= -214 \text{ mol}^{-1}$

Example 8:

For a process A → B, the free energy change at 298 K is 298 kJ mol^{-1}, and ΔH° is given by $\Delta H^\circ = \Delta H_0 + \alpha' T + \beta' T^2 + \gamma' T^3$,

where α', β', γ are constants. Derive an expression for ΔG° as a function of temperature.

Solution:

The change of ΔG° with temperature is best expressed by Gibbs Helmholz equation

$$\left(\frac{\partial(\Delta G^\circ / T)}{\partial T}\right)_P = \frac{\Delta H^\circ}{T^2}$$

$$= -\ \frac{\Delta H_0}{T^2} dT - \frac{\alpha'}{T} dT - \beta' dT - \gamma' T dT$$

Integrating between limits T = 298 K to T = T, we get

$$[\Delta G^\circ / T]_{\Delta G_1^\circ(298)}^{\Delta G_2^\circ(T)}$$

$$= \Delta H_0 \left[\frac{1}{T}\right]_{298}^{T} - \alpha' [\text{In} T]_{298}^{T} - \beta' [T]_{298}^{T} - \frac{1}{2}\gamma' \left[T^2\right]_{298}^{T}$$

or $\dfrac{\Delta G_2^\circ}{T} - \dfrac{\Delta G_1^\circ}{T} = \dfrac{\Delta H^\circ}{T} - \dfrac{\Delta H_0}{298} - \alpha' \text{ In } T + \alpha' \text{ In } 298 - \beta T$

$+\ \beta' \times 298 - \dfrac{1}{2}\gamma' T^2 + \dfrac{\gamma'}{2}(298)^2$

$\because\ \Delta G_1^0 = 298 \text{ kJ} = 298000 \text{ J}$

$$\therefore \Delta G_2^0 = T\left[\frac{298000\text{Jmol}^{-1}}{298\text{ K}} - \frac{\Delta H_0}{298} + \alpha' \text{In } 298 + 298\beta' + \frac{\gamma'}{2}(298)^2\right]$$

$$+ \Delta H_0 - T\alpha' \text{ In } T - \beta' T^2 - \frac{1}{2}\gamma' T^3$$

On substituting the values of various quantities with proper units ΔG^o can be evaluated at any temperature T.

Example 9:

Calculate the free energy change for the process

H_2O (l, 300 K, 1.01325 × 10^5 Pa) → H_2O (v, 300K, 1.01325 × 10^4 Pa) Given that:

$C_{P,m}$ (H_2O, l) = 75.4 J K^{-1} mol^{-1}, $C_{P,m}$ (H_2O, v)

= 33.2 JK^{-1} mol^{-1}

ΔH^o_{vap} (373 K) = 40.85 kJ mol^{-1}

Solution:

The free energy change can be calculated by considering the given process to be split into various reversible steps. $\Delta G = \Delta H - T\Delta S$ equation can be used to calculate the free energy change for isothermal phase transformation and ΔG = RT In P_2/P_1 for isothermal expansion.

Various reversible steps are

(i) H_2O (l, 300 K, 101325 Pa) $\xrightarrow{\text{Heat}}$ H_2O (l, 373 K, 101325 Pa)

$$\Delta H\text{ (i)} = \int_{300}^{373} C_{P,m}(H_2O,l)dT = (75.4\text{ JK}^{-1}\text{mol}^{-1})$$

× (373 K − 300K)

= 5504.2 $Jmol^{-1}$

$$\Delta S\text{ (i)} = C_{P,m}\int d\text{ In}T = (75.4\text{ JK}^{-1}\text{mol}^{-1}) \times 2.303\log\frac{373}{300}$$

= 16.424 $JK^{-1}mol^{-1}$

(ii) H_2O (l, 373 K, 101325 Pa) ⇌ H_2O (v, 373 K, 101325Pa)

ΔH (ii) = 40.85 k J mol^{-1}

$$\Delta S\text{ (ii)} = \frac{\Delta_{vap}}{T_b} = \frac{40850}{373}\text{JK}^{-1}\text{mol}^{-1}$$

$= 109.517\ JK^{-1}mol^{-1}$

(iii) H_2O (v, 373 K, 101325 Pa) $\xrightarrow{\text{Cool}}$ H_2O (v, 300K, 101325 Pa)

$\Delta H\ (iii) = C_{P,m}\ (H_2O,\ v)\ \Delta T = -\ (33.2\ JK^{-1}mol^{-1}) \times (73\ K)$

$= -2423.6\ Jmol^{-1}$

$$\Delta S(iii) = C_{P,\,m}\ (H_2\ O,\ v)\ In\ \frac{300}{373}$$

$$= (33.2JK^{-1}mol^{-1})\ 2.303 \times \log\frac{300}{373}$$

$= -\ 7.232\ JK^{-1}mol^{-1}$

(iv) H_2O (v, 300K, 101325 Pa) $\rightarrow$ H_2O (v, 300K, 10132.5 Pa)

ΔH (iv) = 0 (if the vapours are ideal then ΔH is a function of T only)

$$\Delta S\ (iv) = R\ In\frac{P_1}{P_2} = (8.314\ JK^{-1}mol^{-1}) \times 2.303 \times \log\frac{101325\ Pa}{10132.5\ Pa}$$

$= 19.147\ JK^{-1}mol^{-1}$

For all the steps

$\Delta H = \Delta H\ (i) + \Delta H\ (ii) + \Delta H\ (iii) + \Delta H\ (iv)$

$= (5504.2 + 40850 - 2423.6 + 0) = 43930.6\ Jmol^{-1}$

$\Delta S = \Delta S\ (i) + \Delta S\ (ii) + \Delta S\ (iii) + \Delta S\ (iv)$

$= 16.425 + 109.517 - 7.232 + 19.147 = 137.856\ JK^{-1}mol^{-1}$

$\Delta G = \Delta H - T\Delta S = (43930.6\ J\ mol^{-1}) - (300K)$

$\times\ (137.\ 856\ JK^{-1}\ mol^{-1})$

$=$ **2573.8 J mol⁻¹**

Alternatively the value of ΔG can be evaluated by first considering the Gibbs Helmholtz equation

$$d(\Delta G/T) = -\ \frac{\Delta H}{T^2}\,dT \text{ for the process}$$

(i) H_2O (l, 300K, 101325 Pa) $\rightarrow$ H_2O (v, 300K, 101325 Pa) and the equation $\Delta G = RT\ In\ P_2/P_1$ for the process

(ii) H_2O (v, 300 K, 101325 Pa) $\rightarrow$ H_2O (v, 300 K, 10132.5 Pa)

For step (i) we must know ΔH as a function of temperature, which is given by

$d(\Delta H) = C_P(H_2O, v)\,dT - C_P(H_2O, l)\,dT = [(C_P(v) - C_P(l)]\,dT$

$\therefore \int d(\Delta H) = \Delta C_P \int dT + I_1$

or $\Delta H(T) = \Delta C_P\,T + I_1$

For the given process

$\Delta C_P = 33.2\ JK^{-1}\ mol^{-1}\ 75.4\ JK^{-1} = 42.2\ JK^{-1}\ mol^{-1}$

The integration constant I_1 can be estimated by using

$\Delta H = 40850\ J\ mol^{-1}$ at 373 K.

$\therefore 40850\ J\ mol^{-1} = -(42.\ 2\ JK^{-1}\ mol^{-1} \times 373\ K) + I_1$

or $I_1\ (40850 + 15740.6)\ J\ mol^{-1} = 56590.6\ J\ mol^{-1}$.

With this value of integration constant, the expression for DH (T) may be written as

$\Delta H\ (T) = -\ 42.2\ T + 56590.6$

On substituting this value of ΔH (T) into Gibbs-Helmholtz equation, we get

$$\int d(\Delta G/T) = \int \frac{42.2}{T} dT - \int \frac{56590.6}{T^2} dT + I_2$$

$$\text{or} \quad \frac{\Delta G(i)}{T} = 42.2 \text{ In } T + \frac{56590.6}{T} + I_2$$

Since Δ G = 0 at 373 K for the given process as H_2O (l, 373K) $\rightleftharpoons H_2O$ (v, 373 K) represents a state of equilibrium. Therefore, the integration constant I_2 may be evaluated.

$$0 = 42.2 \times 2.303 \log 373 + \frac{56590.6}{373} + I_2$$

$\therefore I_2 = -\ 401.652\ Jmol^{-1}K^{-1}$

with this value of I_2, the free energy change is given by

$$\frac{\Delta G(i)}{T} = 42.2 \text{ In } T + \frac{56590.6}{T} - 401.652, \text{ At 300 K}$$

$$\Delta G(i) = 300\ [42.2\ JK^{-1}\ mol^{-1} \times 2.303 \log 300 + \frac{56590.6 Jmol^{-1}}{300K} - 401.652\ JK^{-1}\ mol^{-1}]$$

$= 330\ K\ (240.743 + 188.635 - 401.652)\ JK^{-1}mol^{-1}$

= **8317.8 Jmol–1**

The change of free energy for step (ii) is

$$\Delta G\ (ii) = 2.303 \times (8.314\ JK^{-1}\ mol^{-1}) \times (300K) \times \log\frac{10132.5\ Pa}{101325\ Pa}$$

$= -\ 5744.1\ Jmol^{-1}$

$\therefore\ \Delta G = \Delta G\ (i) + \Delta g\ (ii)$

$= (8317.8 - 5744.1)\ Jmol^{-1}$

= **2573.7** mol^{-1}

Here we find that the value of ΔG calculated by the two methods is the same.

(The difference of 0.1 J mol^{-1} in the two calculations is due to certain approximations).

Example 10:

For a given reaction the free energy in Joules is expressed as a function of temperature by the equation

$\Delta G°/J\ mol^{-1}\ 18660 - 14.40\ T \log T - 6.07\ T + 8.24 \times 10^{-3}\ T^2$

Calculate $\Delta G°$ $\Delta H°$ and $\Delta S°$ at 25°C.

Solution:

Using the Gibbs-Helmholtz equation we have

$$\frac{-\Delta H°}{T^2} = \frac{\partial}{\partial T}(\Delta G°/T)_P = \frac{\partial}{\partial T}\left[\frac{18660}{T} - 14.40 \log T\right.$$

$$\left. - 6.07 + 8.24 \times 10^{-3}\ T\right]$$

$$= -\frac{18660}{T^2} - \frac{14.40}{2.303}\frac{\partial \ln T}{\partial T} + 8.24 \times 10^{-3}$$

$$= \frac{-18660}{T^2} - 6.25\frac{1}{T} + 8.24 \times 10^{-3}$$

$$\left(\because \frac{d}{dT}\log T = \frac{1}{2.303} d\ln T = \frac{1}{2.303} \times \frac{1}{T}\right)$$

or $\Delta H° = 18660 + 6.25\ T - 8.24 \times 10^{-3}\ T^2$

$\therefore\ \Delta H°\ (298\ K) =$ **19790.8** $Jmol^{-1}$

At 298 K

$\Delta G^o = 18660 - 14.40 \times 298 \times \log 298 - 6.07 \times 298$

$+ 8.24 \times 10^{-3} \times (298)^2$

$= 18660 - 10617.3 - 1808.8 + 731.7$

= 6965.6 Jmol^{-1}

The relation $\Delta G^o = \Delta H^o - T\Delta S^o$ gives

$$\Delta S^o = \frac{\Delta H^o - \Delta G^o}{T} = \frac{(19790.8 - 6965.6)\text{Jmol}^{-1}}{298\text{ K}}$$

= **43.03** $\text{JK}^{-1}\text{mol}^{-1}$

ΔS^o can also be evaluated by using relation $(\partial \Delta G^o/\partial T)_P = -\Delta S^o$. (At the time of differentiating log T, care must be taken to change it to In T)

$$-\Delta S^o = \left(\frac{\partial \Delta G^o}{\partial T}\right)_P = 0 - 14.40 \log T - \frac{14.40}{2.303} T \times \frac{1}{T}$$

$-6.07 + 2 \times 8.24 \times 10^{-3}\ T$

$\therefore \Delta S^o\ (298\text{ K}) = +\ 14.40 \times 2.4742 + 6.25 + 6.07 - 16.48 \times 10^{-3}$

$\times\ 298$

$= 47.948 - 4.911 = 43.03\ \text{JK}^{-1}\text{mol}^{-1}$

THERMODYNAMIC EQUILIBRIA AND FREE ENERGY FUNCTIONS

The state of equilibrium is represented by a balance of driving and opposing forces, and the system is incapable of undergoing spontaneously any change at this state.

Criteria for equilibria at constant T and V. Under this condition the equilibrium condition can be expressed in terms of Helmholtz free energy A, which is defined by

$A = E - TS$

and $dA = dE - TdS - SdT$

First law gives

$dE = dq + dw = dq - PdV$ if only PV work is considered

$\therefore dA = dq - PdV - TdS - SdT$ ...(16)

At constant temperature dT = 0, and at constant volume dV = 0, therefore equation (16) reduces to

$$(dA)_{T,v} = dq - TdS \quad ...(17)$$

For a reversible change dqrev = TdS, and hence Eq. (17) gives

$$dA_{rev} = 0 \text{ (constant T, V)} \quad ...(18)$$

For finite changes we will have

$$(\Delta A)_{T,v} = 0$$

For an irreversible process the heat change (dq_{irr}) is less than that in a reversible process (dq_{rev}) *i.e.*, dq_{rev} > dqirr and therefore,

$$TdS > dq_{irr}$$

$$\text{or } dq_{irr} - TdS < 0 \quad ...(19)$$

Applying this condition into equation (17) we get

$$dA_{irr} < 0 \quad \text{(At constant T,V)} \quad ...(20)$$

For finite changes

$$(\Delta A)_{T,V} < 0$$

Thus spontaneous processes are accompanied by a decrease of work function and this function does not change at equilibrium. This can be restated in the following form:

"The Helmholtz Free energy decreases in natural (spontaneous) processes. When it reaches a minimum value, the system reaches equilibrium." This statement is obvious from the definition A = E – TS. Since at equilibrium E is minimum and S is maximum, naturally the difference must be a minimum.

Criteria for equilibria at constant T and P. For this case consider the Gibbs free energy function

$$G = H - TS = E + PV - TS$$

$$\text{or } dG = dE + PdV + VdP - TdS - SdT$$

First law of thermodynamics gives

dE = dq + dw = dq – PdV if only PV work is considered.

$$\therefore dG = dq - PdV + PdV + VdP - TdS - SdT$$

$$= dq + VdP - TdS - SdT \quad ...(21)$$

At constant temperature dT = 0, and at constant pressure dP = 0, therefore equation (21) yields

$(dG)T,P = dq - TdS$ ...(22)

Equation (22) is applicable for all types of changes. When the change is reversible $dq = TdS$ and equation (22) gives

$dG_{rev} = 0$ (constant T, P) ...(24)

For finite changes $(\Delta G)_{rev}$ is zero at constant temperature and pressure.

For an irreversible change $dq_{irr} < TdS$ or $dq_{irr} - TdS$ or $dq_{irr} - TdS$ is negative, therefore equation (22) reduces to

$dG_{irr} < 0$ (constant T, P) ...(24)

For finite changes in an irreversible process ΔG is negative at a constant T and P, and at equilibrium the free energy is minimum. This also implies that the free energy in the final state should be lower than that in the initial state. This is in line with daily experiences that the electric current flows from high potential towards lower potential side, heat flows from high temperature body towards a low temperature body. In any chemical reaction the free energy of the reactants should be higher that the free energy of the products which are expected to be formed. This is the reason that ΔG is a measure of driving force and is called **reaction potential**. If Δ G is positive then the change is spontaneous in the reverse direction and forbidden in the direction as represented by equation. When Δ G is zero the reaction is at equilibrium and its properties will not be changed as long as the experimental conditions are unaltered.

Spontaneity and Relative Values of ΔH an TΔS

For spontaneous process it is essential that the free energy change for the system must be negative. Let us examine the tendency of a process in the light of the relative values of the enthalpy and entropy changes under difference conditions. The relation $\Delta G = \Delta H - T\Delta S$ can be used to see what is required of ΔH and ΔS in order that ΔG is negative and the process be spontaneous.

(i) If ΔH is negative and ΔS is positive then the process will certainly be spontaneous since ΔG will be negative.

(ii) If ΔH is positive that is the process is endothermic and ΔS is negative the process will not be spontaneous as ΔG is positive.

(iii) If ΔH and ΔS have same sign (both positive or both negative) the process may or may not be spontaneous depending upon the magnitudes of ΔH, ΔS and T; when

(a) both ΔH and ΔS are positive, for example in the reaction in which bonds are broken, ΔH will predominate at lower temperature and hence ΔG will be positive. As the temperature is raised the term T ΔS becomes larger and the entropy factor becomes more important and at sufficiently high temperature ΔG will be negative and so the process will be spontaneous.

(b) both ΔH and ΔS are negative, for example, the process where bonds are formed, ΔG will be negative at lower temperature and at sufficiently high temperature ΔG will be positive, therefore, the process would not be spontaneous.

(iv) If ΔH = TΔS, then ΔG will be equal to zero and then the reaction will be at equilibrium.

These are summarized in Table 1.

Table 1 : Enthalpy, Entropy Free Energy Changes and the Nature of Reaction

ΔH	ΔS	$\Delta G = \Delta H - T\Delta S$	*Conclusion*
– (exothermic)	+	–	Spontaneous change
– (exothermic)	–	– at low T	Spontaneous change
		+ at high T	Non-spontaneous
+ (endothermic)	+	+ at low T	Non-spontaneous
		– at high T	Spontaneous
+ (endothermic)	–	+	Forbidden in the forward direction

FREE ENERGY OF FORMATION

The free energy of formation of a compound is the free energy change which accompanies the formation of the compound in its standard state form its elements in their standard states, the **free energy of each element in its standard state is assigned a zero value**. The free energy of formation is denoted by ΔGf_f°. The values of ΔG_f° are reported in appendix.

The free energy change for any reaction ΔG° is given by

$$\Delta G^\circ = \Sigma \Delta G_f^\circ \text{ (products)} - \Sigma \Delta G_f^\circ \text{ (reactants)}$$

The values of ΔG_f° for compounds are tabulated and used whenever needed to evaluate ΔG° for any chemical reaction or physical transformations.

The free energy of formation of ions in aqueous solution are calculated from free energy data of the reaction in which the ions are participating. The free energy of formation of the proton in water is arbitrarily taken as zero.

The G_f (ion) obtained in this way is called the **conventional** free energy of formation.

The chemical reactions in which ΔG is negative are called **exoergonic** (*Exo* = out, *ergo* = work) and when ΔG is positive the reaction is said to be **endoergonic**. In exoergonic reactions the ΔG_f° values of products are less than those of reactants. and in endoergonic reactions ΔG_f°(reactants) are less than G_f (products).

The free energy change can also be calculated by using ΔH and ΔS values for reactions with the help of the equation $\Delta G = \Delta H - T\Delta S$. ΔH and ΔS for reaction in turn can be evaluated by the methods discussed in chapter respectively. Since ΔH_f° is **taken as zero for an element we cannot evaluate ΔG_f° for an element or compound by using equation** $\Delta G^\circ = \Delta H^\circ - T\Delta S$.

Example 11:

Calculate the free energy change for the reaction $H_2(g) + Cl_2(g) \rightarrow 2HCl(g)$ *at 25°C by using the bond energy and entropy data:*

DH° (H– H) = 435 kJ mol^{-1}, Dh° (Cl – Cl) = 240 kJ mol^{-1},

DH° (H – Cl) = 430 kJ mol^{-1}; S° (H_2) = 130.59 JK^{-1} mol^{-1},

S° (Cl_2) = 222.95 JK^{-1} mol^{-1}, S° (HCl) = 186.68 JK^{-1}mol^{-1}

Solution:

ΔG° can be evaluated by using the relation $\Delta G^\circ = \Delta H^\circ - TDS^\circ$

ΔH° = energy absorbed – energy relased

= DH° (H–H) + DH° (Cl – Cl) –2 × DH° (H– Cl)

= (435 + 240 –2 × 430) kJ

= – 185 kJ

$\Delta S^\circ = \Sigma S^\circ$ (products) – ΣS° (reactants)

$= 2 \times S^o(HCl) - S^o(H_2) - S^o(Cl_2)$

$= (2 \times 186.68 - 130.59 - 222.95)\ JK^{-1}$

$= 19.82 \times 10^{-3}\ kJ\ K^{-1}$

$\Delta G^o = \Delta H^o - T\Delta S^o$

$= -185\ kJ - (298\ K) \times 19.82 \times 10^{-3}\ kJ\ K^{-1}$

$= -190.9\ kJ$

The reaction is exoergonic by 109.9 kJ.

Example 12:

Hydrazine (N_2H_4) is a potentially valuable rocket fuel, and it is desirable to find a good method for its synthesis. If you were the director of a large industrial research laboratory, would there be any justification in your asking the research officer to look for a catalyst which will make the following reaction proceed at 25°C :

$NH_4NO_3(S) + 3H_2(g) \rightarrow 3H_2(g) + N_2H_4(g)$

Data given are:

Substance	*$NH_4NO_3(s)$*	*$H_2(g)$*	*$H_2O(g)$*	*$N_2H_4(g)$*
ΔH_f^o/kJ mol^{-1}	–365	0	–242	50
S^o/JK^{-1}mol^{-1}	150	130	189	120

Solution:

The given reaction would be feasible if the free energy change ΔG^o is negative. This quantity can be calculated by using the equation

$\Delta G^o = \Delta H^o - T\,\Delta\,S^o$

$\because\ \Delta H^o = 3 \times \Delta H_f^o(H_2O) + H_f^o(N_2H_4) - [(\Delta H_f^o(NH_4NO_3, s)$

$+ 3 \times \Delta H_f^o(H_2)]$

$= (-3 \times 242 + 50)\ kJ - (-365 + 3 \times 0)\ kJ$

$= -311\ kJ$

and $\Delta S = 3 \times S^o(H_2O) + S^o(N_2H_4) - [S^o(NH_4NO_3) + 3 \times S^o(H_2)]$

$= (3 \times 189 + 120)\ JK^{-1} - (150 + 3 \times 130)\ JK^{-1}$

$= 147\ JK^{-1} = 0.147\ kJ\ K^{-1}$

$\therefore \Delta G^o = -311 \text{ kJ} - (298K) \times 0.147 \text{ kJ K}^{-1}$

$= -$ **354.8 kJ**

Since the free energy change calculated for the given reaction comes out to be negative therefore, the reaction should be spontaneous in the forward direction and for the synthesis of hydrazine a suitable catalyst can be used.

Example 13:

Calculate the difference between ΔG and ΔA at 27°C for the reaction

H_2 (g, 1.01×10^5 N m^{-2}) + 1/2 O_2, (g, 1.01×10^5 Nm^{-2}) $\rightarrow$ H_2O (l, 1.01×10^5Nm^{-2}).

Solution:

$\Delta G = \Delta H - T\Delta S$

and $\Delta A = \Delta E - T\Delta S$

$\therefore \Delta G - \Delta A = \Delta H - \Delta E$

ΔH and ΔE for a chemical reaction involving ideal gases are given by

$\Delta H = \Delta E + P\Delta V = \Delta E + \Delta v_g RT$

$\therefore \Delta H - \Delta E = \Delta v_g RT$

For the given reaction $H_2(g) + 1/2\ O_2, (g) \rightarrow H_2O(l)$,

$$\Delta vg = 0 - 1 - \frac{1}{2} = -\frac{3}{2}$$

$\therefore \Delta H - \Delta E = -3/2 \times (8.314 \text{ JK}^{-1} \text{ mol}^{-1}) (300K)$

$= -3714 \text{ J mol}^{-1}$

Hence $\Delta G - \Delta A = -$ **3741 Jmol^{-1}**

Example 14:

The ΔG^o values for the hydrolysis of creatine phosphate (creatine-P) and glucose-6- phosphate (G-6-P) are:

(i) G-6-P + creatine $\rightarrow$ G + creatine-P DG° (iii) = – 12.4 kJ

Solution:

Reversing equation (ii) and the adding it to equation (i), we shall et the result

(iv) Creatine + P → Creatine – P + H_2O ΔG^o (iv) = – ΔG^o (i) = 29.2 kJ

(ii) G-6-P + H_2O → G + P ΔG^o (ii) = – 12.4 kJ

Add

(ii) G-6-P + Creatine → G + Creatine-P ΔG^o (iii) = **16.8kJ**

Example 15:

The oxidation of glucose is a very important reaction in the process of respiration, and we need to know the free energy change for the reaction $C_6H_{12}O_6$ (aq, 0.1m) + $6O_2$ (g, 0.2 atm) → $6CO_2$ (g, 0.0003 atm) + $6H_2O$ (l). Use of following data to calculate ΔG^o for this reaction at 298 K.

Substance	glucose (s)	CO_2(g)	H_2O(l)	O_2(g)
ΔG^o_f/kJ mol^{-1}	–913	–395	–237	0

and $C_6H_{12}O_6$ (s) ⇌ $C_6H_{12}O_6$ (saturated, aq) $\Delta G^o = 0$

A saturated solution of glucose in water is taken approximately 10 molal.

Solution:

In chemistry we are familiar with the combustion of glucose in the presence of oxygen to give CO_2 and H_2O along with energy. For the given reaction we can calculate ΔG^o by realising the fact that chemical equations can be treated as algebraic equations.

(i) $C_6H_{12}O_6$(s) + $6O_2$ (g, 1atm) → $6CO_2$ (g, 1atm) + $6H_2O$(l)

ΔG^o (i) = 6 × ΔG^o_f (CO_2) + 6 × ΔG^o_f (H_2O) – ΔG^o_f ($C_6H_{12}O_6$) – 6 × ΔG^o_f (O_2)

= – 6 × 395 – 6 × 237 + 913 – 0 = – 2879 kJ

ΔG^o for the expansion of 6 moles O_2 from 1 atm to 0.2 atm

(ii) $6O_2$ (g, 1 atm) → $6O_2$ (g, 0.2 atm)

ΔG^o (ii) = nRT In P_2/P_1 = (6 mol) × (8.314 JK^{-1} mol^{-1}) × (298K) × 2.303 log $\frac{0.2}{1}$

= – 23929 J = – 23.929 kJ

ΔG^{o} for the expansion of 6 mol CO_2 from 1 atm to 0.0003 atm

(iii) 6 CO_2 (g, 1 atm) $\rightarrow$ 6CO_2 (g, 0.0003 atm)

$$\Delta G^{o} \text{ (iii)} = 6 \times 8.314 \times 298 \times 2.303 \log \frac{0.0003}{1}$$

= – 120606.07 J – 120.606 kJ

ΔG^{o} for the dilution of glucose solution from 10 molal to 0.1 molal

(iv) glucose (s) $\rightarrow$ glucose (aq, 10 molal) $\Delta G^{o} = 0$

(v) glucose (aq, 10 molal) $\rightarrow$ glucose (aq, 0.1 molal)

ΔG^{o} (v) = nRT In C_2/C_1

Add (iv) and (v)

(vi) glucose (s) $\rightarrow$ glucose (aq, 0.1 molal)

$$\Delta G^{o} \text{ (vi)} = 2.303 \times 8.314 \times 298 \times \log \frac{0.1}{10} = -\ 11411.7 \text{ J}$$

= – 11.41 kJ

On adding equations (i) and (iii), and subtracting equations (ii) and (vi) we get

$C_6H_{12}O_6$ (s) + 6O_2 (g, 1 amd) 6CO_2 (g, 1 atm) – 6O_2 (g, 1 atm) – glucose (s)

$\rightarrow$ 6CO_2 (g, 1 atm) + 6H_2O (l) + 6CO_2 (g, 0.0003 atm) – 6O_2 (g, 0.2 atm) – glucose (aq. 0.1 molal)

$\Delta G^{o} = \Delta G^{o}$ (i) + ΔG^{o} (iii) – ΔG^{o} (ii) – ΔG^{o} (vi)

On rearranging the resulting chemical equation we get

$C_6H_{12}O_6$(aq, 0.1molal) + 6CO_2 (g, 0.0003 atm) + 6H_2O(l)

ΔG^{o} = – 2964.2 kJ

The free energy change is negative, therefore the reaction is spontaneous.

Example 16:

When glucose is burnt in a bomb calorimeter as $C_6H_{12}O_6$ + 6O_2(g) $\rightarrow$ 6CO_2(g) + 6H_2O (g), the reaction is found to be exothermic and ΔE = – 2880 kJ mol^{-1}, the entropy change is found to be 182.4 JK^{-1} mol^{-1}. Calculate the free energy change at 298 K for the process. How much of the energy change can be extracted as heat, and how much can be extracted as work?

Solution:

Since the reaction takes place at constant volume, hence we shall use $\Delta A = \Delta E - T\Delta S$ to evaluate free energy change.

$$\Delta A = -2880 \text{ kJ mol}^{-1} - 298 \text{ K} \times 182.4 \times 10^{-3} \text{ kJK}^{-1} \text{ mol}^{-1}$$

$$= -2934.3 \text{ kJ mol}^{-1}$$

We know that $\Delta A = w$, hence the work done during a reversible oxidation of glucose is – 2934.3 kJ

Since the oxidation takes place at constant volume hence $q_v = \Delta E = -2880 \text{ kJ mol}^{-1}$.

COUPLED REACTIONS

Such reactions have been discussed from two angles of non-equilibrium thermodynamic in chapter.

There are certain important reactions which are endoergonic ($\Delta G > 0$) and as such appear to be thermodynamically forbidden. Nevertheless such reactions can be made to proceed in the desired direction by coupling them with some suitable exoergonic reactions ($\Delta G < 0$). Here by coupling the reaction it is meant that the two reactions are allowed to take place simultaneously. For example, we know the utility of glucose in metabolic processes. But before it can be used in metabolism it must be converted into an activated complex glucose-6 phosphate (G-6-P). This conversion is endoergonic ($\Delta G^o = +12.5 \text{ kJ mol}^{-1}$) and as such cannot take place spontaneously. However, the conversion of adenosine triphosphate (ATP) into adenosine diphosphate (ADP) which is highly exoergonic ($\Delta G^o = -34.1 \text{ kJ mol}^{-1}$) can drive the former reaction to proceed in the desired direction if the two reactions are coupled. This does happen in the metabolic process. The coupling of the two reactions may be represented by the following steps:

(i) Glucose + $H_3PO_4 \rightarrow$ G-6-P ΔG^o (i) = 125 kJ mol^{-1}

(ii) ATP + $H_2O \rightarrow$ ADP + H_3PO_4 ΔG^o (ii) = $-34.1 \text{ kJ mol}^{-1}$

Coupled reaction is

(iii) Glucose + ATP ® G-6-P + ADP

ΔG^o (iii) = ΔG^o (i) + ΔG^o (iii) = -21.7kJ mol^{-1}

The negative free energy change makes the reaction proceed spontaneously.

The other example of the coupled reaction may be considered that of chlorination of titanium oxide in the presence of carbon. The reaction between TiO_2 and Cl_2 is known to be endoergonic ($\Delta G^o = + 162kJ$) and as such the reaction is thermodynamically forbidden. But this reaction can be driven if coupled with the reaction between carbon and oxygen which is exoergonic ($\Delta G^o = - 396$ kJ). The steps are

$TiO_2 + 2Cl_2 \rightarrow TiCl_4 + O_2$ $\quad\quad \Delta G^o = + 162$ kJ

$C + O2 \rightarrow CO_2$ $\quad\quad \Delta G^o = -395$ kJ

Coupled reaction is

$TiO_2 + 2Cl_2 + C \rightarrow TiCl_4 + CO_2$ $\quad\quad \Delta G^o = -233$ kJ

The negative free energy change favours the formation of $TiCl_4$.

On the same line one can argue that the reduction of Al_2O_3 by carbon is not spontaneous at lower temperatures but it becomes spontaneous when carried in the presence of nitrogen gas.

Example 17:

The free energy change for the conversion of malate into fumarate is 3.0 kJ.

Malate $\rightarrow$ *Furmate* ΔG^o (i) = + 3.0 kJ

In metabolism this reaction is coupled with fumarate $\rightarrow$ *asparate, which is exoergonic by 15.5 kJ. Calculate* ΔG^o *for the overall reaction and interpret the result.*

Solution:

The first reaction is endoergonic and the second one is exoergonic. The steps are

(i) Malate $\rightarrow$ Fumarate $\quad\quad \Delta G^o$ (i) = + 3.0 kJ

(ii) Fumarate $\rightarrow$ Asparate $\quad\quad \Delta G^o$ (ii) = – 15.5 kJ

Coupled reaction is

(iiii) Malate $\rightarrow$ Asparate $\quad\quad \Delta G^o$ (ii) = – **12.5 kJ**

The coupling of the reaction drives the reaction (i) in the forward direction because ΔG^o is negative for the coupled reaction. This may also be understood with the help of Le **Chatelier's principle**. In reaction

(ii) fumarate is consumed, the reaction is spontaneous, therefore, this reactant will be needed more. The source of fumarate is reaction (i), hence to meet the demand of the reaction (ii), the reaction (i) must produce fumarate by proceeding in the forward direction.

ELLINGHAM DIAGRAMS AND EXTRACTION OF METALS

The metallurgical operations are carried out at high temperature. Therefore, to predict the feasibility and utility of a given method of extraction of metals we must have sufficient data on ΔG° values for the reactions which take place at higher temperatures. Unfortunately the tabulated ΔG° values for the reactions which take place at higher temperatures. Unfortunately the tabulated ΔG° values refer to 298 K in general. One way to get rid of this problem is to make use of Gibbs-Helmholtz equation, but this would involve long calculations. The convenient and suitable solution to this problem is provided by the Ellingham diagrams.

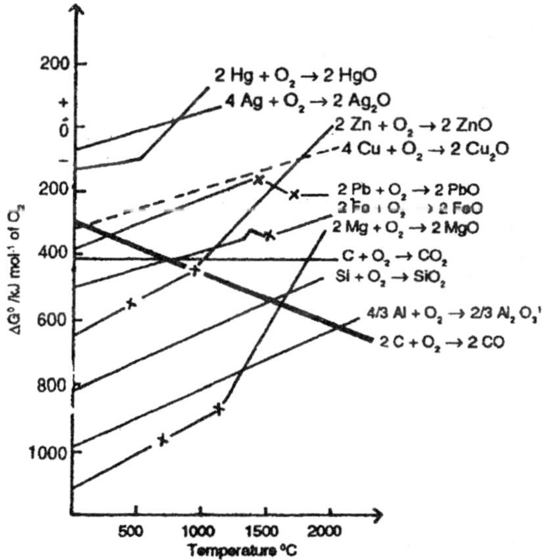

Fig. 4.2 : The Ellingham diagram for the oxide formation. The values of ΔG° for all reactions shown on the diagram involve one mole of oxygen in the oxide formation. $(2x/y)\ M + O_2 \rightarrow (2/y)\ M_xO_y$.

These diagrams are the plots of ΔG° as a function of temperature for several reactions of one class say oxidation of metals. From such diagrams it could be seen at a glance whether a certain metal would reduce the oxide of another. Fig. 4.2 shows a chart for the oxide formation. Similar diagrams are obtained for chlorides, sulphides, nitrides etc.

The important and interesting features of the diagram are :

(i) With the increase of temperature the free energy increases i.e., ΔG^o becomes less negative except for $2C + O_2 \rightarrow 2CO$. This trend in the variation of ΔG^o indicates that the metal oxides become less stable at elevated temperature.

(ii) The lines representing variation of ΔG^o with T, change slope abruptly at each phase transformation like melting or boiling. These are shown by cross (×) on the diagram.

(iii) For the process $2C + O_2 \rightarrow 2CO$, ΔG^o decreases with increase of temperature is raised. This suggests that virtually any metal oxide is reducible by carbon at elevated temperature.

(iv) The free energy change for the reaction $C + O_2 \rightarrow CO_2$ is almost independent of temperature.

(v) The lines for $C + O_2 \rightarrow CO_2$ and $2C + O_2 \rightarrow 2CO$ cross each other at a temperature of 983 K.

Therefore, both CO and C can be used as reducing agents. However, the success of these to be used us a reducing agent is judged by the relative magnitudes of the ΔG^o values for the formation of CO, CO_2 and metal oxide at the temperature concerned i.e., on the ΔG^o values for the following reactions:

$$C + O_2 \rightarrow CO_2$$

$$2C + O_2 \rightarrow 2CO$$

$$2CO + O_2 \rightarrow 2CO_2$$

$$(2x/y)\ M + O_2 \rightarrow (2/y)\ M_xO_y$$

The value of ΔG^o for the formation of a given oxide will be estimated by the entropy change ΔS^o as expressed by the slope of the line $\Delta S^o = -(\partial \Delta G^o/\partial T)_P$, and the enthalpy change ΔH^o. At lower temperature the enthalpy contribution is more important but at higher temperatures ΔS^o will dominate ΔH^o.

For $2C + O_2 \rightarrow 2CO$, the free energy becomes more negative as the temperature increases, this is so because the entropy change ΔS^o increases. This fact is accountable from the reaction $2C + O_2 \rightarrow 2CO$ in which one mole of gaseous O2 disappears and 2 moles of gaseous CO are produced, giving rise to more randomness. That is why CO is more stable than most of the oxides at higher temperature. For metal oxides the free energy becomes more positive as the temperature increases because of the entropy decrease which follows the disappearance of a mole of gaseous oxygen for the reaction under consideration.

The practical utility of the Ellingham diagram may be listed as follows:

(a) oxides of silver and mercury have very small negative ΔG^o and hence can be reduced by heating to temperature where the free energy becomes zero.

(b) Below temperature of 983 K carbon monoxide is the best reducing agent (as the blast furnace reduction of iron oxide).

(c) Above 983 K carbon is the best reducing agent and the cheapest.

(d) CuO can be reduced to Cu by carbon just above room temperature because the free energy of the reaction $2C + O_2 \rightarrow 2CO$ is more negative than that for $4\ Cu + O_2 \rightarrow 2\ Cu2O$.

(e) By the use of carbon as reducing agent the production of silicon is feasible at 1800 K, of magnesium at 1910 K, and of aluminium at about 2350 K. (High temperature reduction is not very economical. Therefore, electrolytic methods are employed for the refining of metals).

PHYSICAL EQUILIBRIA INVOLVING PHASE TRANSITIONS

The Clapeyron Clausius Equation

Clapeyron equation. In a system of a pure substance if several phases are present at equilibrium under a given set of conditions, there is always a possibility of transition of the substance from one phase to another by altering any of the variables of the system. A thermodynamic relation between the change of pressure with change of temperature of a system at equilibrium is called Clapeyron equation. This relationship can be derived as follows:

Let us consider an equilibrium between two phases α and β of a pure substance at a constant temperature and at constant pressure:

Phase (a) $\rightleftharpoons$ phase (b) (at constant T and P)

The free energy change for the process is zero at equilibrium *i.e.*

$\Delta G = G^{\beta} - G^{\alpha} = 0$

$\therefore G^{\alpha} = G^{\beta}$ at equilibrium

Let the temperature and pressure be changed to the new values (T + dT) and (P + dP) in such a manner that the phases a and b are still at equilibrium. Now, if the free energies of the two phases under new set of conditions (T + dT) and (P + dP) are $(G + dG)^{\alpha}$ and $(G + dG)^{\beta}$, then at equilibrium

$(G + dG)^{\alpha} = (G + dG)^{\beta}$

$\because G^{\alpha} = G^{\beta}$

$\therefore (dG)^{\alpha} = (dG)^{\beta}$ ...(25)

for a small change of temperature and pressure. The change of free energy with change of temperature and pressure is expressed as

$dG = -SdT + VdP$ ...(26)

Therefore from equations (25) and (26), we get

$-S^{\alpha}\,dT + V^{\alpha}\,dP = -S^{\beta}\,dT + V^{\beta}\,dP$

or $(S^{\beta} - S^{\alpha})\,dT = (V^{\beta} - V^{\alpha})\,dP$

$$\text{or } \frac{dP}{dT} = \frac{S^{\beta} - S^{\alpha}}{V^{\beta} - V^{\alpha}} = \frac{\Delta S}{\Delta V} \quad ...(27)$$

where $\Delta S = (S^{\beta} - S^{\alpha})$ is the change in entropy and $\Delta V = (V^{\beta} - V^{\alpha})$ is the change in volume for the process. S^{β} and S^{α} and the entropies of the pure substance in the two phases and V^{β} and V^{α} are the corresponding volumes: for 1mole of a pure substance the quantities refer to the molar entropy (S_m) and molar volume (V_m). Further, since at equilibrium $\Delta G = 0$, we have for this transition process, $\Delta S_m = \Delta H_m/T$, where ΔH_m is the change in enthalpy for the reversible transformation T. Substituting this value of ΔS into equation (27) we get,

$$\frac{dP}{dT} = \frac{\Delta H_m}{T\Delta V_m} \quad ...(28)$$

$$\text{or } \Delta H_m = T\frac{dP}{dT}\Delta V_m \quad ...(29)$$

For one gram of the substance equations (28) and (29) can be reduced to

$$\frac{dP}{dT} = \frac{\lambda}{T(V^{\beta} - V^{\alpha})} \qquad ...(30)$$

$$\text{or } \lambda = T\frac{dP}{dT}(V^{\beta} - V^{\alpha})$$

$$\text{where } l = \frac{\Delta H_m}{M} \text{ and } \left(\frac{V^{\beta} - V^{\alpha}}{M}\right) = \left(V^{\beta} - V^{\alpha}\right)$$

M = molecular weight of the substance.

Equations (28) and (30) are called the **Clpeyron equation**. These are general thermodynamic relations applicable to all phase transitions of a pure substance. The equations show that dP/dT is related directly to the enthalpy of transition and inversely to the temperature and the volume change accompanying the transformation.

Equation (28) can be integrated if we know the temperature or pressure dependence of ΔHm and ΔVm. Since such information is usually unaviable, equation is used in differential form as,

$$\frac{\Delta P}{\Delta T} = \frac{P_2 - P_1}{T_2 - T_1} = \frac{\Delta H_m}{T\Delta V_m} \qquad ...(31)$$

where $T = (T_1 + T_2)/2$

The equation is integrated on the assumption that ΔH_m and ΔV_m are constant.

$$\int_{P_1}^{P_2} dP = \frac{\Delta H_m}{\Delta V_m}\int_{T_1}^{T_2}\frac{dT}{T}$$

$$\therefore (P_2 - P_1) = \frac{\Delta H_m}{\Delta V_m} \ln\frac{T_2}{T_1} \qquad ...(32)$$

Equation (32) is sometimes preferred for calculations involving appreciable temperature changes. This form of equation is usually employed for processes involving condensed phases.

APPLICATIONS OF CLAPEYRON EQUATION

Equation (28) can be used to discuss the two-phase equilibria in one component system. Some of its applications are discussed here:

(i) Liquid -vapour equilibria (Vaporization process)

Liquid $\rightleftharpoons$ Vapour

Phase (α) $\rightleftharpoons$ Phase (β)

$\Delta H_m = \Delta H_{vap,m}$ = molar enthalpy of vaporization

$\Delta V = V^{\beta} - V^{\alpha} = (V^{v} - V^{l})$

$$\therefore \quad \frac{dP}{dT} = \frac{\Delta H_{vap,m}}{T\left(V^{v} - V^{l}\right)} \qquad ...(33)$$

Since ΔH_{vap} is positive and $V^v > V^l$ (except at critical point), therefore dP/dT will always be positive i.e., with increase of P, there must be an increase in T, and *vice versa.*

Example 18:

The heat of vaporization of water is 40820 J mol⁻¹, the molar volume of liquid water is 0.019 dm³ mol⁻¹ and the molar volume of steam is 30.199 dm³ mol⁻¹ all at 100°C and 1 atm. What would be the change in the boiling point of water at 100°C if the atmospheric pressure is changed by 1mm?

Solution:

$\Delta H_{vap} = 40820 \text{ J mol}^{-1}$

$V^v = 30.199 \text{ dm}^3 \text{ mol}^{-1}$, $V^l = 0.019 \text{ dm}^3 \text{ mol}^{-1}$,

$T = 100 + 273 = 373 \text{ K}$

$\Delta V_m = V^v - V^l = (30.199 - 0.019) \text{ dm}^3 \text{ mol}^{-1}$

$= 30.18 \text{ dm}^3 \text{ mol}^{-1}$

$= 30.180 \times 10^{-3} \text{ m}^3 \text{ mol}^{-1}$

$$\frac{dP}{dT} = \frac{\Delta H_{vap,m}}{T\Delta V_m} = \frac{40820 \text{Jmol}^{-1}}{(373\text{K}) \times (30.180 \times 10^{-3} \text{m}^3 \text{mol}^{-1})}$$

$= 3626 \text{ J mol}^{-3} \text{ K}^{-1} = 3626 \text{ Nm}^{-2} \text{ K}^{-1}$

$$= 3626 \text{ P}_a\text{K}^{-1} = \frac{3626 \text{P}_a \text{K}^{-1}}{101325 \text{P}_a \text{atm}^{-1}}$$

$= 0.0357 \text{ atm K}^{-1}$

$= (0.0357 \text{ atm K}^{-1}) \times (760 \text{ mm atm}^{-1})$

$= 27.13\ mmK^{-1}$

$\therefore$ dT/dP = **0.0368 K/mm**

= change in the boiling point of water per mm change in atmospheric pressure.

(ii) *Solid-liquid equilibria* (Fusion process).

Solid $\rightleftharpoons$ Liquid

$\Delta H_m = \Delta H_{fus,\ m}$ = molar enthalpy of fusion

ΔV_m = Volume of liquid (V^l_m) – Volume of solid (V^s_m)

Clapeyron equation can be written as

$$\frac{dP}{dT} = \frac{\Delta H_{fus,m}}{T(V^l - V^s)}$$

For the conversion of solid into liquid heat is required and ΔH_{fus} is always positive. But the volume of liquid may be greater or less than that of solid. So that sign of dP/dT will depend on the nature of the system. For example, the volume of liquid water is less than the volume of ice, so ΔV is negative and hence dP/dT is negative. That is, with increase of pressure, the melting point would decrease for ice-water equilibrium. The same is true for bismuth metal. For an overwhelming majority of substances, the quantity dP/dT or dT/dP is positive.

Example 19:

Calculate the change of atmospheric pressure to cause a change of 1°C in the melting point the benzene. Data given $t_f = 5.0°$ C; If = 128 Jg^{-1} the specific volumes of liquid and solid benzene are 1.119 and 1.06 $cm^3\ g^{-1}$ respectively.

Solution:

$$\frac{dP}{dT} = \frac{\lambda_{fus}}{T_{fus}\Delta v_{fus}}$$

$$= \frac{128.6Jg^{-1}}{(278K) \times (0.059cm^3g^{-1})} = 7.84\ J\ Cm^{-3}\ K^{-1}$$

$$= \frac{(7.84) \times 10^6\ Jm^{-3}K^{-1}}{1.01325 \times 10^5\ Nm^{-2}atm^{-1}} = 77.3\ atm\ K{-1}$$

$\therefore$ dP/dT = **77.3 atm K^{-1}**

To cause an increase of 1° C in the melting point of benzene the atmospheric pressure should be changed by 77.3 atm.

Example 20:

Calculate the change in the freezing point of water at 0°C per atm change of pressure. Data at 0°C are given as, heat of fusion of ice is 335 Jg^{-1}, the density of water is 0.9998 g cm^{-3}, and the density of ice is 0.98168 g cm^{-3}.

Solution:

Given data

λ_f = 335 Jg^{-1}, T = 273 K

ρ^l = 0.9998 g cm^{-3}; ρ^s = 0.9168 g cm^{-3}

v^l = 1.0002 cm^3 g^{-1}; v^s = 1.0908 cm^3 g^{-1}

Δv = 1.0002 – 1.0908 = – 0.0906 cm^3 g^{-1}

$$\frac{\Delta P}{\Delta T} = \frac{\lambda_{fus}}{T\Delta v} = -\frac{335 Jg^{-1}}{(273K) \times (0.0906 cm^3 g^{-1})}$$

= – 13.543 J cm–3 K^{-1}

= – 133.7 atm K^{-1}

$$\therefore \quad \frac{\Delta P}{\Delta T} = -\ 0.0075 \text{ K atm}^{-1}$$

Thus an increase in pressure of 1 atm lowers the freezing point by 0.00075. The negative sign indicates that an increase in pressure causes a decrease in temperature.

(iii) *Solid- vapour equilibrium* (sublimation process)

Solid $\rightleftharpoons$ Vapour

$\Delta H_m = \Delta H_{sub}$ = Enthalpy of sublimation. Since ΔH is a state function and the sublimation of a solid is equivalent to the fusion of the solid and vaporization of the resulting liquid at a given temperature, we must have

$$\Delta H_{sub} = \Delta H_{fus} + \Delta H_{vap}$$

Hence, when any two of these quantities are known, the third can be calculated.

ΔV_m = molar volume of vapour phase – molar volume of solid phase. For sublimation process, the Clapeyron equation is written as,

$$\frac{dP}{dT} = \frac{\Delta H_{sub}}{T\Delta V_m}$$

Since ΔH_{sub} is always positive and similar is the change of volume ΔV_m, the solid – vapour equilibria have always a positive slope.

(iv) Equilibrium between two crystalline forms

$$\text{Solid } (\alpha) \rightleftharpoons \text{Solid } (\beta)$$

If α is stable form below the transition point and β is the form stable above the transition point then Clapeyron equation can be written as,

$$\frac{dP}{dT} = \frac{\Delta H_{trans}}{T\left(V_m^{\beta} - V_m^{\alpha}\right)} \qquad \text{...(34)}$$

where ΔH_{trans} = Enthalpy of transition, V_m^{α} and V_m^{β} are the molar volumes of the indicated form. The sign of dP/dT will depend upon the signs of ΔH_{trans} and the densities of the two forms.

Example 21:

Calculate the rate of change of transition temperature with pressure for sulphur. The data given are, transition temperature = 95°C at 1 atm, enthalpy of transition per gram of sulphur is 13.4 J. Monoclinic-sulphur (stable above transition point) has a greater specific volume than that of rhombic sulphur by 0.0126 $cm^3\ g^{-1}$.

Solution:

Given data:

λ_{trans} = 13.4 Jg^{-1} T_t =95.5 + 273 = 368.5 K

Δv = 0.0126 $cm^3\ g^{-1}$

$$\frac{dP}{dT} = \frac{\lambda}{T\Delta v}$$

$$= \frac{13.4 Jg^{-1}}{(368.5K)\,(0.0126 cm^3 g^{-1})}$$

$= 28.5$ atm K^{-1}

$$\therefore \frac{dP}{dT} = 0.0351 \text{ K atm}^{-1}$$

The transition point between the two forms of sulphur should be increased by 0.035° per atmosphere increase of external pressure.

The Clausius-Clapeyron Equation

The Clapeyron equation, when applied to the liquid-vapour equilibria, with the assumption that the volume of the liquid phase is negligible as compared to that of vapour phase which behave as a perfect gas, is called the Clausius-Clapeyron equation. If the temperature is not too near the critical point, then, $V^v >> V^l$ and $V^v_m = RT/P$. With these simplification equation (31) can be written as,

$$\frac{dP}{dT} = \frac{\Delta H_{vap,m}}{TV^v_m} = \frac{P\Delta H_{vap,m}}{RT^2} \qquad ...(35)$$

This differential equation may be employed if the changes in temperature and pressure are small. For example, it is convenient to use the following form in correcting boiling points for barometric fluctuations:

$$\frac{\Delta T}{\Delta P} = \frac{RT^2}{\Delta H_{vap,m} \times P}$$

Integration of Clausius-Clapeyron Equation

Apart from the uses Clapeyron equation discussed earlier, it is also useful to calculate $\Delta H_{vap,m}$ and to study the variation of vapour pressure with temperature.

On rearrangement, equation (35) reads,

$$\frac{dP}{P} = \frac{\Delta H_{vap,m}}{R} \frac{dT}{T^2}$$

Integrating on the assumption that ΔH_{vap} is independent of temperature, we get,

$$\int d \ln P = \frac{\Delta H_{vap,m}}{R} \int T^{-2} dT + \text{Constant}$$

$$\text{or } \ln P = -\frac{\Delta H_{vap,m}}{RT} + C$$

$$\text{or } \log P - \frac{\Delta H_{vap,m}}{2.303RT} + C' \qquad ...(36)$$

C being the integration constant. Equation (36) suggests that a plot of log P versus 1/T should be linear with a slope of $-\Delta H_{vap}/2.303R$. Thus, the heat of vaporization can be calculated, using

$$\Delta H_{vap,m} \text{ J mol}^{-1} = - \text{Slope} \times 2.303 \times 8.314$$

It is more convenient to integrate equation (35) between suitable initial, and final limits to get,

$$\int_{P_1}^{P_2} d \ln P = \frac{\Delta H_{vap,m}}{R} \int_{T_1}^{T_2} T^{-2} dT$$

$$\text{or } \ln P_2 - \ln P_1 = - \frac{\Delta H_{vap,m}}{R} \left[\frac{1}{T_2} - \frac{1}{T_1} \right]$$

$$\text{or } \log \left(\frac{P_2}{P_1} \right) = \frac{\Delta H_{vap,m} (T_2 - T_1)}{2.303RT_2T_1} \qquad ...(37)$$

Equation (37) can be used to calculate the heat of vaporization or the heat of sublimation from the vapour pressure data at two different temperatures. For precise work the variation of ΔH with temperature should also be considered before integrating equation (35).

Example 22:

The vapour pressure of ethylene is given as a function of temperature by the equation

(A) $\log P/\text{(mmHg)} = -\frac{834.13}{T} + 1.75 \log T - 8.375 \times 10^{-3} T +$ 5.3234

Calculate the enthalpy of vaporization of ethylene at its boiling point (–1039° C).

Solution:

Clausius-Clapeyron equation reads

$$\frac{d \ln P}{dT} = \frac{\Delta H_{vap,m}}{RT^2} \qquad ...(i)$$

On transforming the equation (A) into differentiable form, we get,

$$\frac{1}{2.303}\text{In P} = -\frac{834.13}{T} + \frac{1.75}{2.303}\text{ In T} - 8.375 \times 10^{-3}\text{ T} + 5.3224$$

Differentiating with respect to T we have

$$\frac{d\text{ In P}}{dT} = \frac{2.303 \times 834.13}{T^2} + \frac{1.75}{T} - 2.303 \times 8.375 \times 10^{-3} \quad \text{...(ii)}$$

$$\therefore \frac{d\text{ In P}}{dT} = \frac{\Delta H_{vap,m}}{RT^2}$$

$$= \frac{2.303 \times 834.13}{T^2} + \frac{1.75}{T} - 2.303 \times 8.375 \times 10^{-3}$$

and $\Delta H_{vap,m} = R \times 2.303 \times 834.13 + 1.75\text{ RT} + 2.303 \times 8.375 \times 10^{-3} \times RT^2$

$\because$ R = 8.314 J mol^{-1} K^{-1},

and T = – 103.9 + 273.15 = 169.25 K

$\therefore \Delta H_{vap,\,m}$ = **13.840 kJ mol^{-1}**

Example 23:

The temperature dependence of vapour pressure of thiophene is given by

$$\log P/(\text{mmHg}) = -\frac{1246.038}{t + 221.354} + 6.95926.$$

What is the enthalpy of vaporization at 298K.

Solution:

$$\text{In P} = -\frac{2.303 \times 1246.038}{(t + 273.15) - 273.15 + 221.354}$$

$+ 6.95926 \times 2.303$

$$\therefore \frac{d\text{ In P}}{dT} = \frac{2.303 \times 1246.038}{(T - 51.796)^2} = \frac{\Delta H_{vap,m}}{RT^2}$$

$\Delta H_{vap,m}$ (298K) = **34.90 kJ mol^{-1}**

Example 24:

The vapour pressure of decane is 10 torr at 55.7°C and 400 torr at 150.6°C. Calculate $\Delta H_{vap,m}$ and $\Delta S_{vap,m}$ at 100°C, and the normal boiling point.

Solution:

We shall use the integrated form of the Clausius-Clapeyron equation

$$\log (P_2/P_1) = \frac{\Delta H_{vap,m}(T_2 - T_1)}{2.303RT_2T_1}$$

$P_1 = 10$ torr, $T_1 = 55.7 + 273.15 = 328.85$ K

$P_2 = 400$ torr, $T_2 = 150.6 + 273.15 = 423.75$ K

$\Delta H_{vap,m} = [2.303 \times (328.85K)\ (423.75K)\ 5\ (8.314J\ mol^{-1}K^{-1})$

$\times \log (400/10)] \times \dfrac{1}{94.9}$

= 45042 Jmol–1

$\Delta S_{vap,m} = \Delta H_{vap,m}/T = 45042\ J\ mol^{-1}/373\ K =$ **120.75 JK⁻¹mol⁻¹** At the boiling point vapour pressure is 760 torr = P_2 at $T_2 = T_b$, $P_1 = 10$ torr at $T_1 = 328.85$ K. Therefore,

$$\log (760/10) = -\frac{45042 Jmol^{-1}}{2.303 \times 8.314\ JK^{-1}mol^{-1}}\left[\frac{1}{T_b} - \frac{1}{328.85K}\right]$$

$\therefore T_b =$ **446 K.**

Example 25:

An organic compound X melts at 80° C. If the vapour pressure of the liquid X is 10 mm at 85.8° C and 40 mm at 119.3°C, and that of solid X is 1 mm at 52.6°C. Calculate (a) heat of vaporization of liquid, the boiling point and the entropy of vaporization at boiling point, (b) the vapour pressure at the melting point, (c) ΔH_{sub} and ΔH_{fus}, assuming that three phase coexist at 80°C and 1atm, (d) At what temperature the vapour pressure of the solid would be 10^{-5} mm Hg?

Solution:

Given data :

$P_1 = 10$ mm $T_1 = 85.80 + 273.15 = 358.95$ K

P_2 = 40 mm T_2 = 119.30 + 273.15 = 392.45 K

P_s = 1mm T_s = 52.60 + 273.15 = 325.75 K

(a) $\log \frac{P_2}{P_1} = -\frac{\Delta H_{vap,m}}{2.303R}\left[\frac{1}{T_2}-\frac{1}{T_1}\right]$

(i) $\Delta H_{vap,\ m} = \frac{2.303 \times 8.314 \times 392.45 \times 358.95}{33.5} \times \log \frac{40}{10}$

(ii) At boiling point T_b the vapour pressure (P_2) is 1 atm or 760 mm of Hg. Thus,

$$\log \frac{760}{10} = -\frac{48474}{19.147}\left[\frac{1}{T_b}-\frac{1}{358.95}\right]$$

On solving for T_b, we get

$$\frac{1}{T_b} = \frac{1}{358.9} - \frac{19.147}{48474} \times \log 76$$

$$= \frac{1}{358.95} - \frac{36.01}{48474} = 2.7859 \times 10^{-3} - 7.4297 \times 10^{-4}$$

$= 20.423 \times 10^{-4}$

$\therefore T_b = 489.5$ K

(iii) Entropy of vaporization at boiling point can be calculated by

$\Delta G = \Delta H\ T\Delta S = 0$

$$\therefore \Delta S_{vap,m} = \frac{\Delta H_{vap,m}}{T_b} = \frac{48474 mol^{-1}}{489.5 K} = 99.02 \text{ JK}-1 \text{ mol}-1$$

(b) Melting point = T 80.0 + 273.15 = 353.15 K; vapour pressure at melting point $P_m = P_2$ = ?

P_1 = 10 mm at T_1 = 358.95 K

$$\therefore \log \frac{P_2}{P_1} = \frac{48474 \text{ J mol}^{-1}}{2.303R}\left[\frac{1}{353.15 \text{ K}} - \frac{1}{358.95 \text{ K}}\right]$$

$$= -\frac{48474 \text{ J mol}^{-1} \times (5.8\text{K})}{2.303 \times (8.314 \text{Jmol}^{-1}\text{K}^{-1}) \times (353.15\text{K}) \times (358.95\text{K})}$$

$= -0.1158$

Taking antilogarithm we obtain

$$\frac{P_2}{P_1} = 0.766$$

and $P_2 = 0.766 \times P_1 = 7.66$ mm Hg

(c) At 80°C, three phases, *i.e.*, solid, liquid and vapour coexist. Thus, this can be considered triple point. Given that

$P_s = 1\text{mm}\ T_s = 325.75$ K

$P_m = 7.66$ mm $T_m = 353.15$ K

Thus, for solid – vapour equilibria

$$\log \frac{P_m}{P_s} = -\frac{\Delta H_{sub}}{2.303R}\left[\frac{1}{353.15} - \frac{1}{325.75}\right]$$

$$\therefore\ \Delta H_{sub} = \frac{2.303 \times (8.314\text{Jmol}^{-1}\text{K}^{-1}) \times (353.15\text{K}) \times (325.75\text{K})}{27.40\text{K}}$$

$$\times \log \frac{7.66\text{mm}}{1\text{mm}}$$

$= 71082$ J mol^{-1}

$= 71.082$ kJ mol^{-1}

$\because\ \Delta H_{sub} = \Delta H_{vap}$

$\therefore\ \Delta H_{fus} = 71.082$ kJ mol^{-1} – 48.474 kJ mol^{-1}= 22.608 kJ mol^{-1}

(d) Let T_2 be the temperature at which the vapour pressure of solid is 10^{-5} mm Hg. Also $P_1 = 1$ mm at $T_1 = 325.75$ K

$$\log \frac{P_2}{P_1} = -\frac{\Delta H_{sub}}{2.303R}\left(\frac{1}{T_2} - \frac{1}{T_1}\right)$$

On solving for T_2, we get

$$\frac{1}{T_2} = -\frac{2.303 \times R}{\Delta H_{sub}}\log\left(\frac{P_2}{P_1}\right) + \frac{1}{T_1}$$

$$= -\frac{2.303 \times (8.314\text{Jmol}^{-1}\text{K}^{-1})}{71082\text{Jmol}^{-1}}\log\frac{10^{-5}\text{mm}}{1\text{mm}} + \frac{1}{325.7\text{K}}$$

$= 1.3468 \times 10 -3\ K^{-1} + 3.0698 \times 10^{-3}\ K^{-1}$

$= 4.4166 \times 10^{-3}\ K^{-1}$

$\therefore\ T_2 = 226.4\ K$

The vapour pressure of the solid would be less than 10–5 mm of Hg at a temperature below 226 K.

ENTROPY OF VAPORIZATION AND TROUTONS'S RULE

For many liquids the molar entropy of vaporization at the normal boiling point is constant and is approximately equal to 88 JK^{-1} mol^{-1}. This is known as Trouton's Rule and is useful for estimating the molar heat of vaporization of liquids of known boiling point.

The validity of *Trouton*'s Rule for simple liquid be understood with the help of molecular aspect of entropy function, which is a measure of randomness. The entropy of vapour phase is much more greater than that of the liquid phase of the same substance. This leads to the conclusion that there is a difference in the molecular arrangement of the vapour phase and the liquid phase. This difference of the arrangement of the molecules in some substances is almost the same. Since the entropy of vaporization is equal to the molar entropy of vapour phase minus the molar entropy of the liquid phase, the difference of the two entropies must be constant for all those substances for which the extent of disorderness between vapour and liquid phases are identical. However, this rule is not valid for all those liquids which have a tendency to undergo association or dissociation. For example, the entropy of vaporization of water is 108 JK^{-1} which is larger than that predicted by Trouton's Rule. This large value of the entropy of vaporization is attributed to the presence of hydrogen-bonding on the liquid state. Due to H-bonding the freedom of the molecules is considerably restricted leading to a more regular arrangement and hence less entropy as compared with normal liquids. Since association of such molecules is of minor importance in the vapour, therefore, the larger value of entropy of vaporization can be understood.

The entropy of vaporization of acetic acid, with a tendency of forming dimer via hydrogen bonding, is only 62 JK^{-1} mol^{-1}. This value is much less than that predicted by Trouton's Rule. This value in fact is only for 1/2 mol of acetic acid. The tendency of acetic acid to get dimerized in the liquid state persists to an appreciable extent in the vapour state. One mole of liquid in which association is complete in both the liquid and the vapour state would effectively consist of 1/2 mol of dimer and would have half the expected molar entropy, as the entropy is independent of molecular weight as expected from Trouton's Rule. Some values of entropy of vaporization are given in Table 2.

Table 2 : Heat and Entropies of Vaporization of Liquids at their Normal Boiling Points.

Liquid	Boiling Point	ΔH_{vap}	$\Delta S_{vap} = \frac{\Delta H_{vap}}{T_b}$
	°C	$(Jmol^{-1})$	$(JK^{-1}\ mol^{-1})$
Acetic acid (CH_3COOH)	118.2	24390	62
Methane (CH_4)	– 161.4	9270	83
Ethyl ether (C_2H_5 –O –C_2H_5)	34.6	25980	84
Cyclohexane (C_6H_{12})	80.7	30080	85
Benzene (C_6H_6)	80.1	30760	87
Methyl alcohol (CH_3OH)	64.7	35270	104
Ethyl alcohol (C_2H_5OH)	78.5	38570	110

Hildebrand's Rule

It has been shown by Hildebrand that a more consistent set of entropies of vaporization are obtained if they are calculated at temperatures at which the molar volumes of the saturated vapours of the different compounds are the same. However, the abnormal behaviour of hydrogen-bonded liquids is not covered by this rule.

Example 26:

A given liquid obeys Trouton's rule. What would be its vapour pressure at a temperature equal to one third of the normal boiling point T_bK.

Solution:

At the boiling point of a liquid its vapour pressure is equal to 1 atm (= 101325 P_a). To calculate is vapour pressure at some other temperature ($T_2 = 1/3\ T_b$) we can use integrated form of clausius Clapeyron equation In $P_1/P_2 = -(\Delta H/R)(T^{-1}_2 - T^{-1}_1)$; the value of ΔH can be evaluated by Trouton's rule; ΔH_{vap} 88 J $mol^{-1}K^{-1} \times T_b$.

Let P_2 be the vapour pressure of the liquid at $T_2 = 1/3\ T_b$, therefore

$$\log\frac{P_2}{P_1} = -\frac{(88\text{JK}^{-1}\text{mol}^{-1})T_b}{2.303R}\left[\frac{1}{\frac{1}{3}T_b} - \frac{1}{T_b}\right]$$

$$= -\frac{88\text{JK}^{-1}\text{mol}^{-1} \times T_b}{19.147\text{JK}^{-1}\text{mol}^{-1}}\left(\frac{3-1}{T_b}\right)$$

$$= -\frac{88 \times 2}{19.147} = -\ 9.192 = 10.808$$

$\therefore\ P_2/P_1 = 6.4 \times 10^{-10}$

If $P_1 = 1$ atm, then $P_2 = 6.43 \times 10^{-10} \times 1$ atm $= 6.43 \times 10^{-10}$ atm

If $P_1 = 101325$ Pa, then $P_2 = 6.43 \times 10^{-10} \times 101325$ Pa

$= 6.5 \times 10^{-5}$ Pa.

THERMODYNAMIC RELATIONS

The success of thermodynamic predictions are based on the fact that very odd and troublesome experiments can be avoided without lack of precision by using a few master equations. These master equations are the starting point.

$$dE = TdS - PdV$$

$$H = E + PV$$

$$A = E - TS$$

$$G = H - TS = E + PV - TS = A + PV$$

$$C_V = (\partial E/\partial T)_V$$

$$C_P = (\partial H/\partial T)_P$$

With these fundamental equations any desired thermodynamic relations can be deduced by keeping in mind that E, S, H, A, G, T, P, V are all state functions and Euler's theorem of exactness can be applied to the total differential of these functions. [We can emphasize that if any one wants to master the mathematical aspect (Physical aspect will automatically be mastered thereafter) of thermodynamics one must be familiar with these equations.]

Maxwell Relations

(i) Combined form of first and second law is

$$dE = TdS - PdV \quad ...(38)$$

That is, E is a function of entropy and volume, E = f (S, V). The total differential dE is given by

$$dE = \left(\frac{\partial E}{\partial S}\right)_V dS + \left(\frac{\partial E}{\partial V}\right)_S dV \quad ...(39)$$

Comparing the coefficients of dS and dV separately in equation (38) and (39) we get

$$\left(\frac{\partial E}{\partial S}\right)_V = T \quad ...(40)$$

$$\text{and} \left(\frac{\partial E}{\partial V}\right)_S = -P \quad ...(41)$$

On differentiating w.r. to S at constant V, equation (41) gives

$$\frac{\partial}{\partial S}\left[\left(\frac{\partial E}{\partial V}\right)_S\right]_V = -\left(\frac{\partial P}{\partial S}\right)_V \quad ...(43)$$

Since E is a state function and dE is exact differential, the Euler's theorem must be valid, i.e.,

$$\frac{\partial^2 E}{\partial V \partial S} = \frac{\partial^2 E}{\partial S \partial V} \quad ...(44)$$

From equations (42) and (43), we get Maxwell relation

$$\left(\frac{\partial T}{\partial V}\right)_S = -\left(\frac{\partial P}{\partial V}\right)_V \quad ...(45)$$

(ii) Enthalpy is defined by

$H = E + PV$

or $dH = dE + PdV + VdP$

$\because dE + PdV = TdS$

$$\therefore dH = TdS + VdP \quad ...(46)$$

Since H = f (S,P) therefore the total differential dH is given by

$$dH = \left(\frac{\partial H}{\partial S}\right)_P dS + \left(\frac{\partial H}{\partial P}\right)_s dP \qquad ...(47)$$

On comparing the coefficients of dS and dP separately in equations (46) and (47), we get

$$\left(\frac{\partial H}{\partial S}\right)_P = T \qquad ...(48)$$

$$\left(\frac{\partial H}{\partial P}\right)_s = V \qquad ...(49)$$

Differentiation of equation (48) with respect to P at constant S give

$$\frac{\partial}{\partial P}\left\{\left(\frac{\partial H}{\partial S}\right)_P\right\}_s = \left(\frac{\partial T}{\partial P}\right)_s \qquad ...(50)$$

On differentiating w.r.to S at constant P equation (49) yields

$$\frac{\partial}{\partial S}\left\{\left(\frac{\partial H}{\partial P}\right)_s\right\}_P = \left(\frac{\partial V}{\partial S}\right)_P \qquad ...(51)$$

H is a state function, dH is exact differential, therefore,

$$\frac{\partial^2 H}{\partial P \partial S} = \frac{\partial^2 H}{\partial S \partial P} \qquad ...(52)$$

From (50) —(52) we obtain Maxwell relation

$$\left(\frac{\partial T}{\partial P}\right)_s = \left(\frac{\partial V}{\partial S}\right)_s \qquad ...(53)$$

(iii) Helmholtz free energy (work function A) is given by

$A = E - TS$

Differentiation gives

$dA = dE - TdS - SdT$

But the combined first and second laws of thermodynamics reads

$TdS = dE + PdV$

$$\therefore\ dA = -SdT - PdV \qquad ...(54)$$

This equation indicates that A = f (T, V), hence the total differential dA may be expressed as

$$dA = \left(\frac{\partial A}{\partial T}\right)_V dT + \left(\frac{\partial A}{\partial V}\right)_T dV \qquad ...(55)$$

On comparing the coefficient of dT and the coefficient of dV in equations (54) and (55), we have

$$\left(\frac{\partial A}{\partial T}\right)_V = -S \qquad ...(56)$$

$$\left(\frac{\partial A}{\partial V}\right)_T = -P \qquad ...(57)$$

On differentiating w.r. to V at constant T equation (56) gives

$$\frac{\partial}{\partial V}\left\{\left(\frac{\partial A}{\partial T}\right)_V\right\}_T = \left(\frac{\partial S}{\partial V}\right)_T \qquad ...(58)$$

Equation (57), on differentiation w.r. to T at constant V yields

$$\frac{\partial}{\partial T}\left\{\left(\frac{\partial A}{\partial V}\right)_T\right\}_V = -\left(\frac{\partial P}{\partial T}\right)_V \qquad ...(59)$$

Since A is a state function and dA is exact differential hence $\partial^2 A/\partial V\,\partial T = \partial^2 A/\partial T \partial V$. Applying this condition equations (58) and (59) give

$$\left(\frac{\partial S}{\partial V}\right)_T = \left(\frac{\partial P}{\partial T}\right)_V = \alpha/\beta \qquad ...(60)$$

where $\alpha = V^{-1}\,(\partial V/\partial T)_P$ and $\beta = -\,V^{-1}\,(\partial V/\partial P)_T$

(iv) Gibbs free energy G is defined by

G = H – TS

For differential changes we have

dG = dH – TdS – SdT

But H = E + PV

and dH = dE + PdV + VdP

= TdS + VdP ($\because$ TdS = dE + PdV)

Thus, $dG = - SdT + VdP$ ...(61)

At constant T, dT = 0, hence equation (61) reduces to

$(dG)_T = V(dP)_T$

$$\text{or } \left(\frac{\partial G}{\partial P}\right)_T = V \qquad ...(62)$$

At constant P, dP = 0, hence equation (61) yields

$(dG)_P = - S(dT)_P$

$$\text{or} \left(\frac{\partial G}{\partial T}\right)_P = - S \qquad ...(63)$$

On differentiating w.r. to T at constant P equation (62) gives

$$\frac{\partial}{\partial T}\left\{\left(\frac{\partial G}{\partial P}\right)_T\right\}_P = \left(\frac{\partial V}{\partial T}\right)_P \qquad ...(64)$$

When differentiated w.r. to P at constant T equation (63) reduces to

$$\frac{\partial}{\partial P}\left\{\left(\frac{\partial G}{\partial T}\right)_P\right\}_T = -\left(\frac{\partial S}{\partial P}\right)_T \qquad ...(65)$$

Since G is a state function and hence $\partial^2 G/\partial T\,\partial P = \partial^2 G/\partial P \partial T$. With this condition of exactness, equations (64) and (65) give

$$\left(\frac{\partial S}{\partial P}\right)_T = -\left(\frac{\partial V}{\partial T}\right)_P = -\alpha V \qquad ...(66)$$

where $\alpha = V^{-1}(\partial V/\partial T)_P$

Equation (66) is known as Maxwell relation.

Significance of Maxwell Relation. Equations (45), (53), (60) and (66), called Maxwell relations, are very important and helpful in many ways. The use of these relations can reduce the tedious experimental work into simple paper and pencil exercise. For example, the variation of entropy with volume at a constant temperature cannot be studied directly by experimental method as we do not have any apparatus such as entropymeter (like thermometer for temperature of dilatometer for density etc). However, the same can be evaluated by using the values of α, the coefficient of thermal expansion, and b, the compressibility coefficient for any substance by using Maxwell relation $(\partial S/\partial V)T = a/$

b (cf. equation 60). Similarly the knowledge of the value of α of a substance assists us to study the isothermal change in entropy with the change in pressure (cf. equation (66). Maxwell relations are also very useful to deduce many other thermodynamic relations viz., the Clapeyron equation, thermodynamic equation of state etc. which are very important to deduce some more information about the system.

Exercise. If E = f (S, V); H = f (S, P); A = f (T,V) and G = f (T,P), and E, H, A, G are state functions, then by using the definition of E, H, S, A and G along with the combined form of the first and second laws of thermodynamics show that

$$(\partial E/\partial S)_V = T = (\partial H/\partial S)_P$$

$$(\partial E/\partial V)_S = -P = (\partial A/\partial V)_T$$

$$(\partial H/\partial P)_S = V = (\partial g/\partial P)_T$$

$$(\partial G/\partial T)_P = -S = (\partial A/\partial T)_V$$

THERMODYNAMIC EQUATION OF STATE

The relationship between E or H and PVT data is called thermodynamic equation of state.

(i) Combined first and second laws give

$$TdS = dE + PdV \qquad ...(67)$$

For constant T, equation (67) can be written as

$$T\left(\frac{\partial S}{\partial V}\right)_T = \left(\frac{\partial E}{\partial V}\right)_T + P$$

$$\text{Or } \left(\frac{\partial E}{\partial V}\right)_T = T\left(\frac{\partial S}{\partial V}\right)_T - P \qquad ...(68)$$

But dA = – SdT – PdV, gives

$$\left(\frac{\partial S}{\partial V}\right)_T = \left(\frac{\partial P}{\partial T}\right)_V \qquad ...(69)$$

From equations (68) and (69) we get,

$$\left(\frac{\partial E}{\partial V}\right)_T = T\left(\frac{\partial P}{\partial T}\right)_V - P \qquad ...(70)$$

This equation can be employed to study the isothermal variation of internal energy with volume for all types of substances provided we have sufficient P-V-T data or a suitable equation of state relating P-V-T data.

(ii) Change of enthalpy and entropy are related by

$$TdS = dH - VdP \qquad ...(71)$$

For constant temperature Process, we have

$$T\left(\frac{\partial S}{\partial P}\right)_T = \left(\frac{\partial H}{\partial P}\right)_T - V$$

$$\text{or} \left(\frac{\partial H}{\partial P}\right)_T = T\left(\frac{\partial S}{\partial P}\right)_T + V \qquad ...(72)$$

Now the derivative $(\partial S/\partial P)_T$ can be replaced by a measurable derivative using Maxwell relation:

$$G = H - TS = E + PV - TS$$

$$dG = dE + PdV + VdP - TdS - SdT = - SdT + VdP$$

Using Euler's theorem of exactness, we get

$$\left(\frac{\partial S}{\partial P}\right)_T = -\left(\frac{\partial V}{\partial T}\right)_P \qquad ...(73)$$

Substituting equation (73) into equation (72) we get

$$\left(\frac{\partial H}{\partial P}\right)_T = V - T\left(\frac{\partial V}{\partial T}\right)_P \qquad ...(74a)$$

Equation (74a) can also be derived in another way as follows:

G = H – TS gives H = G + TS.

On differentiating with respect to P constant T. we have

$$\left(\frac{\partial H}{\partial P}\right)_T = \left(\frac{\partial G}{\partial P}\right)_T + T\left(\frac{\partial S}{\partial P}\right)_T$$

But dG = – SdT + VdP gives up

$$\left(\frac{\partial G}{\partial P}\right)_T = V,$$

$$\text{and} \left(\frac{\partial S}{\partial P}\right)_T = -\left(\frac{\partial V}{\partial T}\right)_P \quad \text{...(Maxwell relation)}$$

$$\therefore \left(\frac{\partial H}{\partial P}\right)_T = V - T\left(\frac{\partial V}{\partial T}\right)_P = V(1 - \alpha T) \qquad ...(74b)$$

where a = V^{-1} $(\partial V/\partial T)_P$

Equation (74a) may be written as

$$\left(\frac{\partial H}{\partial P}\right)_T = \left\{\frac{\partial(V/T)}{\partial(1/T)}\right\}_P$$

This can be deduced as follows. Multiplying and dividing by T2 the r.h.s. of equation (74a), we get,

$$\left(\frac{\partial H}{\partial P}\right)_T = \left\{\frac{V - T(\partial V/\partial T)_P}{T^2}\right\} T^2$$

The quantity within bracket on r.h.s. of this equation is the differential $-\frac{\partial(V/T)}{\partial T}$. Hence

$$\left(\frac{\partial H}{\partial P}\right)_T = -T^2\left(\frac{\partial(V/T)}{\partial T}\right)_P = -\left(\frac{\partial(V/T)}{T^{-2}\partial T}\right)_P$$

Again $-T^{-2}\,dT = dT^{-1} = d(1/T)$, therefore

$$\left(\frac{\partial H}{\partial P}\right)_T = \left(\frac{\partial(V/T)}{\partial(1/T)}\right)_P \qquad ...(74c)$$

Equation (74a) gives the variation of enthalpy with pressure at constant temperature. If the enthalpy of the system at one pressure is known then we can evaluate its value at some other pressure by using the integrated form of the equation (74a).

$$\int dH = \int V dP - \int T\left(\frac{\partial V}{\partial T}\right)_P dP$$

$$\text{or } H(T, P_2) - H(T, P_1) = \int_{P_1}^{P_2} V dP - T\int_{P_1}^{P_2}\left(\frac{\partial V}{\partial T}\right)_P dP \qquad ...(74b)$$

Here H (T, P_2) is the enthalpy at pressure P_2, and temperature T, and H1 (T, P_1) is the enthalpy at pressure P_1 and temperature T.

The quantities on r. h. s. of this equation can be measured or evaluated by using the PVT data, hence he quantities on left side can be evaluated. More frequently this integrated equation of state is used to test the consistency of the experimental measurement of calorimeter, and dilatometer. That is to say, the enthalpy (change) of a system are measured at different pressures by using a suitable calorimeter, also the change of volume with pressure is evaluated, then the two data are tested by using equation (74d).

(iii) For an ideal gas PV = nRT and PdV + VdP = nRdT

At constant volume dV = 0 and

$$\left(\frac{\partial P}{\partial T}\right)_V = \frac{nR}{V}$$

Substituting this value in equation (70) we see that,

$$\left(\frac{\partial E}{\partial V}\right)_V = \frac{nRT}{V} - P = P - P = 0$$

Again $\left(\frac{\partial V}{\partial T}\right)_P = \frac{nR}{P}$

and hence equation (74a) can be written as

$$\left(\frac{\partial H}{\partial P}\right)_T = V - \frac{nRT}{P} = V - V = 0$$

Thus, for an ideal gas the internal energy and enthalpy are functions of temperature only.

(iv) It can also be shown that for any substance if $(\partial E/\partial V)_T = 0$ and $(\partial H/\partial P)_T = 0$, then the substance must obey the ideal gas equation of state PV = kT. This proved as follows :

Given $(\partial E/\partial V)_T = 0$

But $(\partial E/\partial V)_T = T\,(\partial P/\partial T)_V - P$

$$T\left(\frac{\partial P}{\partial T}\right)_V - P = 0$$

or $\left(\frac{\partial P}{P}\right)_V = \left(\frac{\partial T}{T}\right)_V$

or d In P = d In T

Integration gives In P = In f (V) = In T f (V)

In f (V) is integration constant and is a function of V, its differential will be zero at constant volume.

or P = T f (V) ...(75)

$$\left(\frac{\partial H}{\partial P}\right)_T = V - T\left(\frac{\partial V}{\partial T}\right)_P = 0$$

$\therefore$ d In V = d In T

Integration gives In V = In T + In g (P) = Tg (P)

$\therefore$ V = Tg (P) ...(76)

In g (P) is the integration constant and its differential will become zero at constant pressure.

Comparison of expressions (75) and (76) gives us,

$$T = \frac{V}{g(P)} = \frac{P}{f(V)}$$

or Pg (P) = Vf (V) = constant = k

Since left side is function of pressure and right side depends on volume, therefore,

$$g\ (P) = \frac{k}{P}; f(V) = \frac{k}{V}$$

$$\text{Thus } T = \frac{PV}{k}$$

or PV = kT

However $(\partial E/\partial V)_T = 0$ or $(\partial H/\partial P)_T = 0$ along does not imply that the substance is an ideal gas. For example, for a non-ideal gas with equation of state P = nRT/V –nb, the derivative $(\partial E/\partial V)_T = 0$ but $(\partial H/\partial P)_T$ is not equal to zero. But if both the derivatives $(\partial E/\partial V)_T$ and $(\partial H/\partial P)T$ are equal to zero, the gas must be ideal.

5

Thermochemical Equations

INTRODUCTION

In the first method, the balanced chemical equation representing the actual reaction is written, and the heat change is indicated by mentioning ΔH or ΔE with proper sign to the right side of the equation, *e.g.*, when 1 mole methane burns in the presence of oxygen, 1 mole CO_2 gas and moles H_2O are formed, and 890.3 kj heat is produced measured at constant pressure and 298K.

For completeness, the chemical equations along with the enthalpy change or energy change accompanying the chemical reaction are written in two ways.

This change is represented by the *Thermochemical Equation* as

$$1CH_4\,(g) + 2O_2\,(g) \rightarrow CO_2\,(g) + 2H_2O$$

$$(l)\ \Delta H\ (298K) = -890.3kJ$$

The same equation may be written in thermally balanced manner. Since the reaction is exothermic, 890.3 kJ heat is given to the surroundings. the enthalpies of products are less than those of the reactants by 890.3 kJ. If this amount is added to the product side. the thermally balanced equation is:

$$CH_4\,(g) + 2O_2\,(g) \rightarrow CO_2\,(g) + 2H_2O\,(l) + 890.3\ kJ$$

At present the former method is preferred to the latter. Endothermic reactions may be represented in a similar manner with a changed sign of ΔH or ΔE.

$$N_2\,(g) + O_2\,(g) \rightarrow 2NO\,(g) \quad \Delta H\ (298K) = 180\ kJ$$

$$\text{or } N_2\,(g) + O_2\,(g) \rightarrow 2NO\,(g) - 180\ kJ$$

CONVENTIONS USED IN WRITING THERMOCHEMICAL EQUATIONS

Physical States of Reactants and Products

The physical states of reactants or products have been found to influence the magnitude of heat evolved or absorbed in the reaction.

Therefore, these are indicated by symbols (g), (*l*) and (s) placed after the chemical formulae. These stand for gaseous, liquid and solid states respectively.

However, sometimes confusion arises in representing the physical state of water formed as a chemical reaction. This can be understood by the following reaction at 25°C :

$$H_2(g) + \frac{1}{2}O_2(g) \rightarrow H_2O(l) + 286.2 \text{ kJ}$$

In the above reaction, since experiment is carried out at 25°C, water is obtained in the liquid state.

If the above reaction is carried out at a temperature a little above 100°C under atmospheric pressure, the water produced will remain in the vapour state. The heat evolved in the reaction is 245.5 kJ. The thermochemical equation will now be written as follows :

$$H_2(g) + \frac{1}{2}O_2(g) \rightarrow H_2O(l) + 245.5 \text{ kJ}.$$

Allotropic Forms of Substances

As different allotropic forms of a substance are associated with different amounts of energy, it means that in case of elements showing allotropy, the name of the allotropic form must be mentioned in thermochemical equation.

For example, in a thermochemical equation involving sulphur, we must mention the allotropic forms, rhombic, monoclinic, etc. Thus, in a thermochemical equation, rhombic and monoclinic varieties of sulphur are represented as S_R (rhombic) and SM (monoclinic) respectively.

Reaction in Solutions

When a reaction takes place in a solution, the solvent must be indicated. For example, the symbol (aq) placed after the formula of a substance, reveals that substance is present in large excess of water to ensure that there occurs no heat change on adding further quantity of water.

Heat Changes Involved in Reactions

Two different conventions are used to represent the heat changes of the chemical processes :

(a) *Old Convention :* According to this convention, the heat evolved or adsorbed in a chemical reaction is represented as a part of the equation on the product side. A positive sign (+) is used if heat is given out and negative sign (–) is used if heat is absorbed in a chemical reaction. Thus,

$$A + B \rightarrow C + D + x \text{ kJ}$$

$$A + B \rightarrow C + D - x \text{ kJ}$$

This convention is out of date these days.

(b) *Modern Convention :* This convention is based on thermodynamics. The heat evolved or absorbed in a process may be represented by ΔH, the difference in the total heat contents of the products and reactants. It is not written as a part of equation but is written separately. If heat is absorbed in a chemical reaction, + ΔH is employed and if heat is evolved. Then $-\Delta H$ is employed. For example, the burning of carbon in oxygen to form carbon dioxide was previous represented as

$$C_{\text{graphite}} + O_2(g) \rightarrow CO_2(g) + 393.5 \text{ kJ}$$

At present it is represented as (Old convention)

$$C_{\text{graphite}} + O_2(g) \rightarrow CO_2(g);$$
$$\Delta H = -393.5 \text{ kJ}$$

Temperature and Pressure of Substances

The heat absorbed or evolved in a chemical reaction has been found to depend upon the temperature and pressure at which the reaction is being carried out. These two quantities are, therefore, also in the thermochemical equation. If a reaction is carried out under standard conditions *i.e.*, 25°C or 298 K and 1 atm pressure, the heat change is indicated as ΔH°. The superscript' indicates that pressure is 1 atmosphere and reaction in the standard state.

LAWS OF THERMOCHEMISTRY

For some reactions it is not convenient to measure the heat change in the laboratory. So conventional procedure based on the principle of conservation of energy has been suggested which can be stated as follows:

1. The heat of formation of any compound is equal in magnitude and of opposite sign to the heat of dissociation of that compound at the given temperature and pressure. For example, enthalpy of formation of liquid water from its elements hydrogen and oxygen is – 285.830 kJ mol^{-1} and the enthalpy of dissociation is 285.850 kJ mol^{-1}. Thus the processes can be represented by

   H_2 (g) + 1/2 O_2 (g) → H_2O (l) ΔH (298 K) = – 285.830 kJ

   H_2O (l) → H_2 (g) + 1/2 O_2 (g) ΔH (298 K) = + 285.830 kJ

2. The total enthalpy change of a reaction is the same, regardless of whether the reaction is completed in one step or in several steps. (*Hess's Law of constant heat summation.*) It has been experimentally verified and is also a consequence of the law of conservation of energy. It is of particular utility in calculation of f the heats of reactions which are difficult for practical calorimetric measurements.

Consider a system in which the initial state is represented by the symbol A and the final state by the symbol D and suppose that there are two ways by which the system can change from state A to state D A → D (1st way) direct change.

A → B → C → D (2nd way) stepwise change.

Hess's law required that the enthalpy change accompanying the change from A directly to D be identical to the sum of the enthalpy changes accompanying the changes from A to B to C to D (Fig 5.1)

Thus, $\Delta H_1 = \Delta H_2 + \Delta H_3 + \Delta H_4$

or $H_D - H_A = \Delta H_1 = \Delta H_2 + \Delta H_3 + \Delta H_4$

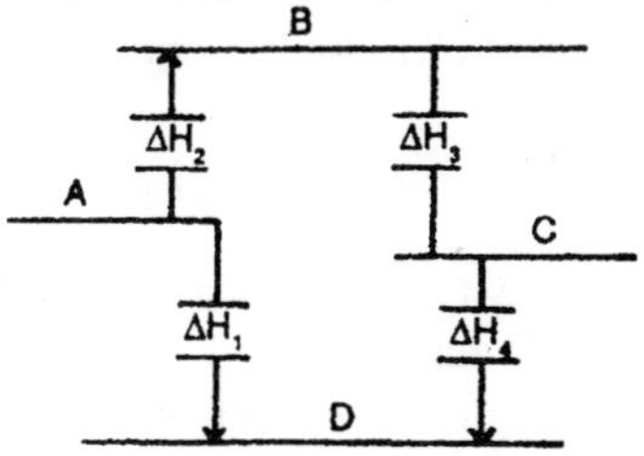

Fig. 5.1 : Hess's law in diagrammatic form.

From the laws of thermochemistry we get the following working rules :

(1) *Chemical equations can be reversed and the sign of DH is changed.*

(2) Chemical equations can be added, subtracted, multiplied, divided by an integer whenever necessary, according to the usual laws of algebra; terms identical in quality and quantity on both sides of the equation may be cancelled.

THE MAXIMUM FLAME TEMPERATURE

When a fuel burns, definite quantities of products are formed and a definite amount of heat is liberated. This quantity of heat may raise the temperature of the products of combustion sufficiently so as to produce a flame. This temperature is termed as the maximum flame temperature, or calorific intensity.

When the experiment is carried out at constant volume the quantity of heat liberated will be ΔE and the maximum temperature is termed explosion temperature. In case the experiment is carried out at constant pressure then heat liberated is ΔH and the maximum temperature to which the products will be raised is termed as the flame temperature. Generally this temperature is termed as the adiabatic maximum flame temperature as it is assumed that there occurs no loss of heat in the process.

Calculation of Maximum Temperature

(1) If we assume that the reaction is very fast and the total heat evolved is used to raise the temperature of the system, then the heat capacities of the species present at the completion of reaction and the heat evolved can be used for calculating the maximum temperature by using Kirchhoff's equation in the following form,

$$\Delta H_{(reaction)} = \int_{T_2}^{T_1} \sum nC_P \; dT \, \Delta H_{(heating)}$$

where it is assumed that $\Delta H_{(heating)} = -\Delta H_{(reaction)}$, as the heat evolved is taken by the products. T_1 represents the initial temperature and T_2 the maximum flame temperature, $\Sigma(nC_P)$ the sum of the heat capacities of all the species present in the system after the reaction gets completed. For constant volume process,

$$\Delta E = \int_{T_2}^{T_1} \sum nC_V \, dT$$

The direct way to know the maximum flame temperature is to carry out the experiment in the laboratory under adiabatic conditions and measure the temperature using standard techniques of temperature measurements.

The calcualted flame temperature has been found to be higher than the observed temperature. This difference has been attributed to a number of complications which are as follows :

(i) The system may not be truly adiabatic.

(ii) In combustion, we generally use excess of air or oxygen, Unreacted air or oxygen is to be heated and therefore a part of the heat would be utilised in heating these gases.

(iii) When flame temperature is to be calculated, heating of the unreacted gases should also be considered.

(iv) At high temperatures, the products might decompose and a part of the heat may be utilised to decompose the product molecules.

(2) If an exothermic reaction is assumed to carry out under adiabatic conditions. the heat evolved would be able to raise the temperature of the system.

It is possible to determine the rise in temperature by considering the reaction to take place in the following two steps. both at constant pressure.

Step I : Reactants (T_o, P) → Products (T_o, P) ΔHT_o

Step II : Products (T_o, P) → Products (T_f, P)

The heat involved in the second step may be given as follows :

$$\Delta H_2 = \int_{T_o}^{T_f} C_P \,(\text{Products})\, dT$$

As the overall reaction refers to the sum of the above two steps, *i.e.*,

Reactants (T_o, P) → Products (T_f, P)

the net heat change is given as follows :

$$\Delta H = \Delta H_{T_o} + \Delta H_2 \quad \text{...(i)}$$

As the reaction is carried out under adiabatic conditions, it is evident that $\Delta H = 0$. Hence Eq. (i) may be put as follows :

$$\Delta H_{To} + \Delta H_2 = 0.$$

or $$\Delta H_2 = -\Delta H_{To} \text{ or}$$

$$\int C_P \text{ (Products) } dT = -\Delta H_{To}$$

If C_Ps are regarded to be independent of temperature, then

$$C_P \text{ (Products) } (T_f - T_o) = -\Delta H_{To} \text{ or } T_f = \frac{-\Delta H_{TO}}{C_p \text{ (products)}} + T_o$$

However, the approximation that C_Ps are independent of temperature has not been correct because T_f is a very large quantity ($\Delta H_{To} \sim$ kJ and CP ~ J) of the order of thousands of kelvin.

If the reaction is carried out in a closed vessel *i.e.*, under the condition of constant volume, then ΔE and C_V replace ΔH and C_P respectively, *i.e.*,

$$T_f = \frac{(-\Delta E_{To})}{C_v \text{ (products)}} + T_o$$

EXPLOSION TEMPERATURE

When the combustion of hydrocarbons is done at constant volume under adiabatic conditions, the maximum temperature attained is termed as the explosion temperature. The explosion temperature is calculated in the same way as that of flame temperature except ΔH is replaced by ΔE. The pressure which the reacting system may acquire at the explosion temperature is termed as the maximum explosion pressure.

BOND ENTHALPIES OR ENERGIES

The bond energy of a particular type of bond in a molecule may be defined as the amount of energy required to dissociate or break one mole of that type present in the compound and separate the resulting atoms or radicals from one another. For example, the bond energy of the H – H bond in H_2 is 433 kJ mol^{-1}. It means that 433 kJ of energy is required to break or dissociate the H – H bond in one mole of hydrogen molecules. Similarly, the bond energies of I – I in I_2 and H – I in HI are 151 kJ mol^{-1} and 229 kJ mol^{-1} respectively.

In case of diatomic molecules the bond energy is a definite quantity as there is only one type of bonds present in these molecules. However, the bond energy becomes a much less definite quantity in case of polyatomic molecules. This is to be expected since in a polyatomic molecule, the energy necessary to rupture a given bond will depend to some extend on the nature of remainder molecule. For example, the energy required to rupture the O – H bond in alcohol would not be expected to the same as the energy needed to rupture same bond in an ionic compound. This is obvious from the fact that alcohols are much less acidic than organic aids. Therefore, it follows that bond energy for a particular bond in polyatomic molecules will vary from compound to compound. However, if there are similar types of bonds present in a ployatomic molecule, the bond energy is equal to the average energy required to break a particular bond in one mole of that compound. For example, C – H bond energy in CH_4 can be calculated as follows :

When methane is formed it elements carbon and hydrogen, 1648.7 kJ of heat is liberated. It means 1648.7 kJ of energy must be supplied to one mole of methane to break four identical C – H bonds, resulting the formation of atomic carbon and atomic hydrogen in the gaseous state. It means that the average C – H bond energy is 1648.7/4 or 412.175 kJ mol^{-1}. This is also the value of heat of formation of C – H bond,

From the above discussion, it is evident that the bond energy or enthalpy should be distinguished carefully from the term bond dissociation energy of enthalpy.

Bond Enthalpy or Energy

Bond enthalpy of a given bond is defined as the average of enthalpies or energy required to dissociate the said bond present in different gaseous compounds into free atoms or radicals in the gaseous state.

Bond Dissociation Enthalpy or Energy

Bond dissociation enthalpy or energy may be defined as the enthalpy or energy required to dissociate a given bond of some specific compound.

The distinction between bond enthalpy and bond dissociation enthalpy may be better understood by a simple example, say of the O – H bond. The enthalpy of dissociation of the O – H bond depends upon the nature of molecular species from which the H atom is getting separated. For example, in the water molecule

$$H_2O(g) \rightarrow H(g) + OH(g) \;;\; \Delta H^\circ = 501.87 \text{ kJ noet}$$

$$OH(g) \rightarrow O(g) + H(g) \;;\; \Delta H^\circ = 423.38 \text{ kJ noet}$$

The bond enthalpy of O – H is defined as the average of these values, *i.e.*,

$$\text{Bond enthalpy} = \frac{501.87 + 423.38}{2} = 462.625 \text{ kJ noet}$$

Hence the bond enthalpy given for any particular pair of atom is the average value of the dissociation enthalpies of the bond for a number of molecules in which the pair of atoms appear.

The bond dissociation energies for the stepwise dissociation of CH_4 are as follows :

$$CH_4(g) \rightarrow CH_3(g) + H(g);\ \Delta H^\circ = 427.0 \text{ kJ}$$

$$CH_3(g) \rightarrow CH_2(g) + H(g);\ \Delta H^\circ = 418.4 \text{ kJ}$$

$$CH_2(g) \rightarrow CH(g) + H(g);\ \Delta H^\circ = 460.2 \text{ kJ}$$

$$CH(g) \rightarrow C(g) + H(g);\ \Delta H^\circ = 343.1 \text{ kJ}$$

The energy for each step is different due to the fact that in each case a different dissociating fragment is involved. The C – H bond enthalpy would be one-fourth of the enthalpy required to break a mole of CH_4 and is 412 kJ.

Calculation of Bond Enthalpy or Energy

It is possible to obtain bond energies from data on heats of combustion and heats of dissociation. Let us illustrate it by calculating the bond enthalpy or energy of C – H. We know,

$$CH_4(g) \rightarrow C(g) + 4H(g)\ \ \epsilon_{C-H} = \frac{\Delta H}{4}$$

The ΔH and hence ϵ_{C-H} for this reaction can be obtained by adding the following equations :

$$CH_4(g) + 2O_2(g) \rightarrow CO_2(g) + 2H_2O(I) \qquad \Delta H^\circ = -890.36\text{kJ}$$

$$CO_2(g) \rightarrow C\ (\text{graphite}) + O_2(g) \qquad \Delta H^\circ = 393.51 \text{ kJ}$$

$$2HO_2(I) \rightarrow 2H_2(g) + O_2(g) \qquad \Delta H^\circ = 571.70 \text{ kJ}$$

$$2H_2(g) \rightarrow 4H(g) \qquad \Delta H^\circ = 871.86 \text{ kJ}$$

$$C\ (\text{graphite}) \rightarrow C(g) \qquad \Delta H^\circ = 716.68 \text{ kJ}$$

$$CH_4(g) \rightarrow C(g) + 4H(g) \qquad \Delta H^\circ = 1663.39 \text{ kJ}$$

$$\text{Thus, at 298.15o k } \epsilon_{C-H} = \frac{1663.39}{4} = 415.85 \text{ kJ}$$

This value of the C – H bond energy does not correspond to the dissociation energy of the carbon-hydrogen bond in methane, which is 426.77 kJ mol^{-1} and refers to the equation

$$CH_4(g) \rightarrow CH_3(g) + H(g)$$

The data on bond energies are used to calculate the approximate heat of formation of a compound of known structure by adding the appropriate bond energies. Wherever direct experimental data are not available, approximate values of enthalpies of reaction can also be obtained by using bond enthalpy data.

Estimation of the Enthalpy of Formation from Bond Enthalpies

Let us calculate the heat of formation of $Se_2CI_2(g)$ from bond enthalpies by involving the following steps ; as the bond energy refers to the dissociation of Se – Se CI gas into the gaseous atoms, the enthalpy change for the formation of this gaseous molecule should be as follows :

$$2Se(g) + 2CI(g) \rightarrow Se_2CI_2(g)$$

$$\Delta H = -[\epsilon Se - Se + 2\ \epsilon Se - CI] = -694.54 \text{ kJ mol}^{-1}.$$

However to estimate the heat of formation it becomes necessary to add two reactions in the above equation, as by definition the heat of formation refers to the elements in their standard states. Therefore, the following enthalpy changes are introduced to convert the elements from their standard states to the gaseous atoms at 298°K

$$CI_2(g) \rightarrow 2CI(g)\ \Delta H = 243.30 \text{ kJ mol}^{-1}$$

$$2\ Se\ (hexagonal) \rightarrow 2\ Se(g) \quad \Delta H = 2 \times 2 \times 202.5 \text{ kJ mol}^{-1}$$

By adding these in the preceding expression, we obtain

$$2\ Se\ (hexagonal) + CI_2(g) \rightarrow Se_2CI_2(g);\ \Delta H_f = -46.22 \text{ kJ mol}^{-1}$$

Estimation of the Heat of Reaction from Bond Enthalpies or Energies

Let us illustrate this by calculating the enthalpy change for the following equation from the bond energy data :

$$C_2H_4(g) + HCI(g) \rightarrow C_2H_5CI(g)$$

The enthalpy change for the above reaction can be calculated as follows :

ΔH = Energy required to break reactants into + Energy released to form products from the gaseous atoms gaseous atoms

$$= [4\ \epsilon_{C-H} + \epsilon_{C=C} + \epsilon_{H-CI}] + [-5\ \epsilon_{C-H} - \epsilon_{C-C} - \epsilon_{C-CI}]$$

$$= (\epsilon_{C=C} + H - CI) - (\epsilon_{C-H} + \epsilon_{C-C} + \epsilon_{C-CI})$$

On substituting the bond enthalpy values in the above equation, we get

$$\Delta H = (615.05kJ + 431.79kJ) - (413.38kJ + 347.69kJ + 328.44kJ)$$
$$= -42.67kJ$$

Bond Energies and Resonance

There occurs a fair agreement between the calculated values of heat of formation obtained from bond enthalpies and any other method. However, there occurs large deviations in case of compounds having alternate single and double bonds. For example, the following reaction.

$$\underset{\text{Benzene}}{C_6H_6(g)} \rightarrow 6C(g) + 6H(g)$$

will require enthalpy of 5368.5 kJ on the basis of bond energies ($3\,\epsilon_{C=C} + 3\epsilon_{C=C} + 6\epsilon_{C-H} = 3 \times 347.69$ kJ $+ 3 \times 615.05$ kJ $+ 6 \times 413.38$ kJ = 5368.5 kJ), while the experimental value is 5351.1 kJ.

This amounts to the fact that benzene is more stable by 166.6 kJ mol^{-1}. This is attributed to resonance, that is, in benzene there occurs no localization of single and double bonds, but the molecule is resonating from one extreme to another.

Factors: The value of the bond energy depends upon the following factors :

(a) *Length of the bond :* In general, it is found that greater the bond length between the two nuclei of the two atoms forming the bonds, the smaller is the bond energy.

Type of bond	H – H	CI – CI	I – I
Bond length (i)	0.74	1.25	2.7
Bond energy (kJ/mol)	433.1	330.5	151.2

(b) *Bond polarity :* Higher is the difference in electronegativity between the combining atoms, greater is the strength of the dipole along the bond and higher is the bond energy.

(c) *Unsaturation :* In general it is observed that with the increase in the number of covalent bonds between the atoms, there occurs an increase in the bond energy but the contribution of each additional bond goes on decreasing.

Bond	*Energy in kJ/mole*
C – C	347.3
C = C	615.0
C = C	811.7

(d) *Resonance effect* : If the bonds in the molecule do not remain fixed but the molecule possesses resonance, the bond energy per bond depends upon the amount of resonance.

(e) *Overlap of Orbitals* : The bond energy depends upon the s character which is determined by extent of overlap of orbitals of the combining atoms.

VARIOUS TYPES OF HEAT CHANGES

Enthalpy of Formation

The enthalpy change when one mole of a pure compound is formed from its elements is called the enthalpy of formation denoted as ΔH_f. When the reacting elements and the product formed are all in their standard state, the enthalpy change accompanying the chemical reaction is called the standard enthalpy of formation denoted as $\Delta H^o{}_f$. For example, the enthalpy of formation of CO_2 at 298 K and 1 atm is equal to the enthalpy change for the reaction C (graphite) + O_2 (g) $\rightarrow$ CO_2(g). This is so because the enthalpy of elements in their standard state is zero by convention.

C (graphite, 1 atm, 298 K) + O_2(g, 1 atm, 298K) $\rightarrow$ CO_2 (g, 1atm, 298K)

$\Delta H^o = -\ 393.5$ kJ mol^{-1}

$\Delta H^o = H^o(CO_2) - H^o(C) - H^o(O_2) = \Delta H^o{}_f(CO_2)$

Since the enthalpy of each element in standard state is zero; therefore, the standard enthalpy of CO_2 is equal to its standard heat of formation i.e.

H° (CO2) = DH°f (CO_2)

$\Delta H^o{}_f$ data for other compounds are given in the *Appendix*.

Heat of Combustion

The enthalpy or energy change when one mole of a substance is burnt in oxygen is defined as the heat of combustion (DH or DE), the product being water and carbon dioxide (usually).

C_2H_6 (g) + 7/2 O_2 (g) → $2CO_2$(g) + $3H_2O$(l) ΔH° (298 K) = – 1560.1 kJ

HEAT OF HYDROGENATION

It is defined as the change in enthalpy accompanying the conversion of one mole of unsaturated organic compound into corresponding saturated compound by reducing the former with hydrogen e.g. when benzene is hydrogenated to cyclohexane, the change in enthalpy is represented by the following equation:

C_6H_6(l) + $3H_2$ (g) → C_6H_{12} (l) ΔH° (298) = –205kJ

The enthalpy change when 1 mole of ethylene is 136.98 kJ at 298K

C_2H_4(g) + H_2(g) → C_2H_6(g) ΔH°(298K) = – 136.98 kJ

Heat of Transition

It is defined as the amount of heat (DE or DH) absorbed or evolved when 1 mole of a substance is transformed from one allotropic form (state) to another allotropic form, e.g., when graphite form of carbon is converted to diamond form, 1.89 kJ of heat is absorbed.

C (graphite) → C (diamond) ΔH° (298K) = 1.89 kJ

Heat of Atomization

It is the amount of heat required to produce gaseous atoms from the stable element. In case of solids heat of atomization is equal to heat of sublimation, *e.g.* the heat of atomization of carbon is 716.68 kJ.

C (crystal, graphite) → C (gas) Δ H° (298K) = 716.68 KJ

and for magnesium.

Mg(c) → Mg (g) ΔH° (298K) = 150.2kJ

In case of homonuclear diatomic gases the heat of atomization is equal to half of heat of dissociation e.g. heat of dissociation of one mole of hydrogen is 435.94 kJ which gives 2 moles of atomic hydrogen. Thus the heat of atomization per mole hydrogen atom is 217.97 kJ

H_2 (g) → 2H (g) ΔH° (298K) = 435.94 kJ

1/2 H_2 (g) → H (g) ΔH° (298K) = 217.97 kJ

and similarly for oxygen

$1/2\ O_2$ (g) $\rightarrow$ O(g) ΔH° (298K) = 249.2 kJ

Heat of reaction from the heats of formation of elements and compounds. The enthalpy change is heat of a reaction which can be calculated provided we know the enthalpies of reactants and products. It is a general convention to take the enthalpies of pure elements in their standard states as zero, and then enthalpy of pure compound is taken as its enthalpy of formation (ΔH°_f) from its elements in their standard state. Thus, the enthalpy change of a reaction ΔH° is given by

$$\Delta H^\circ = \Sigma H^\circ \text{ (products)} - \Sigma H^\circ \text{ (reactants)}$$

$$\text{or } \Delta H^\circ = \Sigma \Delta H^\circ_f \text{ (products)} - \Sigma \Delta H^\circ_f \text{ (reactants)}$$

ΔH°_f values of various compounds are tabulated for ready reference. This tabulation is of great help to compute the enthalpy of reactions. Instead of tabulating the enthalpy of millions of reactions it is convenient of report ΔH°_f values for a few thousands of more important compounds and then use the values to evaluate the heat of any desired reaction with the help of paper and pencil even without performing the actual experiment.

Heat of Vaporization

The enthalpy change accompanying the conversion of one mole of liquid into vapour phase at a given temperature and pressure is called the heat of vaporization per mole of the liquid. For example the heat of vaporization per mole of water at 298 K and 1 atm is 44 kJ.

H_2O(l) $\rightarrow$ H_2O(g) DH° (298 K) = 44.0 kJ

Heat of Fusion

The enthalpy change when one mole of solid is converted to a liquid phase is called the heat of fusion. For example, the heat of fusion of water is 6 kJ mol^{-1}.

H_2O (S) $\rightarrow$ H_2O(1) ΔH (273 K) = 6.0 kJ

Heat of Sublimation

The heat change accompanied with the transformation of one mole of a volatile solid directly into vapour phase is termed as the heat of sublimation. Heat of sublimation is taken as the sum of the heat of fusion and the heat of vaporization, all measured at the same temperature.

$$\Delta H_{sub} = \Delta H_{fus} + \Delta H_{vap}$$

Heat of Solution

When solids or gases are dissolved in liquids or when different liquids are mixed, heat is evolved or absorbed. This thermal change is called the *heat of solution*. The heat of solution is usually reported per mole of the solute but for solutions of liquids in liquids the data are often tabulated as heat change per mole of the solution. The heat of solution depends upon the relative amounts of the substances being mixed and on whether they were initially taken as pure or one of them was introduced into a solution with a certain initial concentration at a given temperature and pressure.

Integral Heat of Solution

The integral heat of solution is the heat change when a known amount of pure solute is added to a known amount of pure solvent under conditions of constant temperature and pressure. The solution process may be represented by a chemical equation such as

$$HCl\ (g) + n\ aq \rightarrow HCl.n\ aq$$

where aq represents the solvent, in this case H_2O, and n represents moles of solvent. The amount of heat evolved in the above reaction depends upon the magnitude of n shown in Fig. 5.2.

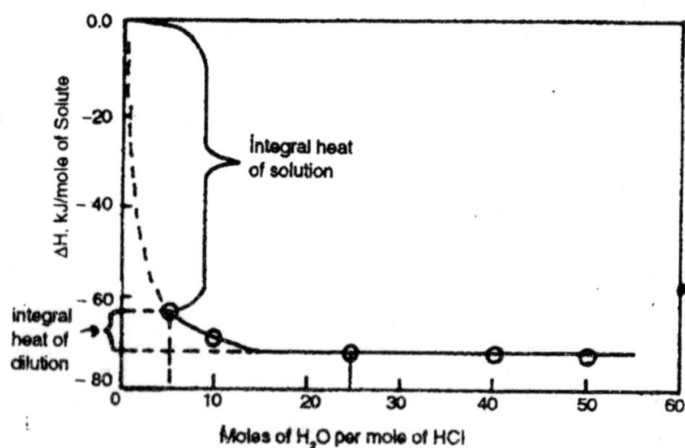

Fig. 5.2 : Variation of heat of solution with the mole of water per mole of HCl at 298 K.

(i) $HCl\ (g) + 5H_2O(l) \rightarrow HCl.5H_2O \quad \Delta H(298\ K) = -\ 63.995\ kJ$

(ii) $HCl\ (g) + 25H_2O(l) \rightarrow HCl.\ 25H_2O \quad \Delta H(298\ K) = -\ 72.177\ kJ$

(iii) HCl (g) + aq → HCl (aq) $\Delta H(298\ K) = -\ 75.076$ kJ

where "$HCl.5H_2O$" represents a solution of 2 mole HCl in 5 moles of water. In each case the heat evolved per given mole of solvent added is an *integral heat of solution* for those conditions. It is the total heat evolved in carrying out that particular process. The term HCl (aq) represents an aqueous solution which is so dilute that further dilution produces no thermal effect. The corresponding Δ H value is called the heat of solution at **infinite dilution** and represents the upper limit of heat released when a substance is dissolved in a large excess of solvent. The integral heats of solution of HCl are plotted versus the number of moles of water per mole of HCl in Fig. 2.

Integral Heat of Dilution

- It is the net change of enthalpy when a solution containing 1 mole of solute at a known concentration (C_1) is diluted by adding more solvent to get a solution of another known concentration (C_2). This quantity is in fact equal to the difference between the integral heats of solution at the two concentrations. For example, when 20 moles water are added to a solution $HCl.5H_2O$ then a solution $HCl.25H_2O$ is made, the accompanying enthalpy change of – 8.182 kJ is the *integral heat of dilution*.

(i) HCl (g) + $5H_2O(l)$ → $HCl.5H_2O$ $\Delta H\ (i) = -\ 63.995\ kJ\ mol^{-1}$

(ii) HCl (g) + 25 $H_2O(l)$→$HCl.25H_2O$ $\Delta H\ (ii) = -72.177\ kJ\ mol^{-1}$

(iii) $HCl.5H_2O$ + 20 H_2O → $HCl.25H_2O$

$\Delta H\ (dilution) = \Delta H\ (ii) - \Delta\ H\ (i) = -\ 8.182\ kJ\ mol^{-1}$

Differential Heat of Solution

It is the change in enthalpy when one mole of a solute is added to such a large amount of solution that the concentration of solution is essentially unchanged. This enthalpy change is also called the partial molar heat of solution.

Similarly, if one mole of a solvent is added to a larger volume of solution without altering the concentration, the accompanying enthalpy change is termed as the *differential* (partial molar) heat of dilution. If Δ H is the enthalpy change when n_2 moles of a solute are mixed with n1 moles of solvent, then

$$\left(\frac{\partial \Delta H}{\partial n_2}\right)_{T.P.n_1} = \Delta H_{2,m} = \text{Differential heat of solution}$$

and $\left(\frac{\partial \Delta H}{\partial n_1}\right)_{T.P.n_2} = \Delta H_{2,m}$ = Differential heat of solution

The total enthalpy change is given by

$\Delta H_1 = n_1 \Delta H_{1,m} + n_2 \Delta H_{2,m}$

If the amount of solvent is kept constant (say 1 kg = 55.55 moles H_2O) and the amount of solute n_2 is varied, then the enthalpy change of solution as a function of molality ($m_2 = n_2/w_1 = n_2$, when w_1 = 1 kg) is given by

$\Delta H = 55.55\ \Delta H_{1,m} + m_2\ \Delta H_{2,m}$

The differential heat of solution depends on the concentration of the solution. It is impractical to measure the differential heat of solution directly, but it can be calculated from the integral heat of solution data. When Δ H value is plotted against the moles of solute m_2 (moles per kg of solvent), the slope of the tangent at a given concentration gives the differential heat of solution $\Delta H_{2,m}$.

Integral heat of solution and heat of dilution can also be calculated from tabulated values of heats of formation in solution. The enthalpy of formation of water is neglected in calculations if there is the same number of moles of water on both sides of the balanced chemical equations. Also, the enthalpy of formation of pure water is used for water aqueous solutions.

Heat of Neutralization

It is change in enthalpy when one mole H^+ ions are completely neutralized by a base in very dilute solutions. The dilute solution is necessary so that on mixing the acid and base the thermal change due to dilution is negligible. The heat of neutralization of hydrochloric acid and sodium hydroxide is –57.0 kJ mol–1 at 25°C and 1 atm pressure. In the case of weak acid and strong base the heat of neutralization is less than 57.0 kJ. The difference in heat of neutralization values of strong acid – strong base, and weak acid – strong base is attributed to the heat of ionization of weak acid.

Heat of neutralization of strong and strong base. Consider the neutralization of aqueous solution of HCl by NaOH. The reaction may be represented by

$H^+ (aq) + Cl^- (aq) + Na^+ (aq) + OH^- (aq)$

$\rightarrow Na^+ (aq) + Cl^- (aq) + H_2O$ (l)

Since Na^+ (aq) and Cl^- (aq) are common on both sides, the neutralization of strong acid and strong base may be represented by

H+ (aq) + OH^- (aq) $\rightarrow H_2O(1)$

Thus heat of nutralization of strong acid and strong base (mono-) is simply the heat of formation of H_2O from H^+ and OH^- ions which is fixed at a given temperature and pressure. This leads to the conclusion that the heat of neutralization of strong monobasic acid with strong base is the same.

Heat of neutralization of a given acid-base pair can experimentally be found by using a calorimeter. The heat of ionization of a weak acid of a base can also be measured by calorimetric method. For example, the heat of ionization of a weak acid (acetic acid) will be equal to the heat of reaction between sodium acetate and hydrochloric acid at the a given temperature. This quantity will also be equal to the difference between the heat of neutralization of NaOH – HCl pair and acetic acid – NaOH pair. This is shown as follows:

(i) CH_3COOH *** CH_3 COO^- + H^+ ΔH (ionization) = ?

(ii) NaOH + HCl $\rightarrow Na^+ Cl^- + H_2O$ ΔH_1 = xkJ

(iii) NaOH + $CH_3COOH \rightarrow CH_3COONa + H_2O$ ΔH_2 = ykJ

Reverse equation (ii) to get (iv)

(iv) $Na^+ Cl^- + H_2O \rightarrow$ NaOH + HCl ΔH_3 = – xkJ

Add (iii) and (iv)

$CH_3COOH + Na^+ + Cl^- \rightarrow CH_3COONa$ + HCl

$\rightarrow CH_3 COO^- + Na^+ + H^+ + Cl^-$

or CH_3 COOH ***$CH_3COO^- + H^+$ ΔH(ion) = $\Delta H_2 + \Delta H_3$

= (y – x) kJ

Heat of Formation of Ions in Solution

The complete thermal data on aqueous solutions of electrolytes can be presented by tabulating heats of formation of ions at unit activity, which can be used to calculate the thermal changes for the reaction in which the ions participate. Since in aqueous solutions the H^+ and OH^- ions are involved in one way or the other, first we shall consider the computation of the standard heats of formation of these ions.

The heat of formation of one mole of water from hydrogen and hydroxyl ions is essentially equal to the heat of neutralization of strong acid and strong base (-57.0 kJ mol^{-1}) and the standard heat of formation of water from its elements H2 and O_2 is -285.8 kJ mol^{-1}. That is

$$H^+ (aq) + OH^- (aq) \rightarrow H_2O(l) \ \Delta H^\circ (298K) = -57.0 \text{ kJ}$$

$$\text{or (i) } H_2O (l) \rightarrow H^+ (aq) + OH^- (aq) \ \Delta H^\circ (298K) = -57.0 \text{ kJ}$$

$$\text{and (ii) } H_2(g) + \frac{1}{2}O_2(g) \rightarrow H_2O(l) \ \Delta H^\circ (298K) = -285.8 \text{ kJ}$$

Adding these equations we get

$$\text{(iii) } H_2 (g) + 1/2\, O_2(g) \rightarrow H^+ (aq) + OH^- (aq) \ \Delta H^\circ (298K) = -228.8 \text{kJ}$$

The sum of the heat of formation of H^+ and OH^- ions is -228.8 kJ mol^{-1}. By convention the standard heat of formation of hydrogen ions in aqueous solution is taken as zero.

$$1/2\ H_2\ (g) \neg\ H^+ (aq) + e^- \ \Delta H^\circ (298\ K) = 0$$

With this convention equation (iii) directly gives the heat of formation of hydroxyl ions,

$$\frac{1}{2}H_2(g) + \frac{1}{2}O_2(g)\ e^- \rightarrow OH^-(aq) \ \Delta H^\circ (298K\ ) = -\ 228.8 \text{ kJ}$$

The standard heat of formation of other ions is solution can be calculated in a similar manner.

KIRCHHOFF'S EQUATION

In 1858 Kirchhoff deduced certain mathematical expressions for defining the variation of heat of reaction with temperature. All these deductions are derived from the first law of thermodynamics . Consider a process,

$$A \rightarrow B$$

Let E_A and H_A be the internal energy and heat content of the system in the initial state A. Similarly, let E_B and H_B be the internal energy and the heat content of the system in the final state B. By the first law of thermodynamics, the change in enthalpy for the given process is given by the following equation

$$\Delta H = H_B - H_A \qquad ...(1)$$

But heat content is a function of temperature and pressure. Therefore, the equation (1) can be differentiated with respect to temperature at constant pressure.

$$\left[\frac{\partial(\Delta H)}{\partial T}\right]_P = \left[\frac{\partial(\Delta_B)}{\partial T}\right]_P - \left[\frac{\partial(H_A)}{\partial T}\right]_P = (C_p)B-(C_P)A$$

$$\left[\because \left(\frac{\partial H}{\partial T}\right)_P C_P\right]$$

or $$\left[\frac{\partial(\Delta H)}{\partial T}\right]_P = \Delta C_P \quad ...(2)$$

where ΔC_p represents the difference in the heat capacities of initial and final state at constant pressure. In order to integrate this equation (2), we have to consider that

(i) ΔC_p is independent of temperature

or (ii) ΔC_p is dependent upon temperature.

Let us take into consideration these cases one by one.

Case I.

When ΔC_p is constant between two temperatures T1 and T_2 : If we consider C_p to be constant and integrate equation (2) between the limits H_1 at T_1 and H_2 and T_2, we get the expression

$$\int_{H_1}^{H_2} d(\Delta H) = \int_{T_1}^{I_2} \Delta C_p \, dT, \text{ or } [\Delta H]_{H_1}^{II_2} = [\Delta C_P]_{T_1}^{T_2}$$

or $$(\Delta H_2 - \Delta H1) = \Delta C_P (T_2 - T_1) \quad ...(3)$$

This relationship is termed as Kirchhoff's equation.

Case II.

When ΔC_P varies with temperature.

It implies that equation (2) can be integrated at constant pressure

$$\int d(\Delta H) = \int \Delta C_P. \, dT$$

or $$\Delta H = \int \Delta C_P. \, dT + K \quad ...(4)$$

where K is a constant of integration. When T = 0. then ΔH_o = K. Hence equation (4) becomes as follows :

$$\Delta H = \Delta H_o + \int \Delta C_P. \, dT \quad ...(4A)$$

where ΔH_o represents the difference in heat contents of products and reactants at the absolute zero.

But $(C_P)_A = \propto_A + \beta_A T + \gamma_A T^2$ ......

$$(C_P)_B = \propto_B + \beta_B T + \gamma_B T^2 \text{ ......}$$

$$\therefore \quad C_P = (C_P)_B - (C_P)_A = (\propto_B + \beta_B T + \gamma_B T^2 \ldots) - (\propto_A + \beta_A T + \gamma_A T^2 + \ldots$$

$$= (\propto_B - \propto_A) + T(\beta_B - \beta_A) + T^2(\gamma B - \gamma A) + \ldots$$

i.e., $$\Delta C_P = \Delta\propto + T\Delta\beta + T^2\,\Delta\gamma + \quad \ldots(5)$$

If we insert this value of ΔC_P in equation (4A) and carry out the integration, we get

$$\Delta H = \Delta H_o + T\,\Delta\propto + \frac{T^2}{2}\Delta\beta + \frac{T^3}{3}\Delta\gamma + \ldots \quad \ldots(6)$$

The values of $\Delta\propto$, $\Delta\beta$, $\Delta\gamma$, ... can be calculated from heat capacity date. Equation (6) is another form of Kirchhoff's equation. This equation can be used to evaluate ΔH_o if the value of ΔH and constant $\Delta\propto$, $\Delta\beta$, $\Delta\gamma$, ... are known.

Consider the following reaction,

$$n_1A + n_2B \rightleftharpoons n_3C + n_4D$$

where n_1 and n_2 be the number of molecules of reactants A and B and n_3 and n_4 be the number of molecules of products C and D.

Now
$$(C_P)_A = n_1(\propto_1 + \beta_1 T + \gamma_1 T^2 + \delta_1 T^3 + \ldots\ldots)$$
$$(C_P)_B = n_2(\propto_2 + \beta_2 T + \gamma_2 T^2 + \delta_2 T^3 + \ldots\ldots)$$
$$(C_P)_C = n_3(\propto_3 + \beta_3 T + \gamma_3 T^2 + \delta_3 T^3 + \ldots\ldots)$$
$$(C_P)_D = n_4(\propto_4 + \beta_4 T + \gamma_4 T^2 + \delta_4 T^3 + \ldots\ldots)$$

The change in heat capacity (ΔC_P) is given as follows :

$$\Delta C_P = [\{(\propto_3 + \propto_4) - (\propto_1 + \propto_2)\}] + \{(\beta_3 + \beta_4) - (\beta_1 + \beta_2)\}T + [\{\gamma_3 + \gamma_4) - (\gamma_1 + \gamma_2)\}\,T^2 + \ldots]\,(n_3 + n_4 - n_1 - n_2)$$

or $$\Delta C_P = (\propto + \beta T + \gamma T^2 + \ldots)\;(n_3 + n_4 - n_1 - n_3)$$

On comparing it with Kirchhoff's equation, we obtain

$$\left[\partial\frac{\partial H}{\partial V}\right]_P = \Delta C_P = \propto + \beta T + \gamma T^2 + \ldots \quad \ldots(6A)$$

On integrating this equation, we obtain

$$\Delta H = \Delta H_o + \propto T + \beta\frac{T^2}{2} + \gamma\frac{T^3}{3} + \ldots \quad \ldots(6B)$$

where ΔH_o is the integration constant and

$\alpha = \alpha_3 + \alpha_4 - \alpha_1 - \alpha_2$, $\beta = \beta_3 + \beta_4 - \beta_1 - \beta_2$,

$\gamma = \gamma_3 + \gamma_4 - \gamma_1 - \gamma_2$, $\delta = \delta_3 + \delta_4 - \delta_1 - \delta_2$,

Equation (6B) is the general expression which defines the variation of ΔH with temperature.

BORN-HABAR CYCLE

When a solution is formed, two processes take place simultaneously. These are the disordering of the crystal lattice resulting into ion formation and the interaction of the ions with the molecules of the solvent. The heat of solution is effectively the combined result of these two processes– heat of solution = crystal energy + solvation energy. The two step process is represented as

$$MX\ (s) \rightarrow M^+\ (g) + X^-\ (g)$$

$$M^+\ (g) + X^-\ (g) + aq \rightarrow M^+\ (aq) + X^-\ (aq)$$

The energy released when one mole of the solid crystal is formed from its constituent ions in gaseous state is termed as the lattice energy, which is equal in magnitude and opposite in sign to the energy required to transform one mole of the crystal to gaseous ions separated by infinite distance. This energy can be calculated using the electron affinity and ionization potential data for the atoms of the constituent elements of the crystal along with the standard heat of formation of the solid substance from its elements. This calculation is easy because all the quantities are taken as state functions. However, one simple way to calculate the lattice energy is to use the **Born-Haber** cycle for a solid MX.

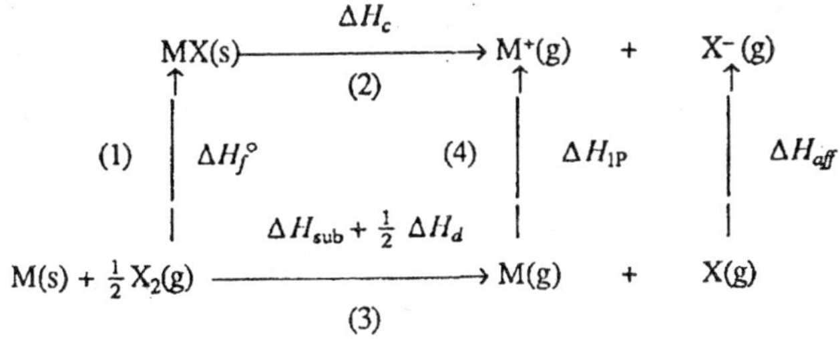

Fig. 5.3

The first step involves the formation of one mole of the solid MX from the elements, for which the heat of formation is ΔH^o_f,

$$M(s) + \frac{1}{2}X_2(g) \rightarrow MX(s)\ \Delta H_f = \text{heat of formation}$$

The second step (2) is the dissociation of the lattice to form the gaseous ions M^+ (g) and X^- (g) for which the energy is ΔH_c

$$MX(s) \rightarrow M^+ (g)\ X^- (g)\ \Delta H_c = \text{crystal energy}$$

Step (3) involves the sublimation of M(s) to M(g) for which ΔHsub is the heat of sublimation, and the dissociation process of 1/2 $X_2 \rightarrow X$ (g) with the heat of dissociation as 1/2 ΔH_d (ΔH_d = heat of dissociation per mole of the substance). The total thermal effect for process (3) is

$$M\ (s) \rightarrow Mg\ (g)\ \Delta H_{sub} = \text{heat of sublimation}$$

$$1/2\ X_2\ (g) \rightarrow X\ (g)\ 1/2\ \Delta H_d = \text{heat of dissociation per mole atom}$$

Step (4) involves the ionization of M(g) to M^+ (g) by supplying ionization energy ΔH_{IP} and the acceptance of electron as X (g) + $e^- \rightarrow$ X^- with electron affinity A or ΔH_{aff}.

$$M\ (g) \rightarrow M^+\ (g) + e^-\ \Delta H_{IP} = \text{Ionization potential}$$

$$X\ (g)\ e^- \rightarrow X^-\ (g)\ \Delta H_{aff} = \text{Electron affinity}$$

Since ΔH is a state function, therefore the thermal changes for step (1) plus step (2) must be equal to those for step (3) and step (4). That is

$$\Delta H^o_f + \Delta H_c = \Delta H_{sub} + 1/2\Delta H_d + \Delta H_{IP} + \Delta H_{aff}$$

$$\text{or}\quad \Delta H_c = \Delta H_{sub} + 1/2\Delta H_d + \Delta H_{IP} + \Delta H_{aff-}\ \Delta H^o_f \qquad ...(1a)$$

$$\text{or}\quad \Delta H^o_f = \Delta H_{sub} + 1/2\ \Delta H_d + \Delta H_{IP} + \Delta H_{aff-}\ \Delta H^o_C$$

(i) The lattice energy or the crystal energy can be calculated by using equation (1) on putting the numerical values with the appropriate signs of the various quantities.

(ii) Any one quantity can be calculated if others are known. The use of this equation along with the heat of solution data is discussed below to calculate the heat of hydration for some ions.

The interaction of gaseous ions with water is called *hydration* and accompanying enthalpy change is called the hydration energy. The hydration energy for lithium ion (radius = 0.060 nm) is larger compared with the hydration energy of Na^+ (radius = 0.095 nm). This suggests that the interaction between Li^+ ion and water dipole is greater (ion⁻ dipole interaction) than the ion – dipole interaction possibility between Na^+ ions and water molecules. This is obvious because there is a size difference between Na^+ ions and Li^+ ions so that the water dipoles are father away from the centre of the sodium ions but nearer to the Li^+ ions, whereas the net ionic charges are the same in both the cases. Therefore water molecules are held less strongly around the sodium ions than around the lithium ions.

The difference in the hydration energies of Li^+ ions and Na^+ ions is about 114 kJ mol^{-1}. However, the crystal energies differ only by about 75 mol^{-1}. The fact that lithium chloride dissolves with the evolution of heat (–35.1kJ) and sodium chloride dissolves with absorption of heat (4.3kJ) must be due to some difference between the lithium ions and sodium ions. This difference is attributed to the fact that the interactions of lithium ions with water produce more heat than the interaction of lithium ions with chloride ions (ion–dipole > ion –ion interaction) whereas the interaction of sodium ions with either water or chloride ions produce about equal enthalpy change.

STANDARD STATES

In the computation of heat of reactions it is a convention to assume that the heat of formation of elements in their standard states is zero. The standard state is taken as 1 atm pressure and a constant temperature. Standard states for various forms of matter are summarized below :

State	*Standard state*
Gas	Ideal gas at 1 atm and the given temperature
Liquid	Pure liquid at 1 atm and the given temperature
Solid	Stable crystalline form at 1 atm and given T (e.g. graphite form of carbon, rhombic form of sulphur).

At standard state the heat of reactions are denoted by ΔE° or ΔH° at given temperature.

Relation between heat of reaction of constant volume (qv) and heat of reaction at constant pressure (qP)

As discussed earlier qv and qP can be taken as ΔE and ΔH respectively. We know that the enthalpy and energy are related by the equation :

$$H = E + PV \Rightarrow \Delta H = \Delta E + P\Delta V$$

Let us consider a general chemical reaction represented by the equation

$aA + bB \rightarrow mM + nN$

The enthalpy change is given by

$\Delta H = \Sigma H \text{ (products)} - \Sigma H \text{ (reactants)}$

$= (mH_M + nHN) - (aHA + bHB)$

$= [mE_M + m\,(PV)_M + nE_N + n\,(PV)_N]$

$- \{aE_A + a(PV)_A + bE_B + b(PV)_B\}$

$= \Delta\, E + P\,[(mV_M - nV_N) - (aV_A + bV_B)]$

$= (\Delta E)P + P\,\Delta V$...If P is constant

where $\Delta E = (mE_M + nE_N) - (aE_A + bE_B)$

$= \Sigma E \text{ (products)} - \Sigma E \text{ (reactants)}$

and $\Delta V = (mV_M + nV_N) - (aV_A + bV_B)$

$= \Sigma V \text{ (products)} - \Sigma V \text{ (reactants)}$

In case the reactants and products being in the solid and liquid state the volume change during the reactions is negligibly small, therefore ΔV can be ignored, and

$\Delta H = (\Delta E)_P = \Delta E$

since $(\Delta E)P$ and are approximately equal.

In the case of gaseous reactants and products ΔV is appreciable and can be evaluated by assuming that the reactants and products obey ideal gas equation $PV = nRT$.

Thus, $\Delta V = \frac{RT}{P}[(m + n) - (a + b)] = \Delta v_{gas}\, RT/P$

Here $\Delta v_{gas} = (m + n)_{gas} - (a + b)_{gas}$

with this value of ΔV, we can write

$\Delta H = \Delta E + \Delta v_{gas}\, RT \Rightarrow qP = qv + \Delta v_{gas}\, RT$

(while calculating Δv, we should not consider the moles of liquid or solid species).

The equation is used to calculate the enthalpy of reaction from the energy of reaction at a given temperature and *vice versa*. If in a reaction $\Delta v_g = 0$, then DH = ΔE.

BOND ENERGIES

Whenever a chemical bond is formed energy is released. Conversely energy is required to rupture a chemical bond. For example, when two gaseous hydrogen atoms combine a molecule of H_2 is formed, and 436 kJ energy is produced per mole of hydrogen molecules. To separate the hydrogen atoms of a hydrogen molecule, therefore, 436 kJ energy per mole must be supplied so that the bonds are broken. This energy is referred to as bond energy. Since the experiments are conveniently carried out at constant pressure the bond energy is commonly called as bond enthalpy. In general, the bond enthalpy of a bond A–B in a gaseous molecule AB is equal to the enthalpy of the reaction represented by

A–B (g) ® A(g) + B (g) DH° (298 K) = DH° (A–B)=x kJ mol–1

Here DH° (A–B) is the bond enthalpy of A–B bond, D gives the flavour of dissociation. The concept of bond energy for diatomic molecules is simpler but the difficulties arise when we consider polyatomic molecules. So it is essential to make a distinction between **bond energy and bond dissociation energy** of a given linkage.

Bond Dissociation Energy. The energy required to break a given bond in some specific compound is defined as the bond dissociation energy. It is a definite quantity. It depends on the presence of other groups in a molecule. For example, the bond dissociation energies of the two O–H linkages in water molecule are different. The energy required to break O–H bond in water is 498 kJ mol^{-1}.

$$H\text{–}O\text{–}H(g) \rightarrow H(g) + OH(g) \quad \Delta H^\circ(298K) = 498 \text{ kJ mol}^{-1}$$

and the energy required to break the O –H bond in hydroxyl group is 430 kJ mol^{-1}

$$O\text{–}H(g) \rightarrow O(g) + H(g) \quad \Delta H^\circ(298K) = 430 \text{kJ mol}^{-1}$$

Thus, the bond dissociation energy of O –H linkage in water molecule is 498kJ whereas that in a hydroxyl group is 430 kJ.

Bond energy. The average of dissociation energies of a given bond in a series of different dissociating species, is termed as bond energy. The O–H bond energy in water molecule is the average of the two bond dissociation energies, in other words

$$DH\ (O\text{–}H) = \frac{498 + 430}{2} = 464 \text{ kJ mol}^{-1}$$

In case of diatomic molecules the bond energy and bond dissociation energies are identical.

Calculation of Bond

The main sources of data for the determination of bond energies are molecular spectroscopy, electron impact studies, and thermochemistry. Here we shall consider only thermochemical methods.

C–H bond energy in methane. Since all the four C–H bonds in methane are equivalent the average of the bond dissociation energies would be equal to the C–H bond energy.

Our aim is to calculate DH°/4 for the reaction CH4(g) ® C(g) + 4H(g) Data given :

(i) C (graphite) $\xrightarrow{\text{Sublimation}}$ C(g) ΔH° (i) = 716.68 kJ

(ii) $\frac{1}{2}H_2$ (g) $\xrightarrow{\text{Dissociation}}$ H(g) ΔH° (ii) = 217.97 kJ

(iii) C (graphite) + $2H_2$ (g) $\rightarrow$ CH_4(g) ΔH° (iii) = 74.75 kJ

Multiply equation (ii) by four, reverse equation (iii), combine these two new equations with equation (i)

(iv) $2H_2$ (g) $\rightarrow$ 4H(g) ΔH° (iv) = 871.88 kJ

(v) CH_4 (g) $\rightarrow$ C (graphite) + 4 H (g) ΔH° (v) = 74.75 kJ

(i) C (graphite) $\rightarrow$ C (g) ΔH° (i) = 716.68 kJ

(vi) CH_4 (g) $\rightarrow$ C (g) + 4 H(g) ΔH° (vi) = 1663.31 kJ

The quantity, 1663.31 kJ, is known as the heat of atomization of methane. This much heat must be supplied to convert a mole of methane into gaseous atoms. This is also known as the **bond dissociation** energies of four equivalent bonds m methane. Thus the C–H bond energy is given by

$$DH^\circ \text{ (C–H)} = \frac{1663.31 \text{kJ mol}^{-1}}{4} = 415.82 \text{ mol}^{-1}$$

The actual value is 413 kJ.

Similarly, C–C, C = C, C ≡ C bond energies can be calculated by using the heats of formation of ethane, ethylene and acetylene respectively.

Finally, by taking methanol, in which the only unknown bond energy is that of the C–O bond we find that the C– O bond energy is of the order of 351 kJ.

DH° values for different bonds are given in the Appendix.

In the calculation of bond energies we have assumed that one type of bond energy calculated for a particular molecule is the same in all other types of molecules. For example, C–H bond energy in methane, ethane, propane etc., will have the same value. Similarly C–C bond energy for ethane is applicable for C–C bond energy in other molecules too.

However, this procedure has got some limitations, e.g., butane exists in two forms as n-butane and iso-butane. Both compounds have three C–C bonds and ten C– H bonds, and from simple bond theory the heats of formation should be the same for both compounds.

Bond energy data = (3C – C + 10 C–H)

$= 3 \times 348 + 10 \times 413 = 5174.0$ kJ

Heat of atomization = Heat of atomization for (4C + 10H)

$= 4 \times 716.7 + 10 \times 217.97 = 5046.5$ kJ

Standard heats of formation of butane from its elements:

ΔH^o_f = (Heats of atomization) – (Bond energies)

$= 5046.5 - 5174.0 = -127.5$ kJ

But the observed ΔH^of for n-butane is – 124.3 kJ and that for isobutane is – 131.2 kJ. There is a difference of about 3 kJ in calculated and observed data. This difference becomes more and more as we consider the still higher aliphatic hydrocarbons, *e.g.*, there is a difference of almost 16 kJ in the standard heats of formations of n-pentane and 2-2 dimethyl propane, although both 4 C – C bonds and 12 C – H bonds.

Apart from the limitations of the bond energy calculations there are certain merits of the procedure in the sense that it puts restrictions on writing incorrect structures of molecules and suggests a suitable form of representing the molecular structures. We shall discuss the merits of thermodynamic calculations by considering the case of benzene.

Structure of Benzene – Resonating structure – Delocalization concept:

We can calculate the standard heats of formation (DH°f) of benzene in the same manner as that for the simpler hydrocarbons. Usually we write the structure of benzene as a ring with three C– C bonds, three C–C bonds and 6 C – H bonds as

Bond energy data

$= 6 \times DH^o\ (C-H) + 3 \times DH^o\ (C-C) + 3 \times DH^o\ (C=C)$

$= (6 \times 413 + 3 \times 348 + 3 \times 610) = 5352 \text{ mol}^{-1}$

Heat of atomization

$= 6 \times \Delta H \text{ [C (graphite)} \rightarrow \text{C (g)]} + 6 \times \Delta H \text{ (1/2 } H_2 \rightarrow \text{H)}$

$= 6 \times 716.7 \text{ kJ mol}^{-1} + 6 \times 217.96 \text{ kJ mol}^{-1}$

$= 5608 \text{ kJ mol}^{-1}$

$\Delta H^o f = 5608 - 5352 = 256 \text{ mol}^{-1}$

The observed ΔH^o_f is only 82.9 kJ mol^{-1}. This shows that benzene molecule is 173 kJ mol^{-1} more stable than predicated by the considered structure. This unexpected stability suggests that the Kekule structure is not a true representative of molecular structure. The real state can be described by a resonance hybrid of the two extreme structures.

The two extreme forms are called "resonance" structures and the difference of the energy of approximately 173 kJ mol–1 is referred to as the resonance stabilization energy. This suggests that the C– C bonds in the benzene ring are neither pure C – C bonds nor pure C = C bonds. All are equivalent and intermediate between single and double bonds. This is shown from C to C bond energy considerations.

Let us suppose that carbon to carbon bonds in the benzene ring are neither pure single nor pure double bonds. Using this approach we can calculate the bond energy of the C linkages.

$C_6H_6 \rightarrow 6C \text{ (graphite)} + 3H_2 \text{ (g)} \qquad \Delta H = -\Delta H^o_f = -82.9 \text{ kJ}$

$6C \text{ (graphite)} \rightarrow 6C \text{ (g)} \qquad \Delta H = 4300.2 \text{ kJ}$

$3H_2 \text{ (g)} \rightarrow 6H \text{ (g)} \qquad \Delta H = 1307.8 \text{ kJ}$

---------- ---------

$C_6H_6 \rightarrow 6C \text{ (g)} + 6H\text{(g)} \qquad \Delta H = 5525.1 \text{ kJ}$

Thus, the energy required to break six C – H bonds and six C to C bond is 5525.1kJ. But the energy required to break six C– H bonds is 2478 kJ. Therefore, the energy of six C to C linkage would be equal to the difference of the two, i.e., (5525.1 –2478.0) = 3047.1 kJ. Each C to C bond in the ring can be assigned a value of 507.8 kJ. Since the energy of pure C – C bond is 348 kJ and for a pure C = C bond, the energy is 610 kJ. This gives an idea that the C to C bonds in benzene have a bond order intermediate between 1 and 2. The bonds between carbon are not localized as double bonds in three places but are spread among all six C– C distances. The electrons involved are delocalized.

The extra stability of the molecule arises from the delocalization of the π-electrons in the ring.

Bond energy data are useful in deciding the nature of bonds in a group with multiple bonds. For example, the comparison of bond energies of C – C and C = C shows that both the bonds in C = C are not identical. If this were so then the C = C bond energy must be just double of the C – C bond energy. However, the valency considerations show that C–C bond is a stable bond of σ-type. Of the two bonds in C = C linkage one is s-type and the other is a weaker p-linkage. This is in line with the thermodynamic calculations which gives a difference of 262 kJ mole–1 between C = C bond and C – C bond energies. This suggests that the bond other than σ-type in C = C bond is weaker. Similarly in C ≡ C linkage the nature of all the three bonds are not identical. one being s-type and the two are π-type bonds.

HEAT CONCEPT OF REACTION

Heat of reaction is defined as the amount of heat absorbed or evolved at a given temperature when the reactants have combined to form the products as represented by the balanced chemical equation. If the heat of-reaction is denoted by q. then according to the first law,

$$q = \Delta E - w$$

where –w is the work done by reacting mixture when it reacts and D E is the energy difference between the reaction products and the reactants.

$$\Delta E = \Sigma E \text{ (products) } \Sigma E \text{ (reactants)}$$

The numerical value of q depends on the manner in which the reaction is performed. For the two methods of conducting chemical reactions in calorimeters, we have

Constant volume : w = 0. and qv = DE; Bomb calorimeter.

Constant pressure : w = – P ΔV and qP = ΔE + P ΔV = ΔH: Open calorimeter.

Thus heat of a reaction at constant volume (qv) is the change of energy (ΔE) accompanying the chemical reaction. The heat of reaction when measured at constant pressure (qp) is equal to the enthalpy change in the reaction given by

$$q_p = \Delta H = \Sigma H \text{ (products) } - \Sigma H \text{ (reactants)}$$

- Constant pressure are much common in chemistry therefore usually whenever we speak of heat of reaction in implies enthalpy change.

Sign conventions. If heat is absorbed by the system ($q > 0$) then the reaction is said to be endothermic and ΔE or ΔH value is given a positive sign. If heat is evolved ($q < 0$) the reaction is said to be **exothermic**, and ΔE or ΔH value is given negative sign.

Endothermic reaction. $\Delta H > 0$, Σ H (Products) > ΣH (reactants). Heat absorbed during the reaction increases the energy contents of the products and hence the enthalpy of the products becomes greater than that of the reactants from which the products are formed.

Exothermic reaction. $\Delta H < 0$, ΣH (products) < ΣH (reactants). Heat is liberated in the reaction and is transferred to the surroundings. Therefore the enthalpy of the products is less than enthalpy of the reactants from which the products are generated. Diagrammatically these may be represented as follows:

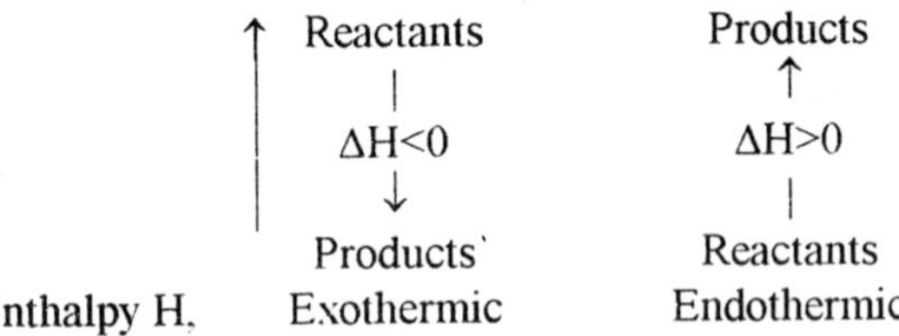

If the reaction is performed at constant temperature, it is indicated by writing TK in parenthesis to the right of Δ H (or Δ E). The state of a product or reactant is indicated by symbols in parentheses after the chemical symbol e. g., (s) or (c) for solid or crystal, (1) for a liquid, (g) for a gas, (aq) for solution in water.

Heat is generally absorbed or evolved whenever reactions take place. a solution is formed, or other changes in the state of matter (fusion, vaporization, etc). occur. Thermochemistry is concerned with the measurement and study of these heat changes.

Heat changes accompanying chemical reactions or physical changes are measured in calorimeters. The system in which the change is to take place is immersed in a bath which is thermally insulated from the rest of the universe. The temperature of the bath is measured before and after and change has occurred. This temperature change is compared with the temperature change produced when a known amount of thermal energy is introduced into the bath, usually by passing an accurately known current at an accurately known voltage for an accurately known time through a resistor immersed in the bath. For precise measurements of heat changes, great care is required in the design and operation of calorimeters.

Thermodynamic principles make the determination of heat changes more simpler even without performing the tedious experiments in the laboratory. This is done by taking advantage of the fact that heat changes under certain conditions (qv = Δ E and qp = ΔH) become state properties. This fact many times helps in the computation of the heat changes for those processes which are otherwise very difficult to be carried out in the laboratory and may require a lot of time and labour.

HEAT OF REACTION OR ENTHALPY OF REACTION

When a chemical reaction is accompanied by evolution of heat, it is termed as exothermic and if it is accompanied by absorption of heat, it is termed as endothernric. *The amount of heat evolved or absorbed during a chemical reaction is called the heat of reaction or enthalpy of reaction.*

From the definition of enthalpy, there always occurs a definite enthalpy change, it follows that it is ΔH, whenever a system undergoes a change of state or a chemical reaction.

Suppose we consider a chemical system having two reactants A and B in the initial state.

Suppose it undergoes a chemical reaction at a constant temperature and pressure, thereby forming the products C and D.

$$A + B \rightarrow C + D$$

Suppose H_1 denotes the enthalpy of the systems in the initial state (*i.e.*, enthalpy of reactants A and B) and H, the enthalpy of the system in the final state at the same temperature and pressure (*i.e.*, enthalpy of products C and D). Then,

$$\Delta H - H_2 - H_1$$

Thus, *at constant pressure and temperature the heat of reaction or enthalpy of reaction is regarded as a measure of the enthalpy chaise, ΔH. In simple words,*

"The heat of reaction or enthalpy of reaction at a given temperature and pressure is the difference between the heat contents of the products and the reactants when the number of gram molecules of substances indicated by the chemical reaction have completely reacted.

Mathematically, it may be put as

ΔH = Heat contents of the products – Heat contents of the reactants.

= Heat contents of 1 mole of C and 1 mole of D – Heat contents of 1 mole of A and 1 mole of B.

ΔH is positive for endothermic reactions and negative for exothermic reaction. For example, the thermochemical equation

$$H_2(g) + \frac{1}{2}O_2(g) \rightarrow 2H_2O(l);\ \Delta H = -285.83\ U$$

indicates that when one mole of gaseous hydrogen combines with half a mole of gaseous oxygen at 298 K at one atmosphere pressure to form 1 mole of liquid water at the same temperature and pressure, 285.83 kJ of heat is given out to the surrounding. It is further assumed that the reaction is fast and proceeds to completion with no side products.

Enthalpy of a Reaction at Constant Pressure and at Constant Volume : The enthalpy of a reaction depends upon the condition under which the reaction is carried out. There are two general conditions under which thermochemical measurements are made. namely :

(i) at constant volume, and

(ii) at constant pressure.

The magnitude of the anthalpy changes in these two conditions are, in general, different.

Heat of reaction or enthalpy of reaction at constant volume (ΔE) : When a reaction is being carried out at a constant volume, the heat change is called the heat of reaction or enthalpy of reaction at constant volume. Example is the combustion of a substance in a bomb calorimeter.

When a reaction is carried out under constant volume by carrying out the reaction in a closed and rigid container, no work is involved and from the first law of thermodynamics, we have

$$\Delta E = qv$$

= Enthalpy of the reaction at constant volume.

Thus, the enthalpy of the reaction at constant volume is exactly equal to the change in the internal energy ΔE of the reacting system. In other words, the thermal change that occurs in a chemical reaction is only due to the difference in the sum of the internal energy of the products and the sum of the internal energy of the reactants.

$$\Delta E = \Sigma E_{products} - \Sigma E_{reactants} \quad ...(1)$$

The importance of Eq. (1) is that the amount of heat absorbed at constant volume can be identified with the change in the thermodynamic quantity.

Heat of reaction or Enthalpy of reaction at constant pressure. ΔH : When pressure is kept constant throughout the reaction, the heat change is then termed as heat of reaction at constant pressure.

At constant pressure, the system is either kept open to the atmosphere or confined with in a vessel on which a constant external pressure is exerted. Under these conditions, the volume of system changes. The thermal change at constant pressure not only involves the change in internal energy of the system but also the work done either in expansion or in contraction of the system. Then, from the first law of thermodynamics, we have

$$q_P = \Delta E + w$$

If w is only pressure volume work, then

$$q_P = \Delta H + P\Delta V = (\Sigma E_P - \Sigma E_R) + P(V_P - V_R)$$

$$= (\Sigma E_P - PV_P) - (\Sigma E_R + PV_R) \quad ...(2)$$

where the subscripts P and R refer to the products and the reactants respectively.

But $H = E + PV$ (By definition)

Therefore, Eq. (2) becomes as follows :

$$q_P = \Sigma H_P - \Sigma H_R = \Delta H \quad ...(3)$$

From the above equation it follows that at constant pressure, the heat of the reaction is exactly equal to the enthalpy change. ΔH. of the reacting system.

Relationship between ΔH and ΔE : In the case of reactions involving solids and liquids, the difference between ΔH and ΔE is almost negligible. Thus, for solids and liquids, $\Delta H \approx \Delta E$. On the other hand, ΔH may not be equal to ΔE in reactions which involve gases. Suppose consider the following reactions :

$$2NH_3(g) \rightarrow N_2(g) + 3H_2(g)$$

2 moles 1 mole 3 moles

$$2H\text{–}2(g) + O_2(g) \rightarrow 2H_2O(g)$$

2 moles 2 moles 1 mole

When these reactions are carried out at constant volume, no work will be done by or on the system as there is neither expansion nor contraction. However, if the reactions are carried out at constant pressure, then the work will be performed by the system in the first case as there occurs an increase in volume

$$(2NH_3 + N_2 + 3H_2)$$

and upon the system in the second case

$$(2H_2 + O_2 \rightarrow 2H_2O)$$

as there occurs a decrease in volume. Consequently, the heat of reaction at constant volume (ΔE) will be more than heat of reaction at constant pressure (ΔH) when contraction occurs during the course of reaction whereas ΔE will be less than ΔH if expansion occurs during the course of reaction. Now consider the following gaseous reaction :

$$aA + bB + ... \rightleftharpoons cC + dD + ...$$

Suppose n_1 = No. of moles of gaseous reactants, *i.e.*, a + b +...

n_2 = No. of moles of gaseous products, *i.e.*, c + d +...

$$\Delta n(g) = (c + d + ...) - (a + b + ...) = (n_2 - n_1),$$

V_1 = volume of the gaseous reactants,

P = the external pressure of the gas, and

V_2 = volume of the gaseous products,

T = temperature of the system on absolute scale,

For an ideal gas,

$$PV = nRT$$

$\therefore$ $PV_1 = nRT$ and $PV_2 = n_2RT$

Now, work done by the gas in expansion or contraction from volume V_1 to V_2 in the reaction

$$aA + bB + ... \rightleftharpoons cC + dD + ...$$

is given by

$$w = \text{pressure} \times \text{change in volume}$$

$$= P(V_2 - V_1) = PV_2 - PV_1$$

$$= n_2RT - n_1Rt\ (n_2 - n_1) = \Delta n(g)RT \quad ...(4)$$

Thus, heat of reaction at constant pressure is given by,

ΔH = heat of reaction at constant volume + work done by or on the gas during change of volume V_1 to V_2

$$= \Delta E + \Delta n(g)RT \quad ...(5)$$

Equation (5) can be used in calculating the value of ΔH from ΔE or vice-versa. A positive value of ΔH or ΔE shows that heat is absorbed while a negative value of ΔH or ΔE shows that heat is evolved or given out by the system to the surroundings,

Case I

When $n_2 > n_1$, $\Delta n(g)$ is positive. Therefore,

$$\Delta H > \Delta E$$

Case II

When $n_2 > n_1$, $\Delta n(g)$ is negative.

Therefore, $\Delta H < \Delta E$

Case III

When there is no change in volume, $n_2 = n_1$. Therefore,

$$\Delta n(g) = 0, \ \Delta H = \Delta E$$

EXOTHERMIC AND ENDOTHERMIC REACTIONS

(a) *Exothermic Reactions* : A reaction which proceeds with the evolution of heat is referred to as an exothermic reaction. Some examples of exothermic reactions are as follows :

$$2Zn(s) + O_2(g) \rightarrow 2ZnO(s) + 693.8 \text{ kJ} \quad ...(1)$$

$$C(s) + O_2(g) \rightarrow CO_2(g) + 393.5 \text{ kJ} \quad ...(2)$$

$$C_2H_2(g) + 2H_2(g) \rightarrow C_2H_6(g) + 314.0 \text{ kJ} \quad ...(3)$$

In all the above reactions, heat is evolved and therefore these are examples of exothermic reactions.

If an exothermic reaction is taking place at constant temperature and constant volume, the energy change is expressed in terms of internal energy changes, *i.e.*, ΔE. Τηυσ,

ΔE = Internal energy of product - Internal energy of reactants = $E_P - E_R$

In exothermic reactions, the total internal energy of the reactants is more than that of products and therefore the excess of internal energy, ΔE, will l appear as heat of reaction or in other words ΔE is negative for exothermic reactions. Thus, equations (1), (2) and (3) in terms of ΔE can be written as :

$$2Zn(s) + O_2(g) \rightarrow 2ZnO(S)\ ;\ \Delta E = -\ 693.8\ kJ$$

$$C(s) + O_2(g) \rightarrow CO_2(g)\ ;\ \Delta E = -\ 393.5\ kJ$$

$$C_2H_2(g) + 2H_2(g) \rightarrow C_2H_6(g)\ ;\ \Delta E = -\ 314.0\ kJ.$$

If an exothermic reaction is taking place at constant temperature and constant pressure, the energy change is expressed in terms of enthalpy change, ΔH, *i.e.*,

$$\Delta H = \text{Enthalpy of products} - \text{Enthalpy of reactants} = H_P - H_R$$

In exothermic reactions, the total heat content of the reactants is more than that of products and, therefore, the excess of heat constant will appear as heat of reaction. In other words ΔH like ΔE is negative. Thus, equations (1), (2) and (3) in terms of ΔH can be written as :

$$2Zn(s) + O_2(g) \rightarrow 2ZnO(S)\ ;\ \Delta H = -\ 693.8\ kJ$$

$$C(s) + O_2(g) \rightarrow CO_2(g)\ ;\ \Delta H = -\ 393.5\ kJ$$

$$C_2H_2(g) + 2H_2(g) \rightarrow C_2H_6(g)\ ;\ \Delta H = -\ 314.0\ kJ$$

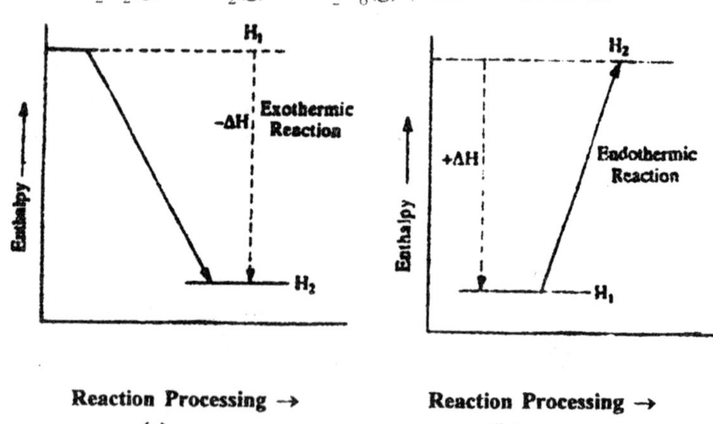

Reaction Processing →
(a)

Reaction Processing →
(b)

Fig. 5.4

As ΔH is negative for exothermic reactions, this is illustrated in Fig. 5.4(a).

(b) *Endothermic Reactions :* A reaction which proceeds with the absorption of heat is referred to as an endothermic reaction. Some examples of endothermic reaction are as follows :

$$2HgO(s) + 180.4 \text{ kJ} \rightarrow 2Hg(l) + O_2(g) \quad ...(1)$$

$$H_2O(g) + C(s) + 131.2 \text{ kJ} \rightarrow CO(g) + H_2(g) \quad ...(2)$$

$$N_2(g) + O_2(g) + 180.5 \text{ kJ} \rightarrow 2NO(g) \quad ...(3)$$

If endothermic reactions are taking place at constant temperature and volume, the total internal energy of the reactants (E_R) is less than the total internal energy of the products (E_P), *i.e.*, ΔE is positive. Thus, equations (1), (2) and (3) can be put as :

$$2HgO(s) \rightarrow Hg(l) + O_2(g); \text{ DH} = 180.4 \text{ kJ}$$

$$H_2O(g) + C(s) \rightarrow CO(g) + H_2(g); \text{ DH} = 131.2 \text{ kJ}$$

$$N_2(g) + O_2(g) \rightarrow 2NO(g); \text{ DE} = 180.5 \text{ kJ.}$$

If endothermic reactions are carried out at constant pressure and constant temperature, then the total heat content of products is more than the total heat content of reactants. In other words, DH is positive. It is represented in Fig. 1.1 (b). In terms of DH, equations (1) (2) and (3) may be put as:

$$2HgO(g) \rightarrow 2Hg(l) + O_2(g); \text{ DE} = 180.4 \text{ kJ}$$

$$H_2O(g) + C(s) \rightarrow CO(g) + H_2(g); \text{ DE} = 131.2 \text{ kJ}$$

$$N_2(g) + O_2(g) \rightarrow 2NO(g); \text{ DH} = 180.5 \text{ kJ}$$

Table 1

Exothermic Reactions	*Endothermic Reactions*
1. Heat is given to the surroundings.	1. Heat is absorbed from the surroundings.
2. The internal energy of reactants is more than that of products, *i.e.*, ΔE is – ve.	2. The internal energy of products is more that of reactants, *i.e.*, ΔE is + ve.
3. The enthalpy of reactants is more than that of Products, *i.e.*, ΔH is –ve.	3. The enthalpy of reactants is less than that of Products, i.e, ΔH is +ve.

Melting of solids and vaporization of liquids are endothermic reactions. Synthesis of proteins in living systems is an endothermic process.

(c) Differences between Exothermic and Endothermic Reactions.

The differences are given in tabular from :

VARIOUS TYPES OF ENTHALPY CHANGES

Enthalpy of a reaction is given different names depending upon the types of reactions. Some of those are described below.

Heat or Enthalpy of Formation

The enthalpy of formation of a compound at the given temperature may be defined as the enthalpy change when one mole of a compound (in its standard state) is formed its elements (in their standard states). It is usually denoted by ΔH°_{f}.

In the above definition, the standard state of any substance is taken as its natural state at 298 K under one atmosphere pressure. Let us consider the following reactions :

$$C(graphite) + O_2(g) \rightarrow CO_2(g);\ \Delta H = \Delta H^{\circ}_{f} = -\ 393.5\ kJ$$

$$C(graphite) + 2H_2(g) \rightarrow CH_4(g);\ \Delta H = \Delta H^{\circ}_{f} = -\ 74.8\ kJ$$

Sometimes, the enthalpy of combustion and enthalpy of formation relate to the same reaction ; for instance, the enthalpy of formation of carbon dioxide is the same as the enthalpy of combustion of graphite.

The compound with negative enthalpy of formations called exothermic compound whereas the compound with positive enthalpy of formations known as endothermic compound. Thus, the formation of water is an exothermic compound whereas the formation of hydrogen iodide is an endothermic compound.

$$H_2(g) + 1/2O_2(g) \rightarrow H_2O(g);\ \Delta H^{\circ}_{f} = -\ 245.5\ kJ$$

$$1/2H_2(g) + 1/2I_2(s) \rightarrow HI(g);\ \Delta H^{\circ}_{f} = +\ 25.94\ kJ$$

The negative enthalpy of formation of a compound means that its heat content is less than that of its elements and hence it is more stable than its elements. The reason is for that the compound with a lower enthalpy (or energy) must be more stable. Therefore, the chemical bonds in H_2O are stronger than those in H_2 or O_2. On the other hand, the positive enthalpy of formation of a compound means that its enthalpy of content is more than that of its elements and hence it is less stable than its elements. Due to this reason, H – I bond is weaker than H – H bond in hydrogen or I – I bond in iodine.

The enthalpy of formation of a compound may be obtained either by direct measurement of ΔH for the reaction or from enthalpies of reactions involving the compound.

Heat or Enthalpy of Combustion

The enthalpy of combustion of a substance at a given temperature may be defined as the enthalpy change accompanying the complete combustion of one mole of the substance in excess of oxygen or air at that temperature.

. The words "complete combustion" in the above definition must be emphasized because it is possible to burn a compound to yield both CO and CO_2. In order to avoid this confusion. The combustion is carried out in a large excess of oxygen or air to yield carbon dioxide and water as the final products. For example, heat of combustion relates to the reaction :

$$C_2H_2(g) + 21/2O_2(g) \rightarrow 2CO_2(g) + H_2O(g);\ \Delta H^\circ = -\ 1258\ kJ$$

and not to the partial oxidation as follows :

$$C_2H_2(g) + 11/2O_2(g) \rightarrow 2CO(g) + H_2O(g);\ \Delta H^\circ - -\ 691kJ$$

The heat of combustion depends upon the physical state of the substances involved. For example, the heat of combustion of hydrogen at 298 K is –285.9 kJ mol^{-1}.

In this case only liquid water is obtained. On the other hand if the combustion is taking place at 313 K or above, the water vapour in this case cannot exist in the liquid state but evaporates to form vapours. The enthalpy of combustion is that case will be –245.5 kJ.

$$H_2(g) + 1/2O_2(g) \rightarrow H_2O(1);\ \Delta H^\circ_{298k} = -\ 285.19\ kJ\ at\ 298\ k$$

$$H_2(g) + 1/2O_2(g) \rightarrow H_2O(g);\ \Delta H^\circ_{373k} = -\ 245.5\ kJ\ at\ 373$$

The enthalpy of combustion at 373 K is lower as compared at 298 K. The reason for this is that small quantity of energy (+ 40 44 kJ) is used up to convert liquid water into its vapours.

$$H_2O(l) \rightarrow H_2O(g);\ \Delta H^\circ - +\ 40.44\ kJ.$$

The enthalpy of combustion of methane is

$CH_4(g) + 2O_2(g) \rightarrow CO_2(g) + 2H_2O(l); \Delta H^\circ = -890.40$ kJ.

Enthalpy of combustion may be used to calculate the enthalpy of formation of a compound provided the enthalpy of formation of other combustion products are known.

For example, it is possible to calculate enthalpy of formation of gaseous methane provided the enthalpies of formation of $CO_2(g)$ and $H_2O(l)$ are known.

$$\Delta H^o = -890.40 = \Delta H^o{}_{CO_2(g)} {}^{+2\Delta H^o}{}_{H_2O(l)} {}^{-\Delta H^o}{}_{CH_4(g)}$$

or $\Delta H^o{}_{CH_4(g)} {}^{=890.40\,+\,\Delta H^o}{}_{CO_2(g)} {}^{+2\Delta H^o}{}_{H_2O(l)}$

$= 890.40 + (-393.50) + 2(-285.83) = -74.76$ kJ mol^{-1}.

Application of Heat Combustion Date : Main applications are :

(i) Calorific value of fuels

(ii) Food values of diets

(iii) Maximum flame temperature

(iv) Calculation of heal of formation

(v) Calculation of heat of reaction.

Heat or Enthalpy of Hydrogenation

It may be defined as the enthalpy change when a mole of an unsaturated organic compound is fully hydrogenated. The enthalpies of hydrogenation of ethylene and benzene are represented as follows .

$CH_2 = CH_2(g) + H_2(g) \rightarrow CH_3 - CH_3(g);$

$\Delta H^\circ = -132.5$ kJ

$C_6H_6(l) + 3H_2(g) \rightarrow C_6H_{12}(l);$

$\Delta H^\circ = -246$ kJ

From the above thermochemical equation, it is to be noted that enthalpy of hydrogenation of benzene is not three times that of enthylene one may expect this due to the presence of three double bonds in benzene and one double bond in enthylene.

The enthalpy of hydrogenation of benzene is less be 151.5kJ mol^{-1} which is attributed to the stabilisation on benzene due to resonance. This method may sometimes be used as the characteristic of stabilisation of aromatic molecules due to resonance.

Heat or Enthalpy of Solution

When a solution is dissolved in a solvent, heat may be either liberated or absorbed. If the heat is absorbed the solution becomes cooler and ΔH is given a positive sign. On the other hand if the heat is evolved, the solution gets warmer and ΔH is given a negative sign.

The enthalpy change per mole of a solute dissolved varies to a considerable extent with dilution and to some extent with the initial temperature of the solvent. Therefore, it becomes essential to express the enthalpy change with reference to the concentration of the solution.

"The enthalpy of solution may be defined as the enthalpy change when one mole of the solute is dissolved in a specified amount of solvent."

The enthalpy of solution can be explained by considering the dissolution of HCI (g) in water at 298 K:

$$HCI\ (g) + 25H_2O(l) \rightarrow HCI.25\ H_2O;\ \Delta H^\circ = -69.5\ kJ$$

$$HCI\ (g) + 40H_2O(l) \rightarrow HCI.40\ H_2O;\ \Delta H^\circ = -73.18\ kJ$$

$$HCI\ (g) + 200H_2O(l) \rightarrow HCI.200H_2O;\ \Delta H^\circ = -74.40\ kJ$$

$$HCI\ (g) + \infty\ H_2O(l) \rightarrow HCI.(aq);\ \Delta H^\circ = -75.17\ kJ$$

From the above equation it means that when one mole of HCI is dissolved in 25 moles of water, 69.5 kJ of heat are evolved. But if the solution is further diluted to such an extent that further dilution would cause no additional heat change, then the equation may be written as :

$$HCI\ (g) + \infty\ H_2O(l) \rightarrow HCI.(aq);\ \Delta H^\circ = -75.17\ kJ\ mol^{-1}$$

where aq. represents a large volume of water. The value of ΔH° correspondents to this reaction is called the enthalpy of solution at infinite dilution. It is the maximum quantity of heat released in the formation of a solution of one mole of HCI(g) in excess of water.

For greater accuracy, two types of heat of solution have been recognised. There are :

(a) *Integral Heat of Enthalpy of Solution :* Integral heat of solution at a particular concentration is the enthalpy change when 1 mole of solute is dissolved in pure solvent to produce a solution of the given concentration. Thus, if 1 mole of solute is dissolved in 500g of water the heat change gives the value of integral heat of solution at the concentration of 2 molal.

While recording integral enthalpies of solution, the usual practice is to state the amount of solvent in which one mole of solute is dissolved. For example :

$$HCI\ (g) + 10H_2O(l) \rightarrow HCI.(10\ H_2O);\ \Delta H^o = -69.5\ kJ.$$

indicates that when one mole of hydrogen chloride gas is dissolved in 10 moles of water, there is an evolutions 65.5 kJ of heat energy.

According to Hess's law, the integral enthalpy of solution is equal to the difference between integral enthalpies of the solution at the two concentrations. For example, if to a solution of 1 mole of hydrogen chloride in 40 moles of water, enough water is added such that

$$HCI.40\ H_2O + aq \rightarrow HCI.(aq)$$

the integral enthalpy change can be calculated as follows :

$$HCI(g) + aq \rightarrow HCI\ ; \quad \Delta H^o = -\ 75.17\ kJ$$

$$HCI(g)\ 40\ H_2O \rightarrow HCI.\ 40H_2O; \quad \Delta H^o = -\ 73.18\ kJ$$

On subtracting, we have

$$HCI.40\ H_2O + aq \rightarrow HCI.(aq); \quad \Delta H^o = -\ 1.99\ kJ$$

If ΔH is plotted against number of solute per 1000 g solvent, the resultant curve is shown in Fig. 5.5. Initially, ΔH increases almost linearly with m but later the increase is not as fast as m. Finally, it reaches a constant value when the solution becomes saturated with respect to solute. If the observers value of enthalpy change (ΔH) is divided by the amount of solute (m) that is dissolved to form a solution of a particular concentration, we get the quantity ΔH/m, which is nothing but the integral enthalpy of solution at the given concentration.

Integral enthalpies of solutions, enthalpies of dilutions and enthalpies of reactions in solutions can be evaluated from the enthalpies of formation in solution. In these calculations, the enthalpy of formation of water is neglected if the same number of moles of water appears on both sides of the balanced thermochemical equation.

(b) *Differential Heat or Enthalpy of Solution* : The differential heat of solution, $DH_{D.s}$ may be defined as the heat of solution of a mole of solute in such a large quantity of solution of known concentration that addition of 1 more mole of solution does not change the concentration appreciably. If on dissolving dm mole of the solute in a finite amount of the solution the heat change is dq, then differential heat of solution will be

$$\Delta H_{D.s} = \delta q/\delta m$$

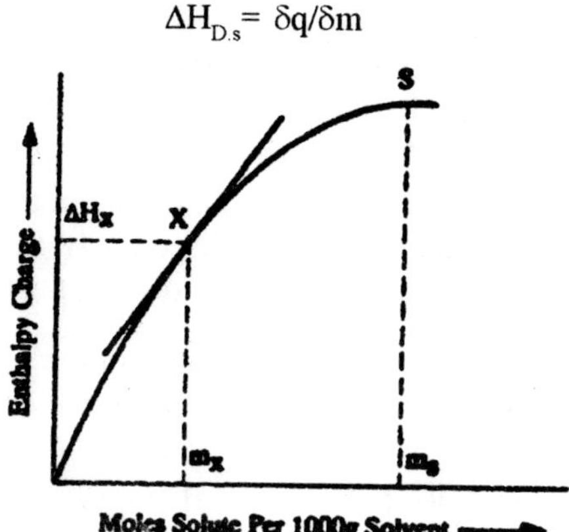

Fig . 5.5 : Variation of enthalpy change when the amount m of solute is dissolved in 1 kg.

From the shape of the curve of Fig. 5.5, it can be seen that the value of the slope will depend upon the concentration of the solution. Thus, the differential enthalpy of a solution besides depending to T and p, will also depend on the amounts of solvent n_1 and solute n_2 that are present in the given solution *i.e.*,

$$d\ (\Delta H)/dm = f(T,\ p\ n_1,\ n_2)$$

Thus, for this reason the concentration of the solution is mentioned while defining the differential enthalpy of solution. The following conclusions may be drawn from the shape of the curve between ΔH and m.

(i) For smaller value of m, the curve is almost linear; thus its slope will have a constant value and will be equal to DH/m. As the latter represents the integral heat of solution, it means that the differential and integral heats of solution are essentially equal for very dilute solutions.

(ii) For higher values on m, the curve is not linear. The reason is that ΔH does not increase as fast as m; thus the slope of the curve decreases as the value of m increases. In other words, the differential heat of solution decreases as the concentration of the solution increases, and becomes zero when the solution gets saturated.

The differential enthalpy of solution can be obtained by plotting ΔH of the solution at various concentrations against the number of moles of solvent associated with a definite quantity of solute, and finding the slope of the curve at a point corresponding to any particular concentration. The value of the slope at a given concentration gives the differential enthalpy of solution at that concentration.

The differential heats of solution and dilution are essentially the partial molar heat of solute and solvent, respectively. Consider a solution having n_1 moles of solvent and n_2 moles of solute. In general,

$$\Delta H = f(T, p\ n_1, n_2)$$

The differential of ΔH is given as follows :

$$d(\Delta H) = \left(\frac{\partial(\Delta H)}{\partial T}\right)_{P, n_1, n_2} dT + \left(\frac{\partial(\Delta H)}{\partial P}\right)_{T, n_1, n_2} dP + \left(\frac{\partial(\Delta H)}{\partial n_1}\right)_{T, P, n_2} dn_1 + \left(\frac{\partial(\Delta H)}{\partial n_2}\right)_{T, P, n_1} dn_2$$

At constant temperature and pressure, the above equation becomes as follows :

$$d(\Delta H) = \left(\frac{\partial(\Delta H)}{\partial n_1}\right)_{T, P, n_2} dn_1 + \left(\frac{\partial(\Delta H)}{\partial n_2}\right)_{T, P, n_1} dn_2$$

or $$d(\Delta H) = \left(\Delta\overline{H}_1\right) dn_1 + \left(\Delta\overline{H}_2\right) dn_2$$

where ΔH_1 and ΔH_2 are termed as partial molar values of heat of solvent and solvent and solute, respectively. It can be seen that ΔH_2 is equal to the slope d $(\Delta H/dn_2)$ of a varve in which the property ΔH is plotted as a function of n_2 for a series of solutions in which n_1 is kept constant. When it is compared with the definition given at the beginning it shows that the differential heat of solution is really the partial molar heat of solution of solute, *i.e.*, ΔH_2. Similarly, the differential heat of dilution is the corresponding partial molar property of the solvent, *i.e.*, $\Delta\overline{H}_1$.

Direct experimental measurement of differential heat of solution is not possible, it can, however, be determined indirectly from the knowledge of integral heat of solution

$$\Delta H_{D.S} = d\ (m\ \Delta H_{I.S})/dm$$

where $m\Delta H_{I.S}$ is the integral heat of solution of the concentration m molal.

As it is obvious from the definitions, the integral and differential heats of solutions approach each other with decreasing the concentration and become idential when the concentration is zero.

Enthalpy of Dilution : If it known than t the enthalpy of solution per mole of a solute gets varied with the concentration of the solution. Therefore, it follows that if a given solution is diluted on the addition of solvent, there occurs noticeable change of enthalpy. This is termed as the enthalpy of dilution.

Enthalpy of dilution may be defined as the change of enthalpy when a solution having 1 mole of solute is diluted from one concentration to another.

Heat or Enthalpy of Neutralization

The enthalpy of neutralization of an acid at a given temperature may be defined as *the enthalpy change accompanying the neutralization of one gram equivalent of a base by an acid in dilute solution at that temperature*. For example, the enthalpy of neutralization one of hydrochloric acid by sodium hydroxide in dilute solution at 298 K is –57.32 kJ.

$$NaOH + HCI \rightarrow NaCI + H_2O$$

As neutralization is the reaction between $H^+(aq)$ form the base, one can write the above equation as :

$$H^+(aq)+CI^-(aq)+Na^+(aq)+OH^-(aq) \rightarrow H_2O(I)+Na^+(aq)+CI^-(aq)$$

or $$H^+(aq) + OH^-(aq) \rightarrow H_2O(I)$$

From this it follows that enthalpy of neutralization should be independent of the nature of acid and the base. This has been found to be true in the case of neutralization of strong acids by strong bases. This is evident form the data given in Table 2.

Table 2 : Enthalpies on Neutralization of Strong Acids by Strong Bases at 298 K

Acid	*Base*	*Enthalpy of Neutralization (kJ per gram equivalent)*
HCI	NaOH	–57.32
HCI	KOH	–57.45
HCI	LiOH	–57.32
HNO_3	NaOH	–57.28

Heat or Enthalpy of Ionisation

For a weak acid or a weak base, the enthalpy of neutralisation is not the same but less than –57.32 kJ. It is because of the fact that these acids or bases are not completely ionised in solution. Some of the heat is consumed in ionising these acids and bases. The heat required to ionise the weak acid or the weak base is known as the enthalpy of ionisation.

It is possible to calculate the enthalpy of ionisation of a weak acid from the enthalpies of neutralisation of strong acid and weak acid by strong bases. For example, in the neutralisation of acetic acid by sodium hydroxide, the enthalpy of neutralisation is – 55.43 kJ mol^{-1}. The enthalpy of ionisation of acetic acid will therefore be equal to +1.89 kJ mol^{-1}. this can seen as follows :

(i) $CH_3COOH(aq) + OH^-(aq) \rightarrow CH_3COO^-(aq) + H_2O(I)$;

$\Delta H^\circ = -\ 55.43$ kJ

(ii) $H^+(aq) + OH^-(aq) \rightarrow H_2O(I)$; $\Delta H = -\ 57.32$ kJ

(iii) $CH_3COOH(aq) \rightarrow CH_3COO^-(aq) + H^+(aq)$;

ΔH° Ionisation = ?

On adding Eqs. (ii) and (iii), we obtain

$$CH_3COOH(aq) + OH^-(aq) \rightarrow CH_3COO^-(aq) + H_2O(I)$$

or $\Delta H^\circ = -57.32 + \Delta H^\circ$ Ionisation

But from Eq. (i), we have

$CH_3COOH(aq) + OH^-(aq) \rightarrow CH_3COO^-(aq) + H_2O(I)$

or $\Delta H^\circ = -55.43$ kJ mol^{-1}

Hence ΔH° Ionisation $= -55.43 + 57.32 = +\ 1.89$ kJ mol.

Enthalpy of Sublimation

The enthalpy of sublimation (ΔH_{sub}) of a substance may be defined as the enthalpy change accompanying the conversion to 1 mole of solid directly into vapour phase at a given temperature below the melting point of the former. Generally, molecular solids like iodine and naphthalene sublime readily. For example, the enthalpy of sublimation of iodine is 62.39 kJ mol^{-1}.

According to Hess's law, the enthalpy of sublimation is equal to sum of enthalpy of fusion (ΔH_f) and enthalpy of vaporization (ΔH_V), *i.e.*,

$$\Delta H_{sub} = \Delta H_f + H_v$$

The above equation is very useful because it can be employed to calculate the enthalpy of sublimation of the enthalpy of fusion and enthalpy of vaporization are known.

Enthalpy of Atomization

This is defined as the enthalpy change required to convert one mole of an element from its normal state at 298 K and one atmosphere pressure into free atoms.

Enthalpies of atomization, are difficult to measure (they are usually obtained from spectroscopic data), but the are useful in relation to bond energies. For example :

$H_2.g = 2H_{atoms}$ $\Delta H = 433$ kJ ...(1)

$O_2.g = 2O_{atoms}$ $\Delta H = 495$ kJ ...(2)

$H_{2g} + 1/2O_{2.g} = H_2O_g$ $\Delta H = -\ 246$ KJ ...(3)

Then (3) – (1) – 1/2 × (2) gives :

$$2H_{atoms} + O_{atoms} = H_2O \quad \Delta H = -\ 926 \text{ kJ}$$

This energy change corresponds to the formation of two oxygen-hydrogen bonds, so that the average bond energy of an oxygen-hydrogen bond in water is 463 kJ.

Heat or Enthalpy of Hydration

The heat or enthalpy of hydration of salt is the change in heat content ΔH when one mole of an anhydrous substance combines with requisite number of moles of water to form the hydrate.

It is obtained from the enthalpies of solutions of hydrous and anhydrous substances. Enthalpy of hydration of copper sulphate can be calculated from the following thermochemical equations :

(i) $CuSO_4.5H_2O + aq \rightarrow CuSO_4\ (aq)$; $\Delta H^\circ = 11.72$ kJ

(ii) $CuSO_4(s) + 5H_2O \rightarrow CuSO_45H_2O\ (s)$; $\Delta H^\circ = x$ kJ

(iii) $CuSO_4(s) + 5H_2O\ (l) + aq \rightarrow CuSO_4\ (aq)$; $\Delta H^\circ = -66.5$ kJ

From the above equations, we get

$$CuSO_4\ (s) + 5H_2O\ (l) \rightarrow CuSO_4\ 5H_2O(s);\ \Delta H^\circ = -78.22 \text{ kJ}.$$

Heat or Enthalpy of Transition

It may be defined as the enthalpy change when one mole of one allotropic form changes to another. It includes transition from solid to liquid (fusion), liquid to vapour (vaporisation), solid to vapour (sublimation) and change from one crystalline form to another crystalline form (polymorphic transition). For example, in the transition of sulphur (rhombic) to sulphur (monoclinic), The enthalpy change is 13.14 kJ.

$$S_{rohmbic} \rightarrow S_{monoclinic}; \quad \Delta H^\circ = 13.14 \text{ kJ}$$

Another example is

$$C_{graphite} \rightarrow C_{diamond}; \quad \Delta H^\circ = 1.90 \text{ kJ}$$

The enthalpy of transition in the above case can be obtained from the enthalpies of combustion of $C_{graphite}$ and $C_{diamond}$.

(i) $C_{graphite} + O_2\,(g) \rightarrow CO_2\,(g); \Delta H^\circ = -393.51$ kJ

(ii) $C_{diamond} + O_2\,(g) \rightarrow CO_2\,(g); \Delta H^\circ = -395.41$ kJ

On subtracting (ii) from (i), we get

$$C_{graphite} \rightarrow C_{diamond}; \quad \Delta H^\circ = 1.90 \text{ kJ}$$

Heat or Enthalpy of Precipitation

It may be defined at the enthalpy change when one mole of a precipitate is formed upon mixing dilute solutions of relevant electrolytes. Some examples are :

$BaCl_2\,(aq) + Na_2SO\,(aq) \rightarrow BaSO_4(s) + 2NaCl\,(aq);$

$\Delta H^\circ = 19.5$ kJ

As the reactants are strong electrolytes, the reaction is essentially ionic in nature and may be written as follows :

$Ba^{2+}\,(aq)\ 2Cl^-(aq) + 2Na^+(aq) + SO_4^{2-}(aq)$

$\rightarrow BaSO_4(s) + 2Na^+\,(aq) + 2Cl^-(aq), \Delta H = -19.50$ kJ

As $Na^+(aq)$ and $Cl^-(aq)$ ions are common on both sides, the reaction, in fact, involves precipitation of $BaSO_4(s)$, *i.e.*,

$Ba^{2+}\,(aq) + SO_4^{2-}(aq) \rightarrow BaSO_4(s); \quad \Delta H = -19.50$ kJ

Another example is as follows :

$AgNO_3\,(aq)\ NaBr\,(aq) \rightarrow AgBr(s) + 2NaNO_3\,(aq);$

$\Delta H^\circ = -17.78$ kJ

Enthalpy of Fusion

The enthalpy of fusion of a substance may be defined as the enthalpy change when 1 mole of the substance is converted from the solid to the liquid phase at its melting point. When 1 mole of ice melts at 273 K, the enthalpy change is approximately 6.6 kJ. This result can be put in the equation form as :

$$H_2O(s) \rightarrow H_2O(I), \qquad \Delta H = 6.0 \text{ kJ at } 273 \text{ K}$$

The enthalpies of fusion of molecular solids (O_2, CCI_4, H_2S, etc.,) are low where as the enthalpies of fusion of ionic solids (NaCI, $MgCI_2$) are large. Thus one may get some idea from the value of fusion, about the nature of the solid and the magnitude of forces acting between the particles constituting the solid.

Enthalpy of Vaporization

The enthalpy of vaporization $\Delta(H_V)$ of a liquid may be defined as the enthalpy change accompanying the conversion of 1 mole of the substances form the liquid to the vapour phase at its boiling point.

In order to cover liquid into its vapours the substance will absorb energy from the surroundings and thus ΔH will have + ve sign. For example, if water is heated at its boiling point, 373 K, it is converted into vapours by absorbing 40.6 kJ mol^{-1} energy from the surroundings. This means ΔH_V of water is 40.6 kJ mol^{-1}.

$$H_2O(l) \rightarrow H_2O(g); \qquad \Delta H_V = +\,40.6 \text{ kJ mol}^{-1}$$

Similarly, ΔH_V for hydrogen and oxygen are 0.904 and 6.7 kJ Mol–1 respectively. If we compare these values with water we realise that the interaction between water molecules is much stronger than that between hydrogen molecule sin the liquid state but less stronger than that between oxygen molecules in the liquid state.

This is quite true because in water there is extensive hydrogen bonding whereas in oxygen and hydrogen there are only weak van der waals forces,

The enthalpy of fusion is very small as compared to the enthalpy of vaporization. The reason for this is that attractive forces in solid and liquid state are much similar (intermolecular distances in solids and liquids are roughly the same) whereas in the gaseous state the attractive forces are much weaker.

INTERCONVERSION OF MATTER AND ENERGY

The interconvertibility of matter and energy is described by the Einstein equation $\Delta E = \Delta mc^2$. Here Δ E is the change of energy and Δm is the change in mass and c is the speed of light. In the language of thermochemistry the above relation means that matter and energy are not conserved separately rather the principle of conservation of energy is broadened to include matter as another form of energy.

STANDARD ENTHALPY CHANGES

As the enthalpy of a reaction depends on temperature, a standard state is chosen to compare the enthalpies or reactions of different substances. A convenient standard state for a substance is the most stable state of aggregation under 1 atmosphere and at the specified temperature, generally at 25°C or 298 K.

For pure solids, liquids and ideal gases, the standard state corresponds to the state of the substances at 1 atm. Pressure and specified temperature.

For the dissolved substances, the standard state of the solute is that concentration which gives unit activity.

The standard enthalpy of a reaction at a temperature T and at a pressure of 1 atm is denoted by $\Delta H°(T)$ and is the value of $\Sigma \Delta H_{products} - \Sigma \Delta H_{reactants}$, at that temperature when all the substances (or elements) in the reaction are in their standard states.

The enthalpy of every element in its standard state is arbitrarily given a zero value. For example, the enthalpy of formation of a mole of $CO_2(g)$ from its elements in the standard state is – 393.5 kg mol^{-1}, *i.e.*,

$$C(graphite) + O_2(g) \rightarrow CO_2(g);\ \Delta H° = -393.5 \text{ kJ}.$$

The value of $\Delta H°_{298K}$ is given as follows:

$$\Delta H°_{298K} = H°CO_2(g) - [H°O_2(g) + H°_{Cgrphite}]$$

In the above equation, H° refers to the standard enthalpy per mol.

As the enthalpies are in their standard states, their heat contents by convention are zero, *i.e.*,

$$H°_{Cgrphite} = H°_{O2(g)} = 0.$$

Hence under these conditions,

$$\Delta H°_{298K} = H°_{CO2(g)} = -393.5 \text{ kJ}.$$

Thus, the standard enthalpy of any compound is the enthalpy of the reaction by which it is formed from its elements in the standard state.

VARIATION OF HEATS OF REACTION WITH TEMPERATURE

Heats of reaction at a given temperature (25°C) and at a given pressure (1 atm) can be calculated using the standard heats of formation given in the International tables. Sometimes it is required to find the heat of reaction at other temperatures and pressures. The variation of heat of reaction with temperature at constant pressure is given by the **Kirchooff's Law**, which can be derived in the following way :

Let us consider the general reaction

$aA + bB \rightarrow mM + nN$

Now the heat of reaction is given by

$\Delta H = (mH_M + nH_N) - (aH_A + bH_B)$

where H_A and H_B are the molar enthalpies of reactants and H_M and H_N and that of the products respectively. Differentiating equation (2) with respect to T at constant P, we get the variation of enthalpy with temperature,

$$\left(\frac{\partial \Delta H}{\partial T}\right)_P = \left[m\left(\frac{\partial H_M}{\partial T}\right)_P + n\left(\frac{\partial H_N}{\partial T}\right)_P\right] - \left[a\left(\frac{\partial H_A}{\partial T}\right)_P + b\left(\frac{\partial H_B}{\partial T}\right)_P\right]$$

$= [MC_{P,m}(M) + n\,C_P, m\,(N)] - [a\,C_{P,m}(A) + b\,C_{P,m}(B)]$

$= \Sigma\, C_P \text{ (products)} - \Sigma\, C_P \text{ (reactants)}$

$= \Delta\, C_P$

or $d\,(\Delta H) = \Delta C_P.\ dT$

Here $C_{P,m} = (\partial H/\partial T)_P$ = molar heat capacity at constant pressure and the quantity $\Delta\, C_P$ is the change in heat capacity in the chemical reaction

$\Delta\, C_P = [mC_{P.m}\,(M) + nC_{P.m}\,(N)] - [aC_{P.m}\,(A) + bC_{P.m}\,(B)]$

The relationship given by equation (1) is known as Kirchhoff's Law. Using this law we can calculate the enthalpy changes at any temperature.

If we have to calculate the heat of reaction at constant volume (D E) then we can derive the Kirchhoff's equation as,

$$\left(\frac{\partial(\Delta E)}{\partial T}\right)_V = DC_V$$

or $\quad d\,(\Delta\, E) = \Delta\, C_V dT$ ...(1)

For small range of temperature, equations (1) and (2) can be integrated taking heat capacities as independent of temperature.

$$\int_1^2 d(\Delta H) = \Delta C_P \int_1^2 dT$$

$\Delta H\,(T_2) - \Delta H\,(T_1) = \Delta C_P\,(T_2 - T_1)$ ...(2)

and $\Delta E\,(T_2) - \Delta\, E\,(T_1) = \Delta\, C_V\,(T_2 - T_1)$ ...(3)

In a reaction if ΔC_P is zero then ΔH is independent of temperature.

VARIATION OF HEAT OF REACTION WITH PRESSURE

Standard heats of formation give values of heat of reaction at a pressure of 1 atm. Sometimes it is necessary to find heats of reaction at other pressures. For this we will make use of the thermodynamic equation of state.

$$\left(\frac{\partial H}{\partial P}\right)_T = V - T\left(\frac{\partial V}{\partial T}\right)_P$$

so that the pressure dependence of the heat of reaction is given by

$$\left(\frac{\partial \Delta H}{\partial P}\right)_P = DV - T\left(\frac{\partial \Delta V}{\partial T}\right)_P \qquad ...(1)$$

where DV is the volume change of the reaction at pressure P.

$\Delta V = \Sigma\, V$ (products) $- \Sigma V$ (reaction)

Differentiation w.r. to T at constant P yields,

$$\left(\frac{\partial \Delta V}{\partial T}\right)_P = \sum\left[\frac{\partial V(\text{products})}{\partial T}\right]_P - \sum\left[\frac{\partial V(\text{reactants})}{\partial T}\right]_P$$

The coefficient of thermal expansion is defined as $\alpha = V^{-1}\,(\partial V/\partial T)_P$

$$\therefore \left(\frac{\partial \Delta V}{\partial T}\right)_P = \Sigma\,\alpha V\,(\text{products}) - \Sigma\,\alpha V\,(\text{reactants})$$

$$= \Delta(\alpha V) \qquad ...(2)$$

From equation (1) and (2), we get

$$\left(\frac{\partial \Delta H}{\partial P}\right)_T = \Delta V - T\Delta\ (\alpha V) \qquad ...(3)$$

If the equations of state of the reactants and products are known ΔV and $\Delta\ (\alpha V)$ will be known functions of the pressures, and equation (3) can be integrated to give

$$\Delta H(P_2) = \Delta H(P_1) + \int_{P_1}^{P_2} [\Delta V - T\Delta(V\alpha)]dP \qquad ...(4)$$

However, if the reactants and products involved in the chemical reaction are ideal gases then the heat of reaction is independent of pressure.

For ideal gases $V = nRT/P$

$$\therefore\ \Delta V = V_{prod} - V_{react} = n_{prod}\frac{RT}{P} - n_{react}\frac{RT}{P}$$

$$\text{or } \Delta V = \Delta v\ \frac{RT}{P}$$

$$\text{and } \left(\frac{\partial \Delta V}{\partial T}\right)_P = \Delta v \frac{R}{P} \qquad ...(5)$$

Putting the expressions from equation (5) into equation (2) we get

$$\left(\frac{\partial \Delta H}{\partial P}\right)_T = \Delta v\ \frac{RT}{P} - T\Delta v\ \frac{R}{P} = 0.$$

THE MAXIMUM FLAME TEMPERATURE

When a fuel burns definite amount of products are formed and a definite quantity of heat is liberated. This quantity of heat may be utilized to raise the temperature of the products of combustion sufficiently to produce a flame. This temperature is called the **maximum flame temperature**, or calorific intensity. If the experiment is performed at constant volume the quantity of heat liberated will be denoted by Δ E and the maximum temperature is called the **explosion temperature**. In case the experiment is performed at constant pressure then heat liberated is denoted by ΔH and the maximum temperature attained is called the flame temperature. Usually this temperature is called the adiabatic maximum flame temperature because it is assumed that no loss of heat in the process takes place.

Calculation of Maximum Temperature

If it is assumed that the reaction is very fast and the total heat evolved is utilized to raise the temperature of the system then the heat capacities of the species present at the completion of reaction and the heat evolved can be used to calculate the maximum temperature with the help of equation

$$\Delta H \text{ (heating)} = \int_{T_1}^{T_2} \Sigma_n C_{P.m} dT = -(\text{reaction}) \qquad ...(1)$$

where we have taken ΔH (heating) = – ΔH(reaction), because the heat evolved taken by the products. T_1 is the initial temperature and T_2 is the maximum flame temperature $\Sigma\ (nC_{P,m})$ is the sum of the heat capacities of all the species present in the system after the reaction is complete. For constant volume process,

$$-\Delta E = \int_{T_1}^{T_2} \Sigma_n C_{V.m} dT \qquad ...(2)$$

The validity of equation (1) os further justified as follows.

Let the reaction be performed at constant temperature and pressure (or constant T.V). Let us assume that we have a reservoir in which the heat evolved in the reaction can be stored.

When the reaction is complete, the stored heat is used to raise the temperature of the products still at constant pressure (or constant volume). When all the heat evolved (stored in the reservoir) has been transferred to the system, the final maximum flame (or explosion) temperature is reached.

This temperature can be measured using standard methods for the measurement of thermal changes for example, thermometer, thermistors, thermopiles etc.

However, this procedure is quite hypothetical, because we do not have any such reservoir in which the heat can be stored and which can be kept completely isolated from the system and the surroudings.

The direct way to know the maximum temperature is to perform the experiment in the laboratory under adiabatic conditions and measure the temperature using standard techniques of temperature measurements.

SOLVED EXAMPLES

Example 1:

For a reaction CO (g) + H_2O (g) $\rightarrow$ CO_2(g) + H_2 (g), ΔH^o (298 K) = – 42.0kJ. The heat capacities of various species are given by

$C_{P, m}$ (CO)/J K^{-1} mol^{-1} = 26.8 + 7.0 × 10^{-3} T

$C_{P, m}$ (H_2O, g)/JK^{-1} mol^{-1} = 30.4 + 9.6 × 10^{-3} T

$C_{P,m}$ (CO_2)/ JK^{-1} mol^{-1} 26.0 – 43.5 × 10^{-3}T

$C_{P, m}$ (H_2)/$JK^{-1}mol^{-1}$ = 29.0 – 0.80 × 10^{-3} T

Calculate (a) ΔH^o (1298K) (b) Δ E^o (1298 K)

Solution:

ΔH^o (1298 K) is given by

$$\Delta H^o \text{ (1298K)} = \Delta H^o \text{ (298K)} + \int_{298}^{1298} \Delta C_P dT$$

Now, ΔC_P = [$C_{P,m}$ (CO_2) + $C_{P,m}$ (H_2)] – [$C_{P,m}$ (CO) + $C_{P,m}$ (H_2O)]

= (26.0 + 29.0) – (26.8 + 30.4) + (43.5 – 0.80)

– (7.0 + 9.6)] × 10^{-3}T

= – 2.2 + 26.1 × 10^{-3} T

$$\therefore \int_{298}^{1298} \Delta C_P dT = 2.2 \int_{298}^{1298} dT + 26.1 \times 10^{-3} \int_{298}^{1298} TdT$$

$$= -2.2 \left[T\right]_{298}^{1298} + 26.1 \times 10\text{–}3 \times \frac{1}{2}\left[T\right]_{298}^{1298}$$

$$= -2.2\ [1298 - 298] + 26.1 \times 10\text{–}3 \times \frac{1}{2}[(1298)^2 - (298)^2]$$

= 18627 J = 18.627 kJ

$\therefore$ Δ H^o (1298K) = – 42.0 kJ + 18.627 kJ = 23.373kJ

Δ 67 (1298 K) = ΔH^o (1298K) – Δv_g RT = ΔH^o (1298K)

= – **23.373 kJ**

Since Δv_g = (1 + 1) – (1 + 1) = 0 for the given reaction.

Example 2:

Using the bond energy data, compute ΔH^o_f for ethyl alcohol (C_2H_5OH). DH^o (C – H) = 413 kJ, D H^o (C – C) = 348 kJ, DH^o (C – O) = 351 kJ, DH^o (O – H) =463kJ.

Solution:

Ethyl alcohol can be represented by the structure

```
      H    H
      |    |
H— C —   C — O — H
      |    |
      H    H
```

Total bond enthalpy = 5 × DH° (C – H) + DH° (C–C)+DH°(C–O) + DH°(O –H)

= 5 × (413 kJ) + 348 kJ + 351 kJ + 363 kJ = 3227 kJ

when 3227 kJ is supplied then 2C (g), 6H (g) and O (g) will be the product given by C_2H_5OH.

Total heat of atomization = 6H (g) + 2C (g) + O (g)

= 6 × 21.797 + 2 × 716.68 + 249.2 = 2990.38 kJ

$\Delta H^\circ f$ (C_2H_5OH) = Enthalpy of atomization – Bond enthalpy

= (2990.38 kJ) – (3227.0 kJ) = – **236.62 kJ**

Heats of Reaction from Bond Energy Data

Bond energy data are used to calculate the heats of gaseous reactions involving substances having only covalent bonds. In using these data, a plus sign in affixed to the bond enthalpies when a bond is broken, since heat is absorbed under such conditions and a minus sign when a bond is formed and heat is evolved.

Example 3:

Using bond energy data calculate the enthalpy change for the reaction:

C_2H_4 (g) + H_2 (g) → C_2H_6 (g)

Solution:

The given reaction can be written as

```
H    H                 H   H
|    |                 |   |
C  =  C + H – H → H – C – C – H
|    |                 |   |
H    H                 H   H
```

In this reaction the four C – H bonds in C_2H_4 are unaffected and a double bond is broken ; in H_2 one H – H bond is broken. In turn, two new C – H bonds and one C – C bond are formed in ethane.

ΔH^o_f = Bond enthalpy in – Bond enthalpy out

= DH° (C = C) + DH° (H – H) 2 × D H° (C – H) – DH° (C – C)

= (610 kJ mol^{-1}) + (436 kJ mol^{-1}) –2 × (413 kJ mol^{-1})

– (348 kJ mol^{-1})

= **–128** kJ mol^{-1}.

Example 4:

Calculate the standard enthalpy of formation of dimethyl ether by using the bond energy data from the appendix and the enthalpy of C (graphite) → C (gas) as 716.7 kJ.

Solution:

The formation of dimethylether may be represented by the equation

(A) 2C (graphite) + $3H_2$(g) + 1/2 O_2 (g) → $H_3C–O–CH_3$ ΔH_f = ?

From bond energy data we get the enthalpy for the following reaction:

(i) H_3C – O – CH_3 (g) → 2C (g) + 6H (g) + O(g)

ΔH = 6DH° (C – H) + 2D H° (C – O)

∴ ΔH = 6 × (413 kJ mol^{-1}) + 2 × (350 kJ mol^{-1}) = 3178 kJ

(ii) $3H_2$ (g) → 2H (g) ΔH = 1307.64 kJ

(iii) 1/2 O_2 (g) → O (g) ΔH = 249.17 kJ

(iv) 2C (graphite) → 2C (g) ΔH = 1433.4 kJ

Reversing eqn.. (i) we get

(v) 2C (g) + 6H (g) + O(g) → $H_3C–O–CH_3$ (g) ΔH

= – 3178 kJ

Adding equations (ii), (iii), (iv) and (v), and cancelling the common terms we get

(iv) 2C (graphite) + $3H_2$ (g) + 1/2 O_2 (g) → H_3 C – O – CH_3 (g)

ΔH (vi) = ΔH (ii) + ΔH (iii) + ΔH (iv) + ΔH (v)

= – 189.79 kJ = DH (A).

Example 5:

Using the data :

$\Delta H^o_f\ (C_2H_6) = -\ 84.5$ kJ

$\Delta H^o_f\ (C_2H_4) = 52.6$ kJ

$\Delta H^o_f\ (C_2H_2) = 226.9$ kJ

$\Delta H^o_f\ (CO_3CHO) = -\ 166.3$ kJ

$\Delta H^o_f\ (H_2O,g) = -\ 241.8$ kJ

$\Delta H^o_f\ (CH_3OH,g) = -\ 201.3$ kJ

$\frac{1}{2}\ DH^o(O_2) = 249.17$ kJ

$\frac{1}{2} DH^o\ (H - H) = 217.97$ kJ

$\Delta H^o C$ (graphite) $\rightarrow$ C(g) = 716.68 kJ

calculate the following bond energies:

(i) DH^o (C –C), (ii) DH^o(C = C),

(iii) DH^o (C ≡ C), (iv) DH^o (O –H)

(v) DH^o (C = O) (vi) DH^o (C –O)

Solution:

C–C bond energy. The energy of a C–C bond can be calculated by applying the method of *previous* section to ethane. Our aim is

$C_2H_6(g) \rightarrow 2C(g) + 6H(g)\ \Delta H^o = ?$

Given data

(i) C_2H_6 (g) $\rightarrow$ 2C (graphite) + $3H_2$ (g) $\Delta H^o = 84.5$ kJ

(ii) 2C (graphite) $\rightarrow$ 2C (g) $\Delta H^o = 1433.4$ kJ

(iii) $3H_2$ (g) $\rightarrow$ 6H (g) $\Delta H^o = 1307.8$ kJ

Add--------------------------

C_2H_6 (g) $\rightarrow$ 2C(g) + 6 H(g) $\Delta H^o = 2825.7$ kJ

Thus 2825.7 kJ is required to break 6C – H bonds and one C–C bond in ethane. The C–C bond energy is equal to 2825.7 kJ less 6 × DH^o (C –H)

$DH^o\ (C - C) = \Delta H^o - 6 \times DH^o\ (C{-}H)$

$= 2825.7 - 6 \times 413 = 347.7$ kJ

C = C bond energy. The of C = C bond can be calculated using the standard enthalpy of formation of ethylene and C– H bond energy data. Here our object would be to study :

$C_2H_4\ (g) \rightarrow 2C\ (g) + 4H\ (g)\ \Delta H^o = ?$

Given data

$C_2H_4\ (g) \rightarrow 2C\ (graphite) + 2H_2\ (g)\ \Delta H^o = -52.6$ kJ

$2C\ (graphite) \rightarrow 2C\ (g) \qquad \Delta H^o = 1433.4$ kJ

$2H_2(g) \rightarrow 4H(g) \qquad \Delta H^o = 871.9$ kJ

Add ------- ---------

$C_2H_4(g) \rightarrow 2C(g) \quad + 4H(g) \qquad \Delta H^o = 2252.7$ kJ

Less 4C – H bonds $\Delta H^o = 1652.0$ kJ

-------- --------

C = C bond DH° (C = C) = **600.7 kJ**

The literature value is slightly greater than this value.

C ° C *bond energy*. Since we already know the C– H bond energy, we can obtain the energy of a C ° C bond by using the ΔH^o_f acetylene and the previous C – H bond energy

$C_2H_2\ (g) \rightarrow 2C\ (graphite) + H_2\ (g)\ \Delta H^o = -226.9$ kJ

$2C\ (graphite) \rightarrow 2C\ (g) \qquad \Delta H^o = 1433.4$ kJ

$H_2\ (g) \rightarrow 2H\ (g)\ \Delta H^o = 435.94$ kJ

Add ---------------- -----------

$C_2H_2(g) \rightarrow 2C\ (g) + 2H(g) \quad \Delta H^o = 1642.44$ kJ

Less 2C – H bonds $2 \times DH^o\ (C\text{–}H) = 826.0$ kJ

-------- --------

C ≡ C bond ΔH° (C ≡ C) = **816.44 kJ**

C = 0 bond energy is calculated using the standard heat of formation of acetaldehyde and the C– C and C – H bond energies. This gives C = O bond energy as 740 kJ mol^{-1}.

The O – H bond energy

$H_2O\ (g) \rightarrow \frac{1}{2}\ O_2\ (g) + H_2(g)\ \Delta H^o = 241.80\ kJ$

$H_2\ (g) \rightarrow 2H\ (g)\ \Delta H^o = 435.94\ kJ$

$\frac{1}{2}\ O_2\ (g) \rightarrow O\ (g)\ \Delta H^o = 249.17 kJ$

Add ------- ----------

$H_2O\ (g) \rightarrow 2H\ (g) + O\ (g) \quad \Delta H^o = \mathbf{926.91\ kJ}$

The quantity, 926.91 kJ is the enthalpy change when the two O – H bonds in H_2O are broken. Therefore, the O – H bond energy is 926.91 kJ/2 = **463.4 kJ**.

Example 6:

The formation of SO_3 from SO_2 and oxygen,

$SO_2\ (g) + 1/2\ \ O_2\ (g) \rightarrow SO_3\ (g)$

is exothermic by 97030 *at* 298 *K and 1 atm measured in a bomb calorimeter. What is the value of* Δ H?

Solution:

The above equation can be written as.

$SO_2\ (g) + 1/2\ O_2\ (g) \rightarrow SO_3\ (g)\ \Delta E\ (298\ K) = -\ 97030\ J.$

We know

$\Delta H = \Delta E + \Delta vRT$

Here $\Delta v = 1 - \left(1 + \frac{1}{2}\right) = -\ \frac{1}{2}$

$\therefore\ \Delta H = -\ 97030 - 1/2\ (8.314)\ (298) = -\ \mathbf{98270\ J}.$

Example 7:

Find ΔE^o *for the following reactions at 1 atm and* 298 K

$C\ (s) + O_2(g) \rightarrow CO_2(g) \quad \Delta H^o\ (298\ K) = -\ 393.5\ kJ\ mol^{-1}$

$C\ (s) + 1/2\ O_2\ (g) \rightarrow CO\ (g) \quad \Delta H^o\ (298\ K) = -\ 110.5\ kJ\ mol^{-1}$

$CH_4(g) + 2O_2(g) \rightarrow CO_2(g) + 2H_2O(l)\ \Delta H^o\ (298\ K)$

$= -\ 890.35\ kJ\ mol^{-1}$

$AgNO_3$ (s) + HCl (g) → AgCl (s) + HNO_3 (l)

ΔH^o (298 K) = – 85.3 kJ mol^{-1}.

Solution:

Since R = 8.314 $JK^{-1}mol^{-1}$

and T = 298 K in the present case

RT = (8.314 JK^{-1} mol^{-1}) × (298K)

= 2477.57 J mol^{-1} = 2.477 kJ mo^{-1}.

The moles of only gaseous species are to be considered

$\Delta H^o = \Delta E^o + \Delta v\ RT$ gives $\Delta E^o = \Delta H^o - \Delta v\ RT$

(i) C(s) + O_2(g) → CO_2 (g) ΔH^o(298K) = – 393.5 kJ mol^{-1}

$\Delta v = 1 - 1 = 0$

∴ ΔH^o (298 K) = ΔE^o (298 K)

= 393.5 KJ mol^{-1}

(ii) C (s) + 1/2 O_2 (g) → CO (g) ΔH^o (298 K)=– 110.5kJ mol^{-1}

$$\Delta v = 1 - \frac{1}{2} = \frac{1}{2}$$

$$\therefore \Delta E^o (298\ K) = -110.5\ kJ\ mol^{-1} - \left(\frac{1}{2}\right) \times (2.477\ kJ\ mol^{-1})$$

= **–111.738 kJ mol^{-1}**

(iii) CH_4 (g) $2O_2$(g) → CO_2 (g) + $2H_2O$(l) ΔH^o (298 K)

= – 890.35 kJ mol^{-1}

$\Delta V = 1 - (1 + 2) = -2$

ΔE^o (298 K) = (– 890.35 kJ mol^{-1}) + 2 × (2. 4 77kj mol^{-1})

= – **885.39 kJ mol^{-1}**

(iv) $AgNO_3$ (s) + HCl (g) → AgCl (s) + HNO_3(l)

ΔH^o (298 K) = – 85.3 kJ

$\Delta v = 0 - 1 = -1$

ΔE^o (298 K) = – (85.3 kJ mol^{-1}) – (–1) × (2.477 kJ mol^{-1})

= – **82.823 kJ mol^{-1}**.

Example 8:

The heat of neutralization of a strong acid with a strong base is – 57.36 kJ mol^{-1}. If the heat of neutralization of a weak acid HA with a strong base be –42.02 kJ mol^{-1} compute the heat of ionization of the weak acid.

Solution:

The neutralization of strong acid with strong base may essentially be represented as

$H^+ (aq) + OH^- (aq) \rightarrow H_2O(l)$ $\Delta H = -\ 57.36$ kJ mol^{-1}

The neutralization of a weak acid with a strong base may be considered to be the resultant of two successive steps.

$HA\ (aq) \rightarrow H+\ (aq) + A^-\ (aq)$ $\Delta H = \Delta H$ (ionization)

$$\frac{H + (aq) + OH^-(aq) \rightarrow H_2O(l)}{HA(aq) + OH^-(aq) \rightarrow H_2O(l) + A^-(aq)} \qquad \Delta H = -\ 57.36 \text{ kJ}$$

$\Delta H = 42.02$ kJ

So that the enthalpy change is

ΔH (neutralization) = ΔH (ionization) – 57.36 kJ mol^{-1}

and ΔH (ionization) = – 42.02 + 57.36

= **15.34** kJ mol^{-1}.

Example 9:

The standard heat of formation of HCl in water at 25°C is –168.0kJ. Calculate the standard heat of formation of chloride ions.

Solution:

(i) $\frac{1}{2}H_2\ (g) + Cl_2(g) + aq \rightarrow H^+\ (aq) + Cl^-\ (aq)$

ΔH° (298K) = –168.0 kJ

(ii) $\frac{1}{2}H_2\ (g) \rightarrow H^+\ (aq) + e^-$ ΔH° (298 K) = 0

Subtracting (ii) from (i) we get

$\frac{1}{2}Cl_2\ (g) + aq + e^- \rightarrow Cl^-\ (aq)$ ΔH° (298 K) = **–168.0kJ**.

Example 10:

Calculate the standard heat of formation of sodium ions in aqueous solution using the standard heat of formation of NaOH in aqueous medium (ΔH^o_f = – 470.7kJ); $\Delta H^o_f (OH^-)$ = – 228.8 kJ.

Solution:

Given data :

(i) $Na(s) + \frac{1}{2} H_2 (g) + \frac{1}{2} O_2 (g) + aq \rightarrow Na^+ (aq) + OH^- (aq)$

$\Delta H^o = -470.7$ kJ

(ii) $\frac{1}{2} H_2(g) + \frac{1}{2} O_2(g) + e^- \rightarrow OH^-(aq)$ $\Delta H^o = -228.8$kJ

Subtracting equation (ii) from (i) we get

$Na(s) + (aq) + e^- \rightarrow Na^+ (aq) + e^-$ $\Delta H^o =$ **–241.9kJ**

Example 11:

When hydrogen is burnt is oxygen, 240.5 kJ heat is liberated per mole of the explosion product measured at 27°C, Calculate the maximum temperature of explosion, $C_{V, m} (H_2O, g) = 24.15\ JK^{-1}\ mol^{-1}$.

Solution:

The reaction can be represented as

$H_2 (g) + \frac{1}{2} O_2 (g) \rightarrow H_2O (g) \Delta E (300K) = -240.5\ kJ\ mol^{-1}$

This amount of heat will be utilized to raise the temperature of the product $H_2O(g)$. Using equation (15) we get

$$240500\ J\ mol^{-1} = \int_{300}^{T_2} (24.15 JK^{-1} mol^{-1}) dT$$

or $240500 = 24.15 (T_2 - 300)$

T_2 = **10258 K**

This calculation is quite approximate, therefore, this temperature would be far from the actual value.

Example 12:

The heat of combustion of methane is – 881.25 kJ mol^{-1} *at 298 ‹ Calculate the maximum theoretical flame temperature when one mole of* CH_4 *is completely burnt in calculated amount of air just sufficient f combustion at a constant pressure. The necessary data of heat capacities and their temperature dependence are also given.*

Data given

$CH_4(g) + 2O_2(g) \rightarrow CO_2(g) + 2H_2O(l)$

ΔH^o (298 K) = – 881.25 kJ mol^{-1}

H_2O (l) $\rightarrow$ $H_2O(g)$ ΔH^o(298 K) = 43.60 kJ mol^{-1}

$C_{P,m}$ $(CO_2)/JK^{-1}mol^{-1} = 26.00 + 43.5 \times 10^{-3}T - 148.3 \times 10^{-7} T^2$

$C_{P,m}$ $(H_2O,g)/JK^{-1}\ mol^{-1} = 30.36 + 9.61 \times 10^{-3}\ T + 11.8 \times 10^{-7}T^2$

$C_{P,m}$ $(N_2)/JK^{-1}\ mol^{-1} = 27.30 + 5.23 \times 10^{-3}\ T - 0.04 \times 10^{-7}T^2$

Solution:

Let T_2 be the maximum theoretical flame temperature to which the products are raised, which include one mole $CO_2(g)$, 2 moles $H_2O(g)$, and 8 moles $N_2(g)$ (present in air associated with 2 moles O_2 used for combustion). Since water is ultimately in the gaseous form at high temperature, that value of ΔH^o os required for the reaction on which H_2O (g) is a product. Thus

(i) CH_4 (g) + $2O_2(g) \rightarrow CO_2$ (g) + $2H_2O(l)$ ΔH^o = –881.25 kJ

(ii) $2H_2O$ (l) $\rightarrow$ $2H_2O(g)$ ΔH^o = + 87.20 kJ

so that, addition of (i) and (ii) gives

CH_4 (g) + $2O_2(g) \rightarrow CO_2$ (g) + $2H_2O$ (g) Δ H^o = – 794.05 kJ

Now, this heat is used to raise the temperature of products after combustion. Equation (14) gives,

$$794050\ J\ mol^{-1} = \int_{298}^{T_2} \Sigma n(C_P)dT \qquad ...(1)$$

Now, $\Sigma nC_{P,m} = C_{P,m}(CO_2) + 2C_{P,m}(H_2O) + 8CP,m\ (N_2)$

$\Sigma\ (nC_{P,m}) = 305.12 + 104.56 \times 10^{-3}\ T - 125 \times 10_{-7}\ T_2$

Therefore, putting the values in equation (1) and carrying out integration we have

$$794050 = \left[30.512T + \frac{1}{2} \times 104.56 \times 10^{-3}T^2 - \frac{1}{3} \times 125 \times 10^{-7}T^3\right]_{298}^{T^2}$$

This equation is solved for T_2. This approximate value of T_2 is found to be 2250 K. The experimental value is about 2150 K. Thus there is a difference of about 100° between theoretical and experimental values. This discrepancy is explained taking into account the various possible reasons.

(i) The process may not be exactly adiabatic and there is exchange of heat between the system and the surroundings.

(ii) In actual practice, excess air is used for combustion process therefore, after combusion some air (O_2 + N_2) is left. Thus the heat evolved will be used to raise the temperature of extra O_2 and N_2 (in air).

(iii) At high temperature attained in the burning of hydrocarbons the products H_2O and CO_2 may get dissociated as

$$H_2O \rightarrow H_2 + \frac{1}{2}O_2$$

$$CO_2 \rightarrow CO + \frac{1}{2}O_2$$

These reactions will involve considerable amount of heat which is not accounted for by the theoretical calculations.

(iv) The combustion may not be complete, hence along with the products some reactants may also be present, and heat evolved will be utilized to raise the temperature of unreacted species.

Exercise. *Calculate the maximum flame temperature when ethane is burnt with twice the amount of air needed for complete combustion to CO_2 and H_2O. The combustion reaction is represented as*

$$C_2H_6\ (g) + \frac{7}{2}O2(g) \rightarrow 2CO_2(g) + 3H_2O(g)$$

$\Delta H°$ (298K) = – 1429.2kJ mol^{-1} **[Ans. 1450 K] [Hint.** Compute

$$\Sigma nC_P = 2C_PCO_2(g) + 3C_PH_2O(g) + 14C_PN_2(g)$$

$$+\left\{\frac{7}{2}C_PO_2(g) + 14C_PN_2(g)\right\}$$

The quantities within braces are for excess air].

Example 13(a):

It is estimated that an average human consumes the equivalents of 10 g glucose per hour. Estimate the power output of the brain in watts. Given that 1 W = 1 JS^{-1};

ΔH^{o}_{f} $(H_2O, l) = -286$ kJ mol^{-1}

ΔH^{o}_{f} $(CO_2\ g) = -394$ kJ mol^{-1}

ΔH^{o}_{f} (*glucose, aq*) $= -1260$ kJ mol^{-1}

Solution:

The heat supplied by 10 g glucose can be calculated if we know its heat of combustion:

$$C_6H_{12}O_6\ (aq) + 6C_2\ (g) \rightarrow 6CO_2\ (g) + 6H_2O\ (l)\ \Delta H = ?$$

$$\Delta H = 6 \times \Delta H^{o}_{f}\ (CO_2) + 6 \times \Delta H^{o}_{f}\ (H_2O) - \Delta H^{o}_{f}\,(C_6H_{12}O_6,\ aq)$$

$$= [(-\ 6 \times 394 - 6 \times 286) + 1260]\ \text{kJ mol}^{-1}$$

$$= -\ 2820\ \text{kJ mol}^{-1}$$

$$\therefore \text{ Heat supplied by 10 g glucose} = -\ 2820\ \text{mol}^{-1} \times \frac{10\text{g}}{180\text{g mol}^{-1}}$$

$$q = -156.66\ \text{kJ per hour}$$

$$= -\ 156666\ \text{J per hour}$$

Power output of the brain = *Heat change per second*

$$= \frac{156666\text{J}}{60 \times 60}\ \text{per second}$$

$$\mathbf{= 43.5\ Js^{-1} = 43.5\ W}$$

Example 13(b):

Calculate the enthalpy change for the process

$$H_2O\ (l, -10°C) \rightarrow H_2O\ (s, -\ 10°C)$$

given that :

$C_{P,m}$ $(H_2O,\ l) = 75.4$ J K^{-1} mol^{-1}

$C_{P,m}$ $(H_2O,s) = 37.2$ J K^{-1} mol^{-1}

H_2O (l, 0°C) $\rightarrow$ H_2O (s, 0° C), $\quad \Delta H = -\ 6008$ J mol^{-1}

Solution:

The given process may be considered to proceed in different stages as

(i) $H_2O\ (l, -10°C) \xrightarrow{\text{Heat}} H_2O\ (l,0°C)$

$$\Delta H\ (i) = \int_{263}^{273} C_{P,m}(H_2O,l)dT = 75.4\ JK^{-1}\ mol^{-1}\ (273K - 263K)$$

$= 754\ J\ mol^{-1}$

(ii) $H_2O\ (l,\ 0°C) \rightarrow H_2O(s,\ 0°C)$

$\Delta H\ (ii) = -\ 6008\ J\ mol^{-1}$

(iii) $H_2O\ (s,\ 0°C) \xrightarrow{Cool} H_2O\ (s,\ -\ 10°C)$

$\Delta\ H\ (iii) = \int_{273}^{263} C_{P,m}(H_2O, s)dT = 37.2\ KJ^{-1}\ mol^{-1}\ (263\ K - 273\ K)$

$= -\ 372\ J\ mol^{-1}$

Addition of equations (i), (ii) and (iii) gives

$H_2O\ (l,\ -\ 10°C) \rightarrow H_2O\ (s, -10°C),\ \Delta H$

$= \Delta\ H\ (i) + \Delta H\ (iii)\ \Delta\ H = -\ 5626\ J\ mol^{-1}$

Example 13(c):

The following reactions might be used to power rockets:

(i) $H_2\ (g) + 1/2\ O_2\ (g) \rightarrow H_2O(g)$ $\quad \Delta H\ (i) = -\ 242\ kJ\ mol^{-1}$

(ii) $CH_3OH\ (l) + 3/2\ O_2\ (g) \rightarrow CO_2\ (g) + 2H_2O\ (g)$

$\Delta H\ (ii) = -\ 640\ kJ\ mol^{-1}$

(iii) $H_2\ (g) + F_2\ (g) \rightarrow 2HF\ (g)$ $\quad \Delta H\ (iii) = -\ 540\ kJ\ mol^{-1}$

Calculate the enthalpy changes for each of these reactions per kg of each second reactant.

Solution:

Moles of reactants can be calculated by using the formula n = mass/ molar mass.

$$n\ (H_2) = \frac{1kg}{0.002\ kg\ mol^{-1}} = 500\ mol$$

$$n\ (O_2) = \frac{1kg}{0.032\ kg\ mol^{-1}} = 31.25\ mol$$

$$n\ (CH_3OH) = \frac{1kg}{0.032\ kg\ mol^{-1}} = 31.25\ mol$$

$$n\ (F_2) = \frac{1kg}{0.038\ kg\ mol^{-1}} = 26.3\ mol$$

The Stoichoiometry of reaction (i) indicates that 31.25 moles O_2 will combine with 62.5 moles H_2 to give 62.5 moles H_2O as

H_2 (g) + 1/2 O_2(g) → H_2O(g), ΔH (i) = – 242 kJ mol^{-1}

62.5 H_2(g) + 31.25 O_2(g) → 62.5 H_2O (g)

ΔH = 62.5 × DH (i) = – **15125 kJ**

The second reaction is written as

20.8 CH_3 OH (l) + 31.25 O_2 (g) → 20.8 CO_2 (g) + 41.6 H_2O(g)

ΔH = 20.8 × ΔH (ii)

= **–13312 kJ**

Third reaction

26.3 H_2 (g) + 26.3 F_2 (g) → 52.6 HF (g)

Δ H = 26.3 × ΔH (iii)

= – **14202 kJ**

Example 14:

An average man weighs about 70 kg and produces about 10^4 kJ of heat per day through metabolic activity. (a) Suppose that a man were an isolated system and that his heat capacity is 4.2 JK^{-1} g^{-1}. If his temperature was 37°C at a given time what would be his temperature 24 hours later?

(b) *Man is actually an open system and the main mechanism of loss of heat is evaporation of water. How much water would need to be evaporated per day to maintain the body temperature at 37° C? ΔH_{vap} (water)* = 2405 J g^{-1} at 37°C.

Solution:

(a) Let T_2 be temperature of the body after 24 hours. Given that C_P = 4.2 JK^{-1} g^{-1},

m = 70 kg = 70 × 10^3 g

ΔH = 10^4 kJ = 10^7 J, T_1 = 310K

$$\Delta H = \int_{T_1}^{T_2} mC_P dT = mC_P(T_2 - T_1)$$

$$\therefore T_2 = \frac{\Delta H + mC_P T_1}{mC_P}$$

$$= \frac{10^7 J + (7 \times 10^4 g) \times (4.2 JK^{-1}g^{-1}) \times (310K)}{(7 \times 10^4 g) \times (4.2 JK^{-1}g^{-1})}$$

= **344 K = 71° C**.

(b) $\because$ 2405 J heat evaporate 1g H_2O

$\therefore$ 10^7 J heat will evaporate $\frac{10^7 J \times 1gH_2O}{2405J}$

= 4158 g H_2O = **4.2 kg H_2O**.

This much water should be evaporated by perspiration to maintain the body temperature.

Example 15:

In an explosion reaction 21000 kJ *are evolved. Calculate the decrease in mass accompanying the reaction.*

Solution:

$\Delta E = \Delta mc^2$ (Einstein equation)

$\Delta E = -21000$ kJ $= -21.0 \times 10^6$ T $= -21.0 \times 10^6$ kg m^2s^{-2}

Speed of light = c = 3.0×10^8 ms^{-1}

$$\therefore \Delta m = \Delta E/c^2 = -\frac{21.0 \times 10^6 kg\ m^2 s^{-2}}{(3.0 \times 10^8 ms^{-1})^2} = -\mathbf{2.3 \times 10^{-10}\ kg}$$

Thus 2.3×10^{-10} kg is the loss of matter which liberates as heat and is transferred to the surroundings.

Exercise. *Energy is emitted by the sun at the rate of 2 ergs per g of the sun per second. The estimated mass of the sun is 2×10^{33} tons. Estimate the decrease in tons and per cent decrease in the mass of the sun per day.*

[**Ans**. 4×10^{17} tons; 2×10^{-14}%. It is interesting to note that if the above calculations were true then after 10^{13} years the sun should vanish.]

Example 16(a):

Compute the heat change in the following reaction

$COCl_2 + 2H_2S \rightarrow 2HCl + H_2O + CS_2$ *given that*

(i) $COCl_2 + H_2S \rightarrow 2HCl + COS$ ΔH° (298) = – 78.705 kJ

(ii) $COS + H_2S \rightarrow H_2O + CS_2$ ΔH° (298) = – 78.705 kJ

Solution:

On adding equation (i) and (ii) and cancelling the identical terms we get the desired result as

$$COCl_2 + H_2S \rightarrow 2HCl + COS \quad \Delta H^\circ_{(i)} = -78.705 \text{ kj}$$

$$\begin{array}{ll} COS + H_2S \rightarrow H_2O + CS_2 & \Delta H^\circ_{(ii)} = +3.420\text{kj} \\ \hline COCl_2 + 2H_2S \rightarrow 2HCl + H_2O + CS_2 & \Delta H^\circ(298) = \mathbf{-75.285kJ} \end{array}$$

Example 16(b):

Calculate the enthalpy change for the following reaction at 25°C.

$$C_2H_6\ (g) + \frac{7}{2}O_2\ (g) \rightarrow 2CO_2(g) + 3H_2O(l)$$

Use the enthalpy of formation of the reactants and products from literature.

Solution:

Enthalpy change for any reaction is given as

$\Delta H^\circ = \Sigma H^\circ_f$ (products) $-\Sigma\Delta H^\circ_f$ (reactants)

For the given reaction

$\Delta H^\circ = 2\Delta H^\circ_f\ (CO_2) + 3\Delta H^\circ_f\ (H_2O) - (\Delta H^\circ_f\ C_2H_6 + 7/2\ \Delta H^\circ_f\ O_2)$

$= -2 \times 393.5 - 3 \times 285.8 - \left(-84.5 + \frac{7}{2} \times 0\right)$

$= -787.0 - 857.4 + 84.5$

$= -$ **1559.9 kJ**.

Example 16(c):

Calculate the heat of formation of benzene from the following data:

$$C_6H_6(l) + \frac{15}{2}\ O_2(g) \rightarrow 6CO_2(g) + 3H_2O(l) \quad \Delta H^\circ = -3303 \text{ kJ}$$

$\Delta H^\circ_f\ (CO_2) = -393.5$ kJ; $\Delta H^\circ_f\ (H_2O) = -285.8$ kJ.

Solution:

For the given reaction

$$\Delta H^\circ = 6 \times \Delta H^\circ_f\ (CO_2) + 3\ \Delta H^\circ_f\ (H_2O) - \Delta H^\circ_f\ (C_6H_6) - \frac{15}{2}\Delta H^\circ_f\ (O_2).$$

$\Delta H^{o}_{f} C_6H_6 = 6 \times \Delta H^{o}_{f} (CO_2) + 3 \times \Delta H^{o}_{f} (H_2O) - \Delta H^{o}$
$= -6 \times 393.5 - 3 \times 285.8 + 3303 =$ **84.6 kJ.**

Example 17:

At 25°C the enthalpy changes for the dissolution of the two forms of glucose and their mutarotation are respectively given by

(i) α-D-*glucose* (s) → α-D-*glucose* (aq) ΔH(i) = 10.72 *kJ*

(ii) β-D-*glucose* (s) → β-D-*glucose* (aq) ΔH(ii) = 4.68 *kJ*

(iii) α-D-*glucose* (aq) → β-D-*glucose* (aq) ΔH(iii) = 1.16 *kJ*

Calculate ΔH *for* α-D-*glucose* (s) → β-D-*glucose* (s) ΔH = ?

Solution:

Reverse equation (ii) and then add this and eqn. (i) and, (iii) cancel the identical terms :

(iv) β-D-*glucose* (aq) → β-D-*glucose* (s) ΔH(iv) = – 4.68 *kJ*

(v) α-D-*glucose* (aq) → β-D-*glucose* (aq) ΔH(iii) = 1.16 *kJ*

$$\text{(i)}\quad \frac{\alpha\text{-D-glucose (s)} \rightarrow \alpha\text{-D-glucose (aq)}}{\alpha\text{-D-glucose (s)} \rightarrow \beta\text{-D glucose (s)}} \quad \frac{\Delta H(i) = 10.72 kJ}{\Delta H = \mathbf{4.88kJ}} \quad \text{Add}$$

Example 18:

Compute the standard heat of formation of methane using the following data:

(i) $CH_4(g) + 2O_2(g) \rightarrow CO_2(g) + 2H_2O(l)$ ΔH° (298K)=– 890.35 kJ

(ii) $H_2(g) + 1/2\ O_2(g) \rightarrow H_2O(l)$ ΔH° (298K) = – 285.84 kJ

(iii) C (*graphite*) + O_2 (g) → CO_2 (g) ΔH° (298K) = – 393.51 kJ

Solution:

Our aim is to get DH° for the following reaction

C (graphite) + $2H_2$ (g) → CH_4(g) ΔH° (298) = ?

Reverse equation (i) and multiply equation (ii) by two.

(iv) CO_2 (g) + $2H_2O$ (l) → CH_4 (g) ΔH° (298K) = – 285.84 kJ

(v) $2H_2$ (g) + O_2(g) → $2H_2O$(l) ΔH° (298K) = – 571.68 kJ

Add equations (iii), (iv) and cancel the common terms on both sides of the equations. This gives

C (graphite) + $2H_2$ (g) → CH_4 (g) ΔH° (298K) = – **74.84 kJ**

Thus the enthalpy change for the formation of methane from its elements is calculated to be 74.84 kJ.

Example 19:

The heat of solution DH at different molalities m of an aqueous solution of a solute X are given below.

m	1.00	1.80	2.82	4.25	6.37	9.95
ΔH(298K)/kJ	–34.2	–56.7	–88.3	–126.0	–174.0	–229.0

Plot Δ H against the molality m, and determine the differential heat of solution of the solute X in a 3-molal solution.

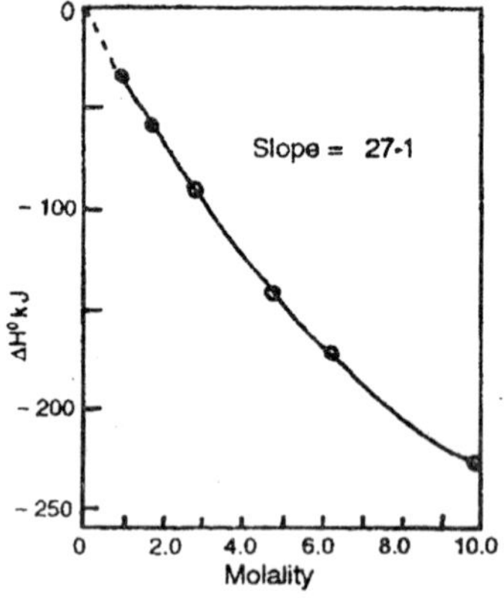

Fig. 5.6

Solution:

ΔH values are plotted against molality as shown in Fig. 5.6. Now draw a tangent on the curve at a point when molality is 3. Extend this tangent to meet the Δ H axis, and find out the slope of the tangent. It is –27.1kJ molal^{-1}. Since the variation of ΔH with molality (m_2) is given by

$$\Delta H = n_1 \, \Delta H_{1, m} + n_2 \, \Delta H_{2, m} = 55.55 \, \Delta H_{1m} + m_2 \, \Delta H_{2, m}$$

The slope of the tangent is equal to $\Delta H_{2, m}$ *i.e.*, $\Delta H_{2, m}$ = **–27.1kJ** molal^{-1}.

Example 20:

Calculate ΔH for the reaction

$HCl.100\ H_2O + NaOH.100\ H_2O \rightarrow NaCl.200\ H_2O + H_2O(l)$

The information given as

(i) HCl (g) + 100 $H_2O \rightarrow$ HCl.100 H_2O ΔH (i) = –75.7 kJ

(ii) NaOH (s) + 100 $H_2O \rightarrow$ NaOH.100H_2O ΔH (ii) = –40.9 kJ

(iii) Nacl (s) + 200 $H_2O \rightarrow$ Nacl.200 H_2O ΔH (iii) = – 4.26 kJ

ΔH_f^{o}'s for HCl (g), NaOH (s), NaCl(s) and H_2O (l) are

–92.3, – 426.7, – 411.0 and – 285.8 kJ *respectively.*

Solution:

The given reaction is

(a) HCl.100 H_2O + NaOH.100 $H_2O \rightarrow$ NaCl.200 H_2O + H_2O(l)

$\Delta H = \Sigma \Delta H^{o}_{f}$ (products) – $\Sigma \Delta H^{o}_{f}$ (reactants)

ΔH for this reaction can be calculated if the ΔH^{o}_{f} data for HCl in 100 H_2O; NaOH in 100 H_2O, NaCl in 200 H_2O and H_2O (l) are known.

From equation (i), we get

ΔH^{o}_{f} (HCl.100 H_2O) = –75.7 –92.3 = – 168.0 kJ

Equation (iii) gives

ΔH^{o}_{f} (NaOH.100 H_2O) = ΔH(ii) + ΔH^{o}_{f} [NaOH(s)]

= – 40.9 – 426.6 = – 467.5 kJ

From equation (iii), we have

ΔH^{o}_{f} (NaCl.200H_2O) = + 4.26 – 411.0 – 406.74 kJ

Also ΔH^{o}_{f} H_2O = – 285.8 kJ

Using these data we have

ΔH(a) = (– 406.74 –285.8) – (– 168.0 – 467.5)

= 692.54 + 635.50 = **–57.04 kJ**.

Example 21(a):

The heat of dissociation per mole of gaseous water at 18°C and 1atm is 241750 J ; calculate its value at 68°C. Data given are

C_P (H_2O) = 33.56; C_P (H_2) = 28.83; C_P (O_2) = 29.12J K^{-1} mol^{-1}

Solution:

The dissociation reaction is,

$H_2O\ (g) \rightarrow H_2\ (g) + 1/2\ O_2\ (g)\ \Delta H^\circ\ (291\ K) = 241750\ J$

$\Delta\ C_P = C_P\ (H_2) + 1/2\ C_P\ (O_2) - C_P\ (H_2O)$

$= (28.83 + 1/2 \times 29.12 - 33.56)\ JK^{-1}\ mol^{-1}$

$= 9.83 J\ K^{-1}\ mol^{-1}$

$\therefore\ \Delta\ C_P \times \Delta T = (9.83\ JK^{-1}\ mol^{-1}) \times (50K)$

$= 491.5\ J\ mol^{-1}$

$\Delta H^\circ\ (341K) = \Delta H^\circ\ (291K) + \Delta\ C_P\ (\Delta\ T)$

$= 241750\ J\ mol^{-1} + 491.5\ J\ mol^{-1}$

$= \mathbf{242241.5\ J\ mol^{-1}}$

For precise calculations we must express the heat capacities of reactants and products as a function of temperature, and then integrate equation (3) $C_{P,m}$ can be expressed as a power series of T in the form

$C_{P,m} = \alpha + \beta T + \gamma T^2$

where α, β, γ are constants for a given substance.

$\therefore \Delta C_P = (m\alpha_M + n\alpha_N) - (a\alpha_A + b\alpha_B) + [(m\ \beta_M + n\beta_N)$

$- (a\beta_A + b\beta_B)\ T + ... + ...$

or $\Delta C_P = \Delta\ \alpha + \Delta\ \beta T + \Delta\gamma T^2$

Equation (3) can be written as

$$d\ (\Delta\ H) = (\Delta\alpha + \Delta\beta T + \Delta\gamma T^2)\ dT \quad ...(7)$$

On integration this equation yields

$$\int d(\Delta H) = \int (\Delta\alpha + \Delta\beta T + \Delta\gamma T^2)\, dT + \text{constant (I)}$$

$$\text{or } \Delta\ H = \Delta\ \alpha T + \frac{1}{2}\Delta\beta T^2 + \frac{1}{3}\ \Delta\gamma T^3 + \text{constant (I)}$$

If T = 0 then

$\Delta\ H = \Delta\ H(T = 0) = \Delta H_0 = \text{constant (I)}$

$$\therefore\ \Delta H = \Delta H_0 + \Delta\alpha T + \frac{1}{2}\ \Delta\beta T^2 + \frac{1}{3}\ \Delta\gamma T^3$$

ΔH_0 is the hypothetical heat of reaction at absolute zero. However, equation (7) can be integrated for suitable range of temperatures T_1 and T_2 to give

$$\int_{T_1}^{T_2} d(\Delta H) = \int_{T_1}^{T_2} (\Delta\alpha + \Delta\beta T + \Delta\gamma T^2) dT$$

$$\text{or} \quad \Delta H (T_2) - \Delta H (T_1) = \Delta\alpha (T_2 - T_1) + 1/2\, \Delta\beta(T_2^2 - T_1^2) + \frac{1}{3}\Delta\gamma (T_2^3 - T_1^3) \qquad ...(8)$$

Equation (8) is used to calculate precisely the enthalpy of reaction at any temperature T_2 provided its value at a temperature T_1 is known. The use of equation (5) or (8) is not restricted only to the chemical reactions. It can suitably be used for the physical transformations as well. For example, in a phase change (vaporization, fusion, sublimation etc). from a phase a to another phase β, the Kirchhoff's equation can be written as

$$d\,(\Delta H_{\alpha \to \beta}) = CP_{\alpha \to \beta}\, dT$$

$$\text{or } \Delta H (T_2) = \Delta H(T_1) + \Delta C_P (T_2 - T_1)$$

where $\Delta H = H\beta - H\alpha$; $\Delta C_P = C_P (b) - C_P (\alpha)$.

Example 21(b):

Examine the reactions tabulated below with the accompanying change in enthalpy :

Process	*Reaction*	*Enthalpy changes (kJ)*
Sublimation :	(i) Li (s) → Li (g)	$\Delta H_{sub} = +161$
	(ii) Na (s) → Na (g)	$\Delta H_{sub} = +109$
Dissociation :	$1/2\ Cl_2 (g) \to Cl (g)$	$1/2\Delta H_d = +122$
Electron-affinity :	$Cl (g) + e \to Cl^- (g)$	$\Delta H_{aff} = -350$
Ionization :	$(i)\ Li (g) \to Li^+ + e$	$\chi H_{IP} = +520$
	$(ii)\ Na (g) \to Na^+ + e$	$\Delta H_{IP} = +496$
Compound	(i) $Li (s) + \frac{1}{2} Cl_2 (g) \to LiCl (s)$	$\Delta H^o_f = -410$
formation :	(ii) $Na (s) + \frac{1}{2} Cl_2(g) \to NaCl(s)$	$\Delta H^o_f = -411$

(a) *Prepare Born-Haber cycles for the separate LiCl and NaCl and calculate their respective lattice energies.*

(b) *Using the following heats of solutions of LiCl and NaCl in water along with the lattice energy data obtained in part* (a), *calculate the heat of hydration of Li^+ and Na^+ ions separately and interpret the results:*

(i) $LiCl (s) + aq \to Li^+ (aq) + Cl^- (aq) + 35.1$ kJ

(ii) $NaCl (s) + aq \to Na^+ (aq) + Cl^-$ 4.3 kJ

Solution:

(a) Born-Haber cyclic is prepared as discussed for general case MX.

For lithium chloride: $LiCl\ (s) \rightarrow Li^+\ (g) + Cl^-\ (g)$

$\Delta H_C = 161 + 122 + 520 + 410$

$= \mathbf{863\ kJ}$

For sodium chloride : $NaCl\ (s) \rightarrow Na^+\ (g) + Cl^+\ (g)$

$\Delta H_C = 109 + 122 + 496 - 350 + 411 = 788\ kJ$

(b) Heat of hydration = Heat of solution – Crystal energy

Solution process : For lithium chloride

$LiCl\ (s) + aq \rightarrow Li^+\ (aq) + Cl^-\ (aq)\ \Delta H = -\ 35.1\ kJ$

Crystal formation process:

$Li^+\ (g) + Cl^-\ (g) \rightarrow LiCl(s)\ \Delta H = -\ 35.1\ kJ$

Adding the two equations, we get

$Li^+\ (g) + Cl^-\ (g) \xrightarrow{aq} Li\ (aq) + Cl^-\ (aq)\ \Delta\ H_{hydration} = -\ 898.1\ kJ$

Similarly for NaCl

$Na^+\ (g) + Cl^-\ (g) \xrightarrow{aq} Na+\ (aq) + Cl^-\ (aq)\ \Delta H_{hydration} = -\ \mathbf{783.7kJ}$

ENCYCLOPAEDIA OF CHEMICAL THERMODYNAMICS

Vol. I

By

K.K. Jain

DISCOVERY PUBLISHING HOUSE PVT. LTD.
NEW DELHI-110 002

Published by:
Namit Wasan
DISCOVERY PUBLISHING HOUSE PVT. LTD.
4383/4B, Ansari Road, Darya Ganj
New Delhi-110 002 (India)
Phone : +91-11-23279245; 23253475; 43596065
E-mail : discoverybooksindia@gmail.com
discoverypublishinghouse@gmail.com
namitwasan9@gmail.com
web : www.discoverypublishinggroup.com

Edition: **2020**

ISBN: 978-81-8356-351-2 (Set)

Encyclopaedia of Chemical Thermodynamics

Printed at:
Infinity Imaging Systems
Delhi

Preface

There are very few books available on chemistry which present the subject matter in a concise manner so that it is interesting and within grasp of B.Sc. (Hons.) and postgraduate students.

Keeping the above approach in mind this book has been planned. It is hoped that it would be useful to chemistry students at all levels. A logical and rigorous approach has been adopted but too much philosophical discussion has been avoided in order to make the material always within the grasp of students.

Each chapter of the book contains a discussion of the subject matter and important mathematical relations followed by suitable solved examples to illustrate the basic principles. At the end of each chapter a set of problems has been given to help the students become familiar with the concepts and principle presented in the text.

Although it has not been possible to give specific reference to contributions which inspired us in writing this book, we do knowledge the help of many who contributed directly or through their writings.

The comments, criticism and suggestions from the readers will be very much appreciated.

Author

CONTENTS

3. Thermochemistry **185**

Introduction, The Concept of Heat of Reaction, Equations of the Thermochemical, Standard States, Laws of Thermochemistry, Various Types of Heat Changes, Heat of Hydrogenation, Born-habar Cycle, Bond Energies, Variation of Heats of Reaction with, Temperature, Variation of Heat of Reaction with Pressure, The Maximum Flame Temperature, Interconversion of Matter and Energy, Experimental Determination of de and dhd, Kirchhoff's Equation, The Maximum Flame Temperature, Explosion Temperature, Law of Thermochemistry, Factors on Which Heat of Reaction or Enthalpy of Reaction Depends, Exothermic and Endothermic Reactions, Standard Enthalpy Changes, Various Types of Enthalpy Changes, Bond Enthalpies or Energies, Conventions Used in Writing Thermochemical Equations, Specific Heat of Solids, Debye Theory, Blackman's Lattice Theory of Specific Heats, Heat of Reaction or Enthalpy of Reaction, Solved Examples.

1

Fundamental Concepts

INTRODUCTION

Thermodynamic deals with energy changes and its relationship with work. It is based on three laws of thermodynamics which are used as axioms just as Newton's laws of motion form the basis of classical mechanics. The first two laws based on facts observed in every day life. The predictions based on these laws have been verified in most cases and so far no case has been reported where the laws break down. The laws can be stated in mathematical form. Hence, thermodynamics is an exact science. The thermodynamic theory can be developed without gaps in the argument using only moderate knowledge of mathematics.

The remarkable feature of thermodynamics is that for the development of the theory, it is not necessary to assume the nature of fundamental particles or mechanism of the phenomena or processes as is done is molecular structure and kinetics. Thermodynamics of a process or partial differentiation and related mathematics without bothering how the process takes place.

Whereas the first law is concerned with the equivalence of different forms of energy, the second law is concerned with the direction of physicochemical change and condition of equilibrium. The two laws deal with changes in thermodynamic variables but the third law attempts to evaluate such variables.

Some of the obvious advantages from the first and second laws of thermodynamics areas follows. Useful relations can be obtained from first law which are valid for both equilibrium and non-equilibrium situations. Further, on the basis of the first and second laws together, one can predict the effect of changes of pressure, temperature and composition on different varieties of physicochemical systems.

We shall be mainly discussing in this text what is called as " Equilibrium Thermodynamics" since most of the thermodynamic predictions are valid only for equilibrium phenomena. There is another branch of thermodynamics called "Non-equilibrium thermodynamics" which has developed into an organized discipline in the course of last few decades. A brief account of this new development.

THERMODYNAMIC TERMS

System

The thermodynamics system is the part of the physical universe to which we direct our attention for the purpose of experimental or theoretical investigation.

Surroundings

The surroundings are the remainder of the universe which can exchange matter of energy with the system. Usually, the surroundings are restricted to a region in the immediate vicinity of the system under investigation.

Boundary

The region or interface separating the system from the surroundings is called the boundary.

Let us consider a gas (or liquid) contained in a cylinder and let the cylinder be placed in a constant temperature bath maintained at 25°C, as shown in Fig. 1.1. In this case the substance (liquid or gas) constitutes the system, the walls of the cylinder are the boundaries and the constant temperature bath is the surrounding.

Since we shall encounter various types of systems as well as walls or boundaries, it will be helpful at this point to define some of the more important type of walls.

(i) *A rigid wall.* A wall whose shape and position are fixed.

(ii) *An impermeable wall.* A wall that prevents passage of matter.

(iii) *A permeable wall.* A wall that allows that passage of matter and consequently, also of energy.

(iv) *An adiabatic wall.* It is wall which, when held rigid will prevent the passage of matter or energy provided no external force field are present (e. g. thermos bottle).

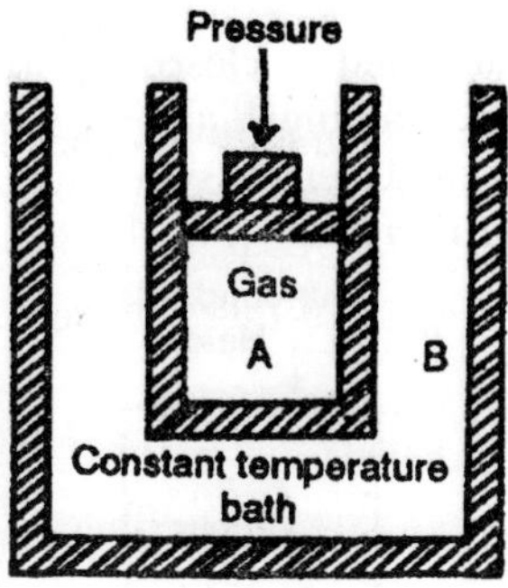

Fig. 1.1

(v) *A diathermic or diathermal wall.* It is a wall which, when held rigid will prevent the passage of matter but allow the passage of energy in the absence of any external force field. Thus it is a wall that is impermeable but not adiabatic. A plain metal container such as an aluminium can in an example of a container whose walls are diathermal.

Depending upon the type of the wall (boundary) of container the following types of systems can be considered.

(i) *An open system.* It is a system enclosed by a permeable wall that can exchange both the energy and the matter with the surroundings across the boundary. Thus $dE \neq 0$, $dm \neq 0$. (dm = change of mass; dE = change of energy).

(ii) *A closed system.* It is a system enclosed by impermeable walls. In system the mass remains constant but the energy may change. Thus $dm = 0$; $dE \neq 0$.

(iii) *An isolated system.* It is a system enclosed by rigid, adiabatic walls and not acted upon by any external force field. Thus in an isolated system neither the energy nor the mass can be exchanged with the surroundings across the boundary and $dE = 0$; $dm = 0$.

Consider a vessel containing a volatile liquid in contact with its vapours. This constitutes an open system as energy (heat) is taken by the system (liquid) for vaporization and the vapours can escape out of the vessel.

Thus system exchanges both the energy and the mass with the surroundings [Fig. 1.2(a)]. If the vapours are not allowed to leave the vessel then the system would be closed one [Fig. 1.2(b)]. However, if the walls of the vessel containing liquid are thermally insulated and the mouth of the vessel is closed then the system can exchange neither energy nor matter with the surroundings [Fig. 1.2(c)].

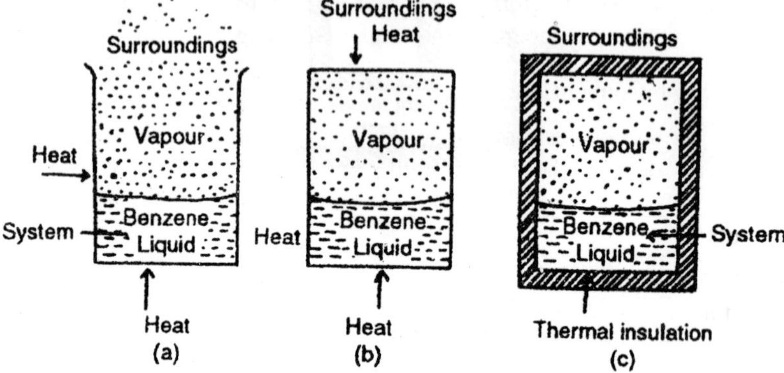

Fig. 1.2 : Types of systems : (a) open, (b) closed and (c) isolated.

THERMODYNAMIC VARIABLES

A system can be defined by measurable properties like temperature (T), pressure (P), volume (V), and the amount (n_i, the number of moles of various species present in the system). These properties are called thermodynamic or state variables.

Table 1 : Common Thermodynamic Variables.

Variable	*Name*	*Defining equation*
P	Pressure	$P = \frac{\text{Force}}{\text{Area}}$
V	Volume	V = Three dimensional space available to system
n	Mole	$n = \frac{\text{Mass of the substance i}}{\text{Molar mass of i}}$
T	Temperature	$T_{\text{triple pt.}}$ of H_2O = 273.16 K
w	Work (external)	w = Force × distance moved = Pressure × area × distance moved = PdV
q	Heat	Heat entering the system

Two additional quantities occur so often that they are considered as thermodynamic variables and are given special symbols, w for work and q for heat. (However, these two are not state properties as would be shown later on). These six variables are central in treatment of energetics of chemical systems from the standpoint of view of thermodynamics. In table 1 these six properties and their definitions are summarized.

The Sign Convention

q or dq denotes the heat added to the system and w or dw denotes the work done on the system. Both q and w are considered to contribute to the internal energy E of the system as expressed by the equation $dE = dq + dw$.

A positive value of q implies that heat has been added to the system, and a negative value of q signifies that heat has been transferred from the system to the surroundings. For example if q = 50 kJ it means that 50 kJ, we can say that system.

A positive value of w means that work has been done on the system whereas the negative magnitude of w means that work has been done by the system on the surroundings. For example, if w = 100 kJ, we say that 100 kJ work has been done on the system, but when w = – 100 kJ it signifies that the system has done 100 kJ of work on the surroundings, in other words we can say that 100 kJ work has been done on the system. Thus, during expansion of a gas the magnitude of w is negative and for compression of the gas w has a positive magnitude. The work of compression is w and the work of expansion is (–w).

It is always convenient to learn the sign of heat change and work done if we think in terms of energy change. Any heat exchange that increases the energy of the system must be positive in magnitude, any work that increases the energy of the system must be considered to have positive magnitude. The q and w which lower the energy should be negative.

Extensive and Intensive Properties

A property is called *extensive* if its value is additive. For example, let us consider a homogeneous system A. Let it be divided into subsystems a_1, a_2, a_3.... etc. Let Y be a property of the system A, and y_1, y_2, y_3...etc. are the properties of the respective subsystems a_1, a_2, a_3... etc. Now Y will be an extensive property provided $Y = y_1 + y_2 + y_3 + ...$ Mass is clearly an extensive property, volume is another example. Extensive property depends on the total amount of the system. Properties are called *intensive* if, when the system is divided into n parts, each part has the same value of the property. That is to say when the property of the system A and that of the subsystems a_1, a_2, a_3... etc. are the same, the property in question is an intensive one. Temperature is an intensive property, because if we take 500 ml water in a beaker kept at room temperature and then divide this amount of water into five parts each containing separately 200 ml, 150 ml, 75 ml, 50 ml and 25 ml, or in any other proportion the temperature of each part of water will be the same. Such properties do not require any specification of the size of sample to which they refer.

The extensive and intensive properties are listed in Table 2.

Table 2

Extensive property	*Intensive property*
Volume	Molar volume
No. of moles	Density
Mass	Refractive index
Energy	Surface tension
Free energy	Viscosity
Entropy	Dielectric constant
Enthalpy	Molar enthalpy
Heat capacity	Molar entropy
	Chemical potential
	Free energy per mole
	Specific heat
	Pressure
	Temperature
	Boiling point; Freezing point etc.

(i) The product, ratio and sum of any intensive properties are intensive. For example if X and Y are arbitrary intensive variables the XY, X/Y, $\partial X/\partial Y$ and X + Y are intensive variables.

(ii) If x and y are arbitrary extensive variables then

(a) (x + y) is an extensive variable, and

(b) x/y and $\partial x/\partial y$ are intensive variables.

(iii) If X is an arbitrary intensive variable and y and arbitrary extensive variable, then yX , y/X and $\partial y/\partial X$ are extensive variables.

Mole fraction is an intensive property, because once a homogeneous mixture is prepared, the relative amount of the constituents are the same in each part of the mixture.

Molarity is an intensive property, because it is same for each part of the homogeneous mixture. Suppose we have prepared 500 mL of 0.1 molar solution of an acid, its molarity is 0.1 Now if we take 10 ml of this solution it is also 0.1 molar or if we take 400 ml of the solution, again it is also 0.1 molar.

Number of mole is an extensive property, because the total number of moles of system will be the sum of the moles of each subsystem. Suppose we take 360 g glucose in a container; the number of moles being 360/180 g mol^{-1} *i.e.*, 2 mole. Now if this quantity is divided as 180 g, 90 g, 54 g, 36 g then moles in each portion would be 1 mole, 0.5 mole, 0.3 mole, 0.2 mole respectively. Again the total mole being 1.0 + 0.5 + 0.3 + 0.2 = 2.0 mole.

Change in state : When the initial and final states of the system are specified then the change in state is completely defined. For example, consider a beaker containing water at 25° C and 1 atm pressure, and heat it to 50° C under 1 atmospheric pressure. Here the change in the temperature of water from 25° C to 50°C is defined as the change in state of the system, because volume, density etc. are accordingly changed. In general, the change in state of a system is illustrated as state I $(P_1, V_1, T_1) \rightarrow$ state II (P_2, V_2, T_2).

THERMODYNAMIC PROCESS

The operation which brings about the changes in the state of the system is called the process. For example, heating is the process in the above example, similarly the various operations like vaporization, freezing, sublimation expansion, contraction by which the state of the system can be changed, are known as processes.

We can discuss following types of processes.

Isothermal process : When the temperature of the system is kept constant during various operations then the process is said to be isothermal. This is attained either by removing heat from the system or by supplying heat to the system. For isothermal process $dT = 0$.

Adiabatic process : There is no exchange of heat between the system and surroundings and q i· ·qual to zero. Thus is adiabatic processes there is always a change in a temperature.

Isobaric process: When the pressure on the system remains constant during the period of the change, the process is said to be isobaric. For an isobaric process dP = 0. For example, conversion of 1 mole of water at 25° C and 1 atm pressure into vapour at 100°C and 1 atm pressure is isobaric process.

Isochoric process : When there is no change in the volume of the system during various operations the change is said to be isochoric. For such changes dV = 0.Combustion of a substance in a bomb calorimeter is an isochoric process.

Cyclic process : If a system having undergone a change returns to initial state then the process is called a cyclic process and the path of the process is called a cycle.

Path : Sequence of steps starting from the initial state to the various intermediate states and ultimately to the final state of the system is called the path. Thus the path of conversion of 1 mole of water at 25°C and 1 atm into vapour at 100°C abd 1 atm can be represented as

$$\underset{\text{(Initial state)}}{1\text{ mole } H_2O\text{ (liq., }25^\circ\text{ C, 1 atm)}} \xrightarrow{\text{Heat}} \underset{\text{(a)}}{H_2O\text{ (liq., }50^\circ\text{C, 1 atm)}} \xrightarrow{\text{Heat}}$$

$$\underset{\text{(b)}}{H_2O\text{ (liq., }75^\circ\text{C, 1 atm)}} \xrightarrow{\text{Heat}} \underset{\text{(c)}}{H_2O\text{ (liq., }100^\circ\text{ C, 1 atm)}}$$

$$\underset{\text{(Final state)}}{H_2O\text{ (vapour, }100^\circ\text{C, 1 atm)}}$$

(a), (b), (c) etc. are the intermediate states and whole of the scheme represents the path of the system.

The final state of a system can be attained either directly or in stages. Accordingly there are two types of paths of a system, consequently two types of processes are recognized. These are :

Reversible process : When the process occurs in such a way that the properties of the system at every instant remain uniform, the process is called reversible. Thus, reversible process is one which proceeds through a succession of equilibrium steps, each of which is an equilibrium state. A reversible process must be carried out at a very slow rate so that enough time is allowed between each change for the properties to come to a constant value. Reversible processes are ideal. In actual practice it is not possible always to perform the experiments under such an ideal situation. Nevertheless the concept of reversibility serves as a means to study the very many processes which are of immense importance to chemists.

Irreversible process : If the change is produced rapidly and the system does not get a chance to attain equilibrium then the process is called irreversible. All natural processes are irreversible and hence *spontaneous*. Once undergone a change the process cannot be reverted without the help of external source and without changing the properties of the surroundings.

Examples of a reversible process are :

(i) Vaporization of a liquid in a closed vessel at a constant temperature is a reversible process. At this temperature the equilibrium pressure of the vapour is equal to the saturation vapour pressure. Now, if the pressure is increased by any amount, the vapour will condense. On the other hand when the pressure is diminished by an infinitesimal amount compared to the equilibrium vapour pressure, the liquid will vaporise. It has to be noted that this does not bring about any change in the surroundings.

(ii) Chemical reaction in a galvanic cell is very nearly a reversible process. The electromotive force (emf) of a cell is measured by a sensitive potentiometer when the driving force of the cell is balanced against an external emf, and the current can be made to flow in one direction or the other by external change of one micro volt without leaving more than a vanishingly small change in the surroundings.

(iii) Chemical reactions of the type A B also come under the class of reversible process.

Examples of a few irreversible processes are – expansion of a gas into vacuum, reaction for the formation of curd from milk, in fact all natural processes are irreversible.

Let us illustrate the significance of the concepts of reversible and irreversible processes for a gas contained in a cylinder which is fitted with a piston.

Let the gas in the cylinder comprise the system and let the piston be loaded with a large number of small weights as shown in Fig. 1.3.

Let the weights be slid off horizontally one at a time, allowing the gas to expand and to work in raising the weights which remain on the piston. As the size of the weights is small and their number is large. the expansion would be reversible for at each level of piston during the reverse process there will be small weight which is exactly at the level of the platform and thus can be placed on the platform without requiring work. In the limit, therefore, as the weights become very small, the reverse process can be accomplished in such a manner that both the system and surrounding are exactly in the same state they were initially.

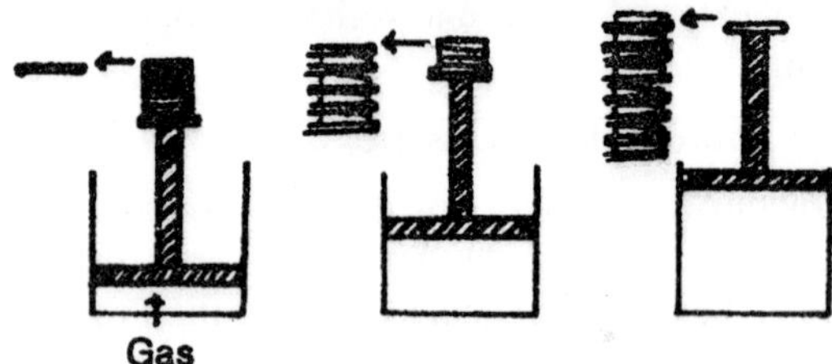

Fig. 1.3 : An Example of a reversible process.

Let us consider another situation. Let the gas in the cylinder be at high pressure restrained by a piston which is secured by pin s depicted in Fig. 1.4. When the pin is removed, the piston is raised and forced abruptly against the stops.

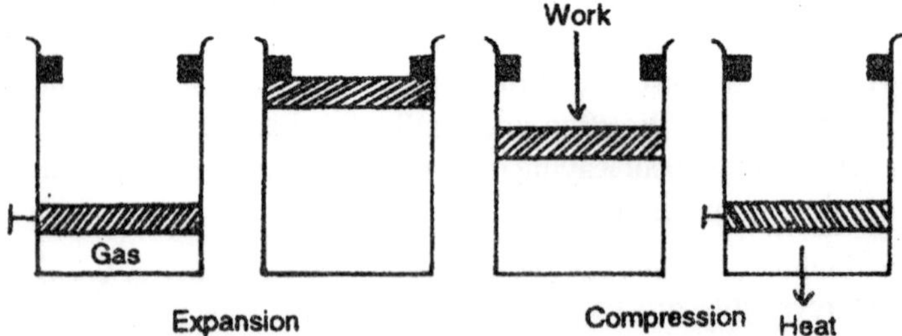

Fig. 1.4 : An example of irreversible process.

Some work is done by the system (gas in he cylinder). since the piston has been raised. If we wish to restore the system to its initial state, a force must be exerted on the piston to compress the gas until the pin could again be inserted in the piston. Since the work done on the gas during compression is greater than the work done by the gas in the expansion process, an amount of heat must be transferred from the system to the surroundings in order that the system is at the same initial state. Thus the system is restored to its initial state, but the surroundings have hanged t_ virtue of the fact that work was required to force the piston iown and heat was transferred to surroundings. Thus the expansion of as under the present condition is an irreversible process because it could ot be reversed without leaving a change in the surroundings.

'HE LAWS OF THERMODYNAMICS

The success of thermodynamics in various fields of science is based on the wide application of the laws of thermodynamics which are as well established, as widely accepted and as free from experimental anomalies as any known scientific laws.

First Law dE = dq + dw

The change in internal energy (dE) in any system is the sum of the heat entering the system (dq) and the work done on the system (dw). The internal energy of an isolated system is constant and dq = 0, dw = 0, dE = 0.

Second Law $dS = dq_{rev}/T$

The change in entropy (dS) during any change in a system is equal to the heat entering the system (when the change is performed reversibly) divided by the absolute temperature T. In all actual processes the entropy of an isolated system always increases ($dS > 0$).

Third Law

The entropy at absolute zero (0K) is zero for any pure crystalline substance *i.e.*, S^0 (0K) = 0.

Zeroth Law

This law states that if two systems A and B are thermal equilibrium with a third system C then A and B will also be in thermal equilibrium when placed together. This law like other laws is based on human experiences. We can explain this law further as follows. Suppose we have three systems A, B and C placed together. A and C are in contact so that the properties of both change and finally the thermal equilibrium is attained and two come at the same temperature; C is in contact with B also then again thermal equilibrium is attained. If this is so then A and B should also be in thermal equilibrium. That is A, B and C should be at the same temperature.

If A (T_A) $\rightleftharpoons$ C (T_C) B (T_B),

then A (T_A) $\rightleftharpoons$ B (T_B), or if $T_A = T_C = T_B$, then $T_A = T_B$.

Thus the law provides a rational definition of temperature.

In thermodynamic studies the temperature's unit is taken as kelvin (K). The kelvin is strictly defined as the fraction 1/273.16 of the temperature interval between the absolute zero (– 273.16°C) and the triple point of water. In practice the temperature in kelvin is obtained by adding 273.15 to the temperature in degree celcius, e.g.,

25°C = 25 + 273. 15 = 298.15 K.

The use of the symbol °K is not recommended. The symbol of temperature in kelvin scale is T.

STATE OF EQUILIBRIUM

It denotes a state of rest, of balance, of absence of change on macroscopic level. An object falls over, rolls about for a time and finally comes to rest; we say that it is then in equilibrium. In thermodynamics the equilibrium must satisfy the following two conditions.

(i) The macroscopic properties such as temperature, pressure, density etc., of the system should not change with time (the submicroscopic properties do change e.g., the Brownion motion is always there even at equilibrium).

(ii) The system should be at rest without the assistance of any external source. In a system in any of the two conditions are not fulfilled then the system will not be at equilibrium. For example, consider a leaky water tank into which a tap is running to maintain a constant water level. The tank is in a condition which is not observably changing from moment to moment i.e., its water level is constant, but its *state is not one of equilibrium* because the level is maintained by external agency. Thus, though it satisfies the first condition, it does not satisfy the second one. Hence it does not represent the equilibrium state. *It is a steady state or stationary state.*

Both the steady and equilibrium state are dynamic in nature. A steady state is dynamic on the macroscopic as well as microscopic level while an equilibrium state is static on macroscopic level but dynamic on *microscopic* level.

A true state of equilibrium in thermodynamics is characterised by mechanical equilibrium, thermal equilibrium and chemical equilibrium.

A system is said to be in a state of mechanical equilibrium when (i) there is no unbalanced force in the interior of a system, and (ii) there is none between the system and its surroundings. When condition (i) is not satisfied the system will undergo a change of state. On the other hand when (ii) is not satisfied, both the system and its surroundings will undergo a change. Such a change is stopped only when the mechanical equilibrium is attained. The condition of mechanical equilibrium is that each part of the system should be at the same pressure. A system is said to be in thermal equilibrium when its each part is at the same temperature and this temperature is identical with that of the surroundings. When these conditions are not satisfied, either the system alone or both the system and the surroundings will undergo a change of state until thermal equilibrium is reached.

The state of chemical equilibrium is characterised by tendency not to undergo spontaneous change of internal structure *i.e.*, the chemical composition of the system must be uniform, with no *net* chemical reaction taking place. In terms of the thermodynamic function the chemical equilibrium is described by the fact that the chemical potential (μ) of each part of the system must be the same.

Types of Equilibrium

Stable equilibrium. A system is in a state of stable equilibrium if after displacement to a new state and release of the constraint causing the displacement, it always returns to the original state. Fig. 1.5 (a) depicts an example of stable equilibrium for a ball resting in a hallow. If the ball is displaced in either direction from its initial position it will try to come to its old position. Such equilibrium is quite common in thermodynamics.

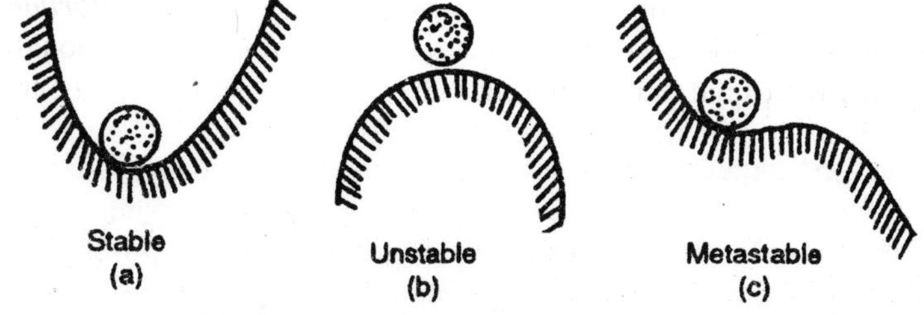

Fig. 1.5 : Types of mechanical equilibrium.

Unstable equilibrium. A system is said to be in a state of unstable equilibrium if, after displacement to a new state and release of the constraint causing the displacement, it never returns to its original state. An example of such a situation is given in Fig. 1.5(b) for a ball resting on a hillock. It may be noted that this is not an equilibrium state.

Metastable equilibrium. It is a state between the stable and unstable equilibria. If the displacement is not too large the state is stable but if the displacement is too large then it is unstable equilibrium. Fig. 1.5(c) depicts an example of metastable equilibrium. A super saturated solution of a super cooled liquid provides an example of the metastable equilibrium.

THERMODYNAMIC FUNCTIONS

The thermodynamic quantities given in Table 1. are quite sufficient to describe the properties of a system. However, new functions of simpler form can be defined, as given in Table 3. These are found to be more useful.

Table 3 : Various Thermodynamic Functions

Definition	*Function*	*Name*
$dE = dq + dw$	E	Internal energy
$dS = \frac{dq_{rev}}{T}$	S	Entropy
$H = E + PV$	H	Enthalpy
$G = H - TS$	G	Gibbs free energy
$A = E - TS$	A	Helmholtz free energy (work function)
$\mu_i = (\partial G/\partial n_i)_{T, P, nj}$ $j \neq i$	μ	Chemical potential
$C_P = (\partial H/\partial T)_P$	C_P	Heat capacity at constant pressure
$C_V = (\partial E/\partial T)_V$	C_V	Heat capacity at constant volume

STATE FUNCTIONS

Thermodynamic functions are state functions. Magnitude of changes in such functions are independent of how the change is accomplished. These depend only upon initial and final states of the system. An analogy of state function can be found in mountain climbing. Suppose that we want to climb a mountain peak 1 km above our lodge. This decision defines our initial (lodge) and final (mountain peak) states. There may be various paths up the mountain but the vertical distance of the peak from the lodge is all the time 1 km. The height of the mountain peak cannot be altered by choosing the different paths though the actual distance travelled and the amount of work and fatigue will be different depending upon the path followed. Vertical distance in this example corresponds to thermodynamic functions, E, H, S, G, μ, T , P, V. These all are state functions. We cannot alter the energy, enthalpy change etc. of the system by choosing different sequence of processes provided the initial and final states for different paths are not altered.

Properties of State Functions–Exact Differentials

State functions give exact differentials. These exact differentials can be integrated between limits, without regard for the actual change that occur as the system moves from one limiting condition to the other. The properties of a state function can be understood if we take a concrete example of volume of definite amount of the substance. It depends upon temperature (T) and pressure (P). Mathematically we can speak that volume is a function of temperature and pressure *i.e.*,

$V = f(T, P)$ (for a definite amount)

If T and P are given a particular value of V is thereupon fixed. If T and P are changed the value of V will accordingly be changed. For this reason T and P are called *independent variables* and V is called the *dependent variable*.

The change in V can be estimated provided the derivatives of the function V with respect to T and P are known. The derivative is the rate of change of dependent variable with the independent variable. For the present case the two derivatives are :

$(\partial V/\partial T)_P$ = rate of change of volume with temperature at constant pressure.

$(\partial V/\partial P)_T$ = rate of change of volume with, pressure at constant temperature.

If the derivative is positive the function V increases with the increase of the independent variable; if the derivative is negative the function decreases with the increase of the independent variable. If the derivative is zero the function is independent of the variables, and the value of the function may be maximum or minimum under that condition.

The change of volume due to infinitesimal change of temperature (dT) at constant pressure is equal to the rate of change of volume with temperature $(\partial V/\partial T)_P$ multiplied by dT, and is given by

$dV = (\partial V/\partial T)_P\ dT$

The change in volume due to infinitesimal change of pressure (dP) at constant temperature is given by

$dV = (\partial V/\partial P)_T\ dP$

If both the temperature and pressure are changed simultaneously then the total change in volume (dV) is given by

$$dV = \left(\frac{\partial V}{\partial T}\right)_P dT + \left(\frac{\partial V}{\partial P}\right)_T dP$$

This is called the total differential of the function (V). This equation was obtained by assuming that *whether we first change temperature then pressure or, first change the pressure then temperature, the final total change in volume is the same*. This is true because V is a *state function* and dV is *an exact differential*. In the language of mathematics the condition of exactness is

$$\frac{\partial}{\partial P}\left[\left(\frac{\partial V}{\partial T}\right)_P\right]_T = \frac{\partial}{\partial T}\left[\left(\frac{\partial V}{\partial P}\right)_T\right]_P$$

or $$\frac{\partial^2 V}{\partial P \partial T} = \frac{\partial^2 V}{\partial T \partial P}$$

This relation is applicable for any state function ϕ.

Let $\phi = f(x, y)$, then total differential $d\phi$ is given by

$$d\phi = \left(\frac{\partial \phi}{\partial x}\right)_y dx + \left(\frac{\partial \phi}{\partial y}\right)_x dy = M(x,y)\ dx + N(x, y)\ dy$$

where $M(x, y) = \left(\dfrac{\partial \phi}{\partial x}\right)_y$ and $N(x, y) = \left(\dfrac{\partial \phi}{\partial y}\right)_x$ ...(1)

If $d\phi$ is an exact differential then

$$\left(\frac{\partial M}{\partial y}\right)_x = \left(\frac{\partial N}{\partial x}\right)_y \qquad ...(2)$$

$$\therefore \frac{\partial^2 \phi}{\partial y dx} = \frac{\partial^2 \phi}{\partial x \partial y} \qquad ...(3)$$

Here ϕ can be considered as any thermodynamic function (E, H, G etc). and (x, y) may be thermodynamic T, P, V etc. The relations given by equations (1, 2) and (1.3) are called the Euler's theorem of exactness. Any function and dY is inexact differential.

Some other properties of derivatives :

(i) They can be turned upside down, the subscript remaining the same.

$$1\Big/\left(\frac{\partial\phi}{\partial x}\right)_y = \left(\frac{\partial x}{\partial\phi}\right)_y$$

However, $1\Big/\left(\frac{\partial^2\phi}{\partial x^2}\right) \neq \frac{\partial x^2}{\partial^2\phi}$

(ii) They can be multiplied together cancelling numerator with denominator

provided the subscripts are the same.

$$\left(\frac{\partial\phi}{\partial x}\right)_y\left(\frac{\partial x}{\partial z}\right)_y = \left(\frac{\partial\phi}{\partial z}\right)_y$$

But $\left(\frac{\partial\phi}{\partial x}\right)_y\left(\frac{\partial x}{\partial z}\right)_t \neq \left(\frac{\partial\phi}{\partial z}\right)$

Example 1:

If pressure, volume and temperature of one mole of a gas are related as $(P + a/V^2)\,V = RT$, *show that* :

(i) *P is a state function*, (ii) *dP is exact differential* (iii) $(\partial P/\partial T)_V\,(\partial T/\partial V)_P\,(\partial V/\partial P)_T + 1 = 0$.

Solution:

dP is an exact differential if $\frac{\partial^2 P}{\partial V\partial T} = \frac{\partial^2 P}{\partial T\partial V}$

Now $P = \frac{RT}{V} - \frac{a}{V^2}$ ...(i)

Differentiating equation (i) with respect to V at constant T, we get

$$\left(\frac{\partial P}{\partial V}\right)_T = -\frac{RT}{V^2} + \frac{2a}{V^3} \qquad \text{...(ii)}$$

Differentiating equation (ii) with respect to T at constant V, we have,

$$\frac{\partial^2 P}{\partial T \partial V} = -\frac{R}{V^2}$$

First differentiate equation (i) with respect to T at constant V and then differentiate the new equation (iii) with respect to V at constant T, to get

$$\left(\frac{\partial P}{\partial T}\right)_V = \frac{R}{V} \qquad \text{...(iii)}$$

$$\frac{\partial^2 V}{\partial V \partial T} = \frac{R}{V^2}$$

Since $\dfrac{\partial^2 V}{\partial T \partial T} = \dfrac{R}{V^2} = \dfrac{\partial^2 P}{\partial V \partial T}$

therefore, dP is an exact differential.

Equation (ii) expresses the change in P with volume at constant temperature, whereas equation (iii) is the change in P with T at constant V. Since the final change in P is reached via equations (ii) and (iii) these may be considered as the intermediate states of the system. The final change in P is the same irrespective of the difference in intermediate changes. So P is the state function. Changes in P depend only on initial and final states of the system.

(iii) See Cyclic rule).

Exercise: *If dE, dH, dS, dA, dG are perfect differentials, what information do you get about E, H, G, S, A? What is the integral of above differentials about a closed and reversible path?*

[**Ans.** (a) E, H, S, G, A are state functions (b) $\phi\ d\phi = 0$]

Example 2:

If X = f (T, V) and V = f (T, P), then show that

(i) $\left(\dfrac{\partial X}{\partial P}\right)_T = \left(\dfrac{\partial X}{\partial V}\right)_T \left(\dfrac{\partial V}{\partial P}\right)_T$

(ii) $\left(\dfrac{\partial X}{\partial T}\right)_P = \left(\dfrac{\partial X}{\partial V}\right)_T \left(\dfrac{\partial V}{\partial T}\right)_P + \left(\dfrac{\partial X}{\partial T}\right)_V$

Solution:

If X and V are state functions then the total differential will be given as

$$dX = \left(\frac{\partial X}{\partial V}\right)_V dT + \left(\frac{\partial X}{\partial V}\right)_T dV \qquad ...(i)$$

$$dV = \left(\frac{\partial V}{\partial T}\right)_P dT + \left(\frac{\partial V}{\partial P}\right)_T dP \qquad ...(ii)$$

Replacing dV from eq. (i) by eq. (ii), we get

$$dX = \left(\frac{\partial X}{\partial T}\right)_V dT + \left(\frac{\partial X}{\partial V}\right)_T \left[\left(\frac{\partial V}{\partial T}\right)_P dT + \left(\frac{\partial V}{\partial P}\right)_T dP\right]$$

$$= \left[\left(\frac{\partial X}{\partial T}\right)_V + \left(\frac{\partial X}{\partial V}\right)_T \left(\frac{\partial V}{\partial T}\right)_P\right] dT + \left(\frac{\partial X}{\partial V}\right)_T \left(\frac{\partial V}{\partial P}\right)_T dP \qquad ...(iii)$$

At constant temperature dT = 0 and eq. (iii) reduces to

$$(dx)_T = \left(\frac{\partial X}{\partial V}\right)_T \left(\frac{\partial V}{\partial P}\right)_T (\partial P)_T$$

or
$$\left(\frac{\partial X}{\partial P}\right)_T = \left(\frac{\partial X}{\partial V}\right)_T \left(\frac{\partial V}{\partial P}\right)_T \qquad ...(iv)$$

At constant pressure dP = 0 and equation (iii) yields.

$$\left(\frac{\partial X}{\partial T}\right)_P = \left(\frac{\partial X}{\partial T}\right)_V + \left(\frac{\partial X}{\partial V}\right)_T \left(\frac{\partial V}{\partial T}\right)_P \qquad ...(v)$$

Cyclic Rule

If P and V are state functions, the expressions for total change in P and V can be used to derive the cyclic rule.

Let P = f (V, T)

$$\therefore\ dP = \left(\frac{\partial P}{\partial V}\right)_T dV + \left(\frac{\partial P}{\partial T}\right)_V dT \qquad ...(4)$$

Let V = f (T, P)

$$\therefore\ dV = \left(\frac{\partial V}{\partial T}\right)_P dT + \left(\frac{\partial V}{\partial P}\right)_T dP \qquad ...(5)$$

Equation (4) for constant pressure (dP = 0) reduces to,

$$\left(\frac{\partial P}{\partial V}\right)_T (\partial V)_P + \left(\frac{\partial P}{\partial T}\right)_V (\partial T)_P = 0$$

Dividing by $(\partial V)P$, we have

$$\left(\frac{\partial P}{\partial V}\right)_T + \left(\frac{\partial P}{\partial T}\right)_V \left(\frac{\partial T}{\partial V}\right)_P = 0$$

Multiplying by $(\partial V/\partial P)_T$ we obtain

$$1 + \left(\frac{\partial P}{\partial T}\right)_V \left(\frac{\partial T}{\partial V}\right)_P \left(\frac{\partial V}{\partial P}\right)_T = 0 \quad ...(6)$$

Equation (5) for constant V (dV = 0) reduces to

$$\left(\frac{\partial V}{\partial T}\right)_P (\partial T)_V + \left(\frac{\partial V}{\partial P}\right)_T (\partial P)_V = 0$$

Dividing by $(\partial P)V$, we have

$$\left(\frac{\partial V}{\partial T}\right)_P \left(\frac{\partial T}{\partial P}\right)_V + \left(\frac{\partial V}{\partial P}\right)_T = 0$$

Multiplying by $(\partial P/\partial V)_T$, we obtain

$$\left(\frac{\partial V}{\partial T}\right)_P \left(\frac{\partial T}{\partial P}\right)_V \left(\frac{\partial P}{\partial V}\right)_T + 1 = 0 \quad ...(7)$$

Equations (6) and (7) express cyclic rules.

COEFFICIENT OF THERMAL EXPANSION (α)

The rate of change of volume with temperature at constant pressure relative to a volume V is thermal expansively α.

$$\alpha = \frac{1}{V}\left(\frac{\partial V}{\partial T}\right)_P$$

a has the dimensions of T^{-1}.

Compressibility Coefficient (β)

The rate of change of volume with pressure at constant temperature relative to a volume is defined as the compressibility of a substance.

$$\beta = -\frac{1}{V}\left(\frac{\partial V}{\partial P}\right)_T$$

The dimensions of β are those of P^{-1}. The expression for b bears a minus sign; this stands for the fact that volume decreases with increase of pressure and *vice versa*.

Relation between α and β

$$V = f(T, P)$$

$$dV = \left(\frac{\partial V}{\partial T}\right)_P dT + \left(\frac{\partial V}{\partial P}\right)_T dP$$

or $dV = \alpha V\, dT - \beta V\, dP$ ...(1)

For a condition of constant volume, $dV = 0$

$\therefore\ \alpha V(\partial T)_V - \beta V - \beta V\,(\partial P)_V = 0$

$$\text{or } \left(\frac{\partial P}{\partial T}\right)_V = \frac{\alpha}{\beta}$$

At constant pressure $dP = 0$ and equation (1) reduces to

$$\left(\frac{\partial V}{\partial T}\right)_P = \alpha V$$

Differentiating it with respect to P at constant T, we get

$$\frac{\partial}{\partial P}\left[\left(\frac{\partial V}{\partial T}\right)_P\right]_T = \alpha \left(\frac{\partial V}{\partial P}\right)_T + V\left(\frac{\partial \alpha}{\partial P}\right)_T$$

$$= -\alpha V\beta + V\left(\frac{\partial \alpha}{\partial P}\right)_T \qquad ...(2)$$

At constant temperature $dT = 0$ and equation (1) transforms to

$$\left(\frac{\partial V}{\partial P}\right)_T = -\beta V$$

Differentiating with respect to T at constant to T constant P we obtain

$$\frac{\partial}{\partial T}\left[\left(\frac{\partial V}{\partial P}\right)_T\right]_P = -\beta \left(\frac{\partial V}{\partial T}\right)_P - V\left(\frac{\partial \beta}{\partial T}\right)_P$$

$$= -\beta V\alpha - V\left(\frac{\partial \beta}{\partial T}\right)_P \qquad ...(10)$$

Since V is a state function and $\partial^2 V/\partial T\,\partial P = \partial^2 V/\partial P\,\partial T$, therefore from equations (2) and (3) we obtain

$$V\left(\frac{\partial \alpha}{\partial P}\right)_T = \left(\frac{\partial \beta}{\partial T}\right)_P$$

or $\left(\frac{\partial \alpha}{\partial P}\right)_T + \left(\frac{\partial \beta}{\partial T}\right)_P = 0$ ...(4)

Example 3:

Calculate a and b for an ideal gas.

Solution:

PV = RT for one mole an ideal gas

Differentiation gives

P dV + V dP = RdT

or $\left(\frac{\partial V}{\partial T}\right)_P = \frac{R}{P}$ [$\because$ at constant P, dP = 0]

$$\therefore \qquad \alpha = \frac{1}{V}\left(\frac{\partial V}{\partial T}\right)_P = \frac{R}{VP} = \frac{1}{T} = T^{-1}$$

Also $\left(\frac{\partial V}{\partial P}\right)_T = -\frac{V}{P}$ for an ideal gas

$$\therefore \qquad \beta = -\frac{1}{V}\left(\frac{\partial V}{\partial P}\right)_T = \frac{V}{VP} = \frac{1}{P} = P^{-1}.$$

Example 4:

(a) *A function ϕ is defined as $\phi(x, y) = x^2 y^3 + x$. Write its partial derivatives and total differential $d\phi$. Test whether $d\phi$ is an exact differential or not.*

Solution:

$\phi(x, y) = x^2y^3 + x$

Partial derivatives :

$$\left(\frac{\partial \phi}{\partial x}\right)_y = 2xy^3 + 1$$

$$\left(\frac{\partial \phi}{\partial y}\right)_x = 3x^2y^2$$

Total differential $d\phi = \left(\frac{\partial \phi}{\partial x}\right)_y dx + \left(\frac{\partial \phi}{\partial y}\right)_x dy$

$= (2xy^3 + 1)\ dx + 3x^2y^2\ dy$

In order that df bean exact differential, we have to prove

$$\frac{\partial^2 \phi}{\partial y \partial x} = \frac{\partial^2 \phi}{\partial x \partial y}$$

Now $\left(\frac{\partial \phi}{\partial x}\right)_y = 2xy^3 + 1$

and $\left(\frac{\partial \phi}{\partial y}\right)_x = 3x^2y^2$ give

$$\frac{\partial}{\partial y}\left[\left(\frac{\partial \phi}{\partial x}\right)_y\right]_x = \frac{\partial}{\partial y}(2xy^3 + 1)_x = 6\ xy^2$$

$$\frac{\partial}{\partial x}\left[\left(\frac{\partial \phi}{\partial y}\right)_x\right]_y = \frac{\partial}{\partial x}\left(3x^2y^2\right)_y$$

$$= 6xy^2 = \frac{\partial}{\partial y}\left[\left(\frac{\partial \phi}{\partial x}\right)_y\right]_x$$

Hence $d\phi$ is an exact differential.

(b) *Prove that in the equation* $dz = (52\ x^3y + 10\ y^5)\ dx + (13\ x^4 + 50\ xy^4)\ dy$, *dz is an exact differential.*

Solution:

If we write the given equation as $dz = M\ dx + N\ dy$, then

$M = 52\ x^3y + 10\ y^5$; $(\partial M/\partial y)_x = 52\ x^3 + 50\ y^4$;

$N = 13\ x^4 + 50\ xy^4$; $(\partial N/\partial x)_y = 52\ x^3 + 50\ x^3 + 50\ y^4$

Since $(\partial M/\partial y)_x = (N/\partial x)_y$, hence dz is exact differential.

Example 5:

A gas obeys the equation of state $\left(P + \frac{a}{V^2}\right)V = RT$. *Derive the cyclic rule for this gas.*

Solution:

The given equation of state can be written as

$$P = \frac{RT}{V} - \frac{a}{V^2} \qquad \text{...(i)}$$

Differentiation gives,

$$dP = \frac{R}{V}dT - \frac{RT}{V^2}dV + \frac{2a}{V^3}dV$$

$$= \frac{R}{V}dT - \left(\frac{RT}{V^2} - \frac{2a}{V^3}\right)dV \qquad \text{...(ii)}$$

As constant volume dV = 0 and hence equation (ii) yields.

$$\left(\frac{\partial P}{\partial T}\right)_V = \frac{R}{V} \qquad \text{...(iii)}$$

When pressure is constant, dP = 0 and equation (ii) reduces to

$$\left(\frac{\partial T}{\partial V}\right) = \left(\frac{RT}{V^2} - \frac{2a}{V^3}\right)\Big/\frac{R}{V} \qquad \text{...(iv)}$$

For constant temperature process, dT = 0 and equation (ii) can be written as

$$\left(\frac{\partial V}{\partial P}\right)_T = -\frac{1}{\left(\frac{RT}{V^2} - \frac{2a}{V^3}\right)} \qquad \text{...(v)}$$

On multiplying equations (iii). (iv) and (v) we get

$$\left(\frac{\partial P}{\partial T}\right)_V \left(\frac{\partial T}{\partial V}\right)_P \left(\frac{\partial V}{\partial P}\right)_T$$

$$= -\left(\frac{R}{V}\right) \times \frac{\left(\frac{RT}{V^2} - \frac{2a}{V^3}\right)}{\left(\frac{R}{V}\right)} \times \frac{1}{\left(\frac{RT}{V^2} - \frac{2a}{V^3}\right)} = -1$$

$$\text{or} \quad \left(\frac{\partial P}{\partial T}\right)_V \left(\frac{\partial T}{\partial V}\right)_P \left(\frac{\partial V}{\partial P}\right)_T + 1 = 0$$

This is cyclic rule.

Example 6:

Derive an expression for the coefficient of thermal expansion (α) *for a gas that obeys van der Waals equation of state.*

Solution:

For one mole of van der Waals gas

$$\left(P+\frac{a}{V^2}\right)(V-b)=RT$$

or $PV^3 - PbV^2 + aV - ab - RTV^2 = 0$

Differentiating with respect to T at constant P, we get,

$$3PV^2\left(\frac{\partial V}{\partial T}\right)_P - 2PbV\left(\frac{\partial V}{\partial P}\right)_P + a\left(\frac{\partial V}{\partial T}\right)_P - 2RTV\left(\frac{\partial V}{\partial T}\right)_P - RV^2 = 0$$

Solving for $\left(\frac{\partial V}{\partial T}\right)_P$, we have

$$\left(\frac{\partial V}{\partial T}\right)_P = \frac{RV^2}{3PV^2-2PbV+a-2RTV} \qquad \text{...(i)}$$

Now α is defined as

$$\alpha = \frac{1}{V}\left(\frac{\partial V}{\partial T}\right)_P \qquad \text{...(ii)}$$

Putting the expression for $(\partial V/\partial T)_P$ from into (ii), we have

$$\alpha = \frac{1}{V}\left[\frac{RV^2}{3PV^2-2PbV+a-2RTV}\right]$$

$$= \frac{R}{3PV-2Pb+\frac{a}{V}-2RT} \qquad \text{...(iii)}$$

van der Waals equation can also be written as

$$RT = PV - Pb + \frac{a}{V} - \frac{ab}{V^2} \qquad \text{...(iv)}$$

Putting the expression for RT into equation (iii), we get

$$\alpha = \frac{R}{3PV-2Pb+\frac{a}{V}-2\left(PV-Pb+\frac{a}{V}-\frac{ab}{V^2}\right)}$$

$$= R\left[PV - \frac{a}{V} + \frac{2ab}{V^2}\right]^{-1}$$

Example 7:

(a) *Derive an expression for* β *for a van der Waals gas.*

Solution :

$$\beta = \frac{1}{V}\left(\frac{\partial V}{\partial P}\right)_T$$

van der Waals equation is written as,

$PV^3 - PbV^2 + aV - ab - RTV^2 = 0$

Differentiating with respect to P at constant T,

$$3PV^2\left(\frac{\partial V}{\partial P}\right)_T + V^3 - 2PbV\left(\frac{\partial V}{\partial P}\right)_T - bV^2 + a\left(\frac{\partial V}{\partial P}\right)_T - 2RTV\left(\frac{\partial V}{\partial P}\right)_T = 0$$

Solving for $\left(\frac{\partial V}{\partial P}\right)_T$, we get

$$\left(\frac{\partial V}{\partial P}\right)_T = \frac{V^2(b-V)}{3PV^2 - 2PbV + a - 2RTV}$$

Also, for van der Waals gas,

$$RTV = PV^2 + a - PbV - \frac{ab}{V}$$

$$\therefore \left(\frac{\partial V}{\partial P}\right)_T = \frac{V^2(b-V)}{3PV^2 - 2PbV + a - 2\left(PV^2 + a - PbV - \frac{ab}{V}\right)}$$

$$= \frac{b-V}{P - a/V^2 + 2ab/V^3}$$

$$\therefore \quad \beta = -\frac{1}{V}\left(\frac{\partial V}{\partial P}\right)_T = -\frac{b-V}{PV - a/V + 2ab/V^2}$$

(b) *Show that volume is a state function for a gas obeying the equation:*

$P = RTV^{-1} - aV^{-2}$.

Solution:

Volume (V) is a state function if dV is exact differential; dV is an exact differential if $(\partial^2V/\partial P\, \partial T) = (\partial^2V/\partial T\, \partial P)$. Differentiating equation $P = RTV^{-1} - aV^{-2}$, we get

$$dp = RV^{-1}dT - RTV^{-2}\, dV + 2aV^{-3}dV$$

$$= RV^{-1}dT + (2aV^{-3} - RTV^{-2})dV$$

$$\therefore (\partial V/\partial P)_T = (2aV^{-3} - RTV^{-2})^{-1} = (aV^{-3} - PV^{-1})^{-1} \quad \text{...(i)}$$

$$\text{and } (\partial V/\partial T)_P = -RV^{-1}/\, (2aV^{-3} - RTV^{-2}) = -R\, [aV^{-2} - P]^{-1} \quad \text{...(ii)}$$

Differentiating equation (i) with respect to T at constant P, we get

$$\frac{\partial^2 V}{\partial T \partial P} = -(aV^{-3}\ PV^{-1})^{-2} \times \left[-3aV^{-4}\left(\frac{\partial V}{\partial T}\right)_P + PV^{-2}\left(\frac{\partial V}{\partial T}\right)_P\right]$$

$$= V^2\, (aV^{-2} - P)^{-2} \times V^{-2}\, (P - 3aV^{-2}) \times R\, (aV^{-2} - P)^{-1}$$

$$= R\, (aV^{-2} - P)^{-3}\, (P - 3aV^{-2}) \quad \text{...(iii)}$$

Differentiating equation (ii) with respect to P at constant T, we have

$$\frac{\partial^2 V}{\partial P \partial T} = R\, (aV{-}2{-}P){-}2 \times \left[-2aV^{-3}(\partial V/\partial P)_T{}^{-1}\right]$$

$$= -R\, (aV^{-2} - P)^{-2} \times [2aV^{-3}(aV^{-3}PV^{-1})^{-1} + 1]$$

$$= R\, (P - 3aV^{-2})\, (aV^{-2} - P)^{-3} \quad \text{...(iv)}$$

Thus $[\partial^2V/\partial T\, \partial P] = [\partial^2V/\partial P\, \partial T]$

HOMOGENEOUS FUNCTION

Any function f (x, y, z) is said to be homogeneous of degree m of the variables x, y, z, if

$$f(\lambda x, \lambda y, \lambda z) = \lambda^m f\ (x, y, z) \quad \text{...(1)}$$

for all positive values of l. This identity implies that an increase in all variables by a factor λ produces an increase in the function by a factor of λ^m. If the degree of homogeneity is unity (m = 1), the operation of multiplying each of the independent variables by a parameter λ produces the same effect as multiplying the overall function by the same factor. All the extensive properties of a system are homogeneous of first degree of moles. Volume is an extensive property and also it is a homogeneous function of first degree of mole, because when we double the moles of the substance its volume is also doubled.

That is for

$V = f(T, P, n)$

keeping T and P constant if n is doubled then the new volume V′ is given by,

$V' = f(T, P, 2n) = 2V$

Thermodynamic functions namely, volume, energy, enthalpy, entropy, free energy and the work function are extensive properties and hence are homogeneous function of the first degree in moles. The intensive properties are homogeneous function of zeroth degree of moles. For example, pressure is an intensive property and hence a homogeneous function of zeroth degree *i.e.*

$P(\lambda n_1, \lambda n_2 ...) = \lambda^{o} P(n_1, n_2 ...)$

Euler's Theorem on Homogeneous Functions

This theorem states that if the function f (x, y, z) is a homogeneous function of degree m then

$$x\frac{\partial f}{\partial x} + y\frac{\partial f}{\partial y} + z\frac{\partial f}{\partial z} = mf(x, y, z) \quad ...(2)$$

and for a first degree function (m = 1), we have

$$x\frac{\partial f}{\partial x} + y\frac{\partial f}{\partial y} + z\frac{\partial f}{\partial z} = f(x, y, z) \quad ...(3)$$

These equations can be generalised for any number of independent variables. This theorem is very useful to deduce the relations involving partial molar quantities.

Example 8:

Prove that the volume of an ideal gas is a homogeneous function of zeroth degree in pressure and temperature.

Solution:

Volume of a fixed amount of a gas is a function of T and P,

$$V = f(T, P) \quad ...(i)$$

Applying the Euler's theorem of homogeneity we have

$$T\frac{\partial V}{\partial T}+P\frac{\partial V}{\partial P} = mV \qquad ...(ii)$$

For an ideal gas of known amount, PV = nRT and

$$\left(\frac{\partial V}{\partial T}\right)_P = \frac{nR}{P},$$

$$\left(\frac{\partial V}{\partial P}\right)_T = -\frac{V}{P}$$

Substituting the expressions for these derivatives into equation (ii) we obtain

$$\frac{nRT}{P} - V = mV$$

$$\text{or} \quad \frac{nRT}{P} - \frac{nRT}{P} = mV = 0 \ (\because V = nRT/P)$$

since V is not zero, therefore m must be zero. Hence for an ideal gas volume is a homogeneous function of the zeroth degree in pressure and temperature *i.e.*,

$$V(\lambda P, \lambda T) = \lambda^{\circ} V(T, P)$$

This is obvious from the relation V = nRT/P also. Suppose we double both the temperature and pressure, the volume of the ideal gas will not be changed.

For other extensive properties we shall have

(i) $A = A(T, V, n_1, n_2)$ ...Helmholtz free energy

$\therefore A(T, \lambda V, \lambda n_1, \lambda n_2) = \lambda A$

since T is an intensive property, therefore, it is not multiplied by λ whereas V being an extensive property is multiplied by λ when the moles (n_i's) are so multiplied.

$S = S(E, V, n_1, n_2)$ ...Entropy

$S(\lambda E, \lambda V, \lambda n_1, \lambda n_2) = \lambda S$

$G = G(T, P, n_1, n_2)$

$G(T, P, \lambda n_1, \lambda n_2) = \lambda G$ ...Gibbs free energy

T and Pare intensive variables and hence are not multiplied by λ.

(iv) $H = H(S, P, n_1, n_2)$ ...Enthalpy

$\therefore$ H (λS, P, λn_1, λn_2) = λH

(v) E = E (S,V, n_1, n_2) ...Intrinsic energy

$\therefore$ E (λS, λV, λn_1, λn_2) = λE

(vi) V = V (T, P, n_1, n_2) ...Volume

$\therefore$ V(T, P, λn_1, λn_2) = λV

The treatment of homogeneity enables us to define the intensive and extensive properties in more precise form. Let us consider a consider a set of intensive variables x, y, z, and a set of extensive variables X, Y, Z, then

(i) The function F (x, y, z,...X, Y, Z,..) is an extensive variable if and only if F (x, y, z, λX, λY, λZ) = λF (x, y, z, X, Y, Z) for arbitrary positive λ.

For example, volume is an extensive property because

V (T, P, λn_1, λn_2) = λV (T, P, n_1, n_2).

(ii) The function f (x, y, z, X, Y,Z) is an intensive property if and only if f (x, y, z, λX, λY, λZ) = f (x, y, z, X, Y, Z) for arbitrary positive λ.

INTEGRATING FACTOR

It is always possible to get an exact differential from an inexact differential if the latter is multiplied by an integrating factor. For example, let us consider an inexact differential dz given by

$$dz = ydx - xdy \qquad ...(1)$$

If dz were exact differential then

$$\left(\frac{\partial y}{\partial y}\right)_x = -\left(\frac{\partial x}{\partial x}\right)_y$$

Equation (15) gives,

$$\left(\frac{\partial z}{\partial x}\right)_y = y\ ;\ \left(\frac{\partial z}{\partial y}\right)_x = -x$$

and $$\frac{\partial^2 z}{\partial y \partial x} = \left(\frac{\partial y}{\partial y}\right)_x = 1; \frac{\partial^2 z}{\partial x \partial y} = -\left(\frac{\partial x}{\partial x}\right)_y = -1$$

obviously dz is inexact differential, since $\partial^2 z/\partial y\ \partial x \neq \partial^2 z/\partial x\ \partial y$

Let $f = x^m y^n$ be the integrating factor, then equation (1) is written as,

$$dz\,(x^m y^n) = y^{n+1}\, x^m\, dx - x^{m+1} y^n\, dy$$

It should be exact differential and the Euler's theorem of exactness would give

$$\frac{\partial}{\partial y}\left[y^{n+1} x^m\right]_x = -\frac{\partial}{\partial x}\left[x^{m+1} y^n\right]_y$$

$$x^m (n+1) y^n = -(m+1) x^m y^n$$

$$(n + m) = -2 \qquad \text{...(2)}$$

There are various values of m and n for which equation 16 is valid, and consequently there are infinite number of integrating factors e.g.,

$n = 0,\ m = -2,$

or $n = 1,\ m = -3.$

Thus one integrating factor is $f = x^2 y^0 = x^2$; with integrating factor equation 15 is now written as

$$f.\,dz = \frac{y}{x^2}dx - \frac{x}{x^2}dy = \frac{y}{x^2}dx - \frac{1}{x}dy$$

Let $M = y/x^2$, and $N = -1/x$

$$\therefore \left(\frac{\partial M}{\partial y}\right)_x = \frac{1}{x^2} \text{ and } \left(\frac{\partial N}{\partial x}\right)_y = 1/x^2$$

That is $\dfrac{\partial^2 z}{\partial y \partial x} = \dfrac{\partial^2 z}{\partial y \partial x} = 1/x^2$

It is seen that dz becomes exact differential if multiplied by (1/x2) and hence $(1/x^2)$ is the *integrating* factor in this case.

Example 9:

Prove that 1/T is an integrating factor for

$$dq = \frac{3}{2}nRdT + \frac{nRT}{V}dV$$

for an ideal gas.

Solution:

Given that $dq = \dfrac{3}{2}nRdT + \dfrac{nRT}{V}dV$

$= M\ dT + N\ dV$

where $M = \frac{3}{2}nR$ and $N = \frac{nRT}{V}$

Now $\left(\frac{\partial M}{\partial V}\right)_T = \frac{\partial}{\partial V}\left(\frac{3}{2}nR\right)_T = 0$

and $\left(\frac{\partial N}{\partial T}\right)_v = \frac{\partial}{\partial T}\left(\frac{nRT}{V}\right) = \frac{nR}{V}$

It is seen that $\left(\frac{\partial M}{\partial V}\right)_T \neq \left(\frac{\partial N}{\partial T}\right)_V$ in the present case.

Let $f = T^{-1}$ be the integrating factor, then

$$f.\ dq = \frac{3}{2}nR\frac{dT}{T} + \frac{nRT}{V}\left(\frac{1}{T}\right)dV$$

$$= \frac{3}{2}nR\frac{dT}{T} + \frac{nR}{V}dV$$

$$= M\ dT + NdV$$

$$M = \frac{3}{2}\frac{nR}{T}, N = \frac{nR}{V}$$

$$\therefore \left(\frac{\partial M}{\partial V}\right)_T = \frac{\partial}{\partial V}\left(\frac{3}{2}\frac{nR}{T}\right)_T = 0;$$

$$\text{and} \left(\frac{\partial N}{\partial T}\right)_V = \frac{\partial}{\partial T}\left(\frac{nR}{V}\right)_V = 0$$

$$\text{Thus} \left(\frac{\partial M}{\partial V}\right)_T = \left(\frac{\partial N}{\partial T}\right)_V = 0$$

Therefore, (dq/T) is exact differential and (1/T) is an integrating factor.

Example 10:

For an ideal gas (PV = nRT) show that (1/T) is an integrating factor for dw = PdV.

Solution:

Given that dw = PdV ...(i)

PV = nRT for ideal gas.

$\therefore$ PdV + VdP = nRdT

and $dV = \frac{nR}{P}dT - \frac{V}{P}dP$ ...(ii)

Substituting the expression for dV in equation (i) we get

$$dw = nRdT - VdP = nRdT - \frac{nRT}{P}dP$$

Now $\frac{\partial}{\partial P}(nR)_T = 0$

and $\frac{\partial}{\partial T}\left(-\frac{nRT}{P}\right)_P = -\frac{nR}{P}$

Therefore, dw is inexact differential.

Let $f = \frac{1}{T}$ be an integrating factor, then

$$f.\ dw = \frac{1}{T}nR\ dT - \frac{nR}{P}dP$$

$$\therefore \frac{\partial}{\partial P}\left[\frac{nR}{T}\right]_T = 0$$

and $\frac{\partial}{\partial T}\left[\frac{nR}{P}\right]_P = 0$

Thus f. $dw = \frac{nR}{T}dT - \frac{nR}{P}dP$ is an exact differential and $\frac{1}{T}$ is and integrating factor.

Example 11:

If E = f(T, V) and dE is an exact differential then show that $\left(\frac{\partial E}{\partial V}\right)_T = T\left(\frac{\partial P}{\partial T}\right)_V - P$, *relationship can be derived by using the relation dq = dE + PdV and taking (1/T) as an integrating factor.*

Solution:

Since E = f (T, V)

Therefore, $dE = \left(\frac{\partial E}{\partial T}\right)_V dT + \left(\frac{\partial E}{\partial V}\right)_T dV$

Given that dq = dE + PdV

Now $dE = \left(\frac{\partial E}{\partial T}\right)_V dT + \left(\frac{\partial E}{\partial V}\right)_T dV$

$$\therefore dq = \left(\frac{\partial E}{\partial T}\right)_V dT + \left(\frac{\partial E}{\partial V}\right)_T + P \Big] dV$$

Let $f = \frac{1}{T}$ be the integrating factor, then f dq must be an exact differential.

$$fdq = \frac{1}{T}\left(\frac{\partial E}{\partial T}\right)_V dT + \frac{1}{T}\left[\left(\frac{\partial E}{\partial V}\right)_T + P\right] dV$$

$= M\ dT + N\ dV$

where, $M = \frac{1}{T}\left(\frac{\partial E}{\partial T}\right)_V$; $N = \frac{1}{T}\left[\left(\frac{\partial E}{\partial T}\right)_T + P\right]$

For the above differential to be exact we must have

$$\left(\frac{\partial M}{\partial V}\right)_T = \left(\frac{\partial N}{\partial T}\right)_V$$

or $\frac{\partial}{\partial V}\left[\frac{1}{T}\left(\frac{\partial E}{\partial T}\right)_V\right]_T = \frac{\partial}{\partial T}\left[\frac{1}{T}\left(\frac{\partial E}{\partial T}\right)_T + \frac{P}{T}\right]_V$

or $\frac{1}{T}\frac{\partial^2 E}{\partial V \partial T} = \frac{1}{T}\frac{\partial^2 E}{\partial T \partial V} - \frac{1}{T^2}\left(\frac{\partial E}{\partial V}\right)_T - \frac{P}{T^2} + \frac{1}{T}\left(\frac{\partial P}{\partial T}\right)_V$ ...(i)

Since dE is an exact differential, therefore,

$\frac{\partial^2 E}{\partial T \partial V} = \frac{\partial^2 E}{\partial V \partial T}$ ...(ii)

From equations (i) and (ii), we get

$$-\frac{1}{T^2}\left(\frac{\partial E}{\partial V}\right)_T - \frac{P}{T^2} + \frac{1}{T}\left(\frac{\partial P}{\partial T}\right)_V = 0$$

or $\left(\frac{\partial E}{\partial V}\right)_T = T\left(\frac{\partial P}{\partial T}\right)_V - P$

Exercise: If H = f (T, P) and dH is an exact differential then prove that

$$\left(\frac{\partial H}{\partial P}\right)_T = V - T\left(\frac{\partial V}{\partial T}\right)_P$$

[**Hint :** Start with dq = dH – V dp taking 1/T as an integrating factor.]

THE SECOND LAW OF THERMODYNAMICS

Relevance of the Second Law

The original statement of the second law of thermodynamics is much concerned with the development of the theoretical aspects of heat engines specially the Carnot's cycle. This has got a very little implication in chemical processes. But, the further developments and various derivations involving the concept of this law are of immense use for chemists. For example, the direction of a particular spontaneous transformation can be ascertained with the help of second law of thermodynamics which is not predicted by first law of thermodynamics.

Spontaneous Process

Most of the natural processes are spontaneous. In fact all human activities and all that happens in the universe involve a significant degree of this concept. For example, the production of wheat from seeds, the formation of clouds in the sky on a sunny day, the fall of rain from clouds, the growth and multiplication of micro-organism when food is digested, are all spontaneous processes.

All those processes which take place of their own accord are called spontaneous process. As such all the spontaneous processes are irreversible in nature. Once having undergone a spontaneous transformation in a particular direction, the process will not take place in the reverse direction spontaneously. The spontaneity however never means that the process is instantaneous. It simply stands that the change is favourable in a particular direction. A few other examples of spontaneous processes are as follows. These will explain as to why do we need a *second law*? A plate dropped on to the floor breaks and the pieces scatter; when we take ice-cream our lips get cold; petroleum vapour and oxygen when sparked react in a car engine producing exhaust gases; water runs downhill; a drop of perfume in a room diffuses quickly to effect the entire atmosphere; aqueous solutions of NaOH and HCl react to produce NaCl (aq) and heat. All these processes are known to proceed in one direction. The question is that why the reverse of the considered processes do not occur?

Can we believe on our eyes to see the pieces of broken plate spontaneously reorganise themselves to reproduce the original plate? Has any one experienced ever hot while taking ice cream? Is it possible exhaust gases CO, CO_2, H_2O can recombine to produce petroleum and oxygen? Is it possible for the dispersed vapours of the perfume to get converted into the droplets and return to the bottle under ordinary conditions? Can water flow of itself upwards? Can we get aqueous solutions of NaOH and HCl on dissolving NaCl in water?

None of these processes is known to be spontaneous as we know by our experiences. However, in the light of the first law of thermodynamics none of these processes seem to be impossible, because in these processes the energy in conserved, converted from one form into another. This implies that the first law of thermodynamics is incapable of predicting the direction of a process. Now the question arises that what law of nature compels the spontaneous processes to proceed in a particular direction? How and why the systems are aware of the directions in which to proceed? What is the driving force that leads the process in a given direction? The answers to all these questions can be provided by considering the second law of thermodynamics.

Statements of the Second Law of Thermodynamics

1. "Heat cannot spontaneously pass from a colder to a warmer body." (R.J.E. Clausius)
2. "It is impossible to transfer heat from a colder system to a warmer without other simultaneous changes occurring in the two systems, or in their environment." (P.S. Epstein)
3. "Every system which is left to itself will, on the average, change toward a condition of maximum probability." (G. N. Lewis)
4. "The state of maximum entropy is the most stable state for an isolated system." (Enrico Fermi).
5. "In any irreversible process the total entropy of all bodies concerned is increased." (G. N. Lewis)
6. "In an adiabatic process the entropy either increases or remains unchanged:

$$\Delta S \geq 0$$

Where the upper sign (>) refers to the irreversible, the lower (=) to the reversible case." (P. S. Epstein)

7. "The entropy function of a system of bodies tends of increase in all physical and chemical processes occurring in nature, if we include in the system all such bodies which are affected by the change." (Saha)
8. "The entropy increases towards a maximum and the energy of the universe in constant." (R.J.E. Clausius)
9. "Gain in information is loss in entropy." (G. N. Lewis)
10. "Entropy is times arrow." (A. Eddington)
11. "Every system left to itself changes rapidly or slowly in such a way to approach a definite final state of rest. No system of its own will change away from the state of equilibrium, except through the influence of external agencies" (G. N. Lewis)
12. " There is a general tendency in nature for energy to pass from more available to less available forms." (J. A. V. Butter)
13. "When an actual process occurs it is impossible to invent a means of restoring every system concerned to its original condition." (G. N. Lewis)
14. "There exists a characteristic thermodynamic function called entropy. The difference of a system in states (1) and (2) is given by the expression

$$S_2 - S_1 = \int_1^2 dq/T$$

over any reversible path.

and $$S_2 - S_1 > \int_1^2 dq/T$$

over any irreversible path.

The entropy is a property of the state only. Its value for an isolated system can never decrease." (R. E. Gibson)

THE CONCEPT OF ENTROPY

The concept of entropy is vital for thermodynamics. It is a Greek word with no equivalent in common language. It was invented by R.J.E. Clausius. He denoted it by the symbol S. **Entropy** stands for trope (a greek word meaning change), the suffix 'en' is written to identify it with *energy*. We have an intuitive feel for the concept of energy; a capacity to do work. However, there is no such intuitive grasp of entropy. Nevertheless, we are quite familiar with the effects the entropy has been used for. Entropy is a measure of randomness or disorderness in a molecular system. More is randomness larger is the value of entropy. More regular is the arrangement lower is the entropy. Entropy is also a measure of unavailable energy. Sometimes entropy is also recognised as an index of exhaustion. Its significance can be appreciated by considering some general observations.

In a cyclic process heat cannot be completely converted into work. The efficiency of working is never hundred per cent. The magnitude of work of irreversible expansion is less than the reversible expansion. Why is it so ? Why the entire heat absorbed cannot be converted into useful work? Wherelse the rest of the unused heat has gone? Why the system's capacity to do work is reduced in an irreversible process? The answers to these questions can partly be given as follows.

When heat is absorbed by a gas, the kinetic energy of the gas molecules will be increased and they will move faster, and collide with (i) each other, (ii) the container and (iii) the face of the piston, thereby the piston will be pushed. Only the third type of collision will bring about the expansions will not bring effective expansion of the gas. Thus some collisions and hence a fraction of heat is not utilized for the expansion of the gas. Certainly the work due to expansion is note exactly equivalent to the amount of heat absorbed. Since the heat goes waste only because of the random motion of the gas molecules, we can correlate the randomness with the unavailable work.

The extent of randomness or the unavailable work is indicated by the entropy of the system. Under an ideal situation, if all the gas molecules were striking the piston perpendicularly then certainly whole of the heat absorbed will be converted into work as shown in Fig. 1.6(a). But this is only possible if there is a regular and only one possible arrangement of the gas molecules. In other situations when the molecules are in random motion, a fraction of heat is converted into such a factor which is a measure of randomness of the molecular system. *So entropy is the measure of randomness of a molecular system.*

This statement is also in line with the observations at the boiling point of a fluid. Under ordinary conditions when a liquid is heated its temperature increases, at the boiling point the temperature remains constant will whole of the liquid is transformed into vapour state with the continuous supply of heat. The question is that why the temperature at the boiling point remains constant? Where is the heat being utilised? The answer can be given if we consider the molecular arrangement in the liquid and vapour state at a given temperature.

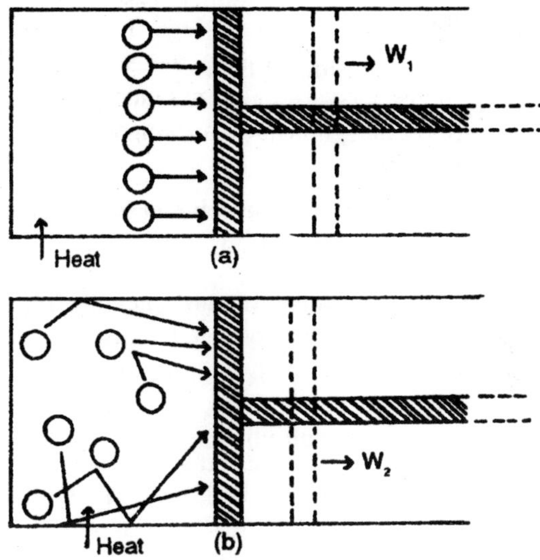

Fig. 1.6 : Conversion of heat work. (a) Ideal case of complete conversion. (b) Partial conversion due to randomness (dotted lines represent the position of the piston after expansion and $w_1 > w_2$).

The molecular arrangement in the liquid state is a bit more regular than the vapour state of the same substance. To disturb the regularity in the liquid state some work must be done. This work done on the system increases the disorderness but not the temperature at the boiling point. Since at the boiling point there is more disorderness in the vapour phase than the liquid, we can say the entropy of the vapour is greater than that of the liquid state of the same substance. Agreeing on the same lines we can say that the entropy of the liquid state is greater than that of the solid state of the same substance. In general, for any substance; S (vapour) > S (liquid) > S (solid) at a given temperature and pressure.

Mathematical Treatment of Entropy Concept

The logical concept of entropy can be given a mathematical foundation. This treatment is essential because we wish to recognise entropy (S) as a thermodynamic function. For this purpose we shall start with the first law of thermodynamics applied to the changes in states of an ideal gas. (Later on this can be extended for other substances also).

Let us consider an ideal gas that is undergoing a reversible change from an initial state T_1, V_1 to a final state T_2, V_2. For infinitesimal changes, the first law of thermodynamics can be written as,

$$dq_{rev} = dE - dw_{rev} \quad ...(i)$$

If only PV work is considered then

$$-dw_{rev} = PdV$$

and $dq_{rev} = dE + PdV$

For an ideal gas $dE = C_V\, dT$, and $P = n\, RT/V$, hence

$$dq_{rev} = Cv\ dT + (nRT/V)\ dV \quad ...(ii)$$

To know the total heat change we must integrate this equation. But we see that the integration is not possible because the second term on right side of the equation involves two variables T and V, whereas we have to carry the integration with respect to V only. However, if both sides of the equation (ii) are multiplied by l/T, then integration can be carried. Thus

$$\frac{dq_{rev}}{T} = C_V \frac{dT}{T} + nR\frac{dV}{V} \quad ...(iii)$$

Integrating equation (iii) between the limits

$V = V_1$. $V = V_2$, $T = T_1$, $T = T_2$, we have

$$\int_1^2 \frac{dq_{rev}}{T} = C_V \ln\left(\frac{T_2}{T_1}\right) + nR\ln\left(\frac{V_2}{V_1}\right) \quad ...(iv)$$

For an ideal gas, $P_1V_1/T_1 = P_2\ V_2/\ T_2$, hence

$$\int_1^2 \frac{dq_{rev}}{T} = C_P \ln\left(\frac{T_2}{T_1}\right) + nR\ln\left(\frac{P_1}{P_2}\right) \quad ...(v)$$

Here $C_P = (C_V + nR)$ = hear capacity at constant pressure. We find that

(i) dq_{rev} is not an exact differential because

$(\partial\ CV/\partial V)_T \neq (\partial n\ RT/\partial T)v$

(ii) (dq_{rev}/T) becomes an exact differential and can be integrated. hence 1/T is the integrating factor.

(iii) Since $dq_{rev}/\ T$ is an exact differential therefore the function whose change is given by dqrev/T must be a state function. Let this state function be S and hence

$dq_{rev}/T = dS$ ...(vi)

Now the exactness of dS and state dependence of S can be shown as follows:

dS is exact differential. From equations (iii) and (vi) we have

$dS = (C_V/T)\, dT + (nR/V)dV$ ...(viii)

At constant T, dT = 0 and

$(\partial S/\partial V)_T = nR/V$

Differentiating with respect to T at constant V we have

$$\frac{\partial^2 S}{\partial T \partial V} = \frac{\partial}{\partial T}(nR/V)_V = 0 \qquad ...(ix)$$

When V is constant dV = 0, and

$$\left(\frac{\partial S}{\partial T}\right)_V = (CV/T) \qquad ...(x)$$

On differentiation with respect to V at constant T, equation (x) gives

$$\frac{\partial^2 S}{\partial V \partial T} = \frac{\partial}{\partial V}(C_V / T)_T = 0 \qquad ...(xi)$$

From equation (ix) and (xi) we see that

$$\frac{\partial^2 S}{\partial T \partial V} = \frac{\partial^2 S}{\partial V \partial T}$$

Hence dS is an exact differential and S is a state function.

S *is a state function,* we have shown that

$$\int_1^2 \frac{dq_{rev}}{T} = C_V \text{ In } (T_2/T_1) + nR \text{ In } (V_2/V_1)$$

The right hand side of this equation depends only on the state variables T_1, V_1 and T_2, V_2, therefore the quantity on left side should also be a function only of state variables. This is

$$\int_1^2 \frac{dq_{rev}}{T} = \int_1^2 dS = S_2(T_2, V_2) - S_1(T_1, V_1) = \Delta S \qquad ...(xii)$$

$\therefore \Delta S = C_V \text{ In } (T_2/T_1) + nR \text{ In } (V_2/V_1)$

This state function S is called entropy function and like other thermodynamic functions the changes in entropy depend only on the state of the system and are independent of the path followed.

Equations (vi) and (xii) are the mathematical statements of the second law of thermodynamics and are used to define the entropy function. Here it must be regarded that this mathematical concept of entropy ($dS = dq_{rev}/T$) holds only for a REVERSIBLE change and change for an irreversible change $dS \neq dq_{irr}/T$.

COMBINED FROM OF THE FIRST AND SECOND LAWS OF THERMODYNAMICS

The mathematical form of the first law is

$$dq = dE - dw = dE + PdV \quad ...(1)$$

(if only PV work is considered).

The mathematical expression for second law is

$$dq_{rev} = TdS \quad ...(2)$$

Equation (1) and (2) give

$$TdS = dE + PdV$$

$$\text{or } dE = TdS - PdV \quad ...(3)$$

This is the combined form of the first and the second laws of thermodynamics. This is very important and useful in the deduction of various other concepts of thermodynamic functions.

Enthalpy and Entropy

Enthalpy is expressed as

$$H = E + PV$$

$$\therefore dH = dE + PdV + VdP$$

Putting $dE + PdV = TdS$ we get

$$dH = TdS + VdP \quad ...(4)$$

Energy as a Function of S and V

Equation (3) clearly indicates that the change in energy depends on the change of entropy (dS) and the change of volume (dV). Thus mathematically E is a function of S and V

$$E = f(S, V)$$

The total differential of E is given as

$$dE = \left(\frac{\partial E}{\partial S}\right)_V dS + \left(\frac{\partial E}{\partial V}\right)_S dV \quad ...(5)$$

Also $dE = TdS - PdV$ ...(6)

Comparing the coefficients of dS and dV separately in equations (5) and (6) we get,

$$\left(\frac{\partial E}{\partial S}\right)_V = T \qquad ...(7)$$

and $$\left(\frac{\partial E}{\partial V}\right)_s = -P \qquad ...(8)$$

Since dE is an exact differential therefore Euler's of exactness when applied with equation (7) gives

$$\left(\frac{\partial T}{\partial V}\right)_s = -\left(\frac{\partial P}{\partial S}\right)_V$$

From equation (3) we see that for constant energy $dE = 0$ and

$$\left(\frac{\partial S}{\partial V}\right)_E = \frac{P}{T}$$

Enthalpy as a Function of S and P

$dH = TdS + VdP$ ...(9)

This equation clearly shows that the change in enthalpy (dH) depends on the changes in entropy (dS) and pressure (dP). Thus

$H = f(S, P)$

$$dH = \left(\frac{\partial H}{\partial S}\right)_P dS + \left(\frac{\partial H}{\partial P}\right)_s dP \qquad ...(10)$$

Comparing the coefficients of dS and dP respectively in equations (10) and (9), we obtain

$$\left(\frac{\partial H}{\partial S}\right)_P = T$$

and $$\left(\frac{\partial H}{\partial S}\right)_s = V \qquad ...(11)$$

ENTROPY CHANGE FOR ISOLATED SYSTEM

For reversible isolated processes the change of entropy is zero whereas for irreversible isolated processes the entropy change is greater than zero. This is shown as follows.

Let us suppose that small changes are made in the system reversibly and irreversibly. The work done by the system on the surrounding in the reversible process $(-dw_{rev})$ i.e.

$$(-dw_{rev}) > (-dw_{irr}) \qquad ...(1)$$

Since energy (E) of the system is a state function and dE is exact differential and must be the same for a given change whether it is performed reversibly or irreversibly. That is $dE_{rev} = dE_{irr} = dE$.

From the first law we have,

$$dE = dq_{rev} + dw_{rev}$$

$$\text{and } dE = d_{irr} + dw_{irr}$$

Since $(-dw_{rev}) > (-dw_{irr})$, R. H. S is positive

$$\therefore\ dq_{rev} > dq_{irr}$$

This is, in any small change a system will absorb more heat under reversible conditions than under irreversible conditions.

On dividing both sides of equation (2) by T, we get

$$\frac{dq_{rev}}{T} > \frac{dq_{irr}}{T} \qquad ...(2)$$

But (dqrev/T) = dS (definition from 2nd law)

$$\therefore\ dS > \frac{dq_{irr}}{T}$$

$$\text{or } dS > \frac{dE - dw_{irr}}{T} \qquad ...(3)$$

$$\text{and } dS = \frac{dE - dw_{rev}}{T} \qquad ...(3b)$$

In an isolated system there is no exchange of energy, matter, heat etc., across the boundary, i.e., dq = 0, dw = 0, dE = 0.

Introducing these conditions into equations (3b) and (3a), we get

$$(dS)_{rev} = 0 \text{ and } (dS)_{irr} > 0$$

$$\text{or } (\Delta s)_{rev} = 0 \text{ and } (\Delta s)_{irr} > 0 \qquad ...(4)$$

The entropy changes expressed by equation (15) indicate as if it were not a state function? But it is. The why ΔS for reversible and irreversible isolated system are different? The explanation is like is like this. An isolated system includes the system under investigation and the surroundings in its immediate vicinity. To calculate any property of an isolated system , therefore, we must account for the properties of both the system and the surroundings. The entropy change for the isolated system includes the changes in the system as well as surroundings, and the total entropy change is called the entropy of the universe.

ΔS (universe) = ΔS (system) + ΔS (surroundings)

For the system $D\,S_{rev}^{syst} = D\,S_{irr}^{syst}$. It is the DS (surrounding) which is different for reversible and irreversible changes.

THERMODYNAMIC RELATIONS BASED ON SECOND LAW

Thermodynamic Equation of State

Any relationship between the thermodynamic function (E or H) and the PVT data is called the thermodynamic equation of state.

(i) **Energy as a function of T and V.** For any reversible process involving only P.V work, the combined form of the first and second laws can be written as

$$TdS = dq_{rev} = dE + PdV \qquad ...(1)$$

E is a state function and dE is exact differential, given as

$$dE = \left(\frac{\partial E}{\partial T}\right)_V dT + \left(\frac{\partial E}{\partial V}\right)_T dV - C_V dT + \left(\frac{\partial E}{\partial T}\right)_V dT + \left(\frac{\partial E}{\partial V}\right)_T dV$$

$C_V = (\partial E/\partial T)_V$ = heat capacity at constant volume

$$\therefore\ dS = \frac{C_V}{T}dT + \frac{1}{T}\left[P + \left(\frac{\partial E}{\partial V}\right)_T\right]dV \qquad ...(2)$$

Entropy is also a state function S = S (T, V) and dS is exact differential given as,

$$dS = \left(\frac{\partial S}{\partial V}\right)_V dT + \left(\frac{\partial S}{\partial V}\right)_T dV \qquad ...(3)$$

Comparing the coefficient of dT and dV separately of equations (2) and (3), we get

$$\left(\frac{\partial S}{\partial T}\right)_V = \frac{C_V}{T} = \frac{1}{T}\left(\frac{\partial E}{\partial T}\right)_V \qquad ...(4)$$

$$\left(\frac{\partial S}{\partial V}\right)_V = \frac{1}{T}\left[P + \left(\frac{\partial E}{\partial V}\right)_T\right] \qquad ...(5)$$

Differentiating equation (4) with respect to V at constant T, we get

$$\left(\frac{\partial^2 S}{\partial V \partial T}\right) = \frac{\partial}{\partial V}\left(\frac{C_V}{T}\right)_T = \frac{1}{T}\frac{\partial}{\partial V}\left[\left(\frac{\partial E}{\partial T}\right)_V\right]_T \quad ...(6)$$

On differentiating equation (5) with respect to T at constant volume, we get

$$\frac{\partial^2 S}{\partial T \partial V} = \frac{\partial}{\partial T}\left[\frac{1}{T}\left\{P + \left(\frac{\partial E}{\partial V}\right)_T\right\}\right]_V$$

$$= -\frac{1}{T^2}\left(\frac{\partial E}{\partial V}\right)_T + \frac{1}{T}\left(\frac{\partial P}{\partial T}\right)_V + \frac{1}{T}\frac{\partial^2 E}{\partial T \partial V} \quad ...(7)$$

Since S is a state function and dS is exact differential, therefore, Euler's theorem of exactness must be valid i.e.,

$$\frac{\partial^2 S}{\partial V \partial T} = \frac{\partial^2 S}{\partial T \partial V}$$

On applying this condition, equations (7) and (6) give

$$\frac{1}{T}\frac{\partial^2 E}{\partial V \partial T} = -\frac{P}{T^2} - \frac{1}{T^2}\left(\frac{\partial E}{\partial V}\right)_T + \frac{1}{T}\left(\frac{\partial P}{\partial T}\right) + \frac{1}{T}\frac{\partial^2 E}{\partial T \partial V} \quad ...(8)$$

Since dE is exact differential

$$\therefore \quad \frac{\partial^2 E}{\partial V \partial T} = \frac{\partial^2 E}{\partial T \partial V}$$

Applying this equality in equation (8), we get

$$\left(\frac{\partial E}{\partial V}\right)_T = T\left(\frac{\partial P}{\partial T}\right)_V - P \quad ...(9)$$

This expression is known as the thermodynamic equation of state and is true for all substances under all conditions and follows directly from the acceptance of the validity of the first and second laws. Equation (9) can be written in another form if we realise that $(\partial P/\partial T)_V = \alpha/\beta$, where $\alpha = V^{-1}(\partial V/\partial T)_P$, and $\beta = -V^{-1}(\partial V/\partial P)_T$.

$$\left(\frac{\partial E}{\partial V}\right)_T = T\frac{\alpha}{\beta} - P$$

Let us apply it to an ideal gas and then to a van der Waals gas.

Thermodynamic equation of state applied to an ideal gas. We have argued that the energy of an ideal gas is independent of volume and have defined an ideal gas as one obeying the two relationships PV = nRT and $(\partial E/\partial V)_T = 0$. With the assistance of the thermodynamic equation of state we can now prove the latter condition.

For an ideal gas PV = nRT

$$\therefore \quad \left(\frac{\partial P}{\partial T}\right)_V = \frac{nR}{V} = \frac{P}{T}$$

On substituting the expression $(\partial P/\partial T)V = PT$ into equation (9) we get,

$$\left(\frac{\partial E}{\partial V}\right)_T = T\left(\frac{\partial P}{\partial T}\right)_V - P = T\frac{P}{T} - P = 0$$

Van der Waals equation. It is given as

$$P = \frac{nRT}{V-nb} - \frac{an^2}{V^2}$$

where a is the attraction coefficient and b is the excluded volume per mole; a measure of bulk of the gas molecules. Using the thermodynamic equation of state we can deduce an expression for the internal pressure $(\partial E/\partial V)_T$ for the gas.

For van der Waals equation

$$\left(\frac{\partial P}{\partial T}\right)_V = \frac{nR}{V-nb}$$

On substituting this expression into thermodynamic equation of state we obtain

$$\left(\frac{\partial E}{\partial V}\right)_T = \frac{nRT}{V-nb} - P = \frac{nRT}{V-nb} - \left(\frac{nRT}{V-nb} - \frac{an^2}{V^2}\right)$$

$$\left(\frac{\partial E}{\partial V}\right)_T = \frac{an^2}{V^2} \qquad ...(10)$$

For one mole of a gas, equation (10) reduces to

$$\left(\frac{\partial E}{\partial V}\right)_T = \frac{a}{V^2}$$

$$\text{or} \quad dE = \frac{a}{V^2}dV$$

Integration gives $\Delta E = -a\left(\frac{1}{V^2} - \frac{1}{V_1}\right)$

The energy of a van der Waals gas is related to the attraction coefficient a. The energy increase with an isothermal expansion of the gas. When the molecules are close to each other, the volume is smaller, the intermolecular forces reduce the energy of the gas.

(ii) **Enthalpy as function of temperature and pressure.**

The enthalpy is defined as

$H = E + PV$

$\therefore\ dH = dE + PdV + VdP$

The combined first and second laws give $dE + PdV = TdS$, therefore

$$TdS = dH - VdP \qquad ...(11)$$

H being a state function H = H (T, P) its total differential is given by

$$dH = \left(\frac{\partial H}{\partial T}\right)_P dT + \left(\frac{\partial H}{\partial P}\right)_T dP = C_p dT + \left(\frac{\partial H}{\partial T}\right)_P dP \qquad ...(12)$$

Here $C_p = (\partial H/\partial T)_P$

Putting this expression into equation (11), we get

$$dS = \frac{C_P}{T} dT + \frac{1}{T}\left[\left(\frac{\partial H}{\partial P}\right)_T - V\right] dP \qquad ...(13)$$

Since S = f (T, P)

$$\therefore\ dS = \left(\frac{\partial S}{\partial T}\right)_P dT + \left(\frac{\partial S}{\partial P}\right)_T dP \qquad ...(14)$$

Comparing the coefficients of dT and dP separately in equations (13) and (14), we get

$$\left(\frac{\partial S}{\partial T}\right)_P = \frac{C_P}{T} = \frac{1}{T}\left(\frac{\partial H}{\partial T}\right)_P \qquad ...(15)$$

$$\left(\frac{\partial S}{\partial P}\right)_T = z\frac{1}{T}\left[\left(\frac{\partial H}{\partial P}\right)_T - V\right] \qquad ...(16)$$

On differentiating equation (15) with respect to P at constant T, we get

$$\frac{\partial^2 S}{\partial P \partial T} = \frac{1}{T}\frac{\partial^2 H}{\partial P \partial T} \quad ...(17)$$

On differentiation w.r. to T at constant P equation (16) yields

$$\frac{\partial^2 S}{\partial T \partial P} = -\frac{1}{T^2}\left[\left(\frac{\partial H}{\partial P}\right)_T - V\right] + \frac{1}{T}\left[\frac{\partial^2 H}{\partial T \partial P} - \left(\frac{\partial V}{\partial T}\right)_P\right] \quad ...(18)$$

Since S and H and state functions, therefore,

$$\frac{\partial^2 S}{\partial T \partial P} = -\frac{\partial^2 S}{\partial T \partial P}, \text{and } \frac{\partial^2 H}{\partial T \partial P} = \frac{\partial^2 H}{\partial P \partial T}$$

Applying this condition, equations (17) and (18) yield

$$\left(\frac{\partial H}{\partial P}\right)_T = V - T\left(\frac{\partial V}{\partial T}\right)_P \quad ...(19)$$

The expression (19) is thermodynamic equation of state.

For ideal gas PV = nRT

$$\therefore \left(\frac{\partial V}{\partial T}\right)_P = \frac{nR}{P} = \frac{V}{T}$$

From equation (34), we have

$$\left(\frac{\partial H}{\partial P}\right)_T = V - T\left(\frac{\partial V}{\partial T}\right)_P = V - T\frac{V}{T} = 0$$

Now we shall consider a few more applications of the results of the results of this section to some simple cases:

(a) **Relation between C_P and C_V**

We developed expressions for the difference between the heat capacities at constant pressure and at constant volume, as

$$C_P - C_V = \left[\left(\frac{\partial E}{\partial V}\right)_T + P\right]\left(\frac{\partial V}{\partial T}\right)_P \quad ...(20)$$

But $\left(\frac{\partial E}{\partial V}\right)_T = T\left(\frac{\partial P}{\partial T}\right)_V - P$ ...(21)

$$\therefore \quad C_P = C_V = T\left(\frac{\partial P}{\partial T}\right)_V \left(\frac{\partial V}{\partial T}\right)_P \quad ...(22)$$

But $(\partial V/\partial T)_P = \alpha V$, where α is the coefficient of thermal expansion. Therefore, equation (22) may be written as,

$$C_P - C_V = TV\alpha\left(\frac{\partial P}{\partial T}\right)_V \qquad ...(23)$$

V is a function of T and P,

$$V = f(T, P)$$

$$\therefore dV = \left(\frac{\partial V}{\partial T}\right)_P dT + \left(\frac{\partial V}{\partial P}\right)_T dP$$

At constant volume, this equation reduces to

$$\left(\frac{\partial V}{\partial T}\right)_P (\partial T)_V + \left(\frac{\partial V}{\partial P}\right)_T (\partial P)_V = 0$$

$$\therefore \left(\frac{\partial P}{\partial T}\right)_V = \frac{(\partial V/\partial T)_P}{(\partial V/\partial P)_T} = \frac{\alpha}{\beta} \qquad ...(24)$$

where β is the compressibility coefficient defined as $b = -V^{-1}(\partial V/\partial P)_T$, and $a = V^{-1}(\partial V/\partial T)_P$

Using equations (23) and (24), We get

$$C_P - C_V = VT\frac{\alpha^2}{\beta}$$

This equation is equally true for solids, liquids and gases.

For water the density is maximum at 3.98° C. and $C_P = C_V$. This can be shown with the help of equation (23) and the definition of density (ρ). The density is defined as mass (m) per unit volume:

$$\rho = m/V,$$

$$\text{hence} \quad \frac{d\rho}{dT} = \frac{d}{dT}(m/V) = -\frac{m}{V^2}(dV/dT)$$

For maximum, the derivative $d\rho/dT$ must be zero, so should be its right side. Since m and V^2 are not zero, therefore dV/dT should be zero. Equation (24) gives $C_P - C_V = 0$, and $C_P = C_V$.

For an ideal gas $\alpha = T^{-1}$, $b = P^{-1}$, therefore $C_P - PV/T = nR$.

(b) **Joule- Thomson coefficient for a van der Waals gas**

Joule-Thomson coefficient is the pressure dependence of temperature under isoenthalpic conditions, and is given as

$$\mu\,\pi = \left(\frac{\partial T}{\partial P}\right)_H = -\frac{1}{C_P}\left(\frac{\partial H}{\partial P}\right)_T \qquad ...(25)$$

From equations (23) and (25), we get,

$$\mu\pi = \frac{1}{C_P}\left[T\left(\frac{\partial V}{\partial T}\right)_P - V\right] = \frac{V}{C_P}(\alpha T - 1)$$

where $\alpha V = (\partial V/\partial T)_P$

The temperature, at which $\mu\pi$ is zero, is called the inversion temperature given by

$T\pi = V(\partial V/\partial T)_P = 1/\alpha$

For one mole, van der Waal's equation of state is written as

$$PV = RT + Pb - \frac{a(V-b)}{V^2}$$

If $V - b = V$ and $V \cong \frac{RT}{P}$ then

$$PV = RT + Pb - \frac{Pa}{RT}$$

and $$V = \frac{RT}{P} + b - \frac{a}{RT}$$

$$\therefore \left(\frac{\partial V}{\partial T}\right)_P = \frac{R}{P} + \frac{a}{RT^2}$$

On substituting the expression for $(\partial V/\partial T)_P$ into (23) we get,

$$\left(\frac{\partial H}{\partial P}\right)_T = V - T\left(\frac{\partial V}{\partial T}\right)_P = V - \frac{RT}{P} - \frac{a}{RT}$$

$$= \left(\frac{RT}{P} + b - \frac{a}{RT}\right) - \frac{RT}{P} - \frac{a}{RT} = b - \frac{2a}{RT}$$

But $(\partial H/\partial P)_T = -C_P\mu\pi$

$$\therefore \mu\pi = \frac{1}{C_P}\left(\frac{2a}{RT} - b\right)$$

For very low temperatures the first term will dominate and the coefficient will be positive, while for higher temperatures the coefficient will be negative. The *inversion temperature* will be $T\pi = \frac{2a}{Rb}$ when mp is zero.

(c) Variation of C_V with Volume at Constant T

$$C_V = \left(\frac{\partial E}{\partial T}\right)_V$$

Differentiating with respect to V at constant T, we get

$$\left(\frac{\partial C_V}{\partial V}\right)_T = \frac{\partial}{\partial V}\left[\left(\frac{\partial E}{\partial T}\right)_V\right] = \frac{\partial}{\partial T}\left[\left(\frac{\partial E}{\partial V}\right)_T\right]_V$$

But $\left(\frac{\partial E}{\partial V}\right)_T = T\left(\frac{\partial P}{\partial T}\right)_V - P$

$$\therefore \quad \left(\frac{\partial C_V}{\partial V}\right)_T = \frac{\partial}{\partial T}\left[T\left(\frac{\partial P}{\partial T}\right)_V - P\right]_V$$

$$= T\frac{\partial^2 P}{\partial T^2} + \left(\frac{\partial P}{\partial T}\right)_V - \left(\frac{\partial P}{\partial T}\right)_V = T\left(\frac{\partial^2 P}{\partial T^2}\right)_V$$

For an ideal gas PV = nRT and $(\partial P/\partial T)_V = nR/V$, therefore

$$\left(\frac{\partial^2 P}{\partial T^2}\right)_V = 0, \text{ and hence}$$

$$\left(\frac{\partial C_V}{\partial V}\right)_T = 0 \text{ for an ideal gas.}$$

For a van der Waals gas $P = \frac{nRT}{V-nb} - \frac{an^2}{V^2}$, and $(\partial P/\partial T)_V = \frac{nR}{V-nb}$.
therefore

$$\left(\frac{\partial^2 P}{\partial T^2}\right)_V = 0, \text{ and hence}$$

$$\left(\frac{\partial C_V}{\partial V}\right)_T = T\left(\frac{\partial^2 P}{\partial T^2}\right)_V = 0 \text{ for a van der Waal's gas.}$$

(d) For **free expansion** of a gas the Joule coefficient μ_J is defined by $\mu_J = (\partial T/\partial V)_E$. With the help of thermodynamic equation of state we can correlate μ_J with some measurable physical properties of a gas. E is a state function and hence

$$dE = \left(\frac{\partial E}{\partial T}\right)_V dT + \left(\frac{\partial E}{\partial V}\right)_T dV = C_V dT + \left(\frac{\partial E}{\partial V}\right)_T dV$$

For constant energy process dE = 0, and

$$C_V\ (\partial T)_E = -\left(\frac{\partial E}{\partial V}\right)_T (\partial V)_E$$

or $$\left(\frac{\partial T}{\partial V}\right)_E = -\frac{1}{C_V}\left(\frac{\partial E}{\partial V}\right)_T$$

But $(\partial T/\partial V)_E = \mu_J$ and $(\partial E/\partial V)_T = T\ (\partial P/\partial T)_V - P$

$$\therefore\ \mu_J = -\frac{1}{C_V}\left[T\left(\frac{\partial P}{\partial T}\right)_V - P\right]$$

Since $(\partial P/\partial T)_V = \alpha/\beta$

$$= \mu_J = \frac{P - \alpha T/\beta}{C_V}$$

For an ideal gas α/β = P/T, therefore $\mu_J = 0$. That is, the free expansion of an ideal gas is isothermal.

Example 1:

A gas obeys the equation of state $P = nRT/V - n^2 RB/V^2$. Show that (i) $\left(\frac{\partial S}{\partial V}\right)_T$ = nR/V; (ii) $(\partial E/\partial V)_T = n^2BR/V^2$, (iii) $(\partial H/\partial V)_T = 2n^2BR/V^2$ *where B is a constant.*

Solution:

(i) $(\partial S/\partial V)_T$ suggests that S = S (T, V), therefore,

$$dS = \left(\frac{\partial S}{\partial T}\right)_V dT + \left(\frac{\partial S}{\partial V}\right)_T dV \qquad \text{...(i)}$$

From the first and second laws we get

$$dS = dq_{rev}/T = \frac{1}{T}(dE + PdV) \qquad \text{...(ii)}$$

But E = E (T,V) and hence

$$dE = \left(\frac{\partial E}{\partial T}\right)_V dT + \left(\frac{\partial E}{\partial V}\right)_T dV = C_V dT + \left(\frac{\partial E}{\partial V}\right)_T dV$$

$$\therefore\ dS = \frac{C_V}{T}dT + \frac{1}{T}\left[\left(\frac{\partial E}{\partial V}\right)_T + P\right]dV$$

Comparing the coefficients of dV in equations (i) and (iii) it is seen that

$$\left(\frac{\partial S}{\partial V}\right)_T = \frac{1}{T}\left[\left(\frac{\partial E}{\partial V}\right)_T + P\right]$$

But $\left(\frac{\partial E}{\partial V}\right)_T = T\left(\frac{\partial P}{\partial T}\right)_V - P$ ...(iii)

$$\therefore \left(\frac{\partial S}{\partial V}\right)_T = \left(\frac{\partial P}{\partial T}\right)_V$$

For the given gas $(\partial P/\partial T)_V = nR/V$

$\therefore \left(\frac{\partial S}{\partial V}\right)_T = nR/V$ ...(iv)

(ii) Putting $(\partial P/\partial T)_V = nR/V$ in thermodynamic equation of state we get

$(\partial E/\partial V)_T = T\frac{nR}{V} - P = n^2RB/V^2$ ...(v)

(iii) $H = E + PV$

$\therefore \left(\frac{\partial H}{\partial V}\right)_T = \left(\frac{\partial E}{\partial V}\right)_T + P + V\left(\frac{\partial P}{\partial V}\right)_T$ ...(vi)

For the given equation $P = nRT/V - n^2 RB/V^2$

$$\left(\frac{\partial P}{\partial V}\right)_T = -nRT/V^2 + 2n^2 RB/V^3$$

$\therefore V\left(\frac{\partial P}{\partial V}\right)_T = -nRT/V + 2n^2RB/V^2$ ...(vii)

Also $(\partial E/\partial V)_T = n^2 RB/V^2$

On substituting the derivatives in equation (vi) we get

$$\left(\frac{\partial H}{\partial V}\right)_T = \frac{n^2RB}{V^2} + P - \frac{nRT}{V} + P - \frac{2n^2RB}{V^2}$$

But $P - \frac{nRT}{V} = \frac{n^2RB}{V^2}$

$$\therefore \left(\frac{\partial H}{\partial V}\right)_T = \frac{2n^2RB}{V^2}$$

Example 2:

A gas obeying the equation of state P (V –B) = RT undergoes a change from the initial state T_1, V_1 to a final state T_2, V_2. Derive an expression for the entropy change of this gas. The variation of heat capacity at constant volume is given by $C_V = a + bT + cT^2$.

Solution:

$$dS = \frac{dE + PdV}{T},\ dE = C_V\, dT + (\partial E/\partial V)_T\, dV \text{ give}$$

$$dS = \frac{C_V}{T}dT + \frac{1}{T}[(\partial E/\partial V)_T + P]dV \qquad \text{...(i)}$$

Thermodynamic equation of state $(\partial E/\partial V)_T + P = T\,(\partial P/\partial T)_V$, when applied, equation (i) yields,

$$dS = \frac{C_V}{T}dT + (\partial P/\partial T)_V\, dV \qquad \text{...(ii)}$$

For the given gas P = RT/(V – B), and

$(\partial P/\partial T)V = R/(V - B)$

Substituting equation (iii) into (ii) we obtain

$$dS = \frac{C_V}{T}dT + R\frac{dV}{V-B} \qquad \text{...(iv)}$$

$$\because\ C_V = a + bT + cT^2$$

$$\therefore\ \frac{C_V}{T} = \frac{a}{T} + b + cT \qquad \text{...(vi)}$$

$$dS = \frac{a}{T}dT + bdT + cTdT + R\frac{dV}{V-B}$$

On integrating for proper limits, we get

$$\int_1^2 dS = a\int_{T_1}^{T_2}\frac{dT}{T} + b\int_{T_1}^{T_2} dT + c\int TdT + R\int_{V_1}^{V_2}\frac{dV}{V-B}$$

$$\therefore\ \Delta S = a \ln\frac{T_2}{T_1} + b\,(T_2 - T_1) + \frac{c}{2}(T_2^2 - T_1^2) + R \ln\frac{V_2 - B}{V_1 - B}$$

Example 3:

(a) *Starting with the fundamental relations prove that*

$$\left(\frac{\partial S}{\partial V}\right)_T = \left(\frac{\partial P}{\partial T}\right)_V = \frac{\alpha}{\beta},\ \text{where } a = V^{-1}(\partial V/\partial T)_P,\ \beta = V^{-1}(\partial V/\partial P)_T$$

Solution:

(a) Combined first and second laws of thermodynamics give

$TdS = dE + PdV$

But $dE = C_V\, dT + \left(\frac{\partial E}{\partial V}\right)_T dV$

$$\therefore\ TdS = C_V\, dT + \left[\left(\frac{\partial E}{\partial V}\right)_T + P\right] dV \quad ...(i)$$

At constant T, dT = 0 and

$$T\left(\frac{\partial S}{\partial V}\right)_T = \left(\frac{\partial E}{\partial V}\right)_T + P \quad ...(ii)$$

Thermodynamic equation of state reads

$$(\partial E/\partial V)T + P = T\left(\frac{\partial P}{\partial T}\right)_V$$

$$\therefore\ T\left(\frac{\partial S}{\partial V}\right)_T = T\left(\frac{\partial P}{\partial T}\right)_V$$

$$\text{or} \quad \left(\frac{\partial S}{\partial V}\right)_T = \left(\frac{\partial P}{\partial T}\right)_V \quad ...(iv)$$

Also V = f (T, P)

$$\therefore\ dV = \left(\frac{\partial V}{\partial T}\right)_P dT + \left(\frac{\partial V}{\partial P}\right)_T dP = \alpha V dT - \beta V dP$$

At constant volume dV = 0 and

$$\left(\frac{\partial P}{\partial T}\right)_V = \frac{\alpha}{\beta} \quad ...(v)$$

From equation (iv) and (v), we get

$$\left(\frac{\partial S}{\partial V}\right)_T = \left(\frac{\partial P}{\partial T}\right)_V = \alpha/\beta$$

(b) *For a given substance a = 1.24 × 10–3 K^{-1} and b = 9.3 × 10–5 atm–1 at 290 K and 1 atm. Assuming a and b to be constant, calculate the change in molar volume which should produce an entropy change of 2.1 JK^{-1} mol^{-1} at 290 K.*

Solution:

(b) We have derived $\left(\frac{\partial S}{\partial V}\right)_T = \left(\frac{\partial P}{\partial T}\right)_V = \alpha/\beta$

or $\frac{\Delta S}{\Delta V} = \alpha/\beta$

$\therefore \Delta V = \Delta S\ \beta/\alpha$

$$= \frac{(2.1\text{JK}^{-1}\text{mol}^{-1}) \times (9.3 \times 10^{-5}\text{atm}^{-1})}{(1.24 \times 10^{-3}\text{K}^{-1})}$$

$$= 0.1583\ \text{mol}^{-1}\ \text{atm}^{-1} = \frac{0.1583\text{Jmol}^{-1}\text{atm}^{-1}}{101.325\text{Jatm}^{-1}\text{dm}^{-3}}$$

$= 1.56 \times 10^{-3}\ \text{dm}^3\ \text{mol}^{-1}$

CALCULATION OF ENTROPY CHANGES

In calculation of entropy changes we must take into consideration the entropy of system and the surroundings. The combination of system and surroundings correspond to an isolated system. The surroundings are usually considered to be the (i) heat reservoir that can add to, or subtract heat from the system, or (ii) mechanical device which can do work on, or accept work from the system. Thus the total entropy change of the isolated system ($\Delta S_{universe}$) is taken as

$$\Delta S_{universe} = \Delta S_{system} + \Delta S_{surroundings}$$

Entropy Change of System (ΔS_{system})

Entropy change of the system in a reversible isothermal change can be calculated by using the relation $dS = dq_{rev}/T$. But this equation cannot be used as such to calculate the entropy of irreversible changes. For this we have to devise the ways to carry the change in reversible ways. This is valid because the entropy is a state function and the changes in entropy of the system are the same irrespective the way the change is brought about.

Entropy Change of the Surroundings ($\Delta S_{surroundings}$)

The entropy changes for the surroundings can be calculated if we know the gain or loss in the heat of the surroundings. For this case, the equation $dS = dq/T$ can be used to calculate $\Delta S_{surroundings}$ because the surroundings are so large that the removal or addition of heat can be taken as reversible processes, as it will not alter the properties of the surroundings to any greater extent. [For analogy we take the case that removal or addition of a few drops of water will not change the water content of ocean, (Surroundings)].

$$\Delta S_{surroundings} = -\frac{q_{system}}{T} = \frac{q_{surroundings}}{T}$$

Now we shall consider the calculation of entropy change for some processes.

Phase changes. As an example of phase changes we may consider the melting of a solid. At melting point, the liquid and solid exist together in equilibrium. The addition of a little heat would melt some of the solid, the removal of a little heat would solidify some of the liquid, but the solid liquid equilibrium will be maintained. This process can be considered as reversible and entropy change can be given by

$$\Delta S_f = \int_1^2 \frac{dq_{rev}}{T_f}$$

Since the temperature is constant during the process, it may be taken outside the integral, giving

$$\Delta S_f = \frac{1}{T_f}\int dq_{rev}$$

The integral is the total heat absorbed which for this constant pressure process is the enthalpy of fusion, so that

$$\Delta S_f = \frac{\Delta H_f}{T_f} \quad ...(1)$$

In general, for any phase transformation taking place at constant temperature and pressure, e.g. [solid $\rightleftharpoons$ liquid (melting); liquid $\rightleftharpoons$ vapour (vaporization); solid $\rightleftharpoons$ vapour (sublimation); solid (α) $\rightleftharpoons$ solid (β) (transition) it must be true that

$$\Delta S_t = \frac{\Delta H_t}{T_t} \quad ...(2)$$

where ΔH_t is the enthalpy change of the transformation and T_t is the constant temperature at which the transformation takes place.

Example 4:

(a) *Calculate the entropy change involved in the conversion of 1mole of ice at 0°C and 1 atm to liquid at 0° C and 1 atm; the enthalpy of fusion per mole of ice is* 6008 J mol^{-1}.

Solution:

The process is reversible.

H_2O (s, 0°C, 1 atm) → H_2O (l, 0° C, 1 atm)

$$\therefore \Delta S_{fus} = \frac{\Delta H_{fus}}{T_m}$$

$$T_m = 273\text{ K};\ \Delta H_{fus} = 6008\text{ J mol}^{-1}$$

$$\therefore \Delta S_{fus} = \frac{6008\text{J mol}^{-1}}{273\text{K}} = 22\ \mathbf{JK^{-1}}\text{ mol}^{-1}$$

This is an increase in the entropy of the system. This means that the entropy of liquid state is greater than that of the solid state of H_2O. Since the randomness in liquid is greater than solid, hence entropy is a measure of randomness.

Example 4:

(b) *Calculate the change of entropy for the process*

H_2O (l, 373 K, 101325 NM^{-2}) → H^2O (v, 373 K, 101325 Nm^{-2})

$\Delta H_{vap} = 40850$ J mol^{-1} at 373 K.

Solution:

The given process is isothermal and isobaric phase transformation.

$$\therefore \Delta S_{vap} = \frac{\Delta H_{vap}}{T_b} = \frac{40850\text{J mol}^{-1}}{373\text{K}} = \mathbf{109.5JK^{-1}mol^{-1}}$$

Since ΔS_{vap} = S(vapour) – S (liquid) = 109.5 JK^{-1} mol^{-1}

∴ S(vapour) = S(liquid) + 109.5 JK^{-1} mol^{-1}.

This observation is in tune with the fact that there is more randomness in the vapour state than the liquid state. Hence entropy is a measure of disorderness.

Example 5:

Evalvate the entropy charges for the following reversible processes:

(a) 1mole H_2O (l, 1atm, 100° C) ⇌ 1 mole H_2O (v, 1 atm, 100°C)

$\Delta H_{vap} = 40850$ J mol^{-1}

(b) 1 mole Sn (a, 1 atm, 13° C) ⇌ 1 mole Sn (β, 1 atm, 13° C)

$\Delta H_{trans} = 2090$ J mol^{-1}

Solution:

(a) $\Delta S_{vap} = \frac{\Delta H_{vap}}{T_b}$

Here $\Delta H_{vap} = 40850 \text{ J mol}^{-1}$

$T_b = 100 + 273 = 373 \text{ K}$

$\therefore \Delta S_{vap} = \frac{40850 \text{Jmol}^{-1}}{373\text{K}} = 109.5 \text{ JK–1 mol}^{-1}$

(b) $\Delta H_{trans} = 2090 \text{ J mol}^{-1}$ and $T_{trans} = 13 + 273 = 286 \text{ K}$

$\therefore \Delta S_{trans} = \frac{2090 \text{Jmol}^{-1}}{286\text{K}} = 7.31 \text{ JK}^{-1}\text{mol}^{-1}$

Entropy Change for Reversible Isothermal Expansion of Ideal Gases

Let the volume of n moles of an ideal gas be changed from V_1 to V_2 at constant temperature. Since for this isothermal expansion, the internal energy does not change, the heat absorbed can be calculated from the first law expression as

$$dq_{rev} = -\,dw_{rev} = PdV = \frac{nRT}{V}dV$$

Therefore, for this process the entropy change of the system can be evaluated as

$$\Delta S = \int_{V_1}^{V_2} \frac{dq_{rev}}{T} = \int_{V_1}^{V_2} nR\left(\frac{dV}{V}\right) = nR\text{In}\left(\frac{V_2}{V_1}\right) \qquad ...(1)$$

Since at constant temperature $P_1/P_2 = V_2/V_1$, we can also write for an isothermal expansion from pressure P_1 to P_2

$$\Delta S = -\,nR \text{ In}\left(\frac{P_2}{P_1}\right)$$

$$= -\,nR \times 2.303 \log\left(\frac{P_2}{P_1}\right) \qquad ...(2)$$

Example 5:

(c) Calculate the entropy change when 1 mole of an ideal gas expands reversibly from an initial volume of 1 dm^3 to a find volume of 10 dm^3 at a constant temperature of 298 K.

Solution:

$$\Delta S = nR \ In \left(\frac{V_2}{V_1}\right) = 2.303 \ nR \log \left(\frac{V_2}{V_1}\right)$$

Here $n = 1$; $V_1 = 1 \ dm^3$, $V_2 = 10 \ dm^3$

$\Delta S = R \ In \ 10 = 2.303 \ R \log 10$

$= 2.303 \times (8.314 \ JK^{-1} \ mol^{-1})$

$19.15 \ JK^{-1} \ mol^{-1}$.

Entropy Change on Heating or Cooling of a Substance

Let us consider n moles of a substance at a temperature T_1. Let its temperature be change to T_2 keeping the volume constant. If this change is brought about reversibly then we can use the defining equation of entropy. Heat change for this process is

$dq_{rev} = C_V dT$

where C_V is the heat capacity at constant volume.

Entropy change is

$$\Delta S = \int_{T_1}^{T_2} \frac{dq_{rev}}{T} = \int_{T_1}^{T_2} \frac{C_V dT}{T}$$

If C_V does not vary with temperature, then

$$\Delta S = C_V \ In \left(\frac{T_2}{T_1}\right) = nC_{V \cdot m} \ In \left(\frac{T_2}{T_1}\right) \quad ...(1)$$

For temperature changes from T_1 to T_2 at constant P.

$$\Delta S = \int_{T_1}^{T_2} C_P \left(\frac{dT}{T}\right) = nC_{P,m} \ In \left(\frac{T_2}{T_1}\right) \quad ...(2)$$

if the heat capacity at con 0

stant pressure (C_P) is taken as independent of temperature. Here $nC_{P,m} = C_P$.

Reversible Adiabatic Changes

If any system undergoes a reversible adiabatic change, then by definition, $dq_{rev} = 0$ at all stages of the process, so that

$\Delta S = q_{rev}/T = 0$

Therefore, reversible adiabatic processes are often called *isentropic* processes. However, in the case of irreversible adiabatic expansion though qirr is zero but ΔS is not zero, because $\Delta S \neq q_{irr}/T$.

The entropy change in a system of an ideal gas being heated or cooled from T_1 to T_2 and expanding from V_1 to V_2, is given by

$$\Delta S = C_V \ In \frac{T_2}{T_1} + nR \ In \frac{V_2}{V_1}$$

This equation is valid for all types of changes which an ideal gas is permitted to undergo. For adiabatic reversible change.

$$\left(\frac{T_2}{T_1}\right) = \left(\frac{V_1}{V_2}\right)^{nR/CV} \quad \text{for an ideal gas}$$

$$\therefore \ In\frac{T_2}{T_1} = \frac{nR}{C_V} In\left(\frac{V_1}{V_2}\right) = \frac{nR}{C_V} In\left(\frac{V_2}{V_1}\right)$$

$$\text{Hence } \Delta S = -\frac{nR}{C_V}.C_V In\frac{V_2}{V_1} + nR.In\frac{V_2}{V_1} = 0$$

That is, any increase in entropy due to expansion is exactly balanced by an equal decrease in entropy due to lowered temperature in an adiabatic reversible expansion of an ideal gas.

$$\frac{T_2}{T_1} = \frac{C_V + (P_2/P_1)nR}{C_P}$$

Therefore, ΔS for irreversible adiabatic change is not zero, this is so because the final temperature (T_2) in adiabatic irreversible change and adiabatic reversible change are different.

Example 6:

0.1kg nitrogen gas at 298 K are held by a piston under 30 atm pressure. The pressure is suddenly released to 10 atm and the gas expands adiabatically. If $C_{V,m}$ *= 20.8* J^{K-1} mol^{-1}*, what is the final temperature and volume? Calculate ΔS system for this expansion. What would be the value of ΔS (surroundings)?*

Solution:

The change is adiabatic (q = 0) but irreversible as the expansion is sudden against a constant pressure of 10 atm. Applying the first law we shall get

$C_V (T_2 - /T_1) = w = - P_2 (V_2 - V_1)$

$P_1V_1/ T_1 = P_2V_2/T_2 = nR$ gives

$C_V (T_2 - T_1) = nR (T_1P_2/P_1 - T_2)$

$$\therefore T_2 = \frac{nC_{V,m} + nRP_2/P_1}{nC_{P,m}} \times T_1$$

$$= \left(\frac{20.8JK^{-1}mol^{-1} + 8.314JK^{-1} \times 10/30}{29.114JK^{-1}mol^{-1}}\right) \times 298 \text{ K}$$

= **241 K**

and $V_2 = P_1V_1T_2/T_1P_2$

$$= \frac{30 \times V_1 \times 241}{298 \times 10} = \mathbf{2.42\ V_1}$$

Entropy change for the system is given by

$$\Delta S\text{system} = nC_{V,m} \ln\frac{T_2}{T_1} + nR\ln\frac{V_2}{V_1}$$

$$= n\left(20.8JK^{-1}mol^{-1} \times 2.303\log\frac{241K}{298K}\right.$$

$$\left. + 8.314\ JK^{-1}\ mol^{-1}\ 2.303 \times \log\frac{2.42V_1}{V_1}\right)$$

$= n (- 4.416\ JK^{-1}mol^{-1} + 7.348\ JK^{-1}\ mol^{-1})$

$= 3.571\ mol \times 2.932\ JK^{-1}\ mol^{-1}$

$= 10.47\ JK^{-1}$

$(\because n\ 0.1/0.028 = 3.571\ mol)$

$\Delta S_{Surroundings} = q_{surr}/T = 0$

Example 7:

Evaluate the change in entropy when 3 moles of a gas are heated from 27° C to 7227° C at a constant pressure of 1 atm. The molar heat capacity of the gas is 23.7 JK^{-1} mol^{-1}.

Solution:

$n = 3$ mol. $T_1 = 300$ K, $T_2 = 100$ K.

$C_{P.m} = 23.7\ J_{K-1}\ mol_{-1}$

$$\Delta S = nC_{P,m} \times 2.303 \log (T_2/T_1)$$
$$= (3 \text{ mol}) \times (23.7 \text{ JK}^{-1} \text{ mol}^{-1}) \times 2.303 \log (100 \text{ K}/300 \text{ K})$$
$$= \mathbf{85.5 \text{ JK}^{-1}}$$

Example 8:

Two blocks of copper metal are of the same size but are at different temperatures T_1 and T_2. These block are brought together and allowed to attain thermal equilibrium i.e. both are allowed to come to the same final temperature T. Show that the entropy change is given by

$$\Delta S = C_P \text{ In} \left[\frac{(T_2 - T_1)^2}{4T_1T_2} + 1 \right]$$

Solution:

Let the metallic block A be at temperature T_1 and block B at temperature T_2 in the initial state. Let $T_2 > T_1$.

Heat lost by block B = C_{PB} $(T_2 - T)$

Heat gained by block A = C_{PA} $(T_2 - T_1)$

At equilibrium

$$C_{PB} (T_2 - T) = C_{PA} (T - T_1)$$

$$\therefore T = \frac{T_1 + T_2}{2} \qquad \because C_{PB} = C_{PA} = C$$

Decrease of entropy of block B

$$= \int_{T_2}^{T} C_P d\text{In}T$$

$$\Delta S_B = C_P \text{ In} \frac{T}{T_2}$$

Increase of entropy of block A

$$\Delta S_A = C_P \text{ In} \frac{T}{T_1}$$

Total entropy change $\Delta S = \Delta S_A + \Delta S_B$

$$= C_P \left[\text{In} \frac{T}{T_1} + \text{In} \frac{T}{T_2} \right]$$

$$= C_P \left[\text{In} \frac{(T_1 + T_2)^2}{4} - \text{In} T_1 T_2 \right]$$

$$= C_P \text{ In } \frac{T_1^2 + T_2^2 + 2T_1T_2}{4T_1T_2}$$

Adding and subtracting $2T_1T_2$ to the In part of the above equation, we get

$$\Delta S = C_P \text{ In} \frac{T_1^2 + T_2^2 + 2T_1T_2 + 4T_1T_2}{4T_1T_2}$$

$$= C_P \text{ In} \left[\frac{(T_2 - T_1)^2}{4T_1T_2} + 1 \right]$$

Now $(T_1 - T_2)^2$ is always positive and hence the quantity within bracket on right side of the above equation is greater than one. Therefore right side is a positive should be spontaneous.

Example 9:

The molar entropy of an ideal gas ($C_{P,m}$ = 20.9 JK^{-1}mol^{-1}) at 298K is 146.0 JK^{-1} mol^{-1}. Find its value at 500 K.

Solution:

We may assume it as a process of heating at constant pressure. Therefore, $dS = dq_{rev}/T = dH/T$ at constant P

$$dS = C_P \frac{dT}{T}$$

On integrating we get

$$\int_{S_1}^{S_2} dS = CP \int_{T_1}^{T_2} d \text{ In } T$$

$$\therefore \text{ S } (T_2) - \text{S } (T_1) = C_P \text{ In } \frac{T_2}{T_1}$$

For the present case

T_1 = 298 K, t_2 = 500 K,

S (298 K) = 146.0JK^{-1} mol^{-1}

$\therefore$ S (500 K) = (146.0 JK^{-1} mol^{-1}) + (20.9 JK^{-1} mol^{-1})

$$\times 2.303 \times \log \frac{500K}{298K}$$

$= 146.0\ JK^{-1}\ mol^{-1} + 10.81\ JK^{-1}\ mol^{-1}$

$= 156.81\ JK^{-1}\ mol^{-1}$

Example 10:

Calculate the entropy change when 10 m^3 of an ideal gas ($C_{P,m}$ = 25R) at 27° C and 1.01 × 10^5 N m^{-2} pressure are heated at constant pressure to 127°C.

Solution:

Given that

$T_1 = 300$ K, $T_2 = 400$ K, $C_{P,m} = 2.5$ R,

$P = 1.01 \times 10^5\ Nm^{-2}$,

Heat absorbed dq = dH = CP. dT = TdS

$$\therefore dS = C_P \frac{dT}{T} = C_P d\ \text{In}\ T$$

$$\int_1^2 dS = \Delta S = C_P \int d\ \text{In}T = C_P\ \text{In}\ \frac{T_2}{T_1} = nC_{P,m} \text{In} \frac{T_2}{T_1}$$

$$\text{Now, } n = \frac{PV}{RT} = \frac{(1.01 \times 10^5 Nm^{-2}) \times 10 \times 10^{-3} m^3}{(8.314 JK^{-1} mol^{-1}) \times (300K)}$$

$$\therefore \Delta S = (0.405\ mol) \times (2.5 \times 8.314\ mol^{-1}) \times 2.303 \log \frac{400}{300}$$

$= 2.422\ JK^{-1}$

Example 11:

1 mole of an ideal gas is allowed to expand isothermally at 27° C until its volume is tripled. Calculate ΔS_{syst} and ΔS_{univ} under the following conditions: (a) The expansion is carried out reversibly, (b) the expansion is a free expansion.

Solution:

(a) Isothermal reversible expansion: entropy change can be calculated by using the defining equation

$dS = dq_{rev}/T$ or $\Delta S = q_{rev}/T$.

In the present case

$q = -w = 2.303\ RT \log V_2/V_1$

$= 2.303 \times 8.314 \times 300 \log 3V_1/V_1$

$= 2740.6\ J\ mol^{-1}$

$\therefore \Delta S_{syst} = q_{rev}/T = 2740.6/300 = 9.135\ JK^{-1}\ mol^{-1}$

$\Delta S_{surr} = -\Delta S_{syst}$; here $\Delta S_{univ} = 0$

(b) In case of free expansion we can not use the equation $dS = dq_{rev}/T$. However, S being a state function, ΔS can be evaluated by

$\Delta S = 2.303\ R \log V_2/V_1 = 9.135\ JK^{-1}\ mol^{-1}$

For surrounding q = 0, and hence $\Delta S_{surr} = 0$

$\Delta S_{univ} = \Delta S_{syst} =$ **9.135 JK⁻¹ mol⁻¹**

Entropy Change for Simultaneous Heating and Expansion of an Ideal Gas

(a) **Temperature and volume are changed**. Let us consider n moles of an ideal gas initially at temperature T_1 and volume V_1. Let its temperature and volume be changed to a final value of T_2 and V_2 respectively. The entropy change can be calculated by devising the following reversible paths for the process.

Path A

Step 1. Expand the gas isothermally at temperature T_1 from volume V_1 to V_2. In this case internal energy change would be zero and the first law will give

$$dq_{rev} = -dw = PdV = \frac{nRT_1}{V} dV$$

$$\therefore \Delta S_1 = \int_{V_1}^{V_2} \frac{dq_{rev}}{T} = \int_{V_1}^{V_2} \frac{nRT_1}{T_1} = nR \ In \left(\frac{V_2}{V_1}\right)$$

Step 2. Heat the gas from T_1 to T_2 at constant volume V_2 then

$$dq_{rev} = C_V dT$$

$$\therefore \Delta S_2 = \int_{T_1}^{T_2} \frac{dq_{rev}}{T} = C_V \int_{T_1}^{T_2} \frac{dT}{T}$$

$$= C_V \ In \left(\frac{T_2}{T_1}\right) = nC_{V,m} \left(\frac{T_2}{T_1}\right)$$

C_V is assumed to be independent of T

The entropy change for the system is

$\Delta S_{system} = \Delta S_1 + \Delta S_2$

$$\Delta S_{system} = nR \ln\left(\frac{V_2}{V_1}\right) + nC_{V,m} \ln\left(\frac{T_2}{T_1}\right) \quad ...(1)$$

Path B

Step 1. Heat the gas at constant Volume V_1 from T_1 to T_2

$$\Delta S_1 = nC_{V,m} \ln\left(\frac{T_2}{T_1}\right)$$

Step 2. Expand the gas isothermally at temperature T_2 from volume V_1 to V_2

$$\Delta S_2 = nR \ln\left(\frac{V_2}{V_1}\right)$$

$$\therefore \Delta S_{system} = \Delta S_1 + \Delta S_2 = nC_{V,m} \ln\left(\frac{T_2}{T_1}\right) + nR \ln\left(\frac{V_2}{V_1}\right) \quad ...(2)$$

Thus we see that the final expression for the entropy change comes out to be the same irrespective of the fact whether we first change the volume and then temperature or the reverse. This suggests that entropy is a state function.

Path C

Step 1. Heat the gas at constant volume from T_1 to an intermediate temperature T.

Step 2. Expand isothermally at T from V_1 to an intermediate volume V′.

Step 3. Heat at constant volume from T′ to T_2.

Step 4. Expand isothermally at T_2 from V′ to V_2.

For step 1: $dq_{rev} = nC_{V,m}\, dT$

$\Delta S_1 = nC_{V,m} \ln (T'/T_1)$

For step 2 : $\Delta S_2 = nR \ln (V'/V_1)$

For step 3 : $\Delta S_3 = nC_{V,m} \ln (T_2/T')$

For step 4 : $\Delta S_4 = nR \ln (V2/V')$

The entropy change for the system is

$$\Delta S_{system} = \Delta S_1 + \Delta S_2 + \Delta S_3 + \Delta S_4$$

$$= nC_{V,m} \ln\left(\frac{T'}{T_1}\right) + nR \ln\left(\frac{V'}{V_1}\right) + nC_{V,m} \ln\left(\frac{T_2}{T'}\right) + nR \ln\left(\frac{V_2}{V'}\right)$$

$$= nC_{V,m} \ln\left(\frac{T_2}{T_1}\right) + nR \ln\left(\frac{V_2}{V_1}\right) \qquad ...(3)$$

(b) **Temperature and pressure are changed.** Let n moles of an ideal gas, initially at temperature T_1 and pressure P_1 be brought to a final temperature T_2 and pressure P_2. We can use the following reversible path between two states.

Step 1. Heat the gas at constant pressure P_1 from T1 to T_2, then

$$dq_{rev} = nC_{P,m}\, dT$$

Step 2. Expand the gas isothermally at temperature T_2 from P_1 to P_2, then

$$dq_{rev} = -dw = PdV$$

For step 1: $\Delta S_1 = \int_{T_1}^{T_2} nC_{P,m} \frac{dT}{T}$

$$= nC_{P,m} \ln\left(\frac{T_2}{T_1}\right)$$

(If C_P is independent of T).

For step 2: $\Delta S_2 = \int_{P_1}^{P_2} \frac{PdV}{T_2} = -\int_{P_1}^{P_2} \frac{VdP}{T_2} = -\int_{P_1}^{P_2} nRd \ln P$

$$= -nR \ln\left(\frac{P_2}{P_1}\right)$$

$$\Delta S = \Delta S_1 + \Delta S_2 = nC_{P,m} \ln\left(\frac{T_2}{T_1}\right) - nR \ln\left(\frac{P_2}{P_1}\right) \qquad ...(4)$$

This result can be transformed into the form of equation (4) by using the following relations:

$$C_{P,m} = C_{V,m} + R$$

$$\frac{P_2}{P_1} = \frac{T_2V_1}{T_1V_2}$$

$$\therefore \Delta S = n(C_{V,m} + R) \ln\left(\frac{T_2}{T_1}\right) - nR \ln\left(\frac{T_2}{T_1}\right) - nR \ln\left(\frac{V_1}{V_2}\right)$$

$$\Delta S = nC_{V,m} \ln\left(\frac{T_2}{T_1}\right) + nR \ln\left(\frac{V_2}{V_1}\right)$$

Entropy change when a van der Waals gas is heated and expanded simultaneously. Let n moles van der Waals gas initially at temperature T_1 and volume V_1 be brought to temperature T_2 and volume V_2. Combined first and second laws give,

$$TdS = dq_{rev} = dE + PdV \quad ...(1)$$

But $E = f(T, V)$

$$\therefore dE = \left(\frac{\partial E}{\partial T}\right)_V dT + \left(\frac{\partial E}{\partial V}\right)_T dV$$

$$= C_V\, dT + \left(\frac{\partial E}{\partial T}\right)_V dV \quad ...(2)$$

From equation (1) and (2), we get

$$dS = \frac{C_V}{T} dT + \frac{1}{T}\left[\left(\frac{\partial E}{\partial V}\right)_T + P\right] dV \quad ...(3)$$

For a van der Waals gas

$$P = \frac{nRT}{V - nb} - \frac{an^2}{V^2}$$

and $\left(\frac{\partial E}{\partial V}\right)_T = \frac{an^2}{V^2}$

$$\therefore \quad dS = \frac{nC_{V,m}}{T} dT + \frac{nR}{V - nb} dV$$

Taking $C_{V,m}$ as independent of T and integrating between suitable limits, we have

$$S_2 - S_1 = \Delta S = nC_{V,m} \ln \frac{T_2}{T_1} + nR \ln\left[\frac{V_2 - nb}{V_1 - nb}\right] \quad ...(4)$$

If the gas expands isothermally then $dT = 0$ and the entropy change is given by

$$\Delta S = nR \text{ In} \left[\frac{V_2 - nb}{V_1 - nb}\right]$$

ΔS can be calculated if b is known. Here we see that the entropy change depends upon the size of the molecules (b).

If $C_{V,m}$ depends upon temperature as given by

$C_{V,m} = \alpha + \beta T + \gamma T_2$, then

$C_{V,m}/T = \alpha/T + \beta + \gamma T$

and $\displaystyle\int_{T_1}^{T_2} \frac{C_{V,m}}{T} dT = \alpha \int_{T_1}^{T_2} \frac{dT}{T} + \beta \int_{T_1}^{T_2} dT + \gamma \int_{T_1}^{T_2} TdT$

$$= \alpha \text{ In } (T_2/T_1) + \beta (T_2 - T_1) + \frac{1}{2}\gamma\left(T_2^2 - T_1^2\right)$$

Therefore, for the n moles of a van der Waals gas the entropy change is given by

$$\Delta S = n\alpha \text{ In } (T_2/T_1) + n\beta (T_2/T_1) + \frac{1}{2}\gamma\left(T_2^2 - T_1^2\right) + nR \text{ In} \left[\frac{V_2 - nb}{V_1 - nb}\right]$$

Similarly, for a gas obeying the equation of state, $\dfrac{PV}{RT} = 1 + \dfrac{B(T)}{V}$, we can calculate ΔS.

We have derived

$$dS = \frac{C_V}{T} dT + \frac{1}{T}\left[\left(\frac{\partial E}{\partial V}\right)_T + P\right] dV$$

and $\left[\dfrac{\partial E}{\partial V}\right]_T = T\left[\dfrac{\partial P}{\partial T}\right]_V - P$

$$\therefore \quad dS = \frac{C_V dT}{T} + \left[\frac{\partial P}{\partial T}\right]_V dV \qquad ...(5)$$

For the given gas

$$\left(\frac{\partial P}{\partial T}\right)_V = \frac{R}{V} + \frac{R}{V^2} B(T) + \frac{RT}{V^2}\left(\frac{dB}{dT}\right)_V \qquad ...(6)$$

Equations (6) and (5) give,

$$dS = \frac{C_V}{T} dT + \left[\frac{R}{V} + \frac{R}{V^2} B(T) + \frac{RT}{V^2}\left(\frac{dB}{dT}\right)_V\right] dV \qquad ...(7)$$

Integrating equation (8) we get,

$$\Delta S = C_V \ln \frac{T_2}{T_1} + R \ln \frac{V_2}{V_1} - RB\left[\frac{1}{V_2} - \frac{1}{V_1}\right] - RT\frac{dB}{dT}\left[\frac{1}{V_2} - \frac{1}{V_1}\right]$$

$$\Delta S = C_V \ln \frac{T_2}{T_1} + R \ln \left[\frac{V_2}{V_1}\right] + R\left[\left(B + T\frac{dB}{dT}\right)\right]\left[\frac{1}{V_1} - \frac{1}{V_2}\right]$$

Thus, it is possible to calculate the entropy change for a gas obeying virial equation of state, provided we know the experimental conditions and the data for B (T).

Example 12:

One mole of an ideal gas is taken through a reversible carnot cycle. Predict the sign of ΔT, ΔP, ΔV, ΔE, ΔH, ΔS in each step without preforming the actual calculations.

Solution:

A reversible Carnot cycle consists of (a) isothermal expansion, (b) adiabatic expansion, (c) isothermal compression, and (d) adiabatic compression.

(a) $\Delta T = 0$ (isothermal); $\Delta V > 0$ (expansion); $\Delta P < 0$ (since $V \propto 1/P$); $\Delta H = 0$, $\Delta E = 0$ since for ideal gas $E = E(T)$ and $H = H(T)$. $\Delta S > q > 0$ for isothermal expansion.

(b) $\Delta V > 0$, (expansion), $\Delta P < 0$, $\Delta E < 0$, $\Delta H < 0$, $\Delta T < 0$ (adiabatic expansion), $\Delta S = 0$ $(q = 0)$

(c) $\Delta T = 0$, $\Delta V < 0$, $\Delta P > 0$, $\Delta E = 0$, $\Delta H = 0$, $\Delta S < 0$.

(d) $\Delta T > 0$, $\Delta V < 0$, $\Delta P > 0$, $\Delta E > 0$, $\Delta H > 0$, $\Delta S < 0$.

Example 13:

Calculate the entropy change when 0.5 dm^3 of an ideal gas ($C_{V,m}$ = 12.6 JK^{-1} mol^{-1}) at 298 K and 1 atm is allowed to expand to double its volume and simultaneous heated to 373 K.

Solution:

ΔS for the system is the same whether the changes take place in one step or in stages. Thus ΔS system is given by

$$\Delta S = nC_{V,m} \ln \frac{T_2}{T_1} + nR \ln \frac{V_2}{V_1}$$

$C_{V, m}$ = 12.6 JK^{-1} mol^{-1}, T_1 = 298K,

T_2 = 373 K, V_1 = 0.5 dm^3,

V_2 = 1.0 dm^3,

P_1 = 1 atm,

$$n = P_1V_1/RT_1 = \frac{[(1atm) \times 0.5\ dm^3]}{(0.082dm^3atmK^{-1}mol^{-1}) \times (298K)}$$

= 0.02 mol

$$\therefore \Delta S = (0.02\ mol) \times (12.6\ JK^{-1}mol^{-1}) \times 2.303 \log \frac{373K}{298K}$$

$$+ (0.02\ mol) \times (8.314JK^{-1}mol^{-1}) \times 2.303 \log \frac{1.0dm^3}{0.5dm^3}$$

= **0.173 JK^{-1}**.

Example 14:

10 g ice at 0° C are added to20 g water at 90° C in a thermally insulated flask of negligible of negligible heat capacity. The heat of fusion of ice is 5980 J mol^{-1}. What is the final temperature, ΔS_{sytem} and $\Delta S_{surroundings}$? $C_{P,\ m}(H_2O, l)$ = 75.42 $JK^{-1}mol^{-1}$.

Solution:

When ice and water are mixed then ice melts and the temperature of the mixture is changed. The change may be considered step by step as follows:

(i) (10/08) mol H_2O (s, 273 K) $\xrightarrow{\text{Melt}}$ (10/18) mol H_2O (1,273K)

$$q\ (i) = n.\ \Delta H_{fus} = 5980 \times \frac{10}{18}\ J = 3322.2\ J$$

The water at 273 K is heated to a temperature T_2K.

(ii) (10/18) H_2O (l, 273 K) $\xrightarrow{\text{Heat}}$ (10/18) H_2O (l,T_2K)

q (ii) = $nC_{P,m}$ (H_2O) (T_2–T_1)

= (10/18) (75.42 JK^{-1} mol^{-1}) (T_2–273)

= 41.9 (T_2 – 273)

(iii) (20/18) mol H_2O (l, 363 K) $\xrightarrow{\text{Cool}}$ (20/18) mol H_2O (l, T_2 K)

q (iii) = (20/18) mol × (75.42 JK^{-1} mol^{-1}) × (T_2 –363)

$= 83.4\ (T_2 - 363)$

$\therefore$ 10g H_2O (s, 0°C) + 20g H_2O (l, 90°C) $\rightarrow$ 30g H_2O (l, T_2K)

Total heat change in the system q = q (i) + q (ii) + q (iii)

Since the change is adiabatic, therefore q = 0, and q (i) + q (ii) + q (iii) = 0

$$\frac{10}{18} \times 5980 + \frac{10}{18} \times 75.42 \times (T_2 - 273) + \frac{20}{18} \times 75.42\ (T_2 - 363) = 0$$

Solving for T_2 we get T_2 = 306 K.

Entropy changes:

$$\Delta S\ (i) = q_1/273 = \frac{5980 \times 10}{273 \times 18} = 12.169\ JK^{-1}$$

$$\Delta S\ (ii) = n_1 C_{P,m}\ \text{In}\ (T_2/T_1)$$

$$= \frac{10}{18} \times 75.42\ \text{In}\ \frac{306}{273} = -\ 4.782\ JK^{-1}$$

$$\Delta S\ (iii) = n_2 C_{P,m}\ \text{In}\ (T_2/T_1)$$

$$= \frac{20}{18} \times 75.42\ \text{In}\ \frac{306}{363} = -\ 14.317\ JK^{-1}$$

$$\Delta S = \Delta S\ (i) + \Delta S\ (ii) + \Delta S\ (iii) = 2.634\ JK^{-1}$$

$\Delta S_{Surrounding}$ = 0, since the process is adiabatic, q_{surr} = 0.

Example 15:

One mole of supercooled water at – 10°C and 1 atm pressure turns into ice at –10°C. Calculate the entropy change in the surroundings and the system and the net entropy change. Heat capacity of water and ice at 1 atm may be taken as constant at 75.42 $JK^{-1}\ mol^{-1}$ and 37.20 $JK^{-1}\ mol^{-1}$ respectively.

ΔH_{fus} (273 K) = 6008 J mol^{-1}.

Solution:

The system is H_2O (l, – 10°C, 1 atm) $\rightarrow$ H_2O (s, –10°C, 1 atm). The given process is not a reversibly one so the entropy change for this cannot be calculated directly. But we can consider a series of reversible steps leading form the liquid water at – 10°C to solid ice at – 10°C.

The various steps are as follows:

Step 1. Heat reversibly the super cooled water at – 10°C to 0°C keeping the pressure constant.

Step 2. Allow the solidification of water at 0°C to ice at 0°C.

Step3. Cool the ice reversibly from 0°C to – 10°C. Thus

Step 1. H_2O (l, –10° C, 1 atm) $\xrightarrow{\text{Heat}}$ H_2O (l, 0° C, 1 atm)

The heat required to raise the temperature of supercooled water by dT is

$$dq_{rev} = C_P \, dT = dH$$

The entropy change in raising the temperature from 263 K to 273 K

$$\Delta S_1 = \int_{263}^{273} \frac{dq_{rev}}{T} = C_P H_2O(l) \ In \frac{273}{263}$$

$$= 2.303 \times (75.42 \ JK^{-1}mol^{-1}) \times \log \frac{273}{263}$$

$$= 2.81 \ JK^{-1} \ mol^{-1}$$

Step 2: H_2O (l, 0°C, 1 atm) → H_2O (s, 0°C, 1atm)

$$\Delta S_2 = -\frac{\Delta H_{fus}}{T}$$

$$= -\frac{6008 Jmol^{-1}}{273K} = -22.0 \ JK^{-1}mol^{-1}$$

Step 3: H_2O (s, 0°C, 1atm) $\xrightarrow{\text{COOl}}$ H_2O (s, – 10°C, 1atm)

Heat change: $dq_{rev} = C_P (H_2O, s) \, dT = dH$

Entropy change : $\Delta S_2 = C_P (H_2O, s) \ In \frac{263}{273}$

$$= 2.303 \times 37.20 \times \log \frac{263}{273}$$

$$= -1.38 \ JK^{-1} \ mol^{-1}$$

On adding the equations of steps 1, 2 and 3, we get

H_2O (l, –10°C, 1 atm) → H_2O (s, – 10°C, 1 atm)

and $\Delta S = \Delta S_1 + \Delta S_2 + \Delta S_3$

$= 2.81 - 22.0 - 1.38 = -$ **20.57** $JK^{-1} \ mol^{-1}$

Here we see that there has been a decrease in entropy of water upon crystallization at –10°C, though the process is irreversible. This example emphasizes the fact that sign of the entropy change for the system plus surroundings, and not merely for any component, is related to irreversibility. Thus we must consider the entropy change for the system as well as that of surroundings. Now ΔS_{surr} can be calculated by considering the surroundings to be large enough so that the exchange of heat can be considered as a reversible process. Thus

$\Delta S_{surr} = \Delta H_{surr}/T = -\Delta H_{system}/T$

$\Delta H_{syst} = \Delta H_1 + \Delta H_2 + \Delta H_3$

ΔH_1 = Heat gained by the system in heating water from 263 K to 273 K

$= C_P (H_2O, l) \times \Delta T = (74.42\ JK^{-1}\ mol^{-1}) \times (10\ K)$

$= 754.2 J\ mol^{-1}$

$\Delta H_2 = -\Delta H. fus = -6008\ mol^{-1}$

ΔH_3 = Heat lost by the system when it is cooled from 273 K to 263 K

$= C_P (H_2O, s) \times \Delta T = -(37.2\ JK^{-1}\ mol^{-1}) \times (10K)$

$= -372.0\ J\ mol^{-1}$

$\Delta H_{syst} = (754.2 - 6008 - 372)\ J\ mol^{-1}$

$= -5626\ J = -\Delta H_{surr}$

$$\Delta S_{surr} = \frac{5626}{263K} = 21.39\ JK{-1}$$

$\Delta S_{universe} = \Delta S_{syst} + \Delta S_{surr} = -20.57\ JK^{-1} + 21.39 JK^{-1}$

$= +0.82\ JK^{-1}$

Example 16:

Illustrate the method to evaluate the entropy change for the following process:

$H_2O(l, 298\ K, 1.01 \times 10^5\ Nm^{-2}) \rightarrow H_2O\ (v, 398\ K, 0.5 \times 10^5\ Nm^{-2})$.

Solution:

The given process is irreversible, therefore, we cannot use the equation $dS = dq_{rev}/T$. However, this system can be split into various reversible systems can be split into various reversible systems for the convenience.

This is a valid procedure since S is a state function. We proceed as follows:

(a) Heat reversibly and isobarically the water from 298 K to 373 K, (b) allow the evaporation of water under reversible isothermal and isobaric condition, (c) heat water vapour reversibly and isobarically from 373 K to 398 K, (d) allow the isothermal and reversible expansion of the water vapour. The processes and the corresponding entropy changes are :

(a) H_2O (l, 298 K, 1.01 × 105 Nm^{-2}) $\xrightarrow{\text{Heat}}$ H_2O (l, 373K, 1.01 × 10^5 N m^{-2})

ΔS (a) = $C_{P,m}$ (H_2O, l) In (373/298)

(b) H_2O (l, 373 K, 1.01 × 10^5 Nm^{-2}) $\rightleftharpoons$ H_2O (v, 373 K, 1.01 × 10^5 N m^{-2})

ΔS (b) = ΔH_{vap}/373

(c) H_2O (v, 373 K, 1.01 × 10^5 Nm^{-2}) $\xrightarrow{\text{Heat}}$ H_2O (v, 398 K, 1.01 × 10^5 N m^{-2})

ΔS (c) = $C_{P,m}$ (H_2O, v) In (398/373)

(d) H_2O (v, 398 K, 1.01 × 10^5 Nm^{-3}) $\xrightarrow{\text{Expand}}$ H_2O (v, 398 K, 0.5 × 10^5 N m^{-2})

$$\Delta S(d) = -\ R\ In\ \frac{0.5 \times 10^5\ Nm^{-2}}{1.01 \times 10^5\ Nm^{-2}}$$

Adding equation (a) to (d) we get,

H_2O (l, 298 K, 1.01 × 10^5 Nm^{-2})→H_2 O (v, 398 K 0.5×10^5 Nm^{-2})

$\Delta S = \Delta S$ (a) + ΔS (b) + ΔS (c) + ΔS (d)

$$= C_{P,m}\ (H_2O,\ l)\ In\ \frac{373}{298} + \frac{\Delta H_{vap}}{373} + C_{P,m}\ (H_2O,\ v)\ In\ \frac{373}{298} + R\ In \frac{1.01}{0.5}$$

On substituting the values of heat capacities of liquid and water vapour, ΔS can be calculated. In the step (d) we assumed the water vapour to behave as an ideal gas.

Example 17:

One mole van der Waals gas is allowed to expand isothermally and reversibly at 273 K from an initial volume of 1.0 dm³ to 50.0 dm³. Calculated the entropy change for the gas. The van der Waals constants are

$a = 6.5$ atm dm^6 mol^{-2}

$b = 0.056$ dm^3 mol^{-1}.

Solution:

Given that the expansion is reversible and isothermal, therefore $\Delta S = q_{rev}/T$ can be used to calculate entropy change.

For 1 mole van der Waals gas

$q_{rev} = RT \ln (V_2 - b)/(V_1 - b)$

$\therefore \Delta S = R \ln (V_2 - b)/(V_1 - b)$

$$= (8.314\ JK^{-1}\ mol^{-1}) \times 2.303 \times \log \frac{(50 - 0.056)}{(1 - 0.056)}$$

$= 32.97\ JK^{-1} mol^{-1}$

Example 18:

Two moles of an ideal gas are expanded isothermally at 298 K from a volume of V to 2.5V. Find the value of ΔS_{gas}, and ΔS_{total} the for following:

(a) Reversible expansion.

(b) Irreversible expansion when 400 J mol^{-1} are less absorbed than in (a).

(c) Free expansion.

Solution:

(a) $\Delta S_{gas} = nR \ln V_2/V_1$

$V_2 = 2.5\ V$, $V_1 = V$,

$n = 2$ mol, hence

$\Delta S_{gas} = 2.303 \times (2\ mol) \times (8.314\ JK^{-1}\ mol^{-1}) \log 2.5$

$= 15.258\ JK^{-1}$

$q_{rev} = nRT \ln V_2/V_1 = 4540.9\ J$

$$\Delta S_{surr} = q_{surr}/T = -q_{syst}/T$$
$$= -4540.9 \text{ J}/298 \text{ K}$$
$$= -15.238 \text{ JK}^{-1}$$
$$\Delta S_{total} = \Delta S_{gas} + \Delta S_{surr}$$
$$= 15.238 - 15.238 = 0$$

(b) $\Delta S_{gas} = 2.303 \text{ nR} \log V_2/V_1 = 15.238 \text{ JK}^{-1}$

$$q_{irr} = q_{rev} - 2 \times 400 \text{ J mol}^{-1} = 4540.9 - 800$$
$$= 3740.9 \text{ J}$$
$$\Delta S_{surr} = q_{surr}/T = -q_{syst}/T$$
$$= -3740.9 \text{ J}/298 \text{ K}$$
$$= -12.533 \text{ JK}^{-1}$$
$$\Delta S_{total} = \Delta S_{gas} + \Delta S_{surr}$$
$$= 15.298 - 12.553 = 2.685 \text{ JK}^{-1}$$

(c) $\Delta S_{gas} = 2.303 \text{ nR} \log V_2/V_1 = 15.238 \text{ JK}^{-1}$

$q = 0$ hence $\Delta S_{surr} = 0$, and

$$\Delta S_{total} = \Delta S_{gas} = 15.238 \text{ JK}^{-1}.$$

Example 19:

Calculate the entropy change for the transformation

12 (s, 1atm, 298 K) $\rightarrow$ I_{-2} (v, 1atm, 457 K)

Given that : ΔH_{fus}, m = 15.68 kJ mol^{-1} at the m.pt 113.6°C

$\Delta H_{vap, m}$ = 25.522 kJ mol^{-1} at the b.pt 184° C.

$C_{P, m}$ (12, s) = 54.684 + 13.431 × 10–4 (t –25) J deg–1 mol–1

$C_{P, m}$ (12, s) = 81.588 JK^{-1} mol^{-1}.

Solution:

Split the process at

(i) J_2 (s, 1 atm, 298 K) $\xrightarrow{\text{heat}}$ I_2 (s, 1 atm, 113.6°C)

$$\Delta S \text{ (i)} = \int_{298}^{386.6} \frac{C_{P,m(I_2,S)}}{T} dT$$

$$C_{P,m} (I_2, S) = 54.684 + 13.431 \times 10^{-4} (t + 273 - 273 - 25)$$

$$\therefore \Delta S\ (i) = 54.684 \int \frac{dT}{T} + 13.4.31 \times 10^{-4} \int \frac{TdT}{T}$$

$$- 13.431 \times 10^{-4} \times 298 \int \frac{dT}{T}$$

$$= (54.684 - 13.431 \times 10^{-4} \times 298)\ \text{In} \frac{386.6}{298}$$

$$+ 13.431 \times 10^{-4}\ (386.6 - 298)$$

$$= 14.248\ JK^{-1}mol^{-1}$$

(ii) I_2 (s, 1atm, 113.6°C) $\xrightarrow{\text{melt}}$ I_2 (l, 1atm, 113.6°C)

$\Delta S\ (ii) = \Delta H_{fus}, m/T_m = 15648\ J\ mol^{-1}/386.6\ K$

= 40.474 JK–1 mol–1

(iii) I_2 (l, 1atm, 113.6°C) $\xrightarrow{\text{heat}}$ I_2 (l, 1atm, 184°C)

$$\Delta S\ (iii) = \int C_{P,m}(I_{2,}\ l) \frac{dT}{t}$$

$= 81.588\ JK^{-1}\ mol^{-1}\ \text{In}\ (457\ K/386.6\ K)$

$= 13.649\ JK^{-1}\ mol^{-1}$

(iv) I_2 (l, 1atm, 457 K) $\xrightarrow{\text{vaporise}}$ I_2 (v, 1atm, 457 K)

$\Delta S\ (iv) = \Delta H_{vap,\ m} / T_b^{\cdot}$

$= 25522\ J\ mol^{-1}/457\ K$

$= 55.846\ JK^{-1}\ mol^{-1}$

Total entropy change for the transformation

$\Delta S = \Delta S\ (i) + \Delta S\ (ii) + \Delta S\ (iii) + \Delta S\ (iv)$

$= 124.218\ JK^{-1}\ mol^{-1}$.

THIRD LAW OF THERMODYNAMICS AND EVALUATION OF ABSOLUTE ENTROPY

So far we have been concerned with the evaluation of entropy changes in different processes under different conditions but can we find the absolute value of entropy of an element or a compound? The thermodynamic relation at hand to evaluate entropy change is

$$dS = \frac{C_P}{T} dT$$

$$\int_{T_1}^{T_2} dS = S(T_2) - S(T_1) = \int_{T_2}^{T_1} \frac{C_P}{T} dT \quad ...(1)$$

$$\text{or } S(T_2) = S(T_1) + \int \frac{C_P}{T} dT \quad ...(2)$$

$S(T_2)$ can be evaluated provided the quantities on R. H. S. of this equation are known. C_P can be measured experimentally over a range of temperature, but the problem remains to find a value of $S(T_1)$. This problem is solved by assigning a zero value of $S(T_1)$ at absolute zero as stated by the **third law of thermodynamics** in the following form:

"Every substance has a finite positive entropy, but at 0K the entropy may become zero, and does so become in case of a perfectly crystalline substance." The statement is confined to pure crystalline solids because theoretical arguments and experimental evidence have shown that the entropy of solutions and supercooled liquids is not zero at 0K. For pure crystalline solids the law has been verified repeatedly, and at present little doubt remains about the general validity of the above statement of the law. Thus as $T \rightarrow 0$, $S(T) \rightarrow 0$. With this convention equation (2) may written as:

$$S^0(T) = \int_0^T \frac{C_P}{T} dT \quad ...(3)$$

$S^0(T)$ is absolute entropy of the substance at a temperature TK. It is a positive quantity. Its value can be calculated if we integrated R.H.S. of equation (53) from $T = 0$ to any desired temperature T. To be able to integrate we must know the values of C_P right from absolute zero to the temperature concerned. The measurement of C_P upto the absolute zero is inconvenient. For this purpose the method of extrapolation is employed. The values of C_P/T are plotted against T and the curve is extended upto $T = 0$. The area under the curve from $T = 0$ to $T = T$ gives the absolute value of entropy of the substance.

In many cases the value of heat capacity down to 0K can be obtained by the use of Debye equation

$$C_V = 3R \left[\frac{4}{5}\pi^4 (T/\theta)^3\right] = a(T/\theta)^3$$

If a substance (X) undergoes phase transformation between $T = 0$ to $T = T$, then all the contributions of the absolute entropy must be considered. For example, the absolute entropy of a gaseous substance (X) at 298 K and 1 atm may be calculated by considering the entropy contributions due to the :

1. heating the substance from $T = 0$ to $T = T_t$, the transition point,
2. transition from one form (α) to another from (β) at $T = T_t$,
3. heating the substance from $T = T_t$ to $T = T_m$, the melting point,
4. melting the substance at a constant temperature $T = T_m$,
5. heating the substance from $T = T_m$ to $T = T_b$, the boiling point,
6. evaporation of the substance at a constant temperature $T = T_b$,
7. heating the substance from $T = T_b$ to $T = 298$ K, the temperature in question.

Various Changes

$$X(T = 0, \alpha) \rightarrow X(Tt, \alpha) \rightleftharpoons X(T_t, \beta) \rightarrow X(T_m, S) \rightleftharpoons X(T_m, l)$$

$$X(298, g) \leftarrow X(T_b, g) \rightleftharpoons X(T_b, l)$$

Entropy Contributions

$$S^0(T) = \int_0^{T_t} C_P(\alpha)\, d\ In\ T + \frac{\Delta H_t}{T_t} + \int_{T_t}^{T_m} C_P(\beta)\, d\ In\ T + \frac{\Delta H_{fus}}{T_m}$$

$$+ \int_{T_m}^{T_b} C_P(l)\, d\ In\ T + \frac{\Delta H_{vap}}{T_b} + \int_{T_b}^{298} C_P(g)\, d\ In\ T$$

The steps involved in the evaluation of entropy for nitrogen are summarized in Table 1.

Table 4 : Evaluation of Entropy of Nitrogen from Heat Capacity

	Contribution	$JK^{-1}mol^{-1}$
1.	Extrapolation from 0 – 10 K and $\int C_p d\ In\ T$ for solid α from 10 to 35.61 K	27.17
2.	Transition, solid $\alpha \rightarrow$ solid β 228.9/35.61	6.43
3.	$\int C_p d\ In\ T$ for solid β from 35.61 to 63.14 K	23.38
4.	Fusion, 720.9/63.14	11.42
5.	$\int C_p d\ In\ T$ for liquid from 63.14 to b.pt.	11.41
6.	Vaporization, 5535.0/77.32	72.13
7.	$\int C_p d\ In\ T$ for liquid from 77.32 to 298 K	39.20

	Entropy of Ideal gas at 298 K and 1atm	191.14

ENTROPY OF REACTION

With the convention that the entropy of each substance is zero at absolute zero, every substance is assigned a positive entropy at any other temperature. These values are tabulated for elements and compounds. For any chemical reaction the change of entropy ΔS can be evaluated by using the relation

$$\Delta S = \Sigma S^0 \text{ (products)} - \Sigma S^0 \text{ (reactants)} \qquad ...(1)$$

For a general reaction of the type $aA + bB \rightarrow mM + nN$, the entropy change is given by

$$\Delta S = [\, mS^0 (M) + nS^0 (N)] - [aS^0 (A) + bS^0 (B)]$$

where S^0, s are the entropies of the given pure substances at a given temperature.

Example 20:

Calculate the standard entropy of formation of ethanol as represented by the following equation

$$2C \text{ (graphite)} + 3H_2 \text{ (g)} + 1/2\ O_2 \text{ (g)} \rightarrow C_2H_5OH \text{ (l)}$$

Solution:

The entropy change for the given reaction is

$$\Delta S^0 = S^0 (C_2H_5OH) - [2S^0 (C \text{ graphite}) + 3S^0 (H_2) + 1/2\ S^0 (O_2)]$$

using the S0 values from the appendix we have

$$\Delta S^0 = (160.7 \text{ JK}^{-1} \text{ mol}^{-1}) - (2 \times 5.7 + 3 \times 130.6 + 1/2 \times 205.1) \text{ JK}^{-1}\text{mol}^{-1}$$

$$= (160.7 - 505.7) \text{ JK}^{-1}\text{mol}^{-1} = -345.0 \text{ JK}^{-1}\text{mol}^{-1}$$

Variation of ΔS^0 with Temperature

If the value of entropy of reaction at one temperature $\Delta S (T_1)$ is known its value at some temperature $\Delta S(T_2)$ can be calculated from the heat capacities of the reaction species.

The calculation is done by using the relation $(\partial S/\partial T)_P = C_P/T$ or $dS = (C_P/T)\, dT = C_P d \ln T$. For the general case

$$\Delta S^0 = mS^0 (M) + nS^0 (N) - aS^0 (A) - bS^0 (B)$$

Differentiating w.r. to T at constant P, we get

$$\left(\frac{\partial \Delta S^0}{\partial T}\right)_P = m\left(\frac{\partial S^0 (M)}{\partial T}\right)_P + n\left(\frac{\partial S^0 (N)}{\partial T}\right)_P$$

$$-a\left(\frac{\partial S^0(A)}{\partial T}\right)_P - b\left(\frac{\partial S^0(B)}{\partial T}\right)_P$$

$$= m\ C_P\ (M)/T + n\ C_P(N)/T - a\ C_P\ (A)/T - b\ C_P\ (B)/T$$

$$= \Delta\ C_P/T \qquad ...(2)$$

where $\Delta C_P = [m\ C_P(M) + n\ C_P\ (N)] - [a\ C_P\ (A) + bC_P\ (B)]$

$= \Sigma C_P$ (products) $- \Sigma C_P$ (reactants)

$$\therefore\ d\ (\Delta S^0) = \Delta C_P \frac{dT}{T} = \Delta C_P d\ \text{In}\ T \qquad ...(3)$$

Integration between proper limits gives

$$\Delta S^0\ (T_2) - \Delta S^0\ (T_1) = \int_{T_1}^{T_2} \Delta C_P d\ \text{In}\ T \qquad ...(4)$$

If the variation of ΔC_P with temperature is known, equation (4) can be integrated. For approximate purposes ΔC_P may be assumed to be independent of T, and integration gives

$$\Delta S^0\ (T_2) - \Delta S^0\ (T_1) = \Delta C_P\ \text{In}\ (T_2/T_1)$$

or $$\Delta S^0\ (T_2) = \Delta S^0\ (T_1) + \Delta C\ 2.303\ \log\ (T_2/T_1) \qquad ...(5)$$

THE CONCEPT OF RESIDUAL ENTROPY

Certain chemical reactions between crystals do not give $\Delta S = 0$ at 0K, which seems to indicate that exceptions to this law exist. Nearly all these exceptional reactions involve either ice, carbon monoxide, nitrous oxide or hydrogen. In the crystalline state, there four substances appear to retain a finite entropy at 0K. Thus any reaction in which one which 1mole of ice is a product gives $\Delta S = 3.3\ JK^{-1}$, at 0K and any reaction in which 1mole of ice is a reactant gives $\Delta S = -3.3\ JK^{-1}$ at 0K.

This shows that if entropies of zero are assigned to the other substances involved, then ice at 0K appears to have a residual entropy of 3.3 JK^{-1}. Similarly, at 0K, CO has a residual entropy of 5.8 $JK^{-1}N_2O$ has a residual entropy of 5.8 JK^{-1} and hydrogen has a residual entropy of 6.2 JK^{-1}. This suggests that the molecules in the crystals of the substances are not in true thermodynamic equilibrium at 0K. The molecular origin of residual entropy is well understood with the help of statistical thermodynamics.

A number of liquids are readily supercooled below their thermodynamic freezing points and can exist in this state even at 0K. For such substance the entropy change of the reaction, Crystal → Liquid, at 0K can be determined and it has been found that liquid has a greater entropy than the crystal. Thus the third law is also not applicable to supercooled liquids e.g. glass.

ENTROPY AT EQUILIBRIUM

Natural processes are accompanied with an increase in the total entropy of the system and its surroundings. This gives us the equilibrium condition when the system cannot undergo any spontaneous change. At this state the entropy of the isolated system is maximum. Mathematically it can be explained with the help of the combined equation of the first and second laws of thermodynamics:

(a) $dS = \frac{dE + PdV}{T}$ for a reversible process

But $dE = C_V\, dT + (\partial E/\partial V)_T\, dV$

$\therefore\ dS = C_V \frac{dT}{T} + \frac{1}{T}[(\partial E/\partial V)_T + P]dV$

In an isolated system dE = 0, dT = 0 any dV = 0, therefore

$(dS)_{E,V} = 0$

For an irreversible change, since $(PdV)_{rev} > (PdV)_{irr}$,

$\therefore\ dS > \frac{dE + PdV_{irr}}{T}$

In an isolated system, dE = 0, dV = 0 and hence

$(dS)_{E,V} > 0$ ...(1)

The combined result may be written as

$(dS)_{E,V} \geq 0$

The state of equilibrium correspond of maximum entropy of the universe.

(b) If the entropy and volume are constant, then dS = 0 dV, and we should have

$(dE)_{S,V} = 0$ for a reversible change.

For an irreversible change $TdS > dE + PdV_{irr}$ or

$dE < TdS - PdV_{irr}$

$\therefore\ (dE)_{S,V} < 0$ ...(2)

The system maintained at constant entropy and constant volume will attain the equilibrium at a state of minimum energy.

(c) At constant entropy and pressure another condition of equilibrium in terms of enthalpy may be given. We have the relation dH = TdS – VdP, at constant S and P, dS = 0, dP = 0

$\therefore (dH)_{S,P} = 0$

For irreversible change $(dH)_{S,P} < 0$.

Thus the state of equilibrium is attained when the energy is minimum and the entropy is maximum. These are the necessary conditions for equilibrium. The sufficient condition of equilibrium is that if any virtual displacement (δ) is made in the variables of the system, then entropy in the new state will be less than the old equilibrium state and δS < 0, and the energy in the new state will be larger than the old equilibrium state and δE > 0. This can be understood by considering the Fig. 1.7(a) and 1.7(b) for a process A $(T_1, V_2, E_1, S_1) \rightarrow (T_2, V_2, E_2, S_2)$.

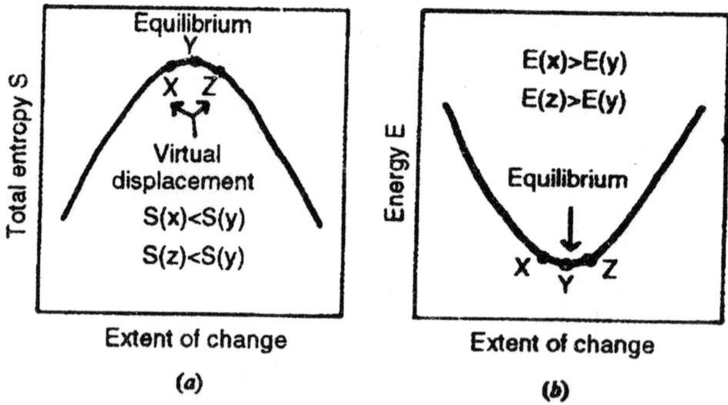

Fig. 1.7 : The condition for equilibrium (a) The entropy is maximum and δS < S < 0; (b) The energy is minimum and dE > 0.

When equilibrium is attained, the entropy is maximum and energy is minimum. Now suppose that by some means, the equilibrium is displaced to new states X or Z. From the figure it is quite clear that the entropy in the state X is less than the entropy in the state Y i.e. S (X) < S(Y) and δS in negative.

Similarly S (Z) < S (Y) and δS < 0. The energy in either of the displaced state is greater than the energy at equilibrium i.e. E (X) > E (Y) and E (Z) > E (Y), therefore δE > 0 is the sufficient condition for equilibrium.

Necessary condition . $(dS)_{E,V} > 0$; $(dE)_{S,V} < 0$.

Sufficient condition. δS < 0; δE > 0.

2

Statistical and Non-Equlibrium Thermodynamics

STATISTICAL THERMODYNAMICS

The Statistical thermodynamics is based on empirical laws in which the variables such as pressure, temperature and energy are regarded as functions of state. However, in the derivation of relations of thermodynamic functions, we ignore the constitution of the material system. But the experimental observations support the laws. The only drawback in classical thermodynamics is that we cannot calculate directly the absolute value off any thermodynamics function of a substance. In order to over come this, a new method of approach has been developed and is known as statistical thermodynamics.

The statement according, to which non-spontaneous process also called negative process cannot be the only result of a complex of processes is not strict are negative processes in microscopic (bulk) system events. Therefore the second law of thermodynamics first law, but a statistical law that is observed with a high degree off accuracy for considerable numbers of molecules and is less applicable, the smaller are the dimensions of the system being studied. So the second law of thermodynamics has statistical nature.

This fact was suggested by L. Bottzmann. He explained in this way the contradiction between the reversibility of mechanical motion and the irreversibility and unidirectional nature of real physical and chemical processes; this unidirectional nature is the result of the molecular structure of the material world. In this way, it was Boltzmann who provided the molecular basis to chemical thermodynamics. This approach is called chemical statistics or statistical thermodynamics and it deals with the applications of mechanics to chemical thermodynamics.

Statistical thermodynamics has bee applied successfully to relate the microscopic properties of the individual molecules (*e.g.* moment of inertia, dipole moment etc.) with macroscopic properties (*e.g.* molar heat capacity, polarization, etc.) of a system having a large number of molecules. In supposed to be composed of innumerable particles in motion. This motion of particles is governed by the laws of dynamics and in fact subject to quantum the number of particles is so large that it is impossible to follow the behaviour of every individual particle Statistical methods can be used to find out the properties of the system, particularly to know what fraction of the molecules has a given energy. It is possible to establish that the laws of thermodynamics are the necessary consequence of the quantum principles, when statistical postulates are used. In this method, we take into account the atomic constitution of matter. The matter is supposed to consist of innumerable particles in motion is governed by the laws of dynamics.

Statistical method can be employed to express the property o the system to find out what fraction of the molecules has a given energy. It is easier to find out the values of entropy, free energy, enthalpy, heat capacity etc., from spectroscopic data by using statistical methods. We shall however, discuss the equilibrium aspects of the system in the succeeding articles. In other words, this is a new approach that deals with the applications of statistical mechanics to thermodynamics. It has been applied successfully to relate the microscopic properties of the individual molecules (moment of inertia, dipole moment, etc.) with macroscopic properties (molar heat capacity, polarization, etc.) of a system having a large number of molecules. Statistical thermodynamics can be used for systems at equilibrium as well as away from equilibrium. We shall limit our discussion to the equilibrium aspects of the system Distinction between quantum mechanics, statistical mechanics and statistical thermodynamics. Statistical mechanics acts as a bridge between thermodynamics and quantum mechanics.

Quantum mechanics provides information about the energy of the molecular system, statistical mechanics tells us about the possible arrangement of the energy among various molecules of the system and introduces the concept of probability and partition functions. Statistical thermodynamics deals. with the relationships between the probability partition functions and thermodynamic properties. In other words. a physical entity generally manifests itself in : (i) A Microscopic, and (ii) A Macroscopic state. The dealt with macroscopic or the bulk state of matter in the preceding chapters.

The macroscopic variables were pressure, volume, temperature, entropy, etc. The macroscopic variables were related mutually on thermodynamical considerations. Further, we assumed that the number of atoms/molecules in them (microscopic details of a macrosystem) were exorbitantly large $\approx 10^{25}$ m^{-3}. The microscopic systems, on the other hand, have very few particles. Now, we know that the actual systems. the macro systems, are formed of the microsystems and so there should be a definite relation between the two.

Classically, if we want to have a complete description of a system we need to know actual velocity (or momentum), location and mutual interaction (if any, with other particles) of every particle. Since every panicle is described by three positions and three momentum coordinates. hence to describe a macrosystem we need a very large number of data. Further, any change in the parameter of an individual particle should be reflected in the properties of the system as a whole. *This poses a question*: How to deal with such a vast data which even a super-computer is not capable to handle, if at all we are able to collect it. If, however, we remember that:

(i) Much of the data and analytical results obtained on micro level are not useful in the study of the macrosystem, and

(ii) The application of classical dynamical principles to a many particle system is theoretically inapplicable, we can easily dispense with the technically impossible and practically useless classical kinetic and thermodynamic methods in favour of new statistical methods.

In the following sections we will see the importance of the concept of probability in the development of the statistical theory. Hence, we shall first consider some elementary theorems on the calculus of probability. In other words, it refers to a class of events of which the particular event is one member. Mathematically, the probability is defined as: When a certain aspect is possible in a ways and is not possible in b ways, then the probability of occurring the aspect is a/(a + b). (Newton defined probability as the ratio of a certain number of aspects, which were possibly observable, to the total number of aspects included in that class.) Here (a + b) is the total number of aspects each of which has the same chance of occurrence. Statistics is applied only to such systems which contain a large number of identical individuals which may be distinguishable by means of some property. In a statistical method, we study the behaviour of aggregates of particles to calculate the average value without investigating the detailed behaviour of the system and finding there from the value for each individual. Statistically, it is sufficient to know the most probable distribution of particles in different accessible states. This branch of physics has seen its development in the hands of Maxwell, Boltzmann, Gibbs, Fermi-Dirac, Bose-Einstein etc.

Now, as we have shown, if an event can occur in a ways, the probability of occurring of this event is a/(a + b), and similarly, if it fails to occur in b ways, then the probability of its failing is b/(a + b). The sum of the two

$$\frac{a}{a+b} + \frac{b}{a+b} = 1$$

corresponds to the certainty of some event, since the event must occur or fail to occur. If b = 0, the probability of occurrence becomes certain but if a = 0, the probability of its failure is certain.

We may define the probability for a certain event occurring m times in n unrestricted (free) attempts as,

$$P = \lim_{n \to \infty} \frac{m}{n}.$$

SOME ENSEMBLES

It would be appropriate to classify the ensembles according of their dependency on macroparameters. Since an ensemble is an assembly of macroscopically identical but mutually independent systems, a change in macroparameter (like volume, pressure, energy, number of molecules, etc.,) may lead to another set of systems:

Canonical Ensemble

In this ensemble all its members are identical and remain in thermal equilibrium with the source (*i.e.*, members of the ensemble can exchange thermal energy with the reservoir). Thus all the members of a canonical ensemble have the same number of particles N, volume V and are at the same temperature T. A canonical ensemble. Its members are shown separated by impermeable but conducting rigid. walls and they are in thermal contact with the source. Gibbs gave a relation.

$$\rho(\varepsilon) = Ae^{-\varepsilon\tau}$$

for density of its phase points. A and T in the relation are constants. The ensemble can be viewed physically as a freezer containing water filled metal-ice cube tray.

Grand Canonical Ensemble

The members of this ensemble can interact thermally, mechanically and also diffusively with each other and with the reservoir. Hence they are at the same temperature of same volume and at same chemical potential such an ensemble. Each member here is separated by flexible, permeable and conducting walls, which help exchange mass and energy between its members and with the surroundings. The constituent molecules in the member systems are thus of different kinds (different chemical species). The density of phase points in such ensemble is

$$\rho(N) = e^{(\Omega + N\mu - \varepsilon)/\tau}$$

where, Ω, μ and τ are constants. A grain of sugar in boiling water can best be viewed as a grand canonical ensemble.

Uniform Ensemble

Those ensembles which have same density in phase space, are called *uniform ensembles.*

Hence, if ΔN represents the number of system, in volume element $d\Gamma$ we have, for a uniform ensemble,

$$\lim_{d\Gamma \to 0} \frac{1}{d\Gamma} \frac{\Delta N}{N} = \text{constant}$$

Microcanonical Ensemble

In these ensembles the energy of their member systems remain constant, *i.e.*, they do not interact with each other or with the surroundings. Thus, they are free from external effects. The ensemble is thus conceptually an idealised one and can be represented by a function that depends only on the energy. An ensemble, here ε and V are constant. Every member is separated from the other by rigid, impermeable and insulated walls. The density of phase points within each member system is uniform and is zero at the walls. Physically, we may view the microcanonical ensemble as a perfect insulating enclosure encompassing several isolated systems. It is suitable for ideal gases.

SOLAR CONSTANT

The source of heat (sun) radiations and it emits heat radiations in all directions. The earth receives only a fraction of the energy emitted by the sun. The atmosphere also absorbs a part of the heat radiations and air, clouds, dust particles etc., in the atmosphere scatter the heat and light radiations falling on them. From the quantity of heat radiations received by the earth, it is possible to estimate the temperature of the sun. Therefore, to determine the value of a constant, called solar constant, certain ideal conditions are taken into consideration. *It is the amount of heat energy (radiation) absorbed per minute by one sq cm of a perfectly black body surface placed at a mean distance of the earth from the sun, in the absence of the atmosphere, the surface being held perpendicular to the sun's rays.*

The instruments used to measure the solar constant are called *pyrheliometers*. The heat energy absorbed by a known area in a fixed time is found with the help of the pyrheliometers. To eliminate the effects of absorption by the atmosphere, the value of the solar constant is found at various altitudes of the sun on the same day under similar sky conditions. If S is the observed solar constant, S_0 the true solar constant and Z the altitude (angular elevation) of the sun, then

$$S = S_0 \, a^{\sec z} \qquad ...(i)$$

$$\log S = \log S_0 + \sec Z \log a \qquad ...(ii)$$

Here a is a constant.

A graph is plotted between log S along the y-axis and sec Z along the x-axis. The graph is a straight line. Produce the graph to meet the y-axis. The intercept on the y-axis gives $\log S_0$.

From the value of $\log S_0$, the value of S_0 the solar constant can be calculated. The value obtained varies between 1.90 and 2.60 calories per sq cm per minute.

Temperature of the Sun

The sun consists of a central hot pardon surrounded by the photosphere. The central portion has a temperature of the order of 10^7 K. The photosphere has a temperature of about 6000 K. This temperature is also called the effective temperature of the sun. Considering the sun as a perfect black body radiator, the temperature of the sun can be calculated.

Let the mean distance of the sun from the earth be R and S the solar constant. Then, the total amount' of heat energy received by the sphere of radius R in one minute = $4\pi R^2 S$.

If r is the radius of the sun, then the amount of heat energy radiated by 1 sq cm surface of the sun in one minute

$$E = \frac{4\pi R^2 S}{4\pi r^2} = \left(\frac{R}{r}\right)^2 \times S$$

Taking $R = 148.48 \times 10^7$ km

$r = 6.928 \times 10^6$ km

The mean value of

$S = 1.94$ cals per cm^2 per minute.

$$E = \left(\frac{148.48\times10^7}{6.928\times10^5}\right)\times\frac{1.94}{60} \text{ cal per second} \quad \text{...(i)}$$

Also $E = \sigma T^4$

But $\sigma = 5.75 \times 10^{-5}$ ergs per cm^2 per second

$$= \frac{5.75\times10^{-5}}{4.2\times10^7} \text{ cal per cm}^2 \text{ per second}$$

$$E = \frac{5.75\times10^{-5}}{4.2\times10^7}.T^4 \quad \text{...(ii)}$$

Equating (i) and (ii)

$$\left(\frac{5.75\times10^{-5}}{4.2\times10^7}\right)T^4 = \left(\frac{148.48\times10^7}{6.928\times10^5}\right)\times\frac{1.94}{60}$$

$$T - 5730 \text{ K.} \quad \text{...(iii)}$$

This temperature gives the effective temperature of the sun acting as a black body radiator. The actual temperature of the sun is higher than this value. The temperature of the sun is usually taken as 6000 K.

Temperature of the sun can also be calculated from Wien's displacement law,

$$\lambda_{max}T = 0.2892$$

The wavelength of the radiations for which the energy is maximum in the spectrum is 4900×10^{-8} cm.

Substituting the value of λ_m, the value of T comes out to be 5902 K. This value is in agreement with the accepted value. Hence, the effective temperature of the sun (photosphere) is about 6000 K.

STEFAN'S LAW (DERIVATION OF MATHEMATICAL)

The fact that blade body radiations exert pressure, similar to a gas, helps in applying thermodynamics to heat radiations.

Let ψ be the energy density of radiations inside a uniform temperature enclosure at temperature T. P is the pressure and V is the volume.

Applying the first law of thermodynamics

$$\delta H = dU + P.dV \qquad ...(i)$$

Applying thermodynamical relation

$$\left(\frac{\partial H}{\partial V}\right)_r = \left(\frac{\partial P}{\partial T}\right)_v \qquad ...(ii)$$

$$\left(\frac{\partial U + P\partial V}{\partial V}\right)_r = T\left(\frac{\partial P}{\partial T}\right)_v$$

$$\left(\frac{\partial U}{\partial V}\right)_r = T\left(\frac{\partial P}{\partial T}\right)_v = P \qquad ...(iii)$$

Now $\qquad U = V\psi$

and $\qquad P = \frac{\psi}{3}$

or $\qquad \left[\frac{\partial U}{\partial V}\right]_r = \psi$

Here ψ is a function of temperature alone.

Substituting these values in equation (iii),

$$\psi = \frac{T}{3}\frac{d\psi}{dT} - \frac{\psi}{3}$$

$$\frac{4\psi}{3} = \frac{T}{3}\frac{d\psi}{dT}$$

or $$\frac{d\psi}{\psi} = 4\frac{dT}{T}$$

Integrating, $\log \psi = 4 \log T + \text{constant}$

or $\qquad \psi = KT^4 \qquad ...(iv)$

Here K is a constant

Also the total rate of emission per unit area of a black body is proportional to the energy density.

$\therefore \qquad R \propto \psi \propto T^4$

$\therefore \qquad R = \sigma T^4 \qquad ...(v)$

where σ is Stefan's constant.

The value of Stefan's constant in C.G.S. system is 5.672×10^{-5} C.G.S. units and in M.K.S. system it is 5.672×10^{-8} M.K.S. units.

COOLING FROM STEFAN'S LAW (DERIVATION OF NEWTON'S LAW)

Stefan's law is applicable for all temperatures of a hot body. But Newton's Law is applicable when the difference of temperature between the hot body and the surrounding is small. Consider a hot body at a temperature T_1 placed in a uniform temperature enclosure at T_1. According to Stefan's law,

$$R = e\sigma(T_1^4 - T_1^4) \qquad ...(i)$$

Here e is the emissivity of the surface of the hot body

$$R = e\sigma(T_1 - T_2)(T_1^3 + T_1^2 T_2 + T_1 T_2^2 + T_2^3)$$

As $(T_1 - T_2)$ is small, T_1 can be taken approximately equal to T_2.

Then,
$$R = e\sigma(T_1 - T_2)(T_2^3 + T_2^3 + T_2^3 + T_2^3)$$

$$R = 4e\sigma T_2^3((T_1 - T_2)$$

Taking $4e\sigma T_2^3 = k$

$$R = k(T_1 - T_2)$$

or

$$R \propto (T_1 - T_2). \qquad ...(ii)$$

This equation represents Newton's law of cooling and is true when the difference of temperature is small.

PHOTON GAS

The interaction of electromagnetic radiations with matter, led to the idea that electromagnetic radiations are composed of discrete energy particles called photons. Each photon has an energy $h\nu$ and momentum h/λ. Here ν is the frequency and λ is the wavelength of the radiations. The electromagnetic radiations trapped in a cavity and in thermal equilibrium with the walls of the cavity are termed as black body radiations. In the equilibrium condition, the black body radiations can be considered as the *photon gas.* It is assumed that the photons do not interact among themselves. The photons interact only with-the atoms of the walls of the cavity. It is further assumed that the photons are indistinguishable and many photons can have the same energy. Photons are taken as bosons and they obey Bose-Einstein statistics.

As the photons can either be emitted or absorbed by the atoms of the walls of the cavity, the number of photons is not constant *i.e.*, the condition $\sum_i dn_i = 0$ is no longer valid.

Due to this reason the value of α is equal to zero in the Bose-Einstein distribution law.

$$n_i = \frac{g_i}{[e^{(\alpha + E_i/kT)}] - 1} \quad \text{...(i)}$$

Here $\alpha = 0$

$$\therefore \quad n_i = \frac{g_i}{[e^{E_i/kT}] - 1} \quad \text{...(ii)}$$

In case, the cavity is large as compared to the wavelength of the radiations, the energy spectrum of the photons is taken to be continuous. In this case, the energy difference between successive allowed energy value is very small. Thus, replacing g_i by $g(E)\ dE$, and n_i by dn

$$dn = \frac{g(E)dE}{\left(e^{E/kT)}\right) - 1} \quad \text{...(iii)}$$

As the energy of a photon, $E = h\nu$, the value of g(E) dE can be taken equal to g (ν) dν. The factor g(ν) dν corresponds to the number of oscillatory modes in the frequency range dν and relating to the energy range dE.

The number of states in a black body radiation in the frequency range ν and ν + dν can be obtained by calculating the spherical volume bound by the spheres of radii $\frac{h(\nu + d\nu)}{c}$ and $\frac{h\nu}{c}$

The volume of the spherical shell

$$= \frac{4}{3}\pi\frac{h^3}{c^3}(\nu + d\nu)^3 - \frac{4}{3}\pi\frac{h^3}{c^3}\nu^3$$

$$= \frac{4}{3}\pi\frac{h^3}{c^3}[\nu^3 + 3\nu^2 d\nu + ... - \nu^3]$$

$$= \frac{4}{3}\pi\frac{h^3}{c^3} \times 3\nu^2 d\nu$$

$$= 4\pi\frac{h^3}{c^3}\nu^2 d\nu \quad \text{...(iv)}$$

The phase space has volume V = h^3 and there are two states of polarization for the radiation. The number of states in the black body radiation in the frequency range ν and ν + dν is given by

$$g(\nu)d\nu = \frac{8\pi V}{c^3}\nu^2 d\nu \quad ...(v)$$

$$dn = \left(\frac{8\pi V\nu^2 d\nu}{c^3}\right)\frac{1}{\left(e^{h\nu/kT}\right)-1} \quad ...(vi)$$

For dn photons in the frequency range ν + dν and ν, the energy is equal to (hν) dn and the energy per unit volume

$$= \frac{(h\nu)dn}{V}$$

Therefore, the energy density distribution for black body radiation is given by

$$E(\nu) = \frac{(h\nu)dn}{V.d\nu} \quad ...(vii)$$

Substituting the value of dn from equation (vi), we get

$$E(\nu) = \frac{8\pi h\nu^2 d\nu.\, h\nu}{c^3 V.d\nu\left[\left(e^{h\nu/kT}\right)-1\right]}$$

$$E(\nu) = \left(\frac{8\pi h\nu^3}{c^3}\right)\left[\frac{1}{\left(e^{h\nu/kT}\right)-1}\right]^{d\nu} \quad ...(viii)$$

Equation (viii) represents Planck's radiation law for black body radiations.

From equation (viii)

$$E(\nu)d\nu = \frac{8\pi h\nu^3}{c^3}\left[\frac{1}{\left(e^{h\nu/kE}\right)-1}\right]^{d\nu}$$

But $\nu = \frac{c}{\lambda}$

or $d\nu = -\frac{c}{\lambda^2}d\lambda$

Neglecting the negative sign, we get

$$\therefore \qquad E(\lambda)d\lambda = \frac{8\pi h\nu^3}{c^3} \times \frac{c}{\lambda^2}\left[\frac{1}{\left(e^{h\nu/kT}\right)-1}\right]d\lambda$$

or

$$E(\lambda) = \frac{8\pi hc}{\lambda^5} \times \frac{c}{\lambda^2}\left[\frac{1}{\left(e^{h\nu/kT}\right)-1}\right] \qquad \text{...(ix)}$$

This gives the energy density for wavelength λ in the spectrum of the black body. Both the equations (viii) and (ix) represent Planck's radiation law for black body radiation.

GIBBS FUNCTION

The Gibbs function G of a system is given by,

$$G = U - TS + PV$$

Consider a system that can do other forms of work, in addition to P.dV work, *e.g.*, a voltaric cell. In the case of a voltaic cell, the electrical work is –E.dI. Similarly for a magnetic material, the magnetic work is –m.dH. In general, the work will be given by P.dV plus a sum of terms each being the product of intensive variable (such as P, E or m) and the differential of an extensive variable (such as dV, dI and dH). In the case of a voltaic cell, the intensive variable is S and differential of extensive variable is dI.

Suppose, in general, in addition to PdV the intensive variable is y and differential of extensive variable is dx. The work done for any reversible process,

$$\delta W = PdV + ydx$$

$$W = \int_{V_1}^{V_2} PdV + \int_{x_1}^{x_2} ydx$$

Take $\int_{x_1}^{x_2} ydx = A$

Consider a process where the system works at constant pressure P_0 and the change in volume is $(V_2 - V_1)$

$$\therefore \qquad \int_{V_1}^{V_2} PdV = P_0\,[V_2 - V_1]$$

$$\therefore \qquad W = P_0[V_2 - V_1] + A \qquad \text{...(i)}$$

For a system that exchanges heat with a reservoir temperature

$$W < (U_1 - U_2) - T_0 (S_1 - S_2) \quad \text{...(ii)}$$

$$P_0[V_2 - V_1] + A < (U_1 - U_2) - T_0 (S_1 - S_2)$$

or $$A < (U_1 - U_2) - T_0 (S_1 - S_2) + P_0[V_1 - V_2] \quad \text{...(iii)}$$

Consider a specific process, where the initial and the final states of the system and the surroundings are at the same temperature (T_0) and pressure (P_0)

$$To = T \text{ and } P_0 = P$$

From equation (iii)

$$A_{P,T} < (U_1 - U_2)_{P,T} - T(S_1 - S_2)_{P,T} + P[V_1 - V_2]_{P,T} \quad \text{...(iv)}$$

But, for the Gibbs function,

$$G = U - TS + PV \quad \text{...(v)}$$

Therefore, for two equilibrium states at the same pressure and temperature,

$$[G_1 - G_2)_{P,T} = (U_1 - U_2)_{P,T} - T[S_1 - S_2]_{P,T} + P[V_1 - V_2]_{P,T} \quad \text{...(vi)}$$

From equations (iv) and (vi)

$$A_{P,T} < (G_1 - G_2)_{P,T} \quad \text{(vii)}$$

Thus, the difference between Gribbs function of a system between two equilibrium states sets the maximum limit to the work in addition to PdV work, provided the initial and the final states are at the same pressure and temperature and the system exchanges heat with a single heat reservoir. The work done will be maximum when the process is reversible. The process in this case will be isothermal–isobaric. If the process is irreversible, work done will be less than the maximum.

STATISTICAL POSTULATES

Following postulates are made for applying the statistical method to physical systems:

(i) Every system has a large number of particles/molecules which are in constant random motion.

(ii) The motion of particles/molecules is described in 2Nf-dimensional phase-space.

(iii) The phase-space is divided into cells.

(iv) Postulate of equal a priori probability holds good.

(v) Probability of a particular configuration is given by the ratio of microstates corresponding to that configuration to total accessible microstates, *i.e.*,

$P_i = \Omega_i/\Sigma\Omega_i$ and the average value of any physical parameter is given by

$$< x > = \Sigma x_i P_i = \frac{\Sigma x_i \Omega_i}{\Sigma \Omega_i}$$

where Qi is the frequency of the configuration.

(vi) The equilibrium state of the system is the state of maximum probability.

(vii) The law of conservation of mass is generally valid whence

$n_1 + n_2 + n_3 + ... = \Sigma n_i = N =$ constant.

(viii) The law of conservation of energy governs the distribution of particles among cells.

Thus, if $\varepsilon_1, \varepsilon_2, \varepsilon_3$ ... denotes energies of each particle in cell 1, 2, 3, etc., then

$E = \varepsilon_1 n_1 + \varepsilon_2 n_2 + \varepsilon_3 n_3 + ... = \Sigma\varepsilon_i n_i =$ constant.

(ix) Every point in the phase-space (cell) represents a microstate, the cell volume being h where/is the degree of freedom.

(x) The number of microstates in the energy interval e and $\varepsilon + d\varepsilon$, per unit volume is represented as $\Omega(\varepsilon)$.

(xi) A collection of accessible microstates form a statistical ensemble.

ENSEMBLE METHOD

Ensemble method is superior to the method of *time average*, in determining average value of a physical quantity. So first we recall the characteristics of an ensemble and its way of representation.

1. It is a group of large number of systems.
2. Every member of an ensemble is an exact miniature copy of the system under test.
3. Every member of an ensemble is subjected to the same experimental conditions.
4. There may be three kinds of interactions between the members of the ensemble:

   (a) Thermal-related to T,

(b) Mechanical-related to V. and

(c) Diffusive-related N or m (*i.e.*, chemical potential).

5. Every member of the ensemble may have a large number of microstates (also called the dynamic state) due to random motion of its particles/molecules/And when the configuration of the member systems is maintained same in terms of generalised coordinates and momenta. the statistical method is expected to yield quite accurate results.

Ensemble Representation

Ensembles can be represented in a Γ-space by the distribution of its phase-points. Generally, the distribution of phase-points in an ensemble is continuous and can be represented by a density function ρ(p. qt).

The number of phase-points in the vicinity of (p. q) in Γ-space is given by p (p. q t) dT where d r = dp dq = volume element of Γ- space.

If ρ (p. q t} is known at time t_1. its value can easily be obtained for some other time t_2. Now. from the probability theory

$$d\rho\ (p,\ q)\ \alpha\ d\Gamma$$

$$\therefore \qquad d\rho\ (p,\ q) = \rho(p,\ q)d\Gamma$$

Here ρ (p. q) is known as the distribution function. The ensemble average of any function R is then given by

$$\langle R \rangle = \int R\ (p,\ q)\ \rho(p,\ q)d\Gamma.$$

Selection of Ensemble

(i) Molecules of real gases are usually formed of 2, 3.... atoms whence the density of phase points is not uniform.

(ii) The density of phase points is non-zero for every range of energy.

(iii) Total number of particles and energy is conserved. Point (i) eliminates use of uniform ensemble. Point (ii) eliminates use of microcanonical ensemble. Point (in) eliminates use of grand canonical ensemble to real gases.

But since the properties of these gases such as N, V, T, ε, H etc., are constant the canonical ensemble finds use in description of real gases for their thermodynamic properties.

ACCESSIBLE MICROSTATES

Let us now count the number of energy states available in a macrosystem. Let E be the total energy of this system which has been divided into small energy intervals δ E so that a sufficient number of quantum stales are available in each interval of δ E. Now, if Ω (E) is the number density of such states in the energy interval E and E + δE then

$$\Omega(E) = \rho(E)\, \delta E \qquad ...(i)$$

where ρ (E) is density of states (the number of states in unit energy interval). Thus

$$\Omega(E) \propto \delta E \qquad ...(ii)$$

Let the value of Ω(E) changes with the magnitude of energy interval. Hence we may consider Ω(E) as a smoothly varying function of E.

Again, if ϕ(E) denotes the number of states with energy up to E and ϕ(E + δE) denotes the number of states with energy up to E + δE then

$$\Omega(E) = \phi(E + \delta E) - \phi(E) \equiv \frac{\delta\phi}{\delta E}\delta E \qquad ...(iii)$$

Comment on ρ (E). Relations (i) and (ii) (possibly) indicate that ρ(E) is a constant. But it is not so. In fact ρ (E) is an increasing function of E. Let us see how! We know that the total number of microstates is equal to the product of thermodynamic probabilities of different macrostates (composite probability is obtained, always by multiplying probabilities for different complexions), *i.e.*,

$$\Omega = \Omega_1\, \Omega_2\, \Omega_3 \ldots\ldots \Omega_i \ldots \Omega_f$$

or $$\Omega = \Pi_1\, \Omega_i\, (i = 1 \text{ to } f)$$

where f is the degree of freedom.

But Ω_i should be somehow related to E, *i.e.*,

$$\Omega \propto E^a \quad \therefore \Omega \propto E^{fa} \qquad ...(iv)$$

usually a $\cong$ 1/2 and/ranges between 10^{22} and 10^{28}. This suggests that for a certain energy interval δE, ρ(E) increases with the value of E.

QUANTUM STATISTICS

Maxwell-Boltzmann distribution, it has been assumed that all the energy levels are accessible to all the particles of the system. However, there may be certain levels prohibited to a certain group of particles. Each energy state is associated with a certain available wave function. The probability of a particular distribution is restricted by the available wave functions of any state. These restrictions are taken into account in quantum statistics. There are two types of quantum statistics

(i) Fermi-Dirac statistics

(ii) Bose-Einstein statistics.

In Fermi-Dirac statistics the particles are assumed to obey Pauli's exclusion principle. They are characterised by antisymmetric wave functions. The particles under, this category are called *fermions*. Protons neutrons and electrons are fermions.

In the case of Bose-Einstein statistics, it is assumed that the particles are not restricted by Pauli's exclusion principle and are characterised by symmetric wave functions. These particles are called *bosons*. It has been experimentally found that all particles having spin zero or integral multiple of 1 are bosons. Helium nuclei and mesons are bosons.

In both the kinds of quantum statistics, the particles are identical and indistinguishable. At high temperature and low pressure, all the three statistics give practically the same result.

STATISTICAL MECHANICS

Statistical Mechanics deals with systems consisting of many particles and the methods employed are to get a collective or macroscopic property of the system without taking into account the individual motion of the particles The word particle is used in a broad sense meaning a fundamental particle *e.g.*, an electron, an atom, a molecule etc. The particle is a well defined and a stable unit of a given physical system. As it is not practically possible to determine the property of each particle individually, a statistical approach is made by using the concept of probability of distribution. This statistical approach helps in determining the bulk or macroscopic property of the system as a whole. The idea of probability does not imply that the particles move in a random way without obeying the laws. For a many particle system, the statistical analysis is valid regarding the applications of probability distribution of particles.

STATISTICAL EQUILIBRIUM

Consider that a thermodynamical isolated system consists of N particles. The energy states available to the particles are E_1, E_2, E_3 etc. These energy states may be quantized or may be continuous and are due to vibrational and rotational energy of the particles.

Suppose, that at any given instant of time, n_1 particles are in state of energy E_1, n_2, particles with energy state E_2 and so on.

The total number of particles in the system is,

$$N = n_1 + n_2 + n_3 + \ldots\ldots = \sum_i n_i \qquad \ldots(i)$$

Here $\qquad i = 1, 2, 3$ etc.

The total energy of the system,

$$U = n_1E_1 + n_2E_2 + n_3E_3 + \ldots$$

$$U = \sum_i n_i E_i \qquad \ldots(ii)$$

Equation (ii) refers to the total energy of a system in which the particles are non-interacting. Here, the energy of each particle depends only on the coordinates of the particle in the system. For an isolated system, the total energy U is constant.

For isolated system.

$$U = \sum_i n_i E_i = \text{constant} \qquad \ldots(iii)$$

Consider a gas having N molecules at a certain temperature and pressure. Its volume, temperature and pressure are kept constant *i.e.*, the system is isolated. The total energy of this system remains constant. But, the molecules of the gas collide with each other and also with the walls of the container. Consequently, the number of molecules change from one energy state to the other energy state. It means that the value of n_1, n_2, n_3 etc., continuously change. It can be reasonably assumed that for each microscopic State of a system of particles, there is a particular most favoured distribution. When this distribution or partition is reached, the system attains statistical equilibrium. For an isolated system, the values of n_1, n_2, n_3 etc., vary only near the values corresponding to the most probable distribution. Hence, the basic problem in statistical thermodynamics is to obtain the most probable distribution law for a given composition of the system.

In practice, three most probable distribution laws are used. They are:

(i) Maxwell-Boltzmann Distribution Law.

(ii) Fermi-Dirac Distribution Law (Statistics).

(iii) Bose-Einstein Distribution Law (Statistics).

MAXWELL-BOLTZMANN DISTRIBUTION LAW

Consider a system that contains a large number of particles that are identical and distinguishable. The identical particles refer to the particles having the same structure. The distinction between the particles is due to their energy states at a given instant. Four particles are in energy state E_1, 2 in state E_2, zero in state E_3 and so on. It is assumed that all the energy states are accessible to each particle. Consequently, it can be assumed that the probability of any particular partition is proportional to the number of different ways in which the particles can be distributed in the existing available energy states so as to produce the desired partition.

To obtain the required distribution as the first particle a in state E_1 can be selected in N ways. The second particle b in state E_1 can be selected in (N – 1) ways and so on. Therefore, the total number of ways in which the first four particles in state E_1 can be selected is given by,

$$N(N-1)(N-2)(N-3) = \frac{N!}{(N-4)!}$$

Moreover, the four particles in state E_1 can be arranged in 4! different orders. For example, *abcd, bcda, cdab* and so on. There are 24 ways. But, it is immaterial for these particles to be arranged in any particular order in state E_1, because they are identical. Thus, the total number of *distinguishable* different ways are,

$$\frac{N!}{4!(N-4)!}$$

In general, if the first state consists of n_1 particles, the distinguishable different ways, for arranging n_1 particles in state E_1, are,

$$p_1 = \frac{N!}{n_1!(N-n_1)!} \qquad ...(i)$$

For the second state E_2, only $(N - n_1)$ particles are available and n_2 particles are in state E_2. The number of distinguishable different ways are,

$$p_2 = \frac{(N-n_1)!}{n_2!(N-n_1-n_2)!} \qquad ...(ii)$$

If this process is continued for all the available states, the total number of distinguishable ways are obtained by multiplying P_1, P_2, P_3 etc.

$$P = p_1 \times p_2 \times p_3 \times \ldots\ldots$$

$$P = \left[\frac{N!}{n_1!(N-n_1)!}\right]\left[\frac{(N-n_1)!}{n_2!(N-n_1-n_2)}\right]\times\ldots.$$

$$P = \frac{N!}{n_1!n_2!n_3!\ldots} \qquad \ldots(iii)$$

The distinguishable ways are,

$$P = \frac{N!}{4!2!0!3!1!} \qquad \ldots(iv)$$

Here 0! is equal to one.

It has been assumed so far that all the available states have the same probability of occupation by the particles. However, it may happen that the states have different intrinsic probabilities say g_i.

For example, a particular energy state may be favourable with more different angular momentum states than the rest and hence it is more likely to be occupied. Taking into account this intrinsic probability factor, the value of P will be different.

If g_i is the probability of locating a particle in a certain energy state E_i, then the probability of locating 2 particles in the same state is $g_i \times g_i = g^2_i$. For n_i particles, the probability is $g_i n_i$. Hence the total probability for a given distribution is given by

$$P = \frac{N!g_1n_1g_2n_2g_3n_3\ldots}{n_1!n_2!n_3!n_4!\ldots} \qquad \ldots(v)$$

Here n_1, n_2, n_3 etc., are the number of particles in states E_1, E_2, E_3, etc. and g_1, g_2, g_3 etc., are the intrinsic probabilities for states E_1, E_2, E_3 etc.

If all the particles are further assumed to be indistinguishable *i.e.*, particles in state E, and particles in state E_4 cannot be distinguished, then N! permutations among the particles themselves and occupying the different states result in the same distribution. The probability in this case is given by

$$P = \frac{1}{N!}\left[\frac{N!g_1n_1g_2n_2g_3n_3\ldots}{n_1!n_2!n_3!\ldots}\right]$$

$$\therefore \qquad P = \frac{g_1n_1g_2n_2g_3n_3\ldots}{n_1!n_2!n_3!\ldots}$$

$$\therefore \qquad P = \prod_{i}^{N} \frac{g_i n_i}{n_1!} \qquad \text{...(vi)}$$

Here Π is the product sign (since probability is the product of such distributions.

The most probable distribution can be obtained by evaluating the maximum value of $\log_e$, P in equation (vi). This should also satisfy the two conditions that

$$\sum_i n_i = N \qquad \text{...(vii)}$$

$$\text{and } \sum_i n_i \; E_i = U \qquad \text{...(vii)}$$

According to Stirling's approximation

$$\log_e x! = x \log_e x - x$$

From equation (vi), applying Stirling's approximation, we get

$$\log_e P = \sum_i (n_i \log_e g_i - \log_e n_i!)$$

$$= \sum_i [n_i \log_e g_i - (n_i \log_e n_i - n_i)]$$

$$= \sum_i (n_i \log_e g_i - n_i \log_e n_i + n_i)$$

$$= \sum_i n_i - \sum_i (n_i \log_e n_i - n_i \log_e g_i)$$

$$\log_e P = N \sum_i n_i \log_e \frac{n_i}{g_i} \qquad \text{...(ix)}$$

Differentiating equation (ix)

$$d(\log_e P) = \sum_i (dn_i) \log_e \frac{n_i}{g_i} - \sum_i n_i d\left(\log_e \frac{n_i}{g_i}\right)$$

$$= \sum_i dn_i \log_e \frac{n_i}{g_i} - \sum_i n_i \frac{dn_i}{n_i}$$

$$= \sum_i dn_i \log_e \frac{n_i}{g_i} - \sum_i dn_i$$

But $\sum_i dn_i = 0$.

$$\therefore \qquad -d(\log_e P) = \sum_i \left[\log_e\left(\frac{n_i}{g_i}\right)\right] dn_i \qquad \text{...(x)}$$

To obtain the maximum value of P,

$$d(\log_e P) = 0$$

$$\therefore \qquad \sum_i \log_e\left(\frac{n_i}{g_i}\right)dn_i = 0 \qquad \text{...(xi)}$$

But $\quad \Sigma dn_i = 0$ ...(xii)

and $\Sigma E_i dn_i = 0$ ...(xiii)

Multiplying (xii) by α and (xiii) by β and adding to equation (xi), we get

$$\sum_i\left[\log_e\left(\frac{n_i}{g_i}\right) + \alpha + \beta E_i\right]dn_i = 0$$

The equilibrium distribution is possible if

$$\log_e \frac{n_i}{g_i} + \alpha + \beta\ Ei = 0 \qquad \text{...(xiv)}$$

or $$\frac{n_i}{g_i} = e^{-\alpha-\beta Ei}$$

or $$n_i = g_i\ e^{-\alpha-\beta Ei} \qquad \text{...(xv)}$$

This gives the maximum probability distribution and a and p are two parameters that depend upon the physical property of the system.

$$N = n_1 + n_2 + n_3 + \ldots$$

$$N = g_1 e^{-\alpha-\beta E1} + g_1 e^{-\alpha-\beta E2} + \ldots.$$

$$N = e^{-\alpha}[g_1 e^{-\beta E1} + g_2 e^{-\beta E2} + \ldots.]$$

$$N = e^{-\alpha}[\sum_i g_i\, e^{-\beta Ei}]$$

Take $[\sum_i g_i\, e^{-\beta Ei} = Z$ ...(xvi)

Here, Z is called the partition function.

$$N = e^{-\alpha}(Z) \qquad \text{....(xvii)}$$

$$e^{-\alpha} = \frac{N}{Z}$$

Substituting this value in equation (vii)

$$n_i = \left(\frac{N}{Z}\right)[g_i e^{-\beta Ei}] \qquad \text{....(xviii)}$$

This equation refers to Maxwell-Boltzmann Distribution Law.

MAXWELL-BOLTZMANN DISTRIBUTION IN TERMS OF TEMPERATURE

The total energy of an isolated system is given by

$$U = \sum_i n_i E_i$$

$$U = n_1E_1 + n_2E_2 + n_3E_3 + ...$$

$$U = (g_1e^{-\alpha-\beta E1}) E_1 + (g_2e^{-\alpha-\beta E2}) E_2 + ...$$

$$U = e^{-\alpha}[g_1E_1e^{-\beta E1} + g_2E_2e^{-\beta E2} + ...]$$

But $$e^{-\alpha} = \frac{N}{Z}$$

$$\therefore \quad U = \frac{N}{Z}[g_1E_1e^{-\beta E1} + g_2E_2e^{-\beta E2} + ...]$$

$$U = \frac{N}{Z}\sum_i g_i E_i e^{-\beta Ei}$$

Here $$Z = \sum_i g_i e^{-\beta Ei}$$

$$\frac{dZ}{d\beta} = \frac{d}{d\beta}\sum_i g_i e^{-\beta Ei} = -\sum_i g_i E_i e^{-\beta Ei}$$

$$\therefore \quad \sum_i g_i E_i e^{-\beta Ei} = \frac{d}{d\beta}\sum_i g_i E_i e^{-\beta Ei}$$

Substituting this value in equation (i)

$$U = -\frac{N}{Z}\frac{d}{d\beta}\sum_i g_i e^{-\beta Ei}$$

$$U = -\frac{N}{Z}\frac{d}{d\beta}$$

$$U = -N\frac{d}{d\beta}[\log_e Z] \quad ...(ii)$$

The average energy of a particle is

$$E_{av} = -\frac{U}{N} = -\frac{d}{d\beta}(\log_e Z) \quad ...(iii)$$

This shows that for a given system, the total energy U, the partition function Z, and the average energy of the particle E_{av}, depend on the parameter β.

Therefore, β may be taken to characterise the internal energy of the system and has the units per joule. If T is the temperature in degrees kelvin, then it is more convenient to represent parameter β as

$$\beta = \frac{1}{kT} \text{ or } KT = \frac{1}{\beta} \quad \text{...(iv)}$$

Here kT has the unit of energy *i.e.*, joule and k is the Boltzmann's constant. Its units are J/K. The value of k is given by

$$k = 1.3805 \times 10^{-23} \text{ J/K}$$

Substituting the value of

$$\beta = \frac{1}{kT}$$

in all the equations, we get

$$Z = \sum_i g_i e^{-(Ei/kT)} \quad \text{...(v)}$$

$$n_i = \frac{N}{Z} \sum_i g_i e^{-(Ei/kT)} \quad \text{...(vi)}$$

This equation represents the Maxwell-Boltzmann distribution law in terms of the temperature of the system.

As $\beta = \frac{1}{kT}$

$$d\beta = -\frac{dT}{kT^2}$$

Substituting this value in equation (ii)

$$\therefore \quad U = (kNT^2)\frac{d}{dT}(\log_e Z) \quad \text{...(vii)}$$

Also, from equation (iii)

$$E_{av} = (kT^2)\frac{d}{dT}(\log_e Z) \quad \text{...(viii)}$$

Equation (viii) gives the relation between the average energy of the particle and its temperature under equilibrium position. Hence, the temperature of a system in statistical equilibrium is the physical quantity related to the average energy of the particle of the system.

FERMI-DIRAC DISTRIBUTION LAW

Consider a system consisting of a large number of particles. It is assumed:

(i) The particles are *identical* and *indistinguishable*, and

(ii) The particles obey exclusion principle. It means that no two particles can have the same dynamical state and the wave function of the whole system must be antisymmetric. The particle satisfying these conditions are called *fermions*. In general, all fundamental particles with spin are fermions.

In quantum statistics the intrinsic probability g_i is governed by the different quantum states relating to a given energy *i.e.*, the degeneracy of the energy state. Each quantum state corresponds to a particle wave function. The Wave functions are determined by each of the possible arrangements of quantum number corresponding to a given energy level. For particles with spin 1/2, in the absence of magnetic forces, each of the particle may be in the energy states with spin +1/2 or −1/2. Hence the intrinsic probability g_i in this case is 2. For motion in a central field, the energy of the particle is independent of the orientation of the orbital angular momentum Due to this, a degeneracy of $(2l + 1)$ is introduced and it is the value of g for that particular energy state. If the particles possess spin, the total degeneracy = $2(2l + 1)$. As no two particles can be in the same energy state having the same quantum number, the instrinsic probabilities g_i's give the maximum number of particles that can be accommodated in a particular energy level without violating the exclusion principle *i.e.*,

$$n_i \leq g_i$$

It means that the n_i value for a given distribution does not exceed the corresponding value of g_i.

Let n_i, be the number of particles for energy level E_i The first particle can be placed in any one of the available g_i states *i.e.*, this particle can be assigned to any of the gi sets of quantum numbers. Thus, the first particle can be distributed in g_i different ways. Similarly, the second particle can be arranged in $(g_i - 1)$ different ways and the process continues.

Thus, the total number of different ways of arranging n_i particles among the available g_i states with energy level E_i is

$$= g_i(g_i - 1)(g_i - 2) \ldots\ldots[g_i - (n_i - 1)]$$

$$= \frac{g_i!}{(g_i - n_i)!} \qquad \ldots(i)$$

Further, if the particles are taken to be indistinguishable, it will not be possible to detect any difference when n_i particles are reshuffled into different states occupied by them in the energy level E_i. Therefore, the total number of different and distinguishable ways is,

$$= \frac{g_i!}{n_i!(g_i - n_i)!} \qquad ...(ii)$$

Therefore, the total number of different and distinguishable ways of getting the distribution n_1, n_2, n_3, etc., among the various energy levels, E_1, E_2, E_3...,etc.. can be obtained by multiplying the various factors.

$$\therefore \qquad P = \frac{g_1!}{n_1!(g_1 - n_1)!} \cdot \frac{g_{2i}!}{n_2!(g_2 - n_2)!} \; ....$$

$$P = \prod_i \frac{g_i!}{n_i!(g_i - n_i)!} \qquad ...(iii)$$

The most probable distribution can by obtained by evaluating the maximum value of $\log_e$, P in equation (iii).

This should also satisfy the condition that

$$\sum_i n_i = N$$

and $\sum_i n_i E_i = U$.

According to Stirling's approximation

$$\log_e x! = x \log_e x - x$$

From equation (iii), applying Stirling's approximation,

$$\log_e P = \sum_i [(g_i \log_e g_i - g_1) - (n_i \log_e n_i - n_i) - [(g_i - n_1) \log_e (g_i - n_1) - (g_i - n_1)]$$

$$\log_e P = \sum_i [g_i \log_e g_i - n_1 - \log_e n_i - (g_i - n_i) \log_e (g_i - n_1)] \qquad ....(iv)$$

Differentiating equation (iv)

$$-d(\log_e P) = \sum_i [\log_e n_i - \log_e (g_i - n_i)] dn_i$$

To obtain the maximum value of P,

$$d(\log_e P) = 0$$

$$\therefore \qquad \sum_i [\log_e n_i - \log_e (g_i - n_i)] dn_i = 0 \qquad ...(v)$$

But $\sum_i dn_i = 0$ ...(vi)

and $\sum_i E_i dn_i = 0$ ...(vii)

Multiplying (vi) by α and (vii) by β and adding to equation (v), we get

$$\sum_i [\log_e n_i - \log_e (g_i - n_1) + \alpha + \beta E_i] dn_i = 0$$

The equilibrium distribution is possible if

$$\therefore \quad \log_e n_i - \log_e (g_i - n_1) + \alpha + \beta E_i = 0 \qquad ...(viii)$$

$$\log_e\left(\frac{n_i}{g_i - n_i}\right) = -\alpha + \beta E_i$$

$$\frac{n_i}{g_i - n_i} = e^{-\alpha - \beta Ei}$$

$$\frac{g_i - n_i}{n_i} = e^{\alpha + \beta Ei}$$

$$\frac{g_i}{n_i} - 1 = e^{\alpha + \beta Ei}$$

$$\frac{g_i}{n_i} = (e^{\alpha + \beta Ei}) + 1$$

$$n_i = \left(\frac{g_i}{e^{\alpha + \beta E_i}}\right) + 1 \qquad ...(ix)$$

Equation (ix) represents the Fermi-Dirac distribution law. The parameter has the same role as in the case of Maxwell-Boltzmann distribution law *i.e.*, for the system consisting of fermions in statistical equilibrium.

$$\beta = \frac{1}{kT}$$

$$\therefore \quad n_i = \frac{g_i}{[e^{(\alpha + E_i/kT)}] + 1}$$

In most cases the value of a is negative and is taken to be equal to

$$\frac{-E_F}{kT}$$

$$n_i = \frac{g_i}{\left[e^{(E_i - E_F)/kT)}\right] + 1}$$

The value of E_F is positive and is independent of temperature.

For T = 0, all the energy states are fully occupied and $n_i = g_i$ All the states with $E > E_F$, are empty *i.e.*, $n_i = 0$.

$$\left[e^{(E_i - E_F)/kT)}\right] = \begin{cases} 0 \text{ for } E_i - E_F < 0 \\ \infty \text{ for } E_i - E_F > 0 \end{cases}$$

In the case of Fermi-Dirac statistics, the accumulation of particles at the ground level is not allowed and at temperature T = 0, the particles occupy the lowest energy levels upto E_F. Here, the energy E_F gives the indication of the maximum energy of the fermions in the system. By is also called Fermi energy. For higher temperatures, the particles, occupy higher energy states greater than E_F. The curves indicate that only those fermions with energies close to E_F can move into unoccupied higher energy states. If $k\theta_F = E_F$, the temperature θ_F is called the Fermi temperature.

MAXWELL-BOLTZMANN DISTRIBUTION AND IDEAL GAS

Most of the gases obey Maxwell-Boltzmann distribution over a wide range of temperature. Consider a gas having mono-atomic molecules. It is assumed that the gas is ideal and has only kinetic energy of translation. Intermolecular attraction is assumed to be absent. If the kinetic energy is not quantized but considered to possess continuous values of energy, then the partition function Z can be written as

$$Z = \int_0^\infty e^{-(E/kT)}\, g(E)dE$$

Here g_i is replaced by g (E) dE

$$\text{But} \quad g(E)dE = \left(\frac{4\pi V(2m^3)^{1/2}}{h^3}\right)E^{1/2}dE$$

Here V is the volume occupied by the gas and h is Planck's constant.

$$Z = \frac{4pV(2m^3)^{1/2}}{h^3}\int_0^\infty E^{1/2}e^{-(E/kT)}dE$$

$$Z = \frac{4\pi V(2m^3)^{1/2}}{h^3}\left[\frac{\sqrt{\pi(kT)^3}}{2}\right]$$

$$Z = \frac{V(2\pi mkT)^{3/2}}{h^3} \quad \text{...(i)}$$

Equation (i) represents the partition function for an ideal mono-atomic gas in terms of the volume and temperature of the gas.

Taking logarithms on both sides

$$\log_e Z = \log_e \left[\frac{(2\pi m)^{3/2} V}{h^3}\right] + \log_e (kT)^{3/2}$$

$$\log_e Z = C + \frac{3}{2}\log_e kT \quad \text{...(ii)}$$

Also $$E_{av} = (kT^2)\frac{d}{dT}(\log_e Z)$$

$$E_{av} = (kT^2)\frac{d}{dT}\left[C + \frac{3}{2}\log_e kT\right]$$

$$= (kT^2)\frac{d}{dT}\left[C + \frac{3}{2}\log_e k + \frac{3}{2}\log_e T\right]$$

$$E_{av} = \frac{3}{2}kT \quad \text{...(iii)}$$

The total energy.

$$U = NE_{av}$$

$$U = N\left(\frac{3}{2}kT\right)$$

$$U = \frac{3}{2}NkT. \quad \text{...(iv)}$$

Equation (iv) shows that the internal energy of an ideal monoatomic gas depends only on its temperature. The same relation does not however. hold good for real gases. In the case of real gases the internal energy is partly potential and partly kinetic. The total energy depends on the volume of the gas.

ELECTRON GAS

Electrons in a metal belong to a most characteristic system of fermions because electrons obey the exclusion principle. For electrons in a metal, the energy levels are grouped in bands. Practically at all temperatures, the lower level energy bands are filled with electrons. The upper level energy bands are only partially filled with electrons. The distribution of electrons is to be considered only in the upper bands called the conduction band.

The zero energy level is taken at the lowest level of the conduction band. It is also assumed that the electrons have free movement within the conductor, provided the energy associated with the electrons is of the order of upper level energy bands.

As the energy of the electron in the conduction band is continuous, the term g_i is replaced by g(E) dE. Here dn electrons have energy in the range E and E + dE. According to Fermi-Dirac distribution law

$$n_i = \frac{g_i}{\left[e^{(E_i - E_F)/kT}\right] + 1} \qquad \text{...(i)}$$

Substituting the value of g_i = g (E) dB in equation (i) and replacing n_i by dn

$$dn = \frac{g(E)dE}{\left[e^{(E_i - E_F)/kT}\right] + 1}$$

As an electron has the spin ±1/2, the total number of states in the sphere is twice. $V/(2\pi)^3$ refers to the translational states per unit volume in the Fermi space Fermi sphere of radius kF has the total number of particles accommodated.

$$n = \frac{2\left[\frac{V}{(2\pi)^3}\right]\left(\frac{4}{3}\pi(kF)^3\right)}{\left[e^{(E_i - E_F)/kT}\right] + 1}$$

Here $$E = \frac{p^2}{2m} \frac{\left(\frac{h}{2\pi}\right)^2 (kF)^2}{2m}$$

$$kF = \frac{p}{h/2\pi} = \frac{2\pi p}{h}$$

$$\therefore \quad n = \frac{2\left[\frac{V}{(2\pi)^3}\right]\left(\frac{4}{3}\pi\left(\frac{2\pi p}{h}\right)^3\right)}{\left[e^{(E_i - E_F)/kT}\right] + 1}$$

But $p = (2mE)^{1/2}$

$$\therefore \qquad n = \frac{2\left[\frac{V}{(2\pi)^3}\right]\left[\left(\frac{4}{3}\pi\left(\frac{2\pi p}{h}\right)^3 (2mE)^{3/2}\right)\right]}{\left[e^{(E_i - E_F)/kT)}\right]+1}$$

$$n = \frac{\left(\frac{8\pi V}{(3h^3}\right)(2m)((2m)^{1/2}E^{3/2}}{\left[e^{(E_i - E_F)/kT)}\right]+1}$$

Differentiating

$$dn = \frac{\left(\frac{8\pi V}{(3h^3}\right)(2m)((2m)^{1/2} \times \frac{3}{2}E^{1/2}dE}{\left[e^{(E_i - E_F)/kT)}\right]+1}$$

$$\frac{dn}{dE} = \frac{\left[\frac{8\pi V(2m^3)^{1/2}}{(3h^3}\right]E^{1/2}}{\left[e^{(E_i - E_F)/kT)}\right]+1} \qquad ...(ii)$$

Equation (ii) represents the energy distribution for free electrons. This is also called Fermi-Dirac formula of free fermions.

At $\qquad T = 0K,$

$$p = (2m\,E_F)^{1/2} \qquad ...(iii)$$

and $N = \left(\frac{2V}{h^3}\right)\left(\frac{4}{3}\pi p^3\right)$

or $$P = \left(\frac{3Nh^3}{8\pi V}\right)^{1/3} \qquad ...(iv)$$

Equating (iii) and (iv)

$$(2mE_F)^{1/2} = \left(\frac{3Nh^3}{8\pi V}\right)^{1/3}$$

Squaring $2mE_F)^{1/2} = \left(\frac{3Nh^3}{8\pi V}\right)^{2/3}$

or $$E_F = \frac{h^2}{8m}\left(\frac{3N}{\pi V}\right)^{2/3} \qquad ...(v)$$

Knowing the value of N/V *i.e.*, the number of free electrons per unit volume. the Fermi energy for electrons in a metal can be obtained.

For silver, the value $\frac{N}{V}$

$$= 5.86 \times 10^{28} \text{ electrons/m}^3$$

$$\therefore \quad E_F = \left(\frac{(6.624\times10^{-34})^2}{8\times9\times10^{-31}}\right)\left(\frac{3\times5.86\times10^{28}}{3.14}\right)^{2/3}$$

$$E_F = 9 \times 10^{-19}\text{J}.$$

But $\quad 1\text{eV} = 1.6 \times 10^{-19}\text{J}$

$$\therefore \quad E_F = \frac{9\times10^{-19}}{1.6\times10^{-19}}$$

$$E_F = 5.625 \text{ eV}.$$

It means that the maximum kinetic energy of free electrons in silver at absolute zero temperature is 5.6 eV.

The Fermi temperature θ_F is given by

$$\theta_F = \frac{E_F}{k}$$

As E_F practically independent of temperature, the value θ_F is fixed for a given metal. The values of θ_F and E_F are given in the following table.

Metals	*E_F (electron Volts)*	*θ_F(K)*
Potassium	2.14	2.4×10^4
Sodium	312	3.7×10^4
Lithium	4.72	6.5×10^4
Silver	5.51	6.4×10^4
Gold	5.64	6.4×10^4
Copper	7.04	8.2×10^4

DISTRIBUTION LAW OF BOSE-EINSTEIN

Bose-Einstein distribution is applied to systems composed of identical and indistinguishable particles that are not restricted by the exclusion principle. In such systems, there is no limit to the number of particles occupying a particular quantum state. These particles are called *bosons*. Their spin is zero or 1. Mesons and helium nuclei are examples of bosons.

The Bose-Einstein statistics values of g_i refer to degeneracy of each energy level. Suppose, that n_i particles are arranged in a row and distributed among g_i quantum states with $(g_i - 1)$ partitions in between.

The total number of possible arrangements of particles and partitions is equal to the total number of permutations of $(n_i + g_i - 1)$ objects in a row. Therefore the total possible ways of arranging n_i particles with $g_i - 1$ partitions

$$= (n_i + g_i - 1)!$$

As the particles are identical and indistinguishable the possible number of distinct arrangements

$$= \frac{(n_i + g_i - 1)!}{n_i!(g_i - 1)!}$$

The total number of distinguishable and distinct ways of arranging N particles in all the available energy states is given by

$$P = \frac{(n_1 + g_1 - 1)!}{n_1!(g_1 - 1)!} \times \frac{(n_2 + g_2 - 1)!}{n_2!(g_2 - 1)!} \times \ldots$$

$$P = \prod_i \frac{(n_i + g_i - 1)!}{n_i!(g_i - 1)!} \quad \ldots(i)$$

The most probable distribution can be obtained by finding the maximum value of $\log_e$. P.

According to Stirling's approximation

$$\log_e x! = x \log_e x - x$$

Also $$\sum_i n_i = N \quad \ldots(ii)$$

and $$\sum_i n_i E_i = U \quad \ldots(iii)$$

From equation (i), applying Stirling's approximation

$$\log_e P = \sum_i [\log_e (n_i + g_i - 1) - \log_e n_i! \log_e (g_i - 1)!]$$

$$\log_e P = \sum_i [(n_i + g_i - 1) \log_e (n_i + g_i - 1) - (n_i + g_i - 1)$$
$$- (n_i \log_e n_i - n_i) - (g_i - 1) \log_e (g_i - 1) - (g_i - 1)]$$

$$\log_e P = \sum_i [(n_i + g_i - 1) \log_e (n_i + g_i - 1)$$
$$- n_i \log_e - n_i - (g_i - 1) \log_e (g_i - 1)] \quad \ldots(iv)$$

The maximum value of P is obtained by taking

$$d\,(\log_e P) = 0.$$

Differentiating equation P.

$$d\,(\log_e P) = \sum_i [\log_e (n_i + g_i - 1)\, dn_i - \log_e n_i - dn_i] = 0$$

$$\therefore \quad -d\,(\log_e P) = \sum_i [-\log_e (n_i + g_i - 1)\, dn_i + \log_e n_i dn_i] = 0$$

or

$$\sum_i [-\log_e (n_i + g_i - 1) + \log_e n_i] dn_i = 0 \quad \text{...(v)}$$

As the total number of particles and total energy are constants, we have,

$$\sum_i dn_i = 0 \quad \text{...(vi)}$$

$$\sum_i E_i dn_i = 0. \quad \text{...(vii)}$$

Multiplying (vi) by α and equation (vii) by β and adding to equation (v), we get

$$\sum_i [-\log_e (n_i + g_i - 1) + \log_e n_i + \alpha + \beta E_i] dn_i = 0$$

$$\therefore \quad -\log_e (n_i + g_i - 1) + \log_e n_i + \alpha + \beta E_i = 0$$

Taking $n_i + g_i$ very large as compared to 1, the quantity 1 can be neglected.

$$\therefore \quad -\log_e (n_i + g_i) + \log_e n_i + \alpha + \beta E_i = 0$$

$$\log_e \left(\frac{n_i}{ng_i + g_i} \right) = -\alpha - \beta E_i$$

$$\frac{n_i}{n_i + g_i} = e^{-\alpha - \beta Ei}$$

$$\frac{n_i + g_i}{n_i} = e^{\alpha + \beta Ei}$$

$$1 + \frac{g_i}{n_i} = e^{\alpha + \beta Ei}$$

$$\frac{g_i}{n_i} = (e^{\alpha + \beta Ei}) - 1$$

$$n_i = \left(\frac{g_i}{e^{\alpha + \beta E_i}} \right) - 1 \quad \text{...(viii)}$$

Equation (ix) represents the Bose-Einstein distribution law.

Taking $$\beta = \frac{1}{kT}$$

$$\therefore \qquad n_i = \frac{g_i}{[e^{(\alpha + E_i/kT)}] - 1}$$

The value of the constant a is governed by the equation $\Sigma n_i = N$. As n_i cannot be negative, α must always have a positive value. The distribution of the particles for different energy levels

SOME DEFINITIONS

It would be appropriate to discuss, at some length, the physical entities involved, in understanding the statistical problems.

A Physical Quantity

Essentially, we have two kinds of physical quantities (a) the intensive quantities that do not depend on the mass of the system, *e.g.*, P and T, and (b) the extensive quantities which depend on the mass of system, *e.g.*, S, V, E, etc.

System

It is a finite region in the space where the experimental body is kept. It Is obligatory for the system to have a finite mass. The system can be a closed one (may have a real physical boundary) or may have an imaginary boundary in space. Hence it can be penetrable or Impenetrable to the transport of energy. The properties of the particles in the system define the properties of the system.

Microstate

Each macrostate can have several microstates associated with it. Microstate is that state of the system where the individuality of the particles is not ignored. Hence the microstates are affected by change in position, momentum, energy (and mutual interaction) of individual particles. These states, however, cannot be identified by any physical experiment. Following table will help in understanding the difference between macro and microstates. We take two boxes A and B and four distinguishable particles a, b, c, d, and investigate their distribution.

It is evident from part B of the table that when the individuality of the particles is ignored, there remains no difference between the number of macrostates and associated microstates. By definition, given above, such a distribution will not give statistical probability or correct probable distribution (last column). Another illustration for the above table is to distribute 4 distinct books on 2 shelves. We have a total of 16 distributions as in part A of the table but if we distribute 4 identical books on same 2 shelves we have only 5 arrangements.

From the table, we see that for the most probable distribution (2, 2) we have 6 microstates. In other words, the macro distribution is most probable when number of microstates involved is maximum. In actual physical problems, we deal with extremely large number of particles as $\approx 10m^{-3}$ whence the number of microstates can be unimaginably large.

Macrostate

It is that state of the system which can be perceived by experiments. It is an isolated state of the system (not influenced by external agencies) and can be defined by certain value of pressure P, temperature T, and volume V. A macrostate can also be defined in terms of its microstructure, *e.g.*, by the specification of number of molecules (or phase-points in each cell of the phase space). Let us consider the distribution of 8 particles in two boxes. We can have (0, 8), (1, 7), (2, 6)... (7, 1) and (8, 0) distributions. These 9 distributions form 9 macrostates. Here we may note that a macrostate does not distinguish individuality of particles. Hence, for n particles we have (n + 1) macrostates when the distribution of particles is restricted in 2 boxes. This number will increase if the number of boxes is increased.

Phase Space

In order to specify the state of the gas from the molecular point of view, we shall know the position and momentum of each of its molecules, *i.e.*, we must specify six quantities x, y, z, p_x, p_y, p_z for each of the molecules. In three-dimensional space, we can easily consider an element of volume dx dy dz. Similarly, we can plot momentum components p_x, p_y and p_z in three mutually perpendicular directions and again consider three-dimensional volume. In other words, let us consider a particle with only a motion of translation. Then, classically we need three positional and three momentum coordinates to describe its state. Just as the space is represented by a three dimensional positional coordinate system, we can also represent momentum space by a three-dimensional momentum coordinate system. The state of the system then can be defined by six-dimensional positional-space and momentum space (phase) system. The two spaces together, defined by 6-dimensional (3 space and 3 momentum) hypothetical coordinate systems, is called phase space. Phase-space is purely a mathematical concept and cannot be depicted by a diagram like. So, phase-space is just a hypothetical space occupied by an infinite set of such six coordinate points. With the passage of time the point P (known as phase-point) traces out a certain trajectory known as phase-path.

As the position and momentum, in translated motion, change with lime, the phase-path may be quite complicated. Such 6-dimensional (6-d) phase space is also known as the μ-space. A large number of 6-d cells, each with a volume of $d\tau = (dx\ dy\ dz)\ (dp_x\ dp_y\ dp_z) = dV.\ dp$, together form the μ-space.

Let us now pass on to a many particle system. Let the system has N particles. Then, to define their state we need 3N positional and 3N momentum coordinates in a 6-d phase-space. We can consider this problem in another way:

A single point in 3-d space represents 3 coordinates; in 6-d space represents 6 coordinates; in n-d space represents n coordinates. Hence in 6N-d space a single point corresponds to 6N coordinates. If N be the number of molecules in a kilogram-mole (or a Mole), the 6N-dimensional phase-space is called r-space (or y-space). We can generalise this representation by assuming that the system has N particles each having f degrees of freedom the Γ- space will then be 2Nf dimensional, having Nf positional coordinates (q_i, i = 1 to Nf) and Nf momentum coordinates (P_i, i = 1 to Nf).

Phase-cells

The concept of phase space enables us to understand the state of micro particles -A their position and energy. When a particle changes its momentum and position with time, there will be infinite points in the phase-space which will express the state of this moving particle. Again in a finite volume of phase space, we have infinite microstates available for the particle. Hence, we can divide the whole of the phase space into finite sized small cells. The volume of the cell, however, cannot indefinitely be small.

Accessible Microstates

Those microstates which follow stipulated conditions (like number of particles is conserved, energy is conserved, system as finite extension) are called the accessible microstates.

The State of Equilibrium

It is the steady state of an isolated system where the values of P, V and T do not change with the passage of time. In such systems, no energy is given to or taken out of it. In statistical terms, equilibrium state corresponds to that macrostate which has maximum microstates. One must, however, remember that an isolated system, in equilibrium, is equally likely to be in any of its accessible microstate.

Events

When an operation is performed or is allowed to take place, the occurrence of certain complexion, configuration or arrangement of the system, at one such trial is called an event. When number of trials is increased the probability of certain event may increase.

STATISTICS APPLIED TO MOLECULES

A monatomic perfect gas. The state of each molecule of the gas in the μ-space is represented by a point (and by a cell in quantum mechanical terms). But, since the gas molecules are in constant motion, their state will be changing continuously in space and, due to collisions (and also mutual interactions), there is a change .in momentum state also. The trajectory of such a molecule constant energy surface given by

$$E = \frac{p_x^2 + p_y^2 + p_z^2}{2m} = V(q_x, q_y, q_z)$$

The real gas molecules, however, are endowed with mutual interactions in addition to usual collisions, hence their state is better represented by a point in y-space. The rate of change of its state, then, depends on the quantum of interactions. The path of representative point in y-space is given by the relation

$$E = \sum_i T_i + \sum_i V_i + \sum_{i,j}{}' U_{ij}$$

where T_i, V_i and U_{ij} refer to kinetic energy, potential energy and energy of interaction of ith particle. The prime on the summation sign suggests that i = j is not allowed (which would mean self-interaction). With this background and remembering that the number of molecules is very large in a macrosystem, we now proceed to investigate the distribution details.

STEFAN'S LAW VERIFICATION

In 1897, Lummer and Pringsheim experimentally verified Stefan's law over a wide range of temperature (100°C to 1,300°C). The apparatus consisted of a black body C. For temperatures between 200°C and 600°C, a hollow copper sphere coated inside with platinum black was used. The fused nitrates of sodium and potassium having a melting point of 219°C were used as the bath surrounding the black body. For temperature between 900°C and 1,300°C, an iron cylinder coated inside with platinum black was used as a black body and it was enclosed in a double walled gas furnace. A thermocouple T was used as a thermometer. A bolometer B was used to measure the intensity of the emitted heat radiations. S_1, S_2 and S_3 were the water-cooling shutters. Another black body A at 100°C was used to standardize the bolometer.

The double walled vessel of the black body A contained boiling water at 100°C. The bolometer B was allowed to face the opening of the black body A and the shutter S_3 raised. The deflections in the galvanometer of the bolometer at various distances were noted and it was found that the deflection was inversely proportional to the square of the distance between the bolometer and the opening of the black body A. Thus, the deflection in the galvanometer was proportional to the intensity of heat radiations.

The shutter S_3 was closed and the bolometer B was allowed to face the opening of the black body C. The shutters of S_1 and S_2 were raised. The bath surrounding the black body was maintained at a constant temperature and the maximum deflection produced in the galvanometer of the bolometer was noted. Thus at various constant temperatures of the black body, corresponding to constant deflections (in the galvanometer of the bolometer) were observed. Then the data was reduced to a common arbitrary unit in terms of the total radiations from the black body A at 100°C.

Let θ be the deflection in the galvanometer, T_1 the temperature of the black body and T_2, the temperature at the entrance of the bolometer. It was found that

$$\theta \propto (T_1^4 - T_2^4)$$

But $\theta \propto R$

$$\therefore \qquad \theta \propto (T_1^4 - T_2^4)$$

This verifies Stefan's law.

Recently Coblentz has verified Stefan's law more accurately. He took an electrically heated black body whose temperature was measured by an accurate thermocouple. An absolute bolometer was used to measure the amount of heat radiations emitted by the black body. He was able to show the correctness of Stefan's law experimentally up to 1,600°C.

MACROSCOPIC SYSTEMS CHARACTERISTIC FEATURES

Let us now discuss some of the characteristic features of the macrosystems. In doing so we will assume that it has a large number of microscopic particles.

Fluctuation

Let the isolated system has N molecules of a perfect gas and that it was left to itself for quite some time so that it is in equilibrium. Further, also consider that the system has been partitioned into two identical (also equal) parts by an imaginary wall, each part A and B has n and n molecules respectively such that n + n = N.

Now, as we have seen, for large values of N, $n \cong n^1 = N/2$

But the molecules of the gas are in a state of constant random motion. So that some of the molecules are leaving portion .A and moving into B and vice-versa. Thus, there is a continuous change in the number of molecules in each portion of the box. Time variation of this change is known as fluctuation. The state of normal fluctuation of an isolated system. Usually, N/2 – n is only a small number. For large deviation, the number of microstates (and hence the probability) reduce drastically. The number of microstates for different values of N and N/2 – n. For the sake of simplicity the deviation N/2– n is represented as a fraction of total number of particles. Value of R ($\cong P_x/P_{max}$) is also shown to arrive at qualitative results;

In actual physical problems the number of molecules are $\sim 10^{24} m^{-3}$ whence even 0.01% deviation also becomes quite improbable. This is the reason for quick diffusion and most equitable distribution of air molecules in the available space.

EQUILIBRIUM STATE OF THE SYSTEM

The macrostate of a gas at a given instant is obtained from the distribution of molecules at that instant. But, if the macrostate of the multimolecular system does not change with time, the isolated system is said to be in equilibrium. Statistically, in the equilibrium state, the probability of the isolated system being in different accessible states does not vary with time. It may be noted that every system has an inherent tendency to approach equilibrium. If, anyhow, a situation (like point t_1

in Fig. 2.1) is created so that the system goes into a higher fluctuation state (lower probability), the motion of the molecules will set in such a way as to bring the system back into normal fluctuation state, with the passage of time. The time taken by the system to attain normal state from its disturbed (excited) state is called the relaxation *time of decay*. It is very easy to perturb the normal state of the system by external effects (by pressure/wind; diffusion of perfume is one such example). But the system will again tend to attain uniform distribution/most probable distribution. To check whether the macrostate is in equilibrium we plot fluctuation versus time curve for a certain period τ. The experiment is repeated again and again. If the curves agree macrostate will be in equilibrium state. In other words, the time average of number of molecules in certain region of space at equilibrium

$$\overline{n(t)} = \frac{1}{\tau}\int_t^{t+\tau} n(t)dt.$$

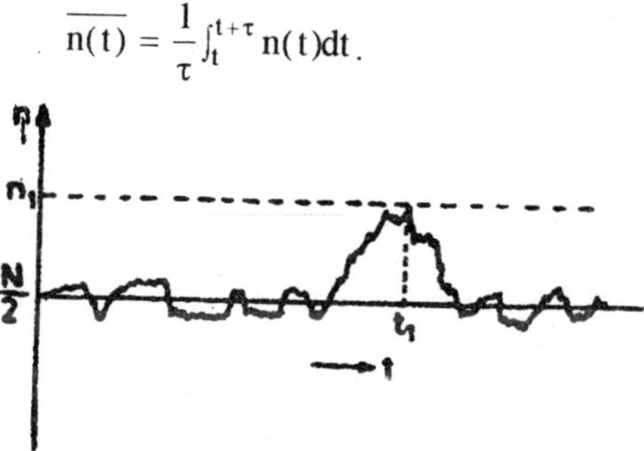

Fig. 2.1

GIBBS ENSEMBLE

The propeties of ensemble was put forth by Gibbs in 1902. In his view, it was a mathematical device which provided us with an opportunity to follow a probabilistic approach in defining the state of the system at a given time. Consider a system A such that a number of similar observations can be taken on it. Further, we assume that the system undergoes exactly similar changes in its state during each observation, *i.e.*, after every set of operations, the system returns back to its initial state (*i.e.*, the system is perfectly reversible). If this be the case, we can define the state of the system in terms of its statistical probability. But this is not practically feasible as:

(i) The number of molecules/particles in the system are so large that it is not possible to have the macrostate exactly the same as it was during the preceding observation,

(ii) Each macrostate has several microstates. We cannot ascertain their identity,

(iii) Each observation may take quite a long time whence for applying statistics, several observations may need several thousand years,

(iv) To manage and process a large number of data, over years together is not possible,

(v) The system may not show reversibility to different operations.

Ensemble was a satisfactory solution to it. An ensemble is a group of N_T identical systems. These N_T systems have the same configuration, energy and accessible states.

In other words, the ensemble has N_T isolated systems each of which are exact copies of the experimental system A Thus, we have a large number of systems that behave exactly in the same manner during a set of operations as the system A Then, if N_r systems out of N_T systems produce a particular result r, we have

$$P_r = N_r/N_T$$

as the probability for that result.

If we want to measure a particular property, say x, of the system

$$< x > = \frac{n_1x_1 + n_2x_2 + ... + n_rx_r +}{n_1 + n_2 + ... + n_r +}$$

or $$< x > = \frac{n_1x_1}{N_T} + \frac{n_2x_2}{N_T} + ... \qquad \because \Sigma n_i = N_T$$

or $$= P_1x_1 + P_2x_2 + ...P_rx_r + ...$$

i.e., $$= \Sigma P_rx_r$$

where x_r is the property of rth group of N_r similar systems and P_r is the probability for that result.

In this unique way Gibbs replaced the method of time averages (observations made on the same system at different times to obtain average) by ensemble average as shown above. The main advantage of the ensemble method is "the observations on all systems can be taken simultaneously under similar conditions", *e.g.*, test of quality of soaps of different brands will be quicker by millions of users.

THEOREMS OF PROBABILITY IN STATISTICAL THERMODYNAMICS

The following are the important probability theorems commonly used in statistical thermodynamics:

(1) The number of ways in which N distinguishable particles can he arranged in order is equal to

$$N!$$

(2) The number of different ways in which N particles can be selected from N distinguishable particles irrespective of the order of selection is equal to

$$\frac{N!}{(N-n)!\,n!}$$

(3) The number of different ways in which n indistinguishable particles can be arranged in g distinguishable states with not more then one particle in each state is equal to

$$\frac{g!}{(g-n)!\,n!}$$

PROBABILITIES CALCULATION

We can extend the theory of probability in order to include the occurrence of two or more independent events. The combined probability in such cases is given by the product of separate probability of each event. Let us examine some cases:

(i) *Tossing of a coin:* On tossing a single coin, there are two possible aspects that may be presented by a fallen coin, a "head" or a "tail". We may assume here that there is no preference for any particular event, so that after making a large number of throws, the ratio of the number of heads to the number of tails is unity. Thus, for a normal coin the probability of each event is 1/2.

(ii) *Tossing of two unlike coins:* Suppose we toss two coins of different denominations or of different sizes, or of different colours, we have four complexions.

1. $Head_1$ $Head_2$
2. $Head_1$ $Tail_2$
3. $Tail_1$, $Tail_2$
4. $Tail_1$ $Head_2$.

Thus, the probability of any one complexion on a single toss will be 1/4 = 1/2.1/2. Or the composite probability is the product of probability of individual events.

(iii) *Tossing of two similar coins*: If two coins cannot be distinguished from each other, then complexions 2 and 4 of (ii) are indistinguishable whence the probability of one head and one tail is twice than that of two heads or two tails.

(iv) *Tossing of two coins in general*: Let us represent occurrence of heads by A_1, A_2 and the occurrence of tails by B_1, B_2 when two coins are tossed. Then the composite event can be represented by $(A_1 + B_1)(A_2 + B_2) = A_1A_1 + A_1B_2 + A_2B_1 + B_1B_2$.

(v) *Tossing of three unlike coins:* With three unlike coins, we have three distinguishable heads A_1, A_2, A_3 and three distinguishable tails, B_1, B_2, B_3. Here we can have eight equally probable complexions $A_1\ A_2\ A_3$. $A_1\ A_2\ B_3$. $A_1\ B_2\ B_3$. $B_1\ A_2\ B_3$, $B_1\ A_2\ B_3$, $B_1\ B_2\ A_3$ and $B_1\ B_2\ B_3$ which are really obtained from the product of $(A_1 + B_1)$. $(A_2 + B_2)$. $(A_3 + B_3)$.

(vi) *Tossing of three similar coins:* In this case we have only four probable states with complexions

1. All heads.
2. Two heads and one tail
3. All tails.
4. One head and two tails.

We can tabulate the result as in (iv) for

$(A + B)(A + B)(A + B) = A_3 + 3A^2B + 3AB^2 + B^3$.

(vii) *Tossing of n coins*: Suppose all the coins are different. Then the possible complexions (also known as probability number) will be $(A_1 + B_1)(A_2 + B_2)$ ..... $(A_n + B_n)$. And, if the coins are all alike the total number of possible complexions will be

$$(A + B)^n = A^n + {}^nC_1A^{n-1}B + {}^nC_2A^{n-2}B^2 + ... {}^nC_rA^{n-r}B^r + ... + {}^nC_nB^n$$

or $$= \sum_{r=0}^{r=n} C_r A^{n-r}B^r$$

or, also $$(A + B)^n = \sum_{r=0}^{r=n} C_r A^rB^{n-r}$$

$$\equiv \sum_{r=0}^{r=n} C_r A^{n-r}B^r$$

where $\quad {}^nC_r = \dfrac{n!}{r!(n-r)!}$ = Number of events with r heads/tails.

Now, since A and B indicate one event each, the total number of events are $(1 + 1)^n = 2^n = {}^nC_1 + {}^nC_2 + ... + {}^nC_n$

∴ The probability of r heads and (n – r) tails to appear is

$$P(r, n-r) = \frac{{}^nC_r}{2^n} = \frac{n!}{r!(n-r)!} \cdot \frac{1}{2^n}$$

Maximum and minimum probability: To find out which of the combination of heads and tails is most likely in a large number of throws of n coins, we have to find the maximum value of ${}^nC_r$. It is known from algebra that the value of ${}^nC_r$ is greatest for r = n/2 when n is even and r = n+1/2 when n is odd.

$$P_{max} = \frac{{}^nC_{n/2}}{2^n} = \frac{n!}{(n/2)!(n/2)} \cdot \frac{1}{2^n}$$

Similarly, the minimum probability, P_{min} obtained for r = 0 or r = n as ${}^nC_n = 1$.

whence $P_{min} = 1/2^n$

(viii) *Tossing of a single coin n times*: When we toss the same coin second, third,... nth time, its behaviour is again unpredictable. It may show a head or a tail each time with 1/2 probability. Hence a single coin tossed n times (and of course, making large number of such attempts, if we want to have satisfactory results) is just equivalent to the case of tossing of n coins as discussed in (vii).

Another equivalent illustration can be: Take 2 boxes, identical in all respects, and throw n balls from a great distance. Assuming that each ball will fall into one or the other box, the distribution of balls in the boxes will be governed by the relation (vii).

(ix) *Tossing of a dice*: If we toss of dice, having six faces, the probability for each face showing up will be 1/6. Since now the total number of aspects is 6 and only one can show up at a time with equal probability. Again, we may take 2, 3,.... n dices and throw them and calculate the composite probability for a certain complexion.

(x) *Tossing of n dices each having m faces*: Here let us assume that dice 1 has faces $a_1, a_2, a_3 ... a_m$; dice 2 has faces $b_1\ b_2\ b_3 ... b_m$;

dice 3 has faces c_1, c_2, c_3 ; etc., then the possible complexions with n dices will be the sum of each aspect

a_1, a_2, a_3... a_m, v_1, b_2, b_3 ... b_m , c_1, c_2, c_3 ... n sets and, if the dice are all alike, we have

$m_{.a}$, $m_{.b}$, $m_{.c}$... n sets.

If the probability for occurring each of the face is same, *i.e.*, a = b = c... = one aspect, we get total number of possible aspects

= m.m.m.... n times = m^n

One can show that the most probable distribution corresponds to equitable distribution of faces *i.e.*, face 1, face 2, face 3, etc., will appear in equal number. Then all the dices showing face 1 have the probability of only $1/m^n$.

A similar illustration to this is : Take *m identical boxes and throw n* balls onto them from a great distance. We will take up this problem later in more detail. As in actual analysis of physical problems we have a large number of particles which are distributed into a large number of (hypothetical) boxes.

(xi) *Probability with weightage*: So far we have considered the case in which the probability of occurring an event and its failure is the same. But there can be cases where the probability of occurring a particular event is greater than that for any other event.

Let a be the probability of occurring an event and b be the probability of its failure, per trial. Then, if the trial is repeated n times, the probability of r favourable events is

$a^r.b^{n-r} = a^r\,(1 - a)^{n-r}$

But there are ${}^nC_r$ ways for r favourable events, therefore, the probability of occurring r events is

$$P(r, n - r) = {}^nC_r\, a^r\,(1 - a)^{n-r}$$

Thus, for example, in a dice a = 1/6

$$\therefore \qquad P(r, n - r) = {}^nC_r \left(\frac{1}{6}\right)^r \left(\frac{5}{6}\right)^{n-r}$$

But, if certain number, say 3, has a = 2/7 then in n tosses the number 3 will occur r times with a probability.

(xii) *Postulate of equal a priori probability:* It states that all accessible microstates corresponding to a macrostate are equally probable.

Thus, the postulate claims that the probability of finding a molecule in one region of phase space is identical with that for any other region of equal volume provided the regions are obeying other conditions of equality (*i.e.*, the regions have extensions of same magnitude and correspond to the same energy). Thus, it follows that the probability of occurrence of a macrostate is proportional to the number of microstates associated with it

i.e., $$P_i \propto \Omega_i$$

where P_i is the probability and Ω_i is the total number of accessible microstates, in the ith set.

SOLVED EXAMPLES

Example 1:

Calculate the work down when 50 g of iron dissolves in hydrochloric acid at 25oC in :

(i) A closed vessel, and

(ii) An open beaker.

(iii) In which case would maximum amount of work be done ?

Solution:

$$\text{Amount of iron} = \frac{50\text{ g}}{55.85\text{ g mol}^{-1}} = \frac{50}{55.85}\text{ mol}$$

(i) In a closed vessel, $\Delta V = 0$. Hence,

$$dw = -PdV = 0$$

that is, no work is involved when iron dissolves in hydrochloric acid in a closed vessel.

(ii) The reaction between iron and hydrochloric acid is

$$Fe + 2HCl \rightarrow FeCl_2 + H_2(g)$$

Thus, the liberated hydrogen gas pushes the atmospheric air and thus involves the work of expansion. In this case, we will have

$$w = -P_{ext}\,\Delta V$$

$$= -P_{ext}\,V_{H2}$$

where V_{H2} is the volume of hydrogen released. According to the ideal gas law, we will have

$$w = - P_{ext} (n_{H2}RT/P_{ext})$$
$$= - n_{H2}RT$$
$$= - (50/55.85) \text{ mol } (8.314 \text{ J K}^{-1} \text{ mol}^{-1}) (298 \text{ K})$$
$$= - 2.2 \text{ kJ}.$$

Example 2:

The opposite faces of a metal plate of 0.2 cm thickness an at a difference of temperature of 109°C and the area of the plate is 200 sq cm. Find the quantity of heat that will flow through the plate in one minute if K = 0.2 CGS units.

Solution:

Here,

$$K = 0.2$$
$$A = 200 \text{ sq cm}$$
$$d = 0.2 \text{ cm}$$
$$(\theta_1 - \theta_2) = 100°C$$
$$t = 60s$$
$$Q = \frac{KA(\theta_1 - \theta_2)t}{d}$$
$$= \frac{0.2 \times 200 \times 100 \times 60}{0.2}$$
$$= 12 \times 10^5 \text{ cal.}$$

Example 3:

A bar of length 30 cm and uniform area of cross-section 5 cm^2 consists of two halves AB of copper and BC of iron welded together at B. The end A is maintained at 200°C and the end C at 0°C. The sides of the bar are thermally insulated. Find the rate of flow of heat along the bar when the steady state is reached. Thermal conductivity of copper if 0.9 and thermal conductivity of iron is 0.12 CGS units.

Solution:

Suppose the temperature at the interface. B is θ after the steady state is reached.

$$\therefore \quad \frac{K_1A(200-\theta)}{d} = \frac{K_2A(\theta-0)}{d}$$
$$0.9(200 - \theta) = 0.12\,\theta$$

$$\theta = 176.5°C.$$

After the steady state is reached, the rate of flow of heat is the same in both the bars.

$$\therefore \quad Q = \frac{K_1 A(\theta - 0)}{d}$$

$$Q = \frac{0.9 \times 5 \times (200 - 176.5)}{15}$$

$$Q = 7.05 \text{ cal/s.}$$

Example 4:

Find the time in which a layer of ice 3 cm thick on the surface of a pond will increase its thickness by 1 mm when the temperature of the surrounding air is –20°C.

Thermal conductivity of ice = 0.005

Latent heat of ice = 80 cal/g

Density of ice at 0°C = 0.91 g/cm³.

Solution:

$$\int dt = \frac{\rho L}{K\theta} \int x dx$$

$$t = \frac{\rho L}{2K\theta} [x_2^2 - x_1^2]$$

Here,

$$\rho = 0.91 \text{ g/cm}^3, \; L = 80 \text{ cal/g}$$

$$K = 0.005, \; \theta = 20°C$$

$$x_1 = 3 \text{ cm}, \; x_2 = 3.1 \text{ cm}$$

$$t = \frac{0.91 \times 80}{2 \times 0.005 \times 20} [(3.1)^2 - (3)^2]$$

$$= 222.04 \text{ s.} = 3 \text{ min } 42 \text{ s approximately.}$$

Example 5:

An ice box is built of wood 1.76 cm thick, lined inside with cork 3 cm thick. If the temperature of the inner surface of the cork is 0°C and that of the outer surface of wood is 12°C, what is the temperature of the interface ? The thermal conductivity of wood and cork are 0.0006 and 0.00012 CGS units respectively.

Solution:

Suppose the temperature of the interface is 9 after the steady state is reached

$$\therefore \qquad \frac{K_1A(12-\theta)}{d_1} = \frac{K_2A(\theta-0)}{d_2}$$

Here $K_1 = 0.0006$,

$K_2 = 0.00012$

$d_1 = 1.75$ cm,

$d_2 = 3$ cm

$$\frac{0.0006(12-\theta)}{1.75} = \frac{0.00012(\theta)}{3}$$

$$\theta = 10.74°C.$$

Example 6(a):

Since $C_V = (\partial q/\partial T)_V$ by definition, one often writes without restriction $dV = C_V dT$. This is generally not true. Explain why.

Solution:

If we take U = f)V, T), then we will have

$$dU = \left(\frac{\partial U}{\partial V}\right)_T dV + \left(\frac{\partial U}{\partial T}\right)_V \cdot dT$$

$$= \left(\frac{\partial U}{\partial V}\right)_T . dV + C_V\, dT$$

Thus, if one writes $dV = C_V\, dT$ without restriction then this statement is not complete. This is applicable for a system in which $(\partial U/\partial V)_T = 0$. An ideal gas is such an example. Alternatively, the system may be undergoing change at constant volume, for which, $(\partial U/\partial V)_T\, dV$ is zero and thus $dV = C_V\, dT$.

Example 6(b):

Is the equation PV^γ = constant valid for all processes ? Explain your answer.

Solution:

The expression PV^γ = constant is valid only for adiabatic reversible volume change (*i.e.*, expansion or compression) involving an ideal gas. For adiabatic process, $dq = 0$.

Hence

$$dU = dw$$

$$C_V dT = -P_{ext}\, dV$$

For revisable process, $P_{ext} = P_{int}$. Now replacing P_{int} = nRT/V and separating the variables followed by integration, we get

$$\frac{Cv}{n}\int\frac{dT}{T} = -R\int\frac{dV}{V}$$

or $$C_{V,m}\ln\frac{T_2}{T_1} = -R\ln\frac{V_2}{V_1}$$

or $$\left(\frac{T_2}{T_1}\right)^{C_{v,m}} = \left(\frac{V_2}{V_1}\right)^{-R}$$

Replacing T_2/T_1 by P_2V_2/P_1V_1 and rearranging, the expression would yield

$$P_2V^{\gamma}_2 = P_1V^{\gamma}_1$$

that is $$PV^{\gamma} = \text{constant.}$$

Example 6(c):

Under what conditions does the adiabatic expansion of gas become a constant energy process ?

Solution:

If the adiabatic expansions done against a vacuum then in this process energy

On comparing the coefficients in Eqs (1) and (2), we have

$$\left(\frac{\partial E}{\partial T}\right)_V = T \qquad ...(1)$$

and $$\left(\frac{\partial E}{\partial V}\right)_S = -P \qquad ...(2)$$

Eq (2) can be written as

$$dE = T\,dS \qquad ...(3)$$

For a given temperature, the above equation on integration becomes as follows :

$$\int dE = \int TdS + \text{Integration constant}$$

$$E = TS + A \qquad ...(4)$$

or $$E = A - TS$$

where A is an integration constant. It is also known as Helmholtz free energy function.

(d) Enthalpy as a Function of S and P

As H is a function of S and P, we can write

$$H = f(S, P)$$

or $$dH = \left(\frac{\partial H}{\partial S}\right)_P dS + \left(\frac{\partial H}{\partial P}\right)_S dP \qquad ...(5)$$

We also know, $$dH = T\,dS + V\,dP \qquad ...(6)$$

On comparing the coefficients in Eqs (5) and (6), we get

$$\left(\frac{\partial H}{\partial S}\right)_P = T \qquad ...(7)$$

$$\left(\frac{\partial H}{\partial P}\right)_S = V \qquad ...(8)$$

When Eq. (7) is integrated, we get a new function

$$\int dH = \int TdS + \text{constant}$$

$$= \int TdS + G$$

or $$H = TS + G$$

or $$G = H - TS$$

where G is an integration constant and is the same as the Gibbs free energy function.

Example 7:

Two moles of an ideal gas at 300 K expand reversibly and isothermally from 4×10^{-2} m^3 to 8×10^{-2} m^3. Calculate the entropy change for the gas. How would you account for the fact that ΔS is not zero although the process is reversible ?

Solution:

If the process is reversible isothermal it means that $\Delta H = 0$, hence the entropy change is given as follows :

$$\Delta S_{sys} = \frac{Q_{rev}}{T} = \frac{W_{rev}}{T} = nR\ \ln\frac{V_2}{V_1}$$

$$= (2 \text{ mol})\ (8.314 \text{ JK}^{-1} \text{ mol}^{-1}) \left(2.303 \log \frac{8 \times 10^{-2}}{4 \times 10^{-2}}\right)$$

$$= 2 \times 8.314 \times 2.303 \log 2 \text{ JK}^{-1}$$

$$= 2 \times 8.314 \times 2.303 \times 0.3010 \text{ JK}^{-1}$$

$$= 11.526579 \text{ JK}^{-1}.$$

The total entropy change (system and its surroundings) will be zero for a reversible process.

Example 8:

Calculate the entropy changes involved in the conversion of 1 mole of ice at 0°C and 1 atm to liquid at 0°C and 1 atm [The enthalpy of fusion per mole of ice is 6008 J mol⁻¹].

Solution:

$$H_2O\ (S) \rightarrow H_2O\ (l)$$

$$\Delta S_{fus} = \frac{\Delta H_{fus}}{T_m}$$

$$= \frac{6008 \text{ J mol}^{-1}}{273 \text{ K}} = 22.007326 \text{ JK}^{-1}$$

Example 9:

Increase in volume of a gas from a given decrease of pressure is more in an isothermal expansion than in an adiabatic expansion. Explain.

Solution:

In an isothermal expansion, temperature remains the same whereas in adiabatic expansion temperature decreases. Since the final pressure is them same, it follows from Charles law ($V \propto T$, p constant) that

$$V_{iso} > V_{adi}$$

Since

$$\Delta V_{iso} = V_{iso} - V_{intial}$$

$$\Delta V_{adi} = V_{adi} - V_{intial}$$

it follows that.

Example 10:

Evaluate the entropy change for the following reversible process.

1 mole Sn ($\propto$*, 13°C* $\rightleftharpoons$ *1 mole Sn (b, 13°C)* *where* $\Delta H_{trans} = 2090\ J\ mol^{-1}$

Solution

$$T_{trans} = 273 + 13 = 286\ K$$

$$\Delta S_{trans} = \frac{\Delta H_{trans}}{T_{trans}}$$

$$= \frac{2090\ J\ mol^{-1}}{286\ K}$$

$$= 7.3076923\ JK^{-1}\ mol^{-1}$$

Example 11:

Calculate the entropy change when two moles of; an ideal gas expands reversibly from an initial volume of 2 dm^3 to a total volume of 20 dm^3 at a constant temperature of 298 K.

Solution:

We know, $\Delta S = nR\ ln\left(\frac{V_2}{V_1}\right)$

$$= 2.303\ nR\ log\left(\frac{V_2}{V_1}\right)$$

$$= 2.303 \times 2 \times 8.314\ J\ JK^{-1}\ mol^{-1}\ log\left(\frac{20dm^3}{2dm^3}\right)$$

$$= 2.303 \times 2 \times 8.314 \times log\ JK^{-1}\ mol^{-1}$$

$$= 38.294284\ JK^{-1}\ mol^{-1}$$

Example 12(a):

The mixing of gases is always accompanied by an increase in entropy. Show that in the formation of a binary mixture of two ideal gases the maximum entropy increase results when $X_1 = X_2 = 0.5$.

Solution:

For a binary :mixture, the entropy per mole of the mixture formed s given by

$$\Delta S\ mixing = -R[X_1\ ln\ X_1 + X_2\ ln\ X_2]$$

$$= -R[X_1\ ln\ X_1 + (1 - X_1)\ ln\ (1 - X_1)]$$

For entropy of mixing to be maximum. the first derivative. $\frac{\delta(\Delta S\,\text{mixing})}{\delta X_1}$ should be zero and the second derivative should be negative. Differentiating, ΔS mixing with respect to X_1 and equating it to zero yields

$$\frac{\delta(\Delta S\,\text{mixing})}{\delta X_1} = -R\left[\ln X_1 + \frac{X_1}{X_1} + \frac{1 - X_1}{1 - X_1}(-1) + (-1)\ln(1 - X_1)\right]$$

$$= 0$$

$$\ln X_1 + 1 - 1 - \ln(1 - X_1) = 0$$

$$\ln \frac{X_1}{1 - X_1} = 0$$

$$X_1 = 1 - X_1$$

$$X_1 = 1/2.$$

Example 12(b):

Under what conditions does the isothermal expansion of a gas become a free expansion process?

Solution:

The isothermal expansion of a gas becomes a free expansion process only when it expands against a vacuum, *i.e.*, $p_{ext} = 0$. Moreover, the gas must be an ideal gas in which case ΔE = 0 at constant temperature.

Example 12(c):

Is every adiabatic system isolated? In what ways can the energy of an adiabatic system change?

Solution:

An isolated system is the one in which no interaction with its surroundings takes place. Neither energy nor matter can be transferred to or from it. An adiabatic system involves no exchange of heat from its surroundings, *i.e.* for such a system dq = 0 and, according to the first law of thermodynamics, dE =dw. Thus, the energy of an adiabatic system can be changed by either doing work on the system or by getting work done by the system.

Example 13:

One mole of a gas is expanded isothermally and reversibly at 27° C form a volume of 5 liters. Calculate q, w, ΔE and ΔH.

Solution:

We know

$$PV = nRT, \qquad n = 1$$

State : Initial

$T_1 = 27 + 273.15 = 300.15$ K

$V_1 = 5$ litres

Final

$T_2 = T_1 = 300.15$K

$V_2 = 10$ litres

(1) Calculation of w :

$$w = \int_{V_1}^{V_2} P\,dV = \int_{V_1}^{V_2} \frac{RT}{V} dV = RT \ln \frac{V_2}{V_1}$$

$$= 2.303 \times 8.314 \times 300 \log (10/5)$$

$$= 2.303 \times 8.314 \times 300 \times \log^2$$

$$= 2.303 \times 8.314 \times 300 \times 0.3010 \quad [\log 2 = 0.3010]$$

$$w = 1729 \text{ J mol}^{-1}$$

(2) Calculation of ΔE :

$$\Delta E = E_2 - E_1$$

For isothermal-reversible expansion of an ideal gas $E_2 = E_1$ and therefore

$$\Delta E = 0.$$

(3) Calculation of q:

$$q = \Delta E + w \text{ (first law)}$$

$$= 0 + w$$

$$q = w = 1729 \text{ J}$$

(4) Calculation of ΔH

$$\Delta H = \Delta E + \Delta (PV) = \Delta E + \Delta(RT)$$

$$0 + 0 = 0$$

Since $\Delta E = 0$ and $\Delta(RT) = 0$; R and T being constant

Example 14:

Calculate the minimum work which must be done to compress 1/2 mole of oxygen at 300 k from a pressure of 2 atm to a pressure of 200 atm.

Solution:

As it is required to calculate the minimum work of compression, the change, is reversible, *i.e.*, the gas gets compressed by gradual increase of pressure so that the external pressure becomes only infinitesimally greater than the pressure of the gas, at each stage. Thus

$$w = \text{work done} = \int_{V_1}^{V_2} PdV$$

$$= 2.303 \text{ n RT} \log \frac{V_2}{V_1}$$

$$= 2.303 \text{ nRT} \log \frac{p_1}{P_2}$$

given date :

$$P_1 = 2 \text{ atm}$$
$$P_2 - 2 \text{ atm}$$
$$T = 300 \text{ K}$$
$$n = \frac{16}{32} = 0.5 \text{mole.}$$

On substituting these values in the above equation, we get

$$w = 2.303 \times 0.5 \times 8.314 \times 300 \log \frac{2}{200}$$

$$= - 5744 \text{ J.}$$

Example 15(a):

Four moles of an ideal gas are held by a piston under 5 atm pressure and at 273.15 K. The pressure is suddenly released to 0.2 atm and the gas is allowed to expand isothermally. Calculate w, q, ΔE and ΔH for the process.

Solution:

As the pressure has been released suddenly, the expansion must be irreversible.

Here n = 4 moles; P_1 = 5 atm; $T_1 = T_2$ = 273.15 K.

(i) Work done irreversibly

$$w = nRT\left\{1-\frac{P_2}{P_1}\right\}$$

$$= 4 \times 8.314 \times 273.15 \left\{1-\frac{0.2}{5}\right\}$$

$$= 4 \times 8.314 \times 273.15 \times 0.96$$

$$= 8720 \text{ J.}$$

(ii) As the ideal gas expands isothermally

$$dT = 0 \text{ and hence } \Delta E = 0;\ \Delta H = 0.$$

(iii) First law gives

$$\Delta E = q - w = 0$$

$$\therefore q = w\ 8720 \text{ J.}$$

Example 15(b):

At low temperature, a nonideal gas follows the relation PV = RT – a/V where a = 0.3636 Nm^4 mole^{-2}. Calculate the work done by one mole of this gas in expanding form 0.224 litre to 22.4 litre at 400 K. Compare this result with the work done in the corresponding expansion of an ideal gas.

Solution:

For an ideal gas, reversible work done

$$= 2.303 \text{ n RT log } \frac{V_2}{V_1}$$

V_1 = 0.224 litres mole^{-1} = 22.4 × 10^{-5} m^3 mole^{-1}

P_2 = 22.4 litres mole^{-1} = 22.4 × 10^{-3} m^3 mole^{-1}

R = 8.314 J K^{-1} mole^{-1}

W_{rew} = (2.303) (1 mole) (8.314 J K^{-1} mole^{-1}) (400 K)

$$\times \log \frac{22.4 \times 10^{-3}}{22.4 \times 10^{-5}}$$

= (2.303) (1 mole) (8.314 J K^{-1} mole^{-1}) (400 K) log 100

= 15320 J mole^{-1}

= 15.32 kJ mole^{-1}.

For the non-ideal gas PV = RT – a/V

$$\text{Therefore, the work done} = \int_{V_1}^{V_2} PdV = \int_{V_1}^{V_2} \left\{ \frac{RT}{V} - \frac{a}{V_2} \right\} dV$$

$$= 2.303 \text{ RT} \log \frac{V_2}{V_1} + a \left\{ \frac{1}{V_2} - \frac{1}{V_1} \right\}$$

The value of first part is from derived eq. (1) as 15.32 kJ mole^{-1} and of another part can be derived as

$$0.3636 \text{ Nm}^4 \text{ mole}^{-2} \left\{ \frac{1}{22.4 \times 10^{-3} \text{ m}^3 \text{ mole}^{-1}} \right.$$

$$\left. - \frac{1}{22.4 \times 10^{-5} \text{ m}^3 \text{ mole}^{-1}} \right\}$$

$$= 1.604 \text{ kJ mole}^{-1}.$$

Thus work done = 15.320 k Joule^{-1} – 1.604 k Joule^{-1}

= 13.716 kJ mole^{-1}.

Example 15(c):

What will be the work done when 65.38 g of zinc dissolves in hydrochloric acid in case of an open beaker and closed beaker at 300K.

Solution:

The process of dissolution of zinc in hydrochloric acid can be given as

$$Zn(s) + 2HCI\ (aq) = ZnCl_2(aq) + H_2(g)$$

From this equation it is clear that for each gram atom of zinc dissolved we will obtain 1 mole of Hydrogen gas. As a result of liberated gas, on the surrounding atmosphere the work performed will be equal to $P\Delta V$. If the initial volume of the system is neglected in comparison to the total volume of the gas produced and the gas shows ideal behaviour, then the work done in the open beaker is

$$W = P\Delta V \approx PV_{H2}$$

$$= nH2\ RT \qquad [\because T = 300 \text{ k } R = 8.314 \text{ Jk}^{-1} \text{ mole}^{-1}]$$

$$= (1 \text{ mole}) (8.314 \text{ JK}^{-1} \text{ mol}^{-1}) (300 \text{ K})$$

$$= 2.4942 \text{ kJ}.$$

If the reaction takes place in a closed beaker there is change in the volume *i.e.*, $\Delta V = 0$ hence work done is zero.

Example 16:

Calculate the work done when one mole of sulphur dioxide gas expands isothermally and reversibly at 300 k from 2.46×10^{-3} m^3 to 24.6×10^{-3} m^3. Assuming that the gas obeys

(a) ideal gas equation and

(b) van der Waals equation

$$\left\{\begin{array}{l} a = 0.6799 \text{ nm}^4 \text{ mole}^{-2} \\ b = 0.564 \times 10^{-3} \text{ mole}^{-1} \end{array}\right\}$$

Solution:

(a) The reversible work done by 1 mole of gas when it expands form

$$2.64 \times 10^{-3}\text{m}^3 \text{ to } 24.6 \times 10^{-3}\text{m}^3 = 2.303 \text{ RT} \log \frac{V_2}{V_1}$$

$$= (2.303)\ (1 \text{ mole})\ (8.314 \text{ J K}^{-1} \text{ mol}^{-1})\ (300 \text{ K}) \times \log 10$$

$$\left\{\because \log\frac{V_2}{V_1} = \log 10\right\}$$

$$= (2.303)\ \ (8.314)\ (300 \text{ K})\ \ \log 10 \text{ J mol}^{-1}$$

$$= 5744 \text{ J mol}^{-1}$$

(b) When the gas obeys van der Waal's equation, work done by 1 mole of gas is

$$= 2.303 \text{ RT} \log \frac{V_2 - b}{V_1 - b} + a\left\{\frac{1}{V_2} - \frac{1}{V_1}\right\}$$

After substituting the values of various quantities and solving, the work done is

$$= 5794 \text{ J mol}^{-1} - 252.4 \text{ J mol}^{-1}$$

$$= 5541.6 \text{ J mol}^{-1}$$

$$= 5.542 \text{ kJ mol}^{-1}.$$

Example 17:

One mole of a van der Walls's gas is allowed to expand isothermally and reversibly from a volume of 1 litre to 50 litre at 0°C. Calculate w, *q, Δ E and ΔH. van der Waal's constants are a = 6.5 atm l^2mole $^{-2}$ = 0.056 litre mole $^{-1}$ and R = 0.082 l atom deg^{-1} $mole^{-1}$.*

Solution:

Here V_1 = 1 litre,

V_2 = 50 litres, n = 1 mole

T = 273 + 0 = 273°k.

R = 0.082 l-atm deg^{-1} $mole^{-1}$

a = 6.5 atm l^2 $mole^{-2}$.

b = 0.056 litre $mole^{-1}$

$$w = 2.303 \text{ nRT} \log\left(\frac{V_2 - nb}{V_1 - nb}\right) + an^2\left(\frac{1}{V_2} - \frac{1}{V_1}\right)$$

$$= 2.303 \times 1 \times 0.082 + 273 \,[\log(50 - 0.056) - \log(1 - 0.056) + 6.5\left(\frac{1}{50} - \frac{1}{1}\right)$$

= 82.47 l atm.

$$\Delta E = -an^2\left(\frac{1}{V_2} - \frac{1}{V_1}\right) = 6.5 \times \frac{49}{50}$$

$$q = 2.303\text{nRT} \log\frac{V_2 - b}{V_1 - b}$$

= ΔE + w = 884 l atm.

Example 18(a):

The van der Waal's constant a and b for hydrogen in litre atmosphere units are 0.246 and 2.67 × 10^{-2} respectively. Calculate the inversion temperature of hydrogen.

Solution:

we know T_i = 2a/Rb

or $$T_i = \frac{2 \times 0.246}{0.0821 \times 0.0267}$$

= 224.5°k = (224.5 − 273) = − 48.5°C.

Example 18(b):

For equation (P _ a/V^2) V = RT, prove that (i) dP is an exact differential, (ii) P is a state function and

$$\text{(iii)} \left(\frac{\partial P}{\partial T}\right)_V \left(\frac{\partial T}{\partial V}\right)_P \left(\frac{\partial V}{\partial T}\right)_T + 1 = 0.$$

Solution:

(i) For dP to be an exact differential, it is be proved that

$$\frac{\partial^2 P}{\partial V \partial T} = \frac{\partial^2 P}{\partial T \partial V} \qquad ...(1)$$

$$(P + a/V^2)\, V = RT$$

or

$$P = \frac{RT}{V} - \frac{a}{V^2} \qquad ...(2)$$

When the above equation is differentiated with respect to V at constant temperature T, we obtain

$$\left(\frac{\partial P}{\partial V}\right)_T = -\frac{RT}{V^2} + \frac{2a}{V^3}$$

When the above equation is differentiated with respect to T at constant V, we obtain

$$\frac{\partial^2 P}{\partial T \partial V} = -\frac{RT}{V^2} \qquad ...(3)$$

When equation (2) is differentiated with respect to T at constant volume V, we obtain

$$\left(\frac{\partial P}{\partial T}\right)_V = \frac{R}{V}$$

When the above equation is further differentiated with respect to V at constant T, we obtain

$$\frac{\partial^2 P}{\partial T \partial V} = -\frac{R}{V^2} \qquad ...(4)$$

From equations (3) and (4), we get

$$\frac{\partial^2 P}{\partial T\, \partial V} = \frac{\partial^2 P}{\partial T\, \partial V} = -\frac{R}{V^2} \qquad ...(5)$$

Thus, we have proved condition, (1). This shows that dP is an exact differential.

(ii) In the said question, first of all a change occurs in volume at constant temperature followed by a change in temperature at constant volume, thereby giving the final equation (3).

Again, a change in temperature is considered at constant volume, followed by a change in volume at constant temperature, thereby giving the final equation (4). Both equations (3) and (4) are same, thereby revealing that the final change in P remains the same irrespective of the difference in intermediate changes. Therefore, P is a state function as changes in P have been found to depend upon the intial and final states of the system.

(iii) Equation (2) is as follows :

$$P = \frac{RT}{V} - \frac{a}{V^2}$$

When the above equation is differentiated, we obtain :

$$dP = \frac{R}{V}\,dT - \frac{RT}{V^2}\,dV + \frac{2a}{V^3}\,dV$$

$$= \frac{R}{V}\,dT - \left(\frac{RT}{V^2} - \frac{2a}{V^3}\right)dV \qquad ...(6)$$

If volume is constant, dV = 0 : equation (6) becomes as follows :

$$dP = \frac{R}{V}\,dT\text{or } \left(\frac{\partial P}{\partial V}\right)_V = \frac{R}{V} \qquad ...(7)$$

If pressure is constant, dP = 0; equation (6) becomes as follow :

$$\frac{R}{V}\,dT - \left(\frac{RT}{V^2} - \frac{2a}{V^3}\right)dV = 0$$

or $$\left(\frac{\partial T}{\partial V}\right)_P = \left(\frac{RT}{V^2} - \frac{2a}{V^3}\right)\Big/\frac{R}{V} \qquad ...(8)$$

If temperature is constant, dT = 0; equation (6) becomes as follows:

$$\left(\frac{\partial T}{\partial V}\right)_T = -\frac{1}{\left(\frac{RT}{V^2} - \frac{2a}{V^3}\right)} \qquad ...(9)$$

On multiplying equations (7), (8) and (9), we obtain

$$\left(\frac{\partial p}{\partial T}\right)_V\left(\frac{\partial T}{\partial V}\right)_P\left(\frac{\partial V}{\partial P}\right)_T = -\frac{R}{V}\times\frac{\left(\frac{RT}{V^2} - \frac{2a}{V^3}\right)}{\left(\frac{RT}{V^2} - \frac{2a}{V^3}\right)}\times\frac{1}{R} = -1$$

or $$\left(\frac{\partial p}{\partial T}\right)_V\left(\frac{\partial T}{\partial V}\right)_P\left(\frac{\partial V}{\partial P}\right)_T + 1 = 0$$

This is cyclic rule.

Example 18(c):

For an ideal gas PV = RT, prove that dP is an exact differential.

Solution:

We know PV = RT

or $$P = \frac{RT}{V} \text{ (R is a constant)} \quad ...(1)$$

On differentiating with respect to T, taking V constant, we obtain

$$\left(\frac{\partial P}{\partial T}\right)_P = \frac{RT}{V} \quad ...(2)$$

Again, on differentiating with respect to V, we obtain

$$\frac{\partial^2 P}{\partial V \partial T} = -\frac{R}{V^2} \quad ..(3)$$

On differentiating (1) with respect to volume, taking T constant, we obtain

$$\left(\frac{\partial P}{\partial V}\right)_T = -\frac{RT}{V^2} \quad ...(4)$$

Again, on differentiating above equation with respect to temperature taking V as constant, we obtain

$$\frac{\partial^2 P}{\partial T \partial V} = -\frac{R}{V^2} \quad ...(5)$$

From equations (3) and (5), we obtain

$$\frac{\partial^2 P}{\partial T \partial V} = \frac{\partial^2 P}{\partial T \partial V}$$

Thus dP is an exact differential.

Example 19:

For an ideal gas PV = RT, prove that DT is an exact differential.

Solution:

$$PV = RT \quad ...(1)$$

In order to prove that dT is a perfect differential, we have to prove that

$$\frac{\partial^2 T}{\partial V \partial P} = \frac{\partial^2 T}{\partial P \partial V}$$

On differentiating Eq. (1), we obtain

$$\left(\frac{\partial T}{\partial V}\right)_P = \frac{P}{R}; \text{i.e., } \frac{\partial^2 T}{\partial P \partial V} = \frac{1}{R}$$

Again $$\left(\frac{\partial T}{\partial P}\right)_V = \frac{V}{R}; \text{i.e., } \frac{\partial^2 T}{\partial V \partial P} = \frac{1}{R}$$

Therefore, $\frac{\partial^2 T}{\partial V \partial P} = \frac{\partial^2 T}{\partial P \partial V}$

Thus, dT is an exact differential.

Example 20:

For an ideal gas (PV = nRT) show that (1/T) is an integrating factor for dw = P dV.

Solution:

$$dw = P\,dV \quad \text{...(1)}$$

$\because$ $$PV = nRT$$

$\therefore$ $$p\,dV + V\,dP = nR\,dT$$

and $$dV = \frac{nR}{P}dT - \frac{V}{P}dP \quad \text{...(2)}$$

On substituting the expression for dV in equation (1), we obtain

$$dw = nR\,dT - V\,dP = nR\,dT - \frac{nRT}{P}dP$$

Now $$\frac{\partial}{\partial P}(nR)_T = 0$$

and $$\frac{\partial}{\partial T}\left(-\frac{nRT}{P}\right)_P = -\frac{nR}{P}$$

Hence, dw is inexact differential.

Suppose f = 1/T be an integrating factor. Then

$$f \,.\, dw = \frac{1}{T}nR\,dT - \frac{nR}{P}dP$$

$\therefore$ $$\frac{\partial}{\partial P}\left[\frac{nR}{T}\right]_T = 0$$

and $$\frac{\partial}{\partial T}\left[\frac{nR}{P}\right]_T = 0$$

Thus f. dw$\frac{nR}{T}$dT – $\frac{nR}{P}$dP is an exact differential while 1/T is an integrating factor.

Example 21:

Calculate the maximum work obtainable by (i) the isothermal expansion and (ii) the adiabatic expansion of 2 mol of on ideal gas initially at 25° C form 10 L to 20 L. Assume Cv = (5/2) R.

Solution:

(i) Maximum work is obtained in reversible expansion. Hence

$$w = -\ nRT\ ln\ \frac{V_2}{V_1}$$

$$= -\ (2\ mol)\ (8.314\ J\ K^{-1}\ mol^{-1})\ (298\ K)\ 2.303\ log\ \frac{20L}{10L}$$

$$= -\ 3435\ J$$

(ii) Firstly, we calculate the final temperature using the expression

$$T_2\ V_2^{\ \gamma-1} = T_1\ V_1^{\gamma-1}$$

where $$\gamma = \frac{C_P}{C_V} = \frac{(7/2)R}{(5/2)R} = \frac{7}{5}$$

Hence, $$T_2 = (298K)\left(\frac{10L}{20L}\right)^{(7/5)-1}$$

$$= (298\ K)\ (0.7578) = 225.8\ K$$

Now. $$W_{adi} = \Delta E$$

$$= n\ C_{V},\ m(T_2 - T_1)$$

$$= (2mol)(2.5 \times 8.314\ JK^{-1}mol^{-1})(225.8K - 298K)$$

$$= -\ 3001\ J.$$

Example 22:

How much heat is required to raise the temperature of 1 mole oxygen from 300K to 1300 K at constant pressure.

$$C_p = 6.095 + 3.325 \times 10^{-3}\ T - 1.017 \times 10^{-6}\ T^2.$$

Solution:

$$H_2 - H_1 \int_{T_1}^{T_2} C_P\ dT$$

$$T_1 = 300 \text{ k}, T_2 = 100 \text{ k}$$

$$\therefore \quad H_{1300} - H_{300}$$

$$= \int_{300}^{1300} \left(6.095 + 3.253 + 10^{-3}\ T - 1.017 \times 10^{-6}\ T^2\right)$$

[Substituting the value of C_p]

$$\because \quad H_{1300} - H_{300} = 6.095\ (1300 - 300) + \frac{3.253 \times 10^{-3}}{2}$$

$$\times\ [(1300)^2 - (300)^2] - 1/3 \times 1.017 \times 10^{-6}\ [(1000)^2 - (300)^2]$$

$$= 6095 + 2602.4 - 735.63$$

$$= 7961.77 \text{ cal/mole.}$$

Example 23(a):

Show that

$$\left(\frac{\partial \alpha}{\partial P}\right)_T + \left(\frac{\partial \beta}{\partial T}\right)_P = 0$$

where α and β are coefficient of thermal expansion and coefficient of compressibility, respectively.

Solution:

Since V is a state function, we can write

$$\frac{\partial^2 V}{\partial T\ \partial P} = \frac{\partial^2 V}{\partial P\ \partial T}$$

or
$$\frac{\partial}{\partial T}(-V\beta)_P = \frac{\partial}{\partial T}(V\alpha)_T$$

Carrying out the differentiation, we get

$$-\beta\left(\frac{\partial V}{\partial T}\right)_P - V\left(\frac{\partial \beta}{\partial T}\right)_P = \alpha\left(\frac{\partial V}{\partial P}\right)_T + V\left(\frac{\partial \alpha}{\partial P}\right)_T$$

that is
$$\left(\frac{\partial \alpha}{\partial P}\right)_T + \left(\frac{\partial \beta}{\partial T}\right)_P = -\frac{\beta}{V}\left(\frac{\partial V}{\partial T}\right)_P - \frac{\alpha}{V}\left(\frac{\partial V}{\partial P}\right)_T$$

$$= -\frac{\beta}{V}(\alpha V) - \frac{\alpha}{V}(-\beta V) = 0$$

Example 23(b):

Show that Joule Thomson coefficient, $\mu_{JT} = 0$ for an ideal gas. Comment on the liquefaction of an ideal gas.

Solution:

By definition, Joule-Thomson coefficient is defined as

$$\mu_{JT} = \left(\frac{\partial T}{\partial P}\right)_H$$

Using the cyclic rule

$$\left(\frac{\partial T}{\partial P}\right)_H \left(\frac{\partial P}{\partial H}\right)_T \left(\frac{\partial H}{\partial T}\right)_P + 1 = 0$$

we get $$\left(\frac{\partial T}{\partial P}\right)_H = -\frac{1}{\left(\frac{\partial P}{\partial H}\right)_T \left(\frac{\partial H}{\partial T}\right)_P} = -\frac{\left(\frac{\partial H}{\partial P}\right)_T}{C_P}$$

Now using, thermodynamic equation of state

$$\left(\frac{\partial T}{\partial P}\right)_H = V - T\left(\frac{\partial V}{\partial T}\right)_P$$

we get $$\mu_{JT} = \frac{V - T\left(\frac{\partial V}{\partial T}\right)_P}{C_P}$$

Now, for an ideal gas

$$V = \frac{nRT}{P}$$

Hence, $$\left(\frac{\partial V}{\partial T}\right)_P = \frac{nR}{P} = \frac{V}{T}$$

Thus $$\mu_{JT} = \frac{V - T\left(\frac{V}{T}\right)}{C_P} = 0$$

The ideal gas cannot be liquefied as there will not occur any change in temperature when the gas undergoes Joule-Thomson experiment.

Example 23(c):

Calculate the work down when 50 g of iron dissolves in hydrochloric acid at 25oC in :

(i) a closed vessel and

(ii) an open beaker.

(iii) In which case would maximum amount of work be done ?

Solution:

$$\text{Amount of iron} = \frac{50\text{ g}}{55.85\text{ g mol}^{-1}} = \frac{50}{55.85}\text{ mol}$$

(i) In a closed vessel, $\Delta V = 0$. Hence,

$$dw = -\,PdV = 0$$

that is, no work is involved when iron dissolves in hydrochloric acid in a closed vessel.

(ii) The reaction between iron and hydrochloric acid is

$$Fe + 2HCl \rightarrow FeCl_2 + H_2(g)$$

Thus, the liberated hydrogen gas pushes the atmospheric air and thus involves the work of expansion. In this case, we will have

$$w = -\,P_{ext}\,\Delta V$$
$$= -\,P_{ext}\,V_{H2}$$

where V_{H2} is the volume of hydrogen released. According to the ideal gas law, we will have

$$w = -\,P_{ext}\,(n_{H2}RT/P_{ext})$$
$$= -\,n_{H2}RT$$
$$= -\,(50/55.85)\text{ mol }(8.314\text{ J K}^{-1}\text{ mol}^{-1})\,(298\text{ K})$$
$$= -\,2.2\text{ kJ}$$

Example 24:

Since $C_V = (\partial q/\partial T)_V$ by definition, one often writes without restriction $dV = C_V dT$. This is generally not true. Explain why.

Solution:

If we take U = f)V, T), then we will have

$$dU = \left(\frac{\partial U}{\partial V}\right)_T dV + \left(\frac{\partial U}{\partial T}\right)_V dT$$
$$= \left(\frac{\partial U}{\partial V}\right)_T .\,dV + C_V\,dT$$

Thus, if one writes $dV = C_V\,dT$ without restriction then this statement is not complete. This is applicable for a system in which $(\partial U/\partial V)_T = 0$.

An ideal gas is such an example. Alternatively, the system may be undergoing change at constant volume, for which, $(\partial U/\partial V)_T\, dV$ is zero and thus $dV = C_V\, dT$.

Example 25(a):

Calculate the work done by the system when 2 mol of an ideal gas expand from 0.01 m^3 to 0.1 m^3 at 300 k isothermally :

(i) into a vacuum, and

(ii) reversibly.

Explain how the surroundings can be restored only in one case and not in the other.

Solution:

(i) When the work is done into a vacuum, the work done by the system is zero since P_{opp} in this case is zero.

(ii) For the reversible isothermal expansion work involved is

$$w = -\, nRT \ln \frac{V_2}{V_1}$$

$$= (2)\,(8.314 \times 300\ J) \ln \frac{0.1}{0.01}$$

$$= -\,2 \times 8.314 \times 300 \times 2.303\ J$$

$$= -\,11488\ J = -\,11.488\ kJ.$$

This surroundings are not disturbed when the expansion is done against vacuum. In case of reversible isothermal expansion, heat is absorbed by the system from the surroundings. Now if we carry out reversible isothermal compression of the system till the original state, then the same quantity or heat is released to the surroundings and thus the latter is restored back to the original state. However, in the first case where expansion is done against vacuum, there will be bet heat received by the surroundings during compression. Hence, in this case, it is not restored back to its original state.

Example 25(b):

Show that PV = nRt, then

(i) $$\frac{\partial^2 P}{\partial n \partial V} = \frac{\partial^2 P}{\partial V \partial n} = -\frac{RT}{V^2}$$

(ii) $\frac{\partial^2 P}{\partial n \partial T} = \frac{\partial^2 P}{\partial T \partial n} = -\frac{R}{V}$

(iii) $\left(\frac{\partial P}{\partial T}\right)_V \left(\frac{\partial T}{\partial V}\right)_P \left(\frac{\partial V}{\partial P}\right)_T = -1$

Solution:

(i) $\because \quad PV = nRT$

$\therefore \quad P = \frac{nRT}{V}$

or $\frac{\partial P}{\partial V} = \frac{\partial}{\partial V}\left(\frac{nRT}{V}\right)$ [keeping T as constant]

$$= nRT \frac{\partial}{\partial V}\left(\frac{1}{V}\right) = -\frac{nRT}{V^2}$$

$$\frac{\partial^2 P}{\partial n \partial V} = \frac{\partial}{\partial n}\left(\frac{\partial P}{\partial V}\right)$$

$= \left(-\frac{nRT}{V^2}\right)$ [Keeping R, T and V as constant]

$$= -\frac{RT}{V^2} \qquad ...(1)$$

Similarly, $PV = nRT$ or $P = \frac{nRT}{V}$

$$\frac{\partial P}{\partial n} = \frac{RT}{V}$$

$\frac{\partial^2 P}{\partial V \partial N} = \frac{\partial}{\partial V}\left(\frac{RT}{V}\right)$ [keeping T and R as constant]

$$= -\frac{RT}{V^2} \qquad ...(2)$$

From equations (1) and (2), we obtain

$$\frac{\partial^2 P}{\partial n \partial V} = \frac{\partial^2 P}{\partial V \partial n} = -\frac{RT}{V^2}$$

(ii) $P = \frac{nRT}{V}$

$\frac{\partial P}{\partial T} = \frac{\partial}{\partial T}\left(\frac{nRT}{V}\right)$ [Keeping n and R as constant]

$$= \frac{nR}{V}$$

$$\frac{\partial^2 P}{\partial n \partial T} = \frac{\partial}{\partial n}\left(\frac{nR}{V}\right) \text{[Keeping V and R as constant]}$$

$$= \frac{R}{V} \qquad ...(3)$$

Again, $P = \frac{nRT}{V}$

$$\frac{\partial P}{\partial n} = \frac{RT}{V} \qquad \text{[Keeping RT and V as constant]}$$

$$\frac{\partial^2 P}{\partial T \partial n} = \frac{\partial}{\partial n}\left(\frac{RT}{V}\right) \qquad \text{[Keeping RT and V as constant]}$$

or $$\frac{\partial^2 P}{\partial T \partial n} = \frac{R}{V} \qquad ...(4)$$

From equations (3) and (4), we obtain

$$\frac{\partial^2 P}{\partial n \partial T} = \frac{\partial^2 P}{\partial T \partial n} = \frac{R}{V}$$

(iii) Again, $P = \frac{nRT}{V}$

$$\left(\frac{\partial P}{\partial T}\right)_V = \frac{nR}{V} \qquad ...(5)$$

Again, $V = \frac{nRT}{P}$

$$\left(\frac{\partial V}{\partial P}\right)_T = -\frac{nRT}{P} \qquad ...(6)$$

Again, $T = \frac{PV}{nR}$

$$\left(\frac{\partial T}{\partial V}\right)_P = -\frac{P}{nR} \qquad ...(7)$$

On multiplying equations (5), (6), and (7), we obtain

$$\left(\frac{\partial P}{\partial T}\right)_V \left(\frac{\partial T}{\partial V}\right)_P \left(\frac{\partial V}{\partial P}\right)_T = \frac{nR}{V} \times -\frac{nRT}{P^2} \times \frac{P}{nR}$$

$$= -\frac{nRT}{PV} = -1\left[\because 1 = \frac{nRT}{PV}\right]$$

Example 25(c):

(i) Calculate ∝ and β for an ideal gas obeying gas equation PV = RT (for one mole).

Solution:

(i) We know

$$PV = RT$$

On differentiating the above equation, we obtain

$$PdV + VdP = RdT \qquad ...(1)$$

At constant pressure dP = 0, the above equation becomes as follows:

$$\left(\frac{\partial V}{\partial T}\right)_P = \frac{R}{V}$$

or
$$\alpha = \frac{1}{V}\left(\frac{\partial V}{\partial T}\right)_P = \frac{R}{VP} = \frac{1}{T} = T^{-1}$$

At constant temperature dT = 0; equation (1) becomes as follows:

$$\left(\frac{\partial V}{\partial P}\right)_T = -\frac{V}{P}$$

or
$$\beta = -\frac{1}{V}\left(\frac{\partial V}{\partial P}\right)_T = \frac{V}{VP} = \frac{1}{P} = P^{-1}$$

(ii) The van der Waal's equation is as follows :

$$\left(P + \frac{a}{V^2}\right)(v - b) = RT$$

or
$$PV^3 - PbV^2 + aV - ab - RTV^2 = 0 \qquad ...(2)$$

On differentiating the above equation with respect to T at constant pressure, we obtain

$$3\,PV^2\left(\frac{\partial V}{\partial T}\right)_P - 2PbV\left(\frac{\partial V}{\partial T}\right)_P + a\left(\frac{\partial V}{\partial T}\right)_P - 2RTV\left(\frac{\partial V}{\partial T}\right)_P - RV^2 = 0$$

or
$$\left(\frac{\partial V}{\partial T}\right)_P = \frac{RV^2}{3PV^2 - 2PbV + a - 2RTV}$$

$$\alpha = \frac{1}{V}\left(\frac{\partial V}{\partial T}\right)_P$$

$$= \frac{1}{V}\left[\frac{RV^2}{3PV^2 - 2PVb + a - 2RTV}\right]$$

$$= \frac{R}{3PV - 2Pb + \frac{a}{V} - 2RT}$$

But the van der Waal's equation may be put as follows :

$$RT = PV - Pb + \frac{a}{V} - \frac{ab}{V^2}$$

$$\therefore \qquad \alpha = \frac{R}{3PV - 2Pb + \frac{a}{V} - 2\left(PV - Pb + \frac{a}{V} - \frac{ab}{V^2}\right)}$$

$$= R\left[PV - \frac{a}{V} + \frac{2ab}{V^2}\right]^{-1}$$

Again, on differentiating equation (2), with respect to P at constant temperature T, we obtain

$$3PV^2\left(\frac{\partial V}{\partial P}\right)_T + V - 2PbV\left(\frac{\partial V}{\partial P}\right)_T$$

$$- bV^2 + a\left(\frac{\partial V}{\partial P}\right)_T - 2RTV\left(\frac{\partial V}{\partial P}\right)_T = 0$$

$$\left(\frac{\partial V}{\partial P}\right)_T = \frac{V^2(b - V)}{3PV^2 - 2PbV + a - 2RTV}$$

But the van der Waal's equation may be put as follows :

$$RTV = PV^2 + a - PbV - ab/V$$

$$\therefore \qquad \left(\frac{\partial V}{\partial P}\right)_T = \frac{V^2(b - V)}{3PV^2 - 2PbV + a - 2\left(PV^2 + a - PbV - \frac{ab}{V}\right)}$$

$$= \frac{b - V}{P - \frac{a}{V^2} + \frac{2ab}{V^3}}$$

Now $$\beta = -\frac{1}{V}\left(\frac{\partial V}{\partial P}\right)_T = -\frac{b-V}{PV - \frac{a}{V} + \frac{2ab}{V^2}}$$

$$= -\frac{V-b}{P - \frac{a}{V} + \frac{2ab}{V^2}}$$

Example 26:

Derive $$C_{P,m} - C_{V,m} = TV_m \frac{\alpha^2}{\beta}$$

where $\propto$ *and* β *are the coefficient of thermal expansion and coefficient of compressibility, respectively.*

Solution:

We start with H = f (T, P) and U = f (T, V) and write their differentials as

$$dH = \left(\frac{\partial H}{\partial T}\right)_P dT + \left(\frac{\partial H}{\partial P}\right)_T dP$$

$$dU = \left(\frac{\partial U}{\partial T}\right)_V dT + \left(\frac{\partial U}{\partial V}\right)_T dV$$

Now since $H = U + PV$

we have $dH = dU + P\,dV + V\,dP$

or $$\left(\frac{\partial H}{\partial T}\right)_P dT + \left(\frac{\partial H}{\partial P}\right)_T dP = \left(\frac{\partial U}{\partial T}\right)_V dT + \left(\frac{\partial U}{\partial V}\right)_T dV + P\,dV + V\,dP$$

Dividing by dT, keeping P constant, we get

$$\left(\frac{\partial H}{\partial T}\right)_P = \left(\frac{\partial U}{\partial T}\right)_V + \left(\frac{\partial U}{\partial V}\right)_T \left(\frac{\partial V}{\partial T}\right)_P + P\left(\frac{\partial V}{\partial T}\right)_P$$

Now using the thermodynamic equation of state

$$\left(\frac{\partial U}{\partial V}\right)_T = T\left(\frac{\partial P}{\partial T}\right)_V - P$$

we get $$\left(\frac{\partial H}{\partial T}\right)_P = \left(\frac{\partial U}{\partial T}\right)_V + T\left(\frac{\partial P}{\partial T}\right)_V \left(\frac{\partial V}{\partial T}\right)_P \qquad ...(1)$$

Now, by definition

$$\alpha = \frac{1}{V}\left(\frac{\partial V}{\partial T}\right)_P \qquad ...(2)$$

$$\beta = -\frac{1}{V}\left(\frac{\partial V}{\partial P}\right)_T \qquad ...(3)$$

Using the cyclic rule

$$\left(\frac{\partial V}{\partial P}\right)_T \left(\frac{\partial P}{\partial T}\right)_V \left(\frac{\partial T}{\partial V}\right)_P + 1 = 0$$

we get $\left(\frac{\partial V}{\partial P}\right)_T = -\left(\frac{\partial V}{\partial T}\right)_P \left(\frac{\partial T}{\partial P}\right)_V$

Hence

$$\beta = \frac{1}{V}\left(\frac{\partial V}{\partial T}\right)_P \cdot \left(\frac{\partial T}{\partial P}\right)_V$$

and $\frac{\alpha}{\beta} = \left(\frac{\partial P}{\partial T}\right)_V$

Hence, Eq (1) becomes

$$C_{P,m} - C_{V,m} = T\left(\frac{\partial P}{\partial T}\right)_V \left(\frac{\partial V}{\partial T}\right)_P \qquad ...(4)$$

With the use of Eqs. (2) and (4), we get

$$C_{P,m} - C_{V,m} = TV\frac{\alpha^2}{\beta}$$

Example 27:

When a bicycle tyre is inflated with hand pump, the temperature of the air inside increases. Explain.

Solution:

Work done on the system is converted into heat which increases the temperature of air inside the tyre.

Example 28:

Can q become a state function ?

Solution:

Heat exchange reversibly at constant volume (or at constant pressure) is equal to the change in internal energy (or enthalpy) of a system. Since the latter is a state function, it follows that in such conditions q also behaves like a state function.

Example 29:

One mole of hydrogen and nine moles of nitrogen are mixed at 291. K and 1 atm pressure. Assuming ideal behaviour for the gases, calculate the entropy of mixing per mole of the mixture formed. Would it make any difference if under similar conditions one mole of hydrogen is mixed with nine moles of oxygen.

Solution:

For ideal gases the entropy of mixing per mole of the mixture is given as follows

$$\Delta S_{mixing} = -R \sum X_i \ln X_i$$

In this case $X_{H2} = 0.1$ and $X_{N2} = 0.9$

$$\Delta Smixing = -R\,[0.1 \ln 0.1 + 0.9 \ln 0.9]$$

$$= -(8.314\ JK^{-1}\ mole^{-1})\ (2.303)\ (0.1)$$

$$[\because R = 8.314\ JK^{-1}\ mol^{-1}]$$

$$\left[\log\frac{1}{10} + 9\log\frac{9}{10}\right]$$

$$= 2.704\quad JK^{-1}\ mol^{-1}$$

When hydrogen is mixed with oxygen, the entropy of mixing would remain unchanged provided oxygen also behaves ideally.

Example 30:

2 moles of an ideal monoatomic gas (C_V = 20.91 J mol^{-1} K^{-1}) are heated from 700 K to 750 K. The volume of the gas changes from 0.08 m^3 to 0.8 m^3. Assuming that the heat capacity remains constant in this temperature range, calculate the entropy change for the system, the surroundings and the universe if the process is carried out (a) reversibly and (b) irreversibly by placing the system in contact with a reservoir at 500 K and allowing the gas to expand against a const. external pressure equal to the final pressure of the gas

Solution:

For the reversible transformation of an ideal gas the entropy change of the system is given as follows :

$$\Delta S_{system} = nC_V \ln \frac{T_2}{T_1} + nR \ln \frac{V_2}{V_1}$$

$$= (2 \text{ mol}) (20.9 \text{ J mol}^{-1} \text{ K}^{-1}) (2.303) \log \left(\frac{750}{700}\right)$$

$$+ (2 \text{ mol}) (8.314 \text{ J mol}^{-1} \text{ K}^{-1}) (2.303) \log \frac{0.80}{0.08}$$

$$= 6.449 + 38.28 = 44.729 \text{ JK}^{-1}.$$

As the process is reversible, the entropy change of the surrounding is as follows.

$$\Delta S_{sur} = - 44.729 \text{ JK}^{-1}$$

and the entropy change of the universe is as follows

$$\Delta S_{univ} = \Delta S_{sys} + \Delta S_{surr} = 0$$

(b) In the irreversible transformation the entropy change of the system will be the same. Heat lost by the surrounding is as

$$- Q = \Delta E - W$$

$$\Delta E = - n C_V (750 - 700)$$

and $$\omega = P_{ext} (V_2 - V_1).$$

Thus according to the final state

$$T_2 = 750 \text{ K}$$

$$V_2 = 0.8 \text{ m}^3$$

and hence the pressure is given at follows

$$P_2 = \frac{nRT_2}{V_2} = \frac{(2 \text{ mol}) \left(8.314 \text{ J mol}^{-1} \text{ K}^{-1}\right) (750 \text{K})}{\left(0.8 \text{ m}^3\right)}$$

$$= 15588.75 \text{ Jm}^{-3}.$$

Now $P_{ext} = P_2 = 15588.75 \text{ Jm}^{-3}$ and hence the heat lost by the surrounding is

$$- Q = - (2 \text{ mol}) (20.9 \text{ J mol}^{-1} \text{ K}^{-1}) (50 \text{ K})$$

$$- (15588.75 \text{ Jm}^{-3}) (0.8 \text{ m}^3 - 0.08 \text{ m}^3)$$

$$= - 2090 \text{ j} - 11223.9$$

$$= 13313.9 \text{ J}.$$

Let us assume that this heat is lost reversibly by the surroundings. Its entropy change is given as

$$\Delta S_{surr} = \frac{13313.9\,j}{500\ K} = 26.6278 JK^{-1}$$

and the entropy change of the universe is

$$\Delta S_{univ} = \Delta S_{sys} + \Delta S_{surr}$$

$$= (44.729 - 26.6278)\ JK^{-1}$$

$$= 18.1012\ JK^{-1}.$$

Example 31:

The temperature of an ideal monoatomic gas is increased from 546 K to 1638 K. Calculate the pressure change in order that the entropy of the gas remains unchanged in the process

Solution:

The entropy change for the system can be explained by the following relation which is as follows

$$\Delta S_{sys} = nC_P \ln\frac{T_2}{T_1}\ nR \ln\frac{P_2}{P_1}$$

$$\Delta S_{sys} = 0,\ \text{thus}$$

$$C_P \ln\frac{T_2}{T_1} = R \ln\frac{P_2}{P_1}$$

$$\frac{5}{2} R \log\frac{T_2}{T_1} = R \log\frac{P_2}{P_1}$$

$$\frac{5}{2} \log\frac{1638\ K}{546\ K} = \log\frac{P_2}{P_1}$$

$$\frac{P_2}{P_1} = 15.60$$

Thus in order to make entropy constant, the pressure of the gas must be increased by 15.60 times its initial value.

Example 32:

Two moles of an ideal gas are allowed to expand isothermally from 0.04 m^3 to 0.4 m^3 at 300 K. Calculate the entropy change for the system, surroundings and universe if the expansion is

(i) reversible and

(ii) irreversible against a constant external pressure of 0.2 atm.

Solution:

$$Q_{rev} = W = n\ RT\ \ln \frac{V_2}{V_1}$$

$$\Delta S_{rev} = \frac{Q_{rev}}{T} = nR\ \ln \frac{V_2}{V_1}$$

$$= (2\ \text{mol})\ (8.314)\ \text{J mol}^{-1}\ \text{K}^{-1}\ (2.303)\ \log \frac{0.40}{0.04}$$

$$= 38.294284\ \text{JK}^{-1}.$$

The surroundings is losing an exactly equivalent amount of heat and hence its entropy decrease is given as follows :

$$\Delta S_{surr} = \frac{(-Q_{rev})}{T} = -19.14\ \text{JK}^{-1}$$

The entropy change of the universe is given as follows :

$$\Delta S_{univ} = \Delta S_{sys} + \Delta S_{surr} = 0$$

(ii) In the irreversible expansion, the entropy change of the system is still the same as in the reversible change, *i.e.*, 19.14 JK^{-1} but heat lost by the surroundings is given by $-P\Delta V$ and consequently the entropy change, if the transfer were reversible, is

$$\Delta S_{surr} = \frac{Q_{surr}}{T} = -\frac{P\Delta V}{T}$$

$$= -\frac{(0.2\,\text{atm.})\left(0.4 - 0.04\,\text{m}^3\right)\left(101325\ \text{Jm}^{-3}\ \text{atm}^{-1}\right)}{300\ \text{K}}$$

$$= -\frac{0.2 \times 0.36 \times 101325}{300}\ \text{JK}^{-1}$$

$$= 24.318\ \text{JK}^{-1}..$$

Example 33:

Calculate the entropy change when 1 mole of perfect gas is allowed to expand at 27°C from a volume of 2 litres to a volume of 20 litres against a constant pressure of one atmosphere.

Solution:

We know

$$dS = C_V \frac{dT}{T} + R \frac{dV}{V}$$

By integrating this equation, we get

$$S_2 - S_1 = C_V \ln \frac{T_2}{T_1} + R \ln \frac{V_2}{V_1}$$

At constant temperature,

$$S_2 - S_1 = R \ln \frac{V_2}{V_1}$$

$$= 2.303\ R \log \frac{V_2}{V_1}$$

$$= 2.303 \times 2 \times \log \frac{20}{2}$$

$$= 2.303 \times 2 \times \log 10$$

$$= 4.606 \text{ cal./ degree/mole.}$$

$$\left[\begin{array}{l} \because R = 2 \text{ cal./deg ree/mole} \\ V_1 = 2 \text{ litres} \\ V_2 = 20 \text{ litres} \end{array}\right]$$

Example 34:

The molar heat capacity at capacity at constant pressure of solid magnesium (0° – 600°C) is expressed by

CP = 6.20 + 1.30 × 10–3 T – 6.80 × 104 T–2

Calculate the increase in entropy when 1 gm atom of metal is heated from 27°C to 127°C at constant pressure.

Solution:

At constant pressure,

$$\Delta S = \int_{T_1}^{T_2} \frac{C_P}{T}\, dT$$

$$\therefore \quad \Delta S = \int_{300}^{400} \frac{\left(6.20 + 1.30 \times 10^{-3}\, T - 6.80 \times 10^4\, T^{-2}\right) dT}{T}$$

$$\int_{300}^{400} 6.20 \frac{dT}{T} + \int_{300}^{400} 1.3 \times 10^{-3} dT - \int_{300}^{400} 6.20 \times 10^{4} \frac{dT}{T^3}$$

$$= 6.20 \ln[\ln T]_{300}^{400} + 1.3 \times 10^{-3} [T]_{300}^{400} - \frac{6.80 \times 10^4}{-2} \left[\frac{1}{T^2}\right]_{300}^{400}$$

$$= 6.20 \ln \frac{400}{300} + 1.3 \text{ Ÿ } 10^{-3} \text{ Ÿ } (400 - 300) + 3.40 \text{ Ÿ } 10^4$$

$$\left[\frac{1}{(400)^2} - \frac{1}{(300)^2}\right] = 1.7487 \text{ cal./degree.}$$

Example 35(a):

A thermostat was maintained at 370.05 K. For nearly an hour 4.2 kJ of heat leaked through the thermostat insulation into a room where the initial temperature of air was 300.05 K. (i) What was the entropy change of the material in the thermostat ? (ii) What was the entropy change of the air in the room ? (iii) Was the process spontaneous? (Material in the thermostat is water).

Solution:

(i) Entropy change of the material in the thermostat

$$-\frac{4.2 \text{ kJ}}{370.05\text{K}} = -11.35 \times 10^{3} \text{ kj K}^{-1}$$

(ii) Entropy change of the air in the room

$$= \frac{4.2 \text{ kJ}}{300.05 \text{ K}} = 14.00 \times 10^{3} \text{ kJ K}^{-1}$$

(iii) Now, $\Delta S_{total} = \Delta S_{thermostat} + \Delta S_{room}$

$$= (-11.35 + 14.00) \text{ JK}^{-1}$$

$$= 2.65 \text{ J K}^{-1}$$

Since ΔStotal is positive, the process of leaking is spontaneous.

Example 35(b):

Calculate ΔS, ΔH and ΔU for the process

$$H_2O(l, 20°C, 1 \text{ atm} \rightarrow H_2O \text{ (g, 250°C, 1 atm)}$$

Given the following data :

$$C_P (l) = 75.6 \text{ J K}^{-1} \text{ mol}^{-1}$$

$$C_P\,(g) = 36.2\ J\ K^{-1}\ mol^{-1}$$

ΔH for vaporization of H_2O at 100°C and 1 atm is equal to 40.85 k J mol^{-1}.

Solution:

The given process may be replaced by the following processes.

(i) H_2O (l, 20°C, 1 atm) → H_2O (l, 100°C, 1 atm)

(ii) H_2O (l, 100°C, 1 atm) → H_2O (g, 100°C, 1 atm)

(iii) H_2O (g, 100°C, 1 atm) → H_2O (g, 250°C, 1 atm)

Thus we have

$$\Delta S = \Delta S_{(1)} + \Delta S_{(ii)}\, \Delta S_{(iii)}$$

$$= C_{p,m}(l)\ \ln\frac{T_2}{T_1} + \frac{\Delta H_{vap}}{T} + C_{p,m}(g)\ \ln\frac{T_2'}{T_1'}$$

$$= \left(75.6\ \ln\frac{373}{293} + \frac{40850}{373} + 36.2\ \ln\frac{523}{373}\right) JK^{-1}\ mol^{-1}$$

$$= (18.25 + 109.52 + 12.24)\ J\ K^{-1}\ mol^{-1}$$

$$= 140.01\ J\ K^{-1}\ mol^{-1}$$

$$\Delta H = \Delta H_{(1)} + \Delta H_{(ii)} + \Delta H_{(iii)}$$

$$= C_{p,m}\ (l)\ (T_2 - T_1) + \Delta H_{vap} + C_{pm}(g)\ (T'_2 - T'_1)$$

$$= (75.6 \times 80 + 40850 + 36.2 \times 150)\ J\ mol^{-1}$$

$$= 52328\ \ J\ mol^{-1}$$

$$\Delta U = \Delta U_{(1)} + \Delta U_{(ii)}\, \Delta U_{(iii)}$$

$$= C_{v,m}\ (l)\ (T_2 - T_1) + \Delta U_{vap} + C_{v,m}(g)\ (T'_2 - T'_1)$$

$$= C_{p,m}(l)(T_2 - T_1) + \{\Delta H_{vap} + \Delta_{V\ g} RT\} + \{C_{p,m}(g) - R\}(T'_2 - T'_1)$$

$$= [75.6 \times 80 + \{40850 - (8.314)\ (373)\}$$

$$+ (36.2 - 8.314)\ (150)]\ J\ mol^{-1}\}$$

$$= (6040 + 37749 + 4183)\ J\ mol^{-1}$$

$$= 47972\ J\ mol^{-1}.$$

Example 35(c):

Describe a process through with the gas can be restored to its initial state in each of the above cases. Explain how the surroundings can be restored only in one ca. e and not in the other.

Solution:

The gas can be restored to its initial state by compressing it isothermally from 5 dm^3 to 1 dm^3. This may be done reversibly or irreversibly. When the compression is done reversibly, the surroundings in the case of (ii) will be restored whereas in case of (i), it cannot be restored whether the compressions is done reversibly or irreversibly. This is because, while on expansion, surroundings remain unchanged whereas during compression, it has to do work on the system and thus suffers changes.

Example 36:

Derive an expression for the entropy change when a van der Waals gas is heated and expanded simultaneously.

Solution:

Taking S = f(T, V), we get

$$dF = \left(\frac{\partial S}{\partial T}\right)_V dT + \left(\frac{\partial S}{\partial V}\right)_T dV$$

$$= \frac{nC_{V,m}}{T} dT + \left(\frac{\partial P}{\partial T}\right)_V dV$$

Now for van der Waal's gas we have

$$P = \frac{n\,RT}{V - nb} - \frac{n^2 a}{V^2}$$

Therefore $$\left(\frac{\partial P}{\partial T}\right)_V = \frac{n\,RT}{V - nb}$$

Thus $$dS = \frac{n\,C_{V,m}}{T} dT + \frac{n\,R}{V - nb} dV$$

Which on integrating within the limit T_1, V_1 and T_2, V_2 gives

$$\Delta S\ nC_{V.m} \ln \frac{T_2}{T_1} + nR \ln \frac{V_2 - nb}{V_1 - nb}$$

Example 37:

For an adiabatic process, dq = 0, If one were to write

$$\Delta S = \int \frac{dq}{T} = \int \frac{0}{T} = 0$$

Then every adiabatic process would be iso-entropic process. Comment.

Solution:

For an adiabatic process, dq – 0, and according to the first law of thermodynamics, dq = dU – dw

Thus, we will have

$$dU = dw$$

The expression of entropy change will be given as

$$\Delta S = \int \frac{dU - dw}{T}$$

$$= \int \frac{dU}{T} + \int \frac{P_{opp}}{T} dV$$

Now, it is only for reversible process the first integral is equal and opposite sign of the second integral and thus $\Delta S = 0$. For an irreversible process, these two integrals do not have equal magnitude and thus sum is not zero. Hence $\Delta S \neq 0$ for an adiabatic irreversible process.

Example 38:

A system consists of 6000 particles distributed in three energy states with equal spacing. The energy of the three states are E_i, = 0, E_2, = x and E_3 = 2x. All the three slates have the same intrinsic probability g. At a certain instant there are 3000 particles in lower level, 2500 in the middle level and 500 in the upper level. Compare the relative probabilities with the distribution obtained, by the transfer of one particle from the middle to the lower level and one particle from the middle to the upper level and the original distribution.

Solution:

Let P_1 and P_2 be the probabilities in the two cases

In the first case,

$$N = 6000,$$

$$n_1 = 3000,$$

$$n_2 = 2500,$$

$$n_3 = 500.$$

$$P_1 = \frac{g^N}{n_1!n_2!n_3!}$$

$$P_1 = \frac{g^{6000}}{3000!2500!500!} \qquad ...(i)$$

In the second case

$$N = 6000.$$

$$n_1 = 3001.$$

$$n_2 = 2498.$$

$$n_3 = 501.$$

$$P_1 = \frac{g^N}{n_1!n_2!n_3!}$$

$$P_1 = \frac{g^{6000}}{3001!2498!501!}$$

Dividing (ii) by (i)

$$\frac{P_2}{P_1} = \frac{3000!2500!500!}{3001!2498!501!}$$

$$\frac{P_2}{P_1} = \frac{2500 \times 2499!}{3001 \times 501!}$$

$$\frac{P_2}{P_1} = 4.157$$

$$\approx 4.2.$$

It means the transfer of one particle from the middle to the upper and one particle from the middle to the lower state has changed the probability by a factor 4.2. This shows that both these distributions are not near the equilibrium state.

Example 39(a):

A thermostat was maintained at 370.05 K. For nearly an hour 4.2 kJ of heat leaked through the thermostat insulation into a room where the initial temperature of air was 300.05 K.

(i) What was the entropy change of the material in the thermostat?

(ii) What was the entropy change of the air in the room ?

(iii) Was the process spontaneous? (Material in the thermostat is water).

Solution:

(i) Entropy change of the material in the thermostat

$$-\frac{4.2 \text{ kJ}}{370.05\text{K}} = -11.35\times 10^3 \text{ kj K}^{-1}$$

(ii) Entropy change of the air in the room

$$= \frac{4.2 \text{ kJ}}{300.05 \text{ K}} = 14.00 \times 10^3 \text{ kJ K}^{-1}$$

(iii) Now

$$\Delta S_{total} = \Delta S_{thermostat} + \Delta S_{room}$$
$$= (-11.35 + 14.00) \text{ JK}^{-1}$$
$$= 2.65 \text{ J K}^{-1}$$

Since ΔStotal is positive. the process of leaking is spontaneous.

Example 39(b):

At low temperature, a nonideal gas follows the relation $PV = RT - a/V$ where $a = 0.3636$ Nm^4 $mole^{-2}$. Calculate the work done by one mole of this gas in expanding form 0.224 litre to 22.4 litre at 400 K. Compare this result with the work done in the corresponding expansion of an ideal gas.

Solution:

For an ideal gas, reversible work done

$$= 2.303 \text{ n RT log } \frac{V_2}{V_1}$$

V_1 = 0.224 litres $mole^{-1}$ = 22.4 × 10^{-5} m^3 $mole^{-1}$

P_2 = 22.4 litres $mole^{-1}$ = 22.4 × 10^{-3} m^3 $mole^{-1}$

R = 8.314 J K^{-1} $mole^{-1}$

$$W_{rew} = (2.303)\ (1 \text{ mole})\ (8.314 \text{ J K}^{-1} \text{ mole}^{-1})\ (400 \text{ K}) \times \log \frac{22.4 \times 10^{-3}}{22.4 \times 10^{-5}}$$

$$= (2.303)\ (1 \text{ mole})\ (8.314 \text{ J K}^{-1} \text{ mole}^{-1})\ (400 \text{ K}) \log 100$$

$$= 15320 \text{ J mole}^{-1}$$

$= 15.32$ kJ mole^{-1}.

For the non-ideal gas PV = RT – a/V

$$\text{Therefore, the work done} = \int_{V_1}^{V_2} PdV = \int_{V_1}^{V_2} \left\{\frac{RT}{V} - \frac{a}{V_2}\right\} dV$$

$$= 2.303\ RT \log \frac{V_2}{V_1} + a\left\{\frac{1}{V_2} - \frac{1}{V_1}\right\}$$

The value of first part is from derived eq. (1) as 15.32 kJ mole^{-1} and of another part can be derived as

$$0.3636\ Nm^4\ mole^{-2} \left\{\frac{1}{22.4 \times 10^{-3}\ m^3\ mole^{-1}} - \frac{1}{22.4 \times 10^{-5}\ m^3\ mole^{-1}}\right\}$$

$= 1.604$ kJ mole^{-1}.

Thus work done = 15.320 k Joule^{-1} – 1.604 k Joule^{-1}

$= 13.716$ kJ mole^{-1}.

Example 39(c):

What will be the work done when 65.38 g of zinc dissolves in hydrochloric acid in case of an open beaker and closed beaker at 300K.

Solution:

The process of dissolution of zinc in hydrochloric acid can be given as

$$Zn(s) + 2HCI\ (aq) = ZnCl_2(aq) + H_2(g)$$

From this equation it is clear that for each gram atom of zinc dissolved we will obtain 1 mole of Hydrogen gas. As a result of liberated gas, on the surrounding atmosphere the work performed will be equal to PΔV. If the initial volume of the system is neglected in comparison to the total volume of the gas produced and the gas shows ideal behaviour, then the work done in the open beaker is

$$W = P\Delta V \approx PV_{H2}$$

$$= nH2\ RT\ [\because T = 300\ k\ R = 8.314\ Jk^{-1}\ mole^{-1}]$$

$$= (1\ mole)\ (8.314\ JK^{-1}\ mol^{-1})\ (300\ K)$$

$$= 2.4942\ kJ.$$

If the reaction takes place in a closed beaker there is change in the volume *i.e.*, $\Delta V = 0$ hence work done is zero.

Example 40:

Calculate the work done when one mole of sulphur dioxide gas expands isothermally and reversibly at 300 k from 2.46×10^{-3} m^3 to 24.6×10^{-3} m^3. Assuming that the gas obeys

(a) ideal gas equation and

(b) van der Waals equation

$$\left\{ \begin{array}{l} a = 0.6799 \text{ nm}^4 \text{ mole}^{-2} \\ b = 0.564 \times 10^{-3} \text{ mole}^{-1} \end{array} \right\}$$

Solution:

(a) The reversible work done by 1 mole of gas when it expands form

$$2.64 \times 10^{-3}\text{m}^3 \text{ to } 24.6 \times 10^{-3}\text{m}^3 = 2.303 \text{ RT} \log \frac{V_2}{V_1}$$

$$= (2.303)\ (1 \text{ mole})\ (8.314 \text{ J K}^{-1} \text{ mol}^{-1})\ (300 \text{ K}) \times \log 10$$

$$\left\{ \because \log \frac{V_2}{V_1} = \log 10 \right\}$$

$$= (2.303)\ \ (8.314)\ (300 \text{ K})\ \ \log 10 \text{ J mol}^{-1}$$

$$= 5744 \text{ J mol}^{-1}$$

(b) When the gas obeys van der Waal's equation, work done by 1 mole of gas is

$$= 2.303 \text{ RT} \log \frac{V_2 - b}{V_1 - b} + a \left\{ \frac{1}{V_2} - \frac{1}{V_1} \right\}$$

After substituting the values of various quantities and solving, the work done is

$$= 5794 \text{ J mol}^{-1} - 252.4 \text{ J mol}^{-1}$$

$$= 5541.6 \text{ J mol}^{-1}$$

$$= 5.542 \text{ kJ mol}^{-1}.$$

Example 41:

One mole of a van der Walls's gas is allowed to expand isothermally and reversibly from a volume of 1 litre to 50 litre at 0°C. Calculate w, *q, Δ E and Δ H. van der Waal's constants are a = 6.5atm l^2mole $^{-2}$ = 0.056 litre mole $^{-1}$ and R = 0.082 l atom deg^{-1} mole^{-1}.*

Solution:

Here V_1 = 1 litre,

V_2 = 50 litres, n = 1 mole

T = 273 + 0 = 273°k,

R = 0.082 l-atm deg^{-1} mole^{-1}

a = 6.5 atm l^2 mole^{-2},

b = 0.056 litre mole^{-1}

$$w = 2.303 \text{ nRT} \log\left(\frac{V_2 - nb}{V_1 - nb}\right) + an^2\left(\frac{1}{V_2} - \frac{1}{V_1}\right)$$

$$= 2.303 \times 1 \times 0.082 + 273\ [\log(50 - 0.056) - \log(1 - 0.056)$$

$$+ 6.5\left(\frac{1}{50} - \frac{1}{1}\right)$$

= 82.47 l atm.

$$\Delta E = -\ an^2\left(\frac{1}{V_2} - \frac{1}{V_1}\right) = 6.5 \times \frac{49}{50}$$

$$q = 2.303 \text{nRT} \log \frac{V_2 - b}{V_1 - b}$$

= ΔE + w = 884 l atm.

Example 42(a):

The van der Waal's constant a and b for hydrogen in litre atmosphere units are 0.246 and 2.67 × 10^{-2} respectively. Calculate the inversion temperature of hydrogen.

Solution:

we know T_i = 2a/Rb

or $$T_i = \frac{2 \times 0.246}{0.0821 \times 0.0267}$$

= 224.5°k = (224.5 – 273)

= – 48.5°C.

Example 42(b):

How much heat is required to raise the temperature of 1 mole oxygen from 300K to 1300 K at constant pressure.

$$C_p = 6.095 + 3.325 \times 10^{-3} T - 1.017 \times 10^{-6} T^2.$$

Solution:

$$H_2 - H_1 \int_{T_1}^{T_2} C_p \, dT$$

$$T_1 = 300 \text{ k}, T_2 = 100 \text{ k}$$

$$\therefore \quad H_{1300} - H_{300}$$

$$= \int_{300}^{1300} \left(6.095 + 3.253 + 10^{-3} \, T - 1.017 \times 10^{-6} \, T^2\right)$$

[Substituting the value of C_p]

$$\because \quad H_{1300} - H_{300} = 6.095 \, (1300 - 300) + \frac{3.253 \times 10^{-3}}{2}$$

$$\times [(1300)^2 - (300)^2] - 1/3 \times 1.017 \times 10^{-6} [(1000)^2 - (300)^2]$$

$$= 6095 + 2602.4 - 735.63$$

$$= 7961.77 \text{ cal/mole.}$$

Example 42(c):

For equation $(P \_ a/V^2) V = RT$, prove that (i) dP is an exact differential, (ii) P is a state function and

$$\text{(iii)} \left(\frac{\partial P}{\partial T}\right)_V \left(\frac{\partial T}{\partial V}\right)_P \left(\frac{\partial V}{\partial T}\right)_T + 1 = 0.$$

Solution:

(i) For dP to be an exact differential, it is be proved that

$$\frac{\partial^2 P}{\partial V \partial T} = \frac{\partial^2 P}{\partial T \partial V} \qquad ...(1)$$

$$(P + a/V^2) \, V = RT$$

or

$$P = \frac{RT}{V} - \frac{a}{V^2} \qquad ...(2)$$

When the above equation is differentiated with respect to V at constant temperature T, we obtain

$$\left(\frac{\partial P}{\partial V}\right)_T = -\frac{RT}{V^2} + \frac{2a}{V^3}$$

When the above equation is differentiated with respect to T at constant V, we obtain

$$\frac{\partial^2 P}{\partial T \partial V} = -\frac{RT}{V^2} \qquad ...(3)$$

When equation (2) is differentiated with respect to T at constant volume V, we obtain

$$\left(\frac{\partial P}{\partial T}\right)_V = \frac{R}{V}$$

When the above equation is further differentiated with respect to V at constant T, we obtain

$$\frac{\partial^2 P}{\partial T \partial V} = -\frac{R}{V^2} \qquad ...(4)$$

From equations (3) and (4), we get

$$\frac{\partial^2 P}{\partial T\, \partial V} = \frac{\partial^2 P}{\partial T\, \partial V} = -\frac{R}{V^2} \qquad ...(5)$$

Thus, we have proved condition, (1). This shows that dP is an exact differential.

(ii) In the said question, first of all a change occurs in volume at constant temperature followed by a change in temperature at constant volume, thereby giving the final equation (3).

Again, a change in temperature is considered at constant volume, followed by a change in volume at constant temperature, thereby giving the final equation (4). Both equations (3) and (4) are same, thereby revealing that the final change in P remains the same irrespective of the difference in intermediate changes. Therefore, P is a state function as changes in P have been found to depend upon the intial and final states of the system.

(iii) Equation (2) is as follows :

$$P = \frac{RT}{V} - \frac{a}{V^2}$$

When the above equation is differentiated, we obtain :

$$dP = \frac{R}{V}\,dT - \frac{RT}{V^2}\,dV + \frac{2a}{V^3}\,dV$$

$$= \frac{R}{V}\,dT - \left(\frac{RT}{V^2} - \frac{2a}{V^3}\right)dV \qquad ...(6)$$

If volume is constant, dV = 0 : equation (6) becomes as follows :

$$dP = \frac{R}{V}\,dT \text{ or } \left(\frac{\partial P}{\partial V}\right)_v = \frac{R}{V} \qquad ...(7)$$

If pressure is constant, dP = 0; equation (6) becomes as follow :

$$\frac{R}{V}\,dT - \left(\frac{RT}{V^2} - \frac{2a}{V^3}\right)dV = 0$$

or
$$\left(\frac{\partial T}{\partial V}\right)_P = \left(\frac{RT}{V^2} - \frac{2a}{V^3}\right)\Big/\frac{R}{V} \qquad ...(8)$$

If temperature is constant, dT = 0; equation (6) becomes as follows:

$$\left(\frac{\partial T}{\partial V}\right)_T = -\frac{1}{\left(\frac{RT}{V^2} - \frac{2a}{V^3}\right)} \qquad ...(9)$$

On multiplying equations (7), (8) and (9), we obtain

$$\left(\frac{\partial p}{\partial T}\right)_V\left(\frac{\partial T}{\partial V}\right)_P\left(\frac{\partial V}{\partial P}\right)_T = -\frac{R}{V} \times \frac{\left(\frac{RT}{V^2} - \frac{2a}{V^3}\right)}{\left(\frac{RT}{V^2} - \frac{2a}{V^3}\right)} \times \frac{1}{R} = -1$$

or
$$\left(\frac{\partial p}{\partial T}\right)_V\left(\frac{\partial T}{\partial V}\right)_P\left(\frac{\partial V}{\partial P}\right)_T + 1 = 0$$

This is cyclic rule.

Example 42(d):

For an ideal gas (PV = nRT) show that (1/T) is an integrating factor for dw = P dV.

Solution:

$$dw = P\,dV \qquad ...(1)$$

$\because$ $$PV = nRT$$

$\therefore$ $$p\,dV + V\,dP = nR\,dT$$

and
$$dV = \frac{nR}{P}dT - \frac{V}{P}\,dP \qquad ...(2)$$

On substituting the expression for dV in equation (1), we obtain

$$dw = nR\,dT - V\,dP$$

$$z = nR\ dT - \frac{nRT}{P} dP$$

Now $\frac{\partial}{\partial P}(nR)_T = 0$

and $\frac{\partial}{\partial T}\left(-\frac{nRT}{P}\right)_P = -\frac{nR}{P}$

Hence, dw is inexact differential.

Suppose f = 1/T be an integrating factor. Then

$$f\ .\ dw = \frac{1}{T} nR\,dT - \frac{nR}{P} dP$$

$$\therefore \quad \frac{\partial}{\partial P}\left[\frac{nR}{T}\right]_T = 0$$

and $\frac{\partial}{\partial T}\left[\frac{nR}{P}\right]_T = 0$

Thus f. dw$\frac{nR}{T}$dT $-$ $\frac{nR}{P}$dP is an exact differential while 1/T is an integrating factor.

Example 43:

The molar heat capacity at capacity at constant pressure of solid magnesium (0° – 600°C) is expressed by

$$CP = 6.20 + 1.30 \times 10 - 3\ T - 6.80 \times 104\ T - 2$$

Calculate the increase in entropy when 1 gm atom of metal is heated from 27°C to 127°C at constant pressure.

Solution:

At constant pressure,

$$\Delta S = \int_{T_1}^{T_2} \frac{C_P}{T} dT$$

$$\therefore \quad \Delta S = \int_{300}^{400} \frac{\left(6.20 + 1.30 \times 10^{-3}\,T - 6.80 \times 10^4\,T^{-2}\right)dT}{T}$$

$$\int_{300}^{400} 6.20\,\frac{dT}{T} + \int_{300}^{400} 1.3 \times 10^{-3} dT - \int_{300}^{400} 6.20 \times 10^4\,\frac{dT}{T^3}$$

$$= 6.20 \ln[\ln T]_{300}^{400} + 1.3 \times 10^{-3} [T]_{300}^{400} - \frac{6.80 \times 10^4}{-2}\left[\frac{1}{T^2}\right]_{300}^{400}$$

$$= 6.20 \ln \frac{400}{300} + 1.3 \times 10^{-3} \times (400 - 300) + 3.40 \times 10^4$$

$$\left[\frac{1}{(400)^2} - \frac{1}{(300)^2}\right]$$

$$= 1.7487 \text{ cal./degree.}$$

Example 44:

Calculate the increase in entropy of three moles of hydrogen as it changes from 300°K atm to 1000°K and 1 atm (C_p = 7 cal./degree/ mole).

Solution:

Entropy change ΔS for n moles

$$= n\, C_P \ln \frac{T_2}{T_1}\; n\, R \ln \frac{P_2}{P_1}$$

$$= 3 \times 7 \times 2.303 \log \frac{1000}{300} - 3 \times 2 \times 2.303 \log \frac{10}{0.1}$$

$$= 25.287996 - 13.818$$

$$= 11.469996 \text{ cal./degree.}$$

Example 45(a):

Calculate ΔS, ΔH and ΔU for the process

$$H_2O(l,\ 20°C,\ 1\ atm \rightarrow H_2O\ (g,\ 250°C,\ 1\ atm)$$

Given the following data :

$$C_P\ (l) = 75.6\ J\ K^{-1}\ mol^{-1}$$

$$C_P\ (g) = 36.2\ J\ K^{-1}\ mol^{-1}$$

ΔH for vaporization of H_2O at 100°C and 1 atm is equal to 40.85 k J mol^{-1}.

Solution:

The given process may be replaced by the following processes.

(i) H_2O (l, 20°C, 1 atm) → H_2O (l, 100°C, 1 atm)

(ii) H_2O (l, 100°C, 1 atm) → H_2O (g, 100°C, 1 atm)

(iii) H_2O (g, 100°C, 1 atm) → H_2O (g, 250°C, 1 atm)

Thus we have

$$\Delta S = \Delta S_{(1)} + \Delta S_{(ii)} \Delta S_{(iii)}$$

$$= C_{p,m}(1) \ln \frac{T_2}{T_1} + \frac{\Delta H_{vap}}{T} + C_{p,m}(g) \ln \frac{T_2'}{T_1'}$$

$$= \left(75.6 \ln \frac{373}{293} + \frac{40850}{373} + 36.2 \ln \frac{523}{373}\right) JK^{-1} \text{ mol}^{-1}$$

$$= (18.25 + 109.52 + 12.24) \text{ J K}^{-1} \text{ mol}^{-1}$$

$$= 140.01 \text{ J K}^{-1} \text{ mol}^{-1}$$

$$\Delta H = \Delta H_{(1)} + \Delta H_{(ii)} + \Delta H_{(iii)}$$

$$= C_{p,m} (l) (T_2 - T_1) + \Delta H_{vap} + C_{pm}(g) (T'_2 - T'_1)$$

$$= (75.6 \times 80 + 40850 + 36.2 \times 150) \text{ J mol}^{-1}$$

$$= 52328 \text{ J mol}^{-1}$$

$$\Delta U = \Delta U_{(1)} + \Delta U_{(ii)} \Delta U_{(iii)}$$

$$= C_{v,m} (l) (T_2 - T_1) + \Delta U_{vap} + C_{v,m}(g) (T'_2 - T'_1)$$

$$= C_{p,m}(l)(T_2-T_1)+\{\Delta H_{vap}+\Delta_{V\,g}RT\}+\{C_{p,m}(g) - R\}(T'_2 - T'_1)$$

$$= [75.6 \times 80 + \{40850 - (8.314)(373)\}$$

$$+ (36.2 - 8.314)(150)] \text{ J mol}^{-1}\}$$

$$= (6040 + 37749 + 4183) \text{ J mol}^{-1}$$

$$= 47972 \text{ J mol}^{-1}.$$

Example 45(b):

For $N = 400$, find the ratios of the probabilities for (300, 100) and (200, 200) distribution in two identical boxes.

Solution:

We have

$$P(300, 100) = \frac{400!}{300!\ 100!} \frac{1}{2^{400}}$$

$$P(200, 200) = \frac{400!}{200!\,200!}\,\frac{1}{2^{400}}$$

$$\frac{P(300, 100)}{P(200, 200)} = \frac{200!\,200}{300!\,100!} \cong 10^{-5}$$

Example 45(c):

Eight similar coins are tossed, find the probability of three heads and t tails.

Solution:

We know $P(3, 5) = \frac{{}^8C_3}{2^8} = \frac{8\times7\times6}{3\times2\times1\times2^8}$

or $P(3, 5) = \frac{7}{32}$

Example 46:

Two bodies at different temperatures are connected by a wire with infinitesimal thermal conductance so that heat flows infinitely slowly from the hot bodies to the cold body until they are at the same temperature. Is this process reversible? Explain.

Solution:

In a reversible process, at any stage, the external condition responsible for the process to occur differs from the internal condition by an infinitesimal amount. The given transfer of heat is not a reversible process as the two bodies are at different temperatures. Moreover, the reversible processes are characterized by the fact that when the system is restored to its original state by traversing the forward sequence of steps in the reverse order, then not only the system but also the surroundings are restored to their original states. In the given process, heat flow from a hot body to a cold body does not disturb the surroundings but if we wish to carry out the reverse process, the surroundings will have to do some work. Hence, the surroundings are not brought back to the original state in the reverse process. Thus, the given process is not a reversible one.

Example 47:

How would the energy of an ideal gas change if it is made to expand into vacuum at constant temperature ?

Solution:

For the expansion against vacuum $P_{opp} = 0$

Hence $w = -P_{opp}\,\Delta V = 0$

Since the expansion is at constant temperature no heat is supplied to the system, *i.e.*, q = 0. Now, from the first law of thermodynamics, we can write

$$\Delta E = q + w = 0$$

that is, the energy of an ideal gas does not change if it is made to expand into vacuum at constant temperature. This also follows from the fact that the energy of an ideal gas depends only on temperature and since temperature remains constant, its energy also remains constant.

Example 48:

Calculate the change of entropy for the process

$$H_2O\ (l,\ 373\ K) \rightarrow H_2O\ (v,\ 373\ K)$$

Here $\Delta H_{vap} = 40850\ J\ mol^{-1}$ *at 373 K.*

Solution:

$$\Delta S_{vap} = \frac{\Delta H_{vap}}{T_b}$$

$$= \frac{40850\ J\,mol^{-1}}{373\ K}$$

$$= 109.51742\ JK^{-1}\ mol^{-1}$$

But $\Delta S_{vap} = S\ (vapour) - S\ (liquid)$

$$= 109.5\ JK^{-1}\ mol^{-1}$$

$\therefore \quad S\ (vapour) = S\ (liquid) + 109.5\ JK^{-1}\ mol^{-1}$

3

Thermochemistry

INTRODUCTION

Heat is generally absorbed or evolved whenever reactions take place, a solution is formed, or other changes in the state of matter (fusion, vaporization, etc). occur. Thermochemistry is concerned with the measurement and study of these heat changes.

Heat changes accompanying chemical reactions or physical changes are measured in calorimeters. The system in which the change is to take place is immersed in a bath which is thermally insulated from the rest of the universe. The temperature of the bath is measured before and after and change has occurred. This temperature change is compared with the temperature change produced when a known amount of thermal energy is introduced into the bath, usually by passing an accurately known current at an accurately known voltage for an accurately known time through a resistor immersed in the bath. For precise measurements of heat changes, great care is required in the design and operation of calorimeters.

Thermodynamic principles make the determination of heat changes more simpler even without performing the tedious experiments in the laboratory. This is done by taking advantage of the fact that heat changes under certain conditions ($q_v = \Delta E$ and $q_p = \Delta H$) become state properties. This fact many times helps in the computation of the heat changes for those processes which are otherwise very difficult to be carried out in the laboratory and may require a lot of time and labour.

THE CONCEPT OF HEAT OF REACTION

Heat of reaction is defined as the amount of heat absorbed or evolved at a given temperature when the reactants have combined to form the products as represented by the balanced chemical equation. If the heat of reaction is denoted by q, then according to the first law,

$q = \Delta E - w$

where –w is the work done by reacting mixture when it reacts and D E is the energy difference between the reaction products and the reactants,

$\Delta E = \Sigma E$ (products) ΣE (reactants)

The numerical value of q depends on the manner in which the reaction is performed. For the two methods of conducting chemical reactions in calorimeters, we have

Constant volume : w = 0, and qv = DE; Bomb calorimeter.

Constant pressure : $w = -P\Delta V$ and $qP = \Delta E + P\Delta V = \Delta H$; Open calorimeter.

Thus heat of a reaction at constant volume (qv) is the change of energy (ΔE) accompanying the chemical reaction. The heat of reaction when measured at constant pressure (qp) is equal to the enthalpy change in the reaction given by

$q_P = \Delta H = \Sigma H$ (products) $- \Sigma H$ (reactants)

Constant pressure are much common in chemistry therefore usually whenever we speak of heat of reaction in implies enthalpy change.

Sign conventions. If heat is absorbed by the system ($q > 0$) then the reaction is said to be endothermic and ΔE or ΔH value is given a positive sign. If heat is evolved ($q < 0$) the reaction is said to be **exothermic**, and ΔE or ΔH value is given negative sign.

Endothermic reaction. $\Delta H > 0$, ΣH (Products) > ΣH (reactants). Heat absorbed during the reaction increases the energy contents of the products and hence the enthalpy of the products becomes greater than that of the reactants from which the products are formed.

Exothermic reaction. $\Delta H < 0$, ΣH (products) < ΣH (reactants). Heat is liberated in the reaction and is transferred to the surroundings. Therefore the enthalpy of the products is less than enthalpy of the reactants from which the products are generated. Diagrammatically these may be represented as follows:

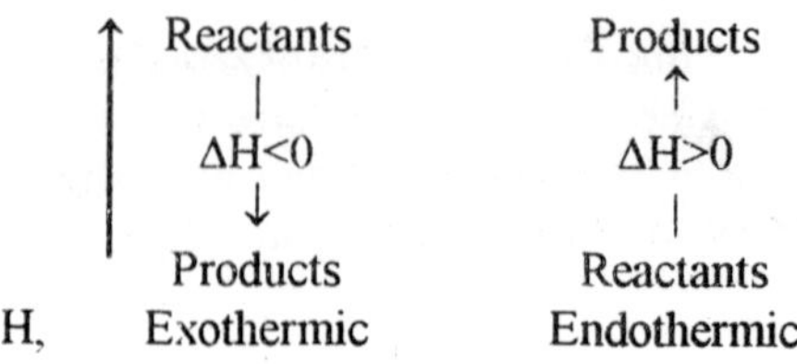

If the reaction is performed at constant temperature, it is indicated by writing TK in parenthesis to the right of Δ H (or Δ E). The state of a product or reactant is indicated by symbols in parentheses after the chemical symbol e. g., (s) or (c) for solid or crystal, (1) for a liquid, (g) for a gas, (aq) for solution in water.

EQUATIONS OF THE THERMOCHEMICAL

For completeness, the chemical equations along with the enthalpy change or energy change accompanying the chemical reaction are written in two ways. In the first method, the balanced chemical equation representing the actual reaction is written, and the heat change is indicated by mentioning ΔH or ΔE with proper sign to the right side of the equation, *e.g.*, when 1 mole methane burns in the presence of oxygen, 1mole CO_2 gas and moles H_3O are formed, and 890.3 kj heat is produced measured at constant pressure and 298K. This change is represented by the *Thermochemical Equation* as

$$1CH_4\ (g) + 2O_2\ (g) \rightarrow CO_2\ (g) + 2H_2O$$

$$(1)\ \Delta\ H\ (298K) = -890.3kJ$$

The same equation may be written in thermally balanced manner. Since the reaction is exothermic, 890.3 kJ heat is given to the surroundings, the enthalpies of products are less than those of the reactants by 890.3 kJ. If this amount is added to the product side, the thermally balanced equation is:

$$CH_4\ (g) + 2O_2\ (g) \rightarrow CO_2\ (g) + 2H_2O\ (1)\ + 890.3\ kJ$$

At present the former method is preferred to the latter. Endothermic reactions may be represented in a similar manner with a changed sign of Δ H or ΔE.

$$N_2\ (g) + O_2\ (g) \rightarrow 2NO\ (g)\quad \Delta H\ (298K) = 180\ kJ$$

$$\text{or } N_2\ (g) + O_2\ (g) \rightarrow 2NO\ (g) - 180\ kJ$$

STANDARD STATES

In the computation of heat of reactions it is a convention to assume that the heat of formation of elements in their standard states is zero. The standard state is taken as 1 atm pressure and a constant temperature. Standard states for various forms of matter are summarized below :

State	*Standard state*
Gas	Ideal gas at 1 atm and the given temperature
Liquid	Pure liquid at 1 atm and the given temperature
Solid	Stable crystalline form at 1 atm and given T . (e.g. graphite form of carbon, rhombic form of sulphur).

At standard state the heat of reactions are denoted by Δ E° or Δ H° at given temperature.

Relation between heat of reaction of constant volume (qv) and heat of reaction at constant pressure (qP)

As discussed earlier qv and qP can be taken as ΔE and Δ H respectively. We know that the enthalpy and energy are related by the equation :

$$H = E + PV \Rightarrow \Delta H = \Delta E + P\Delta V$$

Let us consider a general chemical reaction represented by the equation

$$aA + bB \rightarrow mM + nN$$

The enthalpy change is given by

$$\Delta H = \Sigma H \text{ (products)} - \Sigma H \text{ (reactants)}$$

$$= (mH_M + nHN) - (aHA + bHB)$$

$$= [mE_M + m\,(PV)_M + nE_N + n\,(PV)_N]$$

$$- \{aE_A + a(PV)_A + bE_B + b(PV)_B\}$$

$$= \Delta\, E + P\,[(mV_M - nV_N) - (aV_A + bV_B)]$$

$$= (\Delta E)P + P\,\Delta V \text{...If } P \text{ is constant}$$

where $\Delta E = (mE_M + nE_N) - (aE_A + bE_B)$

$$= \Sigma E \text{ (products)} - \Sigma E \text{ (reactants)}$$

and $\Delta V = (mV_M + nV_N) - (aV_A + bV_B)$

$$= \Sigma V \text{ (products)} - \Sigma V \text{ (reactants)}$$

In case the reactants and products being in the solid and liquid state the volume change during the reactions is negligibly small, therefore ΔV can be ignored, and

$\Delta H = (\Delta E)_P = \Delta E$

since $(\Delta E)P$ and are approximately equal.

In the case of gaseous reactants and products ΔV is appreciable and can be evaluated by assuming that the reactants and products obey ideal gas equation PV = nRT.

Thus, $\Delta V = \frac{RT}{P}[(m + n) - (a + b)] = \Delta v_{gas} RT/P$

Here $\Delta v_{gas} = (m + n)_{gas} - (a + b)_{gas}$

with this value of ΔV, we can write

$\Delta H = \Delta E + \Delta v_{gas} RT \Rightarrow qP = qv + \Delta v_{gas} RT$

(while calculating Δv, we should not consider the moles of liquid or solid species).

The equation is used to calculate the enthalpy of reaction from the energy of reaction at a given temperature and *vice versa*. If in a reaction $\Delta v_g = 0$, then DH = ΔE.

Example 1:

The formation of SO_3 from SO_2 and oxygen,

$SO_2 (g) + 1/2\ O_2 (g) \rightarrow SO_3 (g)$

is exothermic by 97030 *at* 298 *K and 1 atm measured in a bomb calorimeter. What is the value of* Δ H?

Solution:

The above equation can be written as,

$SO_2 (g) + 1/2\ O_2 (g) \rightarrow SO_3 (g)\ \Delta E\ (298\ K) = -97030\ J.$

We know

$\Delta H = \Delta E + \Delta vRT$

Here $\Delta v = 1 - \left(1 + \frac{1}{2}\right) = -\frac{1}{2}$

$\therefore \Delta H = -97030 - 1/2\ (8.314)\ (298) = -$ **98270 J**.

Example 2:

Find ΔE^o *for the following reactions at 1 atm and* 298 K

$C (s) + O_2(g) \rightarrow CO_2(g) \quad \Delta H^o\ (298\ K) = -393.5\ kJ\ mol^{-1}$

$C (s) + 1/2 O_2 (g) \rightarrow CO (g)$ ΔH^o (298 K) = – 110.5 kJ mol^{-1}

$CH_4(g) + 2O_2(g) \rightarrow CO_2(g) + 2H_2O(l)$ ΔH^o (298 K)

= – 890.35 kJ mol^{-1}

$AgNO_3 (s) + HCl (g) \rightarrow AgCl (s) + HNO_3 (l)$

ΔH^o (298 K) = – 85.3 kJ mol^{-1}.

Solution:

Since R = 8.314 $JK^{-1}mol^{-1}$

and T = 298 K in the present case

RT = (8.314 JK^{-1} mol^{-1}) × (298K)

= 2477.57 J mol^{-1} = 2.477 kJ mo^{-1}.

The moles of only gaseous species are to be considered

$\Delta H^o = \Delta E^o + \Delta v\ RT$ gives $\Delta E^o = \Delta H^o - \Delta v\ RT$

(i) $C(s) + O_2(g) \rightarrow CO_2 (g)$ ΔH^o(298K) = – 393.5 kJ mol^{-1}

$\Delta v = 1 - 1 = 0$

$\therefore$ ΔH^o (298 K) = ΔE^o (298 K)

= 393.5 KJ mol^{-1}

(ii) $C (s) + 1/2 O_2 (g) \rightarrow CO (g)$ ΔH^o (298 K)=– 110.5kJ mol^{-1}

$$\Delta v = 1 - \frac{1}{2} = \frac{1}{2}$$

$$\therefore \Delta E^o \text{ (298 K)} = -110.5 \text{ kJ mol}^{-1} - \left(\frac{1}{2}\right) \times (2.477 \text{ kJ mol}^{-1})$$

= **–111.738 kJ mol^{-1}**

(iii) CH_4 (g) $2O_2(g) \rightarrow CO_2 (g) + 2H_2O(l)$ ΔH^o (298 K)

= – 890.35 kJ mol^{-1}

$\Delta V = 1 - (1 + 2) = -2$

Δ E^o (298 K) = (– 890.35 kJ mol^{-1}) + 2 × (2. 4 77kj mol^{-1})

= **– 885.39 kJ mol^{-1}**

(iv) $AgNO_3 (s) + HCl (g) \rightarrow AgCl (s) + HNO_3(l)$

ΔH^o (298 K) = – 85.3 kJ

$\Delta v = 0 - 1 = -1$

ΔE^{o} (298 K) = $-$ (85.3 kJ mol^{-1}) $-$ ($-$1) $\times$ (2.477 kJ mol^{-1})

= $-$ **82.823 kJ mol^{-1}**.

LAWS OF THERMOCHEMISTRY

For some reactions it is not convenient to measure the heat change in the laboratory. So conventional procedure based on the principle of conservation of energy has been suggested which can be stated as follows:

1. The heat of formation of any compound is equal in magnitude and of opposite sign to the heat of dissociation of that compound at the given temperature and pressure.

   For example, enthalpy of formation of liquid water from its elements hydrogen and oxygen is $-$ 285.830 kJ mol^{-1} and the enthalpy of dissociation is 285.850 kJ mol^{-1}. Thus the processes can be represented by

   H_2 (g) + 1/2 O_2 (g) $\rightarrow$ H_2O (l) ΔH (298 K) = $-$ 285.830 kJ

   H_2O (l) $\rightarrow$ H_2 (g) + 1/2 O_2 (g) ΔH (298 K) = + 285.830 kJ

2. The total enthalpy change of a reaction is the same, regardless of whether the reaction is completed in one step or in several steps. (*Hess's Law of constant heat summation.*) It has been experimentally verified and is also a consequence of the law of conservation of energy. It is of particular utility in calculation of f the heats of reactions which are difficult for practical calorimetric measurements.

Consider a system in which the initial state is represented by the symbol A and the final state by the symbol D and suppose that there are two ways by which the system can change from state A to state D A $\rightarrow$ D (1st way) direct change.

A $\rightarrow$ B $\rightarrow$ C $\rightarrow$ D (2nd way) stepwise change.

Hess's law required that the enthalpy change accompanying the change from A directly to D be identical to the sum of the enthalpy changes accompanying the changes from A to B to C to D (Fig 3.1)

Thus, $\Delta H_1 = \Delta H_2 + \Delta H_3 + \Delta H_4$

or $H_D - H_A = \Delta H_1 = \Delta H_2 + \Delta H_3 + \Delta H_4$

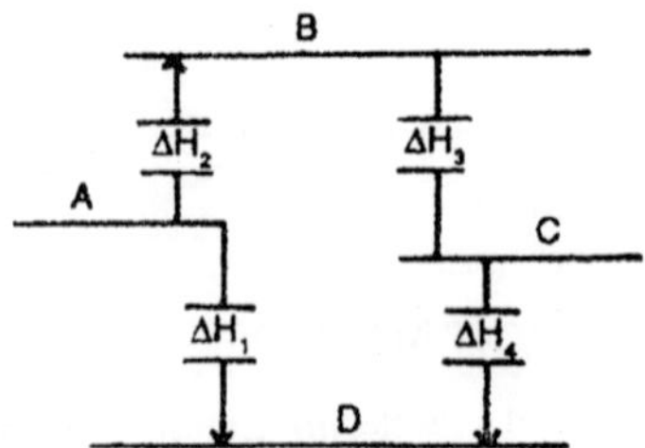

Fig. 3.1 : Hess's law in diagrammatic form.

From the laws of thermochemistry we get the following working rules :

(1) *Chemical equations can be reversed and the sign of DH is changed.*

(2) Chemical equations can be added, subtracted, multiplied, divided by an integer whenever necessary, according to the usual laws of algebra; terms identical in quality and quantity on both sides of the equation may be cancelled.

Example 3:

Compute the heat change in the following reaction

$COCl_2 + 2H_2S \rightarrow 2HCl + H_2O + CS_2$ *given that*

(i) $COCl_2 + H_2S \rightarrow 2HCl + COS$ ΔH^o (298) = – 78.705 kJ

(ii) $COS + H_2S \rightarrow H_2O + CS_2$ ΔH^o (298) = – 78.705 kJ

Solution:

On adding equation (i) and (ii) and cancelling the identical terms we get the desired result as

$COCl_2 + H_2S \rightarrow 2HCl + COS$ $\Delta H^o_{(i)}$ = – 78.705 kj

$$COS + H_2S \rightarrow H_2O + CS_2 \qquad \Delta H^o_{(ii)} = +3.420kj$$

$$COCl_2 + 2H_2S \rightarrow 2HCl + H_2O + CS_2 \qquad \Delta H^o(298) = \mathbf{-75.285kJ}$$

Example 4:

At 25°C the enthalpy changes for the dissolution of the two forms of glucose and their mutarotation are respectively given by

(i) α-D-*glucose* (s) → α-D-*glucose* (aq) ΔH(i) = 10.72 *kJ*

(ii) β-D-*glucose* (s) → β-D-*glucose* (aq) ΔH(ii) = 4.68 *kJ*

(iii) α-D-*glucose* (aq) → β-D-*glucose* (aq) $\Delta H(iii) = 1.16$ *kJ*

Calculate ΔH *for* α-D-*glucose* (s) → β-D-*glucose* (s) ΔH = ?

Solution:

Reverse equation (ii) and then add this and eqn. (i) and, (iii) cancel the identical terms :

(iv) β-D-*glucose* (aq) → β-D-*glucose* (s) $\Delta H(iv) = -4.68$ *kJ*

(v) α-D-*glucose* (aq) → β-D-*glucose* (aq) $\Delta H(iii) = 1.16$ *kJ*

$$\text{(i)}\ \frac{\alpha - \text{D} - \text{glucose (s)} \rightarrow \alpha\text{-D-glucose (aq)}}{\alpha\text{-D-glucose (s)} \rightarrow \beta\text{-D glucose (s)}} \quad \frac{\Delta H(i) = 10.72\text{kJ}}{\Delta H = \mathbf{4.88kJ}}\ \text{Add}$$

Example 5:

Compute the standard heat of formation of methane using the following data:

(i) $CH_4(g) + 2O_2(g) \rightarrow CO_2(g) + 2H_2O(l)$ ΔH° (298K)=– 890.35 kJ

(ii) $H_2(g) + 1/2\ O_2(g) \rightarrow H_2O(l)$ ΔH° (298K) = – 285.84 kJ

(iii) C (*graphite*) + O_2 (g) → CO_2 (g) ΔH° (298K) = – 393.51 kJ

Solution:

Our aim is to get DH° for the following reaction

C (graphite) + $2H_2$ (g) → CH_4(g) ΔH° (298) = ?

Reverse equation (i) and multiply equation (ii) by two.

(iv) CO_2 (g) + $2H_2O$ (l) → CH_4 (g) ΔH° (298K) = – 285.84 kJ

(v) $2H_2$ (g) + O_2(g) → $2H_2O$(l) ΔH° (298K) = – 571.68 kJ

Add equations (iii), (iv) and cancel the common terms on both sides of the equations. This gives

C (graphite) + $2H_2$ (g) → CH_4 (g) ΔH° (298K) = – **74.84 kJ**

Thus the enthalpy change for the formation of methane from its elements is calculated to be 74.84 kJ.

VARIOUS TYPES OF HEAT CHANGES

Enthalpy of Formation

The enthalpy change when one mole of a pure compound is formed from its elements is called the enthalpy of formation denoted as ΔH_f

When the reacting elements and the product formed are all in their standard state, the enthalpy change accompanying the chemical reaction is called the standard enthalpy of formation denoted as $\Delta H^o{}_f$. For example, the enthalpy of formation of CO_2 at 298 K and 1 atm is equal to the enthalpy change for the reaction C (graphite) + O_2 (g) → CO_2(g). This is so because the enthalpy of elements in their standard state is zero by convention.

C (graphite, 1 atm, 298 K) + O_2(g, 1 atm, 298K) → CO_2 (g, 1atm, 298K)

$\Delta H^o = -\ 393.5$ kJ mol^{-1}

$\Delta H^o = H^o\ (CO_2) - H^o\ (C) - H^o\ (O_2) = \Delta H^o{}_f\ (CO_2)$

Since the enthalpy of each element in standard state is zero; therefore, the standard enthalpy of CO_2 is equal to its standard heat of formation i.e.

H° (CO2) = DH°f (CO_2)

$\Delta H^o{}_f$ data for other compounds are given in the *Appendix*.

Heat of Combustion

The enthalpy or energy change when one mole of a substance is burnt in oxygen is defined as the heat of combustion (DH or DE), the product being water and carbon dioxide (usually).

C_2H_6 (g) + 7/2 O_2 (g) → $2CO_2$(g) + $3H_2O$(l) ΔH^o (298 K) = – 1560.1 kJ

HEAT OF HYDROGENATION

It is defined as the change in enthalpy accompanying the conversion of one mole of unsaturated organic compound into corresponding saturated compound by reducing the former with hydrogen e.g. when benzene is hydrogenated to cyclohexane, the change in enthalpy is represented by the following equation:

C_6H_6(l) + $3H_2$ (g) → C_6H_{12} (l) ΔH^o (298) = –205kJ

The enthalpy change when 1 mole of ethylene is 136.98 kJ at 298K

C_2H_4(g) + H_2(g) → C_2H_6(g) ΔH^o(298K) = – 136.98 kJ

Heat of Transition

It is defined as the amount of heat (DE or DH) absorbed or evolved when 1 mole of a substance is transformed from one allotropic form (state) to another allotropic form, e.g., when graphite form of carbon is converted to diamond form, 1.89 kJ of heat is absorbed.

C (graphite) → C (diamond) ΔH° (298K) = 1.89 kJ

Heat of Atomization

It is the amount of heat required to produce gaseous atoms from the stable element. In case of solids heat of atomization is equal to heat of sublimation, *e.g.* the heat of atomization of carbon is 716.68 kJ.

C (crystal, graphite) → C (gas) ΔH° (298K) = 716.68 KJ

and for magnesium,

Mg(c) → Mg (g) ΔH° (298K) = 150.2kJ

In case of homonuclear diatomic gases the heat of atomization is equal to half of heat of dissociation e.g. heat of dissociation of one mole of hydrogen is 435.94 kJ which gives 2 moles of atomic hydrogen. Thus the heat of atomization per mole hydrogen atom is 217.97 kJ

H_2 (g) → 2H (g) ΔH° (298K) = 435.94 kJ

1/2 H_2 (g) → H (g) ΔH° (298K) = 217.97 kJ

and similarly for oxygen

1/2 O_2 (g) → O(g) ΔH° (298K) = 249.2 kJ

Heat of reaction from the heats of formation of elements and compounds. The enthalpy change is heat of a reaction which can be calculated provided we know the enthalpies of reactants and products. It is a general convention to take the enthalpies of pure elements in their standard states as zero, and then enthalpy of pure compound is taken as its enthalpy of formation (ΔH°_f) from its elements in their standard state. Thus, the enthalpy change of a reaction ΔH° is given by

$$\Delta H^\circ = \Sigma H^\circ \text{ (products)} - \Sigma H^\circ \text{ (reactants)}$$

$$\text{or } \Delta H^\circ = \Sigma \Delta H^\circ_f \text{ (products)} - \Sigma \Delta H^\circ_f \text{ (reactants)}$$

ΔH°_f values of various compounds are tabulated for ready reference. This tabulation is of great help to compute the enthalpy of reactions. Instead of tabulating the enthalpy of millions of reactions it is convenient of report ΔH°_f values for a few thousands of more important compounds and then use the values to evaluate the heat of any desired reaction with the help of paper and pencil even without performing the actual experiment.

Example 6:

Calculate the enthalpy change for the following reaction at 25°C.

$$C_2H_6\ (g) + \frac{7}{2}O_2\ (g) \rightarrow 2CO_2(g) + 3H_2O(l)$$

Use the enthalpy of formation of the reactants and products from literature.

Solution:

Enthalpy change for any reaction is given as

$\Delta H^o = \Sigma H^o_f$ (products) $-\Sigma\Delta H^o_f$ (reactants)

For the given reaction

$\Delta H^o = 2\Delta H^o_f\ (CO_2) + 3\Delta H^o_f\ (H_2O) - (\Delta H^o_f\ C_2H_6 + 7/2\ \Delta H^o_f\ O_2)$

$= -2 \times 393.5 - 3 \times 285.8 - \left(-84.5 + \frac{7}{2} \times 0\right)$

$= -787.0 - 857.4 + 84.5$

$= -$ **1559.9 kJ**.

Example 7:

Calculate the heat of formation of benzene from the following data:

$$C_6H_6(l) + \frac{15}{2}\ O_2(g) \rightarrow 6CO_2(g) + 3H_2O(l)\quad \Delta H^o = -3303\ kJ$$

$\Delta H^o_f\ (CO_2) = -393.5$ kJ; $\Delta H^o_f\ (H_2O) = -285.8$ kJ.

Solution:

For the given reaction

$$\Delta H^o = 6 \times \Delta H^o_f\ (CO_2) + 3\ \Delta H^o_f\ (H_2O) - \Delta H^o_f\ (C_6H_6) - \frac{15}{2}\Delta H^o_f\ (O_2).$$

$\Delta H^o_f\ C_6H_6 = 6 \times \Delta H^o_f\ (CO_2) + 3 \times \Delta H^o_f\ (H_2O) - \Delta H^o$

$= -6 \times 393.5 - 3 \times 285.8 + 3303 =$ **84.6 kJ**.

Heat of Vaporization

The enthalpy change accompanying the conversion of one mole of liquid into vapour phase at a given temperature and pressure is called the heat of vaporization per mole of the liquid. For example the heat of vaporization per mole of water at 298 K and 1 atm is 44 kJ.

$H_2O(l) \rightarrow H_2O(g)$ DH° (298 K) = 44.0 kJ

Heat of Fusion

The enthalpy change when one mole of solid is converted to a liquid phase is called the heat of fusion. For example, the heat of fusion of water is 6 kJ mol^{-1}.

H_2O (S) $\rightarrow H_2O(1)$ ΔH (273 K) = 6.0 kJ

Heat of Sublimation

The heat change accompanied with the transformation of one mole of a volatile solid directly into vapour phase is termed as the heat of sublimation. Heat of sublimation is taken as the sum of the heat of fusion and the heat of vaporization, all measured at the same temperature.

$\Delta H_{sub} = \Delta H_{fus} + \Delta H_{vap}$

Heat of Solution

When solids or gases are dissolved in liquids or when different liquids are mixed, heat is evolved or absorbed. This thermal change is called the *heat of solution*. The heat of solution is usually reported per mole of the solute but for solutions of liquids in liquids the data are often tabulated as heat change per mole of the solution. The heat of solution depends upon the relative amounts of the substances being mixed and on whether they were initially taken as pure or one of them was introduced into a solution with a certain initial concentration at a given temperature and pressure.

Integral Heat of Solution

The integral heat of solution is the heat change when a known amount of pure solute is added to a known amount of pure solvent under conditions of constant temperature and pressure. The solution process may be represented by a chemical equation such as

HCl (g) + n aq $\rightarrow$ HCl.n aq

where aq represents the solvent, in this case H_2O, and n represents moles of solvent. The amount of heat evolved in the above reaction depends upon the magnitude of n shown in Fig. 3.2.

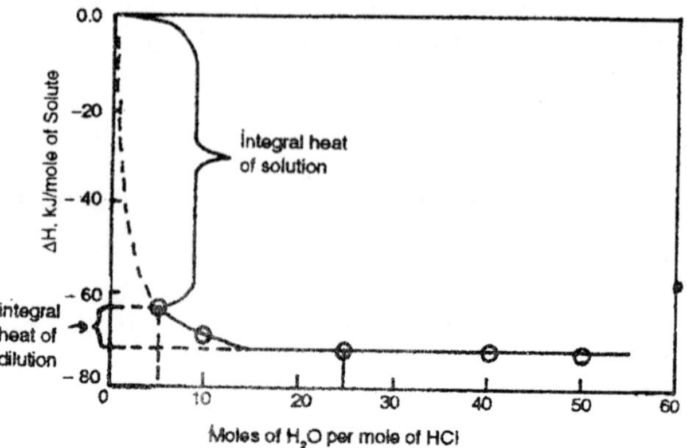

Fig. 3.2 : Variation of heat of solution with the mole of water per mole of HCl at 298 K.

(i) HCl (g) + $5H_2O(l) \rightarrow$ HCl.$5H_2O$ $\Delta H(298\ K) = -\ 63.995$ kJ

(ii) HCl (g) + $25H_2O(l) \rightarrow$ HCl. $25H_2O$ $\Delta H(298\ K) = -\ 72.177$ kJ

(iii) HCl (g) + aq $\rightarrow$ HCl (aq) $\Delta H(298\ K) = -\ 75.076$ kJ

where "HCl.$5H_2O$" represents a solution of 2 mole HCl in 5 moles of water. In each case the heat evolved per given mole of solvent added is an *integral heat of solution* for those conditions. It is the total heat evolved in carrying out that particular process. The term HCl (aq) represents an aqueous solution which is so dilute that further dilution produces no thermal effect. The corresponding Δ H value is called the heat of solution at **infinite dilution** and represents the upper limit of heat released when a substance is dissolved in a large excess of solvent. The integral heats of solution of HCl are plotted versus the number of moles of water per mole of HCl in Fig. 2.

Integral Heat of Dilution

It is the net change of enthalpy when a solution containing 1 mole of solute at a known concentration (C_1) is diluted by adding more solvent to get a solution of another known concentration (C_2). This quantity is in fact equal to the difference between the integral heats of solution at the two concentrations. For example, when 20 moles water are added to a solution HCl.$5H_2O$ then a solution HCl.$25H_2O$ is made, the accompanying enthalpy change of – 8.182 kJ is the *integral heat of dilution.*

(i) HCl (g) + $5H_2O(l) \rightarrow HCl.5H_2O$ ΔH (i) = – 63.995 kJ mol^{-1}

(ii) HCl (g) + 25 $H_2O(l) \rightarrow HCl.25H_2O$ ΔH (ii)=–72.177 kJ mol^{-1}

(iii) $HCl.5H_2O$ + 20 $H_2O \rightarrow HCl.25H_2O$

ΔH (dilution) = ΔH (ii) – Δ H (i) = – 8.182 kJ mol^{-1}

Differential Heat of Solution

It is the change in enthalpy when one mole of a solute is added to such a large amount of solution that the concentration of solution is essentially unchanged. This enthalpy change is also called the partial molar heat of solution.

Similarly, if one mole of a solvent is added to a larger volume of solution without altering the concentration, the accompanying enthalpy change is termed as the *differential* (partial molar) heat of dilution.

If Δ H is the enthalpy change when n_2 moles of a solute are mixed with n1 moles of solvent, then

$$\left(\frac{\partial \Delta H}{\partial n_2}\right)_{T.P.n_1} = \Delta H_{2,m} = \text{Differential heat of solution}$$

and $$\left(\frac{\partial \Delta H}{\partial n_1}\right)_{T.P.n_2} = \Delta H_{2,m} = \text{Differential heat of solution}$$

The total enthalpy change is given by

$$\Delta H_1 = n_1 \Delta H_{1,m} + n_2 \Delta H_{2,m}$$

If the amount of solvent is kept constant (say 1 kg = 55.55 moles H2O) and the amount of solute n_2 is varied, then the enthalpy change of solution as a function of molality ($m_2 = n_2/w_1 = n_2$, when w_1 = 1 kg) is given by

$$\Delta H = 55.55 \Delta H_1, m + m_2 \Delta H_{2,m}$$

The differential heat of solution depends on the concentration of the solution. It is impractical to measure the differential heat of solution directly, but it can be calculated from the integral heat of solution data. When Δ H value is plotted against the moles of solute m_2 (moles per kg of solvent), the slope of the tangent at a given concentration gives the differential heat of solution $\Delta H_{2,m}$.

Integral heat of solution and heat of dilution can also be calculated from tabulated values of heats of formation in solution. The enthalpy of formation of water is neglected in calculations if there is the same number of moles of water on both sides of the balanced chemical equations. Also, the enthalpy of formation of pure water is used for water aqueous solutions.

Example 8:

The heat of solution DH at different molalities m of an aqueous solution of a solute X are given below.

m	1.00	1.80	2.82	4.25	6.37	9.95
ΔH(298K)/kJ	–34.2	–56.7	–88.3	–126.0	–174.0	–229.0

Plot Δ H against the molality m, and determine the differential heat of solution of the solute X in a 3-molal solution.

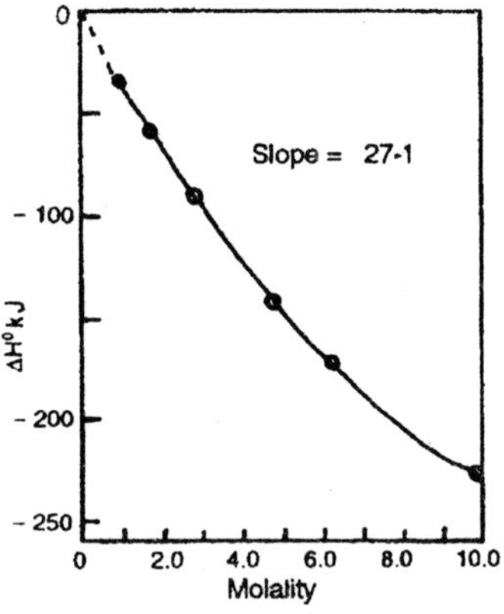

Fig. 3.3

Solution:

ΔH values are plotted against molality as shown in Fig. 3. Now draw a tangent on the curve at a point when molality is 3. Extend this tangent to meet the Δ H axis, and find out the slope of the tangent. It is –27.1kJ molal^{-1}. Since the variation of ΔH with molality (m_2) is given by

$$\Delta H = n_1\ \Delta H_{1\cdot\ m} + n_2\ \Delta H_{2\cdot}\ m = 55.55\ \Delta H_{1m} + m_2\ \Delta H_{2,\ m}$$

The slope of the tangent is equal to $\Delta H_{2\cdot\ m}$ *i.e.*, $\Delta H_{2\cdot\ m}$ = **–27.1kJ** molal^{-1}.

Example 9:

Calculate the integral heat of solution of one mole HCl(g) in 100 H2O (l), using the information that

$\Delta H^o_{f1(HCl)}$ = – 92.3kJ

$\Delta H^o_{f2(HCl \text{ in } 100 H2O)}$ = – 168.0 kJ

Solution:

The dissolution process can be represented by a chemical equation as

$HCl (g) + 100 H_2O \rightarrow HCl.100 H_2O$

$\Delta H^o = \Sigma\Delta H^o_f$ (products) – $\Sigma\Delta H^o_f$ (reactants)

Since, 100 H_2O is on both sides of the above equation, we will not consider the enthalpy of formation of water. Therefore,

ΔH^o = – 168.0 – (–92.3)

= **– 75.7 kJ.**

Example 10:

Calculate ΔH for the reaction

$HCl.100 H_2O + NaOH.100 H_2O \rightarrow NaCl.200 H_2O + H_2O(l)$

The information given as

(i) $HCl (g) + 100 H_2O \rightarrow HCl.100 H_2O$ ΔH (i) = –75.7 kJ

(ii) $NaOH (s) + 100 H_2O \rightarrow NaOH.100H_2O$ ΔH (ii) = –40.9 kJ

(iii) Nacl (s) + 200 H_2O › Nacl.200 H_2O ΔH (iii) – 4.26 kJ

ΔH_f^o's for HCl (g), NaOH (s), NaCl(s) and H_2O (l) are

–92.3, – 426.7, – 411.0 and – 285.8 kJ *respectively.*

Solution:

The given reaction is

(a) $HCl.100 H_2O + NaOH.100 H_2O \rightarrow NaCl.200 H_2O + H_2O(l)$

$\Delta H = \Sigma\Delta H^o_f$ (products) – $\Sigma\Delta H^o_f$ (reactants)

ΔH for this reaction can be calculated if the ΔH^o_f data for HCl in 100 H_2O; NaOH in 100 H_2O, NaCl in 200 H_2O and H_2O (l) are known.

From equation (i), we get

ΔH^o_f, ($HCl.100 H_2O$) = –75.7 –92.3 = – 168.0 kJ

Equation (iii) gives

ΔH^o_f, ($NaOH.100 H_2O$) = ΔH(ii) + ΔH^o_f, [NaOH(s)]

$= - 40.9 - 426.6 = - 467.5$ kJ

From equation (iii), we have

ΔH^o_f (NaCl.200H_2O) = + 4.26 – 411.0 – 406.74 kJ

Also ΔH^o_f H_2O = – 285.8 kJ

Using these data we have

$\Delta H(a) = (- 406.74 - 285.8) - (- 168.0 - 467.5)$

$= 692.54 + 635.50 = $ **–57.04 kJ.**

Heat of Neutralization

It is change in enthalpy when one mole H^+ ions are completely neutralized by a base in very dilute solutions. The dilute solution is necessary so that on mixing the acid and base the thermal change due to dilution is negligible. The heat of neutralization of hydrochloric acid and sodium hydroxide is –57.0 kJ mol–1 at 25°C and 1 atm pressure. In the case of weak acid and strong base the heat of neutralization is less than 57.0 kJ. The difference in heat of neutralization values of strong acid – strong base, and weak acid – strong base is attributed to the heat of ionization of weak acid.

Heat of neutralization of strong and strong base. Consider the neutralization of aqueous solution of HCl by NaOH. The reaction may be represented by

$H^+ (aq) + Cl^- (aq) + Na^+ (aq) + OH^- (aq)$

$\rightarrow Na^+ (aq) + Cl^- (aq) + H_2O$ (l)

Since Na^+ (aq) and Cl^- (aq) are common on both sides, the neutralization of strong acid and strong base may be represented by

H+ (aq) + OH^- (aq) $\rightarrow$ H_2O(1)

Thus heat of nutralization of strong acid and strong base (mono-) is simply the heat of formation of H_2O from H^+ and OH^- ions which is fixed at a given temperature and pressure. This leads to the conclusion that the heat of neutralization of strong monobasic acid with strong base is the same.

Heat of neutralization of a given acid-base pair can experimentally be found by using a calorimeter. The heat of ionization of a weak acid of a base can also be measured by calorimetric method. For example, the heat of ionization of a weak acid (acetic acid) will be equal to the heat of reaction between sodium acetate and hydrochloric acid at the a given temperature. This quantity will also be equal to the difference between the heat of neutralization of NaOH – HCl pair and acetic acid – NaOH pair. This is shown as follows:

(i) CH_3COOH *** $CH_3 \ COO^-$ + H^+ ΔH (ionization) = ?

(ii) NaOH + HCl → $Na^+ \ Cl^-$ + H_2O ΔH_1 = xkJ

(iii) NaOH + CH_3COOH → CH_3COONa + H_2O ΔH_2 = ykJ

Reverse equation (ii) to get (iv)

(iv) $Na^+ \ Cl^-$ + H_2O → NaOH + HCl ΔH_3 = – xkJ

Add (iii) and (iv)

CH_3COOH + Na^+ + Cl^- → CH_3COONa + HCl

→ $CH_3 \ COO^-$ + Na^+ + H^+ + Cl^-

or $CH_3 \ COOH$ ***CH_3COO^- + H^+ ΔH(ion) = $\Delta H_2 + \Delta H_3$

= (y – x) kJ

Example 11:

The heat of neutralization of a strong acid with a strong base is – 57.36 kJ mol⁻¹. If the heat of neutralization of a weak acid HA with a strong base be –42.02 kJ mol⁻¹ compute the heat of ionization of the weak acid.

Solution:

The neutralization of strong acid with strong base may essentially be represented as

H^+ (aq) + OH^- (aq) → $H_2O(l)$ ΔH = – 57.36 kJ mol^{-1}

The neutralization of a weak acid with a strong base may be considered to be the resultant of two successive steps.

HA (aq) → H+ (aq) + A^- (aq) ΔH = ΔH (ionization)

$$\frac{H+(aq) + OH^-(aq) \rightarrow H_2O(l)}{HA(aq) + OH^-(aq) \rightarrow H_2O(l) + A^-(aq)} \qquad \Delta H = -\ 57.36 \text{ kJ}$$

ΔH = 42.02 kJ

So that the enthalpy change is

ΔH (neutralization) = ΔH (ionization) – 57.36 kJ mol^{-1}

and ΔH (ionization) = – 42.02 + 57.36 = **15.34** kJ mol^{-1}.

Heat of Formation of Ions in Solution

The complete thermal data on aqueous solutions of electrolytes can be presented by tabulating heats of formation of ions at unit activity, which can be used to calculate the thermal changes for the reaction in which the ions participate. Since in aqueous solutions the H^+ and OH^- ions are involved in one way or the other, first we shall consider the computation of the standard heats of formation of these ions.

The heat of formation of one mole of water from hydrogen and hydroxyl ions is essentially equal to the heat of neutralization of strong acid and strong base (–57.0 kJ mol^{-1}) and the standard heat of formation of water from its elements H2 and O_2 is –285.8 kJ mol^{-1}. That is

$$H^+ (aq) + OH^- (aq) \rightarrow H_2O(l) \quad \Delta H^\circ (298K) = -57.0 \text{ kJ}$$

or (i) $H_2O (l) \rightarrow H^+ (aq) + OH^- (aq) \quad \Delta H^\circ (298K) = -57.0 \text{ kJ}$

and (ii) $H_2(g) + \frac{1}{2}O_2(g) \rightarrow H_2O(l) \quad \Delta H^\circ (298K) = -285.8 \text{ kJ}$

Adding these equations we get

(iii) $H_2 (g) + 1/2\, O_2(g) \rightarrow H^+ (aq) + OH^- (aq) \quad \Delta H^\circ (298K) = -228.8 \text{kJ}$

The sum of the heat of formation of H^+ and OH^- ions is – 228.8 kJ mol^{-1}. By convention the standard heat of formation of hydrogen ions in aqueous solution is taken as zero.

$$1/2\ H_2 (g) \rightarrow H^+ (aq) + e^- \quad \Delta H^\circ (298 K) = 0$$

With this convention equation (iii) directly gives the heat of formation of hydroxyl ions,

$$\frac{1}{2}H_2(g) + \frac{1}{2}O_2(g)\ e^- \rightarrow OH^-(aq) \quad \Delta H^\circ (298K\ ) = -\ 228.8 \text{ kJ}$$

The standard heat of formation of other ions is solution can be calculated in a similar manner.

Example 12:

The standard heat of formation of HCl in water at 25°C is –168.0kJ. Calculate the standard heat of formation of chloride ions.

Solution:

(i) $\frac{1}{2}H_2 (g) + Cl_2(g) + aq \rightarrow H^+ (aq) + Cl^- (aq)$

ΔH° (298K) = –168.0 kJ

(ii) $\frac{1}{2}H_2$ (g) → H^+ (aq) + e^- ΔH° (298 K) = 0

Subtracting (ii) from (i) we get

$\frac{1}{2}Cl_2$ (g) + aq + e^- → Cl^- (aq) ΔH° (298 K) = **–168.0kJ**.

Example 13:

Calculate the standard heat of formation of sodium ions in aqueous solution using the standard heat of formation of NaOH in aqueous medium (ΔH°_f = – 470.7kJ); ΔH°_f (OH^-) = – 228.8 kJ.

Solution:

Given data :

(i) Na(s) + $\frac{1}{2}$ H_2 (g) + $\frac{1}{2}O_2$ (g) + aq → Na^+ (aq) + OH^- (aq)

ΔH° = – 470.7 kJ

(ii) $\frac{1}{2}$ H_2(g) + $\frac{1}{2}O_2$(g) + e^- → OH^-(aq) ΔH° = –228.8kJ

Subtracting equation (ii) from (i) we get

Na(s) + (aq) + e^- → Na^+ (aq) + e^- ΔH° = **–241.9kJ**

BORN-HABAR CYCLE

When a solution is formed, two processes take place simultaneously. These are the disordering of the crystal lattice resulting into ion formation and the interaction of the ions with the molecules of the solvent. The heat of solution is effectively the combined result of these two processes– heat of solution = crystal energy + solvation energy. The two step process is represented as

MX (s) → M^+ (g) + X^- (g)

M^+ (g) + X^- (g) + aq → M^+ (aq) + X^- (aq)

The energy released when one mole of the solid crystal is formed from its constituent ions in gaseous state is termed as the lattice energy, which is equal in magnitude and opposite in sign to the energy required to transform one mole of the crystal to gaseous ions separated by infinite distance. This energy can be calculated using the electron affinity and ionization potential data for the atoms of the constituent elements of the crystal along with the standard heat of formation of the solid substance from its elements. This calculation is easy because all the quantities are taken as state functions. However, one simple way to calculate the lattice energy is to use the **Born-Haber** cycle for a solid MX.

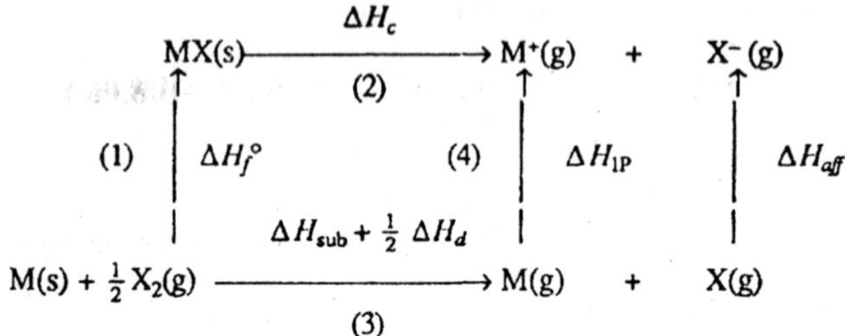

Fig. 3.4

The first step involves the formation of one mole of the solid MX from the elements, for which the heat of formation is $\Delta H^o{}_f$,

$$M(s) + \frac{1}{2}X_2(g) \rightarrow MX(s)\ \Delta H_f = \text{heat of formation}$$

The second step (2) is the dissociation of the lattice to form the gaseous ions M^+ (g) and X^- (g) for which the energy is $\Delta\ H_c$

$$MX(s) \rightarrow M^+\ (g)\ X^-\ (g)\ \Delta H_c = \text{crystal energy}$$

Step (3) involves the sublimation of M(s) to M(g) for which ΔHsub is the heat of sublimation, and the dissociation process of $1/2\ X_2 \rightarrow X$ (g) with the heat of dissociation as $1/2\ \Delta H_d$ (ΔH_d = heat of dissociation per mole of the substance). The total thermal effect for process (3) is

$$M\ (s) \rightarrow Mg\ (g)\ \Delta H_{sub} = \text{heat of sublimation}$$

$$1/2\ X_2\ (g) \rightarrow X\ (g)\ 1/2\ \Delta H_d = \text{heat of dissociation per mole atom}$$

Step (4) involves the ionization of M(g) to M^+ (g) by supplying ionization energy ΔH_{IP} and the acceptance of electron as X (g) + e^- → X^- with electron affinity A or ΔH_{aff}

$$M\ (g) \rightarrow M^+\ (g) + e^-\ \Delta H_{IP} = \text{Ionization potential}$$

$$X\ (g)\ e^- \rightarrow X^-\ (g)\ \Delta H_{aff} = \text{Electron affinity}$$

Since ΔH is a state function, therefore the thermal changes for step (1) plus step (2) must be equal to those for step (3) and step (4). That is

$$\Delta H^o_f + \Delta H_c = \Delta H_{sub} + 1/2\Delta H_d + \Delta H_{IP} + \Delta H_{aff}$$

or $\Delta H_c = \Delta H_{sub} + 1/2\Delta H_d + \Delta H_{IP} + \Delta H_{aff-} \Delta H^o_f$ ...(1a)

or $\Delta H^o_f = \Delta H_{sub} + 1/2\ \Delta H_d + \Delta H_{IP} + \Delta H_{aff-} \Delta H^o_C$

(i) The lattice energy or the crystal energy can be calculated by using equation (1) on putting the numerical values with the appropriate signs of the various quantities.

(ii) Any one quantity can be calculated if others are known. The use of this equation along with the heat of solution data is discussed below to calculate the heat of hydration for some ions.

Example 14:

Examine the reactions tabulated below with the accompanying change in enthalpy :

Process	*Reaction*	*Enthalpy changes (kJ)*
Sublimation :	(i) Li (s) $\rightarrow$ Li (g)	$\Delta H_{sub} = +161$
	(ii) Na (s) $\rightarrow$ Na (g)	$\Delta H_{sub} = +109$
Dissociation :	1/2 Cl (g) $\rightarrow$ Cl (g)	$1/2\Delta H_d = +122$
Electron-affinity :	Cl (g) + $e^- \rightarrow Cl^-$ (g)	$\Delta H_{aff} = -350$
Ionization :	(i) Li (g) $\rightarrow Li^+ + e^-$	$\chi H_{IP} = +520$
	(ii) Na (g) $\rightarrow Na^+ + e^-$	$\Delta H_{IP} = +496$
Compound	(i) Li (s) + $\frac{1}{2}Cl_2$ (g) $\rightarrow$ LiCl (s)	$\Delta H^o_f = -410$
formation :	(ii) Na (s) + $\frac{1}{2}Cl_2$(g) $\rightarrow$ NaCl(s)	$\Delta H^o_f = -411$

(a) *Prepare Born-Haber cycles for the separate LiCl and NaCl and calculate their respective lattice energies.*

(b) *Using the following heats of solutions of LiCl and NaCl in water along with the lattice energy data obtained in part* (a), *calculate the heat of hydration of* Li^+ *and* Na^- *ions separately and interpret the results:*

(i) LiCl (s) + aq $\rightarrow Li^+$ (aq) + Cl– (aq) + 35.1 kJ

(ii) NaCl (s) + aq $\rightarrow Na^+$ (aq) + Cl^- 4.3 kJ

Solution:

(a) Born-Haber cyclic is prepared as discussed for general case MX.

For lithium chloride: $LiCl\ (s) \rightarrow Li^+\ (g) + Cl^-\ (g)$

$\Delta H_C = 161 + 122 + 520 + 410$

$=$ **863 kJ**

For sodium chloride : $NaCl\ (s) \rightarrow Na^+\ (g) + Cl^+\ (g)$

$\Delta H_C = 109 + 122 + 496 - 350 + 411$

$= 788$ kJ

(b) Heat of hydration = Heat of solution – Crystal energy

Solution process : For lithium chloride

$LiCl\ (s) + aq \rightarrow Li^+\ (aq) + Cl^-\ (aq)\ \Delta H = -\ 35.1$ kJ

Crystal formation process:

$Li^+\ (g) + Cl^-\ (g) \rightarrow LiCl(s)\ \Delta H = -\ 35.1$ kJ

Adding the two equations, we get

$Li^+\ (g) + Cl^-\ (g) \xrightarrow{aq} Li\ (aq) + Cl^-\ (aq)\ \Delta\ H_{hydration} = -\ 898.1$ kJ

Similarly for NaCl

$Na^+\ (g) + Cl^-\ (g) \xrightarrow{aq} Na+\ (aq) + Cl^-\ (aq)\ \Delta H_{hydration} = -$ **783.7kJ**

The interaction of gaseous ions with water is called *hydration* and accompanying enthalpy change is called the hydration energy. The hydration energy for lithium ion (radius = 0.060 nm) is larger compared with the hydration energy of Na^+ (radius = 0.095 nm). This suggests that the interaction between Li^+ ion and water dipole is greater (ion– dipole interaction) than the ion – dipole interaction possibility between Na^+ ions and water molecules. This is obvious because there is a size difference between Na^+ ions and Li^+ ions so that the water dipoles are father away from the centre of the sodium ions but nearer to the Li^+ ions, whereas the net ionic charges are the same in both the cases. Therefore water molecules are held less strongly around the sodium ions than around the lithium ions.

The difference in the hydration energies of Li^+ ions and Na^+ ions is about 114 kJ mol^{-1}. However, the crystal energies differ only by about 75 mol^{-1}. The fact that lithium chloride dissolves with the evolution of heat (–35.1kJ) and sodium chloride dissolves with absorption of heat (4.3kJ) must be due to some difference between the lithium ions and sodium ions. This difference is attributed to the fact that the interactions of lithium ions with water produce more heat than the interaction of lithium ions with chloride ions (ion–dipole > ion –ion interaction) whereas the interaction of sodium ions with either water or chloride ions produce about equal enthalpy change.

BOND ENERGIES

Whenever a chemical bond is formed energy is released. Conversely energy is required to rupture a chemical bond. For example, when two gaseous hydrogen atoms combine a molecule of H_2 is formed, and 436 kJ energy is produced per mole of hydrogen molecules. To separate the hydrogen atoms of a hydrogen molecule, therefore, 436 kJ energy per mole must be supplied so that the bonds are broken. This energy is referred to as bond energy. Since the experiments are conveniently carried out at constant pressure the bond energy is commonly called as bond enthalpy. In general, the bond enthalpy of a bond A–B in a gaseous molecule AB is equal to the enthalpy of the reaction represented by

A–B (g) ® A(g) + B (g) DH° (298 K) = DH° (A–B)=x kJ mol–1

Here DH° (A–B) is the bond enthalpy of A–B bond. D gives the flavour of dissociation. The concept of bond energy for diatomic molecules is simpler but the difficulties arise when we consider polyatomic molecules. So it is essential to make a distinction between **bond energy and bond dissociation energy** of a given linkage.

Bond Dissociation Energy. The energy required to break a given bond in some specific compound is defined as the bond dissociation energy. It is a definite quantity. It depends on the presence of other groups in a molecule. For example, the bond dissociation energies of the two O–H linkages in water molecule are different. The energy required to break O–H bond in water is 498 kJ mol^{-1}.

$$H\text{–}O\text{–}H(g) \rightarrow H(g) + OH(g) \quad \Delta H^\circ(298K) = 498 \text{ kJ mol}^{-1}$$

and the energy required to break the O –H bond in hydroxyl group is 430 kJ mol^{-1}

$$O\text{–}H(g) \rightarrow O(g) + H(g) \quad \Delta H^\circ(298K) = 430 \text{kJ mol}^{-1}$$

Thus, the bond dissociation energy of O –H linkage in water molecule is 498kJ whereas that in a hydroxyl group is 430 kJ.

Bond energy. The average of dissociation energies of a given bond in a series of different dissociating species, is termed as bond energy.

The O–H bond energy in water molecule is the average of the two bond dissociation energies, in other words

$$DH\ (O-H) = \frac{498 + 430}{2} = 464\ kJ\ mol^{-1}$$

In case of diatomic molecules the bond energy and bond dissociation energies are identical.

Calculation of Bond

The main sources of data for the determination of bond energies are molecular spectroscopy, electron impact studies, and thermochemistry. Here we shall consider only thermochemical methods.

C–H bond energy in methane. Since all the four C–H bonds in methane are equivalent the average of the bond dissociation energies would be equal to the C–H bond energy.

Our aim is to calculate DH°/4 for the reaction CH4(g) ® C(g) + 4H(g) Data given :

(i) $C\ (graphite) \xrightarrow{\text{Sublimation}} C(g)$ $\quad \Delta H^{\circ}\ (i) = 716.68\ kJ$

(ii) $\frac{1}{2} H_2\ (g) \xrightarrow{\text{Dissociation}} H(g)$ $\quad \Delta H^{\circ}\ (ii) = 217.97\ kJ$

(iii) $C\ (graphite) + 2H_2\ (g) \rightarrow CH_4(g)$ $\Delta H^{\circ}\ (iii) = 74.75\ kJ$

Multiply equation (ii) by four. reverse equation (iii). combine these two new equations with equation (i)

(iv) $2H_2\ (g) \rightarrow 4H(g)$ $\quad \Delta H^{\circ}\ (iv) = 871.88\ kJ$

(v) $CH_4\ (g) \rightarrow C\ (graphite) + 4\ H\ (g)$ $\Delta H^{\circ}\ (v) = 74.75\ kJ$

(i) $C\ (graphite) \rightarrow C\ (g)$ $\quad \Delta H^{\circ}\ (i) = 716.68\ kJ$

(vi) $CH_4\ (g) \rightarrow C\ (g) + 4\ H(g)$ $\quad \Delta H^{\circ}\ (vi) = 1663.31\ kJ$

The quantity, 1663.31 kJ. is known as the heat of atomization of methane. This much heat must be supplied to convert a mole of methane into gaseous atoms. This is also known as the **bond dissociation** energies of four equivalent bonds m methane. Thus the C–H bond energy is given by

$$DH^{\circ}\ (C-H) = \frac{1663.31 kJ\ mol^{-1}}{4} = 415.82\ mol^{-1}$$

The actual value is 413 kJ.

Similarly, C–C, C = C, C ≡ C bond energies can be calculated by using the heats of formation of ethane, ethylene and acetylene respectively.

Example 15:

Using the data :

$\Delta H^o_f\ (C_2H_6) = -\ 84.5$ kJ

$\Delta H^o_f\ (C_2H_4) = 52.6$ kJ

$\Delta H^o_f\ (C_2H_2) = 226.9$ kJ

$\Delta H^o_f\ (CO_3CHO) = -\ 166.3$ kJ

$\Delta H^o_f\ (H_2O,g) = -\ 241.8$ kJ

$\Delta H^o_f\ (CH_3OH,g) = -\ 201.3$ kJ

$\frac{1}{2}\ DH^o(O_2) = 249.17$ kJ

$\frac{1}{2}DH^o\ (H - H) = 217.97$ kJ

ΔH^oC (graphite) $\rightarrow$ C(g) = 716.68 kJ

calculate the following bond energies:

(i) DH^o (C –C), (ii) DH^o(C = C),

(iii) DH^o (C ≡ C), (iv) DH^o (O –H)

(v) DH^o (C = O) (vi) DH^o (C –O)

Solution:

C–C bond energy. The energy of a C–C bond can be calculated by applying the method of *previous* section to ethane. Our aim is

$C_2H_6(g) \rightarrow 2C(g) + 6H(g)\ \Delta H^o = ?$

Given data

(i) C_2H_6 (g) $\rightarrow$ 2C (graphite) + $3H_2$ (g) $\Delta H^o = 84.5$ kJ

(ii) 2C (graphite) $\rightarrow$ 2C (g) $\Delta H^o = 1433.4$ kJ

(iii) $3H_2$ (g) $\rightarrow$ 6H (g) $\Delta H^o = 1307.8$ kJ

Add-------------------------------------

C_2H_6 (g) $\rightarrow$ 2C(g) + 6 H(g) $\Delta H^o = 2825.7$ kJ

Thus 2825.7 kJ is required to break 6C – H bonds and one C–C bond in ethane. The C–C bond energy is equal to 2825.7 kJ less 6 × DH^o (C –H)

$DH^o (C - C) = \Delta H^o - 6 \times DH^o (C-H)$

$= 2825.7 - 6 \times 413 = 347.7$ kJ

C = C bond energy. The of C = C bond can be calculated using the standard enthalpy of formation of ethylene and C– H bond energy data. Here our object would be to study :

$C_2H_4 (g) \rightarrow 2C (g) + 4H (g)$ $\Delta H^o = ?$

Given data

$C_2H_4 (g) \rightarrow 2C$ (graphite)+ $2H_2 (g)$ $\Delta H^o = - 52.6$ kJ

2C (graphite) $\rightarrow$ 2C (g) $\Delta H^o = 1433.4$ kJ

$2H_2(g) \rightarrow 4H(g)$ $\Delta H^o = 871.9$ kJ

Add ------- ---------

$C_2H_4(g) \rightarrow 2C(g) + 4H(g)$ $\Delta H^o = 2252.7$ kJ

Less 4C – H bonds $\Delta H^o = 1652.0$ kJ

-------- --------

C = C bond $DH^o (C = C) =$ **600.7 kJ**

The literature value is slightly greater than this value.

C ° C *bond energy*. Since we already know the C– H bond energy, we can obtain the energy of a C ° C bond by using the ΔH^o_f acetylene and the previous C – H bond energy

$C_2H_2 (g) \rightarrow 2C$ (graphite) + $H_2 (g)$ $\Delta H^o = - 226.9$ kJ

2C (graphite) $\rightarrow$ 2C (g) $\Delta H^o = 1433.4$ kJ

$H_2 (g) \rightarrow 2H (g)$ $\Delta H^o = 435.94$ kJ

Add ---------------- -----------

$C_2H_2 (g) \rightarrow 2C (g) + 2H(g)$ $\Delta H^o = 1642.44$ kJ

Less 2C – H bonds $2 \times DH^o (C- H) = 826.0$ kJ

-------- --------

$C \equiv C$ bond $\Delta H^o (C \equiv C)$ = **816.44 kJ**

C = 0 bond energy is calculated using the standard heat of formation of acetaldehyde and the C– C and C – H bond energies. This gives C = O bond energy as 740 kJ mol^{-1}.

The O – H bond energy

H_2O (g) $\rightarrow \frac{1}{2} O_2$ (g) + H_2(g) ΔH^o = 241.80 kJ

H_2 (g) $\rightarrow$ 2H (g) ΔH^o = 435.94 kJ

$\frac{1}{2} O_2$ (g) $\rightarrow$ O (g) ΔH^o = 249.17kJ

Add ------- ----------

H_2O (g) $\rightarrow$ 2H (g) + O (g) ΔH^o = **926.91 kJ**

The quantity, 926.91 kJ is the enthalpy change when the two O – H bonds in H_2O are broken. Therefore, the O – H bond energy is 926.91 kJ/2 = **463.4 kJ**.

Finally, by taking methanol, in which the only unknown bond energy is that of the C–O bond we find that the C– O bond energy is of the order of 351 kJ.

DH° values for different bonds are given in the Appendix.

In the calculation of bond energies we have assumed that one type of bond energy calculated for a particular molecule is the same in all other types of molecules. For example, C–H bond energy in methane, ethane, propane etc., will have the same value. Similarly C–C bond energy for ethane is applicable for C–C bond energy in other molecules too.

However, this procedure has got some limitations, e.g., butane exists in two forms as n-butane and iso-butane. Both compounds have three C–C bonds and ten C– H bonds, and from simple bond theory the heats of formation should be the same for both compounds.

Bond energy data = (3C – C + 10 C–H)

= 3 × 348 + 10 × 413 = 5174. 0 kJ

Heat of atomization = Heat of atomization for (4C + 10H)

= 4 × 716.7 + 10 × 217.97 = 5046.5 kJ

Standard heats of formation of butane from its elements:

ΔH^o_f = (Heats of atomization) – (Bond energies)

= 5046.5 – 5174. 0 = – 127.5 kJ

But the observed ΔH^of for n-butane is – 124.3 kJ and that for isobutane is – 131.2 kJ. There is a difference of about 3 kJ in calculated and observed data. This difference becomes more and more as we consider the still higher aliphatic hydrocarbons, *e.g.*, there is a difference of almost 16 kJ in the standard heats of formations of n-pentane and 2-2 dimethyl propane, although both 4 C – C bonds and 12 C – H bonds.

Apart from the limitations of the bond energy calculations there are certain merits of the procedure in the sense that it puts restrictions on writing incorrect structures of molecules and suggests a suitable form of representing the molecular structures. We shall discuss the merits of thermodynamic calculations by considering the case of benzene.

Structure of Benzene – Resonating structure – Delocalization concept:

We can calculate the standard heats of formation (DH°f) of benzene in the same manner as that for the simpler hydrocarbons. Usually we write the structure of benzene as a ring with three C– C bonds, three C– C bonds and 6 C – H bonds as

Bond energy data

$= 6 \times DH^{o}\ (C-H) + 3 \times DH^{o}\ (C-C) + 3 \times DH^{o}\ (C=C)$

$= (6 \times 413 + 3 \times 348 + 3 \times 610) = 5352\ mol^{-1}$

Heat of atomization

$= 6 \times \Delta H\ [C\ (graphite) \rightarrow C\ (g)] + 6 \times \Delta H\ (1/2\ H_2 \rightarrow H)$

$= 6 \times 716.7\ kJ\ mol^{-1} + 6 \times 217.96\ kJ\ mol^{-1}$

$= 5608\ kJ\ mol^{-1}$

$\Delta H^{o}f = 5608 - 5352 = 256\ mol^{-1}$

The observed ΔH^{o}_{f} is only 82.9 kJ mol^{-1}. This shows that benzene molecule is 173 kJ mol^{-1} more stable than predicated by the considered structure. This unexpected stability suggests that the Kekule structure is not a true representative of molecular structure. The real state can be described by a resonance hybrid of the two extreme structures.

The two extreme forms are called "resonance" structures and the difference of the energy of approximately 173 kJ mol–1 is referred to as the resonance stabilization energy. This suggests that the C– C bonds in the benzene ring are neither pure C – C bonds nor pure C = C bonds. All are equivalent and intermediate between single and double bonds. This is shown from C to C bond energy considerations.

Let us suppose that carbon to carbon bonds in the benzene ring are neither pure single nor pure double bonds. Using this approach we can calculate the bond energy of the C linkages.

$C_6H_6 \rightarrow 6C \text{ (graphite)} + 3H_2 \text{ (g)} \qquad \Delta H = -\Delta H^\circ_f = -82.9 \text{ kJ}$

$6C \text{ (graphite)} \rightarrow 6C \text{ (g)} \qquad \Delta H = 4300.2 \text{ kJ}$

$3H_2 \text{ (g)} \rightarrow 6H \text{ (g)} \qquad \Delta H = 1307.8 \text{ kJ}$

---------- ---------

$C_6H_6 \rightarrow 6C \text{ (g)} + 6H\text{(g)} \qquad \Delta H = 5525.1 \text{ kJ}$

Thus, the energy required to break six C – H bonds and six C to C bond is 5525.1kJ. But the energy required to break six C– H bonds is 2478 kJ. Therefore, the energy of six C to C linkage would be equal to the difference of the two, i.e., (5525.1 –2478.0) = 3047.1 kJ. Each C to C bond in the ring can be assigned a value of 507.8 kJ. Since the energy of pure C – C bond is 348 kJ and for a pure C = C bond the energy is 610 kJ. This gives an idea that the C to C bonds in benzene have a bond order intermediate between 1 and 2. The bonds between carbon are not localized as double bonds in three places but are spread among all six C– C distances. The electrons involved are delocalized. The extra stability of the molecule arises from the delocalization of the π-electrons in the ring.

Bond energy data are useful in deciding the nature of bonds in a group with multiple bonds. For example, the comparison of bond energies of C – C and C = C shows that both the bonds in C = C are not identical. If this were so then the C = C bond energy must be just double of the C – C bond energy. However, the valency considerations show that C–C bond is a stable bond of σ-type. Of the two bonds in C = C linkage one is s-type and the other is a weaker p-linkage. This is in line with the thermodynamic calculations which gives a difference of 262 kJ mole–1 between C = C bond and C – C bond energies. This suggests that the bond other than σ-type in C = C bond is weaker. Similarly in C ≡ C linkage the nature of all the three bonds are not identical, one being s-type and the two are π-type bonds.

Example 16:

Using the bond energy data, compute ΔH°_f for ethyl alcohol (C_2H_5OH). DH° (C – H) = 413 kJ, D H° (C – C) = 348 kJ, DH° (C – O) = 351 kJ, DH° (O – H) =463kJ.

Solution :

Ethyl alcohol can be represented by the structure

```
      H     H
      |     |
H— C —  C — O — H
      |     |
      H     H
```

Total bond enthalpy = 5 × DH° (C – H) + DH° (C–C)+DH°(C–O) + DH°(O –H)

= 5 × (413 kJ) + 348 kJ + 351 kJ + 363 kJ = 3227 kJ

when 3227 kJ is supplied then 2C (g), 6H (g) and O (g) will be the product given by C_2H_5OH.

Total heat of atomization = 6H (g) + 2C (g) + O (g)

= 6 × 21.797 + 2 × 716.68 + 249.2 = 2990.38 kJ

$\Delta H^{\circ}f$ (C_2H_5OH) = Enthalpy of atomization – Bond enthalpy

= (2990.38 kJ) – (3227.0 kJ) = – **236.62 kJ**

Heats of Reaction from Bond Energy Data

Bond energy data are used to calculate the heats of gaseous reactions involving substances having only covalent bonds. In using these data, a plus sign in affixed to the bond enthalpies when a bond is broken, since heat is absorbed under such conditions and a minus sign when a bond is formed and heat is evolved.

Example 17:

Using bond energy data calculate the enthalpy change for the reaction:

C_2H_4 (g) + H_2 (g) → C_2H_6 (g)

Solution:

The given reaction can be written as

```
H   H          H   H
|   |          |   |
C = C + H – H → H – C – C – H
|   |          |   |
H   H          H   H
```

In this reaction the four C – H bonds in C_2H_4 are unaffected and a double bond is broken ; in H_2 one H – H bond is broken. In turn, two new C – H bonds and one C – C bond are formed in ethane.

ΔH^o_f = Bond enthalpy in – Bond enthalpy out

= DH° (C = C) + DH° (H – H) 2 × D H° (C – H) – DH° (C – C)

= (610 kJ mol^{-1}) + (436 kJ mol^{-1}) –2 × (413 kJ mol^{-1})

– (348 kJ mol^{-1})

= **–128** kJ mol^{-1}.

Example 18:

Calculate the standard enthalpy of formation of dimethyl ether by using the bond energy data from the appendix and the enthalpy of C (graphite) → C (gas) as 716.7 kJ.

Solution:

The formation of dimethylether may be represented by the equation

(A) 2C (graphite) + $3H_2$(g) + 1/2 O_2 (g) → $H_3C–O–CH_3$ ΔH_f = ?

From bond energy data we get the enthalpy for the following reaction:

(i) H_3C – O – CH_3 (g) → 2C (g) + 6H (g) + O(g)

ΔH = 6DH° (C – H) + 2D H° (C – O)

∴ ΔH = 6 × (413 kJ mol^{-1}) + 2 × (350 kJ mol^{-1}) = 3178 kJ

(ii) $3H_2$ (g) → 2H (g) ΔH = 1307.64 kJ

(iii) 1/2 O_2 (g) → O (g) ΔH = 249.17 kJ

(iv) 2C (graphite) → 2C (g) ΔH = 1433.4 kJ

Reversing eqn.. (i) we get

(v) 2C (g) + 6H (g) + O(g) → $H_3C–O–CH_3$ (g) ΔH = – 3178 kJ

Adding equations (ii), (iii), (iv) and (v), and cancelling the common terms we get

(iv) 2C (graphite) + $3H_2$ (g) + 1/2 O_2 (g) → H_3 C – O – CH_3 (g)

ΔH (vi) = ΔH (ii) + ΔH (iii) + ΔH (iv) + ΔH (v)

= – 189.79 kJ = DH (A).

VARIATION OF HEATS OF REACTION WITH TEMPERATURE

Heats of reaction at a given temperature (25°C) and at a given pressure (1 atm) can be calculated using the standard heats of formation given in the International tables. Sometimes it is required to find the heat of reaction at other temperatures and pressures. The variation of heat of reaction with temperature at constant pressure is given by the **Kirchooff's Law**, which can be derived in the following way :

Let us consider the general reaction

$aA + bB \rightarrow mM + nN$

Now the heat of reaction is given by

$\Delta H = (mH_M + nH_N) - (aH_A + bH_B)$

where H_A and H_B are the molar enthalpies of reactants and H_M and H_N and that of the products respectively. Differentiating equation (2) with respect to T at constant P, we get the variation of enthalpy with temperature,

$$\left(\frac{\partial \Delta H}{\partial T}\right)_P = \left[m\left(\frac{\partial H_M}{\partial T}\right)_P + n\left(\frac{\partial H_N}{\partial T}\right)_P\right] - \left[a\left(\frac{\partial H_A}{\partial T}\right)_P + b\left(\frac{\partial H_B}{\partial T}\right)_P\right]$$

$= [MC_{P,m}(M) + n\,C_{P,}\,m\,(N)] - [a\,C_{P,m}(A) + b\,C_{P,m}(B)]$

$= \Sigma\, C_P \text{ (products)} - \Sigma\, C_P \text{ (reactants)}$

$= \Delta C_P$

or $d(\Delta H) = \Delta C_P .\, dT$

Here $C_{P,m} = (\partial H/\partial T)_P$ = molar heat capacity at constant pressure and the quantity ΔC_P is the change in heat capacity in the chemical reaction

$\Delta C_P = [mC_{P.m}(M) + nC_{P.m}(N)] - [aC_{P.m}(A) + bC_{P.m}(B)]$

The relationship given by equation (1) is known as Kirchhoff's Law. Using this law we can calculate the enthalpy changes at any temperature.

If we have to calculate the heat of reaction at constant volume (D E) then we can derive the Kirchhoff's equation as,

$$\left(\frac{\partial(\Delta E)}{\partial T}\right)_V = DC_V \qquad ...(1)$$

or $\quad d(\Delta E) = \Delta C_V dT \qquad ...(2)$

For small range of temperature, equations (1) and (2) can be integrated taking heat capacities as independent of temperature.

$$\int_1^2 d(\Delta H) = \Delta C_P \int_1^2 dT$$

$\Delta H (T_2) - \Delta H (T_1) = \Delta C_P (T_2 - T_1)$ ...(3)

and $\Delta E (T_2) - \Delta E (T_1) = \Delta C_V (T_2 - T_1)$ ...(4)

In a reaction if ΔC_P is zero then ΔH is independent of temperature.

Example 19:

The heat of dissociation per mole of gaseous water at 18°C and 1atm is 241750 J ; calculate its value at 68°C. Data given are

$C_P (H_2O) = 33.56$; $C_P (H_2) = 28.83$; $C_P (O_2) = 29.12 J K^{-1} mol^{-1}$

Solution:

The dissociation reaction is.

$H_2O (g) \rightarrow H_2 (g) + 1/2\ O_2 (g)\ \Delta H^o (291\ K) = 241750\ J$

$\Delta C_P = C_P (H_2) + 1/2\ C_P (O_2) - C_P (H_2O)$

$= (28.83 + 1/2 \times 29.12 - 33.56)\ JK^{-1}\ mol^{-1}$

$= 9.83 J\ K^{-1}\ mol^{-1}$

$\therefore \Delta C_P \times \Delta T = (9.83\ JK^{-1}\ mol^{-1}) \times (50K)$

$= 491.5\ J\ mol^{-1}$

$\Delta H^o (341K) = \Delta H^o (291K) + \Delta C_P (\Delta T)$

$= 241750\ J\ mol^{-1} + 491.5\ J\ mol^{-1}$

$= \mathbf{242241.5\ J\ mol^{-1}}$

For precise calculations we must express the heat capacities of reactants and products as a function of temperature, and then integrate equation (3) $C_{P,m}$ can be expressed as a power series of T in the form

$C_{P,m} = \alpha + \beta T + \gamma T^2$

where α, β, γ are constants for a given substance.

$\therefore \Delta C_P = (m\alpha_M + n\alpha_N) - (a\alpha_A + b\alpha_B) + [(m\ \beta_M + n\beta_N)$

$- (a\beta_A + b\beta_B)\ T + ... + ...$

or $\Delta C_P = \Delta \alpha + \Delta \beta T + \Delta \gamma T^2$

Equation (3) can be written as

$d (\Delta H) = (\Delta\alpha + \Delta\beta T + \Delta\gamma T^2)\ dT$ ...(1)

On integration this equation yields

$$\int d(\Delta H) = \int (\Delta\alpha + \Delta\beta T + \Delta\gamma T^2)\, dT + \text{constant (I)}$$

$$\text{or } \Delta H = \Delta \alpha T + \frac{1}{2}\Delta\beta T^2 + \frac{1}{3}\Delta\gamma T^3 + \text{constant (I)}$$

If T = 0 then

$$\Delta H = \Delta H(T = 0) = \Delta H_0 = \text{constant (I)}$$

$$\therefore \Delta H = \Delta H_0 + \Delta\alpha T + \frac{1}{2}\Delta\beta T^2 + \frac{1}{3}\Delta\gamma T^3$$

ΔH_0 is the hypothetical heat of reaction at absolute zero. However, equation (1) can be integrated for suitable range of temperatures T_1 and T_2 to give

$$\int_{T_1}^{T_2} d(\Delta H) = \int_{T_1}^{T_2} (\Delta\alpha + \Delta\beta T + \Delta\gamma T^2)dT$$

$$\text{or} \quad \Delta H(T_2) - \Delta H(T_1) = \Delta\alpha (T_2 - T_1) + 1/2\, \Delta\beta(T_2^2 - T_1^2) + \frac{1}{3}\Delta\gamma (T_2^3 - T_1^3) \quad ...(2)$$

Equation (2) is used to calculate precisely the enthalpy of reaction at any temperature T_2 provided its value at a temperature T_1 is known. The use of equation (1) or (2) is not restricted only to the chemical reactions. It can suitably be used for the physical transformations as well. For example, in a phase change (vaporization, fusion, sublimation etc). from a phase a to another phase β, the Kirchhoff's equation can be written as

$$d(\Delta H_{\alpha \to \beta}) = CP_{\alpha \to \beta}\, dT$$

$$\text{or } \Delta H(T_2) = \Delta H(T_1) + \Delta C_P (T_2 - T_1)$$

where $\Delta H = H\beta - H\alpha$; $\Delta C_P = C_P(b) - C_P(\alpha)$.

VARIATION OF HEAT OF REACTION WITH PRESSURE

Standard heats of formation give values of heat of reaction at a pressure of 1 atm. Sometimes it is necessary to find heats of reaction at other pressures. For this we will make use of the thermodynamic equation of state.

$$\left(\frac{\partial H}{\partial P}\right)_T = V - T\left(\frac{\partial V}{\partial T}\right)_P$$

so that the pressure dependence of the heat of reaction is given by

$$\left(\frac{\partial \Delta H}{\partial P}\right)_P = DV - T\left(\frac{\partial \Delta V}{\partial T}\right)_P \quad ...(1)$$

where DV is the volume change of the reaction at pressure P.

$\Delta V = \Sigma$ V (products) $- \Sigma$V (reaction)

Differentiation w.r. to T at constant P yields,

$$\left(\frac{\partial \Delta V}{\partial T}\right)_P = \sum\left[\frac{\partial V(\text{products})}{\partial T}\right]_P - \sum\left[\frac{\partial V(\text{reactants})}{\partial T}\right]_P$$

The coefficient of thermal expansion is defined as $\alpha = V^{-1} (\partial V/\partial T)_P$

$$\therefore \left(\frac{\partial \Delta V}{\partial T}\right)_P = \Sigma\ \alpha V\ (\text{products}) - \Sigma\ \alpha V\ (\text{reactants})$$

$$= \Delta(\alpha V) \quad ...(2)$$

From equation (1) and (2), we get

$$\left(\frac{\partial \Delta H}{\partial P}\right)_T = \Delta V - T\Delta\ (\alpha V) \quad ...(3)$$

If the equations of state of the reactants and products are known ΔV and $\Delta\ (\alpha V)$ will be known functions of the pressures, and equation (3) can be integrated to give

$$\Delta H(P_2) = \Delta H(P_1) + \int_{P_1}^{P_2} [\Delta V - T\Delta(V\alpha)]dP \quad ...(4)$$

However, if the reactants and products involved in the chemical reaction are ideal gases then the heat of reaction is independent of pressure.

For ideal gases $V = nRT/P$

$$\therefore \Delta V = V_{prod} - V_{react} = n_{prod}\frac{RT}{P} - n_{react}\frac{RT}{P}$$

$$\text{or } \Delta V = \Delta\nu\ \frac{RT}{P}$$

$$\text{and } \left(\frac{\partial \Delta V}{\partial T}\right)_P = \Delta\nu\frac{R}{P} \quad ...(5)$$

Putting the expressions from equation (5) into we get

$$\left(\frac{\partial \Delta H}{\partial P}\right)_T = \Delta v\ \frac{RT}{P} - T\Delta v\ \frac{R}{P} = 0.$$

Example 20:

For a reaction $CO\ (g) + H_2O\ (g) \rightarrow CO_2(g) + H_2\ (g)$, $\Delta H^o\ (298\ K) = -42.0 kJ$. *The heat capacities of various species are given by*

$C_{P,\ m}\ (CO)/J\ K^{-1}\ mol^{-1} = 26.8 + 7.0 \times 10^{-3}\ T$

$C_{P,\ m}\ (H_2O,\ g)/JK^{-1}\ mol^{-1} = 30.4 + 9.6 \times 10^{-3}\ T$

$C_{P,m}\ (CO_2)/\ JK^{-1}\ mol^{-1}\ 26.0 - 43.5 \times 10^{-3}T$

$C_{P,\ m}\ (H_2)/JK^{-1}mol^{-1} = 29.0 - 0.80 \times 10^{-3}\ T$

Calculate (a) ΔH^o (1298K) (b) $\Delta\ E^o$ (1298 K)

Solution:

ΔH^o (1298 K) is given by

$$\Delta H^o\ (1298K) = \Delta H^o\ (298K) + \int_{298}^{1298} \Delta C_P dT$$

Now, $\Delta C_P = [C_{P,m}\ (CO_2) + C_{P,m}\ (H_2)] - [C_{P,m}\ (CO) + C_{P,m}\ (H_2O)]$

$= (26.0 + 29.0) - (26.8 + 30.4) + (43.5 - 0.80)$

$- (7.0 + 9.6)] \times 10^{-3}T$

$= -2.2 + 26.1 \times 10^{-3}\ T$

$$\therefore \int_{298}^{1298} \Delta C_P dT = 2.2 \int_{298}^{1298} dT + 26.1 \times 10^{-3} \int_{298}^{1298} TdT$$

$$= -2.2\ [T]_{298}^{1298} + 26.1 \times 10-3 \times \frac{1}{2}[T]_{298}^{1298}$$

$$= -2.2\ [1298 - 298] + 26.1 \times 10-3 \times \frac{1}{2}[(1298)^2 - (298)^2]$$

$= 18627$ J

$= 18.627$ kJ

$\therefore\ \Delta\ H^o\ (1298K) = -42.0\ kJ + 18.627\ kJ = 23.373kJ$

Δ 67 (1298 K) $= \Delta H^o\ (1298K) - \Delta v_g\ RT = \Delta H^o\ (1298K)$

$=$ **− 23.373 kJ**

Since $\Delta v_g = (1 + 1) - (1 + 1) = 0$ for the given reaction.

THE MAXIMUM FLAME TEMPERATURE

When a fuel burns definite amount of products are formed and a definite quantity of heat is liberated. This quantity of heat may be utilized to raise the temperature of the products of combustion sufficiently to produce a flame. This temperature is called the *maximum flame temperature*, or calorific intensity. If the experiment is performed at constant volume the quantity of heat liberated will be denoted by Δ E and the maximum temperature is called the *explosion temperature*. In case the experiment is performed at constant pressure then heat liberated is denoted by ΔH and the maximum temperature attained is called the flame temperature. Usually this temperature is called the adiabatic maximum flame temperature because it is assumed that no loss of hea in the process takes place.

Calculation of Maximum Temperature

If it is assumed that the reaction is very fast and the total heat evolved is utilized to raise the temperature of the system then the heat capacitie of the species present at the completion of reaction and the heat evolved can be used to calculate the maximum temperature with the help o equation

$$\Delta H \text{ (heating)} = \int_{T_1}^{T_2} \Sigma_n C_{P.m} dT = -\text{(reaction)} \quad \text{...(1}$$

where we have taken ΔH (heating) = $-$ ΔH(reaction), because the heat evolved taken by the products. T_1 is the initial temperature and T_2 is the maximum flame temperature Σ $(nC_{P,m})$ is the sum of the heat capacities of all the species present in the system after the reaction is complete. For constant volume process,

$$-\Delta E = \int_{T_1}^{T_2} \Sigma_n C_{V.m} dT \quad \text{...(2)}$$

The validity of equation (1) os further justified as follows.

Let the reaction be performed at constant temperature and pressure (or constant T.V). Let us assume that we have a reservoir in which the heat evolved in the reaction can be stored. When the reaction is complete. the stored heat is used to raise the temperature of the products still at constant pressure (or constant volume). When all the heat evolved (stored in the reservoir) has been transferred to the system, the final maximum flame (or explosion) temperature is reached. This temperature can be measured using standard methods for the measurement of thermal changes for example, thermometer, thermistors, thermopiles etc. However, this procedure is quite hypothetical, because we do not have any such reservoir in which the heat can be stored and which can be kept completely isolated from the system and the surroudings. The direct way to know the maximum temperature is to perform the experiment in the laboratory under adiabatic conditions and measure the temperature using standard techniques of temperature measurements.

Example 21:

When hydrogen is burnt is oxygen, 240.5 kJ heat is liberated per mole of the explosion product measured at 27°C, Calculate the maximum temperature of explosion, $C_{V,m}$ $(H_2O, g) = 24.15\ JK^{-1}\ mol^{-1}$.

Solution:

The reaction can be represented as

$$H_2\ (g) + \frac{1}{2}O_2\ (g) \rightarrow H_2O\ (g)\Delta E\ (300K) = -240.5\ kJ\ mol^{-1}$$

This amount of heat will be utilized to raise the temperature of the product $H_2O(g)$. Using equation (15) we get

$$240500\ J\ mol^{-1} = \int_{300}^{T_2} (24.15 JK^{-1} mol^{-1})dT$$

$$\text{or}\quad 240500 = 24.15\ (T_2 - 300)$$

$$T_2 = \mathbf{10258\ K}$$

This calculation is quite approximate, therefore, this temperature would be far from the actual value.

Example 22:

The heat of combustion of methane is – 881.25 kJ mol^{-1} *at 298 K. Calculate the maximum theoretical flame temperature when one mole of* CH_4 *is completely burnt in calculated amount of air just sufficient for combustion at a constant pressure. The necessary data of heat capacities and their temperature dependence are also given.*

Data given

$$CH_4(g) + 2O_2(g) \rightarrow CO_2(g) + 2H_2O(l)$$

$$\Delta H^{\circ}\ (298\ K) = -\ 881.25\ kJ\ mol^{-1}$$

$$H_2O\ (l) \rightarrow H_2O(g) \quad \Delta H^{\circ}(298\ K) = 43.60\ kJ\ mol^{-1}$$

$C_{P,m}$ $(CO_2)/JK^{-1}mol^{-1} = 26.00 + 43.5 \times 10^{-3}T - 148.3 \times 10^{-7} T^2$

$C_{P,m}$ $(H_2O,g)/JK^{-1}$ $mol^{-1} = 30.36 + 9.61 \times 10^{-3} T + 11.8 \times 10^{-7}T^2$

$C_{P,m}$ $(N_2)/JK^{-1}$ $mol^{-1} = 27.30 + 5.23 \times 10^{-3} T - 0.04 \times 10^{-7}T^2$

Solution:

Let T_2 be the maximum theoretical flame temperature to which the products are raised, which include one mole $CO_2(g)$, 2 moles $H_2O(g)$, and 8 moles $N_2(g)$ (present in air associated with 2 moles O_2 used for combustion). Since water is ultimately in the gaseous form at high temperature, that value of ΔH^o os required for the reaction on which H_2O (g) is a product. Thus

(i) CH_4 (g) + $2O_2$(g) → CO_2 (g) + $2H_2O$(l) $\Delta H^o = -881.25$ kJ

(ii) $2H_2O$ (l) → $2H_2O$(g) $\Delta H^o = + 87.20$ kJ

so that, addition of (i) and (ii) gives

CH_4 (g) + $2O_2$(g) → CO_2 (g) + $2H_2O$ (g) $\Delta H^o = - 794.05$ kJ

Now, this heat is used to raise the temperature of products after combustion. Equation (14) gives,

$$794050 \text{ J mol}^{-1} = \int_{298}^{T_2} \Sigma n(C_P)dT \qquad ...(1)$$

Now, $\Sigma nC_{P,m} = C_{P,m}(CO_2) + 2C_{P,m}(H_2O) + 8CP,m\ (N_2)$

$\Sigma\ (nC_{P,m}) = 305.12 + 104.56 \times 10^{-3}\ T - 125 \times 10_{-7}\ T_2$

Therefore, putting the values in equation (17) and carrying out integration we have

$$794050 = \left[30.512T + \frac{1}{2} \times 104.56 \times 10^{-3}T^2 - \frac{1}{3} \times 125 \times 10^{-7}T^3\right]_{298}^{T^2}$$

This equation is solved for T_2. This approximate value of T_2 is found to be 2250 K. The experimental value is about 2150 K. Thus there is a difference of about 100° between theoretical and experimental values. This discrepancy is explained taking into account the various possible reasons.

(i) The process may not be exactly adiabatic and there is exchange of heat between the system and the surroundings.

(ii) In actual practice, excess air is used for combustion process therefore, after combusion some air (O_2 + N_2) is left. Thus the heat evolved will be used to raise the temperature of extra O_2 and N_2 (in air).

(iii) At high temperature attained in the burning of hydrocarbons the products H_2O and CO_2 may get dissociated as

$$H_2O \rightarrow H_2 + \frac{1}{2}O_2$$

$$CO_2 \rightarrow CO + \frac{1}{2}O_2$$

These reactions will involve considerable amount of heat which is not accounted for by the theoretical calculations.

(iv) The combustion may not be complete, hence along with the products some reactants may also be present, and heat evolved will be utilized to raise the temperature of unreacted species.

Exercise. *Calculate the maximum flame temperature when ethane is burnt with twice the amount of air needed for complete combustion to CO_2 and H_2O. The combustion reaction is represented as*

$$C_2H_6\ (g) + \frac{7}{2}O2(g) \rightarrow 2CO_2(g) + 3H_2O(g)$$

ΔH^o (298K) = – 1429.2kJ mol^{-1} **[Ans. 1450 K]**

[**Hint.** Compute

$$\Sigma nC_P = 2C_PCO_2(g) + 3C_PH_2O(g) + 14C_PN_2(g)$$

$$+\left\{\frac{7}{2}C_PO_2(g) + 14C_PN_2(g)\right\}$$

The quantities within braces are for excess air].

Some Interesting Calculations

Example 23:

It is estimated that an average human consumes the equivalents of 10 g glucose per hour. Estimate the power output of the brain in watts. Given that 1 W = 1 JS^{-1};

ΔH^o_f (H_2O, l) = –286 kJ mol^{-1}

ΔH^o_f (CO_2 g) = –394 kJ mol^{-1}

ΔH^o_f (*glucose, aq*) = –1260 kJ mol^{-1}

Solution:

The heat supplied by 10 g glucose can be calculated if we know its heat of combustion:

$C_6H_{12}O_6$ (aq) + $6C_2$ (g) $\rightarrow$ $6CO_2$ (g) + $6H_2O$ (l) ΔH = ?

$\Delta H = 6 \times \Delta H^o_f (CO_2) + 6 \times \Delta H^o_f (H_2O) - \Delta H^o_f (C_6H_{12}O_6, aq)$

$= [(-6 \times 394 - 6 \times 286) + 1260]$ kJ mol^{-1}

$= -2820$ kJ mol^{-1}

$\therefore$ Heat supplied by 10 g glucose $= -2820 \text{ mol}^{-1} \times \dfrac{10g}{180g\ mol^{-1}}$

q = –156.66 kJ per hour

= – 156666 J per hour

Power output of the brain = *Heat change per second*

$= \dfrac{156666J}{60 \times 60}$ per second

= **43.5 Js^{-1} = 43.5 W**

Example 24:

Calculate the enthalpy change for the process

H_2O (l, –10°C) $\rightarrow$ H_2O (s, – 10°C)

given that :

$C_{P,m}$ (H_2O, l) = 75.4 J K^{-1} mol^{-1}

$C_{P,m}$ (H_2O,s) = 37.2 J K^{-1} mol^{-1}

H_2O (l, 0°C) $\rightarrow$ H_2O (s, 0° C), ΔH = – 6008 J mol^{-1}

Solution:

The given process may be considered to proceed in different stages as

(i) H_2O (l, –10°C) $\xrightarrow{\text{Heat}}$ H_2O (l,0°C)

$\Delta H (i) = \int_{263}^{273} C_{P,m}(H_2O, l)dT = 75.4\ JK^{-1}\ mol^{-1}\ (273K - 263K)$

= 754 J mol^{-1}

(ii) H_2O (l, 0°C) $\rightarrow$ H_2O(s, 0°C)

ΔH (ii) = – 6008 J mol^{-1}

(iii) H_2O (s, 0°C) $\xrightarrow{\text{Cool}}$ H_2O (s, – 10°C)

$$\Delta H \text{ (iii)} = \int_{273}^{263} C_{P,m}(H_2O, s)dT = 37.2 \text{ KJ}^{-1} \text{ mol}^{-1} (263 \text{ K} - 273 \text{ K})$$

$= -372 \text{ J mol}^{-1}$

Addition of equations (i), (ii) and (iii) gives

H_2O (l, – 10°C) → H_2O (s,–10°C), ΔH

$= \Delta H \text{ (i)} + \Delta H \text{ (iii)}$ $\Delta H = -5626 \text{ J mol}^{-1}$

Example 25:

The following reactions might be used to power rockets:

(i) H_2 (g) + 1/2 O_2 (g) → H_2O(g) ΔH (i) = – 242 kJ mol^{-1}

(ii) CH_3OH (l) + 3/2 O_2 (g) → CO_2 (g) + 2H_2O (g)

ΔH (ii) = – 640 kJ mol^{-1}

(iii) H_2 (g) + F_2 (g) → 2HF (g) ΔH (iii) = – 540 kJ mol^{-1}

Calculate the enthalpy changes for each of these reactions per kg of each second reactant.

Solution:

Moles of reactants can be calculated by using the formula n = mass/ molar mass.

$$n(H_2) = \frac{1\text{kg}}{0.002 \text{ kg mol}^{-1}} = 500 \text{ mol}$$

$$n(O_2) = \frac{1\text{kg}}{0.032 \text{ kg mol}^{-1}} = 31.25 \text{ mol}$$

$$n(CH_3OH) = \frac{1\text{kg}}{0.032 \text{ kg mol}^{-1}} = 31.25 \text{ mol}$$

$$n(F_2) = \frac{1\text{kg}}{0.038 \text{ kg mol}^{-1}} = 26.3 \text{ mol}$$

The Stoichoiometry of reaction (i) indicates that 31.25 moles O_2 will combine with 62.5 moles H_2 to give 62.5 moles H_2O as

H_2 (g) + 1/2 O_2(g) → H_2O(g), ΔH (i) = – 242 kJ mol^{-1}

62.5 H_2(g) + 31.25 O_2(g) → 62.5 H_2O (g)

$\Delta H = 62.5 \times DH\ (i) = -$ **15125 kJ**

The second reaction is written as

$20.8\ CH_3\ OH\ (l) + 31.25\ O_2\ (g) \rightarrow 20.8\ CO_2\ (g) + 41.6\ H_2O(g)$

$\Delta H = 20.8 \times \Delta H\ (ii)$

= **–13312 kJ**

Third reaction

$26.3\ H_2\ (g) + 26.3\ F_2\ (g) \rightarrow 52.6\ HF\ (g)$

$\Delta H = 26.3 \times \Delta H\ (iii)$

= – **14202 kJ**

Example 26:

An average man weighs about 70 kg and produces about 10^4 kJ of heat per day through metabolic activity. (a) Suppose that a man were an isolated system and that his heat capacity is 4.2 $JK^{-1}\ g^{-1}$. If his temperature was 37°C at a given time what would be his temperature 24 hours later?

(b) *Man is actually an open system and the main mechanism of loss of heat is evaporation of water. How much water would need to be evaporated per day to maintain the body temperature at 37° C? ΔH_{vap} (water)* = 2405 J g^{-1} at 37°C.

Solution:

(a) Let T_2 be temperature of the body after 24 hours. Given that C_P = 4.2 $JK^{-1}\ g^{-1}$.

$m = 70\ kg = 70 \times 10^3\ g$

$\Delta H = 10^4\ kJ = 10^7\ J,\ T_1 = 310K$

$$\Delta H = \int_{T_1}^{T_2} mC_P dT = mC_P(T_2 - T_1)$$

$$\therefore T_2 = \frac{\Delta H + mC_P T_1}{mC_P}$$

$$= \frac{10^7 J + (7 \times 10^4 g) \times (4.2 JK^{-1} g^{-1}) \times (310K)}{(7 \times 10^4 g) \times (4.2 JK^{-1} g^{-1})}$$

= **344 K = 71° C**.

(b) $\because$ 2405 J heat evaporate 1g H_2O

$\therefore$ 10^7 J heat will evaporate $\dfrac{10^7 J \times 1gH_2O}{2405J}$

$= 4158$ g H_2O = **4.2 kg H_2O**.

This much water should be evaporated by perspiration to maintain the body temperature.

INTERCONVERSION OF MATTER AND ENERGY

The interconvertibility of matter and energy is described by the Einstein equation $\Delta E = \Delta mc^2$. Here Δ E is the change of energy and Δm is the change in mass and c is the speed of light. In the language of thermochemistry the above relation means that matter and energy are not conserved separately rather the principle of conservation of energy is broadened to include matter as another form of energy.

Example 27:

In an explosion reaction 21000 kJ *are evolved. Calculate the decrease in mass accompanying the reaction.*

Solution:

$\Delta E = \Delta mc^2$ (Einstein equation)

$\Delta E = -21000$ kJ $= -21.0 \times 10^6$ T $= -21.0 \times 10^6$ kg m^{2}s^{-2}

Speed of light = c = 3.0×10^8 ms^{-1}

$$\therefore \Delta m = \Delta E/c^2 = -\frac{21.0 \times 10^6 \text{kg m}^2\text{s}^{-2}}{(3.0 \times 10^8 \text{ms}^{-1})^2} = -\mathbf{2.3 \times 10^{-10}\ kg}$$

Thus 2.3×10^{-10} kg is the loss of matter which liberates as heat and is transferred to the surroundings.

Exercise. *Energy is emitted by the sun at the rate of 2 ergs per g of the sun per second. The estimated mass of the sun is 2 × 10^{33} tons. Estimate the decrease in tons and per cent decrease in the mass of the sun per day.*

[**Ans.** 4×10^{17} tons; 2×10^{-14}%. It is interesting to note that if the above calculations were true then after 10^{13} years the sun should vanish.]

EXPERIMENTAL DETERMINATION OF ΔE AND ΔHD

Determination of ΔE. If a reaction is allowed to take place at constant temperature and constant volume the internal energy change, ΔE, of the reaction is equal to the heat exchanged with the surroundings. Thus, the external energy change accompanying a chemical reaction can be determined by measuring the heat evolved or absorbed provided the reaction is taking place at constant temperature and constant volume.

Let us now describe the experimental method for determining ΔE. A bomb calorimeter which is mainly used to measure ΔE is shown in Fig. 3.5.

The calorimeter consists of a strong reaction vessel called bomb calorimeter built of heavy steel. For many reactions, the heat is evolved or absorbed provided the reaction is taking place at constant temperature and constant volume.

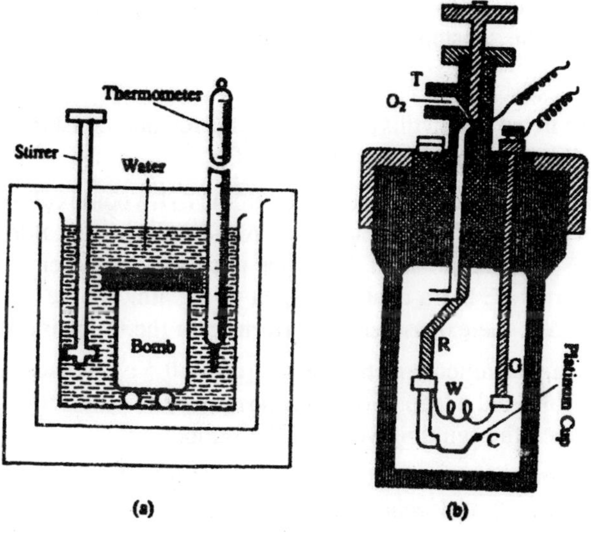

Fig. 3.5

The calorimeter consists of a strong reaction vessel called bomb calorimeter built of heavy steel. For many reactions, the constant volume requirement results in large pressure changes. For this reason the calorimeter is built of heavy steel to withstand high pressure. The bomb calorimeter is placed in the centre of a larger vessel filled with a weighed quantity of water. The larger vessel is also provided with a stirrer and sensitive thermometer. The larger vessel is generally a metal container and is surrounded by an insulating material so that no heat is lost to the surroundings. A weighed sample is placed inside a platinum cap and the temperature of water is noted, The reaction is done inside the platinum cap by passing an electric current through a wire embedded into it. The heat evolved or absorbed due to the reaction changes the temperature of the entire apparatus. The temperature of water is again noted. From the change in temperature and the total water equivalent of the calorimeter, ΔE can be calculated.

The heat capacity of the calorimeter system is known by burning a known weight of a substance of known heat of combustion. For this purpose, usually benzoic acid of highgrade purity is used. Its heat of combustion which has been very carefully measured is taken as – 3226.7 kJ per mole. Let the thermal capacity of the calorimeter system including water be z and θ the rise in temperature produced by burning m of the given substance of molecular weight M. Then, evidently, the heat of combustion of the substance will be equal to

$$z \times \theta \times M/m \text{ Joules.}$$

It may be noted that in this case, the volume remains constant. Hence the heat of the reaction gives a measure of q_v.

Determination of ΔH. In order to determine ΔH, a bomb calorimeter is used as in the case of determination of ΔE. But the bomb calorimeter used for the determination of H is not a closed container as used determination of ΔE but a container open to the atmosphere. In order to determine ΔH, the experiment is performed in the following stages:

1. A weighed amount of the sample (about 0.5 g) is placed in the crucible, and a loop of cotton is positioned so as to act as a fuse between the ignition wire and the sample.
2. The bomb is assembled and charged through the pressure valve to 20 or 30 atmospheres with oxygen.
3. The bomb is then placed in a water calorimeter of known equivalent, and the whole is surrounded by a jacketed vessel which reduces heat losses by radiation.
4. The temperature of the water in the calorimeter is taken (using an accurate thermometer graduated in tenths of a degree) at regular intervals to allow for the application of a cooling correction.
5. The bomb is fired, and the temperature of the water in the calorimeter is taken at intervals, continuing for some minutes after the maximum temperature has been attained.

6. Now the cooling correction is applied. If the heat capacity of the bomb and water calorimeter is known, the heat evolved during the combustion of the sample can be calculated. (A correction is applied for the heat liberated by the combustion of the cotton loop).

From the temperature change and the total water equivalent of the calorimeter, ΔH can be calculated.

In order to convert ΔE into ΔH or vice versa, we have to use the equation

$$\Delta H = \Delta E + \Delta nRT$$

KIRCHHOFF'S EQUATION

In 1858 Kirchhoff deduced certain mathematical expressions for defining the variation of heat of reaction with temperature. All these deductions are derived from the first law of thermodynamics . Consider a process,

$$A \rightarrow B$$

Let E_A and H_A be the internal energy and heat content of the system in the initial state A. Similarly, let E_B and H_B be the internal energy and the heat content of the system in the final state B. By the first law of thermodynamics, the change in enthalpy for the given process is given by the following equation

$$\Delta H = H_B - H_A \qquad \text{...(1)}$$

But heat content is a function of temperature and pressure. Therefore, the equation (1) can be differentiated with respect to temperature at constant pressure.

$$\left[\frac{\partial(\Delta H)}{\partial T}\right]_P = \left[\frac{\partial(\Delta_B)}{\partial T}\right]_P - \left[\frac{\partial(H_A)}{\partial T}\right]_P = (C_p)B - (C_P)A$$

$$\left[\because \left(\frac{\partial H}{\partial T}\right)_P C_P\right]$$

or $$\left[\frac{\partial(\Delta H)}{\partial T}\right]_P = \Delta C_P \qquad \text{...(2)}$$

where ΔC_p represents the difference in the heat capacities of initial and final state at constant pressure. In order to integrate this equation (2), we have to consider that

(i) ΔC_p is independent of temperature

or (ii) ΔC_p is dependent upon temperature.

Let us take into consideration these cases one by one.

Case I.

When ΔC_p is constant between two temperatures T1 and T_2 : If we consider C_p to be constant and integrate equation (2) between the limits H_1 at T_1 and H_2 and T_2, we get the expression

$$\int_{H_1}^{H_2} d(\Delta H) = \int_{T_1}^{T_2} \Delta C_p \, dT, \text{ or } [\Delta H]_{H_1}^{H_2} = [\Delta C_P]_{T_1}^{T_2}$$

or $$(\Delta H_2 - \Delta H1) = \Delta C_P (T_2 - T_1) \qquad ...(3)$$

This relationship is termed as Kirchhoff's equation.

Case II.

When ΔC_P varies with temperature.

It implies that equation (2) can be integrated at constant pressure

$$\int d(\Delta H) = \int \Delta C_P . \, dT$$

or $$\Delta H = \int \Delta C_P . \, dT + K \qquad ...(4)$$

where K is a constant of integration. When T = 0. then $\Delta H_o = K$. Hence equation (4) becomes as follows :

$$\Delta H = \Delta H_o + \int \Delta C_P . \, dT \qquad ...(4A)$$

where ΔH_o represents the difference in heat contents of products and reactants at the absolute zero.

But $(C_P)_A = \propto_A + \beta_A T + \gamma_A T^2$ ......

$$(C_P)_B = \propto_B + \beta_B T + \gamma_B T^2 \ ......$$

$$\therefore \quad C_P = (C_P)_B - (C_P)_A = (\propto_B + \beta_B T + \gamma_B T^2 ...)$$

$$- (\propto_A + \beta_A T + \gamma_A T^2 + ...$$

$$= (\propto_B - \propto_A) + T (\beta_B - \beta_A) + T^2 (\gamma B - \gamma A) + ...$$

i.e., $$\Delta C_P = \Delta\propto + T\Delta\beta + T^2 \Delta\gamma + \qquad ...(5)$$

If we insert this value of ΔC_P in equation (4A) and carry out the integration, we get

$$\Delta H = \Delta H_o + T \, \Delta\propto + \frac{T^2}{2} \Delta\beta + \frac{T^3}{3} \Delta\gamma + ... \qquad ...(6)$$

The values of $\Delta\alpha$, $\Delta\beta$, $\Delta\gamma$, ... can be calculated from heat capacity date. Equation (6) is another form of Kirchhoff's equation. This equation can be used to evaluate ΔH_o if the value of ΔH and constant $\Delta\alpha$, $\Delta\beta$, $\Delta\gamma$, ... are known.

Consider the following reaction,

$$n_1A + n_2B \rightleftharpoons n_3C + n_4D$$

where n_1 and n_2 be the number of molecules of reactants A and B and n_3 and n_4 be the number of molecules of products C and D,

Now
$$(C_P)_A = n_1 (\alpha_1 + \beta_1 T + \gamma_1 T^2 + \delta_1 T^3 + ...)$$
$$(C_P)_B = n_2 (\alpha_2 + \beta_2 T + \gamma_2 T^2 + \delta_2 T^3 + ...)$$
$$(C_P)_C = n_3 (\alpha_3 + \beta_3 T + \gamma_3 T^2 + \delta_3 T^3 + ...)$$
$$(C_P)_D = n_4 (\alpha_4 + \beta_4 T + \gamma_4 T^2 + \delta_4 T^3 + ...)$$

The change in heat capacity (ΔC_P) is given as follows :

$$\Delta C_P = [\{(\alpha_3 + \alpha_4) - (\alpha_1 + \alpha_2)\}] + \{(\beta_3 + \beta_4) - (\beta_1 + \beta_2)\}T$$
$$+[\{\gamma_3 + \gamma_4) - (\gamma_1 + \gamma_2)\} T^2 + ...] (n_3 + n_4 - n_1 - n_2)$$

or $\Delta C_P = (\alpha + \beta T + \gamma T^2 + ...) \ (n_3 + n_4 - n_1 - n_3)$

On comparing it with Kirchhoff's equation, we obtain

$$\left[\partial \frac{\partial H}{\partial V}\right]_P = \Delta C_P = \alpha + \beta T + \gamma T^2 + ... \qquad ...(6A)$$

On integrating this equation, we obtain

$$\Delta H = \Delta H_o + \alpha T + \beta \frac{T^2}{2} + \gamma \frac{T^3}{3} + ... \qquad ...(6B)$$

where ΔH_o is the integration constant and

$\alpha = \alpha_3 + \alpha_4 - \alpha_1 - \alpha_2$, $\beta = \beta_3 + \beta_4 - \beta_1 - \beta_2$,

$\gamma = \gamma_3 + \gamma_4 - \gamma_1 - \gamma_2$, $\delta = \delta_3 + \delta_4 - \delta_1 - \delta_2$,

Equation (6B) is the general expression which defines the variation of ΔH with temperature.

THE MAXIMUM FLAME TEMPERATURE

When a fuel burns, definite quantities of products are formed and a definite amount of heat is liberated. This quantity of heat may raise the temperature of the products of combustion sufficiently so as to produce a flame. This temperature is termed as the maximum flame temperature, or calorific intensity. When the experiment is carried out at constant volume the quantity of heat liberated will be ΔE and the maximum temperature is termed explosion temperature. In case the experiment is carried out at constant pressure then heat liberated is ΔH and the maximum temperature to which the products will be raised is termed as the flame temperature. Generally this temperature is termed as the adiabatic maximum flame temperature as it is assumed that there occurs no loss of heat in the process.

Calculation of Maximum Temperature

(1) If we assume that the reaction is very fast and the total heat evolved is used to raise the temperature of the system, then the heat capacities of the species present at the completion of reaction and the heat evolved can be used for calculating the maximum temperature by using Kirchhoff's equation in the following form,

$$\Delta H_{(reaction)} = \int_{T_2}^{T_1} \sum nC_P \, dT \, \Delta H_{(heating)}$$

where it is assumed that $\Delta H_{(heating)} = -\Delta H_{(reaction)}$, as the heat evolved is taken by the products. T_1 represents the initial temperature and T_2 the maximum flame temperature, $\Sigma(nC_P)$ the sum of the heat capacities of all the species present in the system after the reaction gets completed. For constant volume process,

$$\Delta E = \int_{T_2}^{T_1} \sum nC_V \, dT$$

The direct way to know the maximum flame temperature is to carry out the experiment in the laboratory under adiabatic conditions and measure the temperature using standard techniques of temperature measurements.

The calcualted flame temperature has been found to be higher than the observed temperature. This difference has been attributed to a number of complications which are as follows :

(i) The system may not be truly adiabatic.

(ii) In combustion, we generally use excess of air or oxygen, Unreacted air or oxygen is to be heated and therefore a part of the heat would be utilised in heating these gases.

(iii) When flame temperature is to be calculated, heating of the unreacted gases should also be considered.

(iv) At high temperatures, the products might decompose and a part of the heat may be utilised to decompose the product molecules.

(2) If an exothermic reaction is assumed to carry out under adiabatic conditions, the heat evolved would be able to raise the temperature of the system.

It is possible to determine the rise in temperature by considering the reaction to take place in the following two steps, both at constant pressure.

Step I : Reactants (T_o, P) → Products (T_o, P) ΔHT_o

Step II : Products (T_o, P) → Products (T_f, P)

The heat involved in the second step may be given as follows :

$$\Delta H_2 = \int_{T_o}^{T_f} C_P \text{ (Products) } dT$$

As the overall reaction refers to the sum of the above two steps, *i.e.*,

Reactants (T_o, P) → Products (T_f, P)

the net heat change is given as follows :

$$\Delta H = \Delta H_{To} + \Delta H_2 \qquad ...(i)$$

As the reaction is carried out under adiabatic conditions, it is evident that $\Delta H = 0$. Hence Eq. (i) may be put as follows :

$$\Delta H_{To} + \Delta H_2 = 0.$$

or $$\Delta H_2 = -\Delta H_{To} \text{ or}$$

$$\int C_P \text{ (Products) } dT = -\Delta H_{To}$$

If C_Ps are regarded to be independent of temperature, then

$$C_P \text{ (Products) } (T_f - T_o) = -\Delta H_{To} \text{ or } T_f = \frac{-\Delta H_{TO}}{C_p\text{(products)}} + T_o$$

However, the approximation that C_Ps are independent of temperature has not been correct because T_f is a very large quantity (ΔH_{To} ~ kJ and CP ~ J) of the order of thousands of kelvin.

If the reaction is carried out in a closed vessel *i.e.*, under the condition of constant volume, then ΔE and C_V replace ΔH and C_P respectively, *i.e.*,

$$T_f = \frac{(-\Delta E_{To})}{C_v\text{ (products)}} + T_o$$

EXPLOSION TEMPERATURE

When the combustion of hydrocarbons is done at constant volume under adiabatic conditions, the maximum temperature attained is termed as the explosion temperature. The explosion temperature is calculated in the same way as that of flame temperature except ΔH is replaced by ΔE. The pressure which the reacting system may acquire at the explosion temperature is termed as the maximum explosion pressure.

LAW OF THERMOCHEMISTRY

1. Lavoisier and Laplace's Law. This law was given by Lavoisier and Laplace in 1780 and may be stated as follows :

"The amount of heat supplied to decompose a compound into its elements is equal to the heat of formation of that compound form its elements".

In simple words, the heat of formation of compound is numerically equal to its heat of decomposition of a compound but of opposite sign. Thus, the larger the heat of formation of a compound, the greater will be its stability.

Importance : The importance of this law is that thermochemical equations can be reversed. For example, the equation

$$C(s) + O_2(g) \rightarrow CO_2(g)$$

$$\Delta H^\circ = -\ 296.64\ kJ$$

may be written as

$$CO_2(g) \rightarrow C(s) + O_2(g)$$

$$\Delta H = +\ 296.64\ kJ$$

This law applies not only to reactions involving a compound and its component elements but also to reactions of all other types. Thus, for example,

$$C_2H_5OH + 3O_2 \rightarrow 2CO_2 + 3H_2O$$

$$\Delta H^\circ = -\ 1366.95\ kJ$$

Hence by Lavoisier-Laplace's law, we get

$$2CO_2 + 3H_2O \rightarrow C_2H_5OH + 3O_2$$

$$\Delta H^\circ = +\ 1366.95\ kJ$$

Hess's Law of Constant Heat Summations

Most calculations involving heat changes in thermochemical equations are based on the fundamental law enunciated by Russian chemist G. H. Hess in 1840. This law can be stated in words as follows :

"The enthalpy change of a given chemical reaction is the same whether the process takes place in one or several steps".

In other words, the law is merely a corollary of the law of conservation of energy which implies that heat of reaction does not depend upon the nature of intermediate products but depends only on the initial reactants and final products and is also independent of the path or the manner by which the change is carried out.

Experimental Illustration : In order to bring about experimental illustration, we can carry out the same chemical change in two alternative ways and show that the enthalpy change remains same for both. For example, carbon dioxide may be produced by two process :

First Procedure : The burning of carbon can take place in a single step to form CO_2 with the enthalpy change which is equal to – 393.0 kJ mol^{-1}.

$$C(graphite) + O_2(g) \rightarrow CO_2(g);\ \Delta H = -\ 393\ kJ \qquad ...(1)$$

Second Procedure. It is also possible to carry out the above reactions in two steps. First of all carbon can be converted into carbon monoxide. Then carbon monoxide can be oxidized to carbon dioxide. Thus,

$$C(graphite) + 1/2O_2(g) \rightarrow CO(g);\ \Delta H = -\ 110.5\ kJ \qquad ...(2)$$

$$CO(g) + 1/2O_2(g) \rightarrow CO_2(g);\ \Delta H = -\ 283.2\ kJ \qquad ...(3)$$

On adding equations (2) and (3), we get

$$C(graphite) + O_2(g) \rightarrow CO_2(g);\ \Delta H = -\ 393.7\ kJ \qquad ...(4)$$

From equations (1) and (4), it follows that the enthalpy change in both the procedures is almost same. The small difference can be attributed to experimental error. It also follows from the above discussion that thermochemical equations can be added or subtracted as ordinary algebraic equations.

Theoretical Justification : (Hess's law as a corollary of First Law of Thermodynamics). The law also follows as a mere consequence of the First Law of Thermodynamics. Suppose a substance A can be changed to Z directly.

$$A \rightarrow Z + Q_1 \qquad ...(5)$$

where Q_1 represents the heat evolved in the direct change. If the same change is carried out through intermediate stages,

$$A \rightarrow B + q_1,\ B \rightarrow C + q_2,\ C \rightarrow Z + q_3$$

The total evolution of heat $= q_1 + q_2 + q_3 = Q_2$

By Hess's Law, $Q_1 = Q_2$.

If it is not so, suppose $Q_2 > Q_1$. Then by transforming A to Z through stages and retransforming directly back to A there would occur a gain of heat $= Q_2 - Q_1$. If we repeat this process again and again, an unlimited heat will be produced in an isolated system. This goes against the First Law of Thermodynamics. Hence Q_1 must be equal to Q_2 (Fig. 3.6)

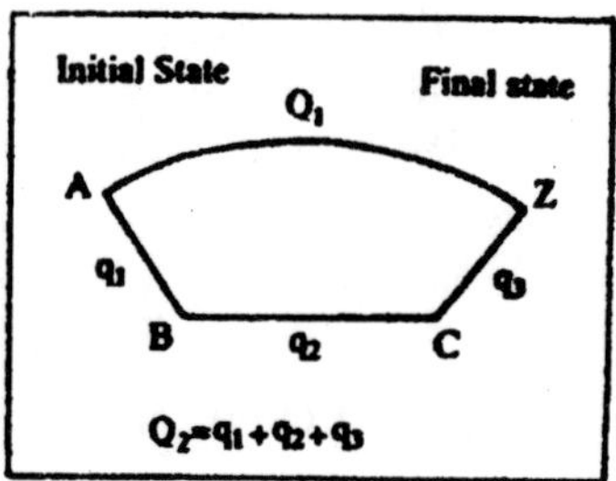

Fig. 3.6 : Theoretical Justification of Hess's law

Applications of Hess's Law

The main practical use of Hess's law lies in the fact that it makes legitimate to add up or subtract thermochemical equations as algebraic equations. This gives rise to the mathematical method of calculating :

(1) Heat of Reaction,

(2) Heat of Transition,

(3) Heat of Formation, and

(4) Lattice Energy of a Crystal.

Let us describe these one by one.

The importance of Eq. (5) is that the amount of heat absorbed at constant volume can be identified with the change in the thermodynamic quantity.

(1) *Calculation of heats of reactions :* By Hess's law it is possible to calculate heats of many reactions which cannot be determined

experimentally. For example it is very difficult to measure the heat evolved when carbon burns in oxygen to form carbon monoxide,

$$C(s) + 1/2O_2(g) \rightarrow CO(g);\ \Delta H = ?$$

From the Hess's law, it is evident that the heat produced in the combustion of carbon to give carbon dioxide is the same, viz., – 393.5 kJ, whether it takes place in a single step as

$$C(s) + O_2(g) \rightarrow CO(g);\ \Delta H = -\ 393.5\ kJ$$

or in two steps as

$$C(s) + 1/2O_2(g) \rightarrow CO(g);\ \Delta H = x\ kJ\ (say)$$

$$CO(g) + 1/2O_2(g) \rightarrow CO_2(g);\ \Delta H = y\ kJ\ (say)$$

Evidently

$$x + y = -\ 393.5\ kJ$$

The x, the heat produced in the conversion of carbon to carbon monoxide cannot be determined but y, the heat change involved in the combination of carbon monoxide and oxygen to give carbon dioxide can be measured and has been found to be – 282.0 kJ

$$x = -\ 393.5 - y$$

$$\text{or } y = \quad 393.5 \quad (\ 282.0) = \quad 111.5\ kJ.$$

Hence the heat evolved during the combination of carbon to give carbon monoxide is – 111.5 kJ.

(2) *Determination of the Heat of Transition* : As transition of an element from one allotropic form to another is an extremely slow process it requires heating of one allotropic form of the element for a long period at constant temperature. Therefore, it is not possible to determine heats of transition of elements experimentally. They are, however, determined indirectly by using Hess's law. Following example will make this clear.

Suppose we intend to determine the heat of transition of sulphur (rhombic) into sulphur (monoclinic), we start with the two allotropic forms separately and measure their heats of combustion.

$$S_{(R)} + O_2(g) \rightarrow SO_2\ ;\ \Delta H = AJ \qquad ...(i)$$

$$S_{(M)} + O_2(g) \rightarrow SO_2\ ;\ \Delta H = BJ \qquad ...(ii)$$

On subtracting (ii) from (i), we get

$$S_{(R)} - S_{(M)} \rightarrow 0 \; ; \; \Delta H = (A - B) \; J$$

$$S_{(R)} - S_{(M)} \; ; \; \Delta H = (A - B) \; J \qquad \text{...(iii)}$$

Equation (iii) expresses the thermochemical equation for the transition of $S_{(R)}$ to $S_{(M)}$ Hence heat of transition $S_{(R)} \rightarrow S_{(M)}$ is (A – B) J.

In this particular case, $\Delta H = 2.5$ kJ.

(3) *Calculation of Heats of Formation* : The heat of formation of those compounds which cannot be determined experimentally can be calculated by using Hess's law. This can be understood by considering the following example:

The enthalpy of formation of methane from graphite and hydrogen cannot be found directly as no chemical reaction occurs between hydrogen and graphite.

(i) $C(\text{graphite}) + 2H_2(g) \rightarrow CH_4(g)$

However, ΔH for the above reaction may be calculated by the help of the following reactions :

(ii) $C(\text{graphite}) + O_2(g) \rightarrow CO_2(g); \; \Delta H_a = -\,393.5$ kJ

(iii) $2H_2(g) + O_2(g) \rightarrow 2H_2O(l); \; \Delta H_b = 2(-285.9 \text{ kJ}) = -\,571.8$ kJ

(iv) $CH_4(g) + 2O_2(g) \rightarrow CO_2(g) + 2H_2O(l); \; \Delta H_c = -\,890.3$ kJ

From the above reactions, the enthalpy change for reaction (i) may be obtained by adding eqquations (ii) and (iii)

(ii) $C(\text{graphite}) + O_2(g) \rightarrow 3CO_2(g); \; \Delta H_a = -\,393.5$ kJ

(iii) $2H_2(g) + O_2(g) \rightarrow 2H_2O(l); \; \Delta H_b = -\,571.8$ kJ

$C(\text{graphite}) + 2O_2(g) + 2H_2(g) \rightarrow CO_2(g) + 2H_2O(l);$

$\Delta H_a + \Delta H_a = -\,965.3$ kJ

On subtracting equation (iv) from the above equation, we obtain

$C(\text{graphite}) + 2O_2(g) \rightarrow CH_4(g);$

$$\Delta H = -\,965.3\text{kJ} + 890.3\text{kJ} = -\,75.0 \text{ kJ}$$

(4) *Lattice Energy of a crystal (Born-Haber Cycle)* : Hess's law can be used to determine lattice energy which may be defined as the amount of heat released when the requisite amounts of ions in the gaseous state combine to yield 1 mole of crystal lattice. An example is the determination of DH_{Lattice} of sodium chloride crystal which corresponds to the reaction.

$$Na^+(g) + Cl^-(g) \rightarrow NaCl\ (s).$$

It is possible to determine the numerical value of $\Delta H_{Lattice}$ for the above reaction directly by using Born-Haber's cycle. The following is the sequence of steps for the formation of NaCI crystals from Na(s) and $Cl_2(g)$

(i) Vaporisation of Na(s)	$Na(s) \rightarrow Na(g)$	ΔH_1
(ii) Ionisation of Na(g)	$Na(g) \rightarrow Na^+(g)+e$	ΔH_2
(iii) Dissociation of chlorine	$1/2\ Cl_2(g) \rightarrow Cl(g)$	ΔH_3
(iv) Formation of Cl^-	$Cl(g) + e \rightarrow Cl^-(g)$	ΔH_4
(v) Condensation of	$Na^+(g)$ and $Cl^-(g)$ $+ Cl-(g) \rightarrow NaCl(s)$	Na+(g) ΔH_5
Net change	$Na(s)+1/2\ Cl_2(g)\ NaCl(s)$	ΔH_6

On applying the Hess's law to the above sequence of steps, we get

$$\Delta H_6 = \Delta H_1 + \Delta H_2 + \Delta H_3 + \Delta H_4 + \Delta H_5$$

Except ΔH_5, it is possible to determine experimentally all of the changes of enthalpy, *i.e.*, ΔH_1, ΔH_2, ΔH_3, ΔH_4, and ΔH_6. Hence ΔH_5 can be determined from the above relation.

FACTORS ON WHICH HEAT OF REACTION OR ENTHALPY OF REACTION DEPENDS

The important factors which influence the heat of a reaction are given as follows :

(i) *Physical States or Reactants or Products* : For example when red and yellow phosphorus burn in oxygen to form P_2O_5, different values of heats of reactions are obtained.

$$4P_{yellow} + 5O_2(g) \rightarrow P_4O_{10}\ ;$$
$$\Delta H = -\ 3024\ kJ$$
$$4P_{red} + 5O_2(g) \rightarrow P_4O_{10};$$
$$\Delta H = -\ 3026\ kJ$$

Thus, there is a difference in the heats of reactions of the two allotropic forms of phosphorus.

(ii) *Temperature* : Heat of reaction has been found to depend upon the temperature at which reaction is occurring. The variation of the heat of reaction which temperature can be ascertained by using Kirchhoff's equation

(iii) *Heats of solution of Reactants and Products :* It is found that different values of heats of reaction are obtained when reactions are carried out with the anhydrous reactants or in solution. For example,

$$H_2S(g) + I_2(g) \rightarrow 2HI(g) + S \; ; \; \Delta H = -72.24 \text{ kJ}$$

$$\underset{\text{(Saturated) solution}}{H_2S(g) + I_2} \rightarrow \underset{\text{Solution}}{2HI + S;} \quad \Delta H = -91.98 \text{ kJ}$$

EXOTHERMIC AND ENDOTHERMIC REACTIONS

(a) *Exothermic Reactions :* A reaction which proceeds with the evolution of heat is referred to as an exothermic reaction. Some examples of exothermic reactions are as follows :

$$2Zn(s) + O_2(g) \rightarrow 2ZnO(s) + 693.8 \text{ kJ} \quad ...(1)$$

$$C(s) + O_2(g) \rightarrow CO_2(g) + 393.5 \text{ kJ} \quad ...(2)$$

$$C_2H_2(g) + 2H_2(g) \rightarrow C_2H_6(g) + 314.0 \text{ kJ} \quad ...(3)$$

In all the above reactions, heat is evolved and therefore these are examples of exothermic reactions.

If an exothermic reaction is taking place at constant temperature and constant volume, the energy change is expressed in terms of internal energy changes, *i.e.*, ΔE. Τηυσ,

ΔE = Internal energy of product - Internal energy of reactants = $E_P - E_R$

In exothermic reactions, the total internal energy of the reactants is more than that of products and therefore the excess of internal energy, ΔE, will l appear as heat of reaction or in other words ΔE is negative for exothermic reactions. Thus, equations (1), (2) and (3) in terms of ΔE can be written as :

$$2Zn(s) + O_2(g) \rightarrow 2ZnO(S) \; ; \; \Delta E = -693.8 \text{ kJ}$$

$$C(s) + O_2(g) \rightarrow CO_2(g) \; ; \; \Delta E = -393.5 \text{ kJ}$$

$$C_2H_2(g) + 2H_2(g) \rightarrow C_2H_6(g) \; ; \; \Delta E = -314.0 \text{ kJ.}$$

If an exothermic reaction is taking place at constant temperature and constant pressure, the energy change is expressed in terms of enthalpy change, ΔH, *i.e.*,

ΔH = Enthalpy of products– Enthalpy of reactants = $H_P - H_R$

In exothermic reactions, the total heat content of the reactants is more than that of products and, therefore, the excess of heat constant will appear as heat of reaction. In other words ΔH like ΔE is negative. Thus, equations (1), (2) and (3) in terms of ΔH can be written as :

$$2Zn(s) + O_2(g) \rightarrow 2ZnO(S) \; ; \; \Delta H = -\,693.8 \text{ kJ}$$

$$C(s) + O_2(g) \rightarrow CO_2(g) \; ; \; \Delta H = -\,393.5 \text{ kJ}$$

$$C_2H_2(g) + 2H_2(g) \rightarrow C_2H_6(g) \; ; \; \Delta H = -\,314.0 \text{ kJ}$$

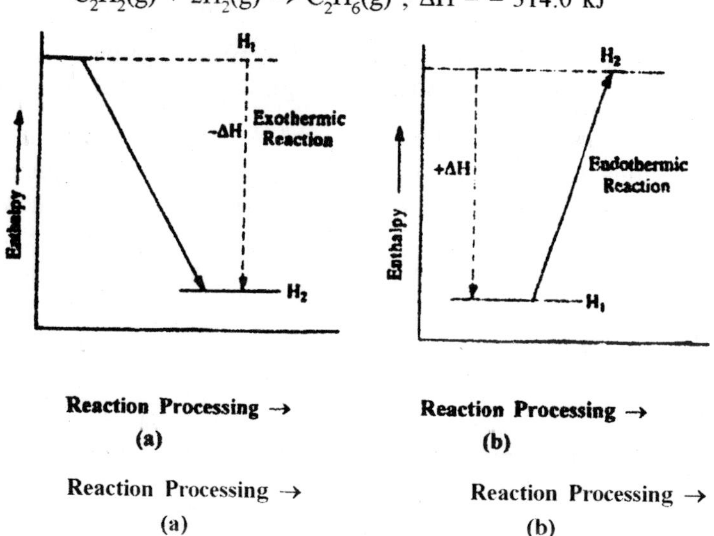

Reaction Processing →
(a)
Reaction Processing →
(b)

Fig. 3.7

As ΔH is negative for exothermic reactions, this is illustrated in Fig. 3.7(a) and (b).

(b) *Endothermic Reactions :* A reaction which proceeds with the absorption of heat is referred to as an endothermic reaction. Some examples of endothermic reaction are as follows :

$$2HgO(s) + 180.4 \text{ kJ} \rightarrow 2Hg(l) + O_2(g) \qquad ...(1)$$

$$H_2O(g) + C(s) + 131.2 \text{ kJ} \rightarrow CO(g) + H_2(g) \qquad ...(2)$$

$$N_2(g) + O_2(g) + 180.5 \text{ kJ} \rightarrow 2NO(g) \qquad ...(3)$$

If endothermic reactions are taking place at constant temperature and volume, the total internal energy of the reactants (E_R) is less than the total internal energy of the products (E_p), *i.e.*, ΔE is positive. Thus, equations (1), (2) and (3) can be put as :

$$2HgO(s) \rightarrow Hg(l) + O_2(g);\ DH = 180.4\ kJ$$

$$H_2O(g) + C(s) \rightarrow CO(g) + H_2(g);\ DH = 131.2\ kJ$$

$$N_2(g) + O_2(g) \rightarrow 2NO(g);\ DE = 180.5\ kJ.$$

If endothermic reactions are carried out at constant pressure and constant temperature, then the total heat content of products is more than the total heat content of reactants. In other words, DH is positive. It is represented. In terms of DH, equations (1) (2) and (3) may be put as:

$$2HgO(g) \rightarrow 2Hg(l) + O_2(g);\ DE = 180.4\ kJ$$

$$H_2O(g) + C(s) \rightarrow CO(g) + H_2(g);\ DE = 131.2\ kJ$$

$$N_2(g) + O_2(g) \rightarrow 2NO(g);\ DH = 180.5\ kJ$$

Melting of solids and vaporization of liquids are endothermic reactions. Synthesis of proteins in living systems is an endothermic process.

(c) Differences between Exothermic and Endothermic Reactions.

The differences are given in tabular from :

Table 1

Exothermic Reactions	*Endothermic Reactions*
1. Heat is given to the surroundings.	1. Heat is absorbed from the surroundings.
2. The internal energy of reactants is more than that of products, *i.e.*, ΔE is − ve.	2. The internal energy of products is more that of reactants, *i.e.*, ΔE is + ve.
3. The enthalpy of reactants is more than that of Products, *i.e.*, ΔH is −ve.	3. The enthalpy of reactants is less than that of Products, i.e, ΔH is +ve.

STANDARD ENTHALPY CHANGES

As the enthalpy of a reaction depends on temperature, a standard state is chosen to compare the enthalpies or reactions of different substances. A convenient standard state for a substance is the most stable state of aggregation under 1 atmosphere and at the specified temperature, generally at 25°C or 298 K.

For pure solids, liquids and ideal gases, the standard state corresponds to the state of the substances at 1 atm. Pressure and specified temperature.

For the dissolved substances, the standard state of the solute is that concentration which gives unit activity.

The standard enthalpy of a reaction at a temperature T and at a pressure of 1 atm is denoted by $\Delta H^\circ(T)$ and is the value of $\Sigma\ \Delta H_{products} - \Sigma\ \Delta H_{reactants}$, at that temperature when all the substances (or elements) in the reaction are in their standard states.

The enthalpy of every element in its standard state is arbitrarily given a zero value. For example, the enthalpy of formation of a mole of $CO_2(g)$ from its elements in the standard state is – 393.5 kg mol^{-1}. *i.e.*,

$$C(graphite) + O_2(g) \rightarrow CO_2(g);\ \Delta H^\circ = -\ 393.5\ kJ.$$

The value of ΔH°_{298K} is given as follows:

$$\Delta H^\circ_{298K} = H^\circ CO_2(g) - [H^\circ O_2(g) + H^\circ_{Cgrphite}]$$

In the above equation, H° refers to the standard enthalpy per mol.

As the enthalpies are in their standard states, their heat contents by convention are zero, *i.e.*,

$$H^\circ_{Cgrphite} = H^\circ_{O2(g)} = 0.$$

Hence under these conditions,

$$\Delta H^\circ_{298K} = H^\circ_{CO2(g)} = -\ 393.5\ kJ.$$

Thus, the standard enthalpy of any compound is the enthalpy of the reaction by which it is formed from its elements in the standard state.

VARIOUS TYPES OF ENTHALPY CHANGES

Enthalpy of a reaction is given different names depending upon the types of reactions. Some of those are described below.

Heat or Enthalpy of Formation

The enthalpy of formation of a compound at the given temperature may be defined as the enthalpy change when one mole of a compound (in its standard state) is formed its elements (in their standard states). It is usually denoted by ΔH°_f.

In the above definition, the standard state of any substance is taken as its natural state at 298 K under one atmosphere pressure. Let us consider the following reactions :

$$C(graphite) + O_2(g) \rightarrow CO_2(g);\ \Delta H = \Delta H^\circ_f = -\ 393.5\ kJ$$

$$C(graphite) + 2H_2(g) \rightarrow CH_4(g);\ \Delta H = \Delta H^\circ_f = -\ 74.8\ kJ$$

Sometimes, the enthalpy of combustion and enthalpy of formation relate to the same reaction ; for instance, the enthalpy of formation of carbon dioxide is the same as the enthalpy of combustion of graphite.

The compound with negative enthalpy of formations called exothermic compound whereas the compound with positive enthalpy of formations known as endothermic compound. Thus, the formation of water is an exothermic compound whereas the formation of hydrogen iodide is an endothermic compound.

$$H_2(g) + 1/2O_2(g) \rightarrow H_2O(g);\ \Delta H^\circ_f = -\ 245.5\ kJ$$

$$1/2H_2(g) + 1/2I_2(s) \rightarrow HI(g);\ \Delta H^\circ_f = +\ 25.94\ kJ$$

The negative enthalpy of formation of a compound means that its heat content is less than that of its elements and hence it is more stable than its elements. The reason is for that the compound with a lower enthalpy (or energy) must be more stable. Therefore, the chemical bonds in H_2O are stronger than those in H_2 or O_2. On the other hand, the positive enthalpy of formation of a compound means that its enthalpy of content is more than that of its elements and hence it is less stable than its elements. Due to this reason, H – I bond is weaker than H – H bond in hydrogen or I – I bond in iodine.

The enthalpy of formation of a compound may be obtained either by direct measurement of ΔH for the reaction or from enthalpies of reactions involving the compound.

Heat or Enthalpy of Combustion

The enthalpy of combustion of a substance at a given temperature may be defined as the enthalpy change accompanying the complete combustion of one mole of the substance in excess of oxygen or air at that temperature.

The words "complete combustion" in the above definition must be emphasized because it is possible to burn a compound to yield both CO and CO_2. In order to avoid this confusion. The combustion is carried out in a large excess of oxygen or air to yield carbon dioxide and water as the final products. For example, heat of combustion relates to the reaction :

$$C_2H_2(g) + 21/2O_2(g) \rightarrow 2CO_2(g) + H_2O(g);\ \Delta H^\circ = -\ 1258\ kJ$$

and not to the partial oxidation as follows :

$$C_2H_2(g) + 11/2O_2(g) \rightarrow 2CO(g) + H_2O(g);\ \Delta H^\circ = -691 kJ.$$

The heat of combustion depends upon the physical state of the substances involved. For example, the heat of combustion of hydrogen at 298 K is −285.9 kJ mol^{-1}.

In this case only liquid water is obtained. On the other hand if the combustion is taking place at 313 K or above, the water vapour in this case cannot exist in the liquid state but evaporates to form vapours. The enthalpy of combustion is that case will be −245.5 kJ.

$$H_2(g) + 1/2O_2(g) \rightarrow H_2O(1);\ \Delta H^\circ_{298k} = -285.19 \text{ kJ at } 298 \text{ k}$$

$$H_2(g) + 1/2O_2(g) \rightarrow H_2O(g);\ \Delta H^\circ_{373k} = -245.5 \text{ kJ at } 373$$

The enthalpy of combustion at 373 K is lower as compared at 298 K. The reason for this is that small quantity of energy (+ 40 .44 kJ) is used up to convert liquid water into its vapours.

$$H_2O(l) \rightarrow H_2O(g);\ \Delta H^\circ - +40.44 \text{ kJ}.$$

The enthalpy of combustion of methane is

$$CH_4(g) + 2O_2(g) \rightarrow CO_2(g) + 2H_2O(l);\ \Delta H^\circ = -890.40 \text{ kJ}.$$

Enthalpy of combustion may be used to calculate the enthalpy of formation of a compound provided the enthalpy of formation of other combustion products are known.

For example, it is possible to calculate enthalpy of formation of gaseous methane provided the enthalpies of formation of $CO_2(g)$ and $H_2O(l)$ are known.

$$\Delta H^\circ = -890.40 = \Delta H^\circ_{CO_2(g)} + 2\Delta H^\circ_{H_2O(l)} - \Delta H^\circ_{CH_4(g)}$$

or $$\Delta H^\circ_{CH_4(g)} = 890.40 + \Delta H^\circ_{CO_2(g)} + 2\Delta H^\circ_{H_2O(l)}$$

$$= 890.40 + (-393.50) + 2(-285.83) = -74.76 \text{ kJ mol}^{-1}.$$

Application of Heat Combustion Date : Main applications are :

(i) Calorific value of fuels

(ii) Food values of diets

(iii) Maximum flame temperature

(iv) Calculation of heal of formation

(v) Calculation of heat of reaction.

Heat or Enthalpy of Hydrogenation

It may be defined as the enthalpy change when a mole of an unsaturated organic compound is fully hydrogenated. The enthalpies of hydrogenation of ethylene and benzene are represented as follows .

$$CH_2 = CH_2(g) + H_2(g) \rightarrow CH_3 - CH_3(g);$$

$$\Delta H^o = -132.5 \text{ kJ}$$

$$C_6H_6(l) + 3H_2(g) \rightarrow C_6H_{12}(l);$$

$$\Delta H^o = -246 \text{ kJ}$$

From the above thermochemical equation, it is to be noted that enthalpy of hydrogenation of benzene is not three times that of enthylene one may expect this due to the presence of three double bonds in benzene and one double bond in enthylene.

The enthalpy of hydrogenation of benzene is less be 151.5kJ mol^{-1} which is attributed to the stabilisation on benzene due to resonance. This method may sometimes be used as the characteristic of stabilisation of aromatic molecules due to resonance.

Heat or Enthalpy of Neutralization

The enthalpy of neutralization of an acid at a given temperature may be defined as *the enthalpy change accompanying the neutralization of one gram equivalent of a base by an acid in dilute solution at that temperature*.

For example, the enthalpy of neutralization one of hydrochloric acid by sodium hydroxide in dilute solution at 298 K is -57.32 kJ.

$$NaOH + HCI \rightarrow NaCI + H_2O$$

As neutralization is the reaction between $H^+(aq)$ form the base, one can write the above equation as :

$$H^+(aq)+CI^-(aq)+Na^+(aq)+OH^-(aq) \rightarrow H_2O(I)+Na^+(aq)+CI^-(aq)$$

or $$H^+(aq) + OH^-(aq) \rightarrow H_2O(I)$$

From this it follows that enthalpy of neutralization should be independent of the nature of acid and the base.

This has been found to be true in the case of neutralization of strong acids by strong bases. This is evident form the data given in Table 2.

Table 2 : Enthalpies on Neutralization of Strong Acids by Strong Bases at 298 K

Acid	*Base*	*Enthalpy of Neutralization (kJ per gram equivalent)*
HCI	NaOH	–57.32
HCI	KOH	–57.45
HCI	LiOH	–57.32
HNO_3	NaOH	–57.28

Heat or Enthalpy of Ionisation

For a weak acid or a weak base, the enthalpy of neutralisation is not the same but less than –57.32 kJ. It is because of the fact that these acids or bases are not completely ionised in solution. Some of the heat is consumed in ionising these acids and bases. The heat required to ionise the weak acid or the weak base is known as the enthalpy of ionisation.

It is possible to calculate the enthalpy of ionisation of a weak acid from the enthalpies of neutralisation of strong acid and weak acid by strong bases. For example, in the neutralisation of acetic acid by sodium hydroxide, the enthalpy of neutralisation is – 55.43 kJ mol^{-1}. The enthalpy of ionisation of acetic acid will therefore be equal to +1.89 kJ mol^{-1}. this can seen as follows :

(i) $CH_3COOH(aq) + OH^-(aq) \rightarrow CH_3COO^-(aq) + H_2O(l)$;
$\Delta H^\circ = -55.43$ kJ

(ii) $H^+(aq) + OH^-(aq) \rightarrow H_2O(l)$; $\Delta H = -57.32$ kJ

(iii) $CH_3COOH(aq) \rightarrow CH_3COO^-(aq) + H^+(aq)$;
ΔH° Ionisation = ?

On adding Eqs. (ii) and (iii), we obtain

$$CH_3COOH(aq) + OH^-(aq) \rightarrow CH_3COO^-(aq) + H_2O(l)$$

or $\Delta H^\circ = -57.32 + \Delta H^\circ$ Ionisation

But from Eq. (i), we have

$CH_3COOH(aq) + OH^-(aq) \rightarrow CH_3COO^-(aq) + H_2O(l)$

or $\Delta H^\circ = -55.43$ kJ mol^{-1}

Hence ΔH° Ionisation = –55.43 + 57.32 = + 1.89 kJ mol.

Heat or Enthalpy of Solution

When a solution is dissolved in a solvent, heat may be either liberated or absorbed. If the heat is absorbed the solution becomes cooler and ΔH is given a positive sign. On the other hand if the heat is evolved, the solution gets warmer and ΔH is given a negative sign.

The enthalpy change per mole of a solute dissolved varies to a considerable extent with dilution and to some extent with the initial temperature of the solvent. Therefore, it becomes essential to express the enthalpy change with reference to the concentration of the solution. "The enthalpy of solution may be defined as the enthalpy change when one mole of the solute is dissolved in a specified amount of solvent."

The enthalpy of solution can be explained by considering the dissolution of HCI (g) in water at 298 K:

$$HCl\,(g) + 25H_2O(l) \rightarrow HCl.25\,H_2O;\ \Delta H^\circ = -69.5\ kJ$$

$$HCl\,(g) + 40H_2O(l) \rightarrow HCl.40\,H_2O;\ \Delta H^\circ = -73.18\ kJ$$

$$HCl\,(g) + 200H_2O(l) \rightarrow HCl.200H_2O;\ \Delta H^\circ = -74.40\ kJ$$

$$HCl\,(g) + \infty\ H_2O(l) \rightarrow HCl.(aq);\ \Delta H^\circ = -75.17\ kJ$$

From the above equation it means that when one mole of HCI is dissolved in 25 moles of water, 69.5 kJ of heat are evolved. But if the solution is further diluted to such an extent that further dilution would cause no additional heat change, then the equation may be written as :

$$HCl\,(g) + \infty\ H_2O(l) \rightarrow HCl.(aq);\ \Delta H^\circ = -75.17\ kJ\ mol^{-1}$$

where aq. represents a large volume of water. The value of ΔH° correspondents to this reaction is called the enthalpy of solution at infinite dilution. It is the maximum quantity of heat released in the formation of a solution of one mole of HCI(g) in excess of water.

For greater accuracy, two types of heat of solution have been recognised. There are :

(a) *Integral Heat of Enthalpy of Solution :* Integral heat of solution at a particular concentration is the enthalpy change when 1 mole of solute is dissolved in pure solvent to produce a solution of the given concentration. Thus, if 1 mole of solute is dissolved in 500g of water the heat change gives the value of integral heat of solution at the concentration of 2 molal.

While recording integral enthalpies of solution, the usual practice is to state the amount of solvent in which one mole of solute is dissolved. For example :

$$HCI\ (g) + 10H_2O(l) \rightarrow HCI.(10\ H_2O);\ \Delta H^\circ = -69.5\ kJ.$$

indicates that when one mole of hydrogen chloride gas is dissolved in 10 moles of water, there is an evolutions 65.5 kJ of heat energy.

According to Hess's law, the integral enthalpy of solution is equal to the difference between integral enthalpies of the solution at the two concentrations. For example, if to a solution of 1 mole of hydrogen chloride in 40 moles of water, enough water is added such that

$$HCI.40\ H_2O + aq \rightarrow HCI.(aq)$$

the integral enthalpy change can be calculated as follows :

$$HCI(g) + aq \rightarrow HCI\ ; \quad \Delta H^\circ = -\ 75.17\ kJ$$

$$HCI(g)\ 40\ H_2O \rightarrow HCI.\ 40H_2O; \quad \Delta H^\circ = -\ 73.18\ kJ$$

On subtracting, we have

$$HCI.40\ H_2O + aq \rightarrow HCI.(aq); \quad \Delta H^\circ = -\ 1.99\ kJ$$

If ΔH is plotted against number of solute per 1000 g solvent, the resultant curve is shown in Fig.3.8. Initially, ΔH increases almost linearly with m but later the increase is not as fast as m. Finally, it reaches a constant value when the solution becomes saturated with respect to solute.

If the observers value of enthalpy change (ΔH) is divided by the amount of solute (m) that is dissolved to form a solution of a particular concentration, we get the quantity $\Delta H/m$, which is nothing but the integral enthalpy of solution at the given concentration.

Integral enthalpies of solutions, enthalpies of dilutions and enthalpies of reactions in solutions can be evaluated from the enthalpies of formation in solution. In these calculations, the enthalpy of formation of water is neglected if the same number of moles of water appears on both sides of the balanced thermochemical equation.

(b) *Differential Heat or Enthalpy of Solution* : The differential heat of solution, $DH_{D.s}$ may be defined as the heat of solution of a mole of solute in such a large quantity of solution of known concentration that addition of 1 more mole of solution does not change the concentration appreciably. If on dissolving dm mole of the solute in a finite amount of the solution the heat change is dq, then differential heat of solution will be

$$\Delta H_{D.s} = \delta q/\delta m$$

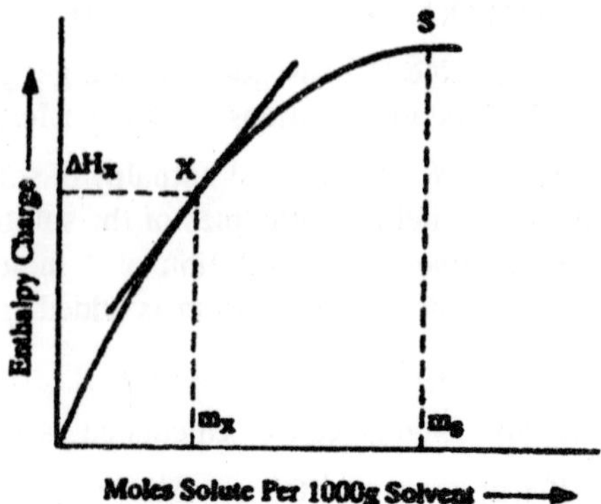

Fig. 3.8 : Variation of enthalpy change when the amount m of solute is dissolved in 1 kg.

From the shape of the curve of Fig. 3.8, it can be seen that the value of the slope will depend upon the concentration of the solution. Thus, the differential enthalpy of a solution besides depending to T and p, will also depend on the amounts of solvent n_1 and solute n_2 that are present in the given solution *i.e.*,

$$d(\Delta H)/dm = f(T, p\ n_1, n_2)$$

Thus, for this reason the concentration of the solution is mentioned while defining the differential enthalpy of solution. The following conclusions may be drawn from the shape of the curve between ΔH and m as shown in Fig. 1.

(i) For smaller value of m, the curve is almost linear; thus its slope will have a constant value and will be equal to DH/m. As the latter represents the integral heat of solution, it means that the differential and integral heats of solution are essentially equal for very dilute solutions.

(ii) For higher values on m, the curve is not linear. The reason is that ΔH does not increase as fast as m; thus the slope of the curve decreases as the value of m increases. In other words, the differential heat of solution decreases as the concentration of the solution increases, and becomes zero when the solution gets saturated.

The differential enthalpy of solution can be obtained by plotting ΔH of the solution at various concentrations against the number of moles of solvent associated with a definite quantity of solute, and finding the slope of the curve at a point corresponding to any particular concentration. The value of the slope at a given concentration gives the differential enthalpy of solution at that concentration.

The differential heats of solution and dilution are essentially the partial molar heat of solute and solvent, respectively. Consider a solution having n_1 moles of solvent and n_2 moles of solute. In general,

$$\Delta H = f(T, p\ n_1, n_2)$$

The differential of ΔH is given as follows :

$$d(\Delta H) = \left(\frac{\partial(\Delta H)}{\partial T}\right)_{P,\, n_1,\, n_2} dT + \left(\frac{\partial(\Delta H)}{\partial P}\right)_{T,\, n_1,\, n_2} dP$$

$$+ \left(\frac{\partial(\Delta H)}{\partial n_1}\right)_{T,P,\, n_2} dn_1 + \left(\frac{\partial(\Delta H)}{\partial n_2}\right)_{T,P,\, n_1} dn_2$$

At constant temperature and pressure, the above equation becomes as follows :

$$d(\Delta H) = \left(\frac{\partial(\Delta H)}{\partial n_1}\right)_{T,P,\, n_2} dn_1 + \left(\frac{\partial(\Delta H)}{\partial n_2}\right)_{T,P,\, n_1} dn_2$$

or $d(\Delta H) = \left(\Delta\overline{H}_1\right) dn_1 + \left(\Delta\overline{H}_2\right) dn_2$

where ΔH_1 and ΔH_2 are termed as partial molar values of heat of solvent and solvent and solute, respectively. It can be seen that ΔH_2 is equal to the slope d $(\Delta H/dn_2)$ of a varve in which the property ΔH is plotted as a function of n_2 for a series of solutions in which n_1 is kept constant. When it is compared with the definition given at the beginning it shows that the differential heat of solution is really the partial molar heat of solution of solute, *i.e.*, ΔH_2. Similarly, the differential heat of dilution is the corresponding partial molar property of the solvent, *i.e.*, $\Delta\overline{H}_1$.

Direct experimental measurement of differential heat of solution is not possible, it can, however, be determined indirectly from the knowledge of integral heat of solution

$$\Delta H_{D.S} = d\ (m\ \Delta H_{I.S})/dm$$

where $m\Delta H_{I.S}$ is the integral heat of solution of the concentration m molal.

As it is obvious from the definitions, the integral and differential heats of solutions approach each other with decreasing the concentration and become idential when the concentration is zero.

Enthalpy of Dilution : If it known than t the enthalpy of solution per mole of a solute gets varied with the concentration of the solution. Therefore, it follows that if a given solution is diluted on the addition of solvent, there occurs noticeable change of enthalpy. This is termed as the enthalpy of dilution.

Enthalpy of dilution may be defined as the change of enthalpy when a solution having 1 mole of solute is diluted from one concentration to another.

Heat or Enthalpy of Hydration

The heat or enthalpy of hydration of salt is the change in heat content ΔH when one mole of an anhydrous substance combines with requisite number of moles of water to form the hydrate. It is obtained from the enthalpies of solutions of hydrous and anhydrous substances. Enthalpy of hydration of copper sulphate can be calculated from the following thermochemical equations :

(i) $CuSO_4.5H_2O + aq \rightarrow CuSO_4\ (aq);\ \Delta H^\circ = 11.72$ kJ

(ii) $CuSO_4(s) + 5H_2O \rightarrow CuSO_45H_2O\ (s);\ \Delta H^\circ = x$ kJ

(iii) $CuSO_4(s) + 5H_2O\ (l) + aq \rightarrow CuSO_4\ (aq);\ \Delta H^\circ = -66.5$ kJ

From the above equations, we get

$$CuSO_4\ (s) + 5H_2O\ (l) \rightarrow CuSO_4\ 5H_2O(s);\ \Delta H^\circ = -78.22 \text{ kJ}.$$

Heat or Enthalpy of Transition

It may be defined as the enthalpy change when one mole of one allotropic form changes to another. It includes transition from solid to liquid (fusion), liquid to vapour (vaporisation), solid to vapour (sublimation) and change from one crystalline form to another crystalline form (polymorphic transition). For example, in the transition of sulphur (rhombic) to sulphur (monoclinic), The enthalpy change is 13.14 kJ.

$$S_{rohmbic} \rightarrow S_{monoclinic};\quad \Delta H^\circ = 13.14 \text{ kJ}$$

Another example is

$$C_{graphite} \rightarrow C_{diamond},\ \Delta H^\circ = 1.90 \text{ kJ}$$

The enthalpy of transition in the above case can be obtained from the enthalpies of combustion of $C_{graphite}$ and $C_{diamond}$

(i) $C_{graphite} + O_2\ (g) \rightarrow CO_2\ (g);\ \Delta H^\circ = -\ 393.51$ kJ

(ii) $C_{diamond} + O_2\ (g) \rightarrow CO_2\ (g):\ \Delta H^\circ = -\ 395.41$ kJ

On subtracting (ii) from (i), we get

$$C_{graphite} \rightarrow C_{diamond}; \qquad \Delta H^\circ = 1.90 \text{ kJ}$$

Heat or Enthalpy of Precipitation

It may be defined at the enthalpy change when one mole of a precipitate is formed upon mixing dilute solutions of relevant electrolytes. Some examples are :

$BaCl_2$ (aq) + Na_2SO (aq) → $BaSO_4$(s) + 2NaCl (aq);

ΔH° = 19.5 kJ

As the reactants are strong electrolytes, the reaction is essentially ionic in nature and may be written as follows :

Ba^{2+} (aq) $2Cl^-$(aq) + $2Na^+$(aq) + SO_4^{2-}(aq)

→ $BaSO_4$(s) + $2Na^+$ (aq) + $2Cl^-$(aq), ΔH = – 19.50 kJ

As Na^+(aq) and Cl^-(aq) ions are common on both sides, the reaction, in fact, involves precipitation of $BaSO_4$(s), *i.e.*,

Ba^{2+} (aq) + SO_4^{2-}(aq) → $BaSO_4$(s); ΔH = – 19.50 kJ

Another example is as follows :

$AgNO_3$ (aq) NaBr (aq) → AgBr(s) + $2NaNO_3$ (aq);

ΔH° = – 17.78 kJ

Enthalpy of Fusion

The enthalpy of fusion of a substance may be defined as the enthalpy change when 1 mole of the substance is converted from the solid to the liquid phase at its melting point. When 1 mole of ice melts at 273 K, the enthalpy change is approximately 6.6 kJ. This result can be put in the equation form as :

$$H_2O(s) \rightarrow H_2O(l), \qquad \Delta H = 6.0 \text{ kJ at } 273 \text{ K}$$

The enthalpies of fusion of molecular solids (O_2, CCl_4, H_2S, etc.) are low where as the enthalpies of fusion of ionic solids (NaCl, $MgCl_2$) are large. Thus one may get some idea from the value of fusion, about the nature of the solid and the magnitude of forces acting between the particles constituting the solid.

Enthalpy of Vaporization

The enthalpy of vaporization $\Delta(H_V)$ of a liquid may be defined as the enthalpy change accompanying the conversion of 1 mole of the substances form the liquid to the vapour phase at its boiling point. In order to cover liquid into its vapours the substance will absorb energy from the surroundings and thus ΔH will have + ve sign. For example, if water is heated at its boiling point, 373 K, it is converted into vapours by absorbing 40.6 kJ mol^{-1} energy from the surroundings. This means ΔH_V of water is 40.6 kJ mol^{-1}.

$$H_2O(l) \rightarrow H_2O(g); \qquad \Delta H_V = +40.6 \text{ kJ mol}^{-1}$$

Similarly, ΔH_V for hydrogen and oxygen are 0.904 and 6.7 kJ Mol–1 respectively. If we compare these values with water we realise that the interaction between water molecules is much stronger than that between hydrogen molecule sin the liquid state but less stronger than that between oxygen molecules in the liquid state. This is quite true because in water there is extensive hydrogen bonding whereas in oxygen and hydrogen there are only weak van der waals forces,

The enthalpy of fusion is very small as compared to the enthalpy of vaporization. The reason for this is that attractive.forces in solid and liquid state are much similar (intermolecular distances in solids and liquids are roughly the same) whereas in the gaseous state the attractive forces are much weaker.

Enthalpy of Sublimation

The enthalpy of sublimation (ΔH_{sub}) of a substance may be defined as the enthalpy change accompanying the conversion to 1 mole of solid directly into vapour phase at a given temperature below the melting point of the former. Generally, molecular solids like iodine and naphthalene sublime readily. For example, the enthalpy of sublimation of iodine is 62.39 kJ mol^{-1}.

According to Hess's law, the enthalpy of sublimation is equal to sum of enthalpy of fusion (ΔH_f) and enthalpy of vaporization (ΔH_V). *i.e.*,

$$\Delta H_{sub} = \Delta H_f + H_v$$

The above equation is very useful because it can be employed to calculate the enthalpy of sublimation of the enthalpy of fusion and enthalpy of vaporization are known.

Enthalpy of Atomization

This is defined as the enthalpy change required to convert one mole of an element from its normal state at 298 K and one atmosphere pressure into free atoms.

Enthalpies of atomization, are difficult to measure (they are usually obtained from spectroscopic data), but the are useful in relation to bond energies. For example :

$H_2.g = 2H_{atoms}$ $\Delta H = 433$ kJ ...(1)

$O_2.g = 2O_{atoms}$ $\Delta H = 495$ kJ ...(2)

$H_{2g} + 1/2O_{2.g} = H_2O_g$ $\Delta H = -246$ KJ ...(3)

Then (3) – (1) – 1/2 × (2) gives :

$2H_{atoms} + O_{atoms} = H_2O$ $\Delta H = -926$ kJ

This energy change corresponds to the formation of two oxygen-hydrogen bonds, so that the average bond energy of an oxygen-hydrogen bond in water is 463 kJ.

BOND ENTHALPIES OR ENERGIES

The bond energy of a particular type of bond in a molecule may be defined as the amount of energy required to dissociate or break one mole of that type present in the compound and separate the resulting atoms or radicals from one another. For example, the bond energy of the H – H bond in H_2 is 433 kJ mol^{-1}. It means that 433 kJ of energy is required to break or dissociate the H – H bond in one mole of hydrogen molecules. Similarly, the bond energies of I – I in I_2 and H – I in HI are 151 kJ mol^{-1} and 229 kJ mol^{-1} respectively.

In case of diatomic molecules the bond energy is a definite quantity as there is only one type of bonds present in these molecules. However, the bond energy becomes a much less definite quantity in case of polyatomic molecules. This is to be expected since in a polyatomic molecule, the energy necessary to rupture a given bond will depend to some extend on the nature of remainder molecule. For example, the energy required to rupture the O – H bond in alcohol would not be expected to the same as the energy needed to rupture same bond in an ionic compound. This is obvious from the fact that alcohols are much less acidic than organic aids. Therefore, it follows that bond energy for a particular bond in polyatomic molecules will vary from compound to compound. However, if there are similar types of bonds present in a ployatomic molecule, the bond energy is equal to the average energy required to break a particular bond in one mole of that compound. For example, C – H bond energy in CH_4 can be calculated as follows :

When methane is formed it elements carbon and hydrogen, 1648.7 kJ of heat is liberated. It means 1648.7 kJ of energy must be supplied to one mole of methane to break four identical C – H bonds, resulting the formation of atomic carbon and atomic hydrogen in the gaseous state. It means that the average C – H bond energy is 1648.7/4 or 412.175 kJ mol^{-1}. This is also the value of heat of formation of C – H bond,

From the above discussion, it is evident that the bond energy or enthalpy should be distinguished carefully from the term bond dissociation energy of enthalpy.

Bond Enthalpy or Energy

Bond enthalpy of a given bond is defined as the average of enthalpies or energy required to dissociate the said bond present in different gaseous compounds into free atoms or radicals in the gaseous state.

Bond Dissociation Enthalpy or Energy

Bond dissociation enthalpy or energy may be defined as the enthalpy or energy required to dissociate a given bond of some specific compound.

The distinction between bond enthalpy and bond dissociation enthalpy may be better understood by a simple example, say of the O – H bond. The enthalpy of dissociation of the O – H bond depends upon the nature of molecular species from which the H atom is getting separated. For example, in the water molecule

$$H_2O(g) \rightarrow H(g) + OH(g)\ ;\ \Delta H^\circ = 501.87 \text{ kJ noet}$$

$$OH(g) \rightarrow O(g) + H(g)\ ;\quad \Delta H^\circ = 423.38 \text{ kJ noet}$$

The bond enthalpy of O – H is defined as the average of these values, *i.e.*,

$$\text{Bond enthalpy} = \frac{501.87 + 423.38}{2} = 462.625 \text{ kJ noet}$$

Hence the bond enthalpy given for any particular pair of atom is the average value of the dissociation enthalpies of the bond for a number of molecules in which the pair of atoms appear.

The bond dissociation energies for the stepwise dissociation of CH_4 are as follows :

$$CH_4(g) \rightarrow CH_3(g) + H(g);\ \Delta H^\circ = 427.0 \text{ kJ}$$

$$CH_3(g) \rightarrow CH_2(g) + H(g);\ \Delta H^\circ = 418.4 \text{ kJ}$$

$$CH_2(g) \rightarrow CH(g) + H(g);\ \Delta H^\circ = 460.2\ kJ$$

$$CH(g) \rightarrow C(g) + H(g);\ \Delta H^\circ = 343.1\ kJ$$

The energy for each step is different due to the fact that in each case a different dissociating fragment is involved. The C – H bond enthalpy would be one-fourth of the enthalpy required to break a mole of CH_4 and is 412 kJ.

Calculation of Bond Enthalpy or Energy

It is possible to obtain bond energies from data on heats of combustion and heats of dissociation. Let us illustrate it by calculating the bond enthalpy or energy of C – H. We know,

$$CH_4(g) \rightarrow C(g) + 4H(g)\ \epsilon_{C-H} = \frac{\Delta H}{4}$$

The ΔH and hence ϵ_{C-H} for this reaction can be obtained by adding the following equations :

$CH_4(g) + 2O_2(g) \rightarrow CO_2(g) + 2H_2O(l)$	$\Delta H^\circ = -890.36 kJ$
$CO_2(g) \rightarrow C\ (graphite) + O_2(g)$	$\Delta H^\circ = 393.51\ kJ$
$2HO_2(l) \rightarrow 2H_2(g) + O_2(g)$	$\Delta H^\circ = 571.70\ kJ$
$2H_2(g) \rightarrow 4H(g)$	$\Delta H^\circ = 871.86\ kJ$
$C\ (graphite) \rightarrow C(g)$	$\Delta H^\circ = 716.68\ kJ$
$CH_4(g) \rightarrow C(g) + 4H(g)$	$\Delta H^\circ = 1663.39\ kJ$

$$\text{Thus, at 298.15o k } \epsilon_{C-H} = \frac{1663.39}{4} = 415.85\ kJ$$

This value of the C – H bond energy does not correspond to the dissociation energy of the carbon-hydrogen bond in methane, which is 426.77 kJ mol^{-1} and refers to the equation

$$CH_4(g) \rightarrow CH_3(g) + H(g)$$

The data on bond energies are used to calculate the approximate heat of formation of a compound of known structure by adding the appropriate bond energies. Wherever direct experimental data are not available, approximate values of enthalpies of reaction can also be obtained by using bond enthalpy data.

Estimation of the Enthalpy of Formation from Bond Enthalpies

Let us calculate the heat of formation of $Se_2Cl_2(g)$ from bond enthalpies by involving the following steps ; as the bond energy refers to the dissociation of Se – Se CI gas into the gaseous atoms, the enthalpy change for the formation of this gaseous molecule should be as follows :

$$2Se(g) + 2CI(g) \rightarrow Se_2CI_2(g)$$

$$\Delta H = -[\in Se - Se + 2 \in Se - CI] = -694.54 \text{ kJ mol}^{-1}.$$

However to estimate the heat of formation it becomes necessary to add two reactions in the above equation, as by definition the heat of formation refers to the elements in their standard states. Therefore, the following enthalpy changes are introduced to convert the elements from their standard states to the gaseous atoms at 298°K

$$CI_2(g) \rightarrow 2CI(g) \ \Delta H = 243.30 \text{ kJ mol}^{-1}$$

$$2 \text{ Se (hexagonal)} \rightarrow 2 \text{ Se(g)} \quad \Delta H = 2 \times 2 \times 202.5 \text{ kJ mol}^{-1}$$

By adding these in the preceding expression, we obtain

$$2 \text{ Se (hexagonal)} + CI_2(g) \rightarrow Se_2CI_2(g);$$

$$\Delta H_f = -46.22 \text{ kJ mol}^{-1}$$

Estimation of the Heat of Reaction from Bond Enthalpies or Energies

Let us illustrate this by calculating the enthalpy change for the following equation from the bond energy data :

$$C_2H_4(g) + HCI(g) \rightarrow C_2H_5CI(g)$$

The enthalpy change for the above reaction can be calculated as follows :

ΔH = Energy required to break reactants into + Energy released to form products from the gaseous atoms gaseous atoms

$$= [4 \in_{C-H} + \in_{C=C} + \in_{H-CI}] + [-5 \in_{C-H} - \in_{C-C} - \in_{C-CI}]$$

$$= (\in_{C=C} + H - CI) - (\in_{C-H} + \in_{C-C} + \in_{C-CI})$$

On substituting the bond enthalpy values in the above equation, we get

$$\Delta H = (615.05kJ + 431.79kJ) - (413.38kJ + 347.69kJ + 328.44kJ)$$

$$= -42.67kJ$$

Bond Energies and Resonance

There occurs a fair agreement between the calculated values of heat of formation obtained from bond enthalpies and any other method.

However, there occurs large deviations in case of compounds having alternate single and double bonds. For example, the following reaction.

$$\underset{\text{Benzene}}{C_6H_6(g)} \rightarrow 6C(g) + 6H(g)$$

will require enthalpy of 5368.5 kJ on the basis of bond energies ($3\,\epsilon_{C=C} + 3\epsilon_{C=C} + 6\epsilon_{C-H} = 3 \times 347.69$ kJ $+ 3 \times 615.05$ kJ $+ 6 \times 413.38$ kJ = 5368.5 kJ), while the experimental value is 5351.1 kJ. This amounts to the fact that benzene is more stable by 166.6 kJ mol^{-1}. This is attributed to resonance, that is, in benzene there occurs no localization of single and double bonds, but the molecule is resonating from one extreme to another.

Factors: The value of the bond energy depends upon the following factors :

(a) *Length of the bond :* In general, it is found that greater the bond length between the two nuclei of the two atoms forming the bonds, the smaller is the bond energy.

Type of bond	H – H	CI – CI	I – I
Bond length (i)	0.74	1.25	2.7
Bond energy (kJ/mol)	433.1	330.5	151.2

(b) *Bond polarity :* Higher is the difference in electronegativity between the combining atoms, greater is the strength of the dipole along the bond and higher is the bond energy.

(c) *Unsaturation :* In general it is observed that with the increase in the number of covalent bonds between the atoms, there occurs an increase in the bond energy but the contribution of each additional bond goes on decreasing.

Bond	*Energy in kJ/mole*
C – C	347.3
C = C	615.0
C = C	811.7

(d) *Resonance effect :* If the bonds in the molecule do not remain fixed but the molecule possesses resonance, the bond energy per bond depends upon the amount of resonance.

(e) *Overlap of Orbitals* : The bond energy depends upon the s character which is determined by extent of overlap of orbitals of the combining atoms.

CONVENTIONS USED IN WRITING THERMOCHEMICAL EQUATIONS

Physical States of Reactants and Products

The physical states of reactants or products have been found to influence the magnitude of heat evolved or absorbed in the reaction.

Therefore, these are indicated by symbols (g), (*l*) and (s) placed after the chemical formulae. These stand for gaseous, liquid and solid states respectively. However, sometimes confusion arises in representing the physical state of water formed as a chemical reaction. This can be understood by the following reaction at 25°C :

$$H_2(g) + \frac{1}{2}O_2(g) \rightarrow H_2O(l) + 286.2 \text{ kJ}$$

In the above reaction, since experiment is carried out at 25°C, water is obtained in the liquid state.

If the above reaction is carried out at a temperature a little above 100°C under atmospheric pressure, the water produced will remain in the vapour state. The heat evolved in the reaction is 245.5 kJ. The thermochemical equation will now be written as follows :

$$H_2(g) + \frac{1}{2}O_2(g) \rightarrow H_2O(l) + 245.5 \text{ kJ}.$$

Allotropic Forms of Substances

As different allotropic forms of a substance are associated with different amounts of energy, it means that in case of elements showing allotropy, the name of the allotropic form must be mentioned in thermochemical equation. For example, in a thermochemical equation involving sulphur, we must mention the allotropic forms, rhombic, monoclinic, etc. Thus, in a thermochemical equation, rhombic and monoclinic varieties of sulphur are represented as S_R (rhombic) and SM (monoclinic) respectively.

Reaction in Solutions

When a reaction takes place in a solution, the solvent must be indicated. For example, the symbol (aq) placed after the formula of a substance, reveals that substance is present in large excess of water to ensure that there occurs no heat change on adding further quantity of water.

Heat Changes Involved in Reactions

Two different conventions are used to represent the heat changes of the chemical processes :

(a) *Old Convention :* According to this convention, the heat evolved or adsorbed in a chemical reaction is represented as a part of the equation on the product side. A positive sign (+) is used if heat is given out and negative sign (–) is used if heat is absorbed in a chemical reaction. Thus,

$$A + B \rightarrow C + D + x\ kJ$$

$$A + B \rightarrow C + D - x\ kJ$$

This convention is out of date these days.

(b) *Modern Convention :* This convention is based on thermodynamics. The heat evolved or absorbed in a process may be represented by ΔH, the difference in the total heat contents of the products and reactants. It is not written as a part of equation but is written separately. If heat is absorbed in a chemical reaction, + ΔH is employed and if heat is evolved. Then $-\Delta H$ is employed. For example, the burning of carbon in oxygen to form carbon dioxide was previous represented as

$$C_{graphite} + O_2(g) \rightarrow CO_2(g) + 393.5\ kJ$$

At present it is represented as (Old convention)

$$C_{graphite} + O_2(g) \rightarrow CO_2(g);$$
$$\Delta H = -393.5\ kJ$$

Temperature and Pressure of Substances

The heat absorbed or evolved in a chemical reaction has been found to depend upon the temperature and pressure at which the reaction is being carried out. These two quantities are, therefore, also in the thermochemical equation.

If a reaction is carried out under standard conditions *i.e.*, 25°C or 298 K and 1 atm pressure, the heat change is indicated as ΔH°. The superscript' indicates that pressure is 1 atmosphere and reaction in the standard state.

SPECIFIC HEAT OF SOLIDS

The specific heat of a substance is defined as the amount of heat required to raise the temperature of 1 gm of the substance through 1°C. The specific heat of a solid increases with rise of temperaturé and tends to zero as the absolute zero of temperature is approached. For example, diamond (carbon) has a specific heat of 0.095 at 0°C, 0.19 at 100°C and 0.459 at 1000°C. The variation of specific heat of solids with temperature has been explained by a number of scientists. Some of the main attempts are outlined below :

Dulong and Petit's Law

The end of eighteenth century has seen great activity in what we will now call theoretical chemistry. *Lavoisier* and *Laplace* had determined the specific heats of a number of metals with considerable accuracy. In 1819, Dulong and Petit, two French scientists, pointed out that, using the accepted atomic weights for many metals and multiplying them by the specific heats, an approximately constant figure was obtained for each metal. The average value of this constant is 6.4. Thus, Dulong and Petit enunciated their law such as :

"At room temperature, the product of the specific heat and the atomic weight of a solid is a constant/or all solid elements, and is approximately equal to 6.4."

or Atomic weight × Specific Heat = –6.4 (approximately).

The product (atomic weight x specific heat) is termed as atomic heat. Hence, the Dulong and Petit's law may be stated in the alternative form :

"The atomic heats of solid elements, at ordinary temperatures, are constant and approximately equal to 6.4

Successes : When this law was applied to different solid substances, it was found that the atomic heat capacities are almost constant at about 6.2 ± 0.4 cal, per degree in spite of the regular increase in the atomic weights from 7 to 200.

Exceptions : The various exceptions to Dulong and Petit's law are as follows:

(i) According to Dulong and Petit's law, the atomic heats of all solids must be about 6.4. Actually this is not true. For example, the elements beryllium, boron, carbon and silicon have exceptionally low atomic heats: the values at ordinary temperatures being 3.5, 2-5, 1-35 (diamond) and 4.7 cal per degree respectively.

(ii) As the atomic weight of an element is constant, the atomic heat must also be constant, *i.e.*, its value should be same at all temperatures. This is contrary to the observed fact.

For example, diamond shows variation of heat capacity with temperature. This is evident from below :

Temperature	–50°C	107°C	185-5°C	260°C	615°C	808°C
Atomic Heat	076	1.135	2.11	3.28	5.33	5.44

In the case of silicon, the specific heat increases rapidly with temperature and reaches a fairly constant value above 200°C The variation in metals like lead and tin is very small.

From the above results, one can draw a very important conclusion that the Dulong and Petit's law would apply to all substances provided the temperature is high enough. However, this view is not correct because the heat capacities of those elements which appear to obey the law at ordinary temperatures increases with increasing temperatures. In some cases these values become as high as 9 cal at their melting points. Thus one can safely say that

"There is no one set of conditions under which the Dulong and Petit's law is applicable to alt elements."

(iii) Measurement of specific heats at low temperatures indicates that atomic heat decreases slowly with fall of temperature and below a certain temperature, characteristic of each element, the heat capacity decreases rapidly, tending finally to the zero value at the absolute zero of temperature. This type of change is not permitted by the law of Dulong and Petit. Thus, we can say that :

"The law of Dulong and Petit\ is entirely accidental and the atomic heat capacity of 6 cal. per degree has a theoretical significance only."

Classical Theory

L. Boltzmann (1871) deduced the value of atomic heat of solid from the kinetic theory on the assumption that :

"An ideal solid is composed of atoms vibrating about their respective equilibrium positions but not interacting with one another in any way"

By the above statement it means that in a solid, atoms are allotted fixed positions about which they simply oscillate, possessing kinetic energy of translational as well as potential energy of attraction. The energy of an oscillating atom (harmonic oscillator) is given by

$$E = \text{kinetic energy} + \text{potential energy}$$

$$= \frac{p^2}{2m} + V_{(r)} = \frac{p^2}{2m} + \frac{1}{2} m\omega_o^2 x^2 \qquad ...(1)$$

where $V_{(r)}$ is equal to $1/2m\omega_o^2x^2$, m is the mass and w_o is the natural frequency of the oscillator. Our next aim is to find out the total energy of the solid due to the various oscillating atoms present in it, *i.e.*, to sum up the energy of all atomic oscillators. The total energy E of a solid is given by

$$E = N \times \overline{E} \qquad ...(2)$$

where E is the mean energy of the atomic oscillator and N, the number of degrees of freedom enjoyed by each atom. From statistical mechanics, it follows that the number of atomic oscillators dN lying between E and E + de at a temperature T is proportional to $e^{-E/KT}$ *i.e.*,

$$dN \propto e^{-E/KT} \text{ or } dN = \text{constant} \times e^{-E/KT}$$

$$= \text{constant} \times e^{-(p2/2m + 1/2mwo2x3)} \qquad ...(3)$$

The mean energy of each oscillator will be given by

$$\overline{E} = \frac{\int_0^\infty E\, dN}{\int_0^\infty dN} \qquad ...(4)$$

Substituting the values of E and dN from equations (1) and (3) in 2), we get

$$\overline{E} = \frac{\int_p \int_x \left(\frac{p^2}{2m} + \frac{1}{2} m\omega_o^2 x^2 \right) e^{-p^2/2mKT} . e^{-m\omega_o^2 x^2/2mKT} dp_x dx}{\int_p \int_x e^{-p^2/2mKT} . e^{-m\omega_o^2 x^2/2mKT} dp_x dx}$$

or

$$\overline{E} = \frac{\int_{-\infty}^{+\infty} \frac{p^2}{2m} e^{-p^2/2mKT} dp_x}{\int_{-\infty}^{+\infty} e^{-p^2/2mKT} dp_x} + \frac{\int_{-\infty}^{+\infty} \frac{1}{2} m\omega_o^2 x^2 e^{-m\omega_o^2 x^2/2mKT} d\, dx}{\int_{-\infty}^{+\infty} e^{-m\omega_o^2 x^2/2mKT} dx} \qquad ...(5)$$

Using the standard integrals

$$\int_0^{\infty} e^{-\alpha u^2}\, du = \frac{1}{2}\left(\frac{\pi}{\alpha}\right)^{1/2}$$

and
$$\int_0^{\infty} u^2 e^{-\alpha u^2}\, du = \frac{1}{4}\left(\frac{\pi}{\alpha^2}\right)^{1/2}$$

we have

$$\frac{\int_{-\infty}^{+\infty} \frac{p^2}{2m} e^{-p^2/2mKT}\, dp_x}{\int_{-\infty}^{+\infty} e^{-p^2/2mKT}\, dp_x} = \frac{2 \times \frac{1}{2m} \times \frac{1}{4}\left[\pi (2mKT)^3\right]^{1/2}}{2 \times \frac{1}{2}\left[\pi\,(2mKT)\right]^{1/2}} = \frac{KT}{2}$$

$$\frac{\int_{-\infty}^{+\infty} \frac{1}{2} m\omega_0^{\,2} x^2 / 2KT\, dx}{\int_{-\infty}^{+\infty} e^{-m\omega_0^{\,2} x^2/2KT}\, dx} = \frac{2 \times \frac{1}{2} \times \frac{1}{4}\left[\pi (2mKT)^3\right]^{1/2}}{2 \times \frac{1}{2}\left[\pi\,(2KT)\right]^{1/2}} = \frac{KT}{2}$$

Substituting these values in equation (5), we get

$$\overline{E} = \frac{KT}{2} + \frac{KT}{2} = KT \qquad ...(6)$$

As an oscillator can vibrate about three mutually perpendicular directions, it means that the mean energy

$$\overline{E} = 3KT \qquad ...(7)$$

If there are N atoms in a mole, the to total energy E is obtained by substituting eq. (7) in (2), we get

$$E = 3NKT = 3RT \qquad ...(8)$$

Differentiating eq. (7) with respect to temperature, we get

$$\left(\frac{dE}{dT}\right)_V = 3R \text{ or } C_V = 3R \text{ because } C_V = \left(\frac{\partial E}{\partial T}\right)_V$$

or $(C_V = 3 \times 1.98 = 5.94$ cal. per gm. atom per A°)

The above value of 5.94 is nearly equal to the value given by the Dulong and Petit's law. This is the derivation of the law of Dulong and Petit from kinetic theory.

Limitation of Classical Theory

It could not explain the discrepancy of the decrease of specific heat at low temperatures which has been observed in case of all solids. The results are also not in accordance with experiment. The classical derivation and its predication of heat capacity to be independent of temperature are completely disproved.

Einstein Theory of Specific Heat

quantum theory of specific heat. Einstein explained the problem by using the conception of quantum theory. He assumed that the atoms in a solid are all independent, and that each atom acts as simple harmonic oscillator with a common frequency. Thus, the energy of each linear oscillator is given by

$$\in = nh\nu, \ n = 0, 1, 2 \dots \dots$$

where n is any positive integer.

The expression for the average energy of an oscillator in quantum theory is quite different from the classical theory. According to Planck's proposal, the linear oscillator vibrates only with integral energy values $0, \in, 2\in, 3\in, \dots \dots$ where $\in$ represents the elementary quantum, *i.e.*, $\in = h\nu$.

Einstein applied the concept of Planck's theory that atomic system can exist with only certain discrete energies. He further provided the key information about the dependence of heat capacities upon temperature. Before giving the details of Einstein, let us examine the general outline of the calculations. The heat capacity at constant volume is given by

$$C_V = \Delta E / \Delta T$$

where ΔE is the change in the total energy of a mole of substance produced by the temperature change, ΔT. We shall calculate ΔE from the expression

$$\Delta E = \Delta N \overline{\in}$$

where N is the number of atoms in a crystal, and $\Delta \overline{\in}$ is the change in the average energy of an atom in the crystal produced by the temperature increase ΔT. Our first objective then is to find an expression that gives us information how the energy $\overline{\in}$ depends on temperature.

Let us consider (as Einstein did) a mole of monoatomic solid which consists of N mass points that can oscillate in the three mutually perpendicular directions with a vibrational frequency ν. According to Planck's quantized energy, an atom oscillating in one direction could have only one set of the energies given by

$$\in_n = nh\nu,\ n = 0,\ 1,\ 2,\ 3,\ \ldots\ \ldots \qquad \ldots(1)$$

where h is Planck's constant and is a proportionality factor between frequency and energy, and n is an integer. Planck's constant has the value 6.6×10^{-27} ergs and the frequency ν is equal to $10^{12}\ sec^{-1}$ for many solids. Although an actual atom in a solid is a three-dimensional oscillator, yet we assume it to be a one-dimensional oscillator and we will correct it for this discrepancy later. At thermal equilibrium, the oscillators in a crystal will be distributed among the various allowed energies according to the Boltzmann distribution law which states that the number N_n of atoms with energy $\in_n$ is related ti the number N_o with energy $\in_o = 0$ by the expression

$$N_n = N_o\ e^{-\in n/KT} = N_o\ e^{-nh\nu/KT} \qquad \ldots(2)$$

The total number of particles, N, is equal to the sum of the number in each energy state,

$$N = N_o + N_1 + N_2 + eN_3 + N_4 + \ldots$$

Substitution of the Boltzmann factor, eq. (2) gives

$$N = N_o + N_o e^{-h\nu/KT} + No\ e^{-2h\nu/KT} + \ldots$$

$$= N_o \sum_{n=0}^{\infty} e^{-nh\nu/KT}$$

We will not calculate the total energy due to the oscillation of the atoms in this one direction only. We can do this by multiplying energy $\in_o$ by the number of particles which have that energy, and adding all these quantities :

$$\Delta E = \in_o N_O + \in_1 N_1 + \in_2 N_2 + \ldots$$

$$= 0N_O + h\nu N_1 + 2h\nu N_2 + \ldots$$

$$= 0 + h\nu N_o e^{-h\nu/KT} + 2h\nu N_o\ e{-}2h\nu/kT + \ldots$$

$$= N_o \sum_{1}^{\infty} nh\nu\ e - nh\nu / kT$$

$$\text{But } \overline{\in} = \frac{\Delta E}{N} = \frac{h\nu \sum ne^{-nh\nu/kT}}{\sum e^{-nh\nu/kT}} \qquad \ldots(3)$$

where $\overline{\in}$ is the average energy of a one-dimensional oscillator. In order to evaluate the value of $\overline{\in}$, the summation in eq. (3) must be done. This can be done if we make the substitution

$$y = e^{-h\nu/KT}\text{ , which gives } e^{-nh\nu/KT} = y^n$$

Then, for the series in the denominator of Eq. (3), we get

$$\sum_{n=1}^{\infty} e^{-nh\nu/kT} = \sum_{n=1}^{\infty} y^n = 1 + y^2 + y^3 + ... = \frac{1}{1-y} \qquad ...(3A)$$

This last step can be verified by algebraic long division of 12 by 1 – y. We treat the series in the number at or of Eq. (3) in the same manner

$$\sum_{1}^{\infty} ne^{-nh\nu/kT} = \sum_{1}^{\infty} ny^n = y\left(1 + 2y + 3y^2 + ...\right) = \frac{y}{(1-y)^2} \qquad ...(3B)$$

Substituting equation (3A) and (3B) in (3), we get

$$\overline{\epsilon} = \frac{h\nu y}{1-y} = \frac{h\nu e^{-h\nu/kT}}{1-e^{-h\nu/kT}}$$

Multiplying numerator and denominator by $e^{h\nu/kT}$ yields

$$\overline{\epsilon} = \frac{h\nu}{e^{h\nu/kT} - 1}$$

This is the average energy of a one-dimension oscillator. The average energy of an atom in a crystal is three times this amount since the atom vibrates in three directions.

Thus, $$\overline{\epsilon}_{atom} = \frac{3h\nu}{e^{h\nu/kT} - 1} \qquad ...(4)$$

We will not compute the heat capacity of the solid at such a high temperature so that $h\nu/kT << 1$. Under these conditions, the denominator of Eq. (4) simplifies considerably if we apply the fundamental properties of the exponential function.

$$e^{h\nu/kT} = 1 + \frac{h\nu}{kT} + \frac{1}{2}\left(\frac{h\nu}{kT}\right)^2 + \frac{1}{6}\left(\frac{h\nu}{kT}\right)^3 ... +$$

But $h\nu/kT << 1$, all terms after the second are very small and may be neglected in comparison with $1 + h\nu/kT$. Then, we get

$$e^{h\nu/kT} \approx 1 + \frac{h\nu}{kT} \qquad ...(5)$$

which when substituted into eq. (4) gives

$$\overline{\epsilon} = \frac{3h\nu}{e^{h\nu/kT} - 1} = \frac{3h\nu}{h\nu / kT} = 3kT \qquad ...(6)$$

Therefore, the total energy of one mole of atoms is given by

$$\Delta E = \overline{N} \in = 3NkT \qquad [\because R = Nk] \qquad ...(7)$$

and the heat capacity is

$$C_V = \Delta E / \Delta T = 3R = 6 \text{ cal/ mole/ deg} \qquad ...(8)$$

The result (eq. 8) is very close to the value used in Dulong and Petit's law. The Einstein's theory thus explains specific heat of solids at high temperatures only. This has been found to be true for a large number of elements. It does not hold true in the case of elements where ν is sufficiently large. In these cases, a very high temperature is needed to attain the classical value. It means that is much depends upon the variable ν. The plot of C_V against T clearly shows that for low values of ν, it holds true.

It is more difficult to explain the behaviour of heat capacity at low temperatures by Einstein theory, because it is not possible to simplify equation (3) appreciably when T is very small. However, if we do simplification of equation (3) at low temperatures by making serious approximations, we observe tat the specific heat of solid decreases as the temperature falls and tends to be zero at absolute zero.

Successes of the Einstein Theory :

1. Einstein explained theoretically that the specific heat (atomic heat) increase with temperature and tends to the values given by Dulong and Petit at high temperatures. For some substances even at ordinary temperatures the condition is satisfied.
2. Specific heat decreases as the temperature falls and tends to be zero at absolute zero.

Limitations of Einstein Theory

1. Einstein theory is only approximate and its limitations are more serious at low temperatures. It was observed that in case of elements like copper, aluminium etc., the atomic heats at low temperatures decrease more rapidly than that predicted by Einstein theory.
2. Another weakness of Einstein theory is that while deducing his theory ν was used. This is obtained empirically and cannot be verified from any other independent physical data.
3. Einstein assumed that the vibrations of a particular atom must be of the same frequency. But the vibrations of a particular atom

must be very complex, because it is under the field of force of other vibrating atoms.

DEBYE THEORY

The discrepancy in theoretical and experimental results at low the temperatures in Einstein model was recognised by Debye (1912). He found that the assumption that all the particles of a crystal vibrate with the same frequency was not entirely sound. He considered the vibrational modes of a crystal as a whole, whereas Einstein considered the vibration of an atom assuming the atom vibrations to be independent of each other. He pointed out by looking at the problem :

(a) That a continuous spectrum could result from modes of vibrations.

(b) and these vibrations are treated as elastic-type vibrations.

Then a continuous solid according to the above conception would execute tow types of waves, viz.

(i) Transverse vibrations with velocity $C_t = \sqrt{(n/\rho)}$

(ii) Longitudinal vibrations with velocity $C_t = \sqrt{(k + 4/3\ n/\rho)}$

Here n is rigidity coefficient and k is the bulk. of modulus. The number of possible vibrational modes in second case enclosed in volume V in frequency internal ν and ν + dν is given by

$$\frac{4\pi\nu^2}{C_l^{\,3}} d_\nu V \qquad ...(1)$$

and the number of modes of vibration in first case would be (in same frequency interval) given by

$$= \frac{8\pi\nu^2 d\nu}{C_l^{\,3}} V \qquad ...(2)$$

because in this case there would be two waves polarised at right angles to each other.

Hence for an elastic sold which can transmit three independent vibrations in any given direction, namely, two transverse vibrations polarised in planes at right angles to each other and a longitudinal wave is given by

$$\int_0^{\nu_m} 4\pi V \left(\frac{1}{C_t^{\,3}} + \frac{2}{C_t^{\,3}}\right) \nu^2 \, d\nu \qquad ...(3)$$

Now, it is very clear that in one gram, *i.e.*, Number of atoms are not vibrating with the same frequency but this is system of 3N oscillators. Hence,

$$3N = \int_0^{\nu_m} 4\pi V \left(\frac{1}{C_t^3} + \frac{2}{C_t^3} \right) \nu^2 \, d\nu \qquad \text{...(3A)}$$

or

$$\nu_m^3 = \frac{9N}{4\pi V \left(\frac{1}{C_t^3} + \frac{2}{C_t^3} \right)} \qquad \text{...(4)}$$

The energy of a harmonic oscillators, according to Planck, is given by

$$\frac{h\nu}{\left(e^{h\nu/kT} - 1\right)}$$

So the energy for 3N harmonic oscillators is given as

$$E = 3N \frac{h\nu}{\left(e^{h\nu/kT} - 1\right)}$$

Now this equation assumes the term from equation (3A)

$$E = 4\pi V \left(\frac{1}{C_t^3} + \frac{2}{C_t^3} \right) \int_0^{\nu_m} \frac{h\nu^3}{\left(e^{h\nu/kT} - 1\right)} d\nu \qquad \text{...(5)}$$

From equation (4), we have

$$E = \frac{9N}{\nu_m^3} \int_0^{\nu_m} \frac{h\nu^3 \, d\nu}{e^{h\nu/kT} - 1} \qquad \text{...(6)}$$

Differentiating the above with respect to temperature, we obtain

$$C_V = \frac{9R}{\nu_m^3} \int_0^{\nu_m} \frac{e^{h\nu/kT}}{\left(e^{h\nu/kT} - 1\right)^2} \left(\frac{h\nu}{kT} \right)^3 \nu^2 \, d\nu \qquad \text{...(7)}$$

Equations (7) is known as Debye heat capacity equation. To evaluate the integral, put

$$\frac{h\nu}{kT} = \Psi \; ; \; \frac{h\nu_m}{k} = \theta \; ; \; \frac{\theta}{T} = \Psi_m = x$$

Substituting these values in equation (15), we get

$$C_V = 3R \left[\frac{12}{x^3} \int_0^x \frac{\Psi^3 \, d\Psi}{e\Psi - 1} - \frac{3x}{e^x - 1} \right] \qquad \text{...(8)}$$

Equation (8) is known as Debye's expression for the atomic heat of a monoatomic solid. The term within the square bracket is a function of x and is generally designated by Debye function, D(x). It s value has been calculated for the values of x which range from to ∞.

Limiting Case :

At moderate and very high temperatures, equation (16) becomes as

$$C_V = 3R\left[1 - \frac{1}{20}\left(\frac{\theta}{T}\right)^2 + \frac{1}{560}\left(\frac{\theta}{T}\right)^4 - \ldots\right] \quad \ldots(9)$$

At sufficiently high temperatures, θ/T becomes small enough for all the terms in the brackets of equation (9) other than the first to be neglected; C_V then equal to the classical value 3R.

(ii) At very low temperatures T is very small and x and Ψ are very large. The second term in equation (8), becomes negligible. Also, the denominator of the first term is also large unless ν is small. At low temperatures, θ/T tends to ∞. Hence, one can replace the upper limit of the first expression of equation (16) be ∞. We have thus

$$C_v = 3R\,\frac{12}{x^3}\int_0^{\infty}\frac{\Psi^3\,d\Psi}{e^{\Psi}-1}$$

$$= 3R\,\frac{12\pi^4}{x^3\,15} = 3R.\frac{12\pi^4}{15}.\left(\frac{T}{\theta}\right)^3 = 77.94 \times 3R\left(\frac{T}{\theta}\right)^3$$

$$= 464.4\left(\frac{T}{\theta}\right)^3 \text{ cal/deg/gm atom.} \quad \ldots(10)$$

As θ is constant for a substance, the specific heat at extremely low temperatures varies as T^3. This statement is known as Debye's – T^3 law. This law has been verified in a number of cases. When t = 0° A, equation (10) becomes as $C_V = 0$

i.e., atomic heat capacity vanishes at absolute zero.

Calculation of Characteristic Temperatures :

we know that

$$\theta = \frac{h\nu_m}{k} = \frac{h}{k}\left\{\frac{9N}{4\pi\nu\left(\frac{2}{C_t^3} + \frac{1}{C_t^3}\right)}\right\}^{\frac{1}{2}} \quad \ldots(11)$$

and $$C_t = \left[\sqrt{\left(k + \frac{4}{3}n\right)}\right] \Big/ \rho \text{ and } C_t = \sqrt{(n/\rho)} \qquad ...(12)$$

$$\therefore \qquad \theta = \frac{h}{k}\left(\frac{9N}{4\pi V}\right)^{1/3} \rho^{1/2} \left[\frac{2}{n^{3/2}} + \frac{2}{(k + 4/3)^{3/2}}\right]^{1/2} \qquad ...(13)$$

where all the letters have their usual significance.

Hence the value of θ can be calculated from equation (13).

The Debye theory predicts that the heat capacity of a solid as a function of temperature should depend only on the characteristic frequency, ν_m. These crystals with closely spaced levels, as a result of weak bonds or heavy atoms, as in lead reach the classical heat capacity value at low temperatures while the crystals with large vibrational spacings reach the value only at high temperatures.

Weaknesses of Debye Theory

(i) This theory does not agree with its theoretical foundation.

(ii) Debye theory cannot be extend to low temperature.

Criticism of Debye Theory

(i) This theory explains quite satisfactorily the aviation of specific heat of solids, but it does not agree with its theoretical functions. It was assumed that :

(ii) When a solid is heated, it is supposed that elastic vibrations like those of continuous medium exist within the body having frequencies from 0 to ν_m, where the spectrum guillotined sharply.

Even if a solid be isotropic, it cannot be treated as continuous since it is composed of atoms separated in space.

(iii) Here the solid is considered to be an isotropic body, but in reality solids are made of definite crystals, either in the form of single crystals or minute crystals brought together by pressure. It has not considered crystalline structure in its derivation. Slater inferred that becomes incorrect due to two reasons :

(a) That the velocity of elastic vibrations depends upon the wavelength of the vibration when it is comparable with atomic distances (as it is for frequencies approaching n_m)

and hence the change of velocity with frequency must be taken into account in deriving.

(b) In an actual solid composed of N atoms, one has to find out the normal or independent modes of vibration of atoms in the solid. This problem was solved by Born by assuming "periodic" boundary conditions or assuming fixed boundaries, the force between atoms to be given by Hooke's law. He showed that there will 3N normal modes of vibration. Thus the total number of modes comes out to be the same as the number of degree of freedom of solid as assumed by Born which is different from Debye even for anisotropic medium.

The Debye model, as also the Einstein's, is applicable to assemblies of harmonic oscillators and cannot, therefore, deal with properties of solids which arise primarily due to anharmonicity of lattice vibrations. Thus to improve upon the Debye theory of specific heat due to lattice vibrations, a more precise vibrational spectrum is required. The problem in its three-dimensional form is complicated. Although to explain the basis and nature of the anharmonic effect, a one-dimensional picture may be constructed in a rather simple manner to understand the problem.

Consider a one-dimensional lattice having a 'grating space' equal to s and containing n identical atoms in it. If the displacement of, say, ith atom from the mean position is x_i the energy E of the lattice is

$$E = \sum_{i-1}^{n} \frac{1}{2} m x_i^2 + \sum_{i-1}^{n} \frac{1}{2} f\left[x_i - x_{i-1}\right]^2 \qquad ...(14)$$

where n represents the number of atoms each of mass m in the lattice and f the force constant. The Lagrangian equation of motion, considering the ith atom, are as follows :

$$mz_i = f[x_{i+1} \; x_i - (x_i - x_{i-1})] = f[x_{i+1} + x_{i-1} - 2x_i], \qquad ...(15)$$

$$i = 1, 2, ..., n$$

If the linear crystal lattice is imagined to be bent into a circle (endless chain) such that the nth atom is next tot the first atom, then the so-called 'cyclic boundary conditions' will be obviously

$$x_i (t) = x_{i+n} (t) \qquad ...(16)$$

where t denotes time motion,

The solutions of Lagrange's equations may be expressed as

$$x_i(t) = Ae^{j(\omega t + i\phi)} \quad ...(17)$$

where $j = (-1)^{1/2}$, A is a constant, $\omega = 2\pi\nu$ and

$$\phi k = \frac{2\pi k}{n}, \; k = 1, 2, ... n \quad ...(18)$$

Hence from eq. (17)

$$x_i(t) = Ae^{j\omega t} e^{2\pi ij/n} \quad ...(19)$$

Now substitute eq (17) in eq. (15)

$$-m\omega^2 = f\left[e^{j\phi} + e^{-j\phi} - 2\right] = f\left[e^{j\phi/2} + e^{-j\phi/2}\right]^2$$

or $\qquad m\omega^2\ 4\ f \sin^2 \phi/2 \quad ...(20)$

And since $\phi = 2\pi k/n$,

$$\omega k = 2\left(\frac{f}{m}\right)^{\frac{1}{2}} \sin\frac{\pi k}{n'} k = 1, 2 ..., n \quad ...(21)$$

and $$V_k = \frac{\omega k}{2\pi} = \frac{1}{\pi}\left(\frac{f}{m}\right)^{1/2} \sin\frac{\pi k}{n} \quad ...(22)$$

Differentiating eq. (22), we get

$$dv_k = \frac{1}{\pi}\left(\frac{f}{m}\right)^{1/2} \sin\frac{\pi k}{n} dk \quad ...(23)$$

Now as $\sin\phi = \sin(\pi - \phi)$, for $\phi \leq \pi/2$; thus there will be two physically distinct values ϕ and $(\pi - \phi)$ which yield the same value of $\sin\phi$, Hence for each value of k such that $1 \leq k \leq n/2$, there will be two distinct normal vibrations, corresponding to $\phi_k = 2\pi k/n$ and $\phi_{n-k} = \frac{2\pi(n-k)}{n}$. But these modes have frequency νk as given by eq. (22), The number of frequencies between ν and $\nu + d\nu$ may also be defined as $\frac{1}{d\nu_k/dk} = g(\nu)$ – the number of vibrational modes, obtained as :

$$g(\nu)\,d\nu = \frac{2n}{\cos(\pi k/n)}(m/f)^{1/2} d\nu \quad ...(23)$$

The additional factor 2 in eq. (23) is present because for every value of k there are two distinct frequencies.

Also denoting

$$\nu_o = \frac{1}{\pi}\left(\frac{f}{m}\right)^{1/2} \quad ...(24)$$

We have $$\nu = \nu_o \sin \frac{\pi k}{n}$$

Such that $$\cos \frac{nk}{n} = \frac{\left(\nu_o{}^2 - \nu^2\right)^{1/2}}{\nu_o} \qquad ...(25)$$

We have, substituting eq. (23) in eq. (24),

$$g(\nu)\, d\nu = \frac{2n}{\pi} \frac{d\nu}{\left(\nu_o{}^2 - \nu^2\right)^{1/2}} \qquad ...(26)$$

Eq. (26) thus gives the frequency spectrum for a non-dimensional linear lattice. This should be compared with the Debye spectrum which for a one-dimensional oscillator, is as :

$$g(\nu)\, d\nu = \nu\, d\nu. \qquad ...(27)$$

Eq. (27) therefore gives an idea of the true picture of the normal vibrations of the lattice and one observes that the spectrum differs considerably from the Debye spectrums.

An interesting calculation, according to Born theory, was given by Thirring. He pointed out that the specific heat at high temperatures could be worked out without finding an explicit form for the spectrum and without any further approximations. This method is to expand the mean energy of an oscillator in powers of $h\nu/kT$ and the result is

$$E - kT \left\{ 1 - \sum_{n-1}^{\infty} (-1)^n \frac{B_n}{(2n)!} \left(\frac{h\nu}{kT} \right)^{2n} \right\}$$

where B_n are the Bernoulli numbers.

The heat capacity of the linear oscillator is then given by

$$C_V = \frac{dE}{dT} = k \left\{ 1 + \sum_{n-1}^{\infty} (-1)^n \frac{B_n (2n-1)}{(2n)!} \left(\frac{h\nu}{kT} \right) 2n \right\}$$

Born and V. Karman theory depends on the calculation of frequency spectrum as a dynamic crystal lattice.

It is expected to apply these theories only to a strictly isotropic body, and to one in which all the atoms are the same. It has been extended by Born and others to apply this theory to crystal and to bodies containing more than one type of atoms.

These theories take into account the fact that body is no continuous but consists of atoms, arranged in a lattice.

BLACKMAN'S LATTICE THEORY OF SPECIFIC HEATS

It is clear that the theory which has been developed so far gives a general account of specific heat of solids except at low temperatures.

From the results obtained by Debye theory, it appears that it is unsatisfactory to treat crystal as continuum. The vibrational spectrum of a crystal lattice can differ very much from that of an elastic solid. Calculation of the vibrational spectrum of a two dimensional lattice by Blackman showed that it differed greatly from the assumed by Debye. Blackman took into account of the interacations between the closed unneighborly particles of the lattice and also between next nearest neighbour particles.

According to this theory specific heats at high and at low temperatures depend on the different combinations of the forces between particles. In high temperature region, the specific heap is controlled by the normal vibrations of the crystals, and the low frequency end of the spectrum plays a negligible part in determining it. On the other hand the specific heat at low temperatures is controlled by the low frequency end of the spectrum.

In order to follow the details of the variation of the specific heat of solid with temperature, spectrum in each case ins recorded.

Comparison with Debye's Theory

(i) Debye theory is remarkable in view of complications occurring in the lattice theory though Debye theory fails miserably at low temperatures.

(ii) the calculation of the vibrational spectrum has thrown new light on the theory of specific heat of solids. The vibrational spectrums is of quite different character from that assumed by Debye.

HEAT OF REACTION OR ENTHALPY OF REACTION

When a chemical reaction is accompanied by evolution of heat, it is termed as exothermic and if it is accompanied by absorption of heat, it is termed as endothernric. *The amount of heat evolved or absorbed during a chemical reaction is called the heat of reaction or enthalpy of reaction.*

From the definition of enthalpy, there always occurs a definite enthalpy change, it follows that it is ΔH, whenever a system undergoes a change of state or a chemical reaction.

Suppose we consider a chemical system having two reactants A and B in the initial state.

Suppose it undergoes a chemical reaction at a constant temperature and pressure, thereby forming the products C and D.

$$A + B \rightarrow C + D$$

Suppose H_1 denotes the enthalpy of the systems in the initial state (*i.e.*, enthalpy of reactants A and B) and H, the enthalpy of the system in the final state at the same temperature and pressure (*i.e.*, enthalpy of products C and D). Then,

$$\Delta H - H_2 - H_1$$

Thus, *at constant pressure and temperature the heat of reaction or enthalpy of reaction is regarded as a measure of the enthalpy chaise, ΔH. In simple words,*

"The heat of reaction or enthalpy of reaction at a given temperature and pressure is the difference between the heat contents of the products and the reactants when the number of gram molecules of substances indicated by the chemical reaction have completely reacted.

Mathematically, it may be put as

ΔH = Heat contents of the products – Heat contents of the reactants.

= Heat contents of 1 mole of C and 1 mole of D – Heat contents of 1 mole of A and 1 mole of B.

ΔH is positive for endothermic reactions and negative for exothermic reaction. For example, the thermochemical equation

$$H_2(g) + \frac{1}{2}O_2(g) \rightarrow 2H_2O(l);\ \Delta H = -285.83\ U$$

indicates that when one mole of gaseous hydrogen combines with half a mole of gaseous oxygen at 298 K at one atmosphere pressure to form 1 mole of liquid water at the same temperature and pressure, 285.83 kJ of heat is given out to the surrounding. It is further assumed that the reaction is fast and proceeds to completion with no side products.

Enthalpy of a Reaction at Constant Pressure and at Constant Volume : The enthalpy of a reaction depends upon the condition under which the reaction is carried out. There are two general conditions under which thermochemical measurements are made, namely :

(i) at constant volume, and

(ii) at constant pressure.

The magnitude of the anthalpy changes in these two conditions are, in general, different.

Heat of reaction or enthalpy of reaction at constant volume (ΔE) : When a reaction is being carried out at a constant volume, the heat change is called the heat of reaction or enthalpy of reaction at constant volume. Example is the combustion of a substance in a bomb calorimeter.

When a reaction is carried out under constant volume by carrying out the reaction in a closed and rigid container, no work is involved and from the first law of thermodynamics, we have

$$\Delta E = qv$$

$$= \text{Enthalpy of the reaction at constant volume.}$$

Thus, the enthalpy of the reaction at constant volume is exactly equal to the change in the internal energy ΔE of the reacting system. In other words, the thermal change that occurs in a chemical reaction is only due to the difference in the sum of the internal energy of the products and the sum of the internal energy of the reactants.

$$\Delta E = \Sigma E_{products} - \Sigma E_{reactants} \qquad ...(1)$$

The importance of Eq. (1) is that the amount of heat absorbed at constant volume can be identified with the change in the thermodynamic quantity.

Heat of reaction or Enthalpy of reaction at constant pressure ΔH : When pressure is kept constant throughout the reaction, the heat change is then termed as heat of reaction at constant pressure.

At constant pressure, the system is either kept open to the atmosphere or confined with in a vessel on which a constant external pressure is exerted. Under these conditions, the volume of system changes. The thermal change at constant pressure not only involves the change in internal energy of the system but also the work done either in expansion or in contraction of the system. Then, from the first law of thermodynamics, we have

$$q_P = \Delta E + w$$

If w is only pressure volume work, then

$$q_P = \Delta H + P\Delta V = (\Sigma E_P - \Sigma E_R) + P\,(V_P - V_R)$$

$$= (\Sigma E_P - PV_P) - (\Sigma E_R + PV_R) \qquad ...(2)$$

where the subscripts P and R refer to the products and the reactants respectively.

But $H = E + PV$ (By definition)

Therefore, Eq. (2) becomes as follows :

$$q_P = \Sigma H_P - \Sigma H_R = \Delta H \qquad ...(3)$$

From the above equation it follows that at constant pressure, the heat of the reaction is exactly equal to the enthalpy change, ΔH, of the reacting system.

Relationship between ΔH and ΔE : In the case of reactions involving solids and liquids, the difference between ΔH and ΔE is almost negligible. Thus, for solids and liquids, $\Delta H \approx \Delta E$. On the other hand, ΔH may not be equal to ΔE in reactions which involve gases. Suppose consider the following reactions :

$$2NH_3(g) \rightarrow N_2(g) + 3H_2(g)$$

2 moles 1 mole 3 moles

$$2H\text{–}2(g) + O_2(g) \rightarrow 2H_2O(g)$$

2 moles 2 moles 1 mole

When these reactions are carried out at constant volume, no work will be done by or on the system as there is neither expansion nor contraction. However, if the reactions are carried out at constant pressure, then the work will be performed by the system in the first case as there occurs an increase in volume

$$(2NH_3 + N_2 + 3H_2)$$

and upon the system in the second case

$$(2H_2 + O_2 \rightarrow 2H_2O)$$

as there occurs a decrease in volume. Consequently, the heat of reaction at constant volume (ΔE) will be more than heat of reaction at constant pressure (ΔH) when contraction occurs during the course of reaction whereas ΔE will be less than ΔH if expansion occurs during the course of reaction. Now consider the following gaseous reaction :

$$aA + bB + ... \rightleftharpoons cC + dD + ...$$

Suppose n_1 = No. of moles of gaseous reactants, *i.e.*, a + b +...

n_2 = No. of moles of gaseous products, *i.e.*, c + d +...

$\Delta n(g) = (c + d +...) - (a + b +...) = (n_2 - n_1)$,

V_1 = volume of the gaseous reactants,

P = the external pressure of the gas, and

V_2 = volume of the gaseous products,

T = temperature of the system on absolute scale,

For an ideal gas,

$$PV = nRT$$

$\therefore \quad PV_1 = nRT \qquad$ and $PV_2 = n_2RT$

Now, work done by the gas in expansion or contraction from volume V_1 to V_2 in the reaction

$$aA + bB + ... \rightleftharpoons cC + dD + ...$$

is given by

w = pressure × change in volume

$= P(V_2 - V_1) = PV_2 - PV_1$

$= n_2RT - n_1Rt\ (n_2 - n_1) = \Delta n(g)RT$ ...(4)

Thus, heat of reaction at constant pressure is given by,

ΔH = heat of reaction at constant volume + work done by or on the gas during change of volume V_1 to V_2

$= \Delta E + \Delta n(g)RT$ ...(5)

Equation (5) can be used in calculating the value of ΔH from ΔE or vice-versa. A positive value of ΔH or ΔE shows that heat is absorbed while a negative value of ΔH or ΔE shows that heat is evolved or given out by the system to the surroundings.

Case I

When $n_2 > n_1$, Δn(g) is positive. Therefore,

$$\Delta H > \Delta E$$

Case II

When $n_2 > n_1$, Δn(g) is negative.

Therefore, $\Delta H < \Delta E$

Case III

When there is no change in volume, $n_2 = n_1$. Therefore,

$$\Delta n(g) = 0,\ \Delta H = \Delta E$$

SOLVED EXAMPLES

Example 1:

For the hypothetical reaction

$$2A(g) \rightarrow A_2(g)$$

$\Delta C_p/J\ k^{-1}\ mol^{-1} = 1.00 + (2.00 \times 10\text{–}3/k)\ T$ *and* $\Delta H^o_{298}k = -5.00$ *kJ* mol^{-1}*. Estimate the temperature at which* $\Delta H^o = 0$ *for this reaction at constant pressure.*

Solution:

We have $$\int_{\Delta H^o}^{0} d(\Delta H) = \int_{298k}^{T} (\Delta C_p)\, dT$$

$$= \int_{298k}^{T} \left[1.00 + \left(\frac{2.00 \times 10^{-3}}{k}\right) T\right] J\ k^{-1}\ mol^{-1}\ dT$$

$$0 - \Delta H^o = (1.00\ J\ k^{-1}\ mol^{-1})(T - 298\ k)$$

$$+ \left(\frac{2.00 \times 10^{-3}}{2k} J\ k^{-1}\ mol^{-1}\right)(T^2 - 298\ k^2)$$

$$5000 = (1.00\ k^{-1})(T - 298\ k) + (10^{-3}\ k^{-2})(T^2 - 298k^2) = 0$$

or $$10^{-3}\left(\frac{T}{K}\right)^2 + \left(\frac{T}{K}\right) - 298^2 \times 10\text{–}3 - 298 - 5000 = 0$$

or $$10^{-3}\left(\frac{T}{K}\right)^2 + \left(\frac{T}{K}\right) - 5209 = 0$$

Solving for T/K, we get

$$\left(\frac{T}{K}\right) = \frac{-1 + \left(1^2 + 4 \times 10^{-3} \times 5209\right)^{1/2}}{2 \times 10^{-3}}$$

$$T = \frac{-1 + 4.67}{2 \times 10^{-3}} k = 1836\ k$$

Example 2:

What are the reasons for the 'calculated maximum adiabatic flame temperature' not being exactly equal to theexperimental value ? What correction would you suggest ?

Solution:

The maximum adiabatic flame temperature is calculated using the expression

$$T_f = \frac{-\Delta H}{C_p\,(\text{products})} + T_0$$

In derving the above expression, it is assumed to C_ps are independent of temperature. However, this is not correct since T_f is a very large quantity (as ΔH is expressed in kJ whereas C_ps are in joule) of the order of thousands of kelvin. Besides this, the above value will be obtained if the combustion process is carried out reversibly, since only then q = ΔH.

But the conditions of experiment are those of irreversible process for which q_p is expected to be less than ΔH. Moreover, it is assumed that Cp (Products) are independent of temperature. To get correct result. one should take into account the variation of heat capacities with temperature.

Exumple 3:

Using bond enthalpies from the following table evaluate ΔH for the isomerization of ethyl alcohol to dimethyl ether :

Bond	*Bond enthalpies at 25°C* *ΔH.kJ mol^{-1}*
H – H	416
O – O	132
O – H	444
C – H	396
C – O	336
C – C	332

Solution:-

We have to calculate the enthalpy change of the reaction

$$CH_3CH_2OH \rightarrow CH_3OCH_3$$

The enthalpy change of the reaction will be given as

$\Delta H = \in(C - C) + 5 \times \in(C - H) + \in(C - O) + \in(O - H) - 6 \times \in(C - H) - 2 \times \in(C - O)$

$= \in(C - C) - \in(C - H) - \in(C - O) + \in(O - H) = (332 - 396 - 336 + 444)\ kJ\ mol^{-1} = 44\ kJ\ mol^{-1}$.

Example 4:

Compute the heat of reaction at 1000°C for

$1/2H_2(g) + 1/2\ Cl_2(g) \rightarrow HCL(g) \quad \Delta H^o_{298k} = 92.236\ k\ J\ mol^{-1}$

$C_p{}^o(H_2, g) = 29.0284 - 0.8355 \times 10^{-3}\ T + 2.0097 \times 10^{-6}\ T^2.$

$C_p{}^o(Cl_2, g) = 31.6555 + 1.0134 \times 10^{-2}\ T - 4.0337 \times 10^{-6}\ T^2.$

$C_p{}^o(HCl, g) = 28.1359 + 1.8074 \times 10^{-2}\ T + 1.5453 \times 10^{-6}\ T^2.$

Solution:

$$\Delta H^o_{1273k} = \Delta H^o_{298k} + \int_{298k}^{1273k} (\Delta C_p)\, dT$$

where $\Delta C_p = \Delta C_p{}^o\ (HCl, g) - 1/2\ C_p{}^o\ (H_2, g) - 1/2C_p{}^o\ (Cl_2, g)$

$= (28.1359 - 1/2 \times 29.0284 - 1/2 \times 31.6555)$

$+ (1.8078 + 1/2 \times 0.8355 - 1/2 \times 10.134) \times 10^{-3}\ T$

$+ (1.5453 - 1/2 \times 2.0097 + 1/2 \times 4.0337) \times 10^{-6}\ T^2$

$= -2.2061 - 2.8415 \times 10^{-3}\ T + 2.5573 \times 10^{-6}\ T^2$

Hence,

$$\Delta H^o{}_{1273k} = \left[\begin{array}{l} -92.236 \times 10^3 + (-2.2061)(1273 - 298) - (2.8415 \times 10^{-3}) \\ \left(\frac{1273^2}{2} - \frac{298^3}{2}\right) + (2.5573 \times 10^{-6})\left(\frac{1273^2}{2} - \frac{298^3}{2}\right) \end{array}\right]$$

$J \cdot mol^{-1}$

$= -92236 - 2151.0 - 2176.2 + 1736.0$

$= -94827\ J\ mol^{-1} = -94.827\ mol^{-1}$

Example 5(a):

The combustion of 1 g of benzene in a bomb calorimeter evolves 41.746 kJ of heat at 25°C, (i) What is ΔU^o for combustion of benzene ? (ii) Calculate heat of formation of benzene.

Solution:

$$\Delta H_f^o\ (CO_2, g) = -\ 393.129\ kJ\ mol^{-1}$$

$$\Delta H_f^o\ (H_2, l) = -\ 285.577\ kJ\ mol^{-1}$$

The molar mass of benzene is 78 g mol^{-1}.

Hence,

$$\Delta U^o = -\ (41.746\ kJ)\left(\frac{78\ g\ mol^{-1}}{1g}\right) = -\ 3256.16\ kJ\ mol^{-1}$$

The chemical equation of combustion is

$$C_6H_6(l) + \frac{15}{2}\ O_2(g) \rightarrow 6CO_2(g) + 3H_2O(l)$$

$$\Delta V_g = 6 - \frac{15}{2} = -\frac{3}{2}$$

Hence, $\Delta H^o = \Delta U^o + (\Delta V_g)\ RT = [(-\ 3256.16 \times 10^3)$

$$+ (\ \ 3/2)\ (8.314 \times 298)]\ J\ mol^{-1}$$

$$= -\ 3259.9 \times 10^3\ J\ mol^{-1}.$$

Finally, we have

$$\Delta H^o = 6\Delta H_f^o + (CO_2, g) + 3\Delta H_f^o\ (H_2O,l) - \Delta H_f^o\ (C_6H_6, l)$$

which gives

$$\Delta H_f^o\ (C_6H_6, l) = 6\Delta H_f^o + (CO_2, g) + 3\Delta H_f^o\ (H_2O,l) - \Delta H_f^o$$

$$= [-\ 6 \times 393.129 - 3 \times 285.577 + 3259.9]\ kJ\ mol^{-1}$$

$$= -\ 44.35\ mol^{-1}$$

Example 5(b):

From the following data, calculate ΔH_f at 298.15 k for (i) $OH^-(aq_l)$ and (ii) $Na^+(aq)$:

ΔH_f^o values is kJ mol^{-1} at 298.15 k for

$H_2O,(l) = -\ 285.8$	$HCl\ (aq) = -\ 167.4$
$NaOH(aq) = -\ 469.6$	$NaCl(aq) = -\ 407.0$

Solution:

(i) $H_2(g) + 1/2\ O_2(g) \rightarrow H_2O(l)$

$\Delta H^\circ = -\ 285.8\ kJ\ mol^{-1}$

(ii) $1/2H_2(g)+1/2\ Cl_2(g)+aq \rightarrow H^+(aq)+Cl^-(aq)$

$\Delta H^\circ = -\ 167.4\ kJ\ mol^{-1}$

(iii) $Na(s) + 1/2\ O_2(g) + 1/2\ H_2(g) + aq \rightarrow Na^+(aq) + OH^-(aq)$

$\Delta H^\circ = -\ 469.6\ kJ\ mol^{-1}$

(iv) $Na(s) + 1/2\ Cl_2(g) + aq \rightarrow Na^+(aq) + Cl^-(aq)$

$\Delta H^\circ = -\ 407.0\ kJ\ mol^{-1}$

Carrying out the manipulation

Eq. (iii) – Eq. (iv) + Eq. (ii)

We get

(v) $1/2O_2(g) + H_2(g) + aq \rightarrow H^+(aq) + OH^-(aq)$

Hence, its enthalpy change becomes

$\Delta H^\circ = \Delta H^\circ_{(m)} - \Delta H^\circ_{(iv)} + \Delta H^\circ_{(ii)}$

$= (-\ 469.6 + 407.0 - 167.4)\ kJ\ mol^{-1}$

$= -\ 230.0\ kJ\ mol^{-1}$

The enthalpy change of Eq. (v) is also given as

$\Delta H = \Delta H_f^\circ(H^+) + \Delta H_f^\circ(OH^-) - 1/2\ \Delta H_f^\circ(O_2) - \Delta H_f^\circ\ (H_2)$

$= 0 + \Delta H_f^\circ(OH^-) - 0 - 0 = \Delta H_f^\circ\ (OH^-)$

Thus $\Delta H_f^\circ\ (OH^-,\ aq) = -\ 230.0\ kJ\ mol^{-1}$

From Eq, (iii), we get

$\Delta H = \Delta H_f^\circ(Na^+) + \Delta H_f^\circ(OH^-) - \Delta H_f^\circ(Na) - 1/2\ \Delta H_f^\circ(O_2) - 1/2\ \Delta H_f^\circ(H_2)$

$-\ 469.6\ \ kJ\ mol^{-1} = \Delta H_f^\circ(Na^+) = -\ 230.0\ kJ\ mol^{-1}$

which gives

$\Delta H_f^\circ(Na^+) = (-\ 469.6 +\ 230.0)\ kJ\ mol^{-1}$

$= -\ 239.6\ kJ\ mol^{-1}$.

Example 5(c):

Calculate ΔH° *for the reaction at 298 K*

$Ba^{2+}\ (aq) + SO_4^{\ 2-} \rightarrow BaSO_4(s)$

Given that

$$\Delta H_f^o\ [Ba^{2+}\ (aq)] = -\ 538.36\ kJ$$
$$\Delta H_f^o\ [SO_4^{2-}\ (aq)] = -\ 907.51\ kJ$$
$$\Delta H_f^o\ [BaSO_4\ (s)] = -\ 1465.24\ kJ$$

Solution:

For the reaction,

$$Ba^{2+}\ (aq) + SO_4^{2-} \rightarrow BaSO_4(s)$$

We have $\Delta H^o = \Delta H_f^o\ [BaSO_4(s)] - \Delta H_f^o\ [Ba^{2+}\ (aq)] - \Delta H_f^o\ [SO_4^{2-}\ (aq)]$

On substituting the values, we get

$\Delta H^o = -\ 1465.24\ kJ - (-\ 538.36\ kJ - 907.51\ kJ) = -\ 19.37\ kJ$

Example 5(d):

Calculate the bond enthalpy for a C – O bond in methanol from the following data :

(i) $C(s)+2H_2(g)+1/2O_2 \rightarrow CH_3OH(g);$ $\Delta H^o = -\ 200\ kJ\ mol^{-1}$

(ii) $C(s) \rightarrow C(g);$ $\Delta H^o = -\ 716.8\ kJ\ mol^{-1}$

(iii) $2H_2(g) \rightarrow 4H(g);$ $\Delta H^o = -872.0\ kJ\ mol^{-1}$

(iv) $1/2O2(g) \rightarrow O(g);$ $\Delta H^o = -249.0\ kJ\ mol^{-1}$

The bond enthalpy for C – H bond is 403 kJ mol^{-1} *and for O–H bond it is 463.6 kJ* mol^{-1}.

Solution:

The enthalpy change for the dissociation of CH_3OH is as follows :

$CH_3OH(g) \rightarrow C(g) + 4H(g) + O(g)$; $\Delta H^o = ?$

On adding (ii), (iii) and (iv), we get

$C(s) + 2H_2(g) + 1/2O_2(g) \rightarrow C(g) + 4H(g) + O(g)$,

$\Delta H^o = 1837.8\ kJ$

On subtracting equation (i) from the above equation, we get

$CH_3OH(g) \rightarrow C(g) + 4H(g) + O(g);$

$\Delta H^o = 2037.8 kJ\ mol^{-1}$

As methanol has three C – H, one C – O bond and one O – H bond, it means that

$$2037.8 \text{ kJ} = 3\,(\Delta H^{o}_{C-H}) + (\Delta H^{o}_{O\,H}) + 1\,(\Delta H^{o}_{C-O})$$

$$= 3\,(413) + (463.6) + ((\Delta H^{o}_{C-O})$$

or $$(\Delta H^{o}_{C-O}) = 335.2 \text{ kJ mol}^{-1}$$

Example 6:

The enthalpy of neutralisation of ammonium hydroxide by hydrochloric acid is 51.46 kJ mol^{-1}. Calculate the enthalpy of ionisation of ammonium hydroxide.

Solution:

(i) $NH_2OH(aq)+H^+(aq) \rightarrow NH_4^+(aq)+H_2O(l)$; $\Delta H^o = -51.46\text{kJ mol}^{-1}$

(ii) $H^+(aq) + OH^-(aq) \rightarrow H_2O(1)$; $\Delta H^o = -57.32\text{kJ mol}^{-1}$

(iii) $NH_2OH(aq) \rightarrow NH_4^+(aq) + OH–(aq)$; $\Delta H^o = \Delta H^o_{\text{Ionisation}}$

On adding (ii) and (iii), we get

$$NH_2OH(aq) + H^+(aq) \rightarrow NH_4^+(aq) + H_2O(1)$$

or $$\Delta H^o = -\,57.32 + \Delta H^o_{\text{Ionisation}}$$

From Eq. (i), we have

$$\Delta H^o = -\,51.46 \text{ kJ mol}^{-1}$$

Hence $(-\,57.32 + \Delta H^o_{\text{Ionisation}})$ kJ mol^{-1} =– 51.46 kJ mol^{-1}

$\Delta H^o_{\text{Ionisation}} = +\,5.86 \text{ kJ mol}^{-1}$.

Example 7:

If E_{C-C} is 34 kJ mole^{-1} and E_{C-H} is 415 kJ mole^{-1}, calculate the heat of formation of propane. The heats of atomization of carbon and hydrogen are 716 kJ mole^{-1} and 433 kJ mole^{-1} respectively.

Solution:

The heat of formation is the sum of the heats of atomization and bond energies. For propane, the heats of atomization as

$$3C_s = 3C_g; \qquad \Delta H = 3 \times 716 = 2148 \text{ kJ}$$

$$4H_{2,g} = 8H_g; \qquad \Delta H = 4 \times 433 = 1732 \text{ kJ}$$

The bond energies are

$$2E_{C-C} = 2 \times -344 = -688 \text{ kJ}$$

$$2E_{C-H} = 8 \times -415 = -3320 \text{ kJ}$$

Adding,

$3C + 4H_2 = C_3H_8$; $\Delta H_f = 2148 + 1732 - 688 - 3320$

$= -128$ kJ mole^{-1}

Example 8:

The heats of formation of PCl_3 and PH_3 are c 306 kJ mole^{-1} and + 8 kJ mole^{-1} respectively, and the heats of atomization of phosphorus, chlorine and hydrogen are given by :

$P_s = P_g$ $\Delta H = 314$ kJ mole^{-1}

$Cl_{2.g} = 2Cl_g$ $\Delta H = 242$ kJ mole^{-1}

$H_{2.g} = 2H_g$ $\Delta H = 433$ kJ mole^{-1}

Calculate E_{P-Cl} and E_{P-H}

Solution:

For phosphorus trichloride, thermochemical equations are :

$P_s = P_g$ $\Delta H_1 = 314$ kJ

$3/2\ Cl_{2g} = 3Cl_g$ $\Delta H_2 = 363$ kJ

$P_g + 3Cl_g = P_{Cl3} = 2H_g$ $\Delta H_3 = ?$

Also,

$P_s + 3/2\ Cl_{2.g} = PCl_3$ $\Delta H_f = -306$ kJ

Equating the two routes leading to the formation of PCl_3,

$$\Delta H_f = \Delta H_1 + \Delta H_2 + \Delta H_3$$

$$-306 = 314 + 363 + \Delta H_3$$

$$\Delta H_3 = -983 \text{ kJ.}$$

This is energy released on the formation of three P – Cl bonds, hence,

$$E_{P-Cl} = \frac{983}{3} = 328 \text{ kJ mole}^{-1}$$

$P_s = P_g$ $\Delta H_1 = 314$ kJ

$3/2\ H_{2,g} = 3H_g$ $\Delta H_2 = 649$ kJ

$$P_2 + 3H_g = PH_3 \qquad \Delta H_3 = ?$$

Also,

$$P_s + 3/2\ H_{2,g} = PH_3 \qquad \Delta H_f = 8\ kJ$$

From which

$$\Delta H_3 = -\ 958\ kJ$$

$$E_{P-H} = \frac{958}{3} = 319.33\ kJ$$

Example 9:

100g of anhydrous $CuSO_4$, when dissolved in excess of water produced 42 kJ of heat. The same amount of $CuSO_4$. $5H_2O$ on dissolving in large excess of water absorbed 4.60 kJ. What is the heat of hydration $CuSO_4$?

Solution:

$$100g\ of\ CuSO_4 = \frac{100}{159.5}\ mole$$

$$Heat\ of\ solution\ of\ CuSO_4\ per\ mole = \frac{42 \times 159.5}{100} = -\ 66.78\ kJ$$

$$100\ g\ of\ CuSO_4.\ 5H_2O = \frac{100}{249.5}\ mole$$

$$Heat\ of\ solution\ of\ CuSO_4.\ 5H_2O\ per\ mole = 4.6 \times \frac{249.5}{100} = 11.48\ kJ$$

Thus,

(i) $CuSO_4.\ 5H_2O + aq \rightarrow CuSO_4(aq) + 66.78\ kJ$

(ii) $CuSO_4.\ 5H_2O + aq \rightarrow CuSO_4(aq) - 11.48\ kJ$

On subtracting (i) from (ii), we get

$$CuSO_4 + 5H_2O \rightarrow CuSO_4.\ 5H_2O + 78.26\ kJ$$

$\because$ Enthalpy of hydration of $CuSO_4$, $\Delta H = -\ 78.26\ kJ$.

Example 10:

Heat of combustion of glucose at constant pressure at 17oC was found to be –651,000 cal. Calculate the heat of combustion of glucose at constant volume considering water to be in the gaseous state.

Solution:

$$\underset{\text{negligible volume}}{C_6H_{12}O_6(s)} + \underset{\text{6 mole}}{6O_2(g)} \rightarrow \underset{\text{6 mole}}{6CO_2(g)} + \underset{\text{6 mole}}{H_2O(g)}$$

$$\Delta n = 12 - 6 = 6$$

$$\Delta H = \Delta E + \Delta nRT$$

$$\Delta E = \Delta H - \Delta nRT = -651000 - 6 \times 2 \times (273 + 17)$$

$$= 651000 - 3480 = -654480 \text{ cal}$$

Hence heat of combustion of glucose at constant volume and at 17°C is – 654,480 cal.

Example 11:

The combustion of methane takes place according to the following equation :

$$CH_4(g) + 2O_2(g) \rightarrow CO_2(g) + 2H_2O(l) \; ; \qquad \Delta H = -890.3 \text{ kJ}$$

(a) How many grams of methane would be required to produce 89.03 kJ of heat of combustion ?

(b) How many grams of carbon dioxide would be formed when 89.03 kJ of heat is evolved ?

Solution:

(a) Molecular weight of $CH_4 = 1 \times 12 + 4 \times 1$

$= 12 + 4 = 16 \qquad [\because C = 12, H = 1]$

$\because$ 890.3 kJ of energy is liberated by the combustion of 1 mole of 16 g of methane.

$\therefore$ 890.3 kJ of energy is liberated by the combustion of

$$16 \times \frac{89.03}{890.3} = 1.6 \text{ g of methane.}$$

= 1.6 g of methane.

(b) Molecular weight of $CO_2 = 1 \times 12 + 2 \times 16$

$= 12 + 32 = 44$

$[\because C = 12, O = 16]$

$\because$ 890.3 kJ of energy is liberated to produce 1 mole or 44 g of CO_2

$\therefore$ 890.3 kJ of energy is liberated to produce

$$44 \times \frac{89.03}{890.3}$$

or 4.4 g of CO_2

= 4.4g of carbon dioxide.

Example 12(a):

Calculate the heat of formation of acetic acid if heat of combustion is 869.0 kJ mol^{-1}. The heats of formation of $CO_2(g)$ and $H_2O(l)$ are – 395.0 kJ mol^{-1} and – 285.0 kJ mol^{-1} respectively.

Solution:

The data given are expressed in the form of three the mochemical equations.

(1) $C(s) + O_2(g) \rightarrow CO_2(g)$; $\Delta H_1 = -395$ kJ

(2) $H_2(g) + 1/2O_2(g) \rightarrow H_2O(l)$; $\Delta H_2 = -285$ kJ

(3) $CH_3COOH(l)+2O_2(g)\rightarrow 2CO_2(g)+2H_2O(l)$; $\Delta H_3 = -869$ kJ

The equation corresponding to the formation of acetic acid from its elements is

$C(s) + 2H_2(g) + O_2(g) \rightarrow CH_3COOH(l)$; $\Delta H = ?$

This equation may be obtained from the three thermochemical equations on multiplying equations (1) and (2) by two and adding and finally subtracting equation (3);

$$2C(s) + 2O_2(g) \rightarrow 2CO_2(g)$$

$$2H_2(g) + O_2(g) \rightarrow 2H_2O(l)$$

Adding $$2C(s) + 3O_2(g) + 2H_2(g) \rightarrow 2CO_2(g) + 2H_2O(l)$$

Subtract Eq. (3) $$CH_3COOH(l) + 2O_2(g) \rightarrow 2CO_2(g) + 2H_2O(l)$$

$$2C(s) + O_2(g) + 2H_2(g) \rightarrow CH_3COOH(l)$$

Rearranging

$$2 \times (1)2 \times \Delta H_1 = -790 \text{ kJ}$$

$$2 \times (2)2 \times \Delta H_2 = -570 \text{ kJ}$$

Adding $$\Delta H' = -1360 \text{ kJ}$$

Subtract Eq. (3) $$\Delta H_3 = -869 \text{ kJ}$$

Giving $$\Delta H = -1360 - (869) = - \text{ kJ}.$$

Example 12(b):

The enthalpy of evaporation of water at 373 K is 40.67 kJ mol^{-1}. What will be the enthalpy of evaporation at 353 K and 393 K if average molar heats at constant pressure in this range for water in liquid and vapour states are 75.312 and 33.89 JK^{-1} mol^{-1} respectively ?

Solution:

$$H_2O(l) \rightleftharpoons H_2O(g)$$

Enthalpy change at 373 K is ΔH_1 = 40.67 kJ mol^{-1}

= 40670 J mol^{-1}

$$\Delta H_2 - \Delta H_1 = \Delta C_p(T_2 - T_1)$$

$$\Delta H_2 = \Delta H_1 + \Delta C_1(T_2 - T_1)$$

$$\Delta C_p = \Delta C_{p,\ H2O(g)} - \Delta C_{p,\ H2O(l)}$$

$$= 33.89 = 75.312 = -41.422.$$

Enthalpy of evaporation at 353 K, ΔH

= 40670 – 41.522(353 – 373)

= 40670 + 41.421 × 20

= 41498.44 J.

Similarly enthalpy of evaporation at 393 K

= 40670 – 41.422(393 – 373)

= 39841.56 J.

Example 12(c):

The heat of formation of one mole of HI from hydrogen and iodine vapour at 25°C is 8000 cal.

$$1/2H_2(g) + 1/2I_2(g) \rightarrow HI(g),\ \Delta H = -8000\ cal.$$

Calculate the heat of formation at 10°C and also the total change in the heat capacity at constant pressure. It is given that the molar heat capacities of hydrogen, iodine and HI vapours are given by the equations :

$C_p = 6.5 + 0.0017T$	*for*	*Hydrogen(g)*
$C_p = 6.5 + 0.0031T$	*for*	*Iodine(g)*
$C_p = 6.5 + 0.0016T$	*for*	*HI(g)*

where T is the absolute temperature.

Solution:

$$\Delta C_p = C_{p(HI)} - 1/2\ [C_{p(HI)} + C_{P(12)}]$$

$$= 6.5\ 0.0016T - 1/2\ [6.5 + 0.0017T + 6.5 + 0.0038T]$$

$$= -0.00115T\ cal.\ deg^{-1}.$$

Here $\Delta H_{298} = \Delta H_{283} + \int_{283}^{298} \Delta C_p \, dT$

where $T_1 = 273 + 10 = 283$ K, $T_2 = 273 + 25 = 298$ K

Also, ΔH_{298} K = – 8000 cal.

$$-8000 = \Delta H_{283} + \int_{283}^{298} (-0.00115)\ T\,dT$$

or $\Delta H_{283} = -8000 + \frac{0.00115}{2}[T_2]_{283}^{298} = -8000 + \frac{0.00115}{2}\left[(298)^2 - (283)^2\right]$

$= -$ 7995 cal (nearly).

Example 13:

How much heat is required to raise the temperature of 1 mole of oxygen from 303 K to 1300 K at constant pressure?

$$C_p = 6.095 + 3.253 \times 10^{-3}\ T - 1.017 \times 10^{-6}\ T^2.$$

Solution:

$$H_2 - H_1 = \int_{T_1}^{T_2} C_p \, dT$$

$$T_1 = 300 \text{ K},\ T_2 = 100\text{K}.$$

$$H_{1300} - H_{300} = \int_{300}^{1300} \left(6.095 + 3.253 \times 10^{-3} T - 1.07 \times 10^{-6}\ T^2\right) dT$$

[Substituting the value of C_p]

$$H_{1300} - H_{300} = 6.095\,(1300 - 300) + \frac{3.253 \times 10^{-3}}{2} \times \left[(1300)^2 - (300)^2\right]$$

$$-\ 1/3 \times 1.017 \times 10\text{–}6\ [(1300)^3 - (300)^3]$$

$$= 6095 + 2602.4 - 735.63 = 7961.77 \text{ cal/mole.}$$

Example 14:

Calculate ΔH at 25°C for the reaction

$$H_2O(g) \rightarrow H_2(g) + 1/2O_2(g)$$

given that ΔH at 18°C is 241.750 kJ mol^{-1} and molar-Heat capacities expressed as Jk^{-1} mole^{-1} are $C_p(H_2)$ = 28.83, $C_p(O)$ = 29.12 $C_p(H_2O)$ = 33.56

Solution:

ΔC_p = Heat capacities of products – Heats capacities of reactants

$= [28.83 + 1/2 \times 29.12] - 33.56 = 43.39 - 33.56 = 9.83 Jk^{-1}mole^{-1}$

$$T_2 - T_1 = (273 + 25) - (273 + 18) = 7 \text{ k}$$

$$\Delta H_2 + \Delta H_1 = \Delta C_p (T_2 - T_1)$$

or $$\Delta H_2 = \Delta H_1 + \Delta C_p (T_2 - T_1) = 241.750 + 9.83 \times 7$$

$$= 310.56 \text{ kJ mol}^{-1}.$$

Example 15:

A student made the following erroneous statement in a laboratory report on bomb calorimetry.

$$\Delta H = \Delta U + p\Delta V$$

Since the bomb calorimetry process is a constant volume one, $\Delta V = 0$ and $\Delta H = \Delta U$. Explain why this argument is incorrect

Solution:

Since, $$H = U + pV$$

we will have $dH = dU + pdV + Vdp$.

For a constant volume process, the above expression is reduced to

$$dH = dU + V\,dp$$

While writing the expression

$$\Delta H = \Delta U + p\,\Delta V$$

the student has not written the expression V dp which makes a significant contribution when the system involves the gaseous phase. Hence, his conclusion $\Delta H = \Delta U$ is incorrect.

Example 16:

H_2 gas is mixed with air at 25°C under a pressure of 1 atmosphere and exploded in a closed vessel. The heat of the reaction

$$H_2(g) + 1/2\, O_2(g) \rightarrow H_2O(g)$$

at constant volume, $\Delta E_{298} = -240.60$ kJ mol^{-1} and C_p values for H_2O vapour and N_2 in the temperature range 298 K and 3,200 K are 39.06 JK^{-1} mol^{-1} and 26.40 JK^{-1} mol^{-1} respectively. Calculate the explosion temperature under adiabatic conditions.

Solution:

For a constant volume process under adiabatic conditions,

$$\Delta E = \Delta E_{heating} + \Delta E_{298} = 0$$

$$\therefore \quad \Delta H_{heating} = -\Delta E_{298}$$

$$= -\int_{298}^{T_f} n \sum C_V \, dT = -240.60 \text{ kJ mol}^{-1} \qquad \text{...(i)}$$

As 2 moles of unreacted N_2 are associated with 1/2 mole of O_2, we have

$$\Sigma nC_V = C_V(H_2O, g) + 2CV\,(N_2, g)$$

$$= (39.06 + 2 \times 26.40) \text{ Jk}^{-1} \text{ mol}^{-1}$$

$$= 91.86 \text{ Jk}^{-1} \text{ mol}^{-1}$$

Hence, from Eq. (i) on integrating, we get

$$(91.86 \text{ Jk}^{-1} \text{ mol}^{-1})(T_f - 298) = 240.600 \text{ J mol}^{-1}$$

$$T_f - 298 = \frac{240600 \text{ jmol}^{-1}}{91.86 \text{ J k}^{-1} \text{ mol}^{-1}} = 2619 \text{ k.}$$

$$T_f = (2619 + 298) \text{ k} = 2917 \text{ k.}$$

Example 17(a):

Estimate the adiabatic flame temperature of a spirit lamp, provided the following data

$$C_2H_5OH(l) + 3O_2(g) \rightarrow 2CO_2(g) + 3H_2O(g)$$

$$\Delta H_{298}\, k = 1370 \text{ kJ mol}^{-1}.$$

$C_{p.m}$ *for (i)* $CO_2(g) = 36.43 \text{ J k}^{-1} \text{ mol}_{-1}$,

(ii) $H_2O(g) = 33.70 \text{ J k}^{-1} \text{ mol}^{-}{}_{1}$ *and*

(iii) $N_2(g) = 294.4 \text{ J k}^{-1} \text{ mol}_{-1}$.

If the above reaction is carried out in air consisting of 80 % N_2 *and 20% of* O_2 *by volume, what would be the final temperature ?*

Solution:

The expression of adiabatic flame temperature is

$$T_f = \frac{-\Delta H}{C_p(\text{products})} = T_o$$

$$= \frac{(1370 \times 10^3 \text{ J mol}^{-1})}{(2 \times 36.43 + 3 \times 33.70)\text{J k}^{-1} \text{ mol}^{-1}} + 298\text{k}$$

$$= 7875 \text{ k} + 298 \text{ k} = 8173 \text{ k}$$

If the reaction is carried out in air, the consumption of 3 mol of O_2 would leave 12 mol of N_2 from air. Hence.

$$T_f = \frac{(1370 \times 10^3 \text{ J mol}^{-1})}{(2 \times 36.43 + 3 \times 33.70 + 12 \times 29.4)\text{J k}^{-1} \text{ mol}^{-1}} + 298\text{k}$$

$$= 2600 \text{ k} + 298 \text{ k} = 2898 \text{ k}$$

Example 17(b):

Use Born-Haber cycle to determine the lattice energy of KCl(s) at 298.15 K. The following data are given at the same temperature.

Enthalpy of formation of KCl	*= – 435.9 k J mol^{-1}*
Ionization energy of K	*= 418.9 k J mol^{-1}*
Electron affinity of Cl	*= 348.6 k J mol^{-1}*
Dissociation enthalpy of Cl_2	*= 242.0 k J mol^{-1}*
Sublimation enthalpy of K	*= 89.8 k J mol^{-1}*

Solution:

We are given that

(1) $K(s) + 1/2\ Cl_2(g) \rightarrow KCl(s)$ $\Delta H_1 = 435.9$ kJ mol^{-1}

(2) $K(g) \rightarrow K^+(g) + e$ $\Delta H_2 = 418.9$ kJ mol^{-1}

(3) $Cl(g) + e \rightarrow Cl^-(g)$ $\Delta H_3 = -348.6$ kJ mol^{-1}

(4) $Cl_2(g) \rightarrow 2Cl(g)$ $\Delta H_4 = 242.0$ kJ mol^{-1}

(5) $K(s) \rightarrow K(g)$ $\Delta H_5 = 89.8$ kJ mol^{-1}

We have to find ΔH of the following equation

(6) $K^+(g) + Cl^-(g) \rightarrow KCl(s)$

Equation (6) can be obtained by the following manipulations.

Eq. (1) – Eq. (5) – Eq. (2) (2) – 1/2 Eq. (4) – Eq. (3)

Accordingly, we have

$$\Delta H_6 = \Delta H_1 - \Delta H_5 - \Delta H_2 - \Delta H_1 .2\ \Delta H_4 - \Delta H_3$$

$\Delta H6 = (-435.9 - 89.8 - 418.9 + 1/2 \times 242.0+348.6)$ kJ mol^{-1}

$= -475.0$ kJ mol^{-1}.

Example 18:

The bond enthalpy of $H_2(g)$ is 436 kJ mol^{-1} and that of $N_2(g)$ is 941.3 kJ mol^{-1}. Calculate the average bond enthalpy of an N – H bond in ammonia $\Delta H_f^o(NH_3) = -46.0$ kJ mol^{-1}.

Solution:

$\because$ (i) $N_2(g) \rightarrow 2N(g)$; $\Delta H^o = 941.3$ kJ mol^{-1}

(ii) $H_2(g) \rightarrow 2H(g)$; $\Delta H^o = 436.0$ kJ mol^{-1}

(iii) $1/2N_2(g) + 3/2H_2(g) \rightarrow NH_3(g)$; $\Delta H^o = -46.0$kJ mol^{-1}

On multiplying equation (i) by 1/2 and equation (ii) by 3/2 and adding, we get

(iv) $1/2N_2(g) + 3/2H_2(g) \rightarrow N(g)+3H(g)$; $\Delta H^o = 1124.6$kJ mol^{-1}

On subtracting equation (iii) from (iv), we get

(v) $NH_3(g) \rightarrow N(g) + 3H(g)$, $\Delta H^o = 1170.6$ kJ mol^{-1}

As there are three N – H bonds, in NH_3, the average bond enthalpy is obtained by dividing the value ΔH^o from (v) by 3. Hence

$\Delta H_{N-H} = 1170.6/3 = 390.2$ kJ mol^{-1}

Example 19:

0.50 f of benzoic acid was subjected to combustion in a bomb calorimeter at 15°C when the temperature of the calorimeter system (including water) was found to rise by 0.55°C. Calculate the heat of combustion of benzoic acid : (i) at constant volume, (ii) at constant pressure. The thermal capacity of the calorimeter system including water was found to be 23.85 kJ.

Solution:

(i) Heat of Combustion at Constant Volume (ΔE)

Weight of benzoic acid, $m = 0.50$ g

Rise in temperature, $\theta = 0.55$°C

Molecular weight of benzoic acid, $M = 122$

$$\left.\begin{array}{l}\text{Thermal capacity}\\ \text{of the calorimeter}\\ \text{system, z}\end{array}\right] = 23.85 \text{ kJ}$$

$$\therefore \left.\begin{array}{l}\text{Heat of combustion}\\ \text{at constant}\\ \text{volume, } \Delta E\end{array}\right] = z \times \theta \times M/m$$

$$= -23.85 \times 0.55 \times \frac{122}{0.50} = -3200.7 \text{ kJ} = -3200.7 \text{ kJ}$$

(ii) Heat of Combustion at constant Pressure (ΔH)

The combustion of benzoic acid at 25°C involves the following reaction :

$C_6H_5COOH\ (s) + 7.5O_2(g) \rightarrow 7CO_2(g) + 3H_2O(l)$

We known that

$\Delta H = \Delta E + \Delta nRT$

In the present case :

$\Delta E = -3200.7$ kJ $\Delta n = 7 - 7.5 = -2.5$ $R = 8.314 \times 10^{-3}$ kJ

$T = 273 + 25 = 298°K$

$\therefore \quad \Delta H = -3200.7 + (\ 0.5) \times 8.314 \times 10^{-3} \times 298 = -3201.9$ kJ

$\therefore$ Heat of Combustion at constant pressure,

$$\Delta H = -3201.9 \text{ kJ}$$

Example 20:

The heats of formation of carbon dioxide and water are 395 kJ and 285 kJ respectively, and the heat of combustion of acetic acid is 869 kJ (all exothermic). Calculate the heat of formation of acetic acid.

Solution:

The date given are expressed in the form of three thermochemical equations :

$C + O_2 = CO_2$	$\Delta H_1 = -395$ kJ (1)
$H_2 + 1/2O_2 = H_2O$	$\Delta H_2 = -285$ kJ (2)
$CH_3COOH + 2O_2 = 2CO_2 + 2H_2O$	$\Delta H_3 = -869$ kJ (3)

The equation corresponding to the formation of acetic acid from its elements is

$$2C + 2H_2 + O_2 = CH_3COOH \qquad \Delta H_f = ?$$

This equation may be obtained from the three thermochemical equations on multiplying equations (1) and (2) by two and adding, finally subtracting equation (3) :

$$2 \times (1) \quad 2C + 2O_2 = 2CO_2$$

$$2 \times (2) \quad 2H_2 + O_2 = 2H_2O$$

Adding $\quad 2C + 3O_2 + 2H_2 = 2CO_2 + 2H_2O$

Subtract (3) $\quad CH_3COOH + 2O_2 = 2CO_2 + 2H_2O$

Giving, after re-arranging $\quad 2C + O_2 + 2H_2 = CH_3COOH$

$2 \times (1) \quad 2 \times \Delta H_1 = -\ 790$ kJ

$2 \times (2) \quad 2 \times \Delta H_2 = -\ 570$ kJ

Adding $\quad \Delta H = -\ 1360$ kJ

Subtract (3) $\Delta H_3 = -\ 896$ kJ

Giving $\quad \Delta H_f = -\ 2360 - (-\ 869) = -\ 491$ kJ

Example 21:

Heats of neutralization (ΔH) of NH_4OH and HF are – 51.5 and 68.6 kJ respectively. Calculate their heats of dissociation.

Solution:

We have

(i) $HCl(aq) + NaOH(aq) \rightarrow NaCl(aq) + H_2O;\ \Delta H = -\ 57.3$ kJ

(ii) $HCl(aq) + NH_4OH(aq) \rightarrow NH_4Cl(aq) + H_2O;\ \Delta H = -\ 51.5$ kJ
(Weak base)

∴ The heat of dissociation of NH4OH,

$$\Delta H = 51.5 - (57.3) = 5.8 \text{ kJ}.$$

Similarly we have

$$HF(aq) + NaOH(aq) \rightarrow NaF + H2O;$$

$$\Delta H = -\ 68.6 \text{ kJ}$$

∴ The heat of dissociation of HF,

$$\Delta H = -68.6 - (57.3) = -11.3 \text{ kJ}.$$

Example 22:

Calculate the heat of reaction at 25°C for the reaction :

$$C_2H_4(g) + H_2(g) \rightarrow C_3H_6(g)$$

given that the heats of combustion of ethylene, hydrogen and ethane at 25oC are 337.0, 68.4 and 373.0 kcal respectively.

Solution:

The thermochemical/equations for the combustion of ethylene, hydrogen and ethane may be written as

(i) $C_2H_4 + 3O_2 \rightarrow 2CO_2 + 2H_2O$ $\Delta H = -337$ kcal

(ii) $H_2 + 1/2O_2 \rightarrow H_2O$ $\Delta H = -68.4$ kcal

(iii) $C_2H_6 + 31/2O_2 \rightarrow 3CO_2 + 3H_2O$ $\Delta H = -337$ kcal

$$\text{(i)} + \text{(ii)} - \text{(iii)} \Rightarrow C_2H_4 + H_2 - C_2H_6$$

Heat of reaction = – 337 – 68.4 + 373 = – 32.3 kcal.

ENCYCLOPAEDIA OF CHEMICAL THERMODYNAMICS

ENCYCLOPAEDIA OF CHEMICAL THERMODYNAMICS

Vol. III

By
K.K. Jain

DISCOVERY PUBLISHING HOUSE PVT. LTD.
NEW DELHI-110 002

Published by:
Namit Wasan

DISCOVERY PUBLISHING HOUSE PVT. LTD.
4383/4B, Ansari Road, Darya Ganj
New Delhi-110 002 (India)
Phone : +91-11-23279245; 23253475; 43596065
E-mail : discoverybooksindia@gmail.com
discoverypublishinghouse@gmail.com
namitwasan9@gmail.com
web : www.discoverypublishinggroup.com

Edition: **2020**

ISBN: 978-81-8356-351-2 (Set)

Encyclopaedia of Chemical Thermodynamics

Printed at:
Infinity Imaging Systems
Delhi

Preface

There are very few books available on chemistry which present the subject matter in a concise manner so that it is interesting and within grasp of B.Sc. (Hons.) and postgraduate students.

Keeping the above approach in mind this book has been planned. It is hoped that it would be useful to chemistry students at all levels. A logical and rigorous approach has been adopted but too much philosophical discussion has been avoided in order to make the material always within the grasp of students.

Each chapter of the book contains a discussion of the subject matter and important mathematical relations followed by suitable solved examples to illustrate the basic principles. At the end of each chapter a set of problems has been given to help the students become familiar with the concepts and principle presented in the text.

Although it has not been possible to give specific reference to contributions which inspired us in writing this book, we do knowledge the help of many who contributed directly or through their writings.

The comments, criticism and suggestions from the readers will be very much appreciated.

Author

CONTENTS

1

Classical Thermodynamics

INTRODUCTION

Classical method can be employed to express the property o the system to find out what fraction of the molecules has a given energy. It is easier to find out the values of entropy, free energy, enthalpy, heat capacity etc., from spectroscopic data by using statistical methods. We shall however, discuss the equilibrium aspects of the system in the succeeding articles. In other words, this is a new approach that deals with the applications of statistical mechanics to thermodynamics. It has been applied successfully to relate the microscopic properties of the individual molecules (moment of inertia, dipole moment, etc.) with macroscopic properties (molar heat capacity, polarization, etc.) of a system having a large number of molecules. Statistical thermodynamics can be used for systems at equilibrium as well as away from equilibrium. We shall limit our discussion to the equilibrium aspects of the system Distinction between quantum mechanics, statistical mechanics and statistical thermodynamics.

Statistical mechanics acts as a bridge between thermodynamics and quantum mechanics. The classical thermodynamics is based on empirical laws in which the variables such as pressure, temperature and energy are regarded as functions of state. However, in the derivation of relations of thermodynamic functions, we ignore the constitution of the material system. But the experimental observations support the laws. The only drawback in classical thermodynamics is that we cannot calculate directly the absolute value off any thermodynamics function of a substance. In order to over come this, a new method of approach has been developed and is known as statistical thermodynamics. The statement according, to which non-spontaneous process also called negative process cannot be the only result of a complex of processes is not strict are negative processes in microscopic (bulk) system events. Therefore the second law of thermodynamics first law, but a statistical law that is observed with a high degree off accuracy for considerable numbers of molecules and is less applicable, the smaller are the dimensions of the system being studied. So the second law of thermodynamics has statistical nature. Classical thermodynamics has bee applied successfully to relate the microscopic properties of the individual molecules (*e.g.* moment of inertia, dipole moment etc.) with macroscopic properties (*e.g.* molar heat capacity, polarization, etc.) of a system having a large number of molecules.

In supposed to be composed of innumerable particles in motion. This motion of particles is governed by the laws of dynamics and in fact subject to quantum the number of particles is so large that it is impossible to follow the behaviour of every individual particle Statistical methods can be used to find out the properties of the system, particularly to know what fraction of the molecules has a given energy.

It is possible to establish that the laws of thermodynamics are the necessary consequence of the quantum principles, when statistical postulates are used. In this method, we take into account the atomic constitution of matter. The matter is supposed to consist of innumerable particles in motion is governed by the laws of dynamics. Quantum mechanics provides information about the energy of the molecular system, statistical mechanics tells us about the possible arrangement of the energy among various molecules of the system and introduces the concept of probability and partition functions. Statistical thermodynamics deals, with the relationships between the probability partition functions and thermodynamic properties. In other words, a physical entity generally manifests itself in :

(i) A Microscopic, and

(ii) A Macroscopic state.

The dealt with macroscopic or the bulk state of matter in the preceding chapters.

The macroscopic variables were pressure, volume, temperature, entropy, etc. The macroscopic variables were related mutually on thermodynamical considerations. Further, we assumed that the number of atoms/molecules in them (microscopic details of a macrosystem) were exorbitantly large $\approx 10^{25}\ m^{-3}$. The microscopic systems, on the other hand, have very few particles. Now, we know that the actual systems, the macro systems, are formed of the microsystems and so there should be a definite relation between the two.

This fact was suggested by L. Bottzmann. He explained in this way the contradiction between the reversibility of mechanical motion and the irreversibility and unidirectional nature of real physical and chemical processes; this unidirectional nature is the result of the molecular structure of the material world. In this way, it was Boltzmann who provided the molecular basis to chemical thermodynamics. This approach is called chemical statistics or statistical thermodynamics and it deals with the applications of mechanics to chemical thermodynamics. If we want to have a complete description of a system we need to know actual velocity (or momentum), location and mutual interaction (if any, with other particles) of every particle. Since every panicle is described by three positions and three momentum coordinates, hence to describe a macrosystem we need a very large number of data. Further, any change in the parameter of an individual particle should be reflected in the properties of the system as a whole. *This poses a question*: How to deal with such a vast data which even a super-computer is not capable to handle, if at all we are able to collect it. If, however, we remember that:

(i) Much of the data and analytical results obtained on micro level are not useful in the study of the macrosystem, and

(ii) The application of classical dynamical principles to a many particle system is theoretically inapplicable, we can easily dispense with the technically impossible and practically useless classical kinetic and thermodynamic methods in favour of new statistical methods.

In the following sections we will see the importance of the concept of probability in the development of the statistical theory. Hence, we shall first consider some elementary theorems on the calculus of probability. In other words, it refers to a class of events of which the particular event is one member. Mathematically, the probability is defined as: When a certain aspect is possible in a ways and is not possible in b ways, then the probability of occurring the aspect is $a/(a + b)$. (Newton defined probability as the ratio of a certain number of aspects, which were possibly observable, to the total number of aspects included in that class.) Here $(a + b)$ is the total number of aspects each of which has the same chance of occurrence. Statistics is applied only to such systems which contain a large number of identical individuals which may be distinguishable by means of some property. In a statistical method, we study the behaviour of aggregates of particles to calculate the average value without investigating the detailed behaviour of the system and finding there from the value for each individual. Statistically, it is sufficient to know the most probable distribution of particles in different accessible states. This branch of physics has seen its development in the hands of Maxwell, Boltzmann, Gibbs, Fermi-Dirac, Bose-Einstein etc.

Now, as we have shown, if an event can occur in a ways, the probability of occurring of this event is a/(a + b), and similarly, if it fails to occur in b ways, then the probability of its failing is b/(a + b). The sum of the two

$$\frac{a}{a+b} + \frac{b}{a+b} = 1$$

corresponds to the certainty of some event, since the event must occur or fail to occur. If b = 0, the probability of occurrence becomes certain but if a = 0, the probability of its failure is certain.

We may define the probability for a certain event occurring m times in n unrestricted (free) attempts as,

$$P = \lim_{n \to \infty} \frac{m}{n}.$$

STATISTICAL MECHANICS

Statistical mechanics deals with systems consisting of many particles and the methods employed are to get a collective or macroscopic property of the system without taking into account the individual motion of the particles The word particle is used in a broad sense meaning a fundamental particle *e.g.*, an electron, an atom, a molecule etc. The particle is a well defined and a stable unit of a given physical system. As it is not practically possible to determine the property of each particle individually, a statistical approach is made by using the concept of probability of distribution. This statistical approach helps in determining the bulk or macroscopic property of the system as a whole. The idea of probability does not imply that the particles move in a random way without obeying the laws. For a many particle system, the statistical analysis is valid regarding the applications of probability distribution of particles.

STATISTICAL POSTULATES

Following postulates are made for applying the statistical method to physical systems:

(i) Every system has a large number of particles/molecules which are in constant random motion.

(ii) The motion of particles/molecules is described in 2Nf-dimensional phase-space.

(iii) The phase-space is divided into cells.

(iv) Postulate of equal a priori probability holds good.

(v) Probability of a particular configuration is given by the ratio of microstates corresponding to that configuration to total accessible microstates, *i.e.*,

$P_i = \Omega_i/\Sigma\Omega_i$ and the average value of any physical parameter is given by

$$< x > = \Sigma x_i P_i = \frac{\Sigma x_i \Omega_i}{\Sigma \Omega_i}$$

where Qi is the frequency of the configuration.

(vi) The equilibrium state of the system is the state of maximum probability.

(vii) The law of conservation of mass is generally valid whence

$n_1 + n_2 + n_3 + ... = \Sigma n_i, = N =$ constant.

(viii) The law of conservation of energy governs the distribution of particles among cells.

Thus, if $\varepsilon_1, \varepsilon_2, \varepsilon_3$ ... denotes energies of each particle in cell 1, 2, 3, etc., then

$E = \varepsilon_1\ n_1, + \varepsilon_2\ n_2 + \varepsilon_3\ n_3 + ... = \Sigma\varepsilon_i n_i =$ constant.

(ix) Every point in the phase-space (cell) represents a microstate, the cell volume being h where/is the degree of freedom.

(x) The number of microstates in the energy interval e and $\varepsilon + d\varepsilon$. per unit volume is represented as $\Omega(\varepsilon)$.

(xi) A collection of accessible microstates form a statistical ensemble.

ENSEMBLE METHOD

Ensemble method is superior to the method of *time average*, in determining average value of a physical quantity. So first we recall the characteristics of an ensemble and its way of representation.

1. It is a group of large number of systems.
2. Every member of an ensemble is an exact miniature copy of the system under test.

3. Every member of an ensemble is subjected to the same experimental conditions.
4. There may be three kinds of interactions between the members of the ensemble:
   (a) Thermal-related to T,
   (b) Mechanical-related to V, and
   (c) Diffusive-related N or m (*i.e.*, chemical potential).
5. Every member of the ensemble may have a large number of microstates (also called the dynamic state) due to random motion of its particles/molecules/And when the configuration of the member systems is maintained same in terms of generalised coordinates and momenta, the statistical method is expected to yield quite accurate results.

Ensemble Representation

Ensembles can be represented in a Γ-space by the distribution of its phase-points. Generally, the distribution of phase-points in an ensemble is continuous and can be represented by a density function ρ(p, qt).

The number of phase-points in the vicinity of (p, q) in Γ-space is given by p (p, q t) dT where d r = dp dq = volume element of Γ- space.

If ρ (p, q t} is known at time t_1, its value can easily be obtained for some other time t_2. Now, from the probability theory

$$d\rho\ (p,\ q)\ \alpha\ d\Gamma$$

$$\therefore \quad d\rho\ (p,\ q) = \rho(p,\ q)d\Gamma$$

Here ρ (p, q) is known as the distribution function. The ensemble average of any function R is then given by

$$\langle R \rangle = \int R\ (p,\ q)\ \rho(p,\ q)d\Gamma.$$

Selection of Ensemble

(i) Molecules of real gases are usually formed of 2, 3.... atoms whence the density of phase points is not uniform,

(ii) The density of phase points is non-zero for every range of energy,

(iii) Total number of particles and energy is conserved. Point (i) eliminates use of uniform ensemble. Point (ii) eliminates use of microcanonical ensemble. Point (in) eliminates use of grand canonical ensemble to real gases.

But since the properties of these gases such as N, V, T, ε, H etc., are constant the canonical ensemble finds use in description of real gases for their thermodynamic properties.

STEFAN'S LAW VERIFICATION

In 1897, Lummer and Pringsheim experimentally verified Stefan's law over a wide range of temperature (100°C to 1,300°C). The apparatus consisted of a black body C. For temperatures between 200°C and 600°C, a hollow copper sphere coated inside with platinum black was used. The fused nitrates of sodium and potassium having a melting point of 219°C were used as the bath surrounding the black body. For temperature between 900°C and 1,300°C, an iron cylinder coated inside with platinum black was used as a black body and it was enclosed in a double walled gas furnace. A thermocouple T was used as a thermometer. A bolometer B was used to measure the intensity of the emitted heat radiations. S_1, S_2 and S_3 were the water-cooling shutters. Another black body A at 100°C was used to standardize the bolometer.

The double walled vessel of the black body A contained boiling water at 100°C. The bolometer B was allowed to face the opening of the black body A and the shutter S_3 raised. The deflections in the galvanometer of the bolometer at various distances were noted and it was found that the deflection was inversely proportional to the square of the distance between the bolometer and the opening of the black body A. Thus, the deflection in the galvanometer was proportional to the intensity of heat radiations.

The shutter S_3 was closed and the bolometer B was allowed to face the opening of the black body C. The shutters of S_1 and S_2 were raised. The bath surrounding the black body was maintained at a constant temperature and the maximum deflection produced in the galvanometer of the bolometer was noted. Thus at various constant temperatures of the black body, corresponding to constant deflections (in the galvanometer of the bolometer) were observed. Then the data was reduced to a common arbitrary unit in terms of the total radiations from the black body A at 100°C. Let θ be the deflection in the galvanometer, T_1 the temperature of the black body and T_2, the temperature at the entrance of the bolometer. It was found that

$$\theta \propto (T_1^4 - T_2^4)$$

But $\theta \propto R$

$$\therefore \qquad \theta \propto (T_1^4 - T_2^4)$$

This verifies Stefan's law.

Recently Coblentz has verified Stefan's law more accurately. He took an electrically heated black body whose temperature was measured by an accurate thermocouple. An absolute bolometer was used to measure the amount of heat radiations emitted by the black body. He was able to show the correctness of Stefan's law experimentally up to 1,600°C.

GIBBS ENSEMBLE

Consider a system a such that a number of similar observations can be taken on it. Further, we assume that the system undergoes exactly similar changes in its state during each observation, *i.e.*, after every set of operations, the system returns back to its initial state (*i.e.*, the system is perfectly reversible). The propeties of ensemble was put forth by Gibbs in 1902. In his view, it was a mathematical device which provided us with an opportunity to follow a probabilistic approach in defining the state of the system at a given time.

If this be the case, we can define the state of the system in terms of its statistical probability. But this is not practically feasible as:

(i) The number of molecules/particles in the system are so large that it is not possible to have the macrostate exactly the same as it was during the preceding observation,

(ii) Each macrostate has several microstates. We cannot ascertain their identity,

(iii) Each observation may take quite a long time whence for applying statistics, several observations may need several thousand years,

(iv) To manage and process a large number of data, over years together is not possible,

(v) The system may not show reversibility to different operations.

Ensemble was a satisfactory solution to it. An ensemble is a group of N_T identical systems. These N_T systems have the same configuration, energy and accessible states.

In other words, the ensemble has N_T isolated systems each of which are exact copies of the experimental system A Thus, we have a large number of systems that behave exactly in the same manner during a set of operations as the system A Then, if N_r systems out of N_T systems produce a particular result r, we have

$$P_r = N_r/N_T$$

as the probability for that result.

If we want to measure a particular property, say x, of the system

$$<x> = \frac{n_1x_1 + n_2x_2 + ... + n_rx_r + }{n_1 + n_2 + ... + n_r + }$$

or $$<x> = \frac{n_1x_1}{N_T} + \frac{n_2x_2}{N_T} + ... \quad \because \Sigma n_i = N_T$$

or $$= P_1x_1 + P_2x_2 + ...P_rx_r + ...$$

i.e., $$= \Sigma P_rx_r$$

where x_r is the property of rth group of N_r similar systems and P_r is the probability for that result.

In this unique way Gibbs replaced the method of time averages (observations made on the same system at different times to obtain average) by ensemble average as shown above. The main advantage of the ensemble method is "the observations on all systems can be taken simultaneously under similar conditions", *e.g.*, test of quality of soaps of different brands will be quicker by millions of users.

MACROSCOPIC SYSTEMS CHARACTERISTIC FEATURES

Let us now discuss some of the characteristic features of the macrosystems. In doing so we will assume that it has a large number of microscopic particles.

Fluctuation

Let the isolated system has N molecules of a perfect gas and that it was left to itself for quite some time so that it is in equilibrium. Further, also consider that the system has been partitioned into two identical (also equal) parts by an imaginary wall, each part A and B has n and n molecules respectively such that n + n = N.

Now, as we have seen, for large values of N, $n \cong n^1 = N/2$

But the molecules of the gas are in a state of constant random motion. So that some of the molecules are leaving portion .A and moving into B and vice-versa. Thus, there is a continuous change in the number of molecules in each portion of the box. Time variation of this change is known as fluctuation. The state of normal fluctuation of an isolated system. Usually, N/2 – n is only a small number. For large deviation, the number of microstates (and hence the probability) reduce drastically. The number of microstates for different values of N and N/2 – n. For the sake of simplicity the deviation N/2-- n is represented as a fraction of total number of particles. Value of R ($\cong P_x/P_{max}$) is also shown to arrive at qualitative results;

In actual physical problems the number of molecules are $\sim 10^{24} m^{-3}$ whence even 0.01% deviation also becomes quite improbable. This is the reason for quick diffusion and most equitable distribution of air molecules in the available space.

THEOREMS OF PROBABILITY IN STATISTICAL THERMODYNAMICS

The following are the important probability theorems commonly used in statistical thermodynamics:

(1) The number of ways in which N distinguishable particles can he arranged in order is equal to

$$N!$$

(2) The number of different ways in which N particles can be selected from N distinguishable particles irrespective of the order of selection is equal to

$$\frac{N!}{(N-n)!\,n!}$$

(3) The number of different ways in which n indistinguishable particles can be arranged in g distinguishable states with not more then one particle in each state is equal to

$$\frac{g!}{(g-n)!\,n!}$$

PROBABILITIES CALCULATION

We can extend the theory of probability in order to include the occurrence of two or more independent events. The combined probability in such cases is given by the product of separate probability of each event. Let us examine some cases:

(i) *Tossing of a coin:* On tossing a single coin, there are two possible aspects that may be presented by a fallen coin, a "head" or a "tail". We may assume here that there is no preference for any particular event, so that after making a large number of throws, the ratio of the number of heads to the number of tails is unity. Thus, for a normal coin the probability of each event is 1/2.

(ii) *Tossing of two unlike coins*: Suppose we toss two coins of different denominations or of different sizes, or of different colours, we have four complexions.

1. $Head_1$ $Head_2$
2. $Head_1$ $Tail_2$
3. $Tail_1$, $Tail_2$
4. $Tail_1$ $Head_2$.

Thus, the probability of any one complexion on a single toss will be 1/4 = 1/2.1/2, Or the composite probability is the product of probability of individual events.

(iii) *Tossing of two similar coins*: If two coins cannot be distinguished from each other, then complexions 2 and 4 of (ii) are indistinguishable whence the probability of one head and one tail is twice than that of two heads or two tails.

(iv) *Tossing of two coins in general*: Let us represent occurrence of heads by A_1, A_2 and the occurrence of tails by B_1, B_2 when two coins are tossed. Then the composite event can be represented by $(A_1 + B_1)(A_2 + B_2) = A_1A_1 + A_1B_2 + A_2B_1 + B_1B_2$.

(v) *Tossing of three unlike coins:* With three unlike coins, we have three distinguishable heads A_1, A_2, A_3 and three distinguishable tails, B_1, B_2, B_3. Here we can have eight equally probable complexions $A_1 A_2 A_3$. $A_1 A_2 B_3$, $A_1 B_2 B_3$, $B_1 A_2 B_3$, $B_1 A_2 B_3$, $B_1 B_2 A_3$ and $B_1 B_2 B_3$ which are really obtained from the product of $(A_1 + B_1)$, $(A_2 + B_2)$, $(A_3 + B_3)$.

(vi) *Tossing of three similar coins:* In this case we have only four probable states with complexions

1. All heads.
2. Two heads and one tail
3. All tails.
4. One head and two tails.

We can tabulate the result as in (iv) for

$$(A + B)(A + B)\ (A + B) = A_3 + 3A^2B + 3AB^2 + B^3.$$

(vii) *Tossing of n coins*: Suppose all the coins are different. Then the possible complexions (also known as probability number) will be $(A_1 + B_1)\ (A_2 + B_2)\ .....\ (A_n + B_n)$. And, if the coins are all alike the total number of possible complexions will be

$$(A + B)^n = A^n + {}^nC_1 A^{n-1} B + {}^nC_2 A^{n-2} B^2 + ... \quad {}^nC_r A^{n-r} B^r + ... + {}^nC_n B^n$$

or $$= \sum_{r=0}^{r=n} C_r A^{n-r} B^r$$

or, also $$(A + B)^n = \sum_{r=0}^{r=n} C_r A^r B^{n-r}$$

$$\equiv \sum_{r=0}^{r=n} C_r A^{n-r} B^r$$

where $${}^nC_r = \frac{n!}{r!(n-r)!} \equiv \text{Number of events with r heads/tails.}$$

Now, since A and B indicate one event each, the total number of events are $(1 + 1)^n = 2^n = {}^nC_1 + {}^nC_2 + ... + {}^nC_n$

∴ The probability of r heads and (n – r) tails to appear is

$$P(r, n-r) = \frac{{}^nC_r}{2^n} = \frac{n!}{r!(n-r)!} \cdot \frac{1}{2^n}$$

Maximum and minimum probability: To find out which of the combination of heads and tails is most likely in a large number of throws of n coins, we have to find the maximum value of ${}^nC_r$. It is known from algebra that the value of ${}^nC_r$ is greatest for r = n/2 when n is even and r = n+1/2 when n is odd.

$$P_{max} = \frac{{}^nC_{n/2}}{2^n} = \frac{n!}{(n/2)!(n/2)} \cdot \frac{1}{2^n}$$

Similarly, the minimum probability, P_{min} obtained for r = 0 or r = n as ${}^nC_n = 1$.

whence $P_{min} = 1/2^n$

(viii) *Tossing of a single coin n times*: When we toss the same coin second, third,... nth time, its behaviour is again unpredictable. It may show a head or a tail each time with 1/2 probability. Hence a single coin tossed n times (and of course, making large number of such attempts, if we want to have satisfactory results) is just equivalent to the case of tossing of n coins as discussed in (vii).

Another equivalent illustration can be: Take 2 boxes, identical in all respects, and throw n balls from a great distance. Assuming that each ball will fall into one or the other box, the distribution of balls in the boxes will be governed by the relation (vii),

(ix) *Tossing of a dice*: If we toss of dice, having six faces, the probability for each face showing up will be 1/6. Since now the total number of aspects is 6 and only one can show up at a time with equal probability. Again, we may take 2, 3,.... n dices and throw them and calculate the composite probability for a certain complexion.

(x) *Tossing of n dices each having m faces*: Here let us assume that dice 1 has faces a_1, a_2, a_3... a_m; dice 2 has faces b_1 b_2 b_3 ... b_m; dice 3 has faces c_1, c_2, c_3 ; etc., then the possible complexions with n dices will be the sum of each aspect

a_1, a_2, a_3... a_m, v_1, b_2, b_3 ... b_m , c_1, c_2, c_3 ... n sets and, if the dice are all alike, we have

$m_{.a}$, $m_{.b}$, $m_{.c}$... n sets.

If the probability for occurring each of the face is same, *i.e.*, a = b = c... = one aspect, we get total number of possible aspects

$= m.m.m.... \text{ n times} = m^n$

One can show that the most probable distribution corresponds to equitable distribution of faces *i.e.*, face 1, face 2, face 3, etc., will appear in equal number. Then all the dices showing face 1 have the probability of only $1/m^n$.

A similar illustration to this is : Take *m identical boxes and throw n* balls onto them from a great distance. We will take up this problem later in more detail. As in actual analysis of physical problems we have a large number of particles which are distributed into a large number of (hypothetical) boxes.

(xi) *Probability with weightage*: So far we have considered the case in which the probability of occurring an event and its failure is the same. But there can be cases where the probability of occurring a particular event is greater than that for any other event.

Let a be the probability of occurring an event and b be the probability of its failure, per trial. Then, if the trial is repeated n times, the probability of r favourable events is

$a^r.b^{n-r} = a^r (1 - a)^{n-r}$

But there are ${}^nC_r$ ways for r favourable events, therefore, the probability of occurring r events is

$$P(r, n - r) = {}^nC_r\, a^r (1 - a)^{n-r}$$

Thus, for example, in a dice a = 1/6

$$\therefore \quad P(r, n - r) = {}^nC_r \left(\frac{1}{6}\right)^r \left(\frac{5}{6}\right)^{n-r}$$

But, if certain number, say 3, has a = 2/7 then in n tosses the number 3 will occur r times with a probability.

(xii) *Postulate of equal a priori probability:* It states that all accessible microstates corresponding to a macrostate are equally probable. Thus, the postulate claims that the probability of finding a molecule in one region of phase space is identical with that for any other region of equal volume provided the regions are obeying other conditions of equality (*i.e.*, the regions have extensions of same magnitude and correspond to the same energy). Thus, it follows that the probability of occurrence of a macrostate is proportional to the number of microstates associated with it

i.e., $$P_i \propto \Omega_i$$

where P_i is the probability and Ω_i is the total number of accessible microstates, in the ith set.

ACCESSIBLE MICROSTATES

Let us now count the number of energy states available in a macrosystem. Let E be the total energy of this system which has been divided into small energy intervals δ E so that a sufficient number of quantum stales are available in each interval of δ E. Now, if Ω (E) is the number density of such states in the energy interval E and E + δE then

$$\Omega(E) = \rho(E)\, \delta E \qquad ...(i)$$

where ρ (E) is density of states (the number of states in unit energy interval). Thus

$$\Omega(E) \propto \delta E \qquad ...(ii)$$

Let the value of Ω(E) changes with the magnitude of energy interval. Hence we may consider Ω(E) as a smoothly varying function of E.

Again, if $\phi(E)$ denotes the number of states with energy up to E and $\phi(E + \delta E)$ denotes the number of states with energy up to E + δE then

$$\Omega(E) = \phi(E + \delta E) - \phi(E) \equiv \frac{\delta\phi}{\delta E}\delta E \qquad \text{...(iii)}$$

Comment on ρ (E). Relations (i) and (ii) (possibly) indicate that ρ(E) is a constant. But it is not so. In fact ρ (E) is an increasing function of E. Let us see how! We know that the total number of microstates is equal to the product of thermodynamic probabilities of different macrostates (composite probability is obtained, always by multiplying probabilities for different complexions), *i.e.*,

$$\Omega = \Omega_1 \Omega_2 \Omega_3 \ldots \Omega_i \ldots \Omega_f$$

or

$$\Omega = \prod_1 \Omega_i \; (i = 1 \text{ to } f)$$

where f is the degree of freedom.

But Ω_i should be somehow related to E, *i.e.*,

$$\Omega \propto E^a \quad \therefore \Omega \propto E^{fa} \qquad \text{...(iv)}$$

usually a ≈ 1/2 and/ranges between 10^{22} and 10^{28}. This suggests that for a certain energy interval δE, ρ(E) increases with the value of E.

STATISTICAL EQUILIBRIUM

Consider that a thermodynamical isolated system consists of N particles. The energy states available to the particles are E_1, E_2, E_3 etc. These energy states may be quantized or may be continuous and are due to vibrational and rotational energy of the particles.

Suppose, that at any given instant of time, n_1 particles are in state of energy E_1, n_2, particles with energy state E_2 and so on.

The total number of particles in the system is,

$$N = n_1 + n_2 + n_3 + \ldots\ldots = \sum_i n_i \qquad \text{...(i)}$$

Here i = 1, 2, 3 etc.

The total energy of the system,

$$U = n_1E_1 + n_2E_2 + n_3E_3 + \ldots$$

$$U = \sum_i n_i E_i \qquad \text{...(ii)}$$

Equation (ii) refers to the total energy of a system in which the particles are non-interacting. Here, the energy of each particle depends only on the coordinates of the particle in the system. For an isolated system, the total energy U is constant.

For isolated system.

$$U = \sum_i n_i E_i = \text{constant} \qquad \text{...(iii)}$$

Consider a gas having N molecules at a certain temperature and pressure. Its volume, temperature and pressure are kept constant *i.e.*, the system is isolated. The total energy of this system remains constant. But, the molecules of the gas collide with each other and also with the walls of the container. Consequently, the number of molecules change from one energy state to the other energy state. It means that the value of n_1, n_2, n_3 etc., continuously change. It can be reasonably assumed that for each microscopic State of a system of particles, there is a particular most favoured distribution. When this distribution or partition is reached, the system attains statistical equilibrium. For an isolated system, the values of n_1, n_2, n_3 etc., vary only near the values corresponding to the most probable distribution. Hence, the basic problem in statistical thermodynamics is to obtain the most probable distribution law for a given composition of the system.

In practice, three most probable distribution laws are used. They are:

(i) Maxwell-Boltzmann Distribution Law.

(ii) Fermi-Dirac Distribution Law (Statistics).

(iii) Bose-Einstein Distribution Law (Statistics).

MAXWELL-BOLTZMANN DISTRIBUTION LAW

Consider a system that contains a large number of particles that are identical and distinguishable. The identical particles refer to the particles having the same structure. The distinction between the particles is due to their energy states at a given instant. Four particles are in energy state E_1, 2 in state E_2, zero in state E_3 and so on. It is assumed that all the energy states are accessible to each particle. Consequently, it can be assumed that the probability of any particular partition is proportional to the number of different ways in which the particles can be distributed in the existing available energy states so as to produce the desired partition.

To obtain the required distribution as the first particle a in state E_1 can be selected in N ways. The second particle b in state E_1 can be selected in (N – 1) ways and so on. Therefore, the total number of ways in which the first four particles in state E_1 can be selected is given by,

$$N(N-1)(N-2)(N-3) = \frac{N!}{(N-4)!}$$

Moreover, the four particles in state E_1 can be arranged in 4! different orders. For example, *abcd, bcda, cdab* and so on, There are 24 ways. But, it is immaterial for these particles to be arranged in any particular order in state E_1, because they are identical. Thus, the total number of *distinguishable* different ways are,

$$\frac{N!}{4!(N-4)!}$$

In general, if the first state consists of n_1 particles, the distinguishable different ways, for arranging n_1 particles in state E_1, are,

$$p_1 = \frac{N!}{n_1!(N-n_1)!} \quad ...(i)$$

For the second state E_2, only $(N - n_1)$ particles are available and n_2 particles are in state E_2. The number of distinguishable different ways are,

$$p_2 = \frac{(N-n_1)!}{n_2!(N-n_1-n_2)!} \quad ...(ii)$$

If this process is continued for all the available states, the total number of distinguishable ways are obtained by multiplying P_1,P_2, P_3 etc.

$$P = p_1 \times p_2 \times p_3 \times \ ......$$

$$P = \left[\frac{N!}{n_1!(N-n_1)!}\right]\left[\frac{(N-n_1)!}{n_2!(N-n_1-n_2)}\right] \times ....$$

$$P = \frac{N!}{n_1!n_2!n_3!...} \quad ...(iii)$$

The distinguishable ways are,

$$P = \frac{N!}{4!2!0!3!1!} \quad ...(iv)$$

Here 0! is equal to one.

It has been assumed so far that all the available states have the same probability of occupation by the particles. However, it may happen that the states have different intrinsic probabilities say g_i.

For example, a particular energy state may be favourable with more different angular momentum states than the rest and hence it is more likely to be occupied. Taking into account this intrinsic probability factor, the value of P will be different.

If g_i is the probability of locating a particle in a certain energy state E_i; then the probability of locating 2 particles in the same state is $g_i \times g_i = g^2_i$. For n_i particles, the probability is $g_i n_i$. Hence the total probability for a given distribution is given by

$$P = \frac{N! g_1 n_1 g_2 n_2 g_3 n_3 \ldots}{n_1! n_2! n_3! n_4! \ldots} \quad \ldots(v)$$

Here n_1, n_2, n_3 etc., are the number of particles in states E_1, E_2, E_3, etc. and g_1, g_2, g_3 etc., are the intrinsic probabilities for states E_1, E_2, E_3 etc.

If all the particles are further assumed to be indistinguishable *i.e.*, particles in state E, and particles in state E_4 cannot be distinguished, then N! permutations among the particles themselves and occupying the different states result in the same distribution. The probability in this case is given by

$$P = \frac{1}{N!}\left[\frac{N! g_1 n_1 g_2 n_2 g_3 n_3 \ldots}{n_1! n_2! n_3! \ldots}\right]$$

$$\therefore \quad P = \frac{g_1 n_1 g_2 n_2 g_3 n_3 \ldots}{n_1! n_2! n_3! \ldots}$$

$$\therefore \quad P = \prod_i^N \frac{g_i n_i}{n_1!} \quad \ldots(vi)$$

Here Π is the product sign (since probability is the product of such distributions.

The most probable distribution can be obtained by evaluating the maximum value of $\log_e$, P in equation (vi). This should also satisfy the two conditions that

$$\sum_i n_i = N \quad \ldots(vii)$$

and $\sum_i n_i E_i = U$ ...(vii)

According to Stirling's approximation

$$\log_e x! = x \log_e x - x$$

From equation (vi), applying Stirling's approximation, we get

$$\log_e P = \sum_i (n_i \log_e g_i - \log_e n_i!)$$

$$= \sum_i [n_i \log_e g_i - (n_i \log_e n_i - n_i)]$$

$$= \sum_i (n_i \log_e g_i - n_i \log_e n_i + n_i)$$

$$= \sum_i n_i - \sum_i (n_i \log_e n_i - n_i \log_e g_i)$$

$$\log_e P = N \sum_i n_i \log_e \frac{n_i}{g_i} \qquad \text{...(ix)}$$

Differentiating equation (ix)

$$d(\log_e P) = \sum_i (dn_i) \log_e \frac{n_i}{g_i} - \sum_i n_i d\left(\log_e \frac{n_i}{g_i}\right)$$

$$= \sum_i dn_i \log_e \frac{n_i}{g_i} - \sum_i n_i \frac{dn_i}{n_i}$$

$$= \sum_i dn_i \log_e \frac{n_i}{g_i} - \sum_i dn_i$$

But $\sum_i dn_i = 0$.

$$\therefore \quad -d(\log_e P) = \sum_i \left[\log_e\left(\frac{n_i}{g_i}\right)\right] dn_i \qquad \text{...(x)}$$

To obtain the maximum value of P,

$$d(\log_e P) = 0$$

$$\therefore \quad \sum_i \log_e\left(\frac{n_i}{g_i}\right) dn_i = 0 \qquad \text{...(xi)}$$

But $\Sigma dn_i = 0$ ...(xii)

and $\Sigma E_i dn_i = 0$ ...(xiii)

Multiplying (xii) by α and (xiii) by β and adding to equation (xi), we get

$$\sum_i \left[\log_e\left(\frac{n_i}{g_i}\right) + \alpha + \beta E_i\right] dn_i = 0$$

The equilibrium distribution is possible if

$$\log_e \frac{n_i}{g_i} + \alpha + \beta Ei = 0 \quad \text{...(xiv)}$$

or $$\frac{n_i}{g_i} = e^{-\alpha-\beta Ei}$$

or $$n_i = g_i\, e^{-\alpha-\beta Ei} \quad \text{...(xv)}$$

This gives the maximum probability distribution and a and p are two parameters that depend upon the physical property of the system.

$$N = n_1 + n_2 + n_3 + ...$$

$$N = g_1 e^{-\alpha-\beta E1} + g_1 e^{-\alpha-\beta E2} + ...$$

$$N = e^{-\alpha}[g_1 e^{-\beta E1} + g_2 e^{-\beta E2} + ...]$$

$$N = e^{-\alpha}[\sum_i g_i\, e^{-\beta Ei}]$$

Take $[\sum_i g_i\, e^{-\beta Ei} = Z$ ...(xvi)

Here, Z is called the partition function.

$$N = e^{-\alpha}(Z) \quad \text{....(xvii)}$$

$$e^{-\alpha} = \frac{N}{Z}$$

Substituting this value in equation (vii)

$$n_i = \left(\frac{N}{Z}\right) [g_i e^{-\beta Ei}] \quad \text{....(xviii)}$$

This equation refers to Maxwell-Boltzmann Distribution Law.

MAXWELL-BOLTZMANN DISTRIBUTION IN TERMS OF TEMPERATURE

The total energy of an isolated system is given by

$$U = \sum_i n_i\, E_i$$

$$U = n_1 E_1 + n_2 E_2 + n_3 E_3 + ...$$

$$U = (g_1 e^{-\alpha-\beta E1})\, E_1 + (g_2 e^{-\alpha-\beta E2})\, E_2 + ...$$

$$U = e^{-\alpha}[g_1 E_1 e^{-\beta E1} + g_2 E_2 e^{-\beta E2} + ...]$$

But $$e^{-\alpha} = \frac{N}{Z}$$

$\therefore$ $$U = \frac{N}{Z}[g_1 E_1 e^{-\beta E1} + g_2 E_2 e^{-\beta E2} + ...]$$

$$U = \frac{N}{Z}\sum_i g_i E_i e^{-\beta E_i}$$

Here $$Z = \sum_i g_i e^{-\beta E_i} \qquad \text{...(i)}$$

$$\frac{dZ}{d\beta} = \frac{d}{d\beta}\sum_i g_i e^{-\beta E_i} = -\sum_i g_i E_i e^{-\beta E_i}$$

$$\therefore \quad \sum_i g_i E_i e^{-\beta E_i} = \frac{d}{d\beta}\sum_i g_i E_i e^{-\beta E_i}$$

Substituting this value in equation (i)

$$U = -\frac{N}{Z}\frac{d}{d\beta}\sum_i g_i e^{-\beta E_i}$$

$$U = -\frac{N}{Z}\frac{d}{d\beta}$$

$$U = -N\frac{d}{d\beta}[\log_e Z] \qquad \text{...(ii)}$$

The average energy of a particle is

$$E_{av} = -\frac{U}{N} = -\frac{d}{d\beta}(\log_e Z) \qquad \text{...(iii)}$$

This shows that for a given system, the total energy U, the partition function Z, and the average energy of the particle E_{av}, depend on the parameter β.

Therefore, β may be taken to characterise the internal energy of the system and has the units per joule. If T is the temperature in degrees kelvin, then it is more convenient to represent parameter β as

$$\beta = \frac{1}{kT} \text{ or } KT = \frac{1}{\beta} \qquad \text{...(iv)}$$

Here kT has the unit of energy *i.e.*, joule and k is the Boltzmann's constant. Its units are J/K. The value of k is given by

$$k = 1.3805 \times 10^{-23} \text{ J/K}$$

Substituting the value of

$$\beta = \frac{1}{kT}$$

in all the equations, we get

$$Z = \sum_i g_i e^{-(E_i/kT)} \qquad \text{...(v)}$$

$$n_i = \frac{N}{Z} \sum_i g_i \, e^{-(E_i/kT)} \quad \text{...(vi)}$$

This equation represents the Maxwell-Boltzmann distribution law in terms of the temperature of the system.

As $\beta = \dfrac{1}{kT}$

$$d\beta = -\frac{dT}{kT^2}$$

Substituting this value in equation (ii)

$$\therefore \quad U = (kNT^2)\frac{d}{dT}(\log_e Z) \quad \text{...(vii)}$$

Also, from equation (iii)

$$E_{av} = (kT^2)\frac{d}{dT}(\log_e Z) \quad \text{...(viii)}$$

Equation (viii) gives the relation between the average energy of the particle and its temperature under equilibrium position. Hence, the temperature of a system in statistical equilibrium is the physical quantity related to the average energy of the particle of the system.

QUANTUM STATISTICS

Maxwell-Boltzmann distribution, it has been assumed that all the energy levels are accessible to all the particles of the system. However, there may be certain levels prohibited to a certain group of particles. Each energy state is associated with a certain available wave function. The probability of a particular distribution is restricted by the available wave functions of any state. These restrictions are taken into account in quantum statistics. There are two types of quantum statistics

(i) Fermi-Dirac statistics

(ii) Bose-Einstein statistics.

In Fermi-Dirac statistics the particles are assumed to obey Pauli's exclusion principle. They are characterised by antisymmetric wave functions. The particles under, this category are called *fermions*. Protons neutrons and electrons are fermions.

In the case of Bose-Einstein statistics, it is assumed that the particles are not restricted by Pauli's exclusion principle and are characterised by symmetric wave functions. These particles are called *bosons*. It has been experimentally found that all particles having spin zero or integral multiple of 1 are bosons. Helium nuclei and mesons are bosons.

In both the kinds of quantum statistics, the particles are identical and indistinguishable. At high temperature and low pressure, all the three statistics give practically the same result.

FERMI-DIRAC DISTRIBUTION LAW

Consider a system consisting of a large number of particles. It is assumed:

(i) The particles are *identical* and *indistinguishable*, and

(ii) The particles obey exclusion principle. It means that no two particles can have the same dynamical state and the wave function of the whole system must be antisymmetric. The particle satisfying these conditions are called *fermions*. In general, all fundamental particles with spin are fermions.

In quantum statistics the intrinsic probability g_i is governed by the different quantum states relating to a given energy *i.e.*, the degeneracy of the energy state. Each quantum state corresponds to a particle wave function. The Wave functions are determined by each of the possible arrangements of quantum number corresponding to a given energy level. For particles with spin 1/2, in the absence of magnetic forces, each of the particle may be in the energy states with spin +1/2 or −1/2. Hence the intrinsic probability g_i in this case is 2. For motion in a central field, the energy of the particle is independent of the orientation of the orbital angular momentum Due to this, a degeneracy of $(2l + 1)$ is introduced and it is the value of g for that particular energy state. If the particles possess spin, the total degeneracy = $2(2l + 1)$. As no two particles can be in the same energy state having the same quantum number, the instrinsic probabilities g_i's give the maximum number of particles that can be accommodated in a particular energy level without violating the exclusion principle *i.e.*,

$$n_i \leq g_i$$

It means that the n_i value for a given distribution does not exceed the corresponding value of g_i.

Let n_i, be the number of particles for energy level E_i The first particle can be placed in any one of the available g_i states *i.e.*, this particle can be assigned to any of the gi sets of quantum numbers. Thus, the first particle can be distributed in g_i different ways. Similarly, the second particle can be arranged in $(g_i - 1)$ different ways and the process continues.

Thus, the total number of different ways of arranging n_i particles among the available g_i states with energy level E_i is

$$= g_i(g_i - 1)(g_i - 2) \ldots\ldots[g_i - (n_i - 1)]$$

$$= \frac{g_i!}{(g_i - n_i)!} \qquad \ldots(i)$$

Further, if the particles are taken to be indistinguishable, it will not be possible to detect any difference when n_i particles are reshuffled into different states occupied by them in the energy level E_i. Therefore, the total number of different and distinguishable ways is,

$$= \frac{g_i!}{n_i!(g_i - n_i)!} \qquad \ldots(ii)$$

Therefore, the total number of different and distinguishable ways of getting the distribution n_1, n_2, n_3, etc., among the various energy levels, E_1, E_2, E_3...,etc.. can be obtained by multiplying the various factors.

$$\therefore \qquad P = \frac{g_1!}{n_1!(g_1 - n_1)!} \cdot \frac{g_{2i}!}{n_2!(g_2 - n_2)!} \ldots.$$

$$P = \prod_i \frac{g_i!}{n_i!(g_i - n_i)!} \qquad \ldots(iii)$$

The most probable distribution can by obtained by evaluating the maximum value of $\log_e$, P in equation (iii).

This should also satisfy the condition that

$$\sum_i n_i = N$$

and $\sum_i n_i E_i = U$.

According to Stirling's approximation

$$\log_e x! = x \log_e x - x$$

From equation (iii), applying Stirling's approximation,

$$\log_e P = \sum_i [(g_i \log_e g_i - g_1) - (n_i \log_e n_i - n_i)$$

$$-[(g_i - n_i)\log_e(g_i - n_i) - (g_i - n_i)]$$

$$\log_e P = \sum_i [g_i \log_e g_i - n_i - \log_e n_i - (g_i - n_i)$$

$$\log_e (g_i - n_i)] \quad \text{...(iv)}$$

Differentiating equation (iv)

$$-d(\log_e P) = \sum_i [\log_e n_i - \log_e (g_i - n_i)]dn_i$$

To obtain the maximum value of P,

$$d(\log_e P) = 0$$

$$\therefore \quad \sum_i [\log_e n_i - \log_e (g_i - n_i)]dn_i = 0 \quad \text{...(v)}$$

But $\sum_i dn_i = 0$ ...(vi)

and $\sum_i E_i dn_i = 0$ ...(vii)

Multiplying (vi) by α and (vii) by β and adding to equation (v), we get

$$\sum_i [\log_e n_i - \log_e (g_i - n_i) + \alpha + \beta E_i]dn_i = 0$$

The equilibrium distribution is possible if

$$\therefore \quad \log_e n_i - \log_e (g_i - n_i) + \alpha + \beta E_i = 0 \quad \text{...(viii)}$$

$$\log_e\left(\frac{n_i}{g_i - n_i}\right) = -\alpha + \beta E_i$$

$$\frac{n_i}{g_i - n_i} = e^{-\alpha - \beta E_i}$$

$$\frac{g_i - n_i}{n_i} = e^{\alpha + \beta E_i}$$

$$\frac{g_i}{n_i} - 1 = e^{\alpha + \beta E_i}$$

$$\frac{g_i}{n_i} = (e^{\alpha + \beta E_i}) + 1$$

$$n_i = \left(\frac{g_i}{e^{\alpha + \beta E_i}}\right) + 1 \quad \text{...(ix)}$$

Equation (ix) represents the Fermi-Dirac distribution law. The parameter has the same role as in the case of Maxwell-Boltzmann distribution law *i.e.*, for the system consisting of fermions in statistical equilibrium.

$$\beta = \frac{1}{kT}$$

$$\therefore \qquad n_i = \frac{g_i}{[e^{(\alpha + E_i/kT)}] + 1}$$

In most cases the value of a is negative and is taken to be equal to

$$\frac{-E_F}{kT}$$

$$n_i = \frac{g_i}{\left[e^{(E_i - E_F)/kT)}\right] + 1}$$

The value of E_F is positive and is independent of temperature.

For T = 0, all the energy states are fully occupied and $n_i = g_i$ All the states with $E > E_F$, are empty *i.e.*, $n_i = 0$.

$$\left[e^{(E_i - E_F)/kT)}\right] = \begin{cases} 0 \text{ for } E_i - E_F < 0 \\ \infty \text{ for } E_i - E_F > 0 \end{cases}$$

In the case of Fermi-Dirac statistics, the accumulation of particles at the ground level is not allowed and at temperature T = 0, the particles occupy the lowest energy levels upto E_F. Here, the energy E_F gives the indication of the maximum energy of the fermions in the system. By is also called Fermi energy. For higher temperatures, the particles, occupy higher energy states greater than E_F. The curves indicate that only those fermions with energies close to E_F can move into unoccupied higher energy states. If $k\theta_F = E_F$, the temperature θ_F is called the Fermi temperature.

MAXWELL-BOLTZMANN DISTRIBUTION AND IDEAL GAS

Most of the gases obey Maxwell-Boltzmann distribution over a wide range of temperature. Consider a gas having mono-atomic molecules. It is assumed that the gas is ideal and has only kinetic energy of translation. Intermolecular attraction is assumed to be absent. If the kinetic energy is not quantized but considered to possess continuous values of energy, then the partition function Z can be written as

$$Z = \int_0^{\infty} e^{-(E/kT)} g(E)dE$$

Here g_i is replaced by g (E) dE

But $\quad g(E)dE = \left(\frac{4\pi V(2m^3)^{1/2}}{h^3}\right)E^{1/2}dE$

Here V is the volume occupied by the gas and h is Planck's constant.

$$Z = \frac{4pV(2m^3)^{1/2}}{h^3}\int_0^\infty E^{1/2}e^{-(E/kT)}dE$$

$$Z = \frac{4\pi V(2m^3)^{1/2}}{h^3}\left[\frac{\sqrt{\pi(kT)^3}}{2}\right]$$

$$Z = \frac{V(2\pi mkT)^{3/2}}{h^3} \qquad \text{...(i)}$$

Equation (i) represents the partition function for an ideal mono-atomic gas in terms of the volume and temperature of the gas.

Taking logarithms on both sides

$$\log_e Z = \log_e\left[\frac{(2\pi m)^{3/2}V}{h^3}\right] + \log_e (kT)^{3/2}$$

$$\log_e Z = C + \frac{3}{2}\log_e kT \qquad \text{...(ii)}$$

Also $\quad E_{av} = (kT^2)\frac{d}{dT}(\log_e Z)$

$$E_{av} = (kT^2)\frac{d}{dT}\left[C + \frac{3}{2}\log_e kT\right]$$

$$= (kT^2)\frac{d}{dT}\left[C + \frac{3}{2}\log_e k + \frac{3}{2}\log_e T\right]$$

$$E_{av} = \frac{3}{2}kT \qquad \text{...(iii)}$$

The total energy.

$$U = NE_{av}$$

$$U = \left(\frac{3}{2}kT\right)$$

$$U = \frac{3}{2}NkT. \qquad \text{...(iv)}$$

Equation (iv) shows that the internal energy of an ideal monoatomic gas depends only on its temperature. The same relation does not however, hold good for real gases. In the case of real gases the internal energy is partly potential and partly kinetic. The total energy depends on the volume of the gas.

ELECTRON GAS

Electrons in a metal belong to a most characteristic system of fermions because electrons obey the exclusion principle. For electrons in a metal, the energy levels are grouped in bands. Practically at all temperatures, the lower level energy bands are filled with electrons. The upper level energy bands are only partially filled with electrons. The distribution of electrons is to be considered only in the upper bands called the conduction band.

The zero energy level is taken at the lowest level of the conduction band. It is also assumed that the electrons have free movement within the conductor, provided the energy associated with the electrons is of the order of upper level energy bands. As the energy of the electron in the conduction band is continuous, the term g_i is replaced by g(E) dE. Here dn electrons have energy in the range E and E + dE. According to Fermi-Dirac distribution law

$$n_i = \frac{g_i}{\left[e^{(E_i - E_F)/kT}\right] + 1} \qquad \text{...(i)}$$

Substituting the value of g_i = g (E) dB in equation (i) and replacing n_i by dn

$$dn = \frac{g(E)dE}{\left[e^{(E_i - E_F)/kT}\right] + 1}$$

As an electron has the spin ±1/2, the total number of states in the sphere is twice. $V/(2\pi)^3$ refers to the translational states per unit volume in the Fermi space Fermi sphere of radius kF has the total number of particles accommodated,

$$n = \frac{2\left[\frac{V}{(2\pi)^3}\right]\left(\frac{4}{3}\pi(kF)^3\right)}{\left[e^{(E_i - E_F)/kT}\right] + 1}$$

Here $$E = \frac{p^2}{2m} \frac{\left(\frac{h}{2\pi}\right)^2 (kF)^2}{2m}$$

$$\therefore \qquad kF = \frac{p}{h/2\pi} = \frac{2\pi p}{h}$$

$$\therefore \qquad n = \frac{2\left[\frac{V}{(2\pi)^3}\right]\left(\frac{4}{3}\pi\left(\frac{2\pi p}{h}\right)^3\right)}{\left[e^{(E_i - E_F)/kT)}\right]+1}$$

But $p = (2mE)^{1/2}$

$$\therefore \qquad n = \frac{2\left[\frac{V}{(2\pi)^3}\right]\left[\left(\frac{4}{3}\pi\left(\frac{2\pi p}{h}\right)^3 (2mE)^{3/2}\right)\right]}{\left[e^{(E_i - E_F)/kT)}\right]+1}$$

$$n = \frac{\left(\frac{8\pi V}{(3h^3}\right)(2m)((2m)^{1/2}E^{3/2}}{\left[e^{(E_i - E_F)/kT)}\right]+1}$$

Differentiating

$$dn = \frac{\left(\frac{8\pi V}{(3h^3}\right)(2m)((2m)^{1/2}\times\frac{3}{2}E^{1/2}dE}{\left[e^{(E_i - E_F)/kT)}\right]+1}$$

$$\frac{dn}{dE} = \frac{\left[\frac{8\pi V(2m^3)^{1/2}}{(3h^3}\right]E^{1/2}}{\left[e^{(E_i - E_F)/kT)}\right]+1} \qquad \text{...(ii)}$$

Equation (ii) represents the energy distribution for free electrons. This is also called Fermi-Dirac formula of free fermions.

At $\qquad T = 0K,$

$$p = (2m\ E_F)^{1/2} \qquad \text{...(iii)}$$

and $N = \left(\frac{2V}{h^3}\right)\left(\frac{4}{3}\pi p^3\right)$

or $$P = \left(\frac{3Nh^3}{8\pi V}\right)^{1/3} \qquad \text{...(iv)}$$

Equating (iii) and (iv)

$$(2mE_F)^{1/2} = \left(\frac{3Nh^3}{8\pi V}\right)^{1/3}$$

Squaring $2mE_F)^{1/2} = \left(\frac{3Nh^3}{8\pi V}\right)^{2/3}$

or $$E_F = \frac{h^2}{8m}\left(\frac{3N}{\pi V}\right)^{2/3} \quad ...($$

Knowing the value of N/V *i.e.*, the number of free electrons per unit volume, the Fermi energy for electrons in a metal can be obtained.

For silver, the value $\frac{N}{V}$

$$= 5.86 \times 10^{28} \text{ electrons/m}^3$$

$$\therefore \quad E_F = \left(\frac{(6.624\times10^{-34})^2}{8\times9\times10^{-31}}\right)\left(\frac{3\times5.86\times10^{28}}{3.14}\right)^{2/3}$$

$$E_F = 9 \times 10^{-19} \text{J}.$$

But $\quad 1\text{eV} = 1.6 \times 10^{-19}\text{J}$

$$\therefore \quad E_F = \frac{9\times10^{-19}}{1.6\times10^{-19}}$$

$$E_F = 5.625 \text{ eV}.$$

It means that the maximum kinetic energy of free electrons in silver at absolute zero temperature is 5.6 eV.

The Fermi temperature θ_F is given by

$$\theta_F = \frac{E_F}{k}$$

As E_F practically independent of temperature, the value θ_F is fixed for a given metal. The values of θ_F and E_F are given in the following table.

Metals	*E_F (electron Volts)*	*θ_F(K)*
Potassium	2.14	2.4×10^4
Sodium	312	3.7×10^4
Lithium	4.72	6.5×10^4
Silver	5.51	6.4×10^4
Gold	5.64	6.4×10^4
Copper	7.04	8.2×10^4

DISTRIBUTION LAW OF BOSE-EINSTEIN

Bose-Einstein distribution is applied to systems composed of identical and indistinguishable particles that are not restricted by the exclusion principle. In such systems, there is no limit to the number of particles occupying a particular quantum state. These particles are called *bosons*. Their spin is zero or 1. Mesons and helium nuclei are examples of bosons.

The Bose-Einstein statistics values of g_i refer to degeneracy of each energy level. Suppose, that n_i particles are arranged in a row and distributed among g_i quantum states with $(g_i - 1)$ partitions in between.

The total number of possible arrangements of particles and partitions is equal to the total number of permutations of $(n_i + g_i - 1)$ objects in a row. Therefore the total possible ways of arranging n_i particles with $g_i - 1$ partitions

$$= (n_i + g_i - 1)!$$

As the particles are identical and indistinguishable the possible number of distinct arrangements

$$= \frac{(n_i + g_i - 1)!}{n_i!(g_i - 1)!}$$

The total number of distinguishable and distinct ways of arranging N particles in all the available energy states is given by

$$P = \frac{(n_1 + g_1 - 1)!}{n_1!(g_1 - 1)!} \times \frac{(n_2 + g_2 - 1)!}{n_2!(g_2 - 1)!} \times \ldots$$

$$P = \prod_i \frac{(n_i + g_i - 1)!}{n_i!(g_i - 1)!} \qquad \ldots(i)$$

The most probable distribution can be obtained by finding the maximum value of $\log_e$. P.

According to Stirling's approximation

$$\log_e x! = x \log_e x - x$$

Also $$\sum_i n_i = N \qquad \ldots(ii)$$

and $$\sum_i n_i E_i = U \qquad \ldots(iii)$$

From equation (i), applying Stirling's approximation

$$\log_e P = \sum_i [\log_e (n_i + g_i - 1) - \log_e n_i!\ \log_e (g_i - 1)!]$$

$$\log_e P = \sum_i [(n_i + g_i - 1)\log_e(n_i + g_i - 1) - (n_i + g_i - 1)$$
$$- (n_i \log_e n_i - n_i) - (g_i - 1)\log_e(g_i - 1) - (g_i - 1)]$$
$$\log_e P = \sum_i [(n_i + g_i - 1)\log_e(n_i + g_i - 1)$$
$$- n_i \log_e - n_i \ -(g_i - 1)\log_e(g_i - 1)] \quad ...(iv)$$

The maximum value of P is obtained by taking

$$d(\log_e P) = 0.$$

Differentiating equation P.

$$d(\log_e P) = \sum_i [\log_e (n_i + g_i - 1)\, dn_i - \log_e n_i - dn_i] = 0$$

$$\therefore \quad -d(\log_e P) = \sum_i [-\log_e (n_i + g_i - 1)\, dn_i + \log_e n_i dn_i] = 0$$

or

$$\sum [-\log_e (n_i + g_i - 1) + \log_e n_i] dn_i = 0 \quad ...(v)$$

As the total number of particles and total energy are constants, we have,

$$\sum_i dn_i = 0 \quad ...(vi)$$

$$\sum_i E_i dn_i = 0 \quad ...(vii)$$

Multiplying (vi) by α and equation (vii) by β and adding to equation (v), we get

$$\sum_i [-\log_e (n_i + g_i - 1) + \log_e n_i + \alpha + \beta E_i] dn_i = 0$$

$$\therefore \quad -\log_e (n_i + g_i - 1) + \log_e n_i + \alpha + \beta E_i = 0$$

Taking $n_i + g_i$ very large as compared to 1, the quantity 1 can be neglected.

$$\therefore \quad -\log_e (n_i + g_i) + \log_e n_i + \alpha + \beta E_i = 0$$

$$\log_e \left(\frac{n_i}{ng_i + g_i} \right) = -\alpha - \beta E_i$$

$$\frac{n_i}{n_i + g_i} = e^{-\alpha - \beta Ei}$$

$$\frac{n_i + g_i}{n_i} = e^{\alpha + \beta Ei}$$

$$1 + \frac{g_i}{n_i} = e^{\alpha + \beta Ei}$$

$$\frac{g_i}{n_i} = (e^{\alpha + \beta Ei}) - 1$$

$$n_i = \left(\frac{g_i}{e^{\alpha+\beta E_i}}\right) - 1 \qquad \text{...(viii)}$$

Equation (ix) represents the Bose-Einstein distribution law.

Taking $$\beta = \frac{1}{kT}$$

$$\therefore \qquad n_i = \frac{g_i}{[e^{(\alpha + E_i/kT)}] - 1}$$

The value of the constant a is governed by the equation $\Sigma n_i = N$. As n_i cannot be negative, α must always have a positive value. The distribution of the particles for different energy levels

SOME DEFINITIONS

It would be appropriate to discuss, at some length, the physical entities involved, in understanding the statistical problems.

A Physical Quantity

Essentially, we have two kinds of physical quantities (a) the intensive quantities that do not depend on the mass of the system, *e.g.*, P and T, and (b) the extensive quantities which depend on the mass of system, *e.g.*, S, V, E, etc.

System

It is a finite region in the space where the experimental body is kept. It Is obligatory for the system to have a finite mass. The system can be a closed one (may have a real physical boundary) or may have an imaginary boundary in space. Hence it can be penetrable or Impenetrable to the transport of energy. The properties of the particles in the system define the properties of the system.

Microstate

Each macrostate can have several microstates associated with it. Microstate is that state of the system where the individuality of the particles is not ignored. Hence the microstates are affected by change in position, momentum, energy (and mutual interaction) of individual particles. These states, however, cannot be identified by any physical experiment. Following table will help in understanding the difference between macro and microstates. We take two boxes A and B and four distinguishable particles a, b, c, d, and investigate their distribution.

It is evident from part B of the table that when the individuality of the particles is ignored, there remains no difference between the number of macrostates and associated microstates. By definition, given above, such a distribution will not give statistical probability or correct probable distribution (last column). Another illustration for the above table is to distribute 4 distinct books on 2 shelves. We have a total of 16 distributions as in part A of the table but if we distribute 4 identical books on same 2 shelves we have only 5 arrangements.

From the table, we see that for the most probable distribution (2, 2) we have 6 microstates. In other words, the macro distribution is most probable when number of microstates involved is maximum. In actual physical problems, we deal with extremely large number of particles as $\approx 10 m^{-3}$ whence the number of microstates can be unimaginably large.

Macrostate

It is that state of the system which can be perceived by experiments. It is an isolated state of the system (not influenced by external agencies) and can be defined by certain value of pressure P, temperature T, and volume V. A macrostate can also be defined in terms of its microstructure, *e.g.*, by the specification of number of molecules (or phase-points in each cell of the phase space). Let us consider the distribution of 8 particles in two boxes. We can have (0, 8), (1, 7), (2, 6)... (7, 1) and (8, 0) distributions. These 9 distributions form 9 macrostates. Here we may note that a macrostate does not distinguish individuality of particles. Hence, for n particles we have (n + 1) macrostates when the distribution of particles is restricted in 2 boxes. This number will increase if the number of boxes is increased.

Phase Space

In order to specify the state of the gas from the molecular point of view, we shall know the position and momentum of each of its molecules, *i.e.*, we must specify six quantities x, y, z, p_x, p_y, p_z for each of the molecules. In three-dimensional space, we can easily consider an element of volume dx dy dz. Similarly, we can plot momentum components p_x, p_y and p_z in three mutually perpendicular directions and again consider three-dimensional volume. In other words, let us consider a particle with only a motion of translation. Then, classically we need three positional and three momentum coordinates to describe its state. Just as the space is represented by a three dimensional positional coordinate system, we can also represent momentum space by a three-dimensional momentum coordinate system. The state of the system then can be defined by six-dimensional positional-space and momentum space (phase) system. The two spaces together, defined by 6-dimensional (3 space and 3 momentum) hypothetical coordinate systems, is called phase space. Phase-space is purely a mathematical concept and cannot be depicted by a diagram like. So, phase-space is just a hypothetical space occupied by an infinite set of such six coordinate points. With the passage of time the point P (known as phase-point) traces out a certain trajectory known as phase-path.

As the position and momentum, in translated motion, change with lime, the phase-path may be quite complicated. Such 6-dimensional (6-d) phase space is also known as the μ-space. A large number of 6-d cells, each with a volume of $d\tau = (dx\ dy\ dz)\ (dp_x\ dp_y\ dp_z) = dV.\ dp$, together form the μ-space.

Let us now pass on to a many particle system. Let the system has N particles. Then, to define their state we need 3N positional and 3N momentum coordinates in a 6-d phase-space. We can consider this problem in another way:

A single point in 3-d space represents 3 coordinates; in 6-d space represents 6 coordinates; in n-d space represents n coordinates. Hence in 6N-d space a single point corresponds to 6N coordinates. If N be the number of molecules in a kilogram-mole (or a Mole), the 6N-dimensional phase-space is called r-space (or y-space). We can generalise this representation by assuming that the system has N particles each having f degrees of freedom the Γ- space will then be 2Nf dimensional, having Nf positional coordinates (q_i, i = 1 to Nf) and Nf momentum coordinates (P_i, i = 1 to Nf).

Phase-cells

The concept of phase space enables us to understand the state of micro particles -A their position and energy. When a particle changes its momentum and position with time, there will be infinite points in the phase-space which will express the state of this moving particle. Again in a finite volume of phase space, we have infinite microstates available for the particle. Hence, we can divide the whole of the phase space into finite sized small cells. The volume of the cell, however, cannot indefinitely be small.

Accessible Microstates

Those microstates which follow stipulated conditions (like number of particles is conserved, energy is conserved, system as finite extension) are called the accessible microstates.

The State of Equilibrium

It is the steady state of an isolated system where the values of P. V and T do not change with the passage of time. In such systems, no energy is given to or taken out of it. In statistical terms, equilibrium state corresponds to that macrostate which has maximum microstates. One must, however, remember that an isolated system, in equilibrium, is equally likely to be in any of its accessible microstate.

Events

When an operation is performed or is allowed to take place, the occurrence of certain complexion, configuration or arrangement of the system, at one such trial is called an event. When number of trials is increased the probability of certain event may increase.

STATISTICS APPLIED TO MOLECULES

A monatomic perfect gas. The state of each molecule of the gas in the μ-space is represented by a point (and by a cell in quantum mechanical terms). But, since the gas molecules are in constant motion, their state will be changing continuously in space and, due to collisions (and also mutual interactions), there is a change .in momentum state also. The trajectory of such a molecule constant energy surface given by

$$E = \frac{p_x^2 + p_y^2 + p_z^2}{2m} = V(q_x, q_y, q_z)$$

The real gas molecules, however, are endowed with mutual interactions in addition to usual collisions, hence their state is better represented by a point in y-space. The rate of change of its state, then, depends on the quantum of interactions. The path of representative point in y-space is given by the relation

$$E = \sum_i T_i + \sum_i V_i + \sum_{i,j}{}' U_{ij}$$

where T_i, V_i and U_{ij} refer to kinetic energy, potential energy and energy of interaction of ith particle. The prime on the summation sign suggests that i = j is not allowed (which would mean self-interaction). With this background and remembering that the number of molecules is very large in a macrosystem, we now proceed to investigate the distribution details.

SOME ENSEMBLES

It would be appropriate to classify the ensembles according of their dependency on macroparameters. Since an ensemble is an assembly of macroscopically identical but mutually independent systems, a change in macroparameter (like volume, pressure, energy, number of molecules, etc.,) may lead to another set of systems:

Canonical Ensemble

In this ensemble all its members are identical and remain in thermal equilibrium with the source (*i.e.*, members of the ensemble can exchange thermal energy with the reservoir). Thus all the members of a canonical ensemble have the same number of particles N, volume V and are at the same temperature T. A canonical ensemble. Its members are shown separated by impermeable but conducting rigid, walls and they are in thermal contact with the source. Gibbs gave a relation,

$$\rho(\varepsilon) = Ae^{-\varepsilon\tau}$$

for density of its phase points. A and T in the relation are constants. The ensemble can be viewed physically as a freezer containing water filled metal-ice cube tray.

Grand Canonical Ensemble

The members of this ensemble can interact thermally, mechanically and also diffusively with each other and with the reservoir. Hence they are at the same temperature of same volume and at same chemical potential such an ensemble. Each member here is separated by flexible, permeable and conducting walls, which help exchange mass and energy between its members and with the surroundings. The constituent molecules in the member systems are thus of different kinds (different chemical species). The density of phase points in such ensemble is

$$\rho(N) = e^{(\Omega+N\mu-\varepsilon)/\tau}$$

where, Ω, μ and τ are constants. A grain of sugar in boiling water can best be viewed as a grand canonical ensemble.

Uniform Ensemble

Those ensembles which have same density in phase space, are called *uniform ensembles*.

Hence, if ΔN represents the number of system, in volume element dΓ we have, for a uniform ensemble,

$$\lim_{d\Gamma \to 0} \frac{1}{d\Gamma} \frac{\Delta N}{N} = \text{constant}$$

Microcanonical Ensemble

In these ensembles the energy of their member systems remain constant, *i.e.*, they do not interact with each other or with the surroundings. Thus, they are free from external effects. The ensemble is thus conceptually an idealised one and can be represented by a function that depends only on the energy. An ensemble, here ε and V are constant. Every member is separated from the other by rigid, impermeable and insulated walls. The density of phase points within each member system is uniform and is zero at the walls. Physically, we may view the microcanonical ensemble as a perfect insulating enclosure encompassing several isolated systems. It is suitable for ideal gases.

SOLAR CONSTANT

The source of heat (sun) radiations and it emits heat radiations in all directions. The earth receives only a fraction of the energy emitted by the sun. The atmosphere also absorbs a part of the heat radiations and air, clouds, dust particles etc., in the atmosphere scatter the heat and light radiations falling on them. From the quantity of heat radiations received by the earth, it is possible to estimate the temperature of the sun. Therefore, to determine the value of a constant, called solar constant, certain ideal conditions are taken into consideration. *It is the amount of heat energy (radiation) absorbed per minute by one sq cm of a perfectly black body surface placed at a mean distance of the earth from the sun, in the absence of the atmosphere, the surface being held perpendicular to the sun's rays.*

The instruments used to measure the solar constant are called *pyrheliometers*. The heat energy absorbed by a known area in a fixed time is found with the help of the pyrheliometers. To eliminate the effects of absorption by the atmosphere, the value of the solar constant is found at various altitudes of the sun on the same day under similar sky conditions. If S is the observed solar constant, S_0 the true solar constant and Z the altitude (angular elevation) of the sun, then

$$S = S_0\, a^{\sec z} \qquad \text{...(i)}$$

$$\log S = \log S_0 + \sec Z \log a \qquad ...(ii)$$

Here a is a constant.

A graph is plotted between log S along the y-axis and sec Z along the x-axis. The graph is a straight line. Produce the graph to meet the y-axis. The intercept on the y-axis gives $\log S_0$.

From the value of $\log S_0$, the value of S_0 the solar constant can be calculated. The value obtained varies between 1.90 and 2.60 calories per sq cm per minute.

Temperature of the Sun

The sun consists of a central hot pardon surrounded by the photosphere. The central portion has a temperature of the order of 10^7 K. The photosphere has a temperature of about 6000 K. This temperature is also called the effective temperature of the sun. Considering the sun as a perfect black body radiator, the temperature of the sun can be calculated.

Let the mean distance of the sun from the earth be R and S the solar constant. Then, the total amount' of heat energy received by the sphere of radius R in one minute = $4\pi R^2 S$.

If r is the radius of the sun, then the amount of heat energy radiated by 1 sq cm surface of the sun in one minute

$$E = \frac{4\pi R^2 S}{4\pi r^2} = \left(\frac{R}{r}\right)^2 \times S$$

Taking $R = 148.48 \times 10^7$ km

$$r = 6.928 \times 10^6 \text{ km}$$

The mean value of

$$S = 1.94 \text{ cals per cm}^2 \text{ per minute.}$$

$$E = \left(\frac{148.48\times10^7}{6.928\times10^5}\right) \times \frac{1.94}{60} \text{ cal per second} \qquad ...(i)$$

Also $E = \sigma T^4$

But $\sigma = 5.75 \times 10^{-5}$ ergs per cm^2 per second

$$= \frac{5.75\times10^{-5}}{4.2\times10^7} \text{ cal per cm}^2 \text{ per second}$$

$$E = \frac{5.75\times10^{-5}}{4.2\times10^7}.T^4 \qquad ...(ii)$$

Equating (i) and (ii)

$$\left(\frac{5.75\times10^{-5}}{4.2\times10^{7}}\right)T^{4} = \left(\frac{148.48\times10^{7}}{6.928\times10^{5}}\right)\times\frac{1.94}{60}$$

$$T = 5730 \text{ K.} \qquad ...(iii)$$

This temperature gives the effective temperature of the sun acting as a black body radiator. The actual temperature of the sun is higher than this value. The temperature of the sun is usually taken as 6000 K.

Temperature of the sun can also be calculated from Wien's displacement law,

$$\lambda_{max}T = 0.2892$$

The wavelength of the radiations for which the energy is maximum in the spectrum is 4900×10^{-8} cm.

Substituting the value of λ_m, the value of T comes out to be 5902 K. This value is in agreement with the accepted value. Hence, the effective temperature of the sun (photosphere) is about 6000 K.

STEFAN'S LAW (DERIVATION OF MATHEMATICAL)

The fact that blade body radiations exert pressure similar to a gas, helps in applying thermodynamics to heat radiations.

Let ψ be the energy density of radiations inside a uniform temperature enclosure at temperature T. P is the pressure and V is the volume.

Applying the first law of thermodynamics

$$\delta H = dU + P.dV \qquad ...(i)$$

Applying thermodynamical relation

$$\left(\frac{\partial H}{\partial V}\right)_r = \left(\frac{\partial P}{\partial T}\right)_v \qquad ...(ii)$$

$$\left(\frac{\partial U + P\partial V}{\partial V}\right)_r = T\left(\frac{\partial P}{\partial T}\right)_v$$

$$\left(\frac{\partial U}{\partial V}\right)_r = T\left(\frac{\partial P}{\partial T}\right)_v = P \qquad ...(iii)$$

Now $$U = V\psi$$

and $$P = \frac{\psi}{3}$$

or $$\left[\frac{\partial U}{\partial V}\right]_r = \psi$$

Here ψ is a function of temperature alone.

Substituting these values in equation (iii),

$$\psi = \frac{T}{3}\frac{d\psi}{dT} - \frac{\psi}{3}$$

$$\frac{4\psi}{3} = \frac{T}{3}\frac{d\psi}{dT}$$

or $$\frac{d\psi}{\psi} = 4\frac{dT}{T}$$

Integrating, $\log \psi = 4 \log T + \text{constant}$

or $$\psi = KT^4 \quad ...(iv)$$

Here K is a constant

Also the total rate of emission per unit area of a black body is proportional to the energy density.

$\therefore$ $$R \propto \psi \propto T^4$$

$\therefore$ $$R = \sigma T^4 \quad ...(v)$$

where σ is Stefan's constant.

The value of Stefan's constant in C.G.S. system is 5.672×10^{-5} C.G.S. units and in M.K.S. system it is 5.672×10^{-8} M.K.S. units.

COOLING FROM STEFAN'S LAW (DERIVATION OF NEWTON'S LAW)

Stefan's law is applicable for all temperatures of a hot body. But Newton's Law is applicable when the difference of temperature between the hot body and the surrounding is small. Consider a hot body at a temperature T_1 placed in a uniform temperature enclosure at T_1. According to Stefan's law,

$$R = e\sigma(T_1^4 - T_1^4) \quad ...(i)$$

Here e is the emissivity of the surface of the hot body

$$R = e\sigma(T_1 - T_2)(T_1^3 + T_1^2T_2 + T_1T_2^2 + T_2^3)$$

As $(T_1 - T_2)$ is small, T_1 can be taken approximately equal to T_2.

Then, $$R = e\sigma(T_1 - T_2)(T_2^3 + T_2^3 + T_2^3 + T_2^3)$$

$$R = 4e\sigma T_2^3((T_1 - T_2)$$

Taking $\quad 4e\sigma T_2^3 = k$

$$R = k(T_1 - T_2)$$

or $\quad R \propto (T_1 - T_2).$ ...(ii)

This equation represents Newton's law of cooling and is true when the difference of temperature is small.

PHOTON GAS

The interaction of electromagnetic radiations with matter, led to the idea that electromagnetic radiations are composed of discrete energy particles called photons. Each photon has an energy $h\nu$ and momentum h/λ. Here ν is the frequency and λ is the wavelength of the radiations. The electromagnetic radiations trapped in a cavity and in thermal equilibrium with the walls of the cavity are termed as black body radiations. In the equilibrium condition, the black body radiations can be considered as the *photon gas.* It is assumed that the photons do not interact among themselves. The photons interact only with-the atoms of the walls of the cavity. It is further assumed that the photons are indistinguishable and many photons can have the same energy. Photons are taken as bosons and they obey Bose-Einstein statistics.

As the photons can either be emitted or absorbed by the atoms of the walls of the cavity, the number of photons is not constant *i.e.*, the condition $\sum_i dn_i = 0$ is no longer valid.

Due to this reason the value of α is equal to zero in the Bose-Einstein distribution law.

$$n_i = \frac{g_i}{[e^{(\alpha + E_i/kT)}] - 1} \quad \text{...(i)}$$

Here $\quad \alpha = 0$

$$\therefore \quad n_i = \frac{g_i}{[e^{E_i/kT}] - 1} \quad \text{...(ii)}$$

In case, the cavity is large as compared to the wavelength of the radiations, the energy spectrum of the photons is taken to be continuous. In this case, the energy difference between successive allowed energy value is very small. Thus, replacing g_i by $g(E)\,dE$, and n_i by dn

$$dn = \frac{g(E)dE}{(e^{E/kT}) - 1} \quad \text{...(iii)}$$

As the energy of a photon, $E = h\nu$, the value of g(E) dE can be taken equal to g (ν) dν. The factor g(ν) dν corresponds to the number of oscillatory modes in the frequency range dν and relating to the energy range dE.

The number of states in a black body radiation in the frequency range ν and ν + dν can be obtained by calculating the spherical volume bound by the spheres of radii $\frac{h(\nu+d\nu)}{c}$ and $\frac{h\nu}{c}$

The volume of the spherical shell

$$= \frac{4}{3}\pi\frac{h^3}{c^3}(\nu+d\nu)^3 - \frac{4}{3}\pi\frac{h^3}{c^3}\nu^3$$

$$= \frac{4}{3}\pi\frac{h^3}{c^3}[\nu^3 + 3\nu^2 d\nu + \ldots - \nu^3]$$

$$= \frac{4}{3}\pi\frac{h^3}{c^3} \times 3\nu^2 d\nu$$

$$= 4\pi \frac{h^3}{c^3} \nu^2 d\nu \qquad \ldots(iv)$$

The phase space has volume $V = h^3$ and there are two states of polarization for the radiation. The number of states in the black body radiation in the frequency range ν and ν + dν is given by

$$g(\nu)d\nu = \frac{8\pi\nu}{c^3}\nu^2 d\nu \qquad \ldots(v)$$

$$dn = \left(\frac{8\pi V\nu^2 d\nu}{c^3}\right)\frac{1}{\left(e^{h\nu/kT}\right)-1} \qquad \ldots(vi)$$

For dn photons in the frequency range ν + dν and ν, the energy is equal to ($h\nu$) dn and the energy per unit volume

$$= \frac{(h\nu)dn}{V}$$

Therefore, the energy density distribution for black body radiation is given by

$$E(\nu) = \frac{(h\nu)dn}{V.d\nu} \qquad \ldots(vii)$$

Substituting the value of dn from equation (vi), we get

$$E(\nu) = \frac{8\pi h\nu^2 d\nu . h\nu}{c^3 V.d\nu\left[\left(e^{h\nu/kT}\right)-1\right]}$$

$$E(\nu) = \left(\frac{8\pi h\nu^3}{c^3}\right)\left[\frac{1}{\left(e^{h\nu/kT}\right)-1}\right]^{d\nu} \qquad \text{...(viii)}$$

Equation (viii) represents Planck's radiation law for black body radiations.

From equation (viii)

$$E(\nu)d\nu = \frac{8\pi h\nu^3}{c^3}\left[\frac{1}{\left(e^{h\nu/kE}\right)-1}\right]^{d\nu}$$

But $\qquad \nu = \dfrac{c}{\lambda}$

or $\qquad d\nu = -\dfrac{c}{\lambda^2}d\lambda$

Neglecting the negative sign, we get

$$\therefore \qquad E(\lambda)d\lambda = \frac{8\pi h\nu^3}{c^3} \times \frac{c}{\lambda^2}\left[\frac{1}{\left(e^{h\nu/kT}\right)-1}\right]^{d\lambda}$$

or

$$E(\lambda) = \frac{8\pi hc}{\lambda^5} \times \frac{c}{\lambda^2}\left[\frac{1}{\left(e^{h\nu/kT}\right)-1}\right] \qquad \text{...(ix)}$$

This gives the energy density for wavelength λ in the spectrum of the black body. Both the equations (viii) and (ix) represent Planck's radiation law for black body radiation.

GIBBS FUNCTION

The Gibbs function G of a system is given by,

$$G = U - TS + PV$$

Consider a system that can do other forms of work, in addition to P.dV work, *e.g.*, a voltaric cell. In the case of a voltaic cell, the electrical work is –E.dI. Similarly for a magnetic material, the magnetic work is –m.dH. In general, the work will be given by P.dV plus a sum of terms each being the product of intensive variable (such as P, E or m) and the differential of an extensive variable (such as dV, dI and dH). In the case of a voltaic cell, the intensive variable is S and differential of extensive variable is dI.

Suppose, in general, in addition to PdV the intensive variable is y and differential of extensive variable is dx. The work done for any reversible process,

$$\delta W = PdV + ydx$$

$$W = \int_{V_1}^{V_2} PdV + \int_{x_1}^{x_2} ydx$$

Take $\int_{x_1}^{x_2} ydx = A$

Consider a process where the system works at constant pressure P_0 and the change in volume is $(V_2 - V_1)$

$$\therefore \quad \int_{V_1}^{V_2} PdV = P_0 [V_2 - V_1]$$

$$\therefore \quad W = P_0[V_2 - V_1] + A \qquad \text{...(i)}$$

For a system that exchanges heat with a reservoir temperature

$$W < (U_1 - U_2) - T_0 (S_1 - S_2) \qquad \text{...(ii)}$$

$$P_0[V_2 - V_1] + A < (U_1 - U_2) - T_0 (S_1 - S_2)$$

or $$A < (U_1 - U_2) - T_0 (S_1 - S_2) + P_0[V_1 - V_2] \qquad \text{...(iii)}$$

Consider a specific process, where the initial and the final states of the system and the surroundings are at the same temperature (T_0) and pressure (P_0)

$$To = T \text{ and } P_0 = P$$

From equation (iii)

$$A_{P,T} < (U_1 - U_2)_{P,T} - T(S_1 - S_2)_{P,T} + P[V_1 - V_2]_{P,T} \qquad \text{...(iv)}$$

But, for the Gibbs function,

$$G = U - TS + PV \qquad \text{...(v)}$$

Therefore, for two equilibrium states at the same pressure and temperature,

$$[G_1 - G_2)_{P,T} = (U_1 - U_2)_{P,T} - T[S_1 - S_2]_{P,T} + P[V_1 - V_2]_{P,T} \qquad \text{...(vi)}$$

From equations (iv) and (vi)

$$A_{P,T} < (G_1 - G_2)_{P,T} \qquad \text{...(vii)}$$

Thus, the difference between Gribbs function of a system between two equilibrium states sets the maximum limit to the work in addition to PdV work, provided the initial and the final states are at the same pressure and temperature and the system exchanges heat with a single heat reservoir. The work done will be maximum when the process is reversible. The process in this case will be isothermal–isobaric. If the process is irreversible, work done will be less than the maximum.

CRITICAL CONSTANTS

(i) ***Critical temperature of a gas (T_c)*** *may be defined as the minimum temperature which must be reached before it can be liquefied by the application of pressure.* A gas cannot be liquefied by pressure above its critical temperature, no matter how high the applied pressure may be. Each gas is having definite critical temperature, *e.g.*, critical temperature of CO_2 is 31.1MC.

(ii) ***Critical pressure of a gas (P_c)*** *may be defined as the minimum pressure necessary for the liquefaction of a gas at critical temperature.*

(iii) ***Critical volume of a gas (V_c)*** *may be defined as the volume occupied b 1 mole of the gas or liquid under critical pressure and under critical temperature.*

CONDITIONS FOR LIQUEFACTION OF GASES

From the Andrew's experiment very useful conditions for the liquefaction of a gas may be stated as :

(i) The temperature must be at or below the critical temperature.

(ii) A suitable pressure must be applied equal to the critical pressure if the temperature is at the critical temperature but becoming less if the temperature is reduced below this value.

van der Waals Equation as an Improvement Over Simple Laws

Van der Waals equation is an improvement over simple laws due to following reasons:

(i) *Van der Waals equation explains the devication from Boyle's law:* Regnault and Amagat studied the behaviour of cases and observed that they do not obey simple gas laws under different conditions of temperature and pressure, more particularly at low temperature and high pressure. They also plotted graphs between the product of pressure and volume ad pressure.and found that PV product was not constant as desired by Bole's law. These deviations from ideal behaviour were explained by van der Waals equations.

(ii) *It also explains critical phenomenon.* This equation satisfactorily explains critical phenomenon, observed in the case of gases. The theoretical isotherm (*i.e.*, the curves between V and P at constant T) obtained from the values of V and the experimental values of 'a' and 'b' from van der Waals equation were found to be quite similar. The theoretical isotherms obtained from simple gas laws are quite different from the experimental isotherms.

PRINCIPLE OF CONTINUITY OF STATES

Although, the transition from liquid to vapour and *vice versa* generally takes place suddenly and the two states can be sharply distinguished from each other as can be see from AB and CD parts of 13.1°C isotherm for carbon dioxide (Fig. 4.11), yet where occurs a possibility of passing from one state of aggregation to the other without involving sudden change.

Hence for further theoretical considerations it becomes essential to show that the liquid stated does not refer to a sharp and discontinuous transition from gaseous state but rather a continuation of the gaseous phase into the region of very strong intermolecular attractions where the volume becomes very small. The idea of a contiguity from liquid to gaseous state will become evident from the following consideration:

The dome-shaped curve depicted b the dashed line in Fig. 4.11 connects the end points of the horizontal portions of the isotherms.

As a horizontal portion represents two phases in equilibrium, all points inside the dome correspond to equilibrium between liquid and vapour. For such points it always becomes possible to draw any sharp distinction between liquid and vapour, because there is a definite boundary surface between the two phases.

For points lying outside the dome only there is only one phase (liquid or vapour) ad here it is not possible to draw any sharp distinction between liquid and vapour. This fact is termed as the *principle of contiguity of states* and can be demonstrated with reference to Fig. 4.11.

Suppose, that carbon dioxide gas under atmospheric pressure is made to heat a temperature above the critical temperature, as to 48°C and that is then subjected to a pressure greater than critical pressure, as, to 100 atmospheres, the system will change volume as shown by the isotherm 48°C.

When the temperature is lowered to 13.1°C while maintaining the pressure at 100 atmospheres, a point on the curve CD will be reached a curve which applies to liquid carbon dioxide. At no point in the process there is any abrupt change of state; the change from gas to liquid will be contiguous. Similarly, it is possible to start with liquid carbon dioxide under the conditions represented, say by the point C.

If we now raise the pressure o the liquid to 10 atmospheres and the temperature to 48°C the condition of the system will be represented by a point on that isotherm; and on reducing the pressure to that of the atmosphere ad then allowing the temperature to fall to 13.1°C, we shall be having carbon dioxide as a gas; and at no time during the whole process any discontinuity of properties will be observed, or an separation into liquid and vapour. The passage from the liquid to the gaseous state would be continuous.

An important conclusion derived from the principle of continuity of states is that one may think of a liquid near the critical point as just a highly compressed gas. If a equation of state can be found which accurately describes the gas close to its critical temperature and pressure that equation should also be able to describe the liquid.

VAN DER WAALS EQUATIONS AND CRITICAL STATE—RELATIONS BETWEEN VAN DER WAALS CONSTANTS AND CRITICAL CONSTANTS

A very good confirmation of the general character of van der Waals equation may be provided b its application to critical phenomenon.

The van der Waals equation may be put as follows:

$$\left(P + \frac{a}{V^2}\right)(V - b) = RT \text{ (putting } = 1) \qquad ...(1)$$

$$PV + \frac{a}{v} - Pb - \frac{ab}{V^2} = RT$$

On multiplying throughout by V^2, we obtain

$$PV^3 + aV - PbV^2 - ab = RTV^2$$

or $$PV^3 + aV - PbV^2 - ab - RTV^2 = 0$$

On arranging in descending powers of V, we get

$$PV^3 - PBV^2 - RTV^2 + aV - ab = 0$$

On dividing b P, we get

$$V^3 - bV^2 - \frac{RTV^2}{P} + \frac{aV}{P} - \frac{ab}{P} = 0$$

$$V^3 - V^2 - \left(b + \frac{RT}{P}\right) + \frac{aV}{P} - \frac{ab}{P} = 0 \qquad ...(2)$$

This is a third degree equation in terms of V (cubic equation in V) and below a certain value of temperature T the equation can be having one of three real roots. With the rise of temperature these roots would approach one another and become identical ultimately. As the temperature gets raised the S-shaped dotted portion of the curve gets reduced to a point. The values for the volume of liquid and the gas approach each other until at this volume of liquid and the gas approach each other until at this point, the three values of V become identical. This point would correspond to the critical temperature T_c and the volume V_c. At the critical point, the three roots of the van der Waals equation would become identical and equal to V_c. But this refers to the value of the three roots of the van der Waals equation at the critical point.

$$(V - V_c)^3 = 0 \qquad ..(3)$$

or $$V^3 - 3V_cV^2 + 3V_c^2V - V_c^3 = 0 \qquad ...(4)$$

On substituting these values in eq. (2), we get,

$$V^3 - V^2\left(b + \frac{RT_c}{P_c}\right) + \frac{aV}{P_c} - \frac{ab}{P_c} = 0 \qquad ...(5)$$

As both eqs. (4) and (5) are identical then the coefficients of the equal powers of V i the two equations may be equated:

$$3V_c = b + \frac{RT_c}{P_c} \qquad ...(6)$$

$$3V_c^2 = \frac{a}{P_c} \qquad ...(7)$$

$$V_c^3 = \frac{ab}{P_c} \qquad ...(8)$$

On dividing Eq. (8) by Eq. (7), we get,

$$\frac{V_c^3}{3V_c^2} = \frac{ab}{P_c} \times \frac{P_c}{a}$$

or $$\frac{V_c}{3} = b$$

$$V_c = 3b \qquad \text{...(9)}$$

Substituting the value of V_c from Eq. (9) into Eq. (7), we obtain

$$\frac{a}{P_c} = 3 \times (3b)^2$$

or $$\frac{a}{P_c} = 27b^2$$

or $$P_c = \frac{a}{27b^2} \qquad \text{...(10)}$$

Substituting the values of V_c and P_c in equation (6), we get

$$3 \times 3b = b + \frac{RT_c}{a/27b^2}$$

$$9b = b + \frac{RT_c}{a/27b^2}$$

or $$8b = RT_c \times \frac{27b^2}{a}$$

or $$T_c = 8b \times \frac{a}{27b^2} \times \frac{1}{R}$$

$$T_c = \frac{8a}{27bR} \qquad \text{...(11)}$$

Derivation of $P_cV_c = \frac{3}{8} RT_c$ from van der Waals Equation and calculation of van der Waals constants 'a' and 'b' in terms of T_c and P_c.

From van der Waals equation, we get

$$V^3 - V^2\left(b + \frac{RT_c}{P_c}\right) + V\frac{a}{P_c} - \frac{ab}{P_c} = 0 \qquad \text{...(12)}$$

and $$V^3 - 3V_cV^2 + 3V_c^2V - V_c^3 = 0 \qquad \text{...(13)}$$

On equating the coefficients of (12), and (13), we get

$$3V_c = \frac{RT_c}{P_c} + b \qquad \text{...(14)}$$

$$3V_c^2 = \frac{a}{P_c} \qquad \text{...(15)}$$

$$V_c^3 = \frac{ab}{P_c} \quad ...(16)$$

On dividing Eq. (16) by (15), we get

$$V_c = 3b \quad \text{or} \quad b = \frac{1}{3}V_c \quad ...(17)$$

On substituting the value of b in (14), we obtain

$$3V_c = \frac{RT_c}{P_c} + \frac{V_c}{3}$$

or $$3V_c - \frac{V_c}{3} = \frac{RT_c}{P_c}$$

or $$\frac{8}{3}V_c = \frac{RT_c}{P_c}$$

or $$P_cV_c = \frac{3}{8}RT_c \quad ...(18)$$

On rearranging Eq. (18), we get

$$V_c = \frac{3}{8}\frac{RT_c}{P_c}$$

On squaring, we obtain

$$V_c^2 = \frac{9}{64}\frac{R^2T_c^2}{P_c^2} \quad ...(19)$$

On substituting the value of $V_c^2 = \dfrac{a}{3P_c}$ from Eq. (15) into Eq. (19), we get,

$$\frac{a}{3P_c} = \frac{9}{64}\frac{R^2T_c^2}{P_c^2}$$

$$a = \frac{27R^2T_c^2}{64P_c} \quad ...(20)$$

On substituting the value of $V_c = 3b$ in Eq. (14), we obtain,

$$9b = \frac{RT_c}{P_c} + b$$

or $$8b = \frac{RT_c}{P_c}$$

or $$b = \frac{RT_c}{8P_c} \quad ...(21)$$

THE PRINCIPLE OF CORRESPONDING STATES

In 1881, van der Waals demonstrated that if the pressure volume and temperature of a gas are expressed in terms of its critical pressure critical volume and critical temperature, then an important generalization called the principle of corresponding states would be obtained.

According to van der Waals equation, we have

$$\left(P + \frac{a}{V^2}\right)(V - b) = RT \qquad ...(1)$$

This equation has constants 'a' and 'b' which cannot be measured directly. Therefore, this equation is not applicable to all gases and in order to make it so, these constants have to be removed.

It has been found experimentally that near the critical point all gases give the same type of isotherms as carbon dioxide, but the actual position of the critical isotherms gets varied with the critical constant of the gas studied. The similarity is great enough to postulate that if for each different gas a suitable scale of pressure, volume and temperature is selected, it will be possible to superimpose all the critical isotherms on one another. If this were done of the critical isotherms it becomes probable that all other isotherms would coincide.

Under such circumstances, there should be one isothermal diagram and one fundamental equation for all gases. Such a scale will be one in which the pressure, volume, and temperature will be expressed as fractions of the critical pressure, volume and temperature.

In order to obtain such an equation, the values of P, V and T must be expressed as corresponding fractions of P_c, V_c and T_c, the critical pressure, volume and temperature,

$$\pi = P/P_c \quad \text{or} \quad P = \pi P_c$$

$$\phi = V/V_c \quad \text{or} \quad V = \phi V_c$$

$$\theta = T/T_c \quad \text{or} \quad T = \theta T_c$$

where π, ϕ and θ are referred to as *reduced pressure, reduced volume and reduced temperature respectively.*

On substituting the values of P, V and T in the van der Waals equation for one mole of an ideal gas, we get

$$\left(\pi P_c + \frac{a}{\phi^2 V_c^2}\right)(\phi V_c - b) = R_q T_c \qquad ...(2)$$

$$P_c = \frac{a}{27b^2}, \quad V_c = 3b$$

$$\text{and} \quad T_c = \frac{8a}{27Rb}$$

On substituting these values in equation (2), we get,

$$\left(\pi\frac{a}{27b^2} + \frac{a}{9\phi^2 b^2}\right)(3\phi\, b - b) = R\theta\frac{8a}{27Rb}$$

or $$\frac{a}{27b^2}\left(\pi + \frac{3}{\phi^2}\right)\times b(3\phi - 1) = \frac{a}{27b}8\theta$$

or $$\frac{a}{27b}\left(\pi + \frac{3}{\phi^2}\right)(3\phi - 1) = \frac{a}{27b}8\theta$$

or $$\left(\pi + \frac{3}{\phi^2}\right)(3\phi - 1) = 8\theta$$

This equation involving neither of the characteristic constants a, b, T_c, P_c and V_c is known as van der Waals reduced equation of state. This equation is applicable to all substances (liquid or gaseous) irrespective of their nature, because it is not involving either of the characteristic constants.

Significance of Law of Corresponding States

It has a great significance in the study of the relationship between physical properties ad chemical constitution of various liquids. As it is known that the pressure has practically no effect on liquids, so while studying the relation between physical properties and chemical constitution the determination of these properties should be performed at the same reduced temperature.

It has been found that the *boiling point of a liquid at atmospheric pressure is almost 2/3rd of its critical temperature* (T_c) *when expressed in absolute scale*. Consequently, boiling point is considered as the *corresponding temperature*. So, various physical properties of liquids such as molecular volumes of various liquids should be compared at their respective boiling points.

In actual practice one cannot use the above reduced equation of state directly. One makes use of the graphs between compressibility factor 'Z' and the reduced pressure at different reduced temperatures. The same graph is applicable to all gases. This can be seen from the following considerations, since $Z = \frac{PV}{RT}$.

Writing this in terms of reduced terms,

We have $$Z = \frac{(\pi P_c)(\phi V_c)}{R(\theta T_c)} = \left(\frac{P_c V_c}{RT_c}\right)\frac{\pi\phi}{\theta} = \frac{3}{8}\frac{\pi\phi}{\theta}$$

Thus, if two gases are having the same reduced temperature and reduced pressure, then their reduced volume will be the same, thus R.H.S. is independent of the as, thus all gases will have the same value of 'Z'. If isotherms are made to plot in terms of the reduced variables, the same curve should be obtained for all gases.

COMPRESSIBILITY FACTOR

It is known that the equation of state, PV = nRT, derived from the postulates of the kinetic theory, *is valid for an ideal gas only*. Real gases tend to obey this equation only approximately ad that too under conditions of low pressure and high temperature.

The higher the pressure and the lower the temperature, the greater would be the deviations from the ideal behaviour. In general, the most easily liquefiable and highly soluble gases exhibit larger deviations. Hence gases like carbon dioxide, sulphur dioxide and ammonia exhibit much larger deviations than hydrogen, oxygen, nitrogen, etc.

It is best to represent the deviations from ideal behaviour i terms of the compressibility factor, Z, which is defined as

$$Z = \frac{PV}{(PV)_{ideal}} = \frac{PV}{nRT} \quad ...(1)$$

For an ideal gas, Z = 1 under all conditions of temperature and pressure. The deviation of Z from unity is, therefore a measure of the imperfection of the gas under consideration.

The graphs plotted for the compressibility factors determined for a number of gases over a range of pressures at a constant temperature (0°C). At extremely low pressures, the value of Z for all the gases are close to unity which implies that the gases behave almost ideally. At very high pressures, all the gases have Z more than unity indicating that the gases are less compressible than an ideal gas. This is attributed to the fact that at high pressures, the molecular repulsive forces are dominant.

It can be seen from Fig. 4.3 that at moderately low pressures, carbon monoxide, methane and ammonia are more compressible than an ideal gas, *i.e.* $Z < 1$. This is attributed to the fact that at low pressures, the long range attractive forces are dominate and favour compression. The compressibility factor Z goes on decreasing with increase in pressure, passes through a minimum at a certain stage and then starts increasing with increase in pressure. The gases would now becomes less compressible than ideal gas, *i.e.*, $Z > 1$. It can be seen from the figure that although carbon monoxide and methane exhibit marked devications from ideal behaviour only at high pressures, yet ammonia exhibits large deviation even at low pressure.

It can be seen that hydrogen and helium at 0°C are less compressible *at all pressures* than the ideal gas, *i.e.*, $Z > 1$. However, if the temperature is sufficiently low (*e.g.*, below – 165°C for hydrogen and below – 240°C for helium) these gases would also give the same type of Z – P plots as are shown by ammonia, carbon monoxide and methane at 0°C. On the other hand, if the temperature becomes sufficiently high, the Z-P plots of ammonia, carbon monoxide and methane will be almost similar to those of hydrogen and helium at 0°C *i.e.*, the value of Z will increase continuously with increasing pressure.

EXPANSIVITY AND COMPRESSIBILITY

When fluids (gases and liquids) are heated, they expand but the expansion of gases is much more than that of liquids. Similarly, they can be compressed. The variation of volume V with temperature T, keeping pressure P constant is termed as coefficient of thermal expansion or the coefficient of isobaric expansion or simple, expansivity, α of the fluid. Thus,

$$\alpha = \frac{1}{V}\left(\frac{\partial V}{\partial T}\right)_P \qquad ...(1)$$

Similarly, the variation of V with P, keeping T constant, is termed *as the coefficient of isothermal compressibility* or, simply, compressibility, β of the fluid, thus,

$$\alpha = -\frac{1}{V}\left(\frac{\partial V}{\partial T}\right)_T \qquad ...(2)$$

Generally, as the volume of a gas gets decreased as pressure increases it means that the quantity $(\partial V/\partial P)_T$ is negative. Hence, a minus sign is included in the definition so as to make β positive. It is to be noted that α has the dimensions of T^{-1} and β those of P^{-1}.

Let us derive the value of α and β for an ideal gas.

For n moles of an ideal gas, we have

$$PV = nRT \quad \text{...(i)}$$

or

$$V = nRT/P \quad \text{...(ii)}$$

On differentiating Eq. (ii) with respect to T at constant P, we get,

$$\left(\frac{\partial V}{\partial T}\right)_P = \frac{nR}{P} = \frac{PV}{TP} = \frac{V}{T}$$

$$\therefore \quad \alpha = \frac{1}{V}\left(\frac{\partial V}{\partial T}\right)_P = \left(\frac{1}{V}\right)\left(\frac{V}{T}\right) = \frac{1}{T}$$

Again, differentiating Eq. (ii) with respect to P at constant T. We get,

$$\left(\frac{\partial V}{\partial P}\right)_T = \frac{nRT}{P^2} = \left(\frac{nRT}{P}\right)\left(-\frac{1}{P}\right) = -\frac{V}{P}$$

$$\therefore \quad \beta = -\frac{1}{V}\left(\frac{\partial V}{\partial P}\right)_T = \left(-\frac{1}{V}\right)\left(-\frac{V}{P}\right) = \frac{1}{P}.$$

POSTULATES OF THE KINETIC MOLECULAR THEORY OF GASES

Kinetic theory of gases is based upon the following assumptions:

(*i*) *All gases are made up of very large umber of minute particles called molecules.*

(*ii*) The molecules are small and separated from one another by large distances so that the actual volume occupied by the molecules is almost negligible as compared to the total volume of the gas.

(*iii*) The molecules are always moving in all directions in straight lines with very high velocities. They keep on colliding against each other (molecular collisions) and against the walls of the vessel at very small intervals of time.

(*iv*) The molecules are spherical and perfectly elastic, and there occurs o loss of kinetic energy due to mutual collisions or collisions with the walls of the container. The kinetic energy may be transferred from one molecule to another but it is not converted into any other form of energy such as heat.

(*v*) The pressure exerted by a gas on the walls of the containing vessel is due to the bombardment of the molecules of the gas on the walls of the containing vessel.

(*vi*) The average kinetic energy of the molecules of gas is directly proportional to the absolute temperature.

(*vii*) There are no attractive forces between molecules or between molecules and the walls of the vessel in which the gas is contained. The molecules move completely independent of one another.

(*viii*) The motion imparted to the molecules by gravity is negligible in comparison to the effect of the continued collisions between them.

(*ix*) The laws of classical mechanics (in particular the Newton's second law of option) are applicable to gaseous molecules.

Justification for the Postulates

The above postulates are sufficiently justified by the following experiences.

1. It is known that gases can be compressed by the application of pressure and can also expand indefinitely filling the whole vessel no matter how large it may be. This reveals that gases are made of very large umber of minute particles separated from each other by large distances. Further these particles occupy very small volumes as compared to the total volume of the gas. In a gas like oxygen, hydrogen or nitrogen, at N.T.P. the volume occupied by the molecules themselves is only 0.014 percent of the total volume. The remaining 99.986 per cent of the volume is merely an empty space.

   [Postulates (*i*) and (*ii*)]

2. Dust particles i air are found to be in constant irregular zig-zag motion termed *Brownian movement*. This is an evidence for the very fast straight line motion of the molecules. [Postulate (*iii*)]

3. Rapid motion of a large number of molecules must involve frequent collisions between then and these collisions must be elastic as otherwise there will occur constant loss of kinetic energy and the molecules must come to stand still which does not happen. [Postulates (*iii*) and (*iv*)]

4. If a gas is contained in a closed vessel, "it is expected that moving molecules will keep on hitting the walls of the vessel continuously thus exerting pressure." [Postulate (*v*)]

5. Molecules are very close in a compressed gas and intermolecular attraction should be high but this attraction must be negligible because even a highly compressed gas expands freely and fills the whole available space.

   [Postulate (*vii*)]

6. Gases intermix freely with other gases in all proportions. Even a heavier gas rises up to intermix with a lighter gas. This indicates that the effect of gravity is negligible i comparison to the effect of the continued collisions between them. [Postulate (*viii*)]

7. On heating, the temperature of gas rises. Heat energy supplied increases the speed of the molecules and their average kinetic energy increases thereby causing the rise of temperature. [Postulate (*vi*)]

Derivation of the Kinetic Equation for Gases

BY taking the postulates of kinetic theory into consideration, it is possible by applying the laws of classical mechanics to derive a mathematical expression which can explain the various gas laws. This expression is known as the *Kinetic Gas Equation*.

The length of each side of the cube = l cm

The total number of molecules = N

The mass of one molecule = m

The selection of the cubical shape of the container is purely a matter of convenience and involves no loss of generality because the pressure of a gas is independent of the shape of the container.

It is regarded that the container are smooth and get oriented perpendicular to the x, y and z-axes of a Cartesian coordinate system. The motion of molecules in the container, at any instant, is totally random.

Although, the molecules are moving in every possible direction the average velocity v of a molecule at any moment can be resolved into three directions X, Y, Z at right angles to one another (Fig. 4.1).

Suppose a single molecule is moving with a velocity v.

Suppose v_x, v_y and v_z be the components of velocity v in the x, y ad z directions parallel to the three edges of the cube.

These three components get related to the resultant velocity v by the expression give below:

$$v = v_x^2 + v_y^2 + v_z^2 \quad ...(1)$$

Suppose the motion of this molecule occurs between the two opposite faces A and B parallel to the x-axis. It strikes the face A at regular intervals.

The velocity of the molecule before it strikes the face A is v_x. But the molecule is perfectly elastic. Therefore it will rebound with the same velocity with the sign changed, *i.e.*, $- v_x$.

$$= 2mv_x \text{ gm. cm/sec} \quad ...(2)$$

In order to strike the same face (A) against, the molecule will have to move to opposite face 'B' and come back. It means that the particle has to travel a distance 2l cm for each successive collision on the same face.

Thus, by traversing 2l cm., one impact is produced. By traversing 1 cm., 1/2l impacts get produced.

∴ By traversing v_x cm, $\frac{v_x}{2l}$ impacts get produced per second.

(v_x is velocity of the molecule that is the distance travelled i one second.)

Hence the number of impacts made by one molecule in one second

$$= \frac{v_x}{2l}$$

∴ Total change ion momentum per second (due to $\frac{v_x}{2l}$ impact) on face A

$$= 2mv_x \times \frac{v_x}{2l}$$

$$= \frac{mv_x^2}{l} \text{ gm cm/sec}^2$$

Total change in momentum per second because of the impacts of the molecule on two opposite faces A and B, along x-axis is therefore, double, *i.e.*, equal to

$$\frac{mv_x^2}{l}+\frac{mv_x^2}{l}=\frac{2mv_x^2}{l} \qquad ...(3)$$

Also the total change of momentum per second due to impact of a single molecule on the two opposite faces along y-axis and two opposite faces along z-axis will be give as follows:

$$\frac{2mv_y^2}{l}, \frac{2mv_z^2}{l} \text{ respectively.}$$

Therefore, the total change in momentum per second o all the six faces of the cube, is equal to

$$=\frac{2mv_x^2}{l}+\frac{2mv_y^2}{l}+\frac{2mv_z^2}{l}$$

or $$=\frac{2m}{l}(v_x^2+v_y^2+v_z^2)$$

or $$=\frac{2mv^2}{l} \text{ gm cm/sec}^2 \qquad ...(4)$$

$$(\because \; v_x^2+v_y^2+v_z^2=v^2)$$

Thus total change of momentum, per second due to one molecule

$$=\frac{2mv^2}{l} \text{ gm cm/sec}^2$$

According to Newton's second law of motion, the rate of momentum is force *i.e.*,

$$\therefore \qquad \text{Force}=\frac{2mv^2}{l} \text{ dynes}$$

The pressure 'P' (force per unit area) is the total force divided by area of six faces of the cube ($6l^2$), *i.e.*,

$$P=\frac{\text{Force}}{6l^2}=\frac{2mv^2}{l}\times\frac{1}{6l^2}=\frac{1}{3}\frac{mv^2}{l^3}$$

or $$P=\frac{1}{3}\frac{mv^2}{l^3} \qquad ...(5)$$

But l^3 is the volume of the cube 'V'

$$P=\frac{1}{3}\frac{mv^2}{V}$$

Hence Pressure exerted by one molecule is

$$P = \frac{1}{3V} mv^2 \qquad ...(6)$$

The total pressure exerted by N different molecules of same mass ad having different velocities $v_1, v_2, v_3 \ldots v_N$ will be as follows:

$$P = P_1 + P_2 + P_3 + \ldots + P_N$$

$$= \frac{1}{3V} mv_1^2 + \frac{1}{3V} mv_2^2 + \frac{1}{3V} mv_3^2 + \ldots + \frac{1}{3V} mv_N^2$$

[from Equation ... (6)]

$$P = \frac{1}{3V} m (v_1^2 + v_2^2 + v_3^2 + \ldots + v_N^2) \qquad ...(7)$$

If we multiply and divide the right had side of equation (7) by , we get

$$P = \frac{mN}{3V} \frac{(v_1^2 + v_2^2 + v_3^2 + \ldots v_N^2)}{N}$$

$$= \frac{mN}{3V} \times (\text{mean square velocity})$$

or $$P = \frac{mN}{3V} \times (\text{root mean square velocity})^2$$

$$= \frac{mNc^2}{3V}$$

where N refers to the total number of molecules and $\sqrt{\bar{c}^2}$, the root mean square velocity give by

$$\sqrt{\bar{c}^2} = \sqrt{\frac{v_1^2 + v_2^2 + v_3^2 + \ldots v_N^2}{N}}$$

$$PV = \frac{1}{3} mN\bar{c}^2 \qquad ...(8)$$

This is known as the Kinetic Gas Equation

It may be expressed as $PV = \frac{1}{3} M\bar{c}^2$ where M is the total mass of the gas.

or $$P = \frac{1}{3} \frac{m}{V} \bar{c}^2 = \frac{1}{3} D\bar{c}^2$$

where D is density of the gas.

Kinetic Energy and Temperature

Translational kinetic energy of a body of mass m with velocity $\sqrt{\bar{c}^2}$ is give by as follows:

$$K.E. = \frac{1}{2}m\bar{c}^2$$

As the velocity of a molecule increases with the rise in temperature, it becomes possible to define mean kinetic energy in terms of temperature of the gas.

Suppose one mole of an ideal gas is under consideration. Then the total number of molecules is N (Avogadro's number).

∴ Kinetic energy per mol of the gas for N particles is given by

$$K.E. = \frac{1}{2}mN\bar{c}^2 \qquad ...(9)$$

The kinetic gas equation (8) for N molecules of a gas is

$$PV = \frac{1}{2}mN\bar{c}^2$$

or $$PV = \frac{2}{3} \times \frac{1}{2}mN\bar{c}^2$$

or $$= \frac{2}{3}K.E. \qquad \left(\because \frac{1}{2}mNc^2 \text{ is the kinetic energy}\right)$$

On comparing this equation with ideal gas equation PV = RT for 1 mole of the gas, we obtain

$$RT = \frac{2}{3}K.E.$$

or $$K.E. = \frac{3}{2}RT$$

or $$K.E. \propto T \qquad ...(10)$$

Hence the kinetic energy of one mole of any gas is directly proportional to its absolute temperature.

This is also the Maxwell's generalisation, which states that "*Kinetic energy of translation of an ideal gas is independent of the nature of the gas, its pressure and depends only upon the temperature.*"

But $\quad K.E. \propto \bar{c}^2$

$\therefore \quad \bar{c}^2 \propto T$

or $\quad \bar{c}^2 \propto \sqrt{T}$ ...(11)

Hence molecular velocity of any gas is proportional to the square root of the absolute temperature. The molecular motion is, therefore, often known as *thermal motion of the molecules.*

At absolute zero, the thermal motion ceases completely and kinetic energy of the gas would become also zero.

Deviation of Gas Laws from the Kinetic Equation

(a) Boyle's Law: *This law states, "Temperature remaining constant, the volume of a give mass of a gas is inversely proportional to the pressure."* mathematically it can be expressed as

$$V \propto \frac{1}{P}$$

$$PV = \text{constant}$$

This law can be derived from the kinetic equation

$$PV = \frac{1}{3} mN\bar{c}^2$$

This can be written as

$$PV = \frac{2}{3} \times \frac{1}{2} mN\bar{c}^2 \qquad ...(i)$$

But $\frac{1}{2} mN\bar{c}^2$ = average kinetic energy of gas which is proportional to the absolute temperature.

i.e. $$\frac{1}{2} mN\bar{c}^2 \propto T \text{ or } \frac{1}{2} mN\bar{c}^2 = KT$$

Substituting in (i) above

$$PV = \frac{2}{3} . KT$$

In this equation $\frac{2}{3}$ KT would be a constant for a give mass of gas at a constant temperature.

$\therefore \quad PV = \text{constant}$

or $$V \propto \frac{1}{P}.$$ This is Boyle's law.

(b) Charle's Law: *According to this law, "The volume of a give mass of a gas is directly proportional to the absolute temperature at a constant pressure."*

We have already derived

$$PV = \frac{2}{3}kT$$

or $$V = \frac{2}{3} \times \frac{k}{P}.T$$

or $$\frac{V}{T} = \frac{2k}{3P}$$ Thus for a give mass of a gas

if pressure P, remains constant, the expression $\frac{2k}{3P}$ is constant.

∴ At constant pressure

$$\frac{V}{T} = K$$

or $$V = K.T.$$

Hence $$V \propto T.$$ This is Charle's law.

(c) Avogadro's Hypothesis: *This law states, "Equal volumes of all gases under similar conditions of temperature ad pressure contain equal number of molecules."*

For any two gases, the kinetic equation can be put as follows:

(i) $$P_1V_1 = \frac{1}{3}m_1N_1\bar{c}_1^2$$

(ii) $$P_2V_2 = \frac{1}{3}m_1N_1\bar{c}_1^2$$

On rewriting:

(i) $$P_1V_1 = \frac{2}{3} \times \frac{1}{2}m_1N_1\bar{c}_1^2$$

(ii) $$P_2V_2 = \frac{2}{3} \times \frac{1}{2}m_1N_1\bar{c}_2^2$$

Since $P_1 = P_2$ and $V_1 = V_2$

$$\therefore \qquad \frac{1}{2}m_1N_1\bar{c}_1^2 = \frac{1}{2}m_1N_1\bar{c}_2^2 \qquad ...(a)$$

Now if the temperature is also constant, the average kinetic energy per molecule must be the same.

or $$\frac{1}{2}m_1\bar{c}_1^2$$

$$= \frac{1}{2}m_2\bar{c}_2^2 \qquad ...(b)$$

On dividing (a) by (b), we have

$$N_1 = N_2$$

This is Avogadro's Law

(d) The Ideal Gas Equation: On Combining Boyle's law, Charles law ad Avogadro's law we find that the volume of a gas depends on the pressure, temperature and number of moles, as follows:

$V \propto 1/P$ (at constant T and n) (Boyle's law)

$V \propto T$(at constant P and n) (Charles' law)

$V \propto n$(at constant T and P) (Avogadro's law)

Thus, V should be proportional to the product of these three terms, *i.e.*,

$$V \propto nT/P$$

or $$V = R(nT/P)$$

or $$PV = RT \qquad ...(i)$$

where R, the proportionality constant, is called the gas constant. Eq. (i) is called the ideal gas equation.

The Gas Constant: As one mole of a ideal gas occupies a volume of 22.414 dm^3 at N.T.P. (0°C ad 1 atmp pressure), we find from Eq. (i) the gas constant

$$R = \frac{PV}{nT} = \frac{(1\text{ atm})(22.414\text{ dm}^3)}{(1\text{ mol})(273.15\text{ K})}$$

$$= 0.08206\text{ dm}^3\text{ atm K}^{-1}\text{ mol}^{-1}$$

In SI units, 1 atom = 1.01325×10^5 N m^{-2}, so that

$$R = \frac{PV}{nT}$$

$$= \frac{(1.03125 \times 10^5\ N\ m^{-2})(22.414 \times 10^{-3}\ m^3)}{(1\ mol)(273.15\ K)}$$

$$= 8.314\ N\ m\ K^{-1}\ mol^{-1}$$

$$= 8.314\ J\ K^{-1}\ mol^{-1} \qquad (\because\ J = Nm)$$

Note: $1\ dm^3 = 1 \times 10^{-3}\ m^3 = 1000\ cm^3 = 1$ litre.

(e) Dalton's Law of partial pressure: This law states, "*Total pressure P of a number of gases i a enclosed space is equal to the sum of the individual (partial) pressures of each gas, provided the gases do not react with one another.*" Thus

$$P = p_1 = p_2 + p_3 + p_4 + \dots + p_N$$

Gas pressure is the force per unit area exerted by the molecules of the gas against the walls of the container.

Suppose N_1 molecules, each of mass m_1 of a gas A, are contained in a vessel of volume V.

The according to the kinetic theory equation the pressure is given by

$$p_a = \frac{m_1 N_1 \bar{c}_1^2}{3V} \qquad \text{...(i)}$$

where $\bar{c}_1$ is the root mean square velocity of the molecules of the gas A.

Now, suppose N_2 molecules, each of mass m_2 of another gas B, are contained i the same vessel at the same temperature and there is no other gas present at that time. The pressure of the gas will b given by

$$p_b = \frac{m_2 N_2 \bar{c}_2^2}{3V} \qquad \text{...(ii)}$$

where $\bar{c}_2$ is root mean square velocity of the molecules of the gas B.

If both the gases are present in the same vessel, the total pressure P will be give by

$$P = \frac{m_1 N_1 \bar{c}_1^2}{3V} + \frac{m_2 N_2 \bar{c}_2^2}{3V} = p_a + p_b$$

Similarly if three, four or ore gases were present, the total pressure P will be given by

$$P = p_a + p_b + p_c + p_d \cdots$$

This is Dalton's law of partial pressures.

(f) Gaham's Law of Gaseous Diffusion: The law may be stated "*Under similar conditions of temperature and pressure, the rates of diffusion of gases are inversely proportional to the square roots of their densities.*"

Mathematically,

$$\frac{r_1}{r_2} = \sqrt{\frac{D_2}{D_1}}$$

According to the kinetic equation

$$PV = \frac{1}{3} m\, N\, \bar{c}^2$$

or

$$\bar{c} = \sqrt{\frac{3PV}{mN}}$$

$$= \sqrt{\frac{3PV}{M}} \qquad \text{[mN = molecular weight of the gas.]}$$

$$= \sqrt{\frac{3P}{D}} \qquad [\because \; D = \frac{M}{V}]$$

$$\bar{c} = \sqrt{\frac{1}{D}} \text{ if P is kept constant.}$$

$\bar{c}$ is the mean velocity of the molecules and rate of diffusion of a gas will be dependent upon it at a particular temperature.

$$r \propto \bar{c}$$

$$r \propto \sqrt{\frac{1}{D}} \text{ if P is kept constant.}$$

$\bar{c}$ is the mean velocity of the molecules and rate of diffusion of a gas will be dependent upon it at a particular temperature.

$$r \propto \bar{c}$$

$$r \propto \sqrt{\frac{1}{D}}$$

This is Graham's Law.

Calculation of Molecular Velocity

In calculating various properties of the gases, one makes use of the following three velocities.

(i) Root mean Square velocity (RMSV): *It may be defined as the square root of the mean of the velocities of the molecules.* If v_1 v_2 ... etc. are the velocities possessed by different molecules, then

$$\sqrt{\bar{c}^2} = \sqrt{\frac{v_1^2 + v_2^2 + \ldots + v_N^2}{N}}$$

The expression to calculate this velocity can be obtained from the kinetic gas equation, according to which

$$PV = \frac{1}{3} mN\bar{c}^2$$

For 1 mole of the gas

$$PV = RT \text{ and } N = NA \text{ thus}$$

$$\sqrt{\bar{c}^2} = \sqrt{\frac{3PV}{mNa}} = \sqrt{\frac{3RT}{mNa}} = \sqrt{\frac{3RT}{M}} = \sqrt{\frac{3kT}{m}} = \sqrt{\frac{3P}{d}}$$

where K is Boltzmann constant k = R/Na

(2) Average Velocity (c): *It is defined as the average of the various velocities possessed by the molecules, i.e., as*

$$\bar{c} = \frac{v_1 + v_2 + \ldots + v_N}{N}$$

It can be calculated by using the expression

$$\bar{c} = \sqrt{\frac{8RT}{\pi M}}$$

$$\bar{c} = \sqrt{\frac{8RT}{\pi M}} = \sqrt{\frac{8kT}{\pi m}}$$

SOLVED EXAMPLES

Example 1:

A system consists of 6000 particles distributed in three energy states with equal spacing. The energy of the three states are $E_1 = 0$, $E_2 = x$ and $E_3 = 2x$. All the three slates have the same intrinsic probability g. At a certain instant there are 3000 particles in lower level, 2500 in the middle level and 500 in the upper level. Compare the relative probabilities with the distribution obtained, by the transfer of one particle from the middle to the lower level and one particle from the middle to the upper level and the original distribution.

Solution:

Let P_1 and P_2 be the probabilities in the two cases

In the first case,

$$N = 6000,$$

$$n_1 = 3000,$$

$$n_2 = 2500,$$

$$n_3 = 500.$$

$$P_1 = \frac{g^N}{n_1!n_2!n_3!}$$

$$P_1 = \frac{g^{6000}}{3000!2500!500!} \qquad ...(i)$$

In the second case

$$N = 6000,$$

$$n_1 = 3001,$$

$$n_2 = 2498,$$

$$n_3 = 501.$$

$$P_1 = \frac{g^N}{n_1!n_2!n_3!}$$

$$P_1 = \frac{g^{6000}}{3001!2498!501!}$$

Dividing (ii) by (i)

$$\frac{P_2}{P_1} = \frac{3000!2500!500!}{3001!2498!501!}$$

$$\frac{P_2}{P_1} = \frac{2500 \times 2499!}{3001 \times 501!}$$

$$\frac{P_2}{P_1} = 4.157$$

$$\approx 4.2.$$

It means the transfer of one particle from the middle to the upper and one particle from the middle to the lower state has changed the probability by a factor 4.2. This shows that both these distributions are not near the equilibrium state.

Example 2:

For an ideal gas (PV = nRT) show that (1/T) is an integrating factor for dw = P dV.

Solution:

$$dw = P\, dV \qquad ...(1)$$

$\because$ $$PV = nRT$$

$\therefore$ $$p\, dV + V\, dP = nR\, dT$$

and $$dV = \frac{nR}{P} dT - \frac{V}{P} dP \qquad ...(2)$$

On substituting the expression for dV in equation (1), we obtain

$$dw = nR\, dT - V\, dP$$

$$z = nR\, dT - \frac{nRT}{P} dP$$

Now $$\frac{\partial}{\partial P}(nR)_T = 0$$

and $$\frac{\partial}{\partial T}\left(-\frac{nRT}{P}\right)_P = -\frac{nR}{P}$$

Hence, dw is inexact differential.

Suppose f = 1/T be an integrating factor. Then

$$f\,.\,dw = \frac{1}{T} nR\, dT - \frac{nR}{P} dP$$

$\therefore$ $$\frac{\partial}{\partial P}\left[\frac{nR}{T}\right]_T = 0$$

and $\frac{\partial}{\partial T}\left[\frac{nR}{P}\right]_T = 0$

Thus f. $dw \frac{nR}{T} dT - \frac{nR}{P} dP$ is an exact differential while 1/T is an integrating factor.

Example 3:

The molar heat capacity at capacity at constant pressure of solid magnesium (0° – 600°C) is expressed by

$$CP = 6.20 + 1.30 \times 10 - 3\ T - 6.80 \times 104\ T - 2$$

Calculate the increase in entropy when 1 gm atom of metal is heated from 27°C to 127°C at constant pressure.

Solution:

At constant pressure,

$$\Delta S = \int_{T_1}^{T_2} \frac{C_P}{T} dT$$

$$\therefore \quad \Delta S = \int_{300}^{400} \frac{\left(6.20 + 1.30 \times 10^{-3} T - 6.80 \times 10^4 T^{-2}\right) dT}{T}$$

$$\int_{300}^{400} 6.20 \frac{dT}{T} + \int_{300}^{400} 1.3 \times 10^{-3} dT - \int_{300}^{400} 6.20 \times 10^4 \frac{dT}{T^3}$$

$$= 6.20 \ln[\ln T]_{300}^{400} + 1.3 \times 10^{-3} [T]_{300}^{400} - \frac{6.80 \times 10^4}{-2}\left[\frac{1}{T^2}\right]_{300}^{400}$$

$$= 6.20 \ln \frac{400}{300} + 1.3 \times 10^{-3} \times (400 - 300) + 3.40 \times 10^4$$

$$\left[\frac{1}{(400)^2} - \frac{1}{(300)^2}\right]$$

$$= 1.7487 \text{ cal./degree.}$$

Example 4:

Calculate the increase in entropy of three moles of hydrogen as it changes from 300°K atm to 1000°K and 1 atm (C_P = 7 cal./degree/ mole).

Solution:

Entropy change ΔS for n moles

$$= n\, C_P \ln \frac{T_2}{T_1}\; n\, R \ln \frac{P_2}{P_1}$$

$$= 3 \times 7 \times 2.303 \log \frac{1000}{300} - 3 \times 2 \times 2.303 \log \frac{10}{0.1}$$

$$= 25.287996 - 13.818$$

$$= 11.469996 \text{ cal./degree.}$$

Example 5:

At low temperature, a nonideal gas follows the relation $PV = RT - a/V$ where $a = 0.3636$ Nm^4 $mole^{-2}$. Calculate the work done by one mole of this gas in expanding form 0.224 litre to 22.4 litre at 400 K. Compare this result with the work done in the corresponding expansion of an ideal gas.

Solution:

For an ideal gas, reversible work done

$$= 2.303 \text{ n RT} \log \frac{V_2}{V_1}$$

$V_1 = 0.224$ litres $mole^{-1} = 22.4 \times 10^{-5}$ m^3 $mole^{-1}$

$P_2 = 22.4$ litres $mole^{-1} = 22.4 \times 10^{-3}$ m^3 $mole^{-1}$

$R = 8.314$ J K^{-1} $mole^{-1}$

$$W_{rew} = (2.303)\,(1 \text{ mole})\,(8.314 \text{ J K}^{-1} \text{ mole}^{-1})\,(400 \text{ K}) \times \log \frac{22.4 \times 10^{-3}}{22.4 \times 10^{-5}}$$

$$= (2.303)\,(1 \text{ mole})\,(8.314 \text{ J K}^{-1} \text{ mole}^{-1})\,(400 \text{ K}) \log 100$$

$$= 15320 \text{ J mole}^{-1}$$

$$= 15.32 \text{ kJ mole}^{-1}.$$

For the non-ideal gas $PV = RT - a/V$

Therefore, the work done $= \int_{V_1}^{V_2} PdV = \int_{V_1}^{V_2} \left\{\frac{RT}{V} - \frac{a}{V_2}\right\} dV$

$$= 2.303 \text{ RT} \log \frac{V_2}{V_1} + a\left\{\frac{1}{V_2} - \frac{1}{V_1}\right\}$$

The value of first part is from derived eq. (1) as 15.32 kJ $mole^{-1}$ and of another part can be derived as

$$0.3636\ Nm^4\ mole^{-2}\left\{\frac{1}{22.4 \times 10^{-3}\ m^3\ mole^{-1}} - \frac{1}{22.4 \times 10^{-5}\ m^3\ mole^{-1}}\right\}$$

$$= 1.604\ kJ\ mole^{-1}.$$

Thus work done = 15.320 k $Joule^{-1}$ – 1.604 k $Joule^{-1}$

= 13.716 kJ $mole^{-1}$.

Example 6:

What will be the work done when 65.38 g of zinc dissolves in hydrochloric acid in case of an open beaker and closed beaker at 300K.

Solution:

The process of dissolution of zinc in hydrochloric acid can be given as

$$Zn(s) + 2HCI\ (aq) = ZnCl_2(aq) + H_2(g)$$

From this equation it is clear that for each gram atom of zinc dissolved we will obtain 1 mole of Hydrogen gas. As a result of liberated gas, on the surrounding atmosphere the work performed will be equal to PΔV. If the initial volume of the system is neglected in comparison to the total volume of the gas produced and the gas shows ideal behaviour, then the work done in the open beaker is

$$W = P\Delta V \approx PV_{H2}$$

$$= nH2\ RT \qquad [\because T = 300\ k\ R = 8.314\ Jk^{-1}\ mole^{-1}]$$

$$= (1\ mole)\ (8.314\ JK^{-1}\ mol^{-1})\ (300\ K)$$

$$= 2.4942\ kJ.$$

If the reaction takes place in a closed beaker there is change in the volume *i.e.*, ΔV = 0 hence work done is zero.

Example 7:

Calculate the work done when one mole of sulphur dioxide gas expands isothermally and reversibly at 300 k from 2.46 × 10^{-3} m^3 to 24.6 × 10^{-3} m^3. Assuming that the gas obeys

(a) ideal gas equation and

(b) van der Waals equation

$$\left\{\begin{array}{l} a = 0.6799\ nm^4\ mole^{-2} \\ b = 0.564 \times 10^{-3}\ mole^{-1} \end{array}\right\}$$

Solution:

(a) The reversible work done by 1 mole of gas when it expands form

$$2.64 \times 10^{-3} m^3 \text{ to } 24.6 \times 10^{-3} m^3 = 2.303\ RT \log \frac{V_2}{V_1}$$

$$= (2.303)\ (1\ mole)\ (8.314\ J\ K^{-1}\ mol^{-1})\ (300\ K) \times \log 10$$

$$\left\{\because \log \frac{V_2}{V_1} = \log 10\right\}$$

$$= (2.303)\ (8.314)\ (300\ K)\ \log 10\ J\ mol^{-1}$$

$$= 5744\ J\ mol^{-1}$$

(b) When the gas obeys van der Waal's equation, work done by 1 mole of gas is

$$= 2.303\ RT \log \frac{V_2 - b}{V_1 - b} + a \left\{\frac{1}{V_2} - \frac{1}{V_1}\right\}$$

After substituting the values of various quantities and solving, the work done is

$$= 5794\ J\ mol^{-1} - 252.4\ J\ mol^{-1}$$

$$= 5541.6\ J\ mol^{-1}$$

$$= 5.542\ kJ\ mol^{-1}.$$

Example 8:

One mole of a van der Walls's gas is allowed to expand isothermally and reversibly from a volume of 1 litre to 50 litre at 0°C. Calculate w q, Δ E and ΔH. van der Waal's constants are a = 6.5 atm $l^2 mole^{-2}$ = 0.056 litre mole $^{-1}$ and R = 0.082 l atom deg^{-1} $mole^{-1}$.

Solution:

Here V_1 = 1 litre,

V_2 = 50 litres, n = 1 mole

T = 273 + 0 = 273°k,

$$R = 0.082 \text{ l-atm deg}^{-1} \text{ mole}^{-1}$$

$$a = 6.5 \text{ atm } l^2 \text{ mole}^{-2},$$

$$b = 0.056 \text{ litre mole}^{-1}$$

$$w = 2.303 \text{ nRT} \log\left(\frac{V_2 - nb}{V_1 - nb}\right) + an^2\left(\frac{1}{V_2} - \frac{1}{V_1}\right)$$

$$= 2.303 \times 1 \times 0.082 + 273 \, [\log(50 - 0.056) - \log(1 - 0.056)$$

$$+ 6.5\left(\frac{1}{50} - \frac{1}{1}\right)$$

$$= 82.47 \text{ l atm.}$$

$$\Delta E = -an^2\left(\frac{1}{V_2} - \frac{1}{V_1}\right) = 6.5 \times \frac{49}{50}$$

$$q = 2.303\text{nRT} \log \frac{V_2 - b}{V_1 - b}$$

$$= \Delta E + w = 884 \text{ l atm.}$$

Example 9:

The van der Waal's constant a and b for hydrogen in litre atmosphere units are 0.246 and 2.67 × 10^{-2} respectively. Calculate the inversion temperature of hydrogen.

Solution:

we know $T_i = 2a/Rb$

or $$T_i = \frac{2 \times 0.246}{0.0821 \times 0.0267}$$

$$= 224.5°k = (224.5 - 273) = -48.5°C.$$

Example 10:

For equation $(P \_ a/V^2)\ V = RT$, prove that (i) dP is an exact differential, (ii) P is a state function and

$$\text{(iii)} \left(\frac{\partial P}{\partial T}\right)_v \left(\frac{\partial T}{\partial V}\right)_P \left(\frac{\partial V}{\partial T}\right)_T + 1 = 0.$$

Solution:

(i) For dP to be an exact differential, it is be proved that

$$\frac{\partial^2 P}{\partial V \partial T} = \frac{\partial^2 P}{\partial T \partial V} \qquad ...(1)$$

$$(P + a/V^2)\, V = RT$$

or
$$P = \frac{RT}{V} - \frac{a}{V^2} \qquad ..\ 2)$$

When the above equation is differentiated with respect to V at constant temperature T, we obtain

$$\left(\frac{\partial P}{\partial V}\right)_T = -\frac{RT}{V^2} + \frac{2a}{V^3}$$

When the above equation is differentiated with respect to T at constant V, we obtain

$$\frac{\partial^2 P}{\partial T \partial V} = -\frac{RT}{V^2} \qquad ...(3)$$

When equation (2) is differentiated with respect to T at constant volume V, we obtain

$$\left(\frac{\partial P}{\partial T}\right)_V = \frac{R}{V}$$

When the above equation is further differentiated with respect to V at constant T, we obtain

$$\frac{\partial^2 P}{\partial T \partial V} = -\frac{R}{V^2} \qquad ...(4)$$

From equations (3) and (4), we get

$$\frac{\partial^2 P}{\partial T\, \partial V} = \frac{\partial^2 P}{\partial T\, \partial V} = -\frac{R}{V^2} \qquad ...(5)$$

Thus, we have proved condition, (1). This shows that dP is an xact differential.

(ii) In the said question, first of all a change occurs in volume at constant temperature followed by a change in temperature at constant volume, thereby giving the final equation (3).

Again, a change in temperature is considered at constant volume, followed by a change in volume at constant temperature, thereby giving the final equation (4). Both equations (3) and (4) are same, thereby revealing that the final change in P remains the same irrespective of the difference in intermediate changes. Therefore, P is a state function as changes in P have been found to depend upon the intial and final states of the system.

(iii) Equation (2) is as follows :

$$P = \frac{RT}{V} - \frac{a}{V^2}$$

When the above equation is differentiated, we obtain :

$$dP = \frac{R}{V}dT - \frac{RT}{V^2}dV + \frac{2a}{V^3}dV$$

$$= \frac{R}{V}dT - \left(\frac{RT}{V^2} - \frac{2a}{V^3}\right)dV \qquad ...(6)$$

If volume is constant, dV = 0 : equation (6) becomes as follows :

$$dP = \frac{R}{V}dT \text{ or } \left(\frac{\partial P}{\partial V}\right)_V = \frac{R}{V} \qquad ...(7)$$

If pressure is constant, dP = 0; equation (6) becomes as follow :

$$\frac{R}{V}dT - \left(\frac{RT}{V^2} - \frac{2a}{V^3}\right)dV = 0$$

or

$$\left(\frac{\partial T}{\partial V}\right)_P = \left(\frac{RT}{V^2} - \frac{2a}{V^3}\right)\Big/\frac{R}{V} \qquad ...(8)$$

If temperature is constant, dT = 0; equation (6) becomes as follows:

$$\left(\frac{\partial T}{\partial V}\right)_T = -\frac{1}{\left(\frac{RT}{V^2} - \frac{2a}{V^3}\right)} \qquad ...(9)$$

On multiplying equations (7), (8) and (9), we obtain

$$\left(\frac{\partial p}{\partial T}\right)_V\left(\frac{\partial T}{\partial V}\right)_P\left(\frac{\partial V}{\partial P}\right)_T = -\frac{R}{V} \times \frac{\left(\frac{RT}{V^2} - \frac{2a}{V^3}\right)}{\left(\frac{RT}{V^2} - \frac{2a}{V^3}\right)} \times \frac{1}{R} = -1$$

or $$\left(\frac{\partial p}{\partial T}\right)_V\left(\frac{\partial T}{\partial V}\right)_P\left(\frac{\partial V}{\partial P}\right)_T + 1 = 0$$

This is cyclic rule.

Example 11:

For an ideal gas PV = RT, prove that dP is an exact differential.

Solution:

We know $PV = RT$

or $$P = \frac{RT}{V} \text{ (R is a constant)} \qquad ...(1)$$

On differentiating with respect to T, taking V constant, we obtain

$$\left(\frac{\partial P}{\partial T}\right)_P = \frac{RT}{V} \quad ...(2)$$

Again, on differentiating with respect to V, we obtain

$$\frac{\partial^2 P}{\partial V \partial T} = -\frac{R}{V^2} \quad ...(3)$$

On differentiating (1) with respect to volume, taking T constant, we obtain

$$\left(\frac{\partial P}{\partial V}\right)_T = -\frac{RT}{V^2} \quad ...(4)$$

Again, on differentiating above equation with respect to temperature taking V as constant, we obtain

$$\frac{\partial^2 P}{\partial T \partial V} = -\frac{R}{V^2} \quad ...(5)$$

From equations (3) and (5), we obtain

$$\frac{\partial^2 P}{\partial T \partial V} = \frac{\partial^2 P}{\partial T \partial V}$$

Thus dP is an exact differential.

Example 12:

For an ideal gas PV = RT, prove that DT is an exact differential.

Solution:

$$PV = RT \quad ...(1)$$

In order to prove that dT is a perfect differential, we have to prove that

$$\frac{\partial^2 T}{\partial V \partial P} = \frac{\partial^2 T}{\partial P \partial V}$$

On differentiating Eq. (1), we obtain

$$\left(\frac{\partial T}{\partial V}\right)_P = \frac{P}{R}; \text{i.e.,} \frac{\partial^2 T}{\partial P \partial V} = \frac{1}{R}$$

Again $$\left(\frac{\partial T}{\partial P}\right)_V = \frac{V}{R}; \text{i.e.,} \frac{\partial^2 T}{\partial V \partial P} = \frac{1}{R}$$

Therefore, $$\frac{\partial^2 T}{\partial V \partial P} = \frac{\partial^2 T}{\partial P \partial V}$$

Thus, dT is an exact differential.

Example 13:

For an ideal gas (PV = nRT) show that (1/T) is an integrating factor for dw = P dV.

Solution:

$$dw = P\,dV \quad ...(1)$$

$\because$ $$PV = nRT$$

$\therefore$ $$p\,dV + V\,dP = nR\,dT$$

and $$dV = \frac{nR}{P}dT - \frac{V}{P}dP \quad ...(2)$$

On substituting the expression for dV in equation (1), we obtain

$$dw = nR\,dT - V\,dP = nR\,dT - \frac{nRT}{P}dP$$

Now $$\frac{\partial}{\partial P}(nR)_T = 0$$

and $$\frac{\partial}{\partial T}\left(-\frac{nRT}{P}\right)_P = -\frac{nR}{P}$$

Hence, dw is inexact differential.

Suppose f = 1/T be an integrating factor. Then

$$f\,.\,dw = \frac{1}{T}nR\,dT - \frac{nR}{P}dP$$

$\therefore$ $$\frac{\partial}{\partial P}\left[\frac{nR}{T}\right]_T = 0$$

and $$\frac{\partial}{\partial T}\left[\frac{nR}{P}\right]_T = 0$$

Thus f. dw $\frac{nR}{T}dT - \frac{nR}{P}dP$ is an exact differential while 1/T is an integrating factor.

Example 14:

A thermostat was maintained at 370.05 K. For nearly an hour 4.2 kJ of heat leaked through the thermostat insulation into a room where the initial temperature of air was 300.05 K. (i) What was the entropy change of the material in the thermostat ? (ii) What was the entropy change of the air in the room ? (iii) Was the process spontaneous? (Material in the thermostat is water).

Solution:

(i) Entropy change of the material in the thermostat

$$-\frac{4.2\ \text{kJ}}{370.05\text{K}} = -11.35\times 10^3\ \text{kj K}^{-1}$$

(ii) Entropy change of the air in the room

$$= \frac{4.2\ \text{kJ}}{300.05\ \text{K}} = 14.00\times 10^3\ \text{kJ K}^{-1}$$

(iii) Now, $\Delta S_{total} = \Delta S_{thermostat} + \Delta S_{room}$

$$= (-11.35 + 14.00)\ \text{JK}^{-1}$$

$$= 2.65\ \text{J K}^{-1}$$

Since ΔStotal is positive, the process of leaking is spontaneous.

Example 15:

Calculate ΔS, ΔH and ΔU for the process

$$H_2O(l,\ 20°C,\ 1\ atm \rightarrow H_2O\ (g,\ 250°C,\ 1\ atm)$$

Given the following data :

$$C_P\ (l) = 75.6\ J\ K^{-1}\ mol^{-1}$$

$$C_P\ (g) = 36.2\ J\ K^{-1}\ mol^{-1}$$

ΔH for vaporization of H_2O at 100°C and 1 atm is equal to 40.8 k J mol^{-1}.

Solution:

The given process may be replaced by the following processes

(i) H_2O (l, 20°C, 1 atm) → H_2O (l, 100°C, 1 atm)

(ii) H_2O (l, 100°C, 1 atm) → H_2O (g, 100°C, 1 atm)

(iii) H_2O (g, 100°C, 1 atm) → H_2O (g, 250°C, 1 atm)

Thus we have

$$\Delta S = \Delta S_{(1)} + \Delta S_{(ii)}\ \Delta S_{(iii)}$$

$$= C_{p,m}(l)\ \ln\frac{T_2}{T_1} + \frac{\Delta H_{vap}}{T} + C_{p,m}(g)\ \ln\frac{T_2'}{T_1'}$$

$$= \left(75.6 \ln \frac{373}{293} + \frac{40850}{373} + 36.2 \ln \frac{523}{373}\right) \text{JK}^{-1} \text{ mol}^{-1}$$

$$= (18.25 + 109.52 + 12.24) \text{ J K}^{-1} \text{ mol}^{-1}$$

$$= 140.01 \text{ J K}^{-1} \text{ mol}^{-1}$$

$$\Delta H = \Delta H_{(1)} + \Delta H_{(ii)} + \Delta H_{(iii)}$$

$$= C_{p,m} (l) (T_2 - T_1) + \Delta H_{vap} + C_{pm}(g) (T'_2 - T'_1)$$

$$= (75.6 \times 80 + 40850 + 36.2 \times 150) \text{ J mol}^{-1}$$

$$= 52328 \text{ J mol}^{-1}$$

$$\Delta U = \Delta U_{(1)} + \Delta U_{(ii)} \Delta U_{(iii)}$$

$$= C_{v,m} (l) (T_2 - T_1) + \Delta U_{vap} + C_{v,m}(g) (T'_2 - T'_1)$$

$$= C_{p,m}(l)(T_2 - T_1) + \{\Delta H_{vap} + \Delta_{V\,g} RT\} + \{C_{p,m}(g) - R\}(T'_2 - T'_1)$$

$$= [75.6 \times 80 + \{40850 - (8.314)(373)\}$$

$$+ (36.2 - 8.314)(150)] \text{ J mol}^{-1}\}$$

$$= (6040 + 37749 + 4183) \text{ J mol}^{-1}$$

$$= 47972 \text{ J mol}^{-1}.$$

Example 16:

Describe a process through with the gas can be restored to its initial state in each of the above cases. Explain how the surroundings can be restored only in one ca. e and not in the other.

Solution:

The gas can be restored to its initial state by compressing it isothermally from 5 dm^3 to 1 dm^3. This may be done reversibly or irreversibly. When the compression is done reversibly, the surroundings in the case of (ii) will be restored whereas in case of (i), it cannot be restored whether the compressions is done reversibly or irreversibly. This is because, while on expansion. surroundings remain unchanged whereas during compression, it has to do work on the system and thus suffers changes.

Example 17:

Derive an expression for the entropy change when a van der Waals gas is heated and expanded simultaneously.

Solution:

Taking S = f(T, V), we get

$$dF = \left(\frac{\partial S}{\partial T}\right)_V dT + \left(\frac{\partial S}{\partial V}\right)_T dV$$

$$= \frac{nC_{V,m}}{T} dT + \left(\frac{\partial P}{\partial T}\right)_V dV$$

Now for van der Waal's gas we have

$$P = \frac{n\,RT}{V - nb} - \frac{n^2 a}{V^2}$$

Therefore $\left(\frac{\partial P}{\partial T}\right)_V = \frac{n\,RT}{V - nb}$

Thus $dS = \frac{n\,C_{V,m}}{T} dT + \frac{n\,R}{V - nb} dV$

Which on integrating within the limit T_1, V_1 and T_2, V_2 gives

$$\Delta S\; nC_{V,m} \ln \frac{T_2}{T_1} + nR \ln \frac{V_2 - nb}{V_1 - nb}$$

Example 18:

For an adiabatic process, dq = 0, If one were to write

$$\Delta S = \int \frac{dq}{T} = \int \frac{0}{T} = 0$$

Then every adiabatic process would be iso-entropic process. Comment.

Solution:

For an adiabatic process, dq – 0, and according to the first law of thermodynamics, dq = dU – dw

Thus, we will have

$$dU = dw$$

The expression of entropy change will be given as

$$\Delta S = \int \frac{dU - dw}{T}$$

$$= \int \frac{dU}{T} + \int \frac{P_{opp}}{T} dV$$

Now, it is only for reversible process the first integral is equal and opposite sign of the second integral and thus $\Delta S = 0$. For an irreversible process, these two integrals do not have equal magnitude and thus sum is not zero. Hence $\Delta S \neq 0$ for an adiabatic irreversible process.

Example 19:

Calculate the maximum work obtainable by (i) the isothermal expansion and (ii) the adiabatic expansion of 2 mol of on ideal gas initially at 25° C form 10 L to 20 L. Assume Cv = (5/2) R.

Solution:

(i) Maximum work is obtained in reversible expansion. Hence

$$w = -\ nRT\ ln\ \frac{V_2}{V_1}$$

$$= -\ (2\ mol)\ (8.314\ J\ K^{-1}\ mol^{-1})\ (298\ K)\ 2.303\ \log\ \frac{20L}{10L}$$

$$= -\ 3435\ J$$

(ii) Firstly, we calculate the final temperature using the expression

$$T_2\ V_2^{\ \gamma-1} = T_1\ V_1^{\gamma-1}$$

where $$\gamma = \frac{C_P}{C_V} = \frac{(7/2)R}{(5/2)R} = \frac{7}{5}$$

Hence, $$T_2 = (298K)\left(\frac{10L}{20L}\right)^{(7/5)-1}$$

$$= (298\ K)\ (0.7578)$$

$$= 225.8\ K$$

Now $$W_{adi} = \Delta E$$

$$= n\ C_{V},\ m(T_2 - T_1)$$

$$= (2mol)(2.5 \times 8.314\ JK^{-1}mol^{-1})(225.8K - 298K)$$

$$= -\ 3001\ J.$$

Example 20:

How much heat is required to raise the temperature of 1 mole oxygen from 300K to 1300 K at constant pressure.

$$C_p = 6.095 + 3.325 \times 10^{-3}\ T - 1.017 \times 10^{-6}\ T^2.$$

Solution:

$$H_2 - H_1 \int_{T_t}^{T_2} C_p \, dT$$

$$T_1 = 300 \text{ k}, \; T_2 = 100 \text{ k}$$

$$\therefore \; H_{1300} - H_{300}$$

$$= \int_{300}^{1300} \left(6.095 + 3.253 + 10^{-3}\,T - 1.017 \times 10^{-6}\,T^2\right)$$

[Substituting the value of C_p]

$$\because \; H_{1300} - H_{300} = 6.095\,(1300 - 300) + \frac{3.253 \times 10^{-3}}{2}$$

$$\times \left[(1300)^2 - (300)^2\right] - 1/3 \times 1.017 \times 10^{-6} \left[(1000)^2 - (300)^2\right]$$

$$= 6095 + 2602.4 - 735.63$$

$$= 7961.77 \text{ cal/mole.}$$

Example 21:

Show that

$$\left(\frac{\partial \alpha}{\partial P}\right)_T + \left(\frac{\partial \beta}{\partial T}\right)_P = 0$$

where α *and* β *are coefficient of thermal expansion and coefficient of compressibility, respectively.*

Solution:

Since V is a state function, we can write

$$\frac{\partial^2 V}{\partial T\,\partial P} = \frac{\partial^2 V}{\partial P\,\partial T}$$

or

$$\frac{\partial}{\partial T}(-V\beta)_P = \frac{\partial}{\partial T}(V\alpha)_T$$

Carrying out the differentiation, we get

$$-\beta\left(\frac{\partial V}{\partial T}\right)_P - V\left(\frac{\partial \beta}{\partial T}\right)_P = \alpha\left(\frac{\partial V}{\partial P}\right)_T + V\left(\frac{\partial \alpha}{\partial P}\right)_T$$

that is

$$\left(\frac{\partial \alpha}{\partial P}\right)_T + \left(\frac{\partial \beta}{\partial T}\right)_P = -\frac{\beta}{V}\left(\frac{\partial V}{\partial T}\right)_P - \frac{\alpha}{V}\left(\frac{\partial V}{\partial P}\right)_T$$

$$= -\frac{\beta}{V}(\alpha V) - \frac{\alpha}{V}(-\beta V) = 0$$

Example 22:

Show that Joule Thomson coefficient, $\mu_{JT} = 0$ for an ideal gas. Comment on the liquefaction of an ideal gas.

Solution:

By definition, Joule-Thomson coefficient is defined as

$$\mu_{JT} = \left(\frac{\partial T}{\partial P}\right)_H$$

Using the cyclic rule

$$\left(\frac{\partial T}{\partial P}\right)_H \left(\frac{\partial P}{\partial H}\right)_T \left(\frac{\partial H}{\partial T}\right)_P + 1 = 0$$

we get $$\left(\frac{\partial T}{\partial P}\right)_H = -\frac{1}{\left(\frac{\partial P}{\partial H}\right)_T \left(\frac{\partial H}{\partial T}\right)_P} = -\frac{\left(\frac{\partial H}{\partial P}\right)_T}{C_P}$$

Now using, thermodynamic equation of state

$$\left(\frac{\partial T}{\partial P}\right)_H = V - T\left(\frac{\partial V}{\partial T}\right)_P$$

we get $$\mu_{JT} = \frac{V - T\left(\frac{\partial V}{\partial T}\right)_P}{C_P}$$

Now, for an ideal gas

$$V = \frac{nRT}{P}$$

Hence, $$\left(\frac{\partial V}{\partial T}\right)_P = \frac{nR}{P} = \frac{V}{T}$$

Thus $$\mu_{JT} = \frac{V - T\left(\frac{V}{T}\right)}{C_P} = 0$$

The ideal gas cannot be liquefied as there will not occur any change in temperature when the gas undergoes Joule-Thomson experiment.

Example 23:

Calculate the work down when 50 g of iron dissolves in hydrochloric acid at 25oC in :

(i) a closed vessel and

(ii) an open beaker.

(iii) In which case would maximum amount of work be done ?

Solution:

$$\text{Amount of iron} = \frac{50 \text{ g}}{55.85 \text{ g mol}^{-1}} = \frac{50}{55.85} \text{ mol}$$

(i) In a closed vessel, $\Delta V = 0$. Hence,

$$dw = -PdV = 0$$

that is, no work is involved when iron dissolves in hydrochloric acid in a closed vessel.

(ii) The reaction between iron and hydrochloric acid is

$$Fe + 2HCl \rightarrow FeCl_2 + H_2(g)$$

Thus, the liberated hydrogen gas pushes the atmospheric air and thus involves the work of expansion. In this case, we will have

$$w = -P_{ext}\,\Delta V$$
$$= -P_{ext}\,V_{H2}$$

where V_{H2} is the volume of hydrogen released. According to the ideal gas law, we will have

$$w = -P_{ext}\,(n_{H2}RT/P_{ext})$$
$$= -n_{H2}RT$$
$$= -(50/55.85)\text{ mol }(8.314\text{ J K}^{-1}\text{ mol}^{-1})\,(298\text{ K})$$
$$= -2.2\text{ kJ}$$

Example 24:

(i) Calculate α and β for an ideal gas obeying gas equation PV = RT (for one mole).

Solution:

(i) We know

$$PV = RT$$

On differentiating the above equation, we obtain

$$PdV + VdP = RdT \qquad ...(1)$$

At constant pressure dP = 0, the above equation becomes as follows:

$$\left(\frac{\partial V}{\partial T}\right)_P = \frac{R}{V}$$

or
$$\alpha = \frac{1}{V}\left(\frac{\partial V}{\partial T}\right)_P = \frac{R}{VP} = \frac{1}{T} = T^{-1}$$

At constant temperature dT = 0; equation (1) becomes as follows:

$$\left(\frac{\partial V}{\partial P}\right)_T = -\frac{V}{P}$$

or
$$\beta = -\frac{1}{V}\left(\frac{\partial V}{\partial P}\right)_T = \frac{V}{VP} = \frac{1}{P} = P^{-1}$$

(ii) The van der Waal's equation is as follows :

$$\left(P + \frac{a}{V^2}\right)(v - b) = RT$$

or
$$PV^3 - PbV^2 + aV - ab - RTV^2 = 0 \qquad ...(2)$$

On differentiating the above equation with respect to T at constant pressure, we obtain

$$3\,PV^2\left(\frac{\partial V}{\partial T}\right)_P - 2PbV\left(\frac{\partial V}{\partial T}\right)_P + a\left(\frac{\partial V}{\partial T}\right)_P$$

$$- 2RTV\left(\frac{\partial V}{\partial T}\right)_P - RV^2 = 0$$

or
$$\left(\frac{\partial V}{\partial T}\right)_P = \frac{RV^2}{3PV^2 - 2PbV + a - 2RTV}$$

$$\alpha = \frac{1}{V}\left(\frac{\partial V}{\partial T}\right)_P$$

$$= \frac{1}{V}\left[\frac{RV^2}{3PV^2 - 2PVb + a - 2RTV}\right]$$

$$= \frac{R}{3PV - 2Pb + \dfrac{a}{V} - 2RT}$$

But the van der Waal's equation may be put as follows :

$$RT = PV - Pb + \frac{a}{V} - \frac{ab}{V^2}$$

$$\therefore \qquad \alpha = \frac{R}{3PV - 2Pb + \frac{a}{V} - 2\left(PV - Pb + \frac{a}{V} - \frac{ab}{V^2}\right)}$$

$$= R\left[PV - \frac{a}{V} + \frac{2ab}{V^2}\right]^{-1}$$

Again, on differentiating equation (2), with respect to P at constant temperature T, we obtain

$$3PV^2\left(\frac{\partial V}{\partial P}\right)_T + V - 2PbV\left(\frac{\partial V}{\partial P}\right)_T$$

$$- bV^2 + a\left(\frac{\partial V}{\partial P}\right)_T - 2RTV\left(\frac{\partial V}{\partial P}\right)_T = 0$$

$$\left(\frac{\partial V}{\partial P}\right)_T = \frac{V^2(b - V)}{3PV^2 - 2PbV + a - 2RTV}$$

But the van der Waal's equation may be put as follows :

$$RTV = PV^2 + a - PbV - ab/V$$

$$\therefore \qquad \left(\frac{\partial V}{\partial P}\right)_T = \frac{V^2(b - V)}{3PV^2 - 2PbV + a - 2\left(PV^2 + a - PbV - \frac{ab}{V}\right)}$$

$$= \frac{b - V}{P - \frac{a}{V^2} + \frac{2ab}{V^3}}$$

Now $\qquad \beta = -\frac{1}{V}\left(\frac{\partial V}{\partial P}\right)_T = -\frac{b - V}{PV - \frac{a}{V} + \frac{2ab}{V^2}}$

$$= -\frac{V - b}{P - \frac{a}{V} + \frac{2ab}{V^2}}$$

Example 25:

Derive $\qquad C_{P,m} - C_{V,m} = TV_m \dfrac{\alpha^2}{\beta}$

where α and β are the coefficient of thermal expansion and coefficient of compressibility, respectively.

Solution:

We start with H = f (T, P) and U = f (T, V) and write their differentials as

$$dH = \left(\frac{\partial H}{\partial T}\right)_P dT + \left(\frac{\partial H}{\partial P}\right)_T dP$$

$$dU = \left(\frac{\partial U}{\partial T}\right)_V dT + \left(\frac{\partial U}{\partial V}\right)_T dV$$

Now since $H = U + PV$

we have $dH = dU + P\,dV + V\,dP$

or $$\left(\frac{\partial H}{\partial T}\right)_P dT + \left(\frac{\partial H}{\partial P}\right)_T dP = \left(\frac{\partial U}{\partial T}\right)_V dT + \left(\frac{\partial U}{\partial V}\right)_T dV + P\,dV + V\,dP$$

Dividing by dT, keeping P constant, we get

$$\left(\frac{\partial H}{\partial T}\right)_P = \left(\frac{\partial U}{\partial T}\right)_V + \left(\frac{\partial U}{\partial V}\right)_T \left(\frac{\partial V}{\partial T}\right)_P + P\left(\frac{\partial V}{\partial T}\right)_P$$

Now using the thermodynamic equation of state

$$\left(\frac{\partial U}{\partial V}\right)_T = T\left(\frac{\partial P}{\partial T}\right)_V - P$$

we get $$\left(\frac{\partial H}{\partial T}\right)_P = \left(\frac{\partial U}{\partial T}\right)_V + T\left(\frac{\partial P}{\partial T}\right)_V \left(\frac{\partial V}{\partial T}\right)_P \quad ...(1)$$

Now, by definition

$$\alpha = \frac{1}{V}\left(\frac{\partial V}{\partial T}\right)_P \quad ...(2)$$

$$\beta = -\frac{1}{V}\left(\frac{\partial V}{\partial P}\right)_T \quad ...(3)$$

Using the cyclic rule

$$\left(\frac{\partial V}{\partial P}\right)_T \left(\frac{\partial P}{\partial T}\right)_V \left(\frac{\partial T}{\partial V}\right)_P + 1 = 0$$

we get $$\left(\frac{\partial V}{\partial P}\right)_T = -\left(\frac{\partial V}{\partial T}\right)_P \left(\frac{\partial T}{\partial P}\right)_V$$

Hence

$$\beta = \frac{1}{V}\left(\frac{\partial V}{\partial T}\right)_P \cdot \left(\frac{\partial T}{\partial P}\right)_V$$

and $$\frac{\alpha}{\beta} = \left(\frac{\partial P}{\partial T}\right)_V$$

Hence, Eq (1) becomes

$$C_{P,m} - C_{V,m} = T\left(\frac{\partial P}{\partial T}\right)_V\left(\frac{\partial V}{\partial T}\right)_P \qquad ...(4)$$

With the use of Eqs. (2) and (4), we get

$$C_{P,m} - C_{V,m} = TV\frac{\alpha^2}{\beta}.$$

Example 26:

When a bicycle tyre is inflated with hand pump, the temperature of the air inside increases. Explain.

Solution:

Work done on the system is converted into heat which increases the temperature of air inside the tyre.

Example 27:

Can q become a state function ?

Solution:

Heat exchange reversibly at constant volume (or at constant pressure) is equal to the change in internal energy (or enthalpy) of a system. Since the latter is a state function, it follows that in such conditions q also behaves like a state function.

Example 28:

One mole of hydrogen and nine moles of nitrogen are mixed at 291. K and I atm pressure. Assuming ideal behaviour for the gases, calculate the entropy of mixing per mole of the mixture formed. Would it make any difference if under similar conditions one mole of hydrogen is mixed with nine moles of oxygen.

Solution:

For ideal gases the entropy of mixing per mole of the mixture is given as follows

$$\Delta S_{mtxing} = -R\sum X_i \ln X_i$$

In this case $X_{H2} = 0.1$ and $X_{N2} = 0.9$

$$\Delta S\text{mixing} = -R\,[0.1 \ln 0.1 + 0.9 \ln 0.9]$$

$$= -(8.314\ JK^{-1}\ mole^{-1})\ (2.303)\ (0.1)$$

$$[\because R = 8.314\ \ JK^{-1}\ mol^{-1}]$$

$$\left[\log\frac{1}{10} + 9\log\frac{9}{10}\right]$$

$$= 2.704\quad JK^{-1}\ mol^{-1}$$

When hydrogen is mixed with oxygen, the entropy of mixing would remain unchanged provided oxygen also behaves ideally.

Example 29:

2 moles of an ideal monoatomic gas ($C_V = 20.91\ J\ mol^{-1}\ K^{-1}$) are heated from 700 K to 750 K. The volume of the gas changes from 0.08 m^3 to 0.8 m^3. Assuming that the heat capacity remains constant in this temperature range, calculate the entropy change for the system, the surroundings and the universe if the process is carried out (a) reversibly and (b) irreversibly by placing the system in contact with a reservoir at 500 K and allowing the gas to expand against a const. external pressure equal to the final pressure of the gas

Solution:

For the reversible transformation of an ideal gas the entropy change of the system is given as follows :

$$\Delta S_{system} = nC_V \ln\frac{T_2}{T_1} + nR\ln\frac{V_2}{V_1}$$

$$= (2\ mol)\ (20.9\ J\ mol^{-1}\ K^{-1})\ (2.303)\ \log\frac{750}{700})$$

$$+ (2\ mol)\ (8.314\ J\ mol^{-1}\ K^{-1})\ (\ 2.303)\ \log\frac{0.80}{0.08}$$

$$= 6.449 + 38.28 = 44.729\ JK^{-1}.$$

As the process is reversible, the entropy change of the surrounding is as follows.

$$\Delta S_{sur} = -\ 44.729\ \ JK^{-1}$$

and the entropy change of the universe is as follows

$$\Delta S_{univ} = \Delta S_{sys} + \Delta S_{surr} = 0$$

(b) In the irreversible transformation the entropy change of the system will be the same. Heat lost by the surrounding is as

$$-Q = \Delta E - W$$

$$\Delta E = -n\, C_V\, (750 - 700)$$

and $$\omega = P_{ext}\, (V_2 - V_1).$$

Thus according to the final state

$$T_2 = 750 \text{ K}$$

$$V_2 = 0.8 \text{ m}^3$$

and hence the pressure is given at follows

$$P_2 = \frac{nRT_2}{V_2} = \frac{(2 \text{ mol})\left(8.314 \text{ J mol}^{-1} \text{ K}^{-1}\right)(750\text{K})}{\left(0.8 \text{ m}^3\right)}$$

$$= 15588.75 \text{ Jm}^{-3}.$$

Now $P_{ext} = P_2 = 15588.75 \text{ Jm}^{-3}$ and hence the heat lost by the surrounding is

$$-Q = -(2 \text{ mol})(20.9 \text{ J mol}^{-1} \text{ K}^{-1})(50 \text{ K})$$

$$-(15588.75 \text{ Jm}^{-3})(0.8 \text{ m}^3 - 0.08 \text{ m}^3)$$

$$= -2090 \text{ j} - 11223.9$$

$$= 13313.9 \text{ J}.$$

Let us assume that this heat is lost reversibly by the surroundings. Its entropy change is given as

$$\Delta S_{surr} = \frac{13313.9 \text{ j}}{500 \text{ K}} = 26.6278 \text{JK}^{-1}$$

and the entropy change of the universe is

$$\Delta S_{univ} = \Delta S_{sys} + \Delta S_{surr}$$

$$= (44.729 - 26.6278) \text{ JK}^{-1}$$

$$= 18.1012 \text{ JK}^{-1}.$$

Example 30:

The temperature of an ideal monoatomic gas is increased from 546 K to 1638 K. Calculate the pressure change in order that the entropy of the gas remains unchanged in the process

Solution:

The entropy change for the system can be explained by the following relation which is as follows

$$\Delta S_{sys} = nC_P \ln\frac{T_2}{T_1}\ nR \ln \frac{P_2}{P_1}$$

$$\Delta S_{sys} = 0, \text{ thus}$$

$$C_P \ln\frac{T_2}{T_1} = R \ln \frac{P_2}{P_1}$$

$$\frac{5}{2} R \log\frac{T_2}{T_1} = R \log \frac{P_2}{P_1}$$

$$\frac{5}{2} \log\frac{1638 \text{ K}}{546 \text{ K}} = \log \frac{P_2}{P_1}$$

$$\frac{P_2}{P_1} = 15.60$$

Thus in order to make entropy constant, the pressure of the gas must be increased by 15.60 times its initial value.

Example 31:

Two moles of an ideal gas are allowed to expand isothermally from 0.04 m^3 to 0.4 m^3 at 300 K. Calculate the entropy change for the system, surroundings and universe if the expansion is

(i) reversible and

(ii) irreversible against a constant external pressure of 0.2 atm.

Solution:

$$Q_{rev} = W = n\ RT \ln \frac{V_2}{V_1}$$

$$\Delta S_{rev} = \frac{Q_{rev}}{T} = nR \ln \frac{V_2}{V_1}$$

$$= (2 \text{ mol})\ (8.314) \text{ J mol}^{-1} \text{ K}^{-1}\ (2.303) \log \frac{0.40}{0.04}$$

$$= 38.294284 \text{ JK}^{-1}.$$

The surroundings is losing an exactly equivalent amount of heat and hence its entropy decrease is given as follows :

$$\Delta S_{surr} = \frac{(-Q_{rev})}{T} = -19.14 \text{ JK}^{-1}$$

The entropy change of the universe is given as follows :

$$\Delta S_{univ} = \Delta S_{sys} + \Delta S_{surr} = 0$$

(ii) In the irreversible expansion, the entropy change of the system is still the same as in the reversible change, *i.e.*, 19.14 JK^{-1} but heat lost by the surroundings is given by $-P\Delta V$ and consequently the entropy change, if the transfer were reversible, is

$$\Delta S_{surr} = \frac{Q_{surr}}{T} = -\frac{P\Delta V}{T}$$

$$= -\frac{(0.2 \text{ atm.})\left(0.4 - 0.04 \text{ m}^3\right)\left(101325 \text{ Jm}^{-3} \text{ atm}^{-1}\right)}{300 \text{ K}}$$

$$= -\frac{0.2 \times 0.36 \times 101325}{300} \text{ JK}^{-1}$$

$= 24.318 \text{ JK}^{-1}$..

Example 32:

Calculate ΔS, ΔH and ΔU for the process

$$H_2O(l,\ 20°C,\ 1 \text{ atm} \rightarrow H_2O\ (g,\ 250°C,\ 1 \text{ atm})$$

Given the following data :

$$C_P\ (l) = 75.6 \text{ J K}^{-1} \text{ mol}^{-1}$$

$$C_P\ (g) = 36.2 \text{ J K}^{-1} \text{ mol}^{-1}$$

ΔH for vaporization of H_2O at 100°C and 1 atm is equal to 40.85 k J mol^{-1}.

Solution:

The given process may be replaced by the following processes.

(i) H_2O (l, 20°C, 1 atm) → H_2O (l, 100°C, 1 atm)

(ii) H_2O (l, 100°C, 1 atm) → H_2O (g, 100°C, 1 atm)

(iii) H_2O (g, 100°C, 1 atm) → H_2O (g, 250°C, 1 atm)

Thus we have

$$\Delta S = \Delta S_{(1)} + \Delta S_{(ii)} \Delta S_{(iii)}$$

$$= C_{p,m}(l) \ln \frac{T_2}{T_1} + \frac{\Delta H_{vap}}{T} + C_{p,m}(g) \ln \frac{T_2'}{T_1'}$$

$$= \left(75.6 \ln \frac{373}{293} + \frac{40850}{373} + 36.2 \ln \frac{523}{373}\right) JK^{-1} \text{ mol}^{-1}$$

$$= (18.25 + 109.52 + 12.24) \text{ J K}^{-1} \text{ mol}^{-1}$$

$$= 140.01 \text{ J K}^{-1} \text{ mol}^{-1}$$

$$\Delta H = \Delta H_{(1)} + \Delta H_{(ii)} + \Delta H_{(iii)}$$

$$= C_{p,m} (l) (T_2 - T_1) + \Delta H_{vap} + C_{pm}(g) (T'_2 - T'_1)$$

$$= (75.6 \times 80 + 40850 + 36.2 \times 150) \text{ J mol}^{-1}$$

$$= 52328 \text{ J mol}^{-1}$$

$$\Delta U = \Delta U_{(1)} + \Delta U_{(ii)} \Delta U_{(iii)}$$

$$= C_{v,m} (l) (T_2 - T_1) + \Delta U_{vap} + C_{v,m}(g) (T'_2 - T'_1)$$

$$= C_{p,m}(l)(T_2 - T_1) + \{\Delta H_{vap} + \Delta_{V\,g} RT\} + \{C_{p,m}(g) - R\}(T'_2 - T'_1)$$

$$= [75.6 \times 80 + \{40850 - (8.314)(373)\}$$

$$+ (36.2 - 8.314)(150)] \text{ J mol}^{-1}\}$$

$$= (6040 + 37749 + 4183) \text{ J mol}^{-1}$$

$$= 47972 \text{ J mol}^{-1}.$$

Example 33:

For N = 400, find the ratios of the probabilities for (300, 100) and (200, 200) distribution in two identical boxes.

Solution:

We have

$$P(300, 100) = \frac{400!}{300!\,100!} \frac{1}{2^{400}}$$

$$P(200, 200) = \frac{400!}{200!\,200!} \frac{1}{2^{400}}$$

$$\frac{P(300, 100)}{P(200, 200)} = \frac{200!\,200}{300!\,100!} \cong 10^{-5}.$$

Example 34:

Eight similar coins are tossed, find the probability of three heads and t tails.

Solution:

We know $$P(3, 5) = \frac{^8C_3}{2^8} = \frac{8\times7\times6}{3\times2\times1\times2^8}$$

or $$P(3, 5) = \frac{7}{32}$$

Example 35:

Two bodies at different temperatures are connected by a wire with infinitesimal thermal conductance so that heat flows infinitely slowly from the hot bodies to the cold body until they are at the same temperature. Is this process reversible? Explain.

Solution:

In a reversible process, at any stage, the external condition responsible for the process to occur differs from the internal condition by an infinitesimal amount. The given transfer of heat is not a reversible process as the two bodies are at different temperatures. Moreover, the reversible processes are characterized by the fact that when the system is restored to its original state by traversing the forward sequence of steps in the reverse order, then not only the system but also the surroundings are restored to their original states. In the given process, heat flow from a hot body to a cold body does not disturb the surroundings but if we wish to carry out the reverse process, the surroundings will have to do some work. Hence, the surroundings are not brought back to the original state in the reverse process. Thus, the given process is not a reversible one.

Example 36:

How would the energy of an ideal gas change if it is made to expand into vacuum at constant temperature ?

Solution:

For the expansion against vacuum $P_{opp} = 0$

Hence $$w = -P_{opp}\,\Delta V = 0$$

Since the expansion is at constant temperature no heat is supplied to the system, *i.e.*, q = 0. Now, from the first law of thermodynamics, we can write

$$\Delta E = q + w = 0$$

that is, the energy of an ideal gas does not change if it is made to expand into vacuum at constant temperature. This also follows from the fact that the energy of an ideal gas depends only on temperature and since temperature remains constant, its energy also remains constant.

Example 37:

Calculate the change of entropy for the process

$$H_2O\ (l,\ 373\ K) \rightarrow H_2O\ (v,\ 373\ K)$$

Here $\Delta H_{vap} = 40850\ J\ mol^{-1}$ *at 373 K.*

Solution:

$$\Delta S_{vap} = \frac{\Delta H_{vap}}{T_b}$$

$$= \frac{40850\ J\ mol^{-1}}{373\ K}$$

$$= 109.51742\ JK^{-1}\ mol^{-1}$$

But ΔS_{vap} = S (vapour) − S (liquid)

$$= 109.5\ JK^{-1}\ mol^{-1}$$

∴ S (vapour) = S (liquid) + 109.5 $JK^{-1}\ mol^{-1}$

Example 38:

A thermostat was maintained at 370.05 K. For nearly an hour 4.2 kJ of heat leaked through the thermostat insulation into a room where the initial temperature of air was 300.05 K.

(i) What was the entropy change of the material in the thermostat?

(ii) What was the entropy change of the air in the room ?

(iii) Was the process spontaneous? (Material in the thermostat is water).

Solution:

(i) Entropy change of the material in the thermostat

$$-\frac{4.2 \text{ kJ}}{370.05\text{K}} = -11.35\times 10^3 \text{ kj K}^{-1}$$

(ii) Entropy change of the air in the room

$$= \frac{4.2 \text{ kJ}}{300.05 \text{ K}} = 14.00 \times 10^3 \text{ kJ K}^{-1}$$

(iii) Now

$$\Delta S_{total} = \Delta S_{thermostat} + \Delta S_{room}$$
$$= (-11.35 + 14.00) \text{ JK}^{-1}$$
$$= 2.65 \text{ J K}^{-1}$$

Since ΔStotal is positive, the process of leaking is spontaneous.

Example 39:

An ice box is built of wood 1.76 cm thick, lined inside with cork 3 cm thick. If the temperature of the inner surface of the cork is 0°C and that of the outer surface of wood is 12°C, what is the temperature of the interface ? The thermal conductivity of wood and cork are 0.0006 and 0.00012 CGS units respectively.

Solution:

Suppose the temperature of the interface is 9 after the steady state is reached

$$\therefore \quad \frac{K_1A(12-\theta)}{d_1} = \frac{K_2A(\theta-0)}{d_2}$$

Here $K_1 = 0.0006$,

$K_2 = 0.00012$

$d_1 = 1.75$ cm,

$d_2 = 3$ cm

$$\frac{0.0006(12-\theta)}{1.75} = \frac{0.00012(\theta)}{3}$$

$$\theta = 10.74°\text{C}.$$

Example 40:

At low temperature, a nonideal gas follows the relation PV = RT – a/V where a = 0.3636 Nm⁴ mole⁻². Calculate the work done by one mole of this gas in expanding form 0.224 litre to 22.4 litre at 400 K.

Compare this result with the work done in the corresponding expansion of an ideal gas.

Solution:

For an ideal gas, reversible work done

$$= 2.303 \text{ n RT log } \frac{V_2}{V_1}$$

V_1 = 0.224 litres mole^{-1} = 22.4 × 10^{-5} m^3 mole^{-1}

P_2 = 22.4 litres mole^{-1} = 22.4 × 10^{-3} m^3 mole^{-1}

R = 8.314 J K^{-1} mole^{-1}

$$W_{rew} = (2.303)\ (1 \text{ mole})\ (8.314 \text{ J K}^{-1} \text{ mole}^{-1})\ (400 \text{ K}) \times \log \frac{22.4 \times 10^{-3}}{22.4 \times 10^{-5}}$$

= (2.303) (1 mole) (8.314 J K^{-1} mole^{-1}) (400 K) log 100

= 15320 J mole^{-1}

= 15.32 kJ mole^{-1}.

For the non-ideal gas PV = RT – a/V

$$\text{Therefore, the work done} = \int_{V_1}^{V_2} P dV = \int_{V_1}^{V_2} \left\{ \frac{RT}{V} - \frac{a}{V_2} \right\} dV$$

$$= 2.303 \text{ RT log } \frac{V_2}{V_1} + a\left\{ \frac{1}{V_2} - \frac{1}{V_1} \right\}$$

The value of first part is from derived eq. (1) as 15.32 kJ mole^{-1} and of another part can be derived as

$$0.3636 \text{ Nm}^4 \text{ mole}^{-2} \left\{ \frac{1}{22.4 \times 10^{-3} \text{ m}^3 \text{ mole}^{-1}} - \frac{1}{22.4 \times 10^{-5} \text{ m}^3 \text{ mole}^{-1}} \right\}$$

= 1.604 kJ mole^{-1}.

Thus work done = 15.320 k Joule^{-1} – 1.604 k Joule^{-1}

= 13.716 kJ mole^{-1}.

Example 41:

What will be the work done when 65.38 g of zinc dissolves in hydrochloric acid in case of an open beaker and closed beaker at 300K.

Solution:

The process of dissolution of zinc in hydrochloric acid can be given as

$$Zn(s) + 2HCI\ (aq) = ZnCl_2(aq) + H_2(g)$$

From this equation it is clear that for each gram atom of zinc dissolved we will obtain 1 mole of Hydrogen gas. As a result of liberated gas, on the surrounding atmosphere the work performed will be equal to PΔV. If the initial volume of the system is neglected in comparison to the total volume of the gas produced and the gas shows ideal behaviour, then the work done in the open beaker is

$$W = P\Delta V \approx PV_{H2}$$

$$= nH2\ RT\ [\because T = 300\ k\ R = 8.314\ Jk^{-1}\ mole^{-1}]$$

$$= (1\ mole)\ (8.314\ JK^{-1}\ mol^{-1})\ (300\ K)$$

$$= 2.4942\ kJ.$$

If the reaction takes place in a closed beaker there is change in the volume *i.e.*, $\Delta V = 0$ hence work done is zero.

Example 42:

Calculate the work done when one mole of sulphur dioxide gas expands isothermally and reversibly at 300 k from 2.46 × 10^{-3} m^3 to 24.6 × 10^{-3} m^3. Assuming that the gas obeys

(a) ideal gas equation and

(b) van der Waals equation

$$\left\{\begin{array}{l} a = 0.6799\ nm^4\ mole^{-2} \\ b = 0.564 \times 10^{-3}\ mole^{-1} \end{array}\right\}$$

Solution:

(a) The reversible work done by 1 mole of gas when it expands form

$$2.64 \times 10^{-3}m^3 \text{ to } 24.6 \times 10^{-3}m^3 = 2.303\ RT \log \frac{V_2}{V_1}$$

$$= (2.303)\ (1\ mole)\ (8.314\ J\ K^{-1}\ mol^{-1})\ (300\ K) \times \log 10$$

$$\left\{\because \log\frac{V_2}{V_1} = \log 10\right\}$$

$$= (2.303)\ \ (8.314)\ (300\ K)\ \ \log 10\ J\ mol^{-1}$$

$= 5744 \text{ J mol}^{-1}$

(b) When the gas obeys van der Waal's equation, work done by 1 mole of gas is

$$= 2.303 \text{ RT log } \frac{V_2 - b}{V_1 - b} + a\left\{\frac{1}{V_2} - \frac{1}{V_1}\right\}$$

After substituting the values of various quantities and solving, the work done is

$$= 5794 \text{ J mol}^{-1} - 252.4 \text{ J mol}^{-1}$$

$$= 5541.6 \text{ J mol}^{-1}$$

$$= 5.542 \text{ kJ mol}^{-1}.$$

Example 43:

One mole of a van der Walls's gas is allowed to expand isothermally and reversibly from a volume of 1 litre to 50 litre at 0°C. Calculate w, q, Δ E and Δ H. van der Waal's constants are a = 6.5atm l²mole⁻² = 0.056 litre mole⁻¹ and R = 0.082 l atom deg⁻¹ mole⁻¹.

Solution:

Here $V_1 = 1$ litre,

$V_2 = 50$ litres, n = 1 mole

T = 273 + 0 = 273°k,

$R = 0.082 \text{ l-atm deg}^{-1} \text{ mole}^{-1}$

$a = 6.5 \text{ atm l}^2 \text{ mole}^{-2}$,

$b = 0.056 \text{ litre mole}^{-1}$

$$w = 2.303 \text{ nRT log}\left(\frac{V_2 - nb}{V_1 - nb}\right) + an^2\left(\frac{1}{V_2} - \frac{1}{V_1}\right)$$

$$= 2.303 \times 1 \times 0.082 + 273 \text{ [log}(50 - 0.056) - \log(1 - 0.056) + 6.5\left(\frac{1}{50} - \frac{1}{1}\right)$$

$= 82.47$ l atm.

$$\Delta E = -an^2\left(\frac{1}{V_2} - \frac{1}{V_1}\right) = 6.5 \times \frac{49}{50}$$

$$q = 2.303nRT \log \frac{V_2 - b}{V_1 - b}$$

$$= \Delta E + w = 884\ 1\ atm.$$

Example 44:

The van der Waal's constant a and b for hydrogen in litre atmosphere units are 0.246 and 2.67 × 10^{-2} respectively. Calculate the inversion temperature of hydrogen.

Solution:

we know $T_i = 2a/Rb$

or $$T_i = \frac{2 \times 0.246}{0.0821 \times 0.0267}$$

$$= 224.5°k = (224.5 - 273)$$

$$= -48.5°C.$$

Example 45:

How much heat is required to raise the temperature of 1 mole oxygen from 300K to 1300 K at constant pressure.

$$C_p = 6.095 + 3.325 \times 10^{-3}\ T - 1.017 \times 10^{-6}\ T^2.$$

Solution:

$$H_2 - H_1 \int_{T_t}^{T_2} C_p\ dT$$

$$T_1 = 300\ k,$$

$$T_2 = 100\ k$$

$\therefore$ $H_{1300} - H_{300}$

$$= \int_{300}^{1300} \left(6.095 + 3.253 + 10^{-3}\ T - 1.017 \times 10^{-6}\ T^2\right)$$

[Substituting the value of C_p]

$\therefore$ $H_{1300} - H_{300} = 6.095\ (1300 - 300) + \frac{3.253 \times 10^{-3}}{2}$

$\times [(1300)^2 - (300)^2] - 1/3 \times 1.017 \times 10^{-6} [(1000)^2 - (300)^2]$

$$= 6095 + 2602.4 - 735.63$$

$$= 7961.77\ cal/mole.$$

Example 46:

For equation $(P - a/V^2)\ V = RT$, prove that (i) dP is an exact differential, (ii) P is a state function and

$$\text{(iii)} \left(\frac{\partial P}{\partial T}\right)_V \left(\frac{\partial T}{\partial V}\right)_P \left(\frac{\partial V}{\partial T}\right)_T + 1 = 0.$$

Solution:

(i) For dP to be an exact differential, it is be proved that

$$\frac{\partial^2 P}{\partial V \partial T} = \frac{\partial^2 P}{\partial T \partial V} \quad \text{...(1)}$$

$$(P + a/V^2)\ V = RT$$

or

$$P = \frac{RT}{V} - \frac{a}{V^2} \quad \text{...(2)}$$

When the above equation is differentiated with respect to 'V' at constant temperature T, we obtain

$$\left(\frac{\partial P}{\partial V}\right)_T = -\frac{RT}{V^2} + \frac{2a}{V^3}$$

When the above equation is differentiated with respect to T at constant V, we obtain

$$\frac{\partial^2 P}{\partial T \partial V} = -\frac{RT}{V^2} \quad \text{...(3)}$$

When equation (2) is differentiated with respect to T at constant volume V, we obtain

$$\left(\frac{\partial P}{\partial T}\right)_V = \frac{R}{V}$$

When the above equation is further differentiated with respect to V at constant T, we obtain

$$\frac{\partial^2 P}{\partial T \partial V} = -\frac{R}{V^2} \quad \text{...(4)}$$

From equations (3) and (4), we get

$$\frac{\partial^2 P}{\partial T\ \partial V} = \frac{\partial^2 P}{\partial T\ \partial V} = -\frac{R}{V^2} \quad \text{...(5)}$$

Thus, we have proved condition, (1). This shows that dP is an exact differential.

(ii) In the said question, first of all a change occurs in volume at constant temperature followed by a change in temperature at constant volume, thereby giving the final equation (3).

Again, a change in temperature is considered at constant volume, followed by a change in volume at constant temperature, thereby giving the final equation (4). Both equations (3) and (4) are same, thereby revealing that the final change in P remains the same irrespective of the difference in intermediate changes. Therefore, P is a state function as changes in P have been found to depend upon the intial and final states of the system.

(iii) Equation (2) is as follows :

$$P = \frac{RT}{V} - \frac{a}{V^2}$$

When the above equation is differentiated, we obtain :

$$dP = \frac{R}{V} dT - \frac{RT}{V^2} dV + \frac{2a}{V^3} dV$$

$$= \frac{R}{V} dT - \left(\frac{RT}{V^2} - \frac{2a}{V^3}\right) dV \qquad \text{...(6)}$$

If volume is constant, dV = 0 : equation (6) becomes as follows :

$$dP = \frac{R}{V} dT \text{or} \left(\frac{\partial P}{\partial V}\right)_V = \frac{R}{V} \qquad \text{...(7)}$$

If pressure is constant, dP = 0; equation (6) becomes as follow :

$$\frac{R}{V} dT - \left(\frac{RT}{V^2} - \frac{2a}{V^3}\right) dV = 0$$

or

$$\left(\frac{\partial T}{\partial V}\right)_P = \left(\frac{RT}{V^2} - \frac{2a}{V^3}\right) \Big/ \frac{R}{V} \qquad \text{...(8)}$$

If temperature is constant, dT = 0; equation (6) becomes as follows:

$$\left(\frac{\partial T}{\partial V}\right)_T = -\frac{1}{\left(\frac{RT}{V^2} - \frac{2a}{V^3}\right)} \qquad \text{...(9)}$$

On multiplying equations (7), (8) and (9), we obtain

$$\left(\frac{\partial p}{\partial T}\right)_V \left(\frac{\partial T}{\partial V}\right)_P \left(\frac{\partial V}{\partial P}\right)_T = -\frac{R}{V} \times \frac{\left(\frac{RT}{V^2} - \frac{2a}{V^3}\right)}{\left(\frac{RT}{V^2} - \frac{2a}{V^3}\right)} \times \frac{1}{R} = -1$$

or $$\left(\frac{\partial p}{\partial T}\right)_V \left(\frac{\partial T}{\partial V}\right)_P \left(\frac{\partial V}{\partial P}\right)_T + 1 = 0$$

This is cyclic rule.

Example 47:

Calculate the work down when 50 g of iron dissolves in hydrochloric acid at 25oC in :

(i) A closed vessel, and

(ii) An open beaker.

(iii) In which case would maximum amount of work be done ?

Solution:

$$\text{Amount of iron} = \frac{50 \text{ g}}{55.85 \text{ g mol}^{-1}} = \frac{50}{55.85} \text{ mol}$$

(i) In a closed vessel, $\Delta V = 0$. Hence,

$$dw = - PdV = 0$$

that is, no work is involved when iron dissolves in hydrochloric acid in a closed vessel.

(ii) The reaction between iron and hydrochloric acid is

$$Fe + 2HCl \rightarrow FeCl_2 + H_2(g)$$

Thus, the liberated hydrogen gas pushes the atmospheric air and thus involves the work of expansion. In this case, we will have

$$w = - P_{ext} \Delta V$$
$$= - P_{ext} V_{H2}$$

where V_{H2} is the volume of hydrogen released. According to the ideal gas law, we will have

$$w = - P_{ext} (n_{H2}RT/P_{ext})$$
$$= - n_{H2}RT$$
$$= - (50/55.85) \text{ mol } (8.314 \text{ J K}^{-1} \text{ mol}^{-1}) (298 \text{ K})$$
$$= - 2.2 \text{ kJ.}$$

Example 48:

The opposite faces of a metal plate of 0.2 cm thickness an at a difference of temperature of 109°C and the area of the plate is 200 sq cm. Find the quantity of heat that will flow through the plate in one minute if K = 0.2 CGS units.

Solution:

Here,

$$K = 0.2$$

$$A = 200 \text{ sq cm}$$

$$d = 0.2 \text{ cm}$$

$$(\theta_1 - \theta_2) = 100°C$$

$$t = 60s$$

$$Q = \frac{KA(\theta_1 - \theta_2)t}{d}$$

$$= \frac{0.2 \times 200 \times 100 \times 60}{0.2}$$

$$= 12 \times 10^5 \text{ cal.}$$

Example 49:

A bar of length 30 cm and uniform area of cross-section 5 cm² consists of two halves AB of copper and BC of iron welded together at B. The end A is maintained at 200°C and the end C at 0°C. The sides of the bar are thermally insulated. Find the rate of flow of heat along the bar when the steady state is reached. Thermal conductivity of copper if 0.9 and thermal conductivity of iron is 0.12 CGS units.

Solution:

Suppose the temperature at the interface. B is θ after the steady state is reached.

$$\therefore \quad \frac{K_1A(200-\theta)}{d} = \frac{K_2A(\theta-0)}{d}$$

$$0.9(200 - \theta) = 0.12\,\theta$$

$$\theta = 176.5°C.$$

After the steady state is reached, the rate of flow of heat is the same in both the bars.

$$\therefore \quad Q = \frac{K_1A(\theta-0)}{d}$$

$$Q = \frac{0.9 \times 5 \times (200 - 176.5)}{15}$$

$$Q = 7.05 \text{ cal/s.}$$

Example 50:

Find the time in which a layer of ice 3 cm thick on the surface of a pond will increase its thickness by 1 mm when the temperature of the surrounding air is –20°C.

Thermal conductivity of ice = 0.005

Latent heat of ice = 80 cal/g

Density of ice at 0°C = 0.91 g/cm³.

Solution:

$$\int dt = \frac{\rho L}{K\theta} \int x dx$$

$$t = \frac{\rho L}{2K\theta} [x_2^2 - x_1^2]$$

Here, $\rho = 0.91 \text{ g/cm}^3, L = 80 \text{ cal/g}$

$K = 0.005, \theta = 20°C$

$x_1 = 3 \text{ cm}, x_2 = 3.1 \text{ cm}$

$$t = \frac{0.91 \times 80}{2 \times 0.005 \times 20} [(3.1)^2 - (3)^2]$$

$= 222.04$ s, $= 3$ min 42 s approximately.

Example 51:

Since $CV = (\partial q/\partial T)\, V$ by definition, one often writes without restriction $dV = C_V dT$. This is generally not true. Explain why.

Solution:

If we take U = f)V, T), then we will have

$$dU = \left(\frac{\partial U}{\partial V}\right)_T dV + \left(\frac{\partial U}{\partial T}\right)_V dT$$

$$= \left(\frac{\partial U}{\partial V}\right)_T . dV + C_V dT$$

Thus, if one writes $dV = C_V dT$ without restriction then this statement is not complete. This is applicable for a system in which $(\partial U/\partial V)_T = 0$.

An ideal gas is such an example. Alternatively, the system may be undergoing change at constant volume, for which, $(\partial U/\partial V)_T\, dV$ is zero and thus $dV = C_V\, dT$.

Example 52:

Is the equation PV^γ = constant valid for all processes ? Explain your answer.

Solution:

The expression PV^γ = constant is valid only for adiabatic reversible volume change (*i.e.*, expansion or compression) involving an ideal gas. For adiabatic process, $dq = 0$.

Hence

$$dU = dw$$

$$C_V dT = -P_{ext}\, dV$$

For revisable process, $P_{ext} = P_{int}$. Now replacing $P_{int} = nRT/V$ and separating the variables followed by integration, we get

$$\frac{Cv}{n}\int \frac{dT}{T} = -R\int \frac{dV}{V}$$

or $$C_{V.m} \ln \frac{T_2}{T_1} = -R \ln \frac{V_2}{V_1}$$

or $$\left(\frac{T_2}{T_1}\right)^{C_{v,m}} = \left(\frac{V_2}{V_1}\right)^{-R}$$

Replacing T_2/T_1 by $P_2\, V_2/P_1\, V_1$ and rearranging, the expression would yield

$$P_2 V^\gamma_2 = P_1\, V^\gamma_1$$

that is $$PV^\gamma = \text{constant.}$$

Example 53:

Under what conditions does the adiabatic expansion of gas become a constant energy process ?

Solution:

If the adiabatic expansions done against a vacuum then in this process energy

On comparing the coefficients in Eqs (1) and (2), we have

$$\left(\frac{\partial E}{\partial T}\right)_V = T \qquad ...(1)$$

and $$\left(\frac{\partial E}{\partial V}\right)_S = -P \qquad ...(2)$$

Eq (2) can be written as

$$dE = T\,dS \qquad ...(3)$$

For a given temperature, the above equation on integration becomes as follows :

$$\int dE = \int TdS + \text{Integration constant}$$

$$E = TS + A \qquad ...(4)$$

or $$E = A - TS$$

where A is an integration constant. It is also known as Helmholtz free energy function.

(d) Enthalpy as a Function of S and P

As H is a function of S and P, we can write

$$H = f(S, P)$$

or $$dH = \left(\frac{\partial H}{\partial S}\right)_P dS + \left(\frac{\partial H}{\partial P}\right)_S dP \qquad ...(5)$$

We also know, $$dH = T\,dS + V\,dP \qquad ...(6)$$

On comparing the coefficients in Eqs (5) and (6), we get

$$\left(\frac{\partial H}{\partial S}\right)_P = T \qquad ...(7)$$

$$\left(\frac{\partial H}{\partial P}\right)_S = V \qquad ...(8)$$

When Eq. (7) is integrated, we get a new function

$$\int dH = \int TdS + \text{constant}$$

$$= \int TdS + G$$

or $$H = TS + G$$

or $$G = H - TS$$

where G is an integration constant and is the same as the Gibbs free energy function.

Example 54:

Two moles of an ideal gas at 300 K expand reversibly and isothermally from 4×10^{-2} m^3 to 8×10^{-2} m^3. Calculate the entropy change for the gas. How would you account for the fact that ΔS is not zero although the process is reversible ?

Solution:

If the process is reversible isothermal it means that $\Delta H = 0$, hence the entropy change is given as follows :

$$\Delta S_{sys} = \frac{Q_{rev}}{T} = \frac{W_{rev}}{T} = nR \ln\frac{V_2}{V_1}$$

$$= (2 \text{ mol})\ (8.314 \text{ JK}^{-1} \text{ mol}^{-1}) \left(2.303 \log \frac{8 \times 10^{-2}}{4 \times 10^{-2}}\right)$$

$$= 2 \times 8.314 \times 2.303 \log 2 \text{ JK}^{-1}$$

$$= 2 \times 8.314 \times 2.303 \times 0.3010 \text{ JK}^{-1}$$

$$= 11.526579 \text{ JK}^{-1}.$$

The total entropy change (system and its surroundings) will be zero for a reversible process.

Example 55:

Calculate the entropy changes involved in the conversion of 1 mole of ice at 0°C and 1 atm to liquid at 0°C and 1 atm [The enthalpy of fusion per mole of ice is 6008 J mol^{-1}].

Solution:

$$H_2O\ (S) \rightarrow H_2O\ (l)$$

$$\Delta S_{fus} = \frac{\Delta H_{fus}}{T_m}$$

$$= \frac{6008 \text{ J mol}^{-1}}{273 \text{ K}}$$

$$= 22.007326 \text{ JK}^{-1}.$$

Example 56:

Increase in volume of a gas from a given decrease of pressure is more in an isothermal expansion than in an adiabatic expansion.

Solution:

In an isothermal expansion, temperature remains the same whereas in adiabatic expansion temperature decreases. Since the final pressure is them same, it follows from Charles law (V ∝ T, p constant) that

$$V_{iso} > V_{adi}$$

Since $\Delta V_{iso} = V_{iso} - V_{intial}$

$$\Delta V_{adi} = V_{adi} - V_{intial}$$

it follows that.

Example 57:

Evaluate the entropy change for the following reversible process.

$$1 \text{ mole Sn } (\alpha, 13°C \rightleftharpoons 1 \text{ mole Sn } (b, 13°C)$$

where $\Delta H_{trans} = 2090 \text{ J mol}^{-1}$

Solution

$$T_{trans} = 273 + 13 = 286 \text{ K}$$

$$\Delta S_{trans} = \frac{\Delta H_{trans}}{T_{trans}}$$

$$= \frac{2090 \text{ J mol}^{-1}}{286 \text{ K}}$$

$$- 7.3076923 \text{ JK}^{-1} \text{ mol}^{-1}$$

Example 58:

Calculate the root mean square velocity of oxygen at N.T.P.

Solution:

For oxygen at N.T.P.

$$P = 76 \text{ cms} = 76 \times 13.6 \times 981 \text{ dynes/sq-cm}$$

$$V = 22.4 \text{ litres} = 22400 \text{ cc.}$$

Substituting the values in the equation:

$$(\text{RMSV}) = \sqrt{\frac{3PV}{M}}$$

we have, $$\text{RMSV} = \sqrt{\frac{3 \times 76 \times 13.6 \times 981 \times 22400}{32}}$$

$$= 46140 \text{ cm/sec.}$$

Type III : *Calculation of molecular velocity when pressure and temperature. (i.e., not at N.T.P.) are given.*

In such calculations, the first step involves the conversion of gram molecular volume (22.4 litres at N.T.P.) to the volume under the different pressure and temperature. Having found the new volume, this value is substituted in the equation:

$$\text{RMSV} = \sqrt{\frac{3PV}{M}}$$

This calculation will become clear from the following example:

Example 59:

Calculate the root mean square velocity of nitrogen at 27°C and 700 mm pressure.

Solution:

(i) Conversion of G.M.V. of new volume

At N.T.P.	At New Conditions
$P_1 = 760$ mm	$P_2 = 700$ mm
$V_1 = 22400$ c.c.	$V_2 = ?$
$T_1 = 273$ °K	$T_2 = 273 + 27 = 300$ °K

By gas equation

$$\frac{P_1V_1}{T_1} = \frac{P_2V_2}{T_2}$$

$$V_2 = \frac{P_1V_1}{T_1} \times \frac{T_2}{P_2}$$

$$= \frac{22400 \times 760}{273} \times \frac{300}{700} = 26720 \text{ c.c.}$$

(ii) Calculating the value of RMSV.

Now $P = 700 \text{ mm} = 70 \text{ cm}$

$$= 70 \times 13.6 \times 981 \text{ dynes/cm}$$

$$V = 26720 \text{ c.c.};\ M = 28 \text{ gm}$$

Substituting the values in the equation:

$$\text{RMSV} = \sqrt{\frac{3PV}{M}}$$

$$\text{RMSV} = \sqrt{\frac{3 \times 70 \times 13.6 \times 981 \times 26720}{28}}$$

$$= 51710 \text{ c/sec.}$$

Type IV : *Calculation of molecular velocity when pressure and density are given.*

$$PV = \frac{1}{3}\text{mnc}^2 = \frac{1}{3}\text{Mc}^2$$

$$\text{RMSV} = \sqrt{\frac{3PV}{M}} = \sqrt{\frac{3P}{D}}$$

where $\frac{V}{M}$ denotes density (D) of the gas in gm/c.c.

Example 60:

Oxygen has a density of 1.429 gm. per litre at N.T.P. Calculate the R.M.S. and A.V. of oxygen molecules.

Solution:

Now RMSV $= \sqrt{\frac{3P}{D}}$

In this case, P = 1 atmosphere

$= 76 \times 13.6 \times 981$ dynes/sq. cm.

D = 1.422 gm/litre

$= \frac{1.429}{1000}$ gm/c.c = 0.001429 gm/c.c.

Substituting the values in the above equation, we have

$$\text{RMSV} = \sqrt{\frac{3 \times 76 \times 13.6 \times 981}{0.001429}} = 46100 \text{ cm/sec}$$

This is the R.M.S. (root mean square mean velocity) of oxygen molecules.

Now A.V. (Average velocity)

$= 0.9213 \times$ R.M.S. velocity

$= 0.9213 \times 46{,}100 = 42458$ cm/sec.

Example 61:

Calculate the entropy change when two moles of; an ideal gas expands reversibly from an initial volume of 2 dm^3 to a total volume of 20 dm^3 at a constant temperature of 298 K.

Solution:

We know, $\Delta S = nR \ln\left(\frac{V_2}{V_1}\right)$

$$= 2.303\ nR \log\left(\frac{V_2}{V_1}\right)$$

$$= 2.303 \times 2 \times 8.314\ J\ JK^{-1}\ mol^{-1} \log\left(\frac{20 dm^3}{2 dm^3}\right)$$

$$= 2.303 \times 2 \times 8.314 \times \log\ JK^{-1}\ mol^{-1}$$

$$= 38.294284\ JK^{-1}\ mol^{-1}$$

Example 62:

The mixing of gases is always accompanied by an increase in entropy. Show that in the formation of a binary mixture of two ideal gases the maximum entropy increase results when $X_1 = X_2 = 0.5$.

Solution:

For a binary ;mixture, the entropy per mole of the mixture formed is given by

$$\Delta S \text{ mixing} = -R[X_1 \ln X_1 + X_2 \ln X_2]$$

$$= -R[X_1 \ln X_1 + (1 - X_1) \ln (1 - X_1)]$$

For entropy of mixing to be maximum, the first derivative, $\frac{\delta(\Delta S \text{ mixing})}{\delta X_1}$ should be zero and the second derivative should be negative. Differentiating, ΔS mixing with respect to X_1 and equating it to zero yields

$$\frac{\delta(\Delta S \text{ mixing})}{\delta X_1} = -R\left[\ln X_1 + \frac{X_1}{X_1} + \frac{1 - X_1}{1 - X_1}(-1) + (-1)\ln(1 - X_1)\right]$$

$$= 0$$

$$\ln X_1 + 1 - 1 - \ln(1 - X_1) = 0$$

$$\ln \frac{X_1}{1 - X_1} = 0$$

$$X_1 = 1 - X_1$$

$$X_1 = 1/2.$$

Example 63:

Under what conditions does the isothermal expansion of a gas become a free expansion process?

Solution:

The isothermal expansion of a gas becomes a free expansion process only when it expands against a vacuum, *i.e.*, $p_{ext} = 0$. Moreover, the gas must be an ideal gas in which case $\Delta E = 0$ at constant temperature.

Example 64:

Is every adiabatic system isolated? In what ways can the energy of an adiabatic system change?

Solution:

An isolated system is the one in which no interaction with its surroundings takes place. Neither energy nor matter can be transferred to or from it. An adiabatic system involves no exchange of heat from its surroundings, *i.e.* for such a system $dq = 0$ and, according to the first law of thermodynamics, $dE = dw$. Thus, the energy of an adiabatic system can be changed by either doing work on the system or by getting work done by the system.

Example 65:

One mole of a gas is expanded isothermally and reversibly at 27° C form a volume of 5 liters. Calculate q, w, ΔE and ΔH.

Solution:

We know

$$PV = nRT, \qquad n = 1$$

State : Initial

$$T_1 = 27 + 273.15 = 300.15 \text{ K}$$

V_1 = 5 litres

Final

$T_2 = T_1$ = 300.15K

V_2 = 10 litres

(1) Calculation of w :

$$w = \int_{V_1}^{V_2} P\,dV = \int_{V_1}^{V_2} \frac{RT}{V} dV = RT\ln\frac{V_2}{V_1}$$

$$= 2.303 \times 8.314 \times 300 \log (10/5)$$

$$= 2.303 \times 8.314 \times 300 \times \log^2$$

$$= 2.303 \times 8.314 \times 300 \times 0.3010 \quad [\log 2 = 0.3010]$$

$$w = 1729 \text{ J mol}^{-1}$$

(2) Calculation of ΔE :

$$\Delta E = E_2 - E_1$$

For isothermal-reversible expansion of an ideal gas $E_2 = E_1$ and therefore

$$\Delta E = 0.$$

(3) Calculation of q:

$$q = \Delta E + w \text{ (first law)}$$

$$= 0 + w$$

$$q = w = 1729 \text{ J}$$

(4) Calculation of ΔH

$$\Delta H = \Delta E + \Delta (PV) = \Delta E + \Delta(RT)$$

$$0 + 0 = 0$$

Since $\Delta E = 0$ and $\Delta(RT) = 0$; R and T being constant

Example 66:

Calculate the minimum work which must be done to compress 1/2 mole of oxygen at 300 k from a pressure of 2 atm to a pressure of 200 atm.

Solution:

As it is required to calculate the minimum work of compression, the change, is reversible, *i.e.*, the gas gets compressed by gradual increase of pressure so that the external pressure becomes only infinitesimally greater than the pressure of the gas, at each stage. Thus

$$w = \text{work done} = \int_{V_1}^{V_2} PdV$$

$$= 2.303 \text{ n RT} \log \frac{V_2}{V_1}$$

$$= 2.303 \text{ nRT} \log \frac{p_1}{P_2}$$

given date :

$P_1 = 2$ atm
$P_2 = 2$ atm
$T = 300$ K
$n = \frac{16}{32} = 0.5$ mole.

On substituting these values in the above equation, we get

$$w = 2.303 \times 0.5 \times 8.314 \times 300 \log \frac{2}{200}$$

$$= -5744 \text{ J.}$$

Example 67:

Four moles of an ideal gas are held by a piston under 5 atm pressure and at 273.15 K. The pressure is suddenly released to 0.2 atm and the gas is allowed to expand isothermally. Calculate w, q, ΔE and ΔH for the process.

Solution:

As the pressure has been released suddenly, the expansion must be irreversible.

Here n = 4 moles; $P_1 = 5$ atm; $T_1 = T_2 = 273.15$ K.

(i) Work done irreversibly

$$w = nRT\left\{1 - \frac{P_2}{P_1}\right\}$$

$$= 4 \times 8.314 \times 273.15 \left\{1 - \frac{0.2}{5}\right\}$$

$$= 4 \times 8.314 \times 273.15 \times 0.96$$

$$= 8720 \text{ J.}$$

(ii) As the ideal gas expands isothermally

$dT = 0$ and hence $\Delta E = 0$; $\Delta H = 0$.

(iii) First law gives

$$\Delta E = q - w = 0$$

$$\therefore\ q = w\ 8720\ J.$$

Example 68:

Calculate the pressure exerted by one mole of carbon dioxide gas in a 1.32 dm³ vessel at 48°C using :

(a) the ideal gas equation and

(b) the van der Waals equation. The van der Waals constants are;

$$a = 3.59\ dm^6\ atm\ mol^{-2}$$

and $$b = 0.0427\ dm^3\ mol^{-1}.$$

Solution:

(a) For an ideal gas, PV = nRT, so that

$$P = \frac{nRT}{V}$$

$$= \frac{(1\ mol)\ (0.08206\ dm^3\ atm\ K^{-1}\ mol^{-1})(321\ K)}{1.32\ dm^3}$$

$$= 19.9\ atm.$$

(b) For a van der Waals as,

$$P = \frac{nRT}{V - nb} - \frac{n^2 a}{V^2}$$

$$= \frac{(1\ mol)\ (0.08206\ dm^3\ atm\ K^{-1}\ mol^{-1})(321\ K)}{(1.32\ dm^3) - (1\ mol)(0.0427\ dm^3\ mol^{-1})}$$

$$= \frac{(1\ mol)^2 (3.59\ dm^6\ atm\ mol^{-2})}{(1.32\ dm^3)^2}$$

$$= 2062\ atm - 2.06\ atm = 18.56\ atm$$

The experimental value of the pressure is 16.40 atm. Thus, van der Waals equation gives a better agreement with experiment than the ideal gas equation.

Example 69:

Two moles of ammonia are enclosed in a five litre flask at 27°C. Calculate the pressure of ammonia using van der Waals equation. For ammonia, $a = 4.17$ *atm. litre*$^{-2}$ *mole*$^{-2}$ *and* $b = 0.0371$ *litre mole*$^{-1}$.

Solution:

V is the volume of n (= 2) moles of the gas = 5 litres.

a = 4.17 atms litre^{-2} mole^{-2} (correction due to inter-molecular attraction).

b = 0.0371 litre mole^{-1} (volume correction)

R = 0.082 litre atms. absolute degree^{-1} mole^{-1}.

$$T = 27 + 273 = 300\text{MK}$$

number of moles (n) = 2

According to van der Waals equation,

$$\left(P + \frac{a}{V^2}\right)(V - b) = Rt, \text{ valid for one mole.}$$

$$\left(P + a\frac{n^2}{V^2}\right)(V - nb) = nRT \text{ valid for n moles.}$$

i.e.
$$P = \frac{nRT}{V - nb} - \frac{an^2}{V^2}$$

Substituting the values of a, b, R, V and T in equation.

$$P = \frac{2 \times 0.082 \times 300}{5 - 2 \times 0.0371} - \frac{4.17 \times 2 \times 2}{5 \times 5}$$

$$= 9.98 - 0.67 = 9.31 \text{ atm.}$$

Example 70:

A vessel of 50 litre capacity contains 10 moles of steam under 5 atmospheres of pressure. Calculate the temperature of the steam using van der Waals equation if for water a = 5.46 atm (litre/mole)^2 and b = 0.031 litre/mole.

Solution:

We know $\left(P + \frac{n^2a}{V^2}\right)(V - nb) = nRT$

i.e.
$$T = \left(P + \frac{n^2a}{V^2}\right)\left(\frac{V - nb}{nR}\right)$$

Here P = 5 atm, n = 10 moles

$$V = 10 \text{ litres}$$

and $$a = 5.46 \text{ atm} \left(\frac{\text{litre}}{\text{mole}}\right)^2$$

and $$b = 0.03/\text{lite/mole.}$$

Substituting the values,

$$T = \left(5 + \frac{(10)^2 \times 5.46}{(50)^2}\right)\left(\frac{50 - 10 \times 0.031}{10 \times 0.08}\right)$$

$$= 311.9°K$$

Example 71:

The critical pressure and critical temperature of a gas obeying van der Waals equation have values 73 atms and 37°C. Calculate constants 'a' and 'b'. The value of the gas constant, R is 0.082 atm degree^{-1}.

Solution:

We have $$T_c = \frac{8a}{27Rb} \quad \text{...(i)}$$

$$P_c = \frac{a}{27b^2} \quad \text{...(ii)}$$

Dividing (i) by (ii) we have,

$$\frac{T_c}{P_c} = \frac{8a}{27Rb} \times \frac{27b^2}{a} = \frac{8b}{R}$$

Now $$T_c = 31° + 273 = 304°K$$

$$P_c = 0.082 \text{ atm degree}^{-1}$$

Hence $$\frac{304}{73} = \frac{8b}{0.082}$$

$\therefore$ $$b = 0.4269$$

Substituting the value of b in (i) *i.e.,*

$$T_c = \frac{8a}{27Rb}, \text{ we get}$$

$$304 = \frac{8a}{27 \times 0.082 \times 0.04269}$$

$$a = 3.591.$$

Example 72:

Calculate the reduced pressure, volume and temperature of CH_4 gas when one mole of gas is contained in a 5 litre flask under 5 atmospheric pressure. What will be the temperature of the as when

Critical temperature $T_c = 190.2°K$

Critical volume $V_c = 0.0988$ *litre/mole and*

Critical pressure $P_c = 45.6$ *atm.*

Solution:

We know,

Reduced pressure $\pi = P/Pc$

Here P = 5 atm $P_c = 45.6$ atm.

Substituting the values, $\pi = 5/45.6 = 0.1096$

Reduced volume $\phi = V/V_c$

Here V = 5 litres and $V_c = 0.0988$ litre/mole

substituting the values $\phi = \dfrac{5}{0.0988} = 50.607$

According to the law of corresponding states,

$$\left(\pi + \frac{3}{\phi^2}\right) \times (3\phi - 1) = 8\theta$$

where θ is the reduced temperature

i.e.

$$\theta = \frac{\left(\pi + \frac{3}{\phi^2}\right) \times (3\phi - 1)}{8}$$

Substituting the values of π and ϕ,

$$\theta = \frac{\left[0.1096 + \frac{3}{(50.607)^2}\right](3 \times 50.607 - 1)}{8}$$

$$= 2.0888$$

We know that reduced temperature

$$\theta = \frac{T}{T_c}$$

$$T = T_c \times \theta$$

Here $T_c = 190.2°K$

Substituting the values,

$$\text{Temperature} = 190.2 \times 2.0888 = 397.3°K.$$

Some Examples are given in Four Types

Type I: *Calculation of velocity when temperature alone is given.*

By kinetic gas equation:

$$PV = \frac{1}{3} mn\bar{c}^2$$

$$(R.M.S.V.) = \sqrt{\frac{3PV}{mn}} = \sqrt{\frac{3PV}{M}} = \sqrt{\frac{3RT}{M}}.$$

M denotes the molecular weight of the gas and $R = 8.3 \times 10^7$ ergs

$$(RMSV) = \sqrt{\frac{3 \times 8.3 \times 10^7 \times T}{M}}$$

$$= 1.58 \times \sqrt{\frac{T}{M}} \times 10^4 \text{ cm/sec.}$$

Example 73:

For a fluid show that $(¶P/¶T)_V = a/b$ *where a is expansivity and b is compressibility of the gas. Hence, show that for an ideal gas.*

$$(\partial P/\partial T)_V = R/V$$

Solution:

As a gas can be defined by any two of the three parameters P, V and T, we can write

$$V = f(T.P)$$

As V is a state function, hence its total differential can be written as follows:

$$dV = \left(\frac{\partial V}{\partial T}\right)_P dT + \left(\frac{\partial V}{\partial P}\right)_T dP \quad ...(1)$$

By definition. $\alpha = \frac{1}{V}\left(\frac{\partial V}{\partial T}\right)_P$ so that $\left(\frac{\partial V}{\partial T}\right)_P = \alpha V$

and $$\beta = -\frac{1}{V}\left(\frac{\partial V}{\partial P}\right)_T \text{ so that} \left(\frac{\partial V}{\partial P}\right) = -\beta V$$

On substituting for α and β in Eq. (i) we get

$$dV = \alpha\, VdT - \beta\, VdP$$

When V is constant, dV = 0. Hence,

$$\alpha\, VdT - \beta\, VdP = 0$$

or $$V(\alpha\, dT - \beta\, dP) = 0$$

But $V \neq 0$, hence

$$\alpha\, dT - \beta\, dP = 0$$

or $$\alpha\, dT = \beta\, dP$$

or $$\left(\frac{\partial P}{\partial T}\right)_V = \frac{\alpha}{\beta}$$

For an ideal gas we know, $\alpha = 1/T$ and $\beta = 1/P$. Hence,

$$\left(\frac{\partial P}{\partial T}\right)_V = \frac{\alpha}{\beta} = \frac{1/T}{1/P} = \frac{P}{T} = \frac{R}{V}$$

This is to be proved.

Example 74:

Calculate the root mean square and average velocity of carbon dioxide at 1000°C.

Solution:

$$(\text{RMSV}) = 1.58 \times \sqrt{\frac{T}{M}} \times 10^4 \text{ cm/sec.}$$

Now T = 273 + 1000 = 1273 K; M = 44

$\therefore$ $$(\text{RMSV}) = 1.58 \times \sqrt{\frac{1273}{44}} \times 10^4 \text{ cm/sec.}$$

$$= 84870 \text{ cm/sec.}$$

This gives the root mean square velocity.

Now average velocity = 0.9213 × R.M.S.V. of CO_2

Average velocity = 0.9213 × 84870

= 78190 cm/sec.

Example 75:

Calculate the entropy change when 1 mole of perfect gas is allowed to expand at 27°C from a volume of 2 litres to a volume of 20 litres against a constant pressure of one atmosphere.

Solution:

We know

$$dS = C_V \frac{dT}{T} + R \frac{dV}{V}$$

By integrating this equation, we get

$$S_2 - S_1 = C_V \ln \frac{T_2}{T_1} + R \ln \frac{V_2}{V_1}$$

At constant temperature,

$$S_2 - S_1 = R \ln \frac{V_2}{V_1}$$

$$= 2.303 \, R \log \frac{V_2}{V_1}$$

$$= 2.303 \times 2 \times \log \frac{20}{2}$$

$$= 2.303 \times 2 \times \log 10$$

$$= 4.606 \text{ cal./ degree/mole.}$$

$$\begin{bmatrix} \because R = 2 \text{ cal./degree/mole} \\ V_1 = 2 \text{ litres} \\ V_2 = 20 \text{ litres} \end{bmatrix}$$

Example 76:

The molar heat capacity at capacity at constant pressure of solid magnesium (0° 600°C) is expressed by

CP = 6.20 + 1.30 × 10–3 T – 6.80 × 104 T–2

Calculate the increase in entropy when 1 gm atom of metal is heated from 27°C to 127°C at constant pressure.

Solution:

At constant pressure,

$$\Delta S = \int_{T_1}^{T_2} \frac{C_P}{T} dT$$

$$\therefore \quad \Delta S = \int_{300}^{400} \frac{\left(6.20 + 1.30 \times 10^{-3} T - 6.80 \times 10^4 T^{-2}\right) dT}{T}$$

$$\int_{300}^{400} 6.20 \frac{dT}{T} + \int_{300}^{400} 1.3 \times 10^{-3} dT - \int_{300}^{400} 6.20 \times 10^4 \frac{dT}{T^3}$$

$$= 6.20 \ln[\ln T]_{300}^{400} + 1.3 \times 10^{-3} [T]_{300}^{400} - \frac{6.80 \times 10^4}{-2} \left[\frac{1}{T^2}\right]_{300}^{400}$$

$$= 6.20 \ln \frac{400}{300} + 1.3 \text{ Ÿ } 10^{-3} \text{ Ÿ } (400 - 300) + 3.40 \text{ Ÿ } 10^4$$

$$\left[\frac{1}{(400)^2} - \frac{1}{(300)^2}\right] = 1.7487 \text{ cal./degree.}$$

Example 77:

Since CV = (∂q/∂T) V by definition, one often writes without restriction dV = C_VdT. This is generally not true. Explain why.

Solution:

If we take U = f)V, T), then we will have

$$dU = \left(\frac{\partial U}{\partial V}\right)_T dV + \left(\frac{\partial U}{\partial T}\right)_V dT$$

$$= \left(\frac{\partial U}{\partial V}\right)_T . dV + C_V dT$$

Thus, if one writes dV = C_V dT without restriction then this statement is not complete. This is applicable for a system in which $(\partial U/\partial V)_T = 0$. An ideal gas is such an example. Alternatively, the system may be undergoing change at constant volume, for which, $(\partial U/\partial V)_T$ dV is zero and thus dV = C_V dT.

Example 78:

Calculate the work done by the system when 2 mol of an ideal gas expand from 0.01 m^3 to 0.1 m^3 at 300 k isothermally :

(i) into a vacuum, and

(ii) reversibly.

Explain how the surroundings can be restored only in one case and not in the other.

Solution:

(i) When the work is done into a vacuum, the work done by the system is zero since P_{opp} in this case is zero.

(ii) For the reversible isothermal expansion work involved is

$$w = -\ nRT\ \ln \frac{V_2}{V_1}$$

$$= (2)\ (8.314 \times 300\ J)\ \ln \frac{0.1}{0.01}$$

$$= -\ 2 \times 8.314 \times 300 \times 2.303\ J$$

$$= -\ 11488\ J = -\ 11.488\ kJ.$$

This surroundings are not disturbed when the expansion is done against vacuum. In case of reversible isothermal expansion, heat is absorbed by the system from the surroundings. Now if we carry out reversible isothermal compression of the system till the original state, then the same quantity or heat is released to the surroundings and thus the latter is restored back to the original state. However, in the first case where expansion is done against vacuum, there will be bet heat received by the surroundings during compression. Hence, in this case, it is not restored back to its original state.

Example 79:

Show that PV = nRt, then

(i) $$\frac{\partial^2 P}{\partial n \partial V} = \frac{\partial^2 P}{\partial V \partial n} = -\frac{RT}{V^2}$$

(ii) $$\frac{\partial^2 P}{\partial n \partial T} = \frac{\partial^2 P}{\partial T \partial n} = -\frac{R}{V}$$

(iii) $$\left(\frac{\partial P}{\partial T}\right)_V \left(\frac{\partial T}{\partial V}\right)_P \left(\frac{\partial V}{\partial P}\right)_T = -1$$

Solution:

(i) $\because$ $$PV = nRT$$

$\therefore$ $$P = \frac{nRT}{V}$$

or $$\frac{\partial P}{\partial V} = \frac{\partial}{\partial V}\left(\frac{nRT}{V}\right)$$ [keeping T as constant]

$$= nRT\ \frac{\partial}{\partial V}\left(\frac{1}{V}\right) = -\frac{nRT}{V^2}$$

$$\frac{\partial^2 P}{\partial n \partial V} = \frac{\partial}{\partial n}\left(\frac{\partial P}{\partial V}\right)$$

$$= \left(-\frac{nRT}{V^2}\right) \quad \text{[Keeping R, T and V as constant]}$$

$$= -\frac{RT}{V^2} \qquad \text{...(1)}$$

Similarly, $PV = nRT$ or $P = \frac{nRT}{V}$

$$\frac{\partial P}{\partial n} = \frac{RT}{V}$$

$$\frac{\partial^2 P}{\partial V \partial N} = \frac{\partial}{\partial V}\left(\frac{RT}{V}\right) \quad \text{[keeping T and R as constant]}$$

$$= -\frac{RT}{V^2} \qquad \text{...(2)}$$

From equations (1) and (2), we obtain

$$\frac{\partial^2 P}{\partial n \partial V} = \frac{\partial^2 P}{\partial V \partial n} = -\frac{RT}{V^2}$$

(ii) $$P = \frac{nRT}{V}$$

$$\frac{\partial P}{\partial T} = \frac{\partial}{\partial T}\left(\frac{nRT}{V}\right) \quad \text{[Keeping n and R as constant]}$$

$$= \frac{nR}{V}$$

$$\frac{\partial^2 P}{\partial n \partial T} = \frac{\partial}{\partial n}\left(\frac{nR}{V}\right) \text{[Keeping V and R as constant]}$$

$$= \frac{R}{V} \qquad \text{...(3)}$$

Again, $$P = \frac{nRT}{V}$$

$$\frac{\partial P}{\partial n} = \frac{RT}{V} \quad \text{[Keeping RT and V as constant]}$$

$$\frac{\partial^2 P}{\partial T \partial n} = \frac{\partial}{\partial n}\left(\frac{RT}{V}\right) \quad \text{[Keeping RT and V as constant]}$$

or $$\frac{\partial^2 P}{\partial T \partial n} = \frac{R}{V} \qquad ...(4)$$

From equations (3) and (4), we obtain

$$\frac{\partial^2 P}{\partial n \partial T} = \frac{\partial^2 P}{\partial T \partial n} = \frac{R}{V}$$

(iii) Again, $$P = \frac{nRT}{V}$$

$$\left(\frac{\partial P}{\partial T}\right)_v = \frac{nR}{V} \qquad ...(5)$$

Again, $$V = \frac{nRT}{P}$$

$$\left(\frac{\partial V}{\partial P}\right)_T = -\frac{nRT}{P} \qquad ...(6)$$

Again, $$T = \frac{PV}{nR}$$

$$\left(\frac{\partial T}{\partial V}\right)_P = -\frac{P}{nR} \qquad ...(7)$$

On multiplying equations (5), (6), and (7), we obtain

$$\left(\frac{\partial P}{\partial T}\right)_V \left(\frac{\partial T}{\partial V}\right)_P \left(\frac{\partial V}{\partial P}\right)_T = \frac{nR}{V} \times -\frac{nRT}{P^2} \times \frac{P}{nR}$$

$$= -\frac{nRT}{PV} = -1 \left[\because 1 = \frac{nRT}{PV}\right]$$

2

Chemical Equilibria

INTRODUCTION

A chemical equilibrium can be defined as a *dynamic state at which the rate of forward and backward reactions are the same and at that instant all the reactants and products are present.* When chemical equilibrium is reached both the forward and backward reactions continue but at the same ate so that the concentrations of the components remain unchanged with time. Chemical equilibrium may be homogeneous or heterogeneous depending upon whether the reaction takes place :

(i) In one phase or

(ii) In two or ore phases.

Examples:

(i) When stream is passed over iron in a closed vessel, it will attain the state of equilibrium and the reactants and products will have a definite concentration at a given temperature.

$$2FE + 4H_2O \rightleftarrows Fe_3O_4 + 4H_2$$

(ii) If a mixture of hydrogen and iodine vapour is passed over platinum heated to 450°C in a closed vessel. the formation of hydrogen iodide and its decomposition into hydrogen and iodine takes place with equal and opposite rate and a state of equilibrium is reached.

$$H_2 + I_2 \rightleftarrows 2HI.$$

Dynamic Nature of Chemical Equilibrium

A reversible reaction must attain the equilibrium state. For instance, if hydrogen and iodine are taken in a reaction vessel, only toward reaction takes place as initially there is no hydrogen iodide. As soon as hydrogen iodide is formed, the reverse reaction starts. As time passes, the concentrations of hydrogen and iodine start decreasing and, therefore, the rate of forward reaction decreases.

However, the rate of reverse reaction increases. As time passes, an equilibrium state may be reached when the rates of toward and reverse reactions become equal. At this stage no further change in consecrations occurs.

Similar results would be obtained if we start with HI. At equilibrium, concentrations of HI, H_2 and I_2 become constant at a certain temperature. This condition is known as chemical equilibrium.

Recognition of Attainment of Equilibrium

It is possible to recognize the attainment of equilibrium by constancy of certain observable properties. For example, the attainment of equilibrium is recognised in the evaporation of water at a given temperature by observing the pressure had become constant. Another example is the dissociation of calcium carbonate in which the attainment of equilibrium is recognised by observing that pressure of carbon dioxide had become constant. In both the examples, it becomes possible to recognise the attainment of equilibrium by observing constancy of pressure.

In some cases, the measurement of concentration of one or more of the reactants or products can be used to recognise the attainment of equilibrium. In order to understand this, we will consider the reaction between hydrogen gas and iodine vapour.

$$H_2(g) + I_2(g) \rightleftarrows 2HI(g)$$

Suppose one mole of hydrogen gas and one mole of iodine vapour are mixed in a flask of 1 litre capacity. Now suppose this flask is allowed to stand in a boiling bath (448°C) for several hours. If the reaction goes to completion, one should expect the formation of two moles of hydrogen iodide. But it is observed that the reaction comes to a stop when only 1.56 moles of hydrogen iodide get formed. This reveals that only 0.78 mole of each of hydrogen and iodide have reacted. Thus, it is evident that 0.22 mole of hydrogen and 0.22 mole of iodine are still present as such without reacting. If the mixture is kept at the same temperature even for a couple of days, there will occur no change in the concentration of either two products or any of the reactants. In this reaction, the attainment of equilibrium is recognised by observing constancy of concentration.

A well known equilibrium reaction between $N_2O_4(g)$ and $NO_2(g)$ is considered,

$$\underset{\text{Almost colourless gas}}{N_2O_4(g)} \rightleftarrows \underset{\text{Deep reddish gas}}{2NO_2}$$

In this reaction, the colour of the gaseous mixture becomes constant at equilibrium.

From the above discussion, it can be concluded that the attainment of equilibrium can be recognised by observing constancy of properties such as pressure, concentration, density of colour whichever may be found to be suitable in a given case.

Characteristics of Chemical Equilibrium

Let us now consider certain important characteristics of chemical equilibrium. These are:

1. Chemical equilibrium, at a given temperature, can be characterised by constancy of certain observable properties like pressure, concentration, density or colour.
2. Chemical equilibrium can be approached from either direction, let us take an example to illustrate this important characteristic of chemical equilibrium. We know that hydrogen gas and iodine vapour react at elevated temperature yielding hydrogen iodide gas, according to the equation.

   $$H_2(g) + I_2(g) \rightarrow 2HI(g)$$

   We know that hydrogen iodide is not a very stable compound. This on heating dissociates into hydrogen and iodine.

   $$2HI(g) \rightarrow H_2(g) + I_2(g)$$

   It means that the hydrogen-iodine reaction is reversible and can be represented as

   $$H_2(g) + I_2(g) \rightleftarrows 2HI(g)$$

   When 1 mole of H_2 is allow to mix with 1 mole of iodine in a closed vessel of one litre capacity and the vessel is allowed to stand for a long time at 448°C (the temperature of boiling sulphur), ultimately, only 1.56 moles of HI, gets formed. There occurs of further increase in the amount of HI even if the reaction is permitted to continue for several days at the same temperature of 448°C. This reveals that only 0.78 mole of each of H_2 and

I_2 has reacted. The composition of the reaction mixture at equilibrium therefore, would be 1.56 moles of Hi, 0.22 mole of H_2 and 0.22 mole of I_2.

Now the reaction is carried out by taking 2 moles of Hi i the same vessel and keeping it again in the both of boiling sulphur at 448°C. After about the same time as before, it will be found that only 0.44 mole of HI gets dissociated into H_2 and I_2. The composition of the reaction mixture, therefore, would be 1.56 moles of HI, 0.22 mole of H_2 and 0.22 mole of I_2 as before.

This reveals that when equilibrium gets attained at a given temperature each reactant and each product is having a fixed concentration and this is independent of the fact whether the reaction is started with the reactants or with the products. Thus, equilibrium can be attained from either side.

3. A catalyst can hasten the approach of equilibrium but fails to alter the state of equilibrium. In other words, it means that the relative concentrations of the products and the reactants remain the same irrespective of the presence or absence of a catalyst. For example,

   $2SO_2(g) + O_2(g) \rightleftarrows 2SO_3(g)$

   This reaction is allowed to take place, generally, at 500°C. The equilibrium in this case would be only attained after a long time. However, if a catalyst (platium or vanadium pentoxide) is add to the system, the equilibrium would be attained fairly rapidly but the equilibrium state is not altered. In other words, the relative amounts of SO_2 (the product) and SO_2 and O_2 (the reactants) at equilibrium would remain the same irrespective of the fact whether a catalyst is added or not in the reaction. Hence the yield of SO_2 would not be improved in any way by using the catalyst.

4. Chemical equilibrium is dynamic in nature because it involves two opposing reactions. One of these reactions takes place from the reactants towards the products and is known as the forward reaction. The other takes place from the products towards the reactants ad is known as the reverse reaction. When equilibrium is attained, there occurs no further change in the concentrations of the products or the reactants. This gives the idea that the reaction has stopped. But this is not the case. Actually, the

opposing reactions, the forward reaction ad the reverse reaction, are taking place simultaneously at equal rates. Hence one reaction completely undoes the effect of the other. In other words, as much of the products get formed in a given time as change back into the reactants i the same time. Hence the chemical equilibrium is dynamic and not static.

LAW OF CHEMICAL EQUILIBRIA AND HETEROGENEOUS EQUILIBRIA

When chemical equilibrium is established in one phase—a mixture of gases, a liquid solution - we have a case of *homogeneous equilibrium*. When more than one phases are involved—for example, gas and solid, or liquid and solid or liquid and gas equilibrium is said to be *heterogeneous*.

If a chemical equilibrium consists of two or more than two phases one of which at least must be a solid or liquid phase. It is said to be heterogeneous equilibrium.

For example, water in contact with its vapour in a closed space forms a two phase (Liquid-Gas) heterogeneous equilibrium. A saturated solution in equilibrium with solute is another familiar instance of *heterogeneous equilibrium*.

Dissociation of red lead can similarly be regarded as a heterogeneous equilibrium consisting of three phases, two of which (Pb_3O_4 ad PbO) are solid and the third (O_2) a gas.

$$2Pb_3O_4(s) \rightleftharpoons 6PbO(s) + O_2(g)$$

For the present we shall concern ourselves with gas-solid reactions which are the ore common.

(i) Thermal Decomposition of $CaCO_3$

A well known example of such a reaction is the dissociation of calcium carbonate:

$$CaCO_3(s) \rightleftharpoons CaO(s) + CO_2(g)$$

where there is one gas and two solids. We could write the equilibrium constant as,

$$K_C = \frac{[CaO][CO_2]}{[CaCO_3]}$$

We now have to decide what is meant by the "concentrations" of the calcium carbonate and calcium oxide. These two solids will form two separate phases, since it is unusual for solids to mix. Any pure solid at a particular temperature and pressure, will contain a constant number of molecules per unit volume, so that the molar concentration of a solid is a constant.

Looking at it another way, we have seen that the vapour pressure of a liquid in a mixture is proportional to its molar concentration; the vapour pressure of pure liquid is constant, at a definite temperature, so that the concentration must be a constant. A solid also has a vapour pressure although it may be too small to be measured.

This vapour pressure is constant provided temperature is constant, irrespective of the amount of solid that may be present. By analog, we can use the vapour pressure of a solid as a measure of the concentration and we shall then arrive at the same result, *i.e.*, the pure solid has a constant concentration. The actual value that is given to these constants does not matter provided that we always use the same values. By convention, the *concentration of a pure liquid or a pure solid is taken as unity*. The actual amount present has no effect on the concentration.

The equilibrium constant for the calcium carbonate reaction now becomes.

$$K_C = [CO_2]$$

and since we are now dealing with the concentration of a gas, it is more convenient to use the pressure:

$$K_P = p_{CO_2}$$

The pressure of the CO_2 is also the total pressure of the system, since it is the only gas present. This is referred to as the *dissociation pressure* and since it is equal to K_P it follows that the dissociation pressure has definite value for any particular temperature.

The pressure of the system is fixed and cannot be altered without upsetting the equilibrium. Experimentally it has been found that the dissociation pressure of CO_2 in equilibrium with solid $CaCO_3$ and solid CaO, at any temperature, is constant.

Similar to homogeneous equilibria, the law of chemical equilibrium is also applicable to heterogeneous equilibria. The application of this law to the heterogeneous equilibria is based of the assumption that the reaction goes on i a heterogeneous system. Consider the above mentioned thermal dissociation of calcium carbonate

$$CaCO_3(s) \rightleftarrows CaO(s) + CO_2(g)$$

We can assume that th reaction here is taking place between calcium carbonate vapour, calcium oxide vapour and carbon dioxide gas, the whole system is thus functioning like a homogenous system.

This assumption the leads us to further conclusion, that the solid $CaCO_3$ and the solid CaO, like other constituents also possess a definite vapour pressure and hence the equilibrium concentrations of the reacting substances can be expressed in terms of their partial pressures as in the case of other homogeneous equilibria. Therefore, p_{CaCO_3}, p_{CaO} and p_{CO_2}, respectively represent the partial pressures of $CaCO_3$, CaO and CO_2, then application of the law of chemical equilibrium gives the expression:

$$K_p = \frac{p_{CaO} \times p_{CO_2}}{p_{CaCO_3}}$$

The partial pressure of a solid (or a pure liquid) is very small but is constant at a particular temperature and remains so as ion as there is any solid (or liquid) present in the system p_{CaCO_3}. and p_{CaO} are thus constant. The equilibrium law equation, thus reduces to.

$$p_{CO_2} = K'_p$$

Thus, when calcium carbonate is heated in a closed vessel at a definite temperature, the pressure of CO_2 formed remains constant irrespective of the amount of calcium carbonate taken.

Therefore at a given temperature, the pressure of CO_2 will be constant. The equilibrium constant K'_p, therefore, appears to be determined only by th pressure of CO_2.

If CO_2 is allowed to escape, its pressure will decrease and further dissociation of calcium carbonate will take place until the original pressure is regarded. That is why in the line kilns, where the CO_2 escapes, the reaction proceeds practically in one direction, *i.e.*, the dissociation of $CaCO_3$.

The equilibrium pressure of CO_2 which is constant at a particular temperature is known as the dissociation pressure of $CaCO_3$. The dissociation pressure at a particular temperature will to change, even if any product of the reaction is added or removed from the system. It changes with the change of temperature.

(ii) Thermal Dissociation of Solid Ammonium Hydrrosulphide

The solid ammonium hydrosulphide, on heating dissociates into two gases, viz., ammonia and hydrogen sulphide $NH_4HS(s) \rightleftharpoons NH_3(g) +$

$H_2S(g)$. It is another example of heterogeneous equilibrium. Since the partial pressure of solid component, *i.e.*, NH_4HS will remain constant at a particular temperature, as ion as there is any solid left in the system th equilibrium constant KP can be expressed as:

$$K_P = \frac{p_{NH_3} \times p_{H_2S}}{p_{NH_4HS}} = \frac{p_{NH_3} \times P_{H_2S}}{\text{constant}}$$

where p_{NH_3} and p_{H_2S} represent the partial pressures of ammonia and hydrogen sulphide. Evidently

$$K'_P = p_{NH_3} \times p_{H_2S}$$

i.e. At any temperature the product of the partial pressures of ammonia and hydrogen sulphide (p_{NH_4HS}) being very small, is neglected), and hence each will be equal to p/2.

i.e. $$p_{NH_3} \times p_{H_2S} = p/2$$

$$K'_P = p_{NH_3} \times p_{H_2S} = p/2 \times p/2 = p^2/4$$

Thus, from the total pressure of the system which is constant at a given temperature, the value K'_P can be determined. If one of the two gases is introduced in the system, from outside at equilibrium, p_{NH_3} will no longer be equal to p_{H_2S}, but the product $p_{NH_3} \times p_{H_2S}$ will still remain constant. Consequently, the dissociation of the solid will be diminished and a certain amount of solid hydrosulphide will be deposited.

Isambart confirmed the above observations by investigating the dissociation equilibrium of ammonium hydrosulphide and ammonium cyanide.

(iii) Iron-Steam Reaction

Another type of heterogeneous equilibrium is the reaction of steam with iron:

$$3Fe(s) + 4H_2O(g) \rightleftarrows Fe_3O_4(s) + 4H_2(g)$$

The equilibrium constant is given by

$$K_c = \frac{[Fe_3O_4][H_2]^4}{[Fe]^3 \times [H_2O]^4}$$

If the active asses of solids are taken constant, then

$$K' = \frac{[H_2]^4}{[H_2O]^4}$$

or $$K'' = \frac{[H_2]}{[H_2O]}$$

or $$K_P = \frac{p_{H_2}}{p_{H_2O}}$$

Experimentally also it has been found that the ratio p_{H_2}/p_{H_2O} is constant at any one temperature and independent of amounts or nature of iron and iron oxide present.

(iv) Water Gas Reaction

In this reaction steam is passed over heated coke:

$$H_2O(g) + C(s) \rightleftarrows CO(g) + H_2(g)$$

for which the equilibrium constant is

$$K_P = \frac{p_{CO} \times p_{H_2}}{p_{H_2O}}$$

as the partial pressure of carbon is taken as unity.

(v) Dehydration Equilibrium of Salt Hydrates

Dehydration of salts is an important class of heterogeneous reactions. In such reactions, the equilibrium constant is totally dependent of the vapour pressure of water. Salt hydrates during dehydration often dissociate in steps to form a number of intermediate hydrates according to the prevailing pressure of moisture is contact with th solid hydrates. Thus copper sulphate pentahydrate on dissociation yields trihydrate, monohydrate and then the anhydrous salt in the above order. Following three equilibria are established:

$$CuSO_4 . 5H_2O \rightleftarrows CuSO_4 . 3H_2O + 2H_2O \quad \text{(1st stage)}$$

$$CuSO_4 . 3H_2O \rightleftarrows CuSO_4 . H_2O + 2H_2O \quad \text{(2nd stage)}$$

$$CuSO_4 . H_2O \rightleftarrows CuSO_4 + H_2O \quad \text{(3rd stage)}$$

However, all the three equilibria do not exist simultaneously. Each one of them has an equilibrium pressure for its existence at a particular temperature. At 25°C the equilibrium vapour pressure for pentahydrate-trihydrate system is 7.8 mm, for the trihydrate-monohydrate, and monohydrate-anhydrous system, the pressure values are 5.6 mm and 0.8 mm respectively. Application of law of chemical equilibrium to th equilibria for th three stages also proves that the equilibrium vapour pressures for three systems are independent of the active masses (or partial pressures) of the solids which are very small and are constant at a particular temperature and remain so, as long as there is any solid present in the system and are arbitrarily fixed as unity. They are, however, different, for different equilibria.

The equilibrium vapour pressure is not that of water vapour or any hydrate. It is the property of the system as a whole and hence, if the system changes, so must the equilibrium vapour pressure.

Thus, the equilibrium for the reaction at the *given temperature* is given by $K_{P_1} = p^{2\,(1)}_{H_2O}$ or $p_{H_2O}(1)m = \sqrt{K_{P_1}}$. So as long as $p_{H_2O}^{(1)}$ is equal to $\sqrt{K_{P_1}}$ both $CuSO_4 . 5H_2O(s)$ and $CuSO_4 . 3H_2O(s)$ will be present. When $P_{H_2O}^{(1)}$ becomes less than $\sqrt{K_{P_1}}$, more and more of $CuSO_4 . 5H_2O(s)$ is converted into $CuSO_4 . H_2O$ till the conversion is complete.

At this stage the third reaction starts for which the equilibrium constant is given by, $K_{P_3} = p^{2\,(3)}_{H_2O}$.

Once again when $p_{H_2O}^{(3)}$ is less than K_{P_3} at a given temperature the dehydration reaction (in 3rd stage) takes place.

If isothermal dehydration of $CuSO_4 . 5H_2O$ at 25°C is carried out, dehydrate is formed according to the 1st stage and the pressure remains constant so long as both penta and trihydrate are i equilibrium. Since the water vapour is being continuously removed, the pentahydrate will completely dissociate into trihydrate. Fig. 2.1 shows all these and subsequent changes in this system on isothermal dehydration. Pressure 7.8 mm as shown in this figure corresponds to complete conversion of pentahydrate into trihydrate.

With the disappearance of pentahydrate, 1st stage equilibrium loses it significance and hence the pressure suddenly drops to 5.6 mm with the appearance of monohydrate. The pressure is constant so long as trihydrate and monohydrate are in equilibrium. Conversion of trihydrate into monohydrate is complete and hence the pressure again drops to 0.8 mm with the appearance of anhydrous copper sulphate.

The pressure would now fall only when monohydrate completely changes over to anhydrous copper sulphate. The stepwise process of dehydration of $CuSO_4 . 5H_2O$ is shown in Fig. 2.1.

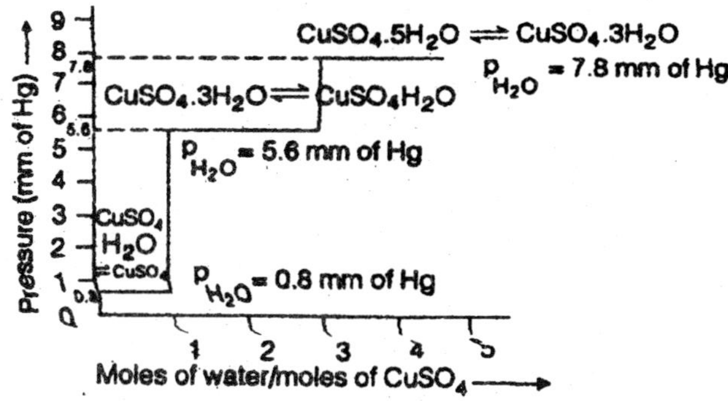

Fig. 2.1 : Dehydration Equilibrium of $CuSO_4 . 5H_2O$ at 25°C.

During summer many salt hydrates *e.g.*, $FeSO_4 . 7H_2O$, $Na_2S_2O_3 . 5H_2O$ get easily dehydrated as K_p at high temperature is large and due to low humidity in air p_{H_2O} is less which are favourable for dehydration. All dehydration reactions can be reversed if the vapour pressure of water beyond a limit determined equilibrium pressure of water is increased.

(vi) Dissociation of Ammoniates of Silver Chloride or Deammination

This also provides another example of heterogeneous equilibria. The deammination of triammoniate of silver chloride $AgCl . 3NH_3$ takes place through stages in the following successive equilibria:

$$\underset{\text{Triammoniate of silver chloride}}{2[AgCl . 3NH_3(s)]} \rightleftharpoons \underset{\text{Sesqui-ammoniate of silver chloride}}{2AgCl . 3NH_3(s) + 3NH_3(g)}$$

$$2AgCl . 3NH_3(s) \rightleftharpoons \underset{\text{Mono-ammoniate of silver chloride}}{2[AgCl . NH_3](s)} + NH_3(g)$$

$$AgCl . NH_3(s) \rightleftharpoons AgCl(s) + NH_3(g)$$

This is similar to dehydration of hydrates and hence the values of the three equilibrium constants, may be represented as

$$Kp_1 = p_3NH_3 \quad \text{or} \quad pNH_3 = 3\sqrt{Kp_1}$$

$$Kp_2 = pNH_3 \quad \text{or} \quad pNH_3 = Kp_2$$

$$Kp_3 = pNH_3 \quad \text{or} \quad pNH_3 = Kp_3$$

During a given stage of demmination of ammoniates, the pressure of ammonia is constant and changes only when one particular stage of deammination is completed and the next stage sets in.

(vii) Liquid-Gas Equilibrium

Action of steam on molten sulphur represents an equilibrium of this type:

$$3S(l) + 2H_2O(g) \rightleftharpoons SO_2(g) + 2H_2S(g)$$

The equilibrium in this case will be given by

$$\frac{(p_{SO_2}) \times (p_{H_2S})^2}{(p_{H_2O})^2} = K'p. \text{ [Sulphur, being pure liquid, } p_s = \text{constant].}$$

(viii) Solid-Liquid Equilibrium

Equilibrium between pure liquid and solid phases only is not possible, as the active masses in these cases have been assumed to be constant. Presence of a gaseous phase or a solution is necessary for any chemical equilibrium to be established.

LE-CHATELIER'S PRINCIPLE : PREDICITING EQUILIBRIUM STATE

This generalisation which predicts the influence on a dynamic equilibrium by altering concentration of species, total pressure and temperature was announced independently by Le-Chatelier in 1885 and Braun in 1886.

This is entirely a general law and applies to physical as well as chemical systems. It may be stated as, *When a system in equilibrium is subjected to a stress, i.e., change of concentration, pressure or temperature, the equilibrium shifts i such a direction as to undo or tends to undo the effect of the change or stress.*

This law is of great help in prediciting the conditions for getting maximum yields of the desired products which will be clear from the following considerations:

(1) The Effect of Changing Concentrations of the Reactants

If any of the components at the equilibrium state is added from outside, its concentration increases ad hence, the equilibrium tends to change in such a way as to counteract the effect of increase of concentration of the added substance. In other words, the equilibrium will shift in the direction which the added substance is being consumed. This is illustrated by the following example:

Let us consider the following reaction in aqueous solution:

$Fe^{3+}(aq) + SCN^-(aq) \rightleftharpoons Fe(SCN)^{2+}(aq)$

If soluble thiocyanate salt is added to the above equilibrium solution containing both Fe^{3+} (aq) and SCN^- the red colour of the complex ion $Fe(SCN)^{2+}$ increases. A new equilibrium is then attained in which more $Fe(SCN)^{2+}$ is present than that was there before the addition of SCN^-. Increasing the concentration of SCN^- ions increases the concentration of the $Fe(SCN)^{2+}$. This is in accord with *Le-Chatelier's principle*. The change imposed on the system was an increase in the concentration of SCN^-. The change can be counteracted in part by some Fe^{3+} and SCN^- ions reacting to form $Fe(SCN)^{2+}$.

The same argument applies to addition of ferric ion from a soluble ferric salt. In each case, the formation of $Fe(SCN)^{2+}$ uses up a portion of the added reactant practically counteracting the change.

So in a chemical equilibrium, increasing the concentrations of the reactants shifts the equilibrium in the direction of the produces whereas increasing the concentrations of the products shifts the equilibrium in the direction of the reactants.

(2) The Effect of Altering Pressure

In most of the system composed of liquids and solids, the effect of pressure on equilibrium is relatively shifts. In gaseous reactions, on the on the hand, the effect of pressure is often extremely important.

Increase of external pressure will cause the equilibrium to shift in the direction which will bring about a lowering of pressure. In a chemical reaction, this means that the equilibrium will shift in the direction which produces the smaller umber of as molecules. This is clear from the following equilibrium:

Example (i) $PCl_5(g) \rightleftharpoons PCl_3(g) + Cl_2(g)$

On increasing the total pressure, the system will now occupy a much smaller volume than it did previously and the total umber of molecules present per unit volume is greater than it was under the original equilibrium conditions. This change can be counteracted in part if some PCl_3 and Cl_2 combine to form PCl_5. The total number of molecules present is reduced (2 molecule unite to form 1 molecule).

Therefore, the increase in pressure shifts the above equilibrium i the reverse direction.

(ii) $N_2(g) + 3H_2(g) \rightleftarrows 2NH_3(g)$

The increase in pressure causes the equilibrium to shift form left to right (forward reaction is favoured).

(iii) $\underset{1\ \text{mole}}{N_2O_4(g)} \rightleftarrows \underset{2\ \text{moles}}{2NO_2(g)}$

In this case also if the pressure is increased the volume of the system decreases ad the total number of moles per unit volume is now more than before. The change can be encountered if the equilibrium shifts in the direction in which total number of molecules decreases. Thus according to *Le-Chatelier's principle*, on applying pressure the system shifts i favour of N_2O_4, *i.e.*, suppresses the dissociation of N_2O_4.

(iv) $\underset{2\ \text{moles}}{2SO_2(g)} + \underset{1\ \text{mole}}{O_2(g)} \rightleftarrows \underset{2\ \text{moles}}{2SO_3(g)}$

The increase in pressure tends to favour equilibrium in the forward direction, *i.e.*, the formation of sulphur trioxide.

(v) $\underset{1\ \text{mole}}{CaCO_3(s)} \rightleftarrows \underset{1\ \text{mole}}{CaCO(s)} + \underset{1\ \text{mole}}{CO_2(g)}$

In this case increase in pressure favours the backward reaction, *i.e.*, suppresses the dissociation of $CaCO_3$.

(vi) $N_2(g) + O_2(g) \rightleftarrows 2NO(g)$

If the pressure on this equilibrium is increased, the gases are compressed to a smaller volume and the concentrations are all increased. Let us suppose that the equilibrium state is changed to favour products, *i.e.*, some N_2 and O_2 molecules react to form a exactly equal number of molecules of NO_2. Since no change takes place in the total number of moles, the proposed change in the equilibrium does not partially reduce the pressure change. According to Le-Chatelier's principle, the processes occur so as to counteract partially imposed change. So no change of the equilibrium state in this reaction is expected when the pressure is altered.

The equilibrium state is not affected by a pressure change for any equilibrium gas mixture where the number of reactant molecules is the same as the number of product molecules in the balanced reaction.

(3) The Effect of Altering Temperature

The shifting of an equilibrium by change of temperature can also be predicted by applying Le-Chatelier's principle. *If the temperature of a system i equilibrium is increased, the equilibrium will shift in the direction which absorbs heat.* Lowering of the temperature will shift the equilibrium in the direction i which heat evolves.

Example: (i) *In the reversible reaction,*

$$N_2(g) + 3H_2(g) \rightleftarrows 2NH_3; \Delta H = -92.38 \text{ kJ}$$

increase in temperature shifts the equilibrium from right to left (absorption of heat), and decrease in temperature will shift it form left to right (evolution of heat).

The equilibrium mixture will contain less ammonia at high temperature, and more at low temperature or formation of ammonia is not favoured at high temperature.

(ii) *In the reaction,*

$$N_2(g) + O_2(g) \rightleftarrows 2NO(g); \Delta H = +180.5 \text{ kJ}$$

increase in temperature will shift the equilibrium from left to right (absorption of heat), and decrease in temperature will shift it from right to left (evolution of heat) or formation of NO is favoured at high temperature.

(iii) *In the reaction,*

$$PCl_5(s) \rightleftarrows PCl_3 + Cl_2(g); \quad \Delta H = -153.72 \text{ kJ}$$

increase in temperature shifts the equilibrium from right to left (absorption of heat). and decrease in temperature will shift it from left to right (evolution of heat). Higher temperature thus favours dissociation of PCl_5.

(iv) *In the reaction,*

$$2SO_2(g) + O_2(g) \rightleftarrows 2SO_3(g); \quad \Delta H = -96.25 \text{ kJ}$$

increase i temperature shifts the equilibrium from right to left (absorption of heat) or favours the backward reaction. A decrease i temperature shifts the equilibrium from left to right (evolution of heat) or favours formation of SO_3.

(v) $CaCO_3(s) \rightleftarrows CaO(s) + CO_{2(g)}$;

$\Delta H = +179.96$ kJ

The decomposition of $CaCO_3$ is accompanied by absorption of heat (endothermic reaction), thus according to *Le-Chatelier's principle*, increase in temperature favours decomposition of $CaCO_3$.

Le-Chatelier's principle finds extensive application in predicting the coditions for getting maximum yield of products in reaction of industrial importance. A few common examples are discussed below.

(i) Formation of Ammoia

(Haber's Process)

$$\underset{1\text{ mole}}{N_2} + \underset{3\text{ moles}}{3H_2} \rightleftharpoons \underset{2\text{ moles}}{2NH_3}; \quad \Delta H = -92.38\text{ kJ}$$

(a) *Effect of temperature:* If temperature is raised, the yield of NH_3 diminishes and the proportion of N_2 and H_2 in the equilibrium mixture increases. On the other hand, when the temperature is lowered the equilibrium shifts i the direction i which heat is produced, *i.e.*, toward reaction. Hence, *low temperature favours the formation of ammonia.* It may be pointed out that at low temperature, the equilibrium is reached very slowly. Even with catalyst, the rate of reaction is not rapid at low temperature. Hence, a compromise is to be made between the low temperature, and time required for equilibrium to be reached. A temperature of 500°C is generally used.

(b) *Effect of pressure:* If at a constant temperature, pressure is increased the total volume will decrease but number of molecules per unit volume will increase. To counteracts that, equilibrium will shift in the direction of decrease of molecules *i.e.*, the forward direction. Hence *high pressure favours the formation of ammonia.*

(c) *Effect of increasing concentrations of reactants:* By adding a quality of one or both of the reactants (N_2 or H_2) to the equilibrium mixture, the reaction will proceed in the direction in which the add components are used up, *i.e.*, the forward direction. Hence, *high concentration of N_2 or H_2 (or both) favours the formation of NH_3.*

(ii) Oxidation of SO_2 to SO_3

(*Contact process for the manufacture of sulphuric acid*).

$$\underset{2\text{ moles}}{2SO_2} + \underset{1\text{ mole}}{O_2} \rightleftharpoons \underset{2\text{ moles}}{2SO_3}; \quad \Delta H = -96.25\text{ kJ}$$

The forward reaction is *exothermic* and is accompanied by a *decrease in number of moles.*

(a) *Effect of temperature :* If temperature is raised, then according to the Le-Chatelier's principle the heat supplied to the mixture constitutes an imposed stress ad reaction will take place to counter this stress, *i.e.*, the resulting reaction is such that it tends to lower the temperature. But a reaction that lowers temperature is endothermic, it is the reverse of the above exothermic reaction.

$$2SO_3 \rightarrow 2SO_2 + O_2 - \text{Heat energy}$$

In other words, as the temperature rises, the yield of SO_2 diminishes and the concentration not the reactants (SO_2 + O_2) increases.

(b) *Effect of concentration :* The addition of more of oxygen imposes a stress. According to Le-Chatelier's principle, to counteract the stress more SO_3 will be formed; thereby decreasing the concentration of SO_2.

(c) *Effect of pressure :* If a constant temperature, the pressures increases, the total volume will decreases but the total umber of molecules per unit volume will increase. To counteract that, equilibrium will shift in the direction of decrease of molecules, *i.e.*, the forward direction.

(iii) *For Nitric of Nitric Oxide*

(*Birkeland and Eyde Process*)

$$\underset{1\ \text{mole}}{N_2(g)} + \underset{1\ \text{mole}}{O_2(g)} \rightleftarrows \underset{2\ \text{mole}}{2NO}; \ \Delta H = 180.5\ \text{kJ}$$

The forward reaction is *endothermic and no change in volume occurs*:

(a) *Effect of temperature :* As the forward reaction is accompanied by a absorption of heat, cosequetly *a increase in temperature will favour the formation of NO.*

(b) *Effect of pressure :* As the toward reaction is accompanied by no change in volume, consequently the pressure will have no effect on the equilibrium.

(c) *Effect of concentration :* The formation of NO will be favoured b an excess of O_2 or N_2 or both.

Applications of Le-Chatelier's Principle to Physical Equilibria

(1) *Effect of pressure on melting point of ice :* At 0°C ice is i equilibrium with water and this equilibrium can be expressed in

the form of an equation which is give below. Ice occupies more volume than corresponding amount of water at the same temperature.

$$\underset{\text{more volume}}{\text{ice}} \rightleftharpoons \underset{\text{less volume}}{\text{water}}$$

If this system is contained in a cylinder with a piston of which the temperature is 0°C and if pressure be increased then according to Le-Chatelier's principle the system will adjust in the direction of the lesser volume, *i.e.*, more ice will melt. This means the melting point of ice is lowered by increasing the pressure and *vice versa*.

(2) *Effect of temperature on melting of ice*

$$\underset{\text{more volume}}{\text{ice}} \rightleftharpoons \underset{\text{less volume}}{\text{water}} \quad - \text{Q joules or } \Delta H = +\text{ve}$$

The forward reaction is accompanied by absorption of heat. Hence increase of temperature favours the formation of water.

(3) *Effect of pressure on solubility of salts :* We come across two type of salts. Those which *dissolve with a decrease in volume* [*i.e.*, the volume of solution is less than the volume of the solid and volume of the solvent] their solubility according to this principle will increase with increase of pressure and *vice versa*.

And the second type is of those salts which dissolve leading to an increase in volume, their solubility will decease with increase of pressure and *vice versa*.

(4) *Effect of temperature on solubility of salts :* The dissolution of salts like KCl, NH_4Cl etc..

$$NH_4Cl + aq \rightleftharpoons NH_4Cl(aq) - \text{Q joules or } \Delta H = +\text{ve}$$

is accompanied by absorption of heat and hence the forward reaction or solubility of these salts will increase with the increase of temperature.

On the other had, solubility of substances like NaOH, KOH, $Ca(OH)_2$ etc.

$$NaOH + aq \rightleftharpoons NaOH\ (aq) + \text{Q joules or } \Delta H = -\text{ve}$$

which are accompanied by evolution of heat, their solubility will decrease with the rise of temperature.

(5) *Effect of heat and pressure on vapourisation of water*

$$\text{Water (l)} \rightleftharpoons \text{Water vapour} - \text{Q joules or } \Delta H = +\text{ve}$$

The forward reaction is accompanied by a large increase of volume and absorption of heat. Hence an increase in pressure will favour the backward reaction, *i.e.*, liquefaction, while increase of temperature will favour the forward reaction, *i.e.*, vapourisation.

THERMODYNAMIC TREATMENT OF LE CHATELIER'S PRINCIPLE

Let us consider the following general chemical reaction:

$$aA + bB \rightleftarrows mM + nN$$

As the free energy change, ΔG for the reaction depends upon the temperature, pressure and the extent of the reaction ζ, it implies that

$$\Delta G = f(T, P, \zeta) \qquad ...(1)$$

As ΔG is a state function, the total change in the reaction free energy with T, P and ζ can be put as follows:

$$\text{or } d(\Delta G) = \left(\frac{\partial(\Delta G)}{\partial T}\right)_{P,\zeta} dT + \left(\frac{\partial(\Delta G)}{\partial P}\right)_{T,\zeta} dP + \left(\frac{\partial(\Delta G)}{\partial \zeta}\right)_{T,P} d\zeta \qquad ...(2)$$

$$\text{But, } \Delta G = \left(\frac{\partial G}{\partial \zeta}\right)_{T,P} = G' \text{ (By definition)} \qquad ...(3)$$

Therefore, eq. (2) can be put as follows:

$$\therefore \quad d(\Delta G) = \frac{\partial}{\partial T}\left\{\left(\frac{\partial G}{\partial \zeta}\right)_{T,P}\right\}_{P,\zeta} dT + \frac{\partial}{\partial T}\left\{\left(\frac{\partial G}{\partial \zeta}\right)_{T,P}\right\}_{T,\zeta} dP + \frac{\partial}{\partial \zeta}\left(\frac{\partial G}{\partial \zeta}\right) d\zeta \qquad ...(4)$$

From the mathematical properties of state functions, it is possible to express $d(\Delta G)$ as follows:

$$\therefore \; d(\Delta G) = \frac{\partial}{\partial \zeta}\left\{\left(\frac{\partial G}{\partial T}\right)_{P,\zeta}\right\} dT + \frac{\partial}{\partial \zeta}\left\{\left(\frac{\partial G}{\partial P}\right)_{T,\zeta}\right\} dT + \left(\frac{\partial^2 G}{\partial \zeta^2}\right) d\zeta \qquad ...(5)$$

But $(\partial G/\partial T)_P = -S$ and $(\partial G/\partial P)_T = V$. Therefore, Eq. (2) may be written as follows:

$$\therefore \; d(\Delta G) = -\left(\frac{\partial S}{\partial \zeta}\right)_{T,P} dT + \left(\frac{\partial V}{\partial \zeta}\right)_{T,P} dP + \left(\frac{\partial^2 G}{\partial \zeta^2}\right)_{T,P} d\zeta \qquad ...(6)$$

The quantity $(\partial S/\partial \zeta)_{T, P} = \Delta S$ = Rate of change of entropy with the progress variable, ζ *i.e.*, this quantity refers to the entropy change of the reaction and the quantity $(\partial V/\partial \zeta)_{T, P} = \Delta V$ refers to the volume change of the reaction. Thus, Eq. (5) can be put as follows:

$$d(\Delta G) = -\Delta S\,dT + \Delta V\,dP + \left(\frac{\partial^2 G}{\partial \zeta^2}\right)_{T, P} d\zeta \qquad ...(7)$$

If a case is considered where T, P and get changed in such a manner that equilibrium is maintained, then, ΔG would be minimum and $d(\Delta G) = 0$, so that

or $$-\Delta S\,dT + \Delta V\,dP + G''d\zeta = 0 \qquad ...(8)$$

where $G'' = (\partial^2 G/\partial \zeta^2)_{T, P}$ is positive because G is minimum at equilibrium.

At equilibrium, $\Delta G = 0$ and $\Delta S = \Delta H/T$, so that

$$-(\Delta H/T)dT + \Delta V dP + G''d\zeta = 0 \qquad ...(9)$$

Let us now consider two cases.

Case A: Variation of ζ with T at constant P at equilibrium. If P is constant, $dP = 0$, so that Eq. (9) becomes as follows:

$$-(\Delta H/T)(dT)_P + G''d\zeta = 0 \qquad ...(10)$$

$$(\partial \zeta/\partial T)_P = \Delta H/(TG'')$$

As G'' is a positive quantity, the sign of $(\partial \zeta/\partial T)_P$ would depend on the sign of ΔH. Remember that $(\Delta H = H_{products} - H_{reactants})$.

(i) For isothermic reactions, ΔH is negative $(H_{products} < H_{reactants})$. Hence, $(\partial \zeta/\partial T)_P$ would be negative.

If dT is positive (which is the case when temperature gets increased), t~en $d\zeta$ would be negative, *i.e.*, ζ should decrease with increase in T and the concentration of the reactant would increase in the direction of high enthalpy.

When dT is negative (*i.e.*, when temperature gets decreased), $d\zeta$ should be positive, *i.e.*, ζ should increase with decrease in T and concentration of the product would get increased in the direction of low enthalpy.

(ii) For endothermic reactions, ΔH is positive $(H_{products} < H_{reactants})$. Hence $(\partial \zeta/\partial T)_P$ would be positive.

When dT is negative, (*i.e.*, when the temperature is lowered), then dζ too, must be negative) so than ζ should decrease with decrease in T and reactants with lower enthalpy are formed.

If dT is positive (*i.e.*, when the temperature is increased), then dζ should also be positive and ζ should increase causing the formation of more products with higher enthalpy.

Hence, it can be concluded that an increase in temperature would shift the equilibrium towards the high enthalpy side and a decrease in temperature would shift the equilibrium towards the low enthalpy side. This statement has been similar to a statement of Le Chatelier's principle.

Case B. Variation of ζ with P at constant T at equilibrium : When T is constant, dT = 0, so that Eq. (9) would become as follows:

$$\Delta V(dP)_T + G''(d\zeta) = 0 \qquad \text{...(11)}$$

$$(\partial\zeta/\partial P)_T = \Delta V/G'' \qquad \text{...(12)}$$

As G'' is positive, the sign of $(\partial\zeta/\partial T)_P$ would depend upon that of ΔV.

(i) When $\Delta V = 0$ (*i.e.*, $V_{products} = V_{reactants}$) then ζ would be independent of P at constant T.

(ii) When ΔV is positive (*i.e.*, $V_{products} > V_{reactants}$) the RHS of Eq. (12) would be negative so that $(\partial\zeta/\partial T)_P$ will also be negative.

Increase in pressure (dP > 0) will decrease ζ(dζ > 0) so that equilibrium would shift towards the reactant side. On the other hand, decrease in pressure (dP < 0) will increase ζ(dζ > 0) therefore equilibrium would shift towards the product side.

(iii) When ΔV is negative (*i.e.* $V_{products} < V_{reactants}$) then the RHS of Eq. 12 would be positive so that $(\partial\zeta/\partial P)_T$ would also be positive.

Increase in pressure (dP > 0) will increase ζ(dζ > 0) so that equilibrium will get shifted towards the product side. On the other hand decrease in pressure (dP < 0) will decrease ζ(dζ < 0), so that equilibrium would be shifting towards the reactant side.

Thus it can be concluded that an increase in pressure would shift the equilibrium towards the low volume side of the reaction, whereas a decease in pressure would shift the equilibrium towards the high volume side. This is another statement of Le Chatelier's Principle.

DE DONDERS CONCEPT OF DEGREE OF ADVANCEMENT OF A REACTION

In order to understand this concept, *i.e.*, the progress of a reaction, we shall consider the hypothetical reaction of the following type which is proceeding in a closed vessel.

$$aA + bB \rightleftarrows cC + dD.$$

As this reaction is being carried out in a closed vessel, it means that the total mass of the system, evidently, will remain constant although the number of moles of each reactant and product will go on changing with time. If ζ represents the progress or degree of advancement of the reaction then $d\zeta$ will represent the small increase in the extent of the reaction that takes place i any given interval of time. Due to this, the amount of A decreases by a $d\zeta$ and that of B by b $d\zeta$ while the amount of C increases by cd ζ and that of D increased by d $d\zeta$. Mathematically, these changes may be represented as follows:

$$dn_A = -ad\zeta \qquad dn_B = -bd\zeta$$

$$dn_C = -cd\zeta \qquad dn_D = -dd\zeta$$

If the temperature and pressure of the system are assumed to be constant, the change in Gibbs free energy for this small progress of this reaction will be as follows:

$$(dG)_{T,P} = \sum_{I} \mu_i dn_i$$

$$= (c\mu_C + d\mu_D - a\mu_A - b\mu_B)d\zeta$$

or

$$\left(\frac{\partial G}{\partial z}\right)_{T,P} = (c\mu_C + d\mu_D - a\mu_A - b\mu_B)$$

$$= \sum_{I} v_i \mu_i$$

where v_i is the difference between the stoichiometric coefficients of the products and those of the reactants, *i.e.*, (c + d) – (a + b).

De Donder introduced the term $\left(\frac{\partial G}{\partial \zeta}\right)_{T,P}$ which may be defined as the change of *Gibbs free energy per unit change in the extent of the reaction*. Three cases may arise:

(i) If the value of this term is negative, the reaction will proceed from left to right.

(ii) If the value of this term is positive, the reaction will proceed from right to left.

(iii) If the value of this term is zero, th system will be in a state of equilibrium.

DERIVATION OF LAW OF MASS ACTION

From Kinetic Theory of Gases

According to kinetic considerations the number of reactions per second between any tow gases is proportional to the number of collisions of molecules of both the gases taking place per second.

Consider a homogeneous gaseous reaction

$$A + B \rightarrow \text{Products}$$

Suppose a unit volume contains one molecule of each A and B, the probability of collision at any instant is $1 \times 1 = 1$.

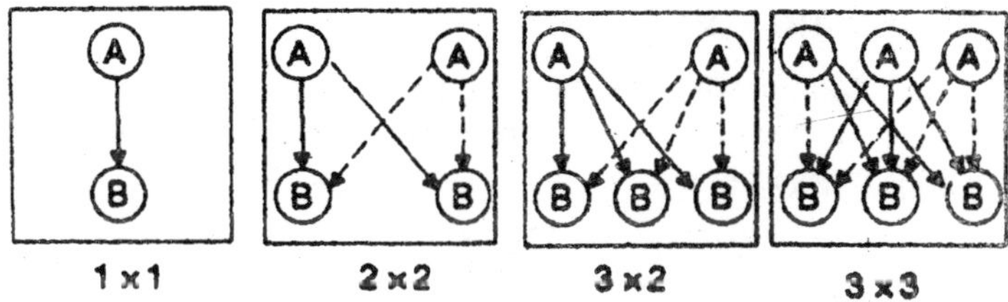

Fig. 2.2 : Collision of reacting molecules.

Where the number of molecules of A and B are two each then the possibility of collision is equal to $2 \times 2 = 4$. Suppose, there are two molecules of A and three molecules of B in it volume, the number of chances of collisions per unit time is $2 \times 3 = 6$. Similarly, if there are three molecules of each A and B per unit volume, the chances of collision will be $3 \times 3 = 9$.

Thus, chances of collisions are equal to the product of umber of molecules present per unit volume.

$\therefore$ Rate of reaction $\propto$ product of molecules of A and B per unit volume.

$$\propto [A][B]$$

This is exactly in accordance with the statement of the law of mass action.

REVERSIBLE REACTIONS

A reaction is said to be reversible if the composition of the reaction mixture on the approach of equilibrium at a given temperature is the same irrespective of the initial state of the system, *i.e.*, irrespective of the fact whether we start with the reactants or the products. A reversible reaction never tends to completion.

Some examples of reversible reactions are as follows:

$$Fe + 4H_2O \rightleftarrows Fe_3O_4 + 4H_2$$

$$N_2 + O_2 \rightleftarrows 2NO$$

$$H_2 + I_2 \rightleftarrows 2HI$$

The extent of reversibility of a reaction is governed by the nature of the reaction and the experimental conditions of temperature, concentrations of substances and pressure on the system, if gases are involved. Most of the reaction that we know are reversible in nature.

IRREVERSIBLE REACTIONS

A reaction in which the product do not interact to form the original reactants is called an irreversible reaction. Some examples of irreversible reactions are as follows:

$$2KClO_3 \rightarrow KCl + 3O_2$$

$$NaCl + AgNO_3 \rightarrow AgCl + NaNO_3$$

Irreversible reactions proceed to completion in one direction. Irreversible reactions have their equilibrium positions almost on one extreme.

Most of the precipitation reactions are irreversible.

EXPERIMENTAL PROOF FOR DYAMICAL EQUILIBRIUM

Now, it has been possible to establish the dynamic nature of chemical equilibrium experimentally in several cases. This has been demonstrated in the case involving hydrogen-iodine-hydrogen iodide reaction.

$$H_2(g) + I_2(g) \rightleftarrows 2HI(g)$$

The equilibrium of this reaction is attained at 448°C. As soon as this equilibrium is attained, a radioactive isotope of iodine is introduced into the reaction mixture. After sometime, it will be found that HI contains radioactive iodine whereas the relative amounts of HI. H_2 and I_2 do not get changed. This reveals that although no change is taking place in the relative amounts of the reactants and products, yet chemical reaction is taking place from left to right. But there occurs no increase i the amount of Hi. Therefore, it means that the reaction is taking place from left to right as well at the same rate. Thus, it means that although the system has been in equilibrium, yet the two opposing reactions have been proceeding and the equilibrium conditions are being maintained by a dynamic balance between the two reactions.

LIMITATIONS OF THE EQUATION FOR CHEMICAL EQUILIBRIUM

The equation for representing chemical equilibrium falls to reveal how far the reaction would proceed from left to right or right to left before the equilibrium is attained. Suppose the following equilibrium is considered:

$$N_2O_4(g) \rightleftharpoons 2NO_2(g).$$

The above equation does not reveal what fraction of N_2O_4 gets dissociated into NO_2. If the temperature of the reaction is raised by 10°C, there occurs a considerable increase in the dissociation of N_2O_4 but even then the equation for the reaction remains the same.

The equation used for representing the chemical equilibrium fails to reveal how long it takes for a reaction to attain equilibrium. This can be understood by considering the equilibrium of water vapour with hydrogen and oxygen

$$H_2O(g) \rightleftharpoons H_2(g) + \frac{1}{2}O_2(g)$$

Water is a highly stable compound. Its degree of dissociation is appreciably small. Even at 2000°C, its dissociation is mainly 0.6%. Thus, the equilibrium is lying towards much to the left. However, it has been verified experimentally that a mixture of hydrogen and oxygen remains as such for a sufficient long time without undergoing reaction to form water and establish the above equilibrium. The equation for the equilibrium fails to reveal that it requires so long to the equilibrium.

If an electric arc is struck in the hydrogen-oxygen mixture, the reaction starts instantaneously and the equilibrium is attained within minute or so. The equation of the chemical equilibrium fails to reveal the rate at which a chemical system comes to equilibrium and the conditions under which the equilibrium can be attained rapidly.

LOW OF MASS ACTION

A qualitative relation connecting the concentrations of the reactants and products with the amount of chemical change produced known as the low of mass action was proposed in 1867 by Guldbeg and Waage. This law can be stated as follows:

"*The rate at which a substance reacts a proportional to the active mass and the rate of a chemical reaction is proportional to the product of active masses of the reactants.*"

By the term active mass, it means the *molecular concentration.* In gases and solutions, it means the umber of gram molecules (*i.e.*, moles) present per litre. In the case of solids or pure liquids the active mass may be regarded as constant. *i.e.*, for sake of simplicity, it is taken as unity.

Active mass of a substance is expressed by enclosing the symbol or the formula of a substance i square brackets *i.e.*, [A] or by the symbol C at the base of which the symbol of the substance is placed. *i.e.*, C_A

Let us consider the following reversible reaction taking place at constant temperature:

$$A + B \rightleftarrows C + D$$

According to the law of mass action, the velocity of forward reaction: v_1 is given by

$$v_1 \propto C_A + C_B$$

$$v_1 \propto k_1 C_A + C_B \qquad ...(1)$$

where C_A and C_B are the molecular concentrations of A and B and k_1 is a constant which may be called the *velocity constant of the forward reaction*. The concentrations of A and B go on decreasing with the progress of the reaction and hence the velocity of the forward reaction gradually decreases with time as shown graphically in Fig. 2.3.

Since the reaction is reversible, the products as soon as they accumulate, start reacting to reform the reactants. The velocity of backward reaction is given by

$$v_2 = k_2 + C_C + C_D \qquad ...(2)$$

where C_C ad C_D are the molecular concentrations of C and D and k_2 is the velocity constant of the backward reaction. As the concentrations of the products gradually increase, the velocity of the backward reaction will also steadily increase, as shown in Fig. 2.3.

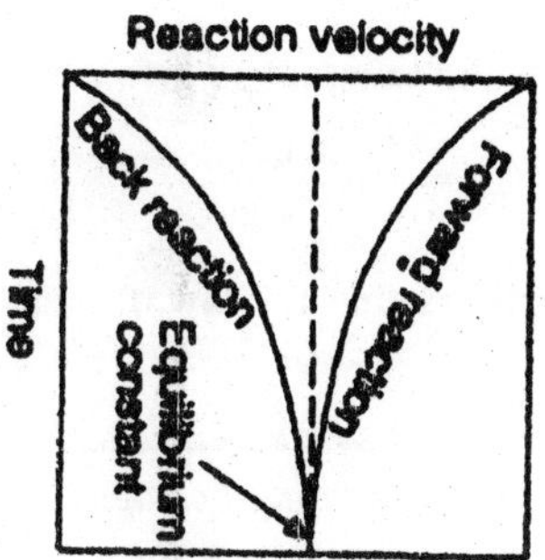

Fig. 2.3

Since the commencement of the reaction (zero time), the concentrations of the products gradually increase, the velocity of the backward reaction will also steadily increase. Ultimately a stage called *dynamic equilibrium* is reached when the velocity of toward reaction becomes equal to that of backward reaction, *i.e.*,

$$v_1 = v_2 \qquad ...(3)$$

At this stage the number of products molecules produced per second in the forward reaction would be equal to the number of product molecules which would disappear in the backward reaction.

Equilibrium Constant

On substituting Eqs. (1) and (2) in Eq. (3), we get

$$k_1 CAcB = k_2 C_C C_D$$

or

$$\frac{k_1}{k_2} = K_C = \frac{C_C C_D}{C_A C_B} \qquad ...(4)$$

where K_C is a constant, called the equilibrium constants. It is defined as the ratio of velocity constants of two opposing reactions.

Eq. (4) is known as the equilibrium law equation.

When the reaction is of the type $2A \rightleftarrows C + D$, one may write

$$2A \rightleftarrows C + D$$

Hence the equilibrium constant is given by

$$k_C = \frac{C_C \times C_D}{C_A C_A} = \frac{C_C \times C_D}{C_A^2}$$

A general equation for a reversible reaction can be written as

$$aA + bB + ... \rightarrow cC + dD + ...$$

Hence the equilibrium constant is given by

$$k_C = \frac{C_C^c \times C_D^d \times ...}{C_A^a \times C_B^b \times ...} \quad ...(5)$$

From Eq. (5), it follows that the law of mass action may be stated in another way

"At a give temperature, the rate of a reaction is proportional to the molar concentration of arch reacting species raised to the power of the number of its molecules taking part in the reaction as represented in the stoichiometic equation describing the reaction."

In Eq. (5), when the concentrations are expressed in moles per litre, the equilibrium concentration is written as K_C.

In the case of gaseous reactions, the concentrations may be put proportional to the partial pressures. Hence the gaseous reactions, Eq. (5) may be put in the following form:

$$K_P = \frac{P_C^c \times P_C^d \times ...}{P_A^a \times P_B^b \times ...} \quad ...(6)$$

where K_P is another equilibrium constant. The partial pressures are generally expressed in atmospheres.

Relationship between K_P, and K_C. For an ideal gas,

$$P_iV = nRT$$

or

$$P_i = n\frac{RT}{V} = C_iRT \quad ...(7)$$

where C is the concentration. On substituting the concentrations of various species from Eq. (7), in Eq. (6), we get

$$K_P = \frac{(C_CRT)^c \times (C_DRT)^d \times ...}{(C_ART)^a \times (C_BRT)^b \times ...}$$

$$K_p = \frac{C_C^c \times C_D^d \times ...}{C_A^a \times C_B^b \times ...} \times (RT)^{(c+d)-(a+b)}$$

$$K_p = K_C \times RT^{\Delta n}$$

where

$$\Delta n = (c + d + ...) - (a + b + ...)$$

Equilibrium Constant : The equilibrium constant is defined as simply the ratio of the velocity constants of two opposing reactions.

OR

The equilibrium constant is the product of the concentrations of the reacting species, each concentration raised to the power which is the stoichiometric coefficient of the substance in the chemical equation.

In general, for the reaction represented by the equation,

$$aA + bB + cC \rightleftarrows lL + mM + nN$$

the equilibrium constant is give by

$$K_C = \frac{[K_1]}{[K_2]} = \frac{[L]^l[M]^m[N]^n}{[A]^a[B]^b[C]^c}$$

The equilibrium constant for the reaction $2SO_2 + O_2 \rightleftarrows 2SO_3$ is as follows:

$$K_c = \frac{[SO_3]^2}{[SO_2]^2[O_2]} = \frac{(\text{litre/mol})}{(\text{litre/mol})(\text{litre/mol})^2}$$

It means that the equilibrium constant for this reaction is constant.

Factors Influencing Equilibrium Constant

The various characteristics or the reactors that influence the equilibrium constants are as follows :

1. *Change in Temperature :* The equilibrium constant has a definite value for every chemical reaction at a given temperature. The value of equilibrium constant varies with temperature. It has also the same value (at a give temperature) irrespective of the pressure on the system or the concentrations of the reactants and products.
2. *The mode of Expressing the Reaction :* If a reaction is represented in two opposite ways, the value of equilibrium constant in each case would be the reciprocal of the equilibrium constant i th other case For example,

   $$A + B \rightleftarrows C + D, \quad K = \frac{C_c \times C_d}{C_A \times C_B}$$

   $$C + D \rightleftarrows A + B, \quad K_1 = \frac{C_A \times C_B}{C_C \times C_D} = \frac{1}{K}$$

   It means that if the equilibrium constant is large in one direction, it will be smaller in the other direction.
3. *Change of Concentration or Partial Pressure Units :* The units of equilibrium constants K_p and K_C are

   $K_p = (\text{Atm})^{\Delta n}$.

$$K_C = (\text{Mol dm}^{-3})^{\Delta n}$$

For reactions in which Δn is zero, *i.e.*, the number of moles of products is same as that of reactants, K has no units. For reactions in which Δn is not zero, any change in units expressing concentration or partial presents of the involved substances will alter the equilibrium constant.

4. *Representation of Stoichiometric Equation* : The value of equilibrium constant depends upon the stoichiometric equation representing the reversible reaction. If it is expressed in different ways, the value of K is affected. The formation of SO_3 from SO_2 and O_2 may be represented by either of the following stoichiometric equations.

$$2SO_2 + O_2 \rightleftarrows 2SO_3 \quad ...(i)$$

$$SO_2 + \frac{1}{2} \rightleftarrows SO_3 \quad ...(ii)$$

K_1 from reaction (i),

$$K_1 = \frac{C_{SO_3^2}}{C_{SO_2} \times C_{O2}} \quad ...(iii)$$

K_2 from reaction (ii),

$$K_2 = \frac{C_{SO_3}}{C_{SO_2} \times C_{O2}^{1/2}} \quad ...(iv)$$

From (iii) and (iv), we have

$$K_2 = \sqrt{K_1}$$

Significance of Equilibrium Constant

The equilibrium constant is regarded as a convenient measure of the extent to which a reaction goes before equilibrium is established. In order to illustrate the statement, we consider the following reaction:

$$2H_2O(g) \rightleftarrows 2H_2(g) + O_2(g);$$

$$K = 1.35 \times 10^{-11} \text{ mol l}^{-1} \text{ at } 1073 \text{ K}$$

As the value of K is very small, it means that there will be very small quantities of H_2 and O_2 before equilibrium is established. Thus, the reverse reaction is almost complete. Due to this reaction, a mixture of H_2 and O_2 is dangerously explosive.

Let us now consider another reaction

$$2O_3(g) \rightleftharpoons 3O_2(g);$$

$$K = 10^{55} \text{ mol } l^{-1} \text{ at } 298 \text{ K}$$

because K is very large, the reaction mixture at equilibrium will contain a high proportion of oxygen and reaction will be nearly complete at equilibrium, *i.e.*, O_2 is more stable than O_3.

EXPERIMENTAL VERIFICATION OF LAW OF MASS ACTION

Synthesis of Hydrogen Iodide

The synthesis of hydrogen iodide can be represented as follows:

$$H_2 + I_2 \rightleftharpoons 2HI.$$

Bodnstin selected this reaction for the experimental verification of law of mass action because it proceeds very slowly at ordinary temperatures but rapidly at the boiling point of sulphur (444°C) to establish the equilibrium. H_2 and I_2 mixtures of varying compositions are filled in glass bulbs and then sealed. These glass bulbs are heated in the vapour of boiling sulphur (Fig. 2.4) for sufficient time to attain equilibrium.

The sealed glass bulbs are then suddenly cooled to the room temperature to fix the equilibrium. The bulbs are then broken i the NaOH solution which absorbs the HI formed and I_2 and leaves H_2 free. Thus, the concentration of HI and I_2 are measured.

Suppose 'a' moles of H_2 and 'b' moles of iodine be heated in a sealed glass bulb having a volume V litres in a thermostat till equilibrium is established. If the concentration of HI formed after analysis be 2x, then according to the above reaction 2x moles of HI will be obtained from x moles of each H_2 and I_2.

The equilibrium concentration per litre of various reactants and product may be but as follows:

$$[HI] = 2x/V$$

$$[H_2] = (a - x)/V$$

$$[I_2] = (v - x)/V$$

Applying the law of mass action to the synthesis of HI

$$H_2 + I_2 \rightleftarrows 2HI$$

$$K = \frac{[HI]^2}{[H_2][I_2]}$$

Substituting the values of various terms

$$K = \frac{(2x/V)^2}{\left(\frac{a-X}{V}\right)\left(\frac{b-x}{V}\right)} = \frac{4x^2}{(a-x)(b-x)} \quad ...(1)$$

The two unknown factor sin the above equation are K and x and therefore. the law of mass action can be verified in two ways:

(a) The value of x is obtained by analysis and substituted in equation (1). If the value of K is always the same with different compositions of H_2 and I_2 mixtures the law of mass action is correct.

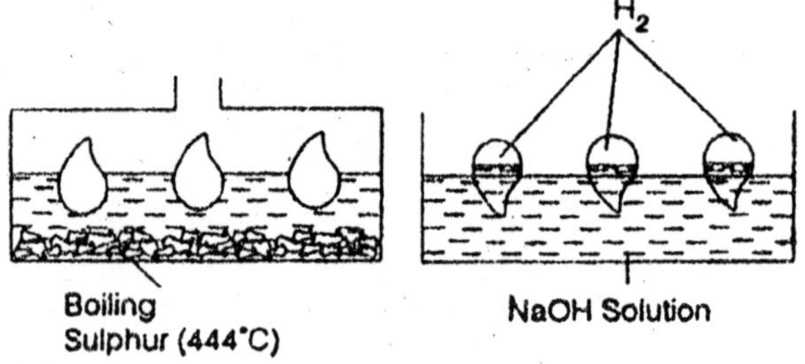

Fig. 2.4 : Experimental verification of law of mass action.

(b) From equation (1)

$$K = \frac{4x^2}{(a-x)(b-x)}$$

or $$ab - ax - bx + x^2 = \frac{4x^2}{K}$$

or $$x^2\left(1 - \frac{4}{K}\right) - x(a+b) + ab = 0$$

∴ $$x = (a+b) \pm \frac{\sqrt{(a+b)^2 - 4(1-4/K)(ab)}}{2(1-4K)}$$

From the values of K calculated from accurate observation, x can be calculated and if it agrees with the experimental values of x, th law is valid.

THERMODYNAMIC DERIVATION OF LAW OF MASS ACTION

Van't Hoff derived law of mass action thermodynamically by using an equilibrium box in which all the systems wherein equilibrium with each other at constant temperature (Fig. 2.5).

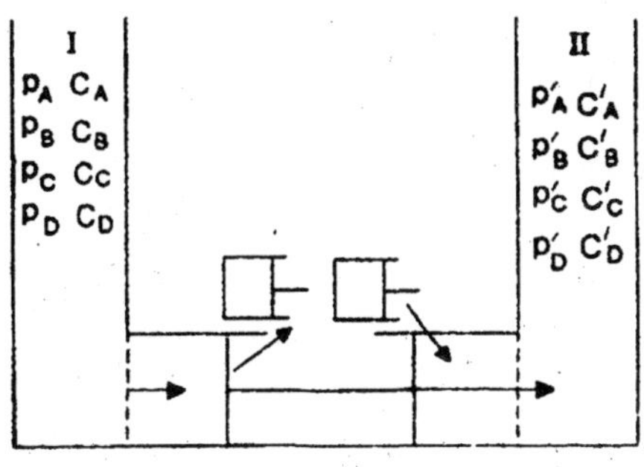

Fig. 2.5

This box is simply hypotectical and its properties are as follows:

(i) The walls of the box are rigid having indefinite dimensions.

(ii) The walls can be suitable made permeable to various components.

(iii) The various components present i, the box are always in equilibrium with each other.

(iv) If a system undergoes the chemical change inside the box, the work done is zero.

(v) The box is assumed to have large dimensions so that there occurs no change in concentration or in partial pressure of any component, when a small amount of these is added or removed from it.

(vi) When the working substances are gases, they are considered to be ideal ones.

(vii) The box is kept at constant temperature.

Consider the following chemical change which is taking place at constant temperature.

$$A + B \rightleftarrows C + D$$

Consider two equilibrium boxes I and II in which components A, B, C, and D are always in equilibrium.

For box No (I) C_A, C_B, C_C and C_D represent the concentrations of A, B, C and D and p_A, p_B, p_C and p_D are corresponding partial pressures, similarly, C'_A, C'_B, C'_C and C'_D represent the concentrations of A, B, C and D and p_A, p_B, p_C and p_D the corresponding pressures in the box No. (II). Suppose one mole of the substance A gets transferred from (I) to box (II). This can be achieved if it is assumed that th walls of the two boxes are preamble to A.

Now calculate the work done for the transfer of one mole of A which involves the following steps:

(a) One mole of A is removed from box No. I by moving the piston infinitesimally slowly outwards so that the partial pressure of A (*i.e.*, p_A) remains constant. The amount of work done is $P_V V_A$ where V_A represents the molar volume under pressure p_A.

(b) As the partial pressure of A is different in two boxes the gas removed from box I is changed isothermally and irreversibly from p_A and p'_A by working the piston, the amount of work done is

$$RT \ln \frac{p_A}{p'_A}$$

(c) The gas is now ready for putting it in box II. This will be put in box II by compressing the as infinitesimally slowly by work in the piston so that the partial pressure of A in box No. II (*i.e.* p'_A) remains constant. The work done in this process $p'_A V'_A$ where V'_A represents the molar volume of A under a partial pressure p'_A. Hence the total work done w_1 in transferring one mole of A reversibly and isothermally from box (I) to (II) is

$$w_1 = P_A V_A + RT \ln \frac{p_A}{p'_A} - p'_A V'_A$$

The gas is ideal. $p_A V_A = p'_A V'_A$

Hence
$$w_1 = RT \ln \frac{p_A}{p'_A} \quad \text{...(1)}$$

But the partial pressures are proportional to the molar concentrations. Therefore the amount of work done is

$$w_1 = RT \ln \frac{C_A}{C'_A} \quad \text{...(2)}$$

Similarly, the work done w_2 in transferring one mole B of from box I to box (II) is as follows:

$$w_2 = RT \ln \frac{p_B}{p'_B} \quad ...(3)$$

$$w_2 = RT \ln \frac{C_B}{C'_B} \quad ...(4)$$

The concentrations of A and B do not undergo any change in box (I) as proportionate amounts of C and D will react to form A and B to maintain equilibrium.

Now one mole of C is removed from box (II) to box (I), then the amount of work done is

$$w_3 = RT \ln \frac{p'_C}{p_C} \quad ...(5)$$

$$w_3 = RT \ln \frac{C'_C}{C_C} \quad ...(6)$$

Also one mole of D is removed from box (II) to box (I). Then, the amount of work done w_4 is given by

$$w_4 = RT \ln \frac{p'_D}{p_D} \quad ...(7)$$

$$w_4 = RT \ln \frac{C'_D}{C_D} \quad ...(8)$$

When C and D are transferred to box (I), they interact to form necessary amounts of A and B to maintain the equilibrium.

This process can be considered as an isothermal cyclic one because

1. Both the boxes have regained their initial states and
2. The process is carried out at constant temperature.

For such an isothermal cyclic process, total amount of work done is zero, *i.e.*

$$w_1 + w_2 + w_3 + w_4 = 0$$

On substituting eqs. (1), (3), (5) and (7) in (9), we obtain

$$RT \ln \frac{p_A}{p'_A} + RT \ln \frac{p_B}{p'_B} + RT \ln \frac{p'_C}{p_C} + RT \ln \frac{p'_D}{p_D} = 0$$

or $$\ln \frac{p_A}{p'_A} + \ln \frac{p_B}{p'_B} = \ln \frac{p_C}{p'_C} + \ln \frac{p_D}{p'_D}$$

$$\ln \frac{p_A \times p_B}{p'_B \times p'_B} = \ln \frac{p_C \times p_D}{p'_C \times p'_D}$$

or $$\frac{p_A \times p_B}{p'_B \times p'_B} = \frac{p_C \times p_D}{p'_C \times p'_D}$$

or $$\frac{p'_C \times p'_D}{p'_A \times p'_B} = \frac{p_C \times p_D}{p_A \times p_B} = K_P \quad ...(10)$$

Similarly

$$\frac{C_C \times C_D}{C_A \times C_B} = \frac{C'_C \times C'_D}{C'_A \times C'_B} = K_C \quad ...(11)$$

[Using equations (2), (4), (6) and (1)]

Thus the equations (10) and (11) are the two forms of the law of mass action.

DERIVATION OF LAW OF MASS ACTION FROM CHEMICAL POTENTIAL

Consider a reaction

$$pA + qB + ... \rightleftharpoons rC + mD + ...$$

Suppose the concentrations of the reactions A, B ... at the start of reaction are C_A, C_B ... respectively. Let the concentrations of the products C and C, ∴ at the end of reaction be C_C and C_D ... respectively. The free energies per mole (*i.e.*, chemical potentials) of the various substances at temperature T will be denoted by μ_A, μ_B, μ_C, μ_D ... respectively.

We know that the free energy of a system consist in of several components at constant temperature and pressure is give by

$$G = n_1\mu_1 + n_2\mu_2 + ... \quad ...(1)$$

where n_1, n_2 ... are the number of moles of the various components and m_1, m_2, ... are their respective chemical potentials. Hence one may write

Free energy of products = $r\mu_C + m\mu_D$

Free energy of reactants = $p\mu_A + q\mu_B$

∴ $$\Delta G = (r\mu_C + m\mu_D) - (p\mu_A + q\mu_B) \quad ...(2)$$

The chemical potential of a substance in any state is give by the equation

$$\mu = \mu^o + RT \ln a$$

where μ^o is chemical potential in the standard state of unit activity and μ the chemical potential at activity a.

Replacing the values of chemical potentials of the products and the reactants in equation (2), we get

$$\Delta G = [r(\mu_C^o + RT \ln a_C) + m(m_D^o + RT \ln a_D)] + ...$$

$$-[p(\mu_A^o + RT \ln a_A) + q(\mu_B^o + RT \ln a_B) + ...]$$

Rearranging,

$$\Delta G = [(r\mu_C^o + m\mu_D^o + ...) - (p\mu_A^o + q\mu_B^o + ...)]$$

$$+ RT \ln \frac{(a_C)^r (a_D)^m ...}{(a_A)^p (a_B)^q ...} \quad ...(4)$$

The μ^o terms refer to standard states of unit activity of the substance concerned. The first term in square brackets on the right side of equation (4), therefore, represents the increase in free energy of the reactants when these at unit activity react to form products also at the unit activity. If this energy increase is represented by ΔG^o, equation (4) may be written as

$$\Delta G = \Delta G^o + RT \ln \frac{(a_C)^r (a_D)^m ...}{(a_A)^p (a_B)^q ...} \quad ...(5)$$

$$= \Delta G^o + RT \ln x \quad ...(6)$$

where x stands for reaction quotients of activities of the products and reactants, viz.,

$$x = \frac{(a_C)^r (a_D)^m ...}{(a_A)^p (a_B)^q ...}.$$

Equation (6) is commonly known as the reaction isotherm.

If the above reaction is in equilibrium state, then ΔG will be zero. Hence from equation (5),

$$\Delta G^o = -RT \ln \left(\frac{(a_C)^r (a_D)^m}{(a_A)^p (a_B)^q} \right) \quad ...(7)$$

Since ΔG^o represents standard free energy change, *i.e.*, when the reactants as well as products are in their standard states of unit activity it must be constant at a given temperature. Consequently the expression on the right side of equation (7) must be constant as well. Since R is a gas constant, it follows, therefore, that if the temperature T is constant, then

$$\left(\frac{(a_C)^r(a_D)^m}{(a_A)^p(a_B)^q}\right) = \text{Constant} = K_a \quad ...(8)$$

where K_a is the equilibrium constant of the reaction. Equation (8) is the mathematical statement of the law of mass action.

Substituting equation (8) in (7), we get

$$\Delta G^o = -RT \text{ in } K_a \quad ...(9)$$

This is another important relation.

VAN'T HOFF'S REACTION ISOTHERM

This can be deduced by either of the following methods:

1. Free Energy Change Method

In the earlier deduction it was assumed that the system remains in equilibrium and hence the change in free energy is zero. Now let us proceed to calculate the change in free energy when a chemical reaction is taking place out at constant temperature from some arbitrary concentrations of reactants to some other arbitrary concentrations of the products. Let us assume a reaction involving for gases. A, B, C and D

$$A + B \rightleftarrows C + D \quad ...(1)$$

I this reaction amounts of A and B are decreasing while those of C and D are increasing. Subsequently, the free energies of A and B are decreasing while those C and D are increasing.

The free energy of the substance A per mole at temperature T is give as follows:

$$G_A = G_A^o + RT \ln p_A$$

where p_A represents the pressure of A and G_A^o represents the free energy at some standard state (p = 1) and is known as standard free energy.

Similarly, the free energies of B, C and D are as follows:

$$G_B = G_A^o + RT \ln p_B$$

$$G_C = G_C^o + RT \ln p_C$$

$$G_D = G_D^o + RT \ln p_D$$

By definition, the free energies of the reaction (ΔG) is

$$\Delta G = (\Delta G_C + G_D) + (G_A + G_B)$$

Hence $$\Delta G = (G^o_C + RT \ln p_C + G^o_D + RT \ln p_D) - (G^o_A + RT \ln p_A + G^o_B + RT \ln p_B)$$

$$= (G^o_C + G^o_D - G^o_A - G^o_B) + RT \ln \frac{p_C \times p_D}{p_A \times p_B}$$

$$= \Delta G^o + RT \ln \frac{p_C \times p_D}{p_A \times p_B} \quad ...(2)$$

where ΔG^o represents standard free energy of reaction.

But at equilibrium $\Delta G = 0$

Therefore, $\Delta G^o + RT \ln \left[\frac{p_C \times p_D}{p_A \times p_B}\right] = 0$

where pressure are equilibrium pressures.

or $$\Delta G^o + RT \ln K_P = 0$$

or $$\Delta G^o = -RT \ln K_P$$

where $$K_P = \frac{p_C \times p_D}{p_A \times p_B}$$

On substituting this value of ΔG^o in equation (2), we get

$$\Delta G = -RT \ln K_P + RT \ln \frac{p_C \times p_D}{p_A \times p_B}$$

or $$-\Delta G = RT \ln K_P - RT \ln \frac{p_C \times p_D}{p_A \times p_B} \quad ...(3)$$

Now consider the general reaction

$$n_1A + n_2B + ... \rightleftharpoons^* n_3C + n_4D + ...$$

By following the same procedure, one may write

$$-\Delta G = RT \ln K_P + RT \ln \frac{(p_C)^{n_3} \times (p_D)^{n_4}}{(p_A)^{n_1} \times (p_B)^{n_2}} \quad ...(4)$$

$$-\Delta G = RT \ln K_P - RT \Sigma \ln p \quad ...(5)$$

Thus $$-\Delta G = RT \ln K_c - RT \Sigma \ln c \quad [\because \; p \propto c] \quad ...(6)$$

where K_c is equilibrium constant. Equations (5) and (6) are known as Van't Hoff isotherm, because the reaction is taking place at constant temperature. This relation is very important because at may be used to show the direction in which a reaction tends to proceed.

2. The Van't Hoff's Equilibrium Box Method

It is a theoretical device which has been used successfully for calculating the free energy change in a chemical reaction taking place at constant temperature from arbitrarily chosen starting concentration of the reactants up to some of other arbitrarily chosen concentrations of the reactants. Consider the following chemical reaction.

$$n_1A + n_2B \rightarrow n_3C + n_4D \qquad ...(1)$$

Let there be four reservoirs (Fig. 2.6) each of indefinite dimensions, one for each of the components A, B, C and D. Let the concentration, pressure and volume of A in the reservoir I be C_A, P_A and V_A, respectively. Similarly let C_B, P_B and V_B, C_C, P_C and VC and C_D, P_D and V_D be the corresponding concentrations, pressures and volumes of components B, C and D i the reservoirs II, I and IV respectively. Besides these vessels, there is a central reservoir which is an equilibrium box V having A, B C and D where their concentrations, partial pressures and molar volumes are $(C_A)_e$, $(P_A)_e$ and $(V_A)_e$, $(C_B)_e$, $(P_B)_e$ and $(V_B)_e$ $(C_C)_e$, $(P_C)_e$ $(P_C)_e$ and $(C_D)_e$, $(P_D)_e$ and $(V_D)_e$ respectively. Let us assume the temperature to be T. Now transfer n_1 moles of A from reservoir I into the equilibrium box V. This can be reversibly done by the following three stage process:

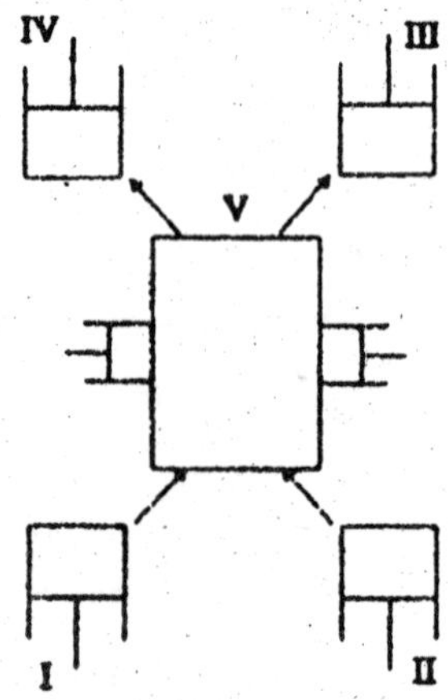

Fig. 2.6

(1) The n_1 moles of A at pressure P_A and temperature T are taken out by slowly moving a piston of reservoir I. Thus, the necessary amount of A will come out through a wall which is permeable to A. It will involve work done by the gas, equivalent to n_1. P_A. V_A. *i.e.*

$$(\text{work done})_1 = n_1(P_A V_A) \qquad ...(2)$$

(2) Before the gas has been introduced into the equilibrium reservoir V, the pressure of A is changed from P_A to $(P_A)_e$. The work done in the process for n_1 moles is

$$(\text{work done})_2 = n_1 RT \ln \frac{P_A}{(P_A)_e} \qquad ...(3)$$

(3) In the last stage, the gas is compressed at a pressure of $(P_A)_e$ into the equilibrium box. This involves work done which is equal to

$$(\text{work done})_3 = -n_1(P_A)_e(V_A)_e \qquad ...(4)$$

The total work done is, therefore, equal to sum of the equations (2), (3) and (4), *i.e.*,

$$w_1 = (\text{work done})_1 + (\text{work done})_2 + (\text{work done})_3$$

or $$w_1 = n_1 P_A V_A + n_1 RT \ln \frac{P_A}{(P_A)_e} - n_1 (P_A)_e (V_A)_e \qquad ...(5)$$

But the gas is ideal and the process is isothermal. Therefore,

$$n_1 P_A v_A = n_1(P_A)_e(V_A)_e \qquad ...(6)$$

On substituting equation (6) in (5), we obtain

$$w_1 = n_1 P_A V_A + n_1 RT \ln \frac{P_A}{(P_A)_e} - n_1 P_A V_A$$

$$= n_1 RT \ln \frac{P_A}{(p_A)_e} \qquad ...(7)$$

Similarly, if n_2 moles of B are transferred from reservoir II to the equilibrium box V the work done is given as follows:

$$w_2 = n_2 RT \ln \frac{p_B}{(p_B)_e} \qquad ...(8)$$

When n_1 moles of A and n_2 moles of B have bee introduced inside the equilibrium box they will react to form equilibrium concentrations of C and D. We have now to remove n_3 moles of C and n_4 moles of D from the equilibrium box V into the respective reservoirs III and IV. The work w_3 ad w_4 involved in these processes are as follows:

$$w_3 = n_3 RT \ln \frac{(p_C)_e}{P_C} \qquad ...(9)$$

$$w_4 = n_4 RT \ln \frac{(p_D)_e}{P_D} \qquad ...(10)$$

Total work done in this process is given as follows:

$$W = w_1 + w_2 + w_3 + w_4$$

$$W = n_1RT \ln \frac{P_A}{(P_A)_e} + n_2RT \ln \frac{P_B}{(P_B)_e} + n_3RT \ln \frac{(P_C)_e}{P_C} + n_4RT \ln \frac{(P_D)_e}{P_D} \quad ...(11)$$

The work W is derived from the free energy decrease of the system *i.e.*

$$W = -\Delta G \quad ...(12)$$

From equations (11) and (12), we have

$$-\Delta G = n_1RT \ln \frac{P_A}{(P_A)_e} + n_2RT \ln \frac{P_B}{(P_B)_e} + n_3RT \ln \frac{(P_C)_e}{P_C} + n_4RT \ln \frac{(P_D)_e}{P_D} \quad ...(13)$$

$$-\Delta G = RT \ln \frac{(P_C)_e^{n_3}(P_D)_e^{n_4}}{(P_A)_e^{n_1}(P_B)_e^{n_2}} - RT \ln \frac{(P_C)^{n_3}(P_D)^{n_4}}{(P_A)^{n_1}(P_B)^{n_2}}$$

$$= RT \ln K_p - RT \Sigma n \ln P$$

Where $K_p = \dfrac{(P_C)_e^{n_3}(P_D)_e^{n_4}}{(P_A)_e^{n_1}(P_B)_e^{n_2}}$

and $\Sigma \ln n\, P = \ln(P_C)\, n_3 + \ln (P_D) n_4 - (\ln P_A)\, n_1 - (\ln P_B) n_2$

If, however, the work done is expressed in terms of C instead of P we would get

$$-\Delta G = RT \ln \frac{(C_C)_e^{n_3}(C_D)_e^{n_4}}{(C_A)_e^{n_1}(C_B)_e^{n_2}} - RT \ln \frac{(C_C)^{n_3}(C_D)^{n_4}}{(C_A)^{n_1}(C_B)^{n_2}}$$

$$= RT \ln K_c - RT \Sigma n \ln C \quad ...(15)$$

where the term $\Sigma n \ln C$ has a similar significance as $\Sigma n \ln P$ already referred to.

VAN'T HOFF'S REACTION ISOCHORE

Van't Hoff's reaction isochore deals with the variation of equilibrium constant with temperature. According to Van't Hoff's reaction isotherm, the change in free energy is given as follows:

$$\Delta G = RT \ln K_c - RT \Sigma \ln C \quad ...(1)$$

The Van't Hoff's isochore can be derived by differentiating equation (1) of Van't Hoff's reaction isotherm with respect to temperature at constant volume. But at constant volume the change in free energy is not given by ΔG but by ΔA. Therefore, equation (1) modifies to

$$-\Delta A = RT \ln K_c - RT \Sigma \ln c \quad ...(2)$$

On differentiating it with respect to temperature at constant volume, we obtain

$$\left(\frac{\partial(-\Delta A)}{\partial T}\right)_V = RT \frac{d}{dT}(\ln K_c) + R \ln K_c$$

$$-RT \frac{d}{dT}(\Sigma n \ln c) + R \ln c$$

$RT \frac{d}{dT} \Sigma_n \ln c = 0$ because this arbitrary concentration is not considered as a function of temperature. Thus,

$$\left(\frac{\partial(-\Delta A)}{\partial T}\right)_V = RT \frac{d}{dT}(\ln K_c) + R \ln K_c - R \Sigma n \ln c$$

or $$-\frac{\partial(\Delta A)}{\partial T} = RT \frac{d}{dT}(\ln K_c) + R \ln K_c - R \Sigma n \ln c$$

or $$-\frac{\partial(\Delta A)}{\partial T} = RT \frac{d}{dT}(\ln K_c) + R \ln K_c - R \Sigma n \ln c$$

or $$-T\frac{\partial(\Delta A)}{\partial T} = RT^2 \frac{d}{dT}(\ln K_c) + R \ln K_c - R \Sigma n \ln c$$

$$= RT^2 \frac{d}{dT}(\ln K_c) - \Delta A \quad ...(3)$$

[use equation (2)]

The Gibb's Helmholtz equation is as follows:

$$\Delta A = \Delta E + T\left[\frac{\partial(AA)}{\partial T}\right]_V$$

or $$-T\left[\frac{\partial(AA)}{\partial T}\right]_V = \Delta E - \Delta A \quad ...(4)$$

On comparison of equation (3) with (2), we get

$$\Delta E = RT^2 \frac{d}{dT} \ln K_c$$

or $$\frac{\Delta E}{RT^2} = \frac{d}{dT} \ln K_c \quad ...(5)$$

But $q_v = \Delta E$ where q_v is the heat absorbed by the system at constant volume. Hence we get

$$\frac{d}{dT}\ln K_c = \frac{q_v}{RT^2} \quad ...(6)$$

It is the required equation.

Integrated Form of the Van't Hoff Isochore

The integrated form of this can be obtained by integrating equation (5) between the temperatures T_1 and T_2 then

$$\int_1^2 d\ln K_c = \int_{T_1}^{T_2} \frac{\Delta E}{RT^2} dT$$

$$\frac{K_{c_2}}{K_{c_1}} = \frac{\Delta E}{R}\left[\frac{1}{T_1} - \frac{1}{T_2}\right] \quad ...(7)$$

This equation is useful for calculating the value of equilibrium constant at any temperature provided equilibrium constant at one temperature and ΔE are given. Even ΔE can also be calculated when K_{c_1} and K_{c_2} at two temperature T_1 and T_2 are known.

VAN'T HOFF REACTION ISOBAR

Some confusion exists in the literature about the reaction isobar. Some prefer to call it by the name of Van't Hoff reaction Isotherm. We will now deduce this equation by using the following relation.

$$K_P = K_c(RT)^{\Delta n}$$

where Δn represents change in number of moles during the reaction.

On taking logarithms of both sides, we get

$$\ln K_P = \ln K_c + \Delta n \ln RT$$

or

$$\ln K_P = \ln K_c + \Delta n \ln R + \Delta n \ln T$$

On differentiating this equation with respect to temperature we obtain

$$\frac{d}{dT}(\ln K_P) = \frac{d}{dT}(\ln K_c) + \frac{d}{dT}(\Delta n \ln R) + \frac{d}{dT}(\Delta n \ln T)$$

But $\frac{d}{dT}[\Delta n \ln R] = 0$, R is a constant

$$\therefore \quad \frac{d}{dT}(\ln K_P) = \frac{d}{dT}(\ln K_c) + \frac{\Delta n}{T} \qquad \left(\because \frac{d}{dT}\ln T = \frac{1}{T}\right)$$

But $\frac{d}{dT}(\ln K_c) = \frac{q_v}{RT^2}$ [From reaction isochore]

$$\therefore \quad \frac{d}{dT}(\ln K_c) = \frac{q_v}{RT^2} + \frac{\Delta n}{T} = \frac{q_v + \Delta n\,RT}{RT^2} \qquad ...(1)$$

For an ideal as,

$$P\Delta V = \Delta n\,RT$$

Therefore, equation (1) becomes as follows:

$$\frac{d}{dT}(\ln K_P) = \frac{q_v + P\Delta V}{RT^2}$$

But $q_p = q_v + P\Delta V$, where q_p represents the heat absorbed by the system at constant pressure. Hence,

$$\frac{d}{dT}(\ln K_P) = \frac{q_p}{RT^2} \qquad ...(2)$$

We also know

$$\Delta H = q_p$$

$$\therefore \quad \frac{d}{dT}(\ln K_P) = \frac{\Delta H}{RT^2} \qquad ...(3)$$

This equation is known as Van't Hoff's reaction isobar. By integrating equation (3), we obtain

$$\int_1^2 d\ln K_P = \int_{T_1}^{T_2} \frac{\Delta H}{RT^2}\,dH$$

$$\frac{K_{P_2}}{K_{P_1}} = \frac{\Delta H}{R}\left[\frac{1}{T_1} - \frac{1}{T_2}\right]$$

Knowing the equilibrium constant K_{P_2} at temperature T_2 it is possible to calculate the equilibrium constant K_{P_1} at temperature T_1 and *vice versa*, provided the heat of reaction ΔH is known. Alternatively if the equilibrium constants of a reaction at two temperatures are known, the heat of reaction (ΔH) can be calculated.

APPLICATIONS OF LAW OF MASS ACTION

1. Dissociation of Hydrogen iodide

This is first type (*i.e.*, $K_P = K_c$) of a homogeneous gaseous reversible reaction as in this case the total number of molecules on both sides is the same.

	$2HI \rightleftharpoons H_2$	$+ \; I_2$	$-$ Heat
Initial	1 mole	0	0
At equilibrium	$(1-x)$ mole	$x/2$ mole	$x/2$ mole
Concentration	$\frac{(1-x)}{V}$	$\frac{x}{2V}$	$\frac{x}{2V}$

(where V = volume of vessel)

Let us start with '1' g mole of HI enclosed in a vessel of volume V litres. Let 'x' mole of HI decompose at the equilibrium to yield x/2 mole of each H_2 and I_2. Then the equilibrium molar concentrations are:

$$[HI] = \frac{(1-x)}{V};$$

$$[H_2] = \frac{x/2}{V}; \quad [I_2] = \frac{x/2}{V}$$

$\therefore$ From the Law of Mass action, we get

$$K_P = K_C = \frac{k_f}{k_b} = \frac{[H_2][I_2]}{[HI]} = \frac{\frac{x}{2V} \cdot \frac{x}{2V}}{\left[\frac{1-x}{V}\right]^2} = \frac{x^2}{4(1-x^2)}$$

(1) *Effect of pressure change,* From the above equation it is clear that K_c is *independent* of total volume 'V' of the system. A change of pressure alters the total volume and as K_c is independent of 'V' consequently *change of pressure will not alter the final state of equilibrium.*

(2) *Effect of temperature,* With the rise and fall of temperature the velocity constants k_f and k_b vary differently. It has been found that k_f increases more rapidly than k_b with the increase of temperature. So, with the increase of temperature $K_c = k_f/k_b$ will increase. Evidently, the concentration of products (*i.e.* H_2 and I_2) will increase and that of reactant HI will decrease. *Hence, the temperature will favour the dissociation of HI.*

(3) *Effect of adding products,* Alternative equation for equilibrium constant K_P in terms of partial pressures, will be:

$$K_P = \frac{P_{H_2} \times P_{I_2}}{P_{HI}^2}$$

On adding of any product, say H_2 the value of P_{H_2} will increase. In order to keep the value of K_P constant, the value of P_{I_2} must decrease and of P_{HI} must increase. In other words, *adding of H_2 will suppress the dissociation of HI.* Similar will be the consequences if I_2 is added.

(4) *Effect of adding an inert gas:* Addition of an inert gas like argon to the equilibrium mixture will increase the total pressure of the reactants without change the partial pressures of H_2, I_2 or HI. Hence, *the addition of an inert gas will have no effect on the degree of dissociation of HI.*

2. Dissociation of PCl_5

It is a second type (*i.e.*, $K_P \neq K_c$) of a homogeneous gaseous reaction as the total number of molecules of the reactants *change* as a result of chemical reaction.

$$PCl_5 \rightleftharpoons^{*} PCl_3 + Cl_2$$

	PCl_5	PCl_3	Cl_2
Initial	1 mole	0	0
At equilibrium	(1 − x) mole	x mole	x mole
		= 1 − x + x + x = 1 + x mole	
Concentration	$\frac{(1-x)}{V}$	$\frac{x}{V}$	$\frac{x}{V}$ where V = Volume
Partial pressure	$\frac{(1-x)}{(1+x)}P$	$\frac{xP}{(1+x)}$	$\frac{xP}{(1+x)}$ where P = Pressure

Let us start with '1' g mole of PCl_5 enclosed in a vessel of volume 'V' litres. Let 'x' mole of PCl_5 decompose at the equilibrium to yield x mole of each PCl_3 and Cl_2. Then, equilibrium molar concentrations are:

$$[PCl_5] = \frac{1-x}{V}$$

$$[PCl_3] = [Cl_2] = \frac{x}{V}$$

∴ Applying Law of Mass Action we get

$$\{K_c = \frac{k_f}{k_b} = \frac{[PCl_3][Cl_2]}{[PCl_5]} = \frac{x^2}{(1-x)V} \quad ...(i)$$

The expression for K_P can be deduced as follows:

Total No. of moles at equilibrium = 1 − x + x + x = (1 + x). Let P = Total pressure of the system,

Then $$P_{PCl_5} = \frac{1-x}{1+x}P; \quad P_{PCl_3} = \frac{xP}{1+x};$$

$$P_{PCl_2} = \frac{xP}{1+x}$$

$$\therefore \quad K_P = \frac{p_{PCl_3} \times p_{Cl_2}}{p_{PCl_5}}$$

$$= \frac{x^2P}{(1-x)(1+x)} = \frac{x^2P}{(1-x^2)} \quad \text{...(ii)}$$

(1) *Effect of pressure change:* In the expression (ii) for K_P, p is in the numerator. Hence, it follows if P is increased, the value of 'x' must decrease i order to keep K_P constant. Evidently increase in pressure suppresses the dissociation of PCl_5.

(2) *Effect of temperature:* It has been found that *k_f increases more rapidly than k_b with the rise of temperature*. Thus, it is evident that with the increase of temperature the value of $K = k_f/k_b$ increases, *i.e.*, the concentration of products increases and that of reactant PCl_5 decreases. *Hence, high temperature favours the dissociation of PCl_5.*

(3) *Effect of adding products at constant volume:* We have:

$$K_P = \frac{p_{PCl_3} \times p_{Cl_2}}{p_{PCl_5}}$$

Addition of any product, say Cl_2 will result in the increase in the value of PCl_3. In order to keep K_P constants, the value of PCl_3 will decrease and that of PCl_5 would decrease. Evidently, *addition of Cl_2 suppresses the dissociation of PCl_5*. Similarly consequences will happen by addin PCl_3.

(4) *Effect of adding an inert gas at constant volume:* The addition of an inert gas will increase the total pressure P, since K_P is proportional to P. Hence, in order to keep K_P constant the increase of P by the addition of inert gas will decrease the value of x. Evidently, *addition of an inert as will support the dissociation of PCl_5.*

3. Dissociation of N_2O_4

This is a second type (*i.e.*, $K_P \neq K_C$) of homogeneous gaseous reaction as the number of molecules alters as a result of reaction.

	$N_2O_4 \rightleftharpoons$*	$2NO_2$
Initial	1 mole	0
At equilibrium	$(1-x)$ mole	2x mole

Total = 1 – x + 2 = 1 + x moles

Concentration $\frac{(1-x)}{V}$ $\frac{2x}{V}$ where V = Volume

Partial pressure $\frac{(1-x)}{(1+x)}P$ $\frac{2xP}{(1+x)}$ where P = Pressure

Let us start with 1 mole of N_2O_4 enclosed in a vessel of capacity 'V' litres. Let at equilibrium 'x' mole of N_2O_4 is dissociated into 2x moles of NO_2. Then, equilibrium concentrations are:

$$[N_2O_4] = \frac{1-x}{V};$$

$$[NO_2] = \frac{2x}{V}$$

∴ Applying Law of Mass Action, we get

$$K_C = \frac{k_f}{k_b} = \frac{[NO_2]^2}{[N_2O_4]} = \frac{4x^2}{(1-x)V} \quad ...(i)$$

The expression for K_P can be calculated as follows:

Total No. of moles at equilibrium = 1 – x + 2x = 1 + x. Let total pressure of the system = P.

Then,
$$p_{N_2O_4} = \frac{(1-x)}{(1+x)}P;$$

$$p_{NO_2} = \frac{2xp}{1+x}$$

$$\therefore \quad K_P = \frac{(p_{NO_2})^2}{p_{N_2O_4}} = \frac{4x^2P}{(1-x^2)} \quad ...(ii)$$

(1) *Effect of pressure change:* From the expression (ii) it is clear that when P increases x must decrease and *virsa* in order to keep K_P constant. In other words, *increase of pressure suppresses the dissociation of N_2O_4.*

(2) *Effect of adding product at constant volume:* We have

$$K_P = p^2_{NO_2} / p_{N_2O_4}$$

If NO_2 is added to the equilibrium mixture in a close vessel, the partial pressure p_{NO} will increase. In order to keep K_P constant, the value of $p_{N_2O_4}$ must increase. Hence, *dissociation of N_2O_4 will be suppressed by the addition of NO_2.*

(3) *Effect of change of temperature:* It has been found that with the *rise of temperature k_f increases more rapidly than k_b, i.e., K increases* or the concentration of product NO_2 will increase. Hence, *increase in temperature will favour the formation of NO_2, i.e., dissociation of N_2O_4.*

(4) *Effect of adding an inert gas at constant volume:* The addition of an inert gas will increase the total pressure P, since K_P, is directly proportional to P. Hence, in order to keep K_P constant, the increase of total pressure P by the addition of inert gas will decrease the value of 'x'. Evidently, *the addition of an inert gas at constant volume will suppress the dissociation of N_2O_4.*

4. Synthesis of Ammonia

This is a second type (*i.e.*, $K_P \neq K_C$) of homogeneous gaseous reaction as the total number of molecules *changes* as a result of reaction

$$N_2 + 3H_2 \rightleftharpoons^{*} 2NH_3$$

	N_2	$3H_2$	$2NH_3$
Initial	1 mole	3 moles	0
At equilibrium	$(1-x)$ mole	$(3-3x)$ moles	$2x$ mole

$$= 1 - x + 3 - 3x + 2x = 4 - 2x$$

Concentration	$\frac{(1-x)}{V}$	$\frac{(3-3x)}{V}$	$\frac{2x}{V}$

where V = Volume

Partial pressure	$\frac{(1-x)P}{(4-2x)}$	$\frac{(3-3x)P}{(4-2x)}$	$\frac{2xP}{(4-2x)}$

where P = Pressure

Let us start with 1 mole of N_2 and 3 moles of H_2 in a closed vessel of volume 'V' litres. Let 'x' mole of N_2 and 3x moles of H_2 are used up at equilibrium. Then, equilibrium concentrations are:

$$[N_2] = \frac{1-x}{V}; \quad [H_2] = \frac{3(1-x)}{V}; \quad [NH_3] = \frac{2x}{V}$$

$\therefore$ Applying law of Mass Action, we get

$$K_C = \frac{k_f}{k_b} = \frac{[NH_3]^2}{[N_2][H_2]^3} = \frac{\left[\frac{2x}{V}\right]^2}{\left[\frac{1-x}{V}\right]\left[\frac{3(1-x)}{V}\right]^3} = \frac{4x^2V}{27(1-x)^4} \quad \text{...(i)}$$

The expression for K_P can be calculated as follows:

Total No. of molecules in the system at equilibrium

$$= 1 - x + 3 - 3x + 2x = 4 - 2x.$$

Let total pressure of system = P.

Then $p_{N_2} = \frac{1-x}{4-2x}P;$

$$p_{H_2} = \frac{3-3x}{4-2x}P;$$

$$p_{NH_3} = \frac{2xP}{4-2x}$$

$$\therefore \quad K_P = \frac{(p_{NH_3})^2}{p_{N_2} \times (p_{H_2})^3}$$

$$\frac{\left[\frac{2xP}{4-2x}\right]^2}{\left[\frac{1-x}{4-2x}P\right]\left[\frac{(3-3x)P}{4-2x}\right]^3} = \frac{16x^2(2-x)^2}{27P^2(1-x)^4} \quad ...(ii)$$

1. *Effect of pressure change:* As K_P is inversely proportional to P^2, so an increase in P will cause a decrease in x. In other words, *increase of pressure increases the formation of NH_3.*

2. *Effect of temperature change:* It has bee found that *with the rise of temperature k_b increases more rapidly than k_f.* Thus, with the increase in temperature the backward reaction is favoured. Hence, at high temperature the dissociation of NH_3 will occur. Conversely *low temperature favours the formation of NH_3.*

3. *Effect of adding reactants at constant volume:* We have:

$$K_P = \frac{(p_{NH_3})^2}{p_{N_2} \times (p_{H_2})^3} m$$

Addition of any reactant, say N_2, will result ion the increase in the value of p_{N_2}. In order to keep K_P constant, the value of p_{NH_3} will increase and that of p_{H_2} will decrease. Evidently, addition of *N_2 favours the formation of NH_3.* Similar consequences will happen by adding H_2.

4. *Effect of adding an inert gas at constant volume:* The addition of an inert gas will decrease the total pressure P, since K_P is inversely proportional to P.

Hence, in order to keep K_P constant the increase of P by addition of inert gas will increase the value of x. Evidently, *addition of an inert gas will favour the formation of NH_3.*

5. Formation of Nitric Oxide

This is a first type (*i.e.*, $K_P = K_C$) of a homogeneous reaction as the total number of molecules remains the same as a result of the reaction:

	N_2 +	O_2 $\rightleftarrows$*	2NHO – Heat
Initial	a mole	b moles	0
At equilibrium	(a – x) mole	(b – x) moles	2x mole

= 1 – x + 3 – 3x + 2x = 4 – 2x

Concentration	$\frac{(a-x)}{V}$	$\frac{b-x}{V}$	$\frac{2x}{V}$

(where V = Volume of vessel)

Let us start with 'a' moles of N_2 and 'b' moles of O_2 in a closed vessel of capacity 'V' litres.

Let at equilibrium 'x' moles each of N_2 and O_2 are used up to form O_2 moles of NO. Then, the equilibrium concentrations are:

$$[N_2] = (a - x)/V\ ;$$

$$[O_2] = (b - x)/V;$$

$$[NO] = 2x/V$$

∴ Applying Law of Mass Action, we get

$$K_P = K_C = \frac{[NO]^2}{[N_2][O_2]} = \frac{4x^2}{(a-x)(b-x)}$$

1. *Effect of pressure change:* As V ∝ 1/P and K_C is independent of V. Therefore, K_C is independent of P. Hence, *the equilibrium concentrations of reaction remain uneffected by change of pressure.*

2. *Effect of adding reactants:* We have

$$K_P = \frac{p_{NO}^2}{p_{N_2} \times p_{O_2}}$$

of adding a reactant, say N_2 the partial pressure of N_2 will increase. In order to keep K_p constant, the value of p_{O_2} will decrease and that of p_{NO} will increase, *i.e.*, more N_2 and O_2 will react to form NO.

In other words, addition of N_2 will favour the formation of NO. Similar effects will be caused by the addition of oxygen.

3. *Effect of temperature change:* It has been found that with the rise of *temperature k_f increases more rapidly than k_b i.e.* with the increase of temperature the about of product, *i.e.*, NO, will increase.

Hence *formation of nitric oxide is favoured by high temperature.*

6. Esterification of Ethyl Alcohol by Acetic Acid

This is a reaction in solution and is represented by the equation.

	CH_3COOH +	C_2H_5OH $\rightleftarrows$	$CH_3COOC_2H_5$ +	H_2O
Initial	a mole	b moles	0	0
At equilibrium	(a − x) mole	(b − x) moles	x mole	x mole
Concentration	$\frac{(a-x)}{V}$	$\frac{b-x}{V}$	$\frac{x}{V}$	$\frac{x}{V}$

Let us start with 'a' moles of acid and 'b' moles of alcohol. Let 'x' moles each of ester and water are formed at equilibrium. If V is thee total volume of the solution mixture, then equilibrium molar concentrations are:

$$[CH_3COOH] = (a-x)/V;$$

$$[C_2H_5OH] = \frac{b-x}{V};$$

$$[CH_3COOC_2H_3] = x/V \quad [H_2O] = x/V$$

∴ Applying Law of Mass Action, we get

$$K_C = \frac{k_f}{k_b}$$

$$= \frac{[CH_3COOC_2H_5][H_2O]}{[CH_3COOH][C_2H_5OH]}$$

$$= \frac{x/V \times x/V}{\frac{a-x}{V} \times \frac{b-x}{V}} = \frac{x^2}{(a-x)(b-x)}$$

SOLVED EXAMPLES

Example 1:

When ethyl alcohol and acetic acid are mixed i equivalent proportions equilibrium is attained when 2/3rds of the acid and alcohol are consumed. How much ester will be present in the equilibrium mixture if we start with 240 g of acetic acid and 138 g of alcohol?

Solution:

For reaction:	CH_3COOH +	C_2H_5OH ⇄	$CH_3COOC_2H_5$	$+H_2O$
Initial	a moles	b moles	0	0
At equilibrium	(a – x) moles	(b – x) moles	x mole	x mole
Concentration	(a – x)/V	(b – x)/V	x/V	x/V

$$\text{We have } K_C = \frac{[x/V]\times[x/V]}{[(a-x)/V][(b-x)/V]} = \frac{x^2}{(a-x)(b-x)}$$

In first case if a = 1, b = 1, x = 2/3

$$K_C = \frac{(2/3)^2}{\left(1-\frac{1}{3}\right)\left(1-\frac{2}{3}\right)} = \frac{[2/3]\times[2/3]}{[1/3]\times[1/3]} = 4.0$$

In second case a = 240 g/60 = 4 moles (∵ mole wt of acid = 6)
and b = 138 g/46 = 3 mole (∵ mole wt. of alcohol = 46)

$$\therefore \quad K_C = 4 = \frac{x^2}{(4-x)(3-x)}$$

or $\quad x^2 = 4(12 - 7x + x^2) = 48 - 28x + 4x^2$

or $3x^2 - 28x + 36 = 0$

$$\therefore \quad x = \frac{28 \pm \sqrt{(28)^2 - 4\times3\times36}}{2\times3} = 2.26$$

as the second value is not admissible. Hence, the amount of ester present at equilibrium

$$= 2.26 \text{ mole} = 2.26 \times 88 = 198.88 \text{ g.}$$

Example 2:

Phosphorus pentachloride is 82% dissociated into phosphorus trichloride and chlorine, all gaseous, at 250°C and 1 atmosphere. Calculate the equilibrium constant of dissociation

Solution:

For reaction:	PCl_5	PCl_3 +	Cl_2
Initial	1 moles	0	0
At equilibrium	(1 – x) mole	x mole	x mole

= 1 – x + x + x = 1 + x mole

Partial pressure	$\frac{(1-x)}{(1+x)}P$	$\frac{x.P}{(1+x)}$	$\frac{x.P}{(1+x)}$

$$\text{We have } K_P = \frac{p_{Cl_2} \times p_{PCl_3}}{p_{PCl_5}} = \frac{\frac{x.P}{1+x} \times \frac{x.P}{1+x}}{\frac{(1-x)P}{(1+x)}} = \frac{x^2P}{(1-x^2)}$$

Substituting x = 0.82; P = 1 atmosphere, we et

$$\therefore \quad K_P = \frac{(0.82)^2 \times 1}{(1+0.82)(1-0.82)}$$

$$= \frac{0.82 \times 0.82}{1.82 \times 0.18} = 2.577 \text{ atmosphere.}$$

Example 3:

The equilibrium constant for the reaction $N_2O_4 \rightleftarrows 2NO_2$ at 500°C is found to be 640. If the pressure of the gas mixture is 180 mm, calculate the percentage dissociation of N_2O_4. At what pressure will it be half dissociated?

Solution:

For reaction:	$N_2O_4 \rightleftarrows$	$2NO_2$
Initial	1 moles	0
At equilibrium	(1 – x) mole	2x mole = 1 – x + 2x = (1 + x) mole
Partial pressure	$\frac{(1-x)}{(1+x)}P$	$\frac{2xP}{(1+x)}$

$$\text{We have} \quad K_P = \frac{(p_{NO_2})^2}{p_{N_2O_4}} = \frac{\left[\frac{2x.P}{(1+x)}\right]^2}{\frac{(1-x)P}{(1+x)}} = \frac{4x^2P}{(1-x^2)}$$

Substituting K_P = 640; P = 180/760 = 9/38 atmospheres, we get

$$\therefore \qquad 640 = \frac{4x^2 \times 9}{(1 - x^2) \times 38}$$

or $$36x^2 = 640 \times 38 - 640 \times 38x^2$$

or $$24.356x^2 = 24,320$$

or $$x = 0.9992 \text{ or } 99.92\% \text{ dissociated at } 500°C.$$

Let N_2O_4 is half-dissociated at P, *i.e.*, x = 0.5

$$640 = K_P = \frac{4x^2P}{(1 - x^2)} = \frac{4 \times (0.5)^2 \times P}{(1 + 0.5)(1 - 0.5)} = \frac{P}{0.75}$$

$$P = 640 \times 0.75 = 480 \text{ atmosphere.}$$

Example 4:

When a mixture of nitrogen and hydrogen in the volume ration of 1 : 3 allowed to attain equilibrium at 400°C and 1000 atm. 0.2491 mole fraction of ammonia is formed. Calculate the equilibrium constant for the above reaction at the experimental temperature and pressure.

Solution:

For reaction:	N_2	+	$3H_2 \rightleftharpoons$	$2NH_3$
Initial	1 mole		3 moles	0
At equilibrium	(1 – x)		(3 – 3x)	2x
			= 1 – x + 3 – 3x + 2x = 4 – 2x mole	
Partial pressure	$\frac{(1-x)P}{(4-2x)}$		$\frac{(3-3x)P}{(4-2x)}$	$\frac{2xP}{(4-2x)}$

We have: $$K_p = \frac{(p_{NH_3})^2}{p_{N_2} \times (p_{H_2})^3}$$

$$= \frac{\left[\frac{2xP}{(4-2x)}\right]^2}{\left[\frac{(1-x)P}{(4-2x)}\right]\left[\frac{(3-3x)P}{(4-2x)}\right]^3} = \frac{16x^2(2-x)^2}{27P^2(1-x)^4}$$

Here P = 1000 atmospheres and mole fraction of NH_3 = 0.2491 = 2x(4 – 2x), we get

$$\therefore \qquad 2x = 0.9964 - 0.4982x$$

or $$2.4982x = 0.9964$$

or $$x = 0.9964/2.4982 = 0.4$$

then, $$K_p = \frac{16 \times (0.4)^2 \times (1.6)^2}{27 \times (1000)^2 \times (0.6)^2}$$

$$= 1.617 \times 10^{-6} \text{ atmospheres}^{-2}$$

Example 5:

In an experiment, at a total pressure of 10 atmospheres and 400°C, i the equilibrium mixture $2NH_3 \rightleftarrows N_2 + 3NH_2$ the ammonia was found to have dissociated to the extent of 96%. Calculate K_P for the reaction.

Solution:

For reaction:	NH_3	$\rightleftarrows$	N_2 +	$3H_2$
Initial	1 mole		0	0
At equilibrium	$(1 - x)$		$x/2$	$3x/2$

$$= 1 - x + x/2 + 3x/2 = 1 + x$$

Partial pressure	$\frac{(1-x)P}{(4-2x)}$	$\frac{x.P}{2(1+x)}$	$\frac{3x.P}{2(1+x)}$

We have $$K_p = \frac{(p_{N_2})(p_{H_2})^3}{(pNH_3)}$$

$$= \frac{\frac{xP}{1(1+x)}\left[\frac{3xP}{2(1+x)}\right]^3}{(1-x)/(1+x)P} = \frac{27x^4P^2}{16(1-x)^2(1+x)^2}$$

Substituting $x = 0.96$ and $P = 10$ atmosphere, we get

$$\therefore \quad K_p = \frac{27(0.96)^4(10)^2}{16(1-0.96)^2(1+0.96)^2} = 23,320.$$

Example 6:

Calculate K_P and K_C at 1 atmosphere and at 25°C for the reaction.

$$N_2O_4 = 2NO_2$$

$$(\Delta G^o = 1.1 \text{ K cal.})$$

Solution:

From equation, we have

$$\Delta G^o = -RT \ln K_P$$

Here $$\Delta G^o = 1.1 \text{ K cal} = 1100 \text{ cals.}$$

$$R = 2 \text{ cals.}$$

$$T = 253 + 25 = 298 \text{ K}$$

$$\therefore \quad 1100 = -2 \times 298 \ln K_P$$

$$K_P = 0.1579 \text{ atm.}$$

The relation between K_P and K_C is

$$K_P = K_C(RT)^{\Delta n}$$

where $\Delta n = 2 - 1 = 1$

$$\therefore \quad K_C = \frac{K_P}{RT}$$

Here $K_P = 0.1579$ atm., R = 0.082 litre-atm/degree

$$\therefore \quad K_C = \frac{0.1579}{0.082 \times 298} = 6.461 \times 10^{-3} \text{ mole/litre}$$

Example 7:

The equilibrium constant for this reaction $N_2 + 3H_2 \rightleftarrows 2NH_3$ at 400°C is 1.064×10^{-4} and at 500°C is 0.144×10^{-4}. Calculate the mean heat of formation of 1 mole of ammonia from its elements in the given range of temp.

Solution:

We have

$$\ln \frac{K_{C_2}}{K_{C_1}} = \frac{q_v}{R}\left(\frac{1}{T_1} - \frac{1}{T_2}\right)$$

$$2.303 \log \frac{K_{C_2}}{K_{C_1}} = \frac{q_v}{R}\left(\frac{T_2 - T_1}{T_1 T_2}\right)$$

or

$$\log \frac{0.144 \times 10^{-4}}{1.64 \times 10^{-4}} = \frac{q_v}{1.987 \times 2.303} \times \left(\frac{773 - 673}{773 \times 673}\right)$$

or

$$q_v = -25.150 \text{ cal.}$$

Thus the heat of reaction

$$N_2 + 3H_2 \rightleftarrows 2NH_3$$

is -25150 cal.

In simple words, heat of formation of 2 moles of NH_3 is -25.150 Therefore, heat of formation of mole of NH_3

$$= -\frac{1}{2} \times 25150 = -12575 \text{ cal.}$$

Example 8:

Prove that

$$\left(\frac{\partial \ln K_P}{\partial T}\right) = \frac{\Delta H^\circ}{RT^2}$$

when $\Delta G^\circ = -RT \ln K_P$

Solution:

$$\Delta G^\circ = -RT \ln K_P \quad ...(1)$$

On differentiating with respect to temperature at constant pressure, we get

$$\left(\frac{\partial \Delta G^\circ}{\partial T}\right)_P = -R \ln K_P - RT\left(\frac{\partial \ln K_P}{\partial T}\right)_P$$

On multiplying both sides by T, we get

$$\left(\frac{\partial \Delta G^\circ}{\partial T}\right)_P = -RT \ln K_P - RT^2\left(\frac{\partial \ln K_P}{\partial T}\right)$$

$$= \Delta G^\circ - RT^2\left(\frac{\partial \ln K_P}{\partial T}\right)_P \quad \text{[use (1)] ...(2)]}$$

Gibbs-Helmholtz equation for substances in their standard states may be written as follows:

$$\Delta G^\circ = \Delta H^\circ + T\left(\frac{\partial \Delta G^\circ}{\partial T}\right)_P$$

$$T\left(\frac{\partial \Delta G^\circ}{\partial T}\right)_P = \Delta H^\circ - \Delta G^\circ \quad ...(3)$$

From Eqs. (2) and (3), we have

$$\Delta H^\circ - \Delta G^\circ = -\Delta G^\circ + RT^2\left(\frac{\partial \ln K_P}{\partial T}\right)$$

$$\frac{\Delta H^\circ}{RT^2} = \left(\frac{\partial \ln K_P}{\partial T}\right)$$

Example 9:

The equilibrium constant of the reaction

$$2SO_2 + O_2 \rightleftarrows 2SO_3$$

at 528°C is 98.0 and at 680°C is 10.5 and find the heat of reaction.

Solution:

According to the integrated form of Van't Hoffs Isochore equation,

$$\log_{10} K_{p2} - \log_{10} K_{p1} = \frac{\Delta H}{2.303R}\left(\frac{1}{T_1} - \frac{1}{T_2}\right) = \frac{\Delta H}{2.303R}\left(\frac{T_2 - T_1}{T_1 T_2}\right)$$

where K_p is the equilibrium constant, ΔH is the heat of reaction.

$$K_{p2} = 10.5, \quad K_{p1} = 98.0$$

$$T_1 = 528 + 273 = 801 \text{ K},$$

$$T_2 = 680 + 283 = 953 \text{ K}$$

$$\log_{10} 10.5 - \log_{10} 98 = \frac{\Delta H}{2.303 \times 1.987}\left\{\frac{953 - 801}{953 \times 801}\right\}$$

$$(\log_{10} 10.5 - \log_{10} 98) = \left\{\frac{953 \times 801}{953 - 801}\right\} = \frac{\Delta H}{2.303 \times 1.987}$$

or

$$\Delta H = (2.303 \times 1.987)(\log_{10} 10.5 - \log_{10} 98)\left\{\frac{953 \times 801}{953 - 801}\right\}$$

$$= \frac{2.303 \times 1.987 \times (-0.97)801 \times 953}{152}$$

$$= -22290 \text{ cal.}$$

Example 10:

The equilibrium constant K for the reaction

$$N_2 + 3H_2 \rightleftarrows 2NH_3$$

is 1.064×10^{-4} at 400°C and 0.144×10^{-4} at 500°C. Calculate the mean heat of formation of mole of ammonia from its elements in this temperature range.

Solution:

$$\log_{10} K_{p2} - \log K_{p1} = \frac{\Delta H}{2.303 R}\left(\frac{T_2 - T_1}{T_1 T_2}\right)$$

$$\log_{10} 0.144 \times 10^{-4} - \log_{10} 1.64 \times 10^{4} = \frac{\Delta H}{2.303 \times 1.988}\left\{\frac{773 - 673}{773 \times 673}\right\}$$

as $T_1 = 400 + 273 = 673$ K and $T_2 = 500 + 273 = 773$ K

or $$(\log_{10} 0.144 \times 10^{-4} - \log_{10} 1.64 \times 10^{-4})\frac{673 \times 773}{773 - 673} = \frac{\Delta H}{2.303 \times 1.987}$$

or $\Delta H = 2.303 \times 1.987 \times (\log_{10} 0.144 \times 10^{-4} - \log_{10} 1.64$

$$\times 10^{-4} \times \frac{673 \times 773}{100}$$

$$= \frac{2.303 \times 1.987 \times (-1.05648) \times 773 \times 673}{100} \text{ cal.}$$

$$\Delta H = -25150 \text{ cal.}$$

But this is the heat of reaction for the formation of 2 g moles of ammonia, hence heat at of formation of one g molecule.

$$= -\frac{25150}{2} = -12575 \text{ cal}$$

Negative sign shows that heat is evolved in this reaction.

Example 11:

At 1000 k, water vapour at 1 atm has bee found to be dissociated into hydrogen and oxygen to the extent of 3 × 10^{-7}%. Calculate the free energy decrease of the system, assuming ideal behaviour.

Solution:

$$2H_2O \rightleftarrows 2H_2 + O_2$$

$$K_p = \frac{(p_{H_2}O)^2}{(p_{H_2})^2 (p_{O_2})}$$

If the pressure is 1 atm then

$$p_{H_2} = 3.0 \times 10^{-7}, \; p_{O_2} = \frac{1}{2}(3.0 \times 10^{-7})$$

$$= 1.5 \times 10^{-7}$$

$$\because \quad K_p = \frac{1}{(3.0 \times 10^{-7})^2 (1.5 \times 10^{-7})}$$

$$= \frac{1}{1.35 \times 10^{-20}} = 7.4074 \times 10^{19}$$

$$\therefore \quad -\Delta G = RT \ln K_p = 2.303 \, RT \log K_p$$

$$= 2.303 \times 8.314 \times 100 \times \log 7.4074 \times 10^{19}$$

$$= 380447.3 \text{ J} = 380.4473 \text{ kJ}$$

Example 12:

At 375 K and total pressure of 1 atom sulphuryl chloride is dissociated according to the equation.

$$SO_2Cl_2 \rightleftharpoons SO_2 + Cl_2$$

to the extent of 91.2%. Find K_P and the et work by this reaction at this temp.

Solution:

$$SO_2Cl_2 \rightleftharpoons SO_2 + Cl_2$$

is
$$K_P = \frac{\alpha^2 P}{1-\alpha^2}$$

or
$$K_P = \frac{\alpha^2 P}{(1-\alpha)(1+\alpha)}$$

$$= \frac{(0.912)^2 \times 1}{(1-0.912)(1+0.912)} = 4.944$$

Net work w = et decrease in free energy = $-\Delta G$

But
$$-\Delta G = RT \ln K_P$$

$$= 2.303 \times 375 \times 8.314 \times \log 4.944$$

$$= 4983.6063 \text{ J}$$

Example 13:

The equilibrium constant of the reaction

$$2SO_2 + O_2 \rightleftharpoons 2SO_3$$

at 528°C is 98 and at 680°C is 10.5. Find the heat of reaction.

Solution:

According to the integrated form of Vant Hoff's Isochore equation.

$$\log_{10} K_{P_2} - \log_{10} K_{P_1} = \frac{\Delta H}{2.303}\left(\frac{1}{T_1} - \frac{1}{T_2}\right) = \frac{\Delta H}{2.303}\left(\frac{T_2 - T_1}{T_1 T_2}\right)$$

where K_P is the equilibrium constant, ΔH is the heat of reaction.

$$K_{P_1} = 10.5,\ K_{P_2} = 98.0$$

$$T_1 = 528 + 273 = 801 \text{ K},$$

$$T_2 = 680 + 273 = 953 \text{ K}$$

$$\therefore \log_{10} 10.5 - \log_{10} 98 = \frac{\Delta H}{2.303 \times 8.314}\left\{\frac{953 - 801}{953 \times 801}\right\}$$

$$(\log_{10} 10.5 - \log_{10} 98) \times \left\{\frac{953 \times 801}{953 - 801}\right\} = \frac{\Delta H}{2.303 \times 8.314}$$

$$\Delta H = (2.303 \times 8.314)\ (\log_{10} 98 - \log_{10} 10.5) \left\{\frac{953 \times 801}{953 - 801}\right\} \text{J}$$

$$= \frac{2.303 \times 8.314 \times (-0.97) \times 953 \times 801}{152} \text{J}$$

$$\Delta H = -93273.332 \text{ J.}$$

Example 14:

The temperature dependence of the equilibrium constant of the isomerization reaction D ⇄ is give by the relation

$$K_P = 4.814 - \frac{2059 \text{ K}}{\text{T}}$$

where T is in kelvin. Calculate the entropy of isomerization at 25°C and ΔG^o and ΔH^o at the same temperature.

Solution:

At 25°C (= 298 K), we will have

In $\quad K_P = 4.814 - \dfrac{2059 \text{ K}}{298 \text{ K}} = -2.095$

Now $\quad \Delta G^\circ = -RT \ln K_P$

$$= -(8.314 \text{ JK}^{-1} \text{ mol}^{-1})(298 \text{ K})(-2.095)$$

$$= 5190.5 \text{ J mol}^{-1}$$

Now the differentiation of ln K_P would give the expression

$$d \ln K_P = -(2059 \text{ K})d(1/T)$$

or $\quad \dfrac{d \ln K_P}{d(1/T)} = -(2059 \text{ K})$

Hence $\quad -\dfrac{\Delta H^\circ}{R} = -(2059 \text{ K})$

or $\quad \Delta H^\circ = (2059 \text{ K})(8.314 \text{ J K}^{-1} \text{ mol}^{-1})$

$$= 17118.5 \text{ J mol}^{-1}$$

Finally $\quad \Delta S^\circ = \dfrac{\Delta H^\circ - \Delta G^\circ}{T}$

$$= \frac{(17118.5 - 5190.5) \text{ J mol}^{-1}}{298 \text{ K}}$$

$$= 40.03 \text{ J K}^{-1} \text{ mol}^{-1}$$

Example 15:

At 250°C, $PCl_5(g)$ dissociates according to

$$PCl_5(g) \rightleftarrows PCl_3(g) + Cl_2(g)$$

to the extent of 80% under a total pressure of 1 atm. Find (i) K_P (ii) the total volume of the system and (iii) the concentration of each component in mol/litre if 1 mol of PCl_5 was taken initially.

Solution:

At equilibrium we will have

$$PCl_5(g) \rightleftarrows PCl_3(g) + Cl_2(g)$$

$$1\text{ mol} - 0.8\text{ mol} \qquad 0.8\text{ mol} \qquad 0.8\text{ mol}$$

Total amount of species = 1.8 mol.

$$\text{Partial pressure of } PCl_5 = \frac{0.2}{1.8} \times 1\text{ atm}$$

$$\text{Partial pressure of } PCl_3 = \frac{0.8}{1.8} \times 1\text{ atm}$$

$$\text{Partial pressure of } Cl_2 = \frac{0.8}{1.8} \times 1\text{ atm}$$

Now $$K_P = \frac{p_{PCl_3} p_{Cl_2}}{p_{PCl_5}} = \frac{(0.8\text{ atm}/1.8)^2}{(0.2\text{ atm}/1.8)} = 1.778\text{ atm}$$

(ii) The total volume of the system is given as

$$V = \frac{nRT}{p}$$

$$= \frac{(1.8\text{ mol})(0.082\text{ atm LK}^{-1}\text{ mol}^{-1})(523\text{ K})}{(1.8\text{ mol})}$$

$$= 42.89\text{ L}$$

(iii) Concentrations of the species are as follows:

$$[PCl_5] = \frac{0.2\text{ mol}}{42.89\text{ L}}$$

$$= 4.663 \times 10^{-3}\text{ mol L}^{-1}$$

$$[PCl_3] = [Cl_2] = \frac{1.8\text{ mol}}{42.89\text{ L}}$$

$$= 4.20 \times 10^{-2}\text{ mol L}^{-1}$$

Example 16:

The equilibrium constant for the reaction $H_2 + I_2 \rightleftarrows 2HI$ at 450°C is found to be 64. If 6 moles of hydrogen are mixed with 3 moles of iodine ion a one-litre vessel at this temperature, what will be the concentration of each of three components when the equilibrium is attained?

Solution:

For reaction:	H_2 +	$I_2 \rightleftarrows$*	2HI
Initial	6 moles	3 moles	0
At equilibrium	(6 – x) moles	(3 – x) moles	2x moles
Concentration	(6 – x)/1 moles litre^{-1}	(3 – x)/1 moles litre^{-1}	2x/1 moles litre^{-1}

$$\therefore \quad K_C = \frac{[HI]^2}{[H_2][I_2]}$$

$$= \frac{(2x)^2}{(6-x)(3-x)} = \frac{4x^2}{18-9x+x^2} = 64$$

or $$\frac{x^2}{18-9x+x^2} = 16$$

or $$x^2 = 16 \times 18 - 16 \times 9x + 16x^2$$

or $$15x^2 - 144x + 288 = 0$$

or $$5x^2 - 48x + 96 = 0$$

$$x = \frac{48 \pm \sqrt{(48)^2 - 4 \times 5 \times 96}}{2 \times 15}$$

$$= \frac{48 \pm \sqrt{384}}{30} = \frac{48 \pm 19.75}{30} = 2.258 \text{ or } 0.942$$

Hence, composition of equilibrium mixture will be

H_2 = 6 – 2.258 (0.942) = 3.742 or 5.058 mole/litre

I_2 = 3 – 2.258 (0.942) = 3.742 or 2.058 mole/litre

HI = 2 × 2.258 (0.942) = 4.516 or 1.884 mole/litre

3

Phase Rule

DERIVATION OF PHASE RULE

J. Willard Gibbs (1876) deduced a *mathematical* relation between the number of degrees of freedom (F), the number of components (C) and the number of phases (P) in a heterogeneous system at equilibrium. The relation is as follows :

$$F = C - P + 2$$

It was first derived by Gibbs from thermodynamic considerations and is known as *Gibbs phase rule*. Let us consider a system of 'C' components distributed between 'P' phases and let each phase contain all the 'C' components. To define the composition of each phase it is required to specify (C – 1) concentration terms and hence the concentration of the remaining components will be defined by the difference. As there are 'P' phases the total number of variables for concentration alone is P (C – 1).

As the process is being carried out at a constant temperature and pressure, it is required to add two more variables. Hence the total number of variables is becoming equal to P(C – 1) + 2.

We will now determine how many equations involving these variables are there when all the phases are in equilibrium. This can be shown in the following manner :

Consider a heterogeneous system of two phases (a and b) containing three components (1, 2, 3) in equilibrium. The chemical potential (μ) of these components in two phases can be represented as :

$$\mu_1 (a), \mu_2 (a), \mu_3 (a), \mu_1 (b), \mu_2 (b), \mu_3 (b)$$

As we are considering the closed system in equilibrium at a definite temperature, and pressure, the *Gibbs-Duhem* equation can be applied to it.

Thus, we get $\Sigma\mu \, dn = 0.$

Suppose a small amount δn_1 is transferred. under equilibrium conditions from phase a to b and therefore:

$$-\mu_1 (a) \, dn_1 + \mu_1 (b) \, dn_1 = 0$$

or $\mu_1 (a) = \mu_1 (6)$

This equation reveals that the equilibrium is between two phases when the chemical potential of a particular component is the same in each phase. This can be applied to all the components. It, therefore, follows that

$$\mu_2 (a) = \mu_2 (b); \; \mu_3 (a) = \mu_3 (b); \text{ etc.}$$

Now, imagine the system is having three phases. Therefore, the chemical potential of a particular component must be same in all the phases. *i.e.*,

$$\mu_1(a) = \mu_1(b) = \mu_1(c)$$

Hence the number of independent equations defining the equilibrium between three phases are two only. In general, the above conditions also hold good for 'P' phases and 'C' components to give a set of following relations :

$$\mu_1(a) = \mu_1(\&) = \mu_1(c) = \ldots = \mu_1(P)$$

$$\mu_1(a) = \mu_2 (b) = \mu_2(c) = \ldots = \mu_2 (P)$$

$$\ldots \quad \ldots \quad \ldots \quad \ldots \quad \ldots$$

$$\ldots \quad \ldots \quad \ldots \quad \ldots \quad \ldots$$

$$\mu_c(a) = \mu_c(b) = \mu_c(c) = \ldots = \mu_c(P)$$

These equations constitute C(P – 1) independent equations. Thus the degree of freedom F must be equal to

$$F = [P(C - 1) + 2] - [C(P - 1)]$$

$$= C - P + 2$$

This is the phase rule of Gibbs.

Discussion : If a component is missing from a particular phase, however, the number of concentration variables is decreased by one.

But at the same time the number of possible equations is also decreased by one. Hence the value of (C–P), and therefore F. remains the same whether each constituent is present in every phase or not. This means that the phase rule is not restricted by the assumption made, and is generally valid under all conditions of distribution provided that equilibrium exists in the system.

Advantages of Phase Role

(i) It gives a simple method of classifying equilibrium states of systems.

(ii) It confirms that the different systems having the same number of degrees of freedom behave in like manner.

(iii) It predicts the behaviour of systems when subjected to changes in the variables such as pressure, temperature and volume.

(iv) The phase rule is applicable to macroscopic systems. Therefore, it is not necessary to take into account about their molecular structures.

(v) It is applicable to physical as well as to chemical phase reactions.

(vi) Phase rule takes no account of nature of the reactants or products in phase reactions.

(vii) Phase rule predicts that a number of substances would remain in equilibrium or not in equilibrium if some of the substances have been transformed into the new substances.

Limitations of the Phase Role

(i) As the phase rule is applicable to heterogeneous systems in equilibrium, it is therefore of no use for such systems which are slow in reaching the equilibrium state.

(ii) As the phase rule is applicable to a single equilibrium state, it never tells about the number of other equilibrium possible in the system.

(iii) In phase rule various variables are temperature, pressure and composition. This phase rule does not consider the electric and magnetic influences. If such variables are considered, the factor 2 of the phase rule has to be adjusted accordingly.

(iv) All the phases in the system must be present under the same pressure, temperature and gravitational force.

(v) No liquid or solid phases should be finely divided otherwise their vapour pressures will differ from their normal values.

ONE-COMPONENT SYSTEMS

When one component is the smallest number by means of which the composition of each phase is expressed, it is known as a one component system.

Maximum Number of Phases : When F is minimum, P becomes maximum. The minimum number of degrees of freedom possible in a system is zero. Consider an equilibrium state of a one component system having zero degree of freedom. According to the phase rule,

we have $P = C - F + 2$ ...(1)

or $P = 1 - 0 + 2 = 3$ ...(2)

Thus, the *maximum number of phases in any equilibrium state of one component system having zero degree of freedom will be three*. It cannot be more then three.

Maximum Number of F : The maximum value of F is expected in the case when P is minimum. As the minimum number of phases in any system is one, it follows from phase rule equation that the maximum degree of freedom in one component system are two.

$$F = C - P + 2 = 1 - 1 + 2 = 2$$

THE WATER SYSTEM

It is one component system as H_2O is the only chemical compound involved. Water exists in three possible phases, namely, solid, liquid and vapour. The phase diagram of this system is given in Fig. 3.1 which is obtained by plotting pressure—temperature curves.

The diagram consists of:

(i) *Curve :* Three curves are OA, OB and OC.

(ii) *Triple point :* The above three curves meet at the point 0 which is known as triple point.

(iii) *Areas :* Three curves divide the diagram into three areas AOC, AOB and BOC.

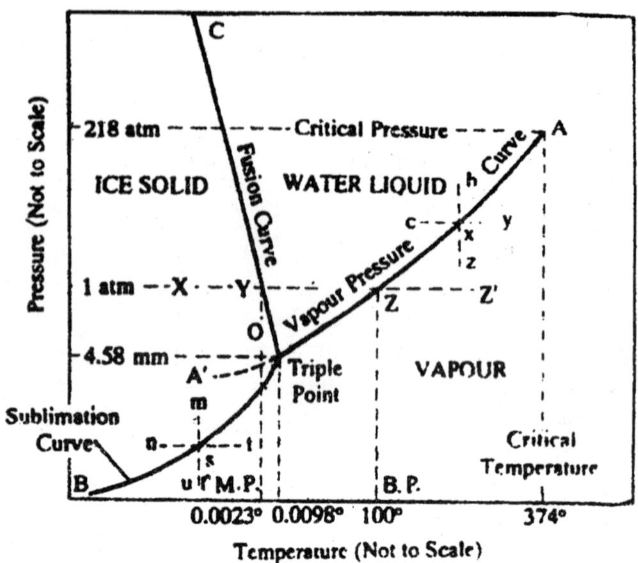

Fig. 3.1 : Water System.

Description of the Phase Diagram

Curves

AO : This curve is known as the *vapour pressure curve* because it gives the vapour pressure of water at different temperatures. The curve starts from O which is the freezing point of water and ends at A, the critical temperature of water (374°C). Beyond the point A, *i.e.*, critical temperature, the two phases liquid water and vapour merge into each other. From this curve it is seen that for any given temperature, there exists a fixed value of pressure. Similarly for each vapour pressure, temperature has also a fixed value. Thus, the degree of freedom on any point of this curve is one or it is univariant. This is also predicted by the phase rule.

$$F = C - P + 2 = 1 - 2 + 2 = 1$$

The number of phases is taken as two because liquid water and vapour exist along the curve OA.

OB : It is the *vapour pressure carve* of ordinary ice and is known as the sublimation curve. Along this curve, solid ice is in equilibrium with its vapour. Thus, this curve shows the variation of vapour pressure of ice at various temperatures.

The curve starts at O and ends at B, *i.e.*, absolute zero (–273°C). At this temperature, no vapour can exist and, therefore only ice is left. But on the other points of the curve OB, ice is in equilibrium with vapour. Thus, there are two phases. Applying the phase rule, we have

$$F = C - P + 2 = 1 - 2 + 2 = 1$$

Thus, the system has one degree of freedom or is univariant. It means that each vapour pressure term can be maintained only at a fixed temperature. This is found to be so.

OC : It is the *melting point or fusion curve* of ice as it indicates the effect of pressure on melting point of ice. The inclination of the curve OC towards the pressure axis, *i.e.*, y-axis indicates that the melting point of ice is lowered by increase of pressure. At any point on the curve OC two phases ice and liquid water are in equilibrium. Applying the phase rule, we get

$$F = C - P + 2 - 1 - 2 + 2 \text{ or } F = 1$$

Thus, the system is univariant. It means that for any given pressure, melting point must have one definite value. This has been found in actual practice.

Triple point O : The three curves OA, OB and OC meet at a point O which is known as *triple point*. At this point all the three phases namely, ice, water and vapour co-exist. Thus, the value of P is three. Applying the phase rule to this point,

$$F = C - P + 2 = 1 - 3 + 2 \text{ or } F = 0$$

Thus, the degree of freedom at triple point is zero. It means that three phases can co-exist in equilibrium only at a definite temperature and pressure which correspond to the point 0. The values of pressure and temperature at the point O are 4.5 mm and 0.0075°C. If pressure and temperature are varied from the value for this point, three different cases may arise :

(i) If pressure is raised without changing the temperature, only liquid phase will be there. Other phases ice and vapour will be converted into the liquid.

(ii) If pressure is lowered without changing the temperature, only vapour phase will be there. Two phases ice and liquid will also be converted into the vapour.

(iii) *If pressure and temperature are varied together, one or the other curve (i.e., OA or OB or OC) will be followed depending upon values of pressure and temperature.*

Areas, AOB, BOC and COA : The areas between the curves OA OB and OC arc AOB, BOC and COA. In each area a single phase *i.e.*, solid ice, water and vapour is of stable existence Thus,

(i) in the area AOB, only vapour exists.

(ii) in the area AOC, only water exists.

(iii) and in the area BOC, only ice exists.

All these areas are bivariant. *i.e.*, in order to locate any point in the area, the temperature and pressure must be defined. This also follows from the phase rule,

$$F = C - P + 2 = 1 - 1 + 2 = 2$$

Metastable Equilibrium : Sometimes it is possible to cool water below its freezing point without the separation of solid ice. Thus, the curve AO can be extended to A' by cooling below its freezing point without separating ice. Along the curve OA' the liquid is in metastable equilibrium with the vapour. Therefore, this curve (OA') is said to be in metastable equilibrium. As soon as a small ice particle is kept in contact with the supercooled liquid, it at once changes into the solid ice and the curve merges in OB.

From the phase diagram, it is seen that the curve OB lies below the curve OA'. Thus the metastable system processes a higher vapour pressure than the stable system if same temperature is considered.

Effect of Change of Temperature and Pressure on Equilibrium

The real significance of phase diagram can be understood only if we follow the changes that are occurring on changing the pressure or temperature of the system. Suppose we are interested to study the effect of heating ice under a pressure of 1 atmosphere and at a temperature represented by the point X in the Fig. 3.2. At the point X. there is only one phase, *i.e.*, ice and therefore this point is said to be bivariant (F = 1 – 1 + 2 = 2). Due to the bivariant nature of X. The temperature can have any value at the same pressure. Therefore, heating the ice slowly at constant pressure will cause the system to move along XY. Up to point Y there is only ice phase. Beyond the Y. fusion sets in, *i.e.*, the liquid phase also appears. It means that at point beyond Y there are two phases. *i.e.*, ice and water hence the system has one degree of freedom ($F = 1 - 2 + 2 = 1$). Due to one degree of freedom the temperature cannot alter without the alteration of pressure. The effect of continued heating will result the conversion of ice into water. As soon as the point Z is reached. there occurs complete conversion of solid into liquid. At point Z. there is only liquid phase and the system is bivariant. Beyond the point Z. vapourisation sets in. It means that the number of phases will become two and the system has only one degree of freedom. The continued heating will at change the temperature because the pressure is being kept constant. The only effect of further heating will be to convert more and more of the liquid into vapour. The temperature will rise along the line ZZ'.

Effect of Change of Temperature and Pressure on Equilibrium

It is very interesting to see the effect of heating and applying pressure on the system at the triple point 0 where all the three phases are coexisting. If we do the heating at the triple point 0, there occurs the melting of more and more of the solid into liquid but there occurs no increase in temperature or pressure until the whole of ice has completely converted into liquid water. When this is completed. the system will have now two phases. *i.e.*, liquid and vapour. This means that the system will change from nonvariant to univariant. Therefore. further heating will result a rise in temperature. causing the equilibrium to shift along the curve OA.

Now if we apply the pressure to the system at the triple point. no change in pressure or temperature will occur as long as all the three phases are present. The only effect of applying pressure will be to convert vapour into liquid or solid phase. Ultimately vapour phase will disappear completely and convert into liquid and solid phases. and the further application of pressure will cause increase of pressure with change of temperature along the curve OC.

TWO COMPONENT SYSTEMS

When two components denote the smallest number by means of which composition of each phase can be expressed, it is known as a two component system.

Maximum number of Phases : When F is minimum. P becomes maximum. The minimum number of degrees of freedom possible in any system is zero. Consider an equilibrium state of a two component system having zero degree of freedom. According to the phase rule, we have

$$F = C - P + 2 \qquad ...(1)$$

$$0 = 2 - P + 2 \text{ or } P = 4$$

Thus, the maximum number of phases in any equilibrium state of a two component system having zero degree of freedom will be four.

Maximum number of F : The maximum value of F is expected in the case when P is minimum.

As the minimum number of phases in any system is one, it follows from equation (1) that maximum degree of freedom in a two component system are three.

$$F = C - P + 2 = 2 - 1 + 2 \text{ or } F = 3$$

Phase Diagram, As the value of F is three, it means that a phase is to be defined by three variables, *i.e.,*. pressure, temperature and composition. Pressure, temperature and composition could be represented by three co-ordinate axes at right angles, yielding models for two component systems. But the space models cannot be conveniently represented on paper.

In order to have simple phase diagram for two component systems, it is a usual practice to choose any two of three variables for graphic representation, assuming third to be constant. In this manner, we can have

(i) Pressure-temperature diagram (P–T)

(ii) Temperature-composition diagram (T–C)

(iii) Composition-pressure diagram (C–P)

In most of the studies, it is the usual practice to keep the pressure constant and thus, an T–C diagram is the one which is most often met.

Reduced phase rule : In all the above P–T. T–C and C–P diagrams, one of the variables is fixed. Thus the degree of freedom of the system is reduced by one. Therefore the phase rule equation (1) for the component system is written as

$$F - 1 = C - P + 1$$

$$F = C - P + 1 \qquad ...(2)$$

where F' gives the remaining degrees of freedom of the system. Equation (2) is known as *reduced phase rule*.

TYPES OF TWO COMPONENT SYSTEMS

We will only discuss the solid-liquid systems of two components.

Solid-liquid Systems of two Components

Condensed systems : The effect of pressure on the equilibrium between liquids and solids is very small. Therefore, the effect of pressure on such systems may be disregarded. Therefore, the experiments in liquid-solid equilibria are usually conducted under constant atmospheric pressure. *"Such a system in which only solid and liquid phase are considered and the experiments are carried out under atmospheric pressure, is called a condensed system."*

The graphical representation of condensed systems of two-components is simplified as there are only two variables, *i.e.*, temperature and composition, keeping pressure to be constant and these can be represented on a plane paper in the case of a condensed system, variables gets fixed. In other words, one of the degrees of freedom of the system is reduced by one. Therefore, the equation (1) becomes as:

$$F - 1 = C - P + 1$$

or $$F = C - P + 1 \quad ...(3)$$

where F' gives the remaining degrees of freedom of the system. Equation (3) is known as *condensed phase rule* which is similar to the equation (2). Solid-liquid equilibria of number of types are known. But the following types are more important:

Type I. Two components are completely miscible m liquid phase and the solid phase consist of pure components. Examples are : (a) Lead-antimony, (b) Lead-silver, (c) Potassium iodide-water system.

Type II. The components enter into chemical combination giving rise to one or more compounds. Such types of two-component systems are of two kinds :

IInd (A). Two components form a compound with congruent melting point. Examples are :

(i) Zinc-magnesium system, (ii) $FeCl_3$–water system.

IInd (B). Two components form a compound with incongruent melting point. Examples are :

(i) Picric acid-benzene system, (ii) Gold-antimony system.

(ii) Sodium sulphate-water system.

Simple Eutectic System or Solid-liquid Equilibrium of Type I.

Consider a two-component system having two components A and B. These components are completely miscible in liquid state. *They do not form any compound.* The phase equilibrium diagram for this type is shown in Fig. 3.2.

The phase diagram (3) consists of

(i) Curves. AC and CB

(ii) Points. A, B and C

(iii) *Areas* : ADC, D'CB, OO'D'D and they are above the curve ACB.

Description

(i) *Point A* : It represents the melting point of the pure component A.

(ii) *Point B* : It represents the melting point of the pure component B.

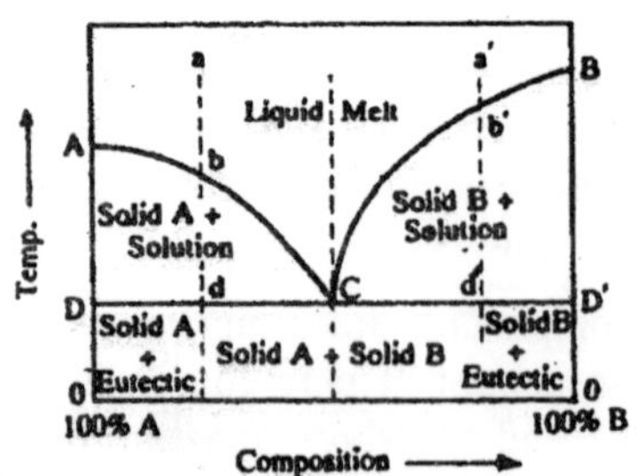

Fig. 3.2 : Solid-liquid system (Type I)

(iii) *Curve AC* : On gradual addition of small amounts of B to A, the freezing point of A is lowered along this curve AC. Therefore, this curve AC is known as the freezing point of the component A.

Along this curve AC, the solid A is in equilibrium with the solution of the component B in A (liquid) at different temperatures. Thus, the number of phases along this curve AC is two. As measurements are being made at constant atmospheric pressure, it follows from the condensed phase rule (3),

$$F = C - P + 1 = 2 - 2 + 1 = 1$$

Thus any point on the curve AC has one degree of freedom or is univariant. It means that only composition varies along the curve AC.

Curve BC : As increasing quantities of A are added to B, the freezing point of B falls along the curve BC. Along this curve, the solid B is in equilibrium with the solution of the component A in B at different temperatures. Again the number of phases along the curve BC like the curve AC is two. Again, applying the condensed phase rule equation, we get

$$F' = C - P + 1 = 2 - 2 + 1 = 1$$

Thus, any point on the curve BC is univariant, *i.e.*, only composition varies along BC.

Point C : At the point C, where the two curves AC and BC intersect, obviously both the sides A and B must be in equilibrium with the liquid phase. As three phases co-exist at this point, it follows from the modified phase rule (3).

$$F' = C - P + 1 = 2 - 3 + 1 = 0$$

Thus, the system at the point C is non-variant, *i.e.*, has no degree of freedom. This means that there is only one temperature at which the liquid phase can be in equilibrium with both solids A and B, and that the composition of the liquid phase must also be definite. From the figure, it is seen that the point C is the lowest temperature at which the liquid can exist. Therefore C has been called the eutectic point because it represents the lowest melting point of any mixture of solids A and B.

Area Above ABC : In this area the two components A and B exist as a homogeneous liquid solution because the temperature in it is above the melting point of any mixture. It means that there is only one phase in the area. Therefore, applying the condensed phase rule into this area, we obtain

$$F' = C - P + 1 = 2 - 1 + 1 s = 2$$

Thus, the system in the area above curve ACB is bivariant. It means that temperature and composition are required to define any point in the area above curve ACB.

Area OO'D'D : In this area only solid can exist because the liquid phase can not exist below the eutectic temperature.

Area ADC : Any point in the area ADC represents equilibrium between solid A and liquid. The composition of the phases in equilibrium can he obtained by drawing a horizontal line through the particular point. This line, where it cuts the curve AD, gives the nature of the solid and the AC to give the composition of the liquid.

Area CD'B : The above explanation is also true for this area.

Cooling of liquid mixtures : It is possible to predict the behaviour of any system on heating or cooling by using equilibrium diagram (3). This type of study is of special importance in the study of alloys in metallurgy.

Consider a liquid mixture of components represented by a point a. Allow it to cool at constant pressure. This fall in temperature continues without any change of composition till the point b on the curve AC is reached. At the point b, solid A separates out at a temperature t_1 which corresponds to the point b. Thus, the system at b involves two phases. *i.e.*, solid A and liquid solution.

Applying the condensed phase rule to this point b. it gives

$$F' = C - P + = 2 - 2 + 1 - 1$$

Thus, the system at point b is univariant. It means that there will be a fall in temperaıure only when the composition, of liquid phase changes. Therefore, if the solid A keeps on separating out along the curve bC and the solution becomes relatively richer in B. At the point y solid A is in equilibrium with its mixture of composition y at a temperature t_2 which corresponds to this point. Thus, in the area ACD solid A is in equilibrium with solution of different compositions depending upon different temperatures.

When the point d is reached which corresponds to the eutectic point C, the second solid B also starts separating out. Now the system will have three phases. Again, applying the phase rule to this system, we obtain

$$F' = C - P + 1 = 2 - 3 + 1 \text{ or } F' = 0$$

Thus, the system is univariant. It implies that on cooling further, the solid A and solid B separate out in the fixed ratio so that the composition of the liquid mixture remains constant as required at the point C. Here temperature does not change. This continues till the whole solution phase has been completely solidified. Thus, the system in the area OO'D'D consists of mixture of solid A and solid B. In that area, the system becomes univariant,

$$F' = C - P + 1 = 2 - 2 + 1$$

or $$F' = 1.$$

It means that on further cooling the temperature will only fall in the area BD'DA keeping the composition of two solids as constant. Similarly, one can explain if one considers a liquid of composition a' on the right side of eutectic point. Examples are : (i) Animony-lead system, (ii) Silicon-aluminium system. (Hi) Bismuth-cadmium system, (iv) KCl–$CaCl_2$ system, (v) Benzene-methyl chloride system, (vi) KI-H_2O system, (vii) Lead-silver system.

We will now discuss some examples of type 1 of two-component involving solid-liquid equilibria.

Lead-silver system : It is a two component system. The two metals lead and silver are completely miscible in the liquid state and do not form any compound. Therefore, the equilibrium diagram of this system is similar to that shown in the figure (3.2). The special features of lead-silver system are demonstrated in Fig. 3.3.

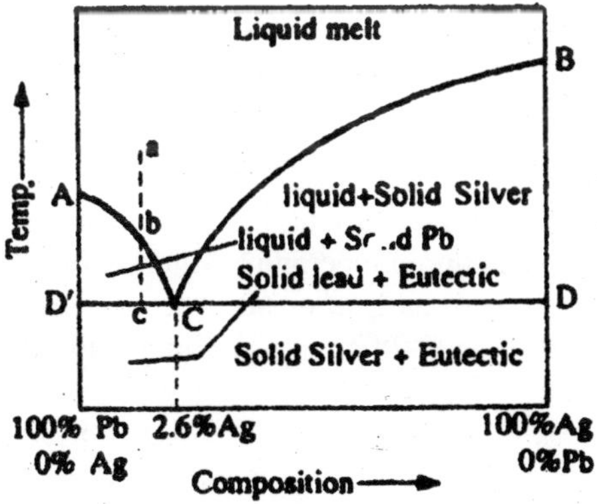

Fig. 3.3 : Lead-Silver System.

As the experiment is being carried out at constant pressure, the vapour phase is not considered and the condensed phase rule is applied.

$$F' - C - P + 1$$

Curve AC : A is the melting point of lead (327°). When increasing quantities of pure silver are added to lead metal, the freezing point of lead lowers along the curve AC. Therefore, AC is the freezing point curve of lead.

Along the curve AC, the solid lead is in equilibrium with the liquid melt. Thus, the number of phases along the curve is two. As all measurements are made at atmospheric pressure, on applying condensed phase rule, F' = C – P + 1 = 2 – 2 + 1 = 1

Therefore, the system is univariant. It means only composition varies along the curve AC.

Curve BC : Point B represents the melting point of pure silver (961°). When increasing quantities of pure lead are added, the freezing of silver lowers along the curve BC. Therefore, this curve may be called the *freezing point* curve of silver. Again, the solid silver is in equilibrium with the liquid melt along the curve BC. Thus, there are two phases. Applying the condensed phase rule equation, we obtain

$$F' = C - P + 1 = 2 - 2 + 1 = 1.$$

Again, the curve BC like AC is univariant and only composition varies along this.

Eutectic point C : The two curves AC and BC intersect at the eutectic point C. This point, being common to both the curves, represents the condition under which three phases lead, solid silver and liquid melt co-exist. The degree of freedom here is zero which also follows from reduced phase rule :

$$F' = C - P + 1 = 2 - 3 + 1 = 0$$

Thus, the point C is invariant, *i.e.*, has no degree of freedom. This point C (308°C) lies at a temperature which is lower than the melting points of silver as well as lead metal. The composition of an alloy of lead (97.6%) and silver (2.4%) corresponding to the point C is also fixed. Therefore, the point C is called the *eutectic point i.e., lowest temperature at which the liquid melt can exist. The temperature corresponding to this point is known as eutectic temperature and composition to this point is known as eutectic composition.*

Area above curves AC and BC : In this area, silver and lead are present as a homogeneous liquid solution. Thus, there is only one phase in this area. Applying the condensed rule to any point in this area, we obtain :

$$F' = C - P + 1 = 2 - 1 + 1 = 2$$

Thus, the system in this area is bivariant. It means that both temperature and composition are required to define any point in this area.

Area ACD' : In this area, solid lead is in equilibrium with the liquid melt.

Area BCD : In this area, solid silver is in equilibrium with the liquid melt,

Significance of lead-silver system : The phase diagram is utilised in the separation of silver from lead in Pattinson's Process for desilverisation of lead.

Suppose a represents the molten argentiferous lead containing a small amount of silver in it. Allow it to cool. Only temperature of the liquid melt falls with no change in concentration till the point b on the curve AC is reached. On further cooling lead begins to crystallise out and the solution becomes richer in silver. Further cooling will continue along the curve bC. Lead continues to separate out and is constantly removed by means of ladels. The liquid melt continues to be richer and richer in silver till the point C is reached where an alloy containing 2.6% of silver is obtained. This is then subjected to cupellation process.

The above process of increasing the concentration of silver in argentiferous lead is known as *Pattinson's process.*

Potassium Iodide-water System

The potassium iodide-water system is a typical example of a two-component system which forms a eutectic mixture. Hence its phase diagram is similar to the Fig. 3.4. The various features of this system are illustrated in the T–C diagram, pressure being considered as one *atm* pressure. In the above diagram, the melting point of potassium iodide is not shown because it is very high.

Curve AC : Point A represents the freezing point of water at 1 atm pressure. This temperature is 0°C.

When increasing quantities of potassium iodide are added, the freezing point of water is lowered along the curve AC. Thus, AC may be called the freezing point curve of water.

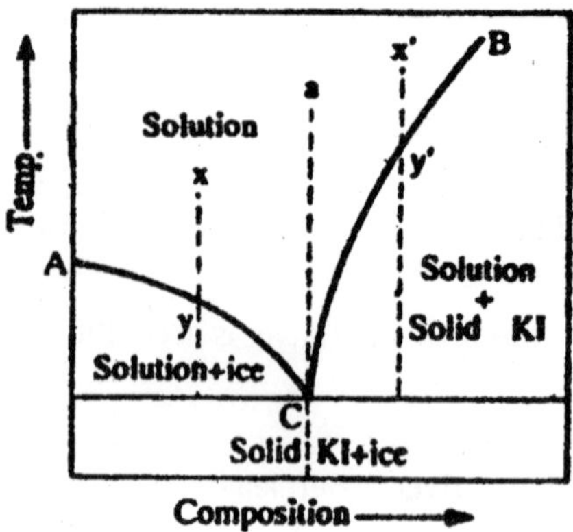

Fig. 3.4 : KI-H_2O System.

Along the curve AC, the solution of KI in H_2O is in contact with ice. Thus, there are two phases. Applying the condensed phase rule to any point on this area.

$$F' = C - P + 1 = 2 - 2 + 1$$

or $$F' = 1.$$

Thus, the system is univariant, *i.e.*, has one degree of freedom. It means that there will be a definite composition of the solution corresponding to each temperature. When further quantities of potassium iodide are added, the lowering of freezing point continues along the curve AC till the point C is reached. Here the solution becomes saturated with respect to solid potassium iodide and freezes at a constant temperature (–23°C) and fixed composition (52% KI).

Curve BC : This is known as the solubility curve of potassium iodide in water at different temperatures. The solubility increases with rise of temperature and becomes maximum at the point B which is the boiling point of the saturated solution. Allow a saturated solution of potassium iodide at B to cool. The solution will freeze along the curve BC. During this, solid KI crystallises out.

This crystallisation of KI continues till the temperature and composition corresponding to the point C is reached. At this point, the solution freezes as a whole with fixed composition.

Along the curve BC, solid KI is in equilibrium with solution.

Again, applying the condensed phase rule to this system, we obtain

$$F' = C - P + 1 = 2 - 2 + 1 \text{ or } F' = 1$$

Thus the system is univariant, *i.e.*, has no degree of freedom. The steep rise of curve BC indicates that the solubility of KI increases slowly with the rise of temperature.

Point C : It is the eutectic or cryohydric point. At this point the two curves AC and BC intersect. This point corresponds to a definite temperature (–23°C) and definite composition (527% KI + 487% ice).

At this point C, ice, solution and solid KI are in equilibrium. This point has no degree of freedom which is evident from phase rule equation.

$$F' = C - P + 1 = 2 - 3 + 1 \text{ or } F' = 0$$

Thus, C is a non-variant point.

Area above AOB : If the point lies above the curve BC. It means that the dilute solution of KI is there. But the area above the curve BC represents the concentrated solution of potassium iodide.

Cooling : Three cases may arise :

(i) Consider a dilute solution represented by the point x. On cooling, the temperature will lower along the curve xy without any change in composition.

This continues till the point y is reached. At this point ice separates out. Here the system will become univariant and further cooling will take place along the curve yC. This continued till the point C is reached. At this point ice and K.I freeze as a whole with fixed composition, *i.e.*, eutectic mixture.

(ii) Consider a concentrated solution represented by the point x'. Allow it to cool. This cooling continues without any change in composition along the curve x'y'. At the point y', solid KI separates out.

Further cooling will proceed along the curve y'C until the point C is reached. At this point, ice and K[freeze as a whole to give the eutectic mixture [KI = 52% and ice = 48%].

(iii) Consider a solution which is represented by the point a. This point lies vertically above the eutectic point C. When the solution is allowed to cool, the temperature will lower along the curve aC until the eutectic point C is reached. At this point, ice and potassium iodide separate out simultaneously.

This mixture of potassium iodide and ice deposited at the eutectic point is known as *cryohydrate*.

Type IInd A.

Solid-liquid equilibrium of type IInd A :

A compound is said to hive a congruent melting point when it melts sharply at a constant temperature into a liquid of the same composition as that of solid from which it is derived.

Suppose A and B are two components and AB is the compound formed by their chemical combination. The compound AB has a sharp congruent melting point. The phase diagram for such a phase system is shown in Figure (3.5).

Curve AC : Point A represents the melting point of pure component A. By gradual addition of B to A, the freezing point of A is lowered along the curve AC. Thus, the curve AC represents the freezing point curve of the component A. Along this curve, the solid A is in equilibrium with the solution of the component B in A. Therefore, the number of phases along AC is two. Applying the condensed phase rule to the system, we obtain

$$F' = C - P + 1 = 2 - 2 + 1 = 1$$

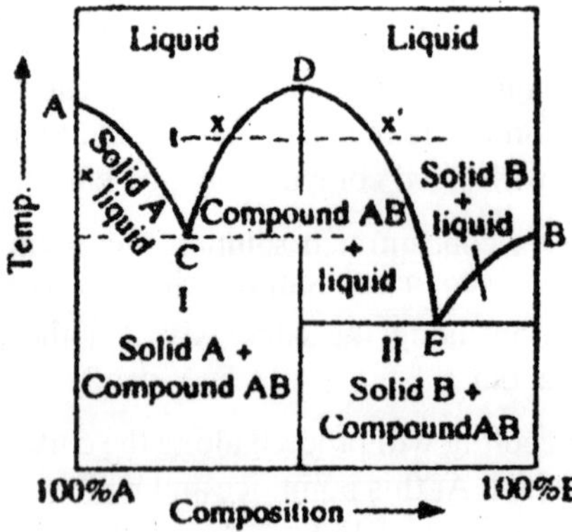

Fig. 3.5 : Solid-liquid equilibrium (Type II)

Thus, the system is univariant, *i.e.*, only the composition varies along AB.

Curve BE : Point B represents the melting point of pure component B. When A is added to B, the freezing point B is lowered along the curve BE. The curve is known as the *freezing point curve* of the component B.

Along this curve, the solid B is in equilibrium with the solution of the component A in B. Thus, the number of phases along BE is two. Applying the condensed phase rule to this system,

$$F' = C - P + 1 = 2 - 2 + 1$$

or $$F' = 1.$$

Thus, the system is univariant, *i.e.*, only the composition varies along BE.

Curve CDE : This curve is known as *freezing point carve* of the solid compound AB. Along this curve, the solid compound AB is in equilibrium with the liquid phase at different temperatures. The point D is known as *congruent melting* point of the compound because both the liquid and solid phases have the same composition at the point D. It means that two-component system becomes one component at this temperature as both the solid and liquid phases contain the same compound AB. Applying the phase rule to this point D, we get

$$F' = C - P + 1 = 1 - 2 + 1 = 0$$

Thus, the point D is non-variant : It means that this point D represents a definite melting point A.

Eutectic point C : The two curves AC and CDE intersect at the point C. At this point, solid A and compound AB are in equilibrium with liquid phase. Therefore, the number of phases is three. Now, applying the phase rule equation to this point C, we obtain

$$F' - C - P + 1 = 2 - 3 + 1 = 0$$

Thus, the system at C is invariant, *i.e.*, has no degree of freedom. It means that solid A, solid compound AB and liquid can only exist at a definite temperature and the composition of the liquid phase must also be definite.

Eutectic point E : The two-curves BE and CDE intersect at the point E where solid B, solid compound AB are in equilibrium with liquid phase. Again, the point E is invariant, *i.e.*, has no degree of freedom. The horizontal line passing through the eutectic point C and E is known as solid because only phase exists along it.

Areas I and II : The names of the various phases in areas I and II are shown in the Fig. 3.14.

General Discussion

(i) In Fig. 3.14 it is shown that the melting point (congruent) lies above the melting points of A and B. But it is not so. Certain cases are known in which the congruent melting point of the compound AB may be above, below or between those of two components A and B.

(ii) In Fig. 3.14 it is shown that the compound AB contains equimolecular amounts of A and B. This is not always so.

(iii) Draw a line XX_1. It means that the liquid phase has two compositions X and X, in equilibrium with the same solid AB at a certain temperature, say, t. In simple words it means that the compound AB has two solubilities at the same temperature* This paradox is explained by the diagram (3.5). AB is made up of two parts with DD' as the dividing line. Thus, the left half of the diagram will be a two component system of solid A and compound AB and the right half of the diagram shows the two component system involving solid B and compound AB. Also, the curve DC will represent the freezing point curve of solid compound AB when A is added to it. Similarly, the curve DE will represent the freezing point curve of compound AB when B is added to it.

(iv) When two components form more than one compound, the equilibrium diagram will have curve analogous to CDE for each compound.

Examples are:

(i) Zinc-magnesium system,

(ii) Aluminium-magnesium system,

(ii) Mercury-thorium system,

(iv) Gold-tin system,

(v) Phenol-aniline system,

(vi) Ferric chloride-water system.

We will now discuss some examples.

Zinc-Magnesium System : It is an example of a two component system in which two metals form a compound having congruent melting point. The compound is an alloy having the formula $Mg(Zn)_2$ with a true melting point 590°C which lies between the melting points of two metals. The composition of $Mg(Zn)_2$ by weight is

$$Mg = 79\%$$

and Zn = 21%

The melting point of zinc is 420°C and that of magnesium is 650°C. The melting point of $Mg(Zn)_2$ is 590°C. The compound is fairly stable and it melts without change of composition.

As the experiment is being carried out at constant pressure, vapour phase therefore, is not considered and the condensed phase rule will be applied :

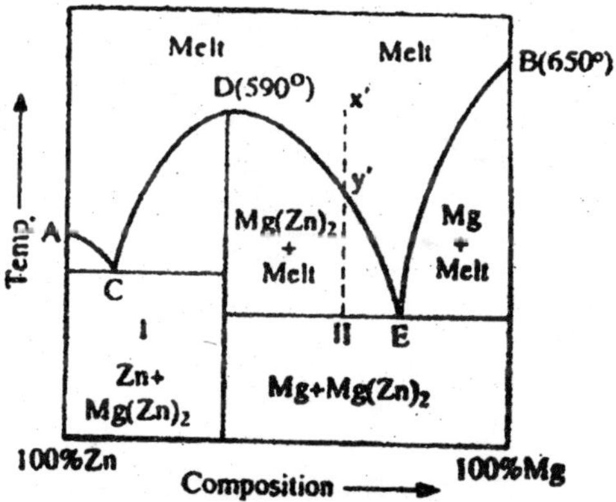

Fig. 3.6 : Zn—Mg system

$$F' = C - P + 1$$

The equilibrium diagram of such a system is represented in Fig. 3.6.

Point A : It represents the melting point of pure zinc (M.P. of zinc = 420°C],

Curve AC : By gradual addition of Mg to Zn, the melting point of zinc is lowered along the curve AC. Along this curve the solid zinc is in equilibrium with melt (solution). Thus, the number of phases is two. Applying the condensed phase rule equation to this curve, we get

$$F' = C - P + 1 = 2 - 2 + 1 = 1$$

Thus, the system is univariant, *i.e.*, only the composition varies along the curve.

Point B : This represents the melting point of pure magnesium [M.P. = 65°C].

Curve BE : When zinc is added to Mg the melting point of Mg is lowered along the curve BE. Along the curve BE, the solid Mg is in equilibrium with melt at different temperatures and hence is regarded as the freezing or melting point curve of magnesium. Again, only composition varies along this curve which is evident from phase rule,

$$F' = C - P + 1 = 2 - 2 + 1 = 1 \qquad [\because P=2]$$

Curve DC : When further quantities of zinc are added, the melting point of the compound $Mg(Zn)_2$ lowers along this curve. Thus, the curve represents the lowering of melting point of compound $Mg(Zn)_2$ when zinc is added.

Curve DE : This curve represents the lowering of melting point of compound $Mg(Zn)_2$ when magnesium metal is added to it.

Point D : At this point, the liquid and solid have the same composition, *i.e.*, $Mg(Zn)_2$ and melt at this point without decomposition. Thus, the temperature corresponding to the point D is regarded as-the congruent melting point of $Mg(Zn)_2$.

Point C : The two curves AC and CD intersect at the point C. Thus, there will be three phases at this point. These three phases are solid zinc, compound $Mg(Zn)_2$ and melt. This point is invariant which is evident from the phase rule.

$$F' = C - P + 1 = 2 - 3 + 1 = 0$$

The point C is known as eutectic point.

Point E : This is the point where two curves BE and DE intersect at the point E. Thus, there are three phases which are in equilibrium at this point. These three phases are solid Mg, compound [$Mg(Zn)_2$] and melt. Again, it is invariant and is known as eutectic point.

Cooling : Consider a melt of composition represented by a point x. Allow it to cool along the line xy. When it reaches y, the solid $Mg(Zn)_2$ separates out from the melt. At this point there are two phases and hence the system becomes univariant.

$$F' = C - P + 1 = 2 - 2 + 1 = 1$$

It means that the change in composition will also take place if the temperature is further lowered beyond the point y. This cooling will, therefore, take place along yC. At the point C, solid zinc will separate out. Thus, there are three phases at this point and hence it is nonvariant.

$$F' = C - P + 1 = 2 - 3 + 1 = 0$$

It means that there will not be any change in temperature as well as composition at this point if cooling is continued and as long as all the three phases are present. Similarly, consider a liquid melt of composition represented by a point x'. Allow it to cool. The temperature will, fall along x'y'. At point y'. solid $Mg(Zn)_2$ separates. Further cooling will therefore, take place along y 'E'. At the point E. magnesium also separates out.

Ferric Chloride-water System

It is an interesting example in which ferric chloride and water form a number of compounds. All compounds have congruent melting points.

There are in all 8 phases in the system out of which there are four stable hydrates.

Phases	*Composition*
(i) $Fe_2Cl_6.12H_2O$	$Fe_2Cl_6 + 12H_2O$
(ii) $Fe_2Cl_6.7H_2O$	$Fe_2Cl_6 + 7H_2O$
(iii) $Fe_2Cl_6.5H_2O$	$Fe_2Cl_6 + 5H_2O$
(iv) $Fe_2Cl_6.4H_2O$	$Fe_2Cl_6 + 4H_2O$
(v) Anhydrous Fe_2Cl_6	$Fe_2Cl_6 + H_2O$
(vi) Solution	$Fe_2Cl_6 + xH_2O$
(vii) Ice	$0\ Fe_2Cl_6 + xH_2O$
(viii) Vapour	$0\ Fe_2Cl_6 + xH_2O$

As the composions of all the above phases can be represented by Fe_2Cl_6 and H_2O it is therefore. regarded as a two component system. The complete diagram of this system is shown in the Fig. (3.7). As all the measurements are being carried out at constant pressure, vapour phase is, therefore, not considered and the condensed phase rule will be applied.

$$F' = C - P + 1.$$

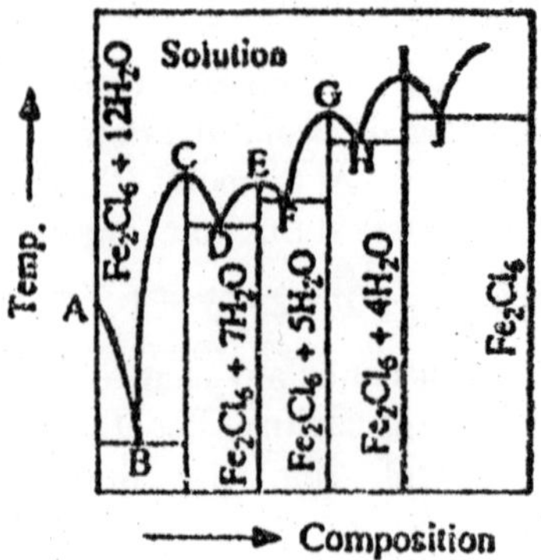

Fig. 3.7 : $FeCl_3$ – H–2–O system.

DEUTERIUM OXIDE SYSTEM

Deuterium oxide occurs in ordinary water in the ratio of 1 : 4000. This is prepared from ordinary water by fractional electrolysis. The chemical properties of H_2O and D_2O are more or less similar but there is marked difference in the physical properties of D_2O and H_2O. The phase diagram of heavy water (deuterium oxide) is shown in Fig. 3.8. This diagram is Just similar to the water system but has some differences which are as follows :

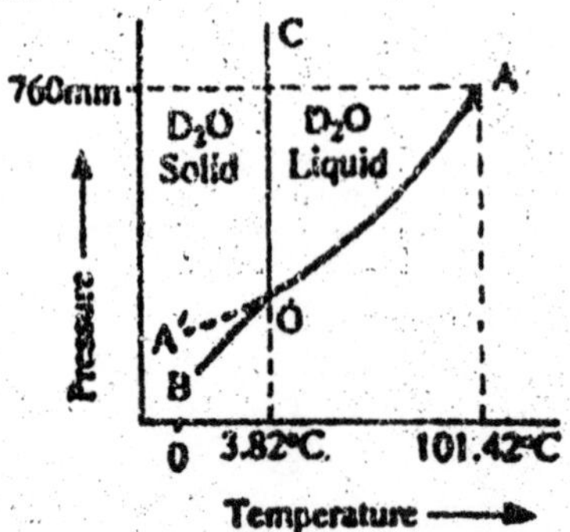

Fig. 3.8 : Deuterium oxide system.

(i) Triple point in heavy water system is 3.82°C but not 0.0075°C as in H_2O system.

(ii) Vapour pressure of the liquid heavy water becomes equal to 760mm of Hg at 111.42°C whereas in case of H_2O system it becomes at 100°C.

(iii) On freezing, heavy water expands. The melting point of solid heavy water is lowered by 0.0069°C for each atmospheric pressure applied. Hence, the curve OC is sloping towards the left in D_2O system but this is less than the corresponding curve for H_2O.

The equilibrium diagram of heavy water contains three curves OA, OB and OC. These three curves divide the phase diagram into three regions AOC, BOC and AOB. Point 0 is the triple point. All the curves, areas and triple points can be explained on exactly the same lines as in water system. The salient features of heavy water system are outlined in Table 3.1.

Table 3.1 : Salient Feature of D_2O System

Curves/Reglong/ Triple point	*Name*	*Phase*	*Degree of Freedom*
Curve OA	Vaporisation curve of D_2O liquid	$D_2O(l) \rightleftharpoons D_2O(v)$	1
Curve OB	Vaporisation curve of D_2O solid	$D_2O(s) \rightleftharpoons D_2O(v)$	1
Curve OC	Fusion curve of D_2O solid	$D_2O(s) \rightleftharpoons D_2O(l)$	1
Region AOB	–	D_2O (vapour)	2
Region AOC	–	D_2O (liquid)	2
Region BOC		D_2O (solid)	2
Point 0	Triple point	$D_2O(s) \rightleftharpoons D_2O(l) \rightleftharpoons D_2O(v)$	0
Curve OA'	Metastable vaporisation curve of liquid D_2O	$D_2O(l) \rightleftharpoons D_2O$ (v)	2

ICE SYSTEM

Tammann ann Bridgmann **observed that** water under high pressure exists in seven different crystalline **forms** of ice which are shown in the phase diagram Fig. 3.9. The seven **forms** of ice are ice I, ice II, ice III, ice IV, ice V, ice VI and ice VII. Ice I is ordinary ice, whereas the other forms are different from ice I in density, crystal structure, and other physical properties and are stable at higher pressure with the exception of IV which is not shown on the diagram because its existence though indicated was not confirmed.

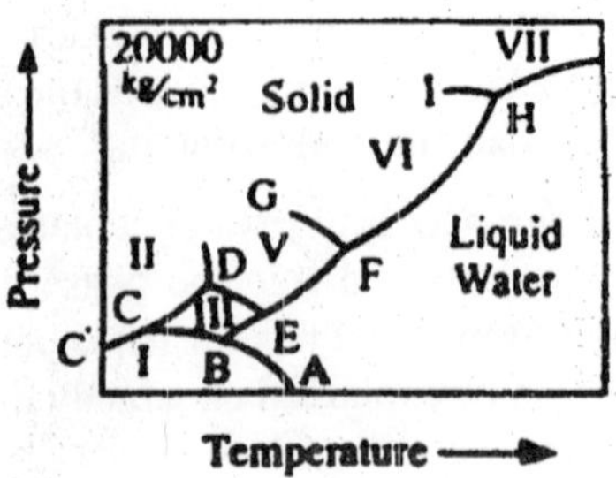

Fig. 3.9 : Ice System

There are ten stable curves in the system, viz., AB, BC. CD, BE, DE, EF. FG. FH. HI and HJ.

A is the normal triple point as it exists in water system. At A, ice I, liquid and vapour are in equilibrium. The degree of freedom of this point is one which follows from the phase rule :

$$F = C - P + 2 = 1 - 3 + 2 = 0$$

This point A is stable at 0.0075°C at a corresponding pressure of 4.58 mm. of Hg.

When the pressure is increased. the melting point of ice I falls until the triple point B is reached which corresponds to –22°C at 2110 kg cm^2. At the point. ice I, ice III and liquid are in equilibrium. The degree of freedom of this point is zero. If pressure is further applied to the state corresponding to point B, a value of –34°C at pressure of 2170 kg. cm^2 corresponding to point C is reached. At this triple point C. three phases in equilibrium are ice I, ice II and ice III.

Further increase in pressure results in the transformation of ice III into ice V, ice VI and ice VII. Thus, there are Seven triple points in this system which are all non-variant. Only three forms of ice viz. ice I, ice VI and ice VII are stable above 0°C.

In fact, ice VII is solid at the normal boiling point of water (100°C at a pressure of 22400 kg cm^2). It is remarkable that melting of ice is quite hot. The melting point curve beyond the ice VII curve has been extended to 190°C and pressure of 40,000 kg. cm^2.

The various salient features of ice system are given in Table 3.2.

Table 3.2 : Salient Features of the Ice System

Triple point	*Temperature°C*	*Pressure*	*Phases*
A	+00075°	4.58 mm of Hg	Ice I ⇌ liquid ⇌ Vapour
B	–22°	2,110 kg cm^2	Ice I ⇌ Ice III ⇌ liquid
C	–34.7°	2,170 kg cm^2	Ice I ⇌ Ice II ⇌ Ice III
D	–24.3°	3,500 kg cm^2	Ice II ⇌ Ice III ⇌ Ice V
E	–17°	3,520 kg cm^2	Ice II ⇌ Ice V ⇌ Liquid
P	+01°	6,375 kg cm^2	Ice V ⇌ Ice VI ⇌ Liquid
G Liquid	+81.6"	22,400 kg cm^2	Ice VI ⇌ Ice VII ⇌

THE SULPHUR SYSTEM

It is a one component system consisting of four phases :

Rhombic sulphur S_R..............Solid phase

Monoclinic sulphur S_M. ....Solid phase

Liquid sulphur S_L,................. Liquid phase

Vapour sulphur S_V.................Vapour phase.

The phase diagram of sulphur system is represented.

The diagram consists of :

(i) *Curves* AB, BC, CD, BE, CE and EG and for metastable curves.

(ii) *Areas* : The above curves divide the diagram into 4 areas ABEG, ABCD, GECD and BEC.

(iii) *Stable triple points* : Three stable triple points are B, C and E.

(iv) *Metastable point* : F is metastable point.

Description

Curve AB : This curve is the *sublimation* or *vapour pressure* curve of rhombic sulphur (S_R). This gives the variation of vapour pressure of S_R at different temperatures. Thus, along this curve, two phases rhombic sulphur and vapour are in equilibrium with each other. It follows from the phase rule.

$$F = C - P + 2 = 1 - 2 + 2 = 1.$$

Thus, the system is univariant. It means that there can be one vapour pressure value at one temperature.

Point B : At the point B, rhombic sulphur changes into monoclinic sulphur. This is known as transition temperature. Thus, there are three phases (two solids and one vapour) in equilibrium at this point (B) and it is non-variant.

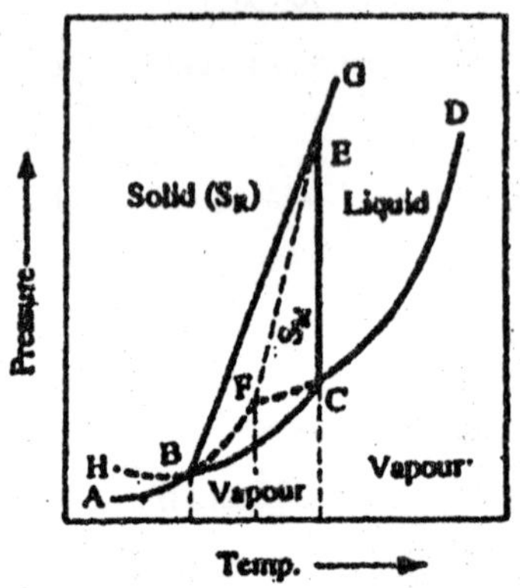

Fig. 3.10 : Sulphur system.

$$F = C - P + 2 = 1 - 3 + 2 = 0$$

Thus, B is a triple point.

Curve BC : It is the *vapour pressure* or *sublimation* curve for monoclinic sulphur. It gives the variation of vapour pressure of S_M (monoclinic sulphur) at different temperatures. At any point on this curve BC, the two phases SM and Sy are in equilibrium. The system is univariant. *i.e.*, it has one degree of freedom.

$$F = C - P + 2 = 1 - 2 + 2 = 1$$

Point C : It represents the *melting point* of monoclinic sulphur (120°). This is another triple point. At this point three phases (S_M. S_L, and S_V) are in equilibrium. Thus, this is a non-variant point.

Curve CD : It is the vapour pressure curve of liquid sulphur. This curve starts at point C which is melting point of S_M and terminate at point D, the critical temperature. Beyond this temperature there is one phase S_V.

Along this curve SL and, Sy are in equilibrium. Thus the system is univariant since at any point on the curve for the fixed value of pressure, temperature has a fixed value. This is also evident from the phase rule,

$$F = C - P + 2 = 1 - 2 + 2 \text{ or } F - 1$$

Curve BE : This is known as the *transition curve* of S_R to S_M. It gives the effect of pressure on the transition point. Both phases (S_R and S_M) are solids. The system is monovariant. This curve terminates at E beyond which SM disappears. As the transition point is raised With increase of pressure, the curve BE, therefore, slopes away from the pressure axis.

Curve CE : It is the melting or fusion curve for monoclinic sulphur. It shows the effect of pressure on melting point of S_M which is raised by increase of pressure, the curve CE, therefore slopes slightly away from the pressure axis.

At any point of the curve, the two phases S_M and S_L are in equilibrium with each other. This also has one degree of freedom.

Point E : The two curves BE and CE meet at the point E. Thus, it is another triple point which is nonvariant. At this point, three phases (S_R, S_M and S_L) are in equilibrium.

Curve EG : It is the fashion, curve for rhombic sulphur. Along this curve S_R and S_L are in equilibrium. Therefore, the number of phases is one and hence the system is monovariant.

Metastable Equilibrium

Curve BF : If the temperature of rhombic sulphur at 95.5°C is allowed to rise quickly, the transition of S_R or S_M does not take place at B But this curve is extended to F, the melting point of S_R. The curve BF so obtained is the metastable curve of rhombic sulphur. If system is disturbed slightly, the curve BE merges into BC.

Curve CF : If liquid sulphur is allowed to cool very carefully the solid phase will not separate out at C. Thus, the curve CD can be extended to F by supercooling the liquid carefully. Thus, the curve CF represents metastable equilibrium between S_L and S_V.

Curve EF : It is the metastable fusion curve of rhombic sulphur. Along this curve, rhombic sulphur is in metastable equilibrium with liquid sulphur. This curve gives the effect of pressure on the melting point of rhombic sulphur in metastable state.

Point F : It is the metastable triple point where S_R. S_L and S_V co-exist in the equilibrium ; like B, C and E, it is non-variant.

Areas. There are four areas :

(i) Area ABEG : It contains only rhombic sulphur.

(ii) Area ABCD : It contains only vapour sulphur.

(iii) Area GECD : It contains on ly liquid sulphur.

(iv) Area BEC : It contains only monoclinic sulphur.

Thus, all the above areas contain one phase only. It means that there are two degrees of freedom which is also evident from the phase rule equation.

$$F = C - P + 2 = 1 - 1 + 2 \text{ or } F = 2$$

It means that both pressure and temperature are required to define any point in all the above areas.

GIBBS EQUATIONS

The phase rule of Gibbs may be expressed in the simple form :

$$P + F = C + 2 \qquad ...(1)$$

where P is the number of *phases*, F is the number of *degrees of freedom* and C is the number of *components* in a system at equilibrium. Equation (1) does not involve the mass of phase as it has no influence on the state of equilibrium. In words, the phase rule (Eq. 1) may be stated as:

"In a heterogeneous system in equilibrium, the number of degrees of freedom plus the number of phases are equal to the number of components plus 2."

Before deriving the phase rule, it is necessary to explain the meaning of phase, component and degree of freedom. *Phase* : As stated earlier, a heterogeneous system consists of different parts of each of which is homogeneous in itself and is separated from others by bounding surfaces or interfaces. The different parts of heterogeneous system can be separated by mechanical methods. *These homogeneous, physically distinct and mechanical separable parts of the heterogeneous system in equilibrium are called phases. A phase may be gas, liquid or solid.*

Examples : (i) In a system consisting of ice, water and vapour, there are three phases; ice (solid phase), water (liquid phase) and vapour (gas phase). Each phase is homogeneous in itself and can be separated from the other phases by mechanical methods.

(ii) *All gases are completely miscible :* Therefore, a gaseous mixture irrespective of the number of gases present, consists of one phase only.

(iii) Two or more liquids which are miscible with one another constitute a single phase as there is no bounding surfaces separating the different liquids. Example is the alcohol and water which exist in one phase only as two are completely miscible. As there will be vapour in contact with the liquid, there will be two phases, *i.e.*, liquid and gas in miscible liquids.

(iv) A solid mixture consists of many as phases as are solids present. A heterogeneous equilibrium,

$$CaCO_3 \rightleftharpoons CaO + CO_2$$

involves three phases, solid ($CaCO_3$), solid (CaO) and gas (CO_2).

Thus, there are three phases, two solids and one gaseous.

Component : A component is an element or a compound present in a system. The concentration of this can undergo variation independently. But in the phase rule the component does not stand for the total number of constituents of the system. This term, is however, defined as :

"The number of components of a system at equilibrium is defined as the minimum number of molecular species in terms of which the composition of all the phases may be expressed quantitatively:

The meaning of this term will be made clear by considering some actual cases.

(i) Consider a water system consisting of the three phases :

$$\text{Ice} \rightleftharpoons \text{water} \rightleftharpoons \text{Vapour.}$$

Thus, water system is one component system as each phase is essentially a different physical form of the same chemical compound represented by H_2O.

(ii) Consider the dissociation of $CaCO_3$ by heat

$$CaCO_3 \rightleftharpoons CaO + CO_2$$

This equilibrium consists of three phases, two solids and one gas. At first sight, it seems that it is a three component system. This cannot be so as three constituents are not independent of each other as demanded by the definition. For example :

$$CaCO_3 = CaO + CO_2$$

or $$CaO = CaCO_3 - CO_2$$

or $$CO_2 = CaCO_3 - CaO.$$

Thus, the composition of all the three phases can be expressed by taking two of the constituents only. Hence it forms a two component system. For example :

(a) By taking CaO and CO_2 as the components

Phase	Component
$CaCO_3$	$= CaO + CO_2$
CaO	$= CaO + 0\ CO_2$
COa	$= 0\ CaO + CO_2$

(b) By taking $CaCO_3$ and CO_2 as components

Phase	Component
$CaCO_3$	$= CaCO_3 + 0\ CO_2$
CaO	$= CaCO_3 - CO_2$
CO_2	$= 0\ CaCO_3 + CO_2$

(c) By taking $CaCO_3$ and CaO as components

Phase	Component
$CaCO_3$	$= CaCO_3 + 0\ CaO$
CaO	$= 0\ CaCO_3 + CaO$
CO_2	$= CaCO_3 - CaO$

(iii) Consider the dissociation of ammonium chloride in a closed vessel which is in equilibrium with its products, ammonia and hydrochloric acid

$$\underset{\text{solid}}{NH_4Cl} \rightleftharpoons \underset{\text{gas}}{NH_4Cl} \rightleftharpoons NH_3 + HCl$$

The system is considered as a one-component system because the composition of both the solid and gaseous phases can be stated in terms of NH_4Cl only. This is shown as below:

Phase	*Component*
Solid	NH_4Cl
Gaseous	$x.NH_3 + x.HCl$ or $x\ NH_4Cl$

If a small amount of NH_3 or HCl is added to the equilibrium mixture then the number of components would become two. Suppose excess of hydrochloric acid is added, then the gaseous mixture will contain x moles of NH_3 + x moles of HCl + y moles of HCl. Tlherefore, the composition of two phases of ammonium chloride may be expressed as

Phase	*Component*
Solid NH_4Cl	NH_4Cl
Gas	.t.NHa +X.HCl –{ y.HCl
	$= x\ NH_4Cl + y\ HCl$

Thus, the minimum number of components by means of which the composition of both phases can be expressed arc the two components NH_4Cl and HCl. Thus, it is a two *component system*. Similarly it behaves as a two component system if ammonia is added to the equilibrium mixture.

Degree of freedom or variance of the system : The degree of freedom or variance of a system is defined as the least number of variable factors such as temperature, pressure and concentration which should be arbitrarily fixed in order to define the system completely.

(i) A gaseous mixture, say of carbon dioxide and nitrogen, is completely defined when three variables temperature, pressure and concentration are specified. Then, the system is said to have three degrees of freedom or to be trivariant.

(ii) If one consider a system consisting of unsaturated water vapour, two variables pressure and temperature are to be specified to define the system because pressure may have any value at a particular temperature. This system is said to have two degrees of freedom or bivariant.

PHASE REACTIONS

There are certain physical or chemical reactions which are accompanied by the appearance or disappearance of a phase. Such reactions are known as phase reactions.

A phase reaction is generally indicated in Fig. 3.11.

From Fig. 3.11, it follows that the examples of phase reactions are as follows:

(i) Change of solid to liquid.

(ii) Change of liquid to solid.

(iii) Change of liquid to gas.

(iv) Change of solid to gas.

(v) Change of gas to liquid.

(vi) Change of gas to solid.

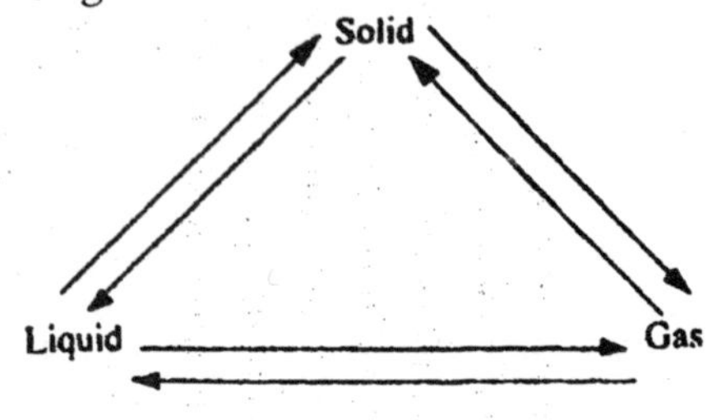

Fig. 3.11

From the above examples, it is evident that all processes of vaporisation, sublimation, freezing, melting, solidification, crystallisation, and precipitation are phase reactions.

All chemical processes may not be phase reactions, *e.g.*, the combination of gaseous ammonia and gaseous hydrogen chloride to form solid ammonium chloride is a phase reaction because the. reactants and products are in different phases. But the reaction between gaseous hydrogen and gaseous chlorine to form gaseous hydrogen chloride cannot be regarded as a phase reaction because all the, substances are in one phase.

CONDITIONS FOR EQUILIBRIUM BETWEEN PHASES

In order for each phase to be in equilibrium in a multiphase system, the following equilibrium conditions are to be fulfilled :

(a) *Thermal Equilibrium :* It means that the temperature of all the phases must be identical, otherwise there would occur the flow of heat from one phase to another. One can prove it thermodynamically as follows :

Suppose we consider two phases α and β at temperature T^{α} and T^{β}. Suppose S^{α} and S^{β} are the entropies of the two phases. Suppose the heat, dq, is transferred from phase α to β at equilibrium. Due to this heat transfer, the change of entropy of the system at equilibrium will be as follows :

$$dS = dS^{\alpha} + dS^{\beta} = 0 \quad ...(1)$$

We know, $$dS^{\alpha} = -\frac{dq}{T\alpha} \text{ and } dS^{\beta} = \frac{dq}{T\beta} \quad ...(2)$$

In equation (2), –ve and +ve signs are used to signify the heat loss and heat gain by the phases α and β respective ly.

On substituting equation (2) in (1), we get

$$-\frac{dq}{T\alpha} + \frac{dq}{T\beta} = 0$$

or $$T\alpha = T\beta \quad ...(3)$$

From the equation (3), *it follows that the temperature of all the phases in equilibrium is same.*

(b) *Mechanical Equilibrium :* By this equilibrium it is meant that all the phases in equilibrium must be under the same pressure. It is not so, one phase would increase its volume at the expense of another. This may be deduced as follows :

Suppose there occurs an expansion in volume, dV, of α-phase at the expense of β-phase. Then at constant temperature, the change in Helmholtz free energy will become as follows :

$$dA = dA^{\alpha} + dA^{\beta} = 0 \text{ (At equilibrium)} \quad ...(4)$$

we know $$dA^{\alpha} = P^{\alpha}\, dv \text{ and } dA^{\beta} = -P^{\beta}\, dv \quad ..(5)$$

In equilibrium (5), positive sign indicates increase in volume of α-phase whereas negative sign indicates decrease in volume of β-phase.

On substituting equation (5) in (4), we get

$$P^{\alpha} = P^{\beta} \quad ...(6)$$

From equation (1), it follows *that the phases in equilibrium are under the same pressure.*

(c) *Chemical Equilibrium :* By this equilibrium, it is meant that the chemical potential of any components is same in all the phases at equilibrium.

Let us consider a closed system of P phases indicated by α, β, γ, ... having u total of C components indicated by 1, 2, 3, ... P in equilibrium and each at the same pressure and temperature. We know that the Gibbs tree energy (G) of each phase is a function of temperature, pressure and composition, *i.e.*,

$$G^{\alpha} = f(T, P, n_i)^{\alpha} \qquad ...(7)$$

$$G^{\alpha} = f(T, P, n_i)^{\beta} \qquad ...(8)$$

$$\ldots \quad \ldots \quad \ldots\ldots$$

$$G^{P} = f(T, P, n_i)^{P} \qquad ...(9)$$

where i = 1, 2, 3, 4, ..., i, C The free energy change of the entire system will be the sum of change of free energies of each phase.

$$dG = dG^{\alpha} + + dG^{\beta} + dG^{v} + ...$$

For a multicomponent system, we know

$$dG = -S\,dT + V\,dp + \Sigma\mu_i\,dn$$

At constant temperature and pressure, dT = dP = 0 and therefore the above equation becomes as

$$(dG)r, p = \Sigma\mu_i\,dn_i$$

If there occurs a transfer of mass from one phase to another, we can write

$$dG = \sum_{i=1}^{0} \mu_i^{\alpha}\,dn_i^{\alpha} + \Sigma\mu_i^{\beta}\,dn_i^{\beta} + \Sigma\mu_i^{y}\,dn_i^{y} + ... \qquad ...(10)$$

If we are considering a closed system at equilibrium (dG = 0), we can write

$$\Sigma\mu_i^{\alpha}\,dn_i^{\alpha} + \Sigma\mu_i^{\beta}\,dn_i^{\beta} + \Sigma\mu_i^{y}\,dn_i^{y} + ... = 0$$

At equilibrium total mass of each component is constant because we are considering a closed system. Therefore, we can write

$$\left.\begin{array}{l} dn_1^\alpha + dn_1^\beta + dn_1^y + ... + dn_1^p = 0 \\ dn_2^\alpha + dn_2^\beta + dn_2^y + ... + dn_2^p = 0 \\ ... \quad ... \quad ... \quad ... \quad ... \\ dn_i^\alpha + dn_i^\beta + dn_i^y + ... + dn_i^p = 0 \\ ... \quad ... \quad ... \quad ... \quad ... \\ dn_c^\alpha + dn_c^\beta + dn_c^y + ... + dn_c^p = 0 \end{array}\right\} \quad ...(11)$$

For all the equations in expression (11) to be zero for all possible variations of n, subject to the restrictions posed by equation (10), it is required that

$$\left.\begin{array}{l} \mu_1^\alpha = \mu_1^\beta = \mu_1^y = ... = \mu_1^p \\ \mu_2^\alpha = \mu_2^\beta = \mu_2^y = ... = \mu_2^p \\ ... \quad ... \quad ... \quad ... \\ \mu_c^\alpha = \mu_c^\beta = \mu_c^y = ... = \mu_c^p \end{array}\right\} \quad ...(12)$$

From expression (12), it is evident that for any component i in the system, the value of chemical potential (α_i) must be same in every phase provided the system is in equilibrium at constant T and P.

EXPERIMENTAL DETERMINATION OF TRANSITION POINT

The various methods are as follows :

1. *The Dilatometric Method* : This method is based on the simple principle that there occurs a change in volume due to the transition of one form into another because each form has a different density.

The determination is carried out in an apparatus known as *dilatometer* as shown in Fig. 3.12. It consists of a wide tube having a graduated capillary (in mm) side-tube A. The substance whose transition temperature is to be determined is filled in the lower half of the tube whereas the rest of the tube is filled with an inert solvent such as xylene. The stopper carrying the thermometer is inserted into the tube. The dilatometer is now kept in a thermostat. The level of the xylene in the capillary is noted. The temperature of the thermostat is increased degree by degree. As long as no change is taking place in the solid, the rise in the xylene level in the capillary is regular and uniform. As soon as the transition point is reached, there occurs a sudden change in volume of the solid which causes an abrupt rise or fall in the level.

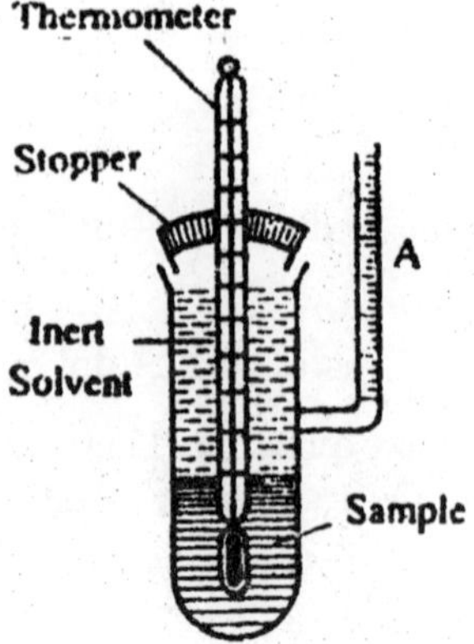

Fig. 3.12

The same thing happens on cooling when the transition point is again reached. But the heating curve and cooling curve do not coincide as there is always a lag. Usually, a mean of the temperatures is taken as the transition temperature. This is shown in Fig. 3.13.

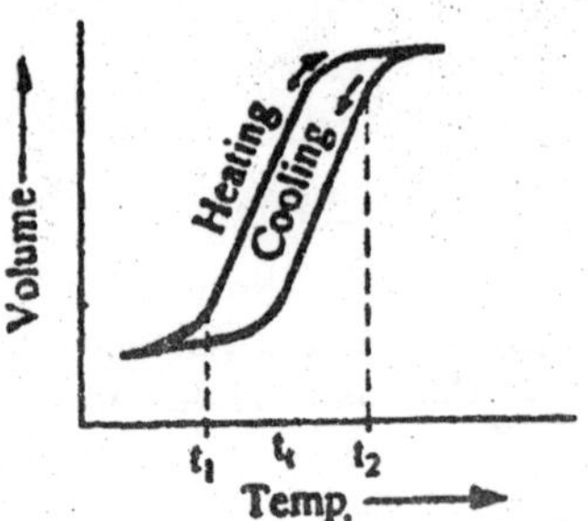

Fig. 3.13

2. *The Solubility Method* : Each form possesses a different solubility at a given temperature. But both the forms have the same solubility at the transition temperature.

   In this method, the solubility curves for both the forms are plotted against temperature (Fig. 3.14). In this figure, transition temperature corresponds to that point at which the two curves intersect each other.

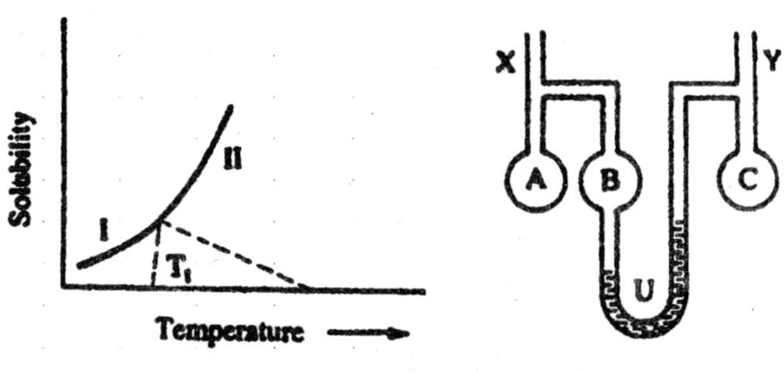

Fig. 3.14 Fig. 3.15

3. *The Vapour Pressure Method :* The principle of this method is that the vapour pressure of the two forms becomes equal at the transition temperature. Bremer and Frowein employed an apparatus shown in Fig. 3.15 known as tensiometer to measure the vapour pressure. It consists of a U-tube containing mercury. The two limbs of U-tube are connected to bulbs B and C. The bulb B is fused to another bulb A which has side-let X. Similarly the bulb C has side-let Y.

In bulb A, water is kept and then the bulb is sealed at X. The pure hydrated substance is kept in the bulb C. An exhaust pump is connected to Y and then the pump is started to reduce the pressure in the apparatus by removing all the air from it. Now the bulb C is also sealed at Y. Note the mercury level in the U-tube. After this, the bulb C is heated to start dehydration of the hydrate by keeping the apparatus in a thermostat. When the equilibrium is attained, the difference in the mercury level is accurately read out with the help of a travelling microscope. Then, the value of p_1 can be obtained by applying the following equation:

$$p_1 - p_2 = \text{Difference in mercury level}$$

where p_1 = Decomposition pressure of the substance kept in the bulb C, and

p_2 = Saturation pressure due to the water present in bulb A.

Various values of pi are obtained for different temperatures and are plotted against temperature to get a graph shown in Fig. 3.16. The sharp break in the curve corresponds to the transition temperature.

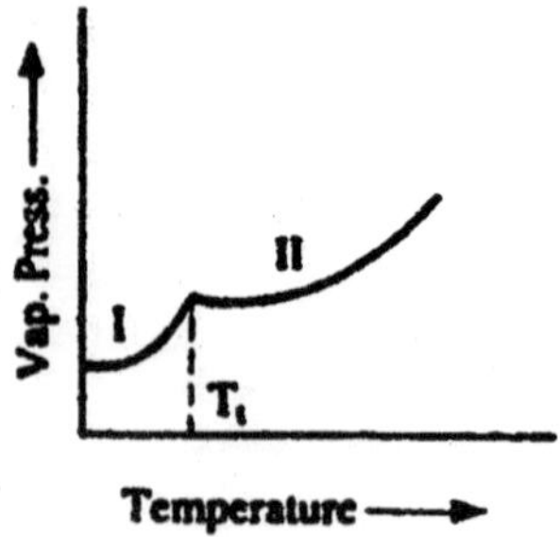

Fig. 3.16

4. *Thermometric Method :* The principle of this method is that each form possesses a different energy content. If a substance which can undergo transition is heated, its temperature rises gradually until the transition temperature is reached at which the temperature remains constant until its transition into the other form is complete.

BUFFER SOLUTIONS OR BUFFERS

We know that the pH of neutral water is 7. If we add 1 cc of 1N HCl to 1 litre of water, the $[H^+]$ concentration in the solution becomes 10^{-3} gm ions and the pH of the solution becomes 3. Similarly, the addition of 1 cc of 1N NaOH in a litre of pure water changes the pH of the solution to 11.

Thus, we see that the pH of water changes by large amounts on the addition of small quantity of a strong acid or a strong base. But for certain biological and analytical purposes, we require solution whose pH should not change on keeping or on addition of small amounts of a strong acid or a strong base.

Such solutions which oppose the change in their pH on the addition of small amount of an acid or a base are called buffer solutions. Some characteristics of buffer solution are:

(i) Their pH should not change on keeping.

(ii) Their pH should remain same on dilution.

(iii) Their pH should remain same on the addition of small amounts of strong acid and base.

Action of Buffers

A buffer solution invariably consists of a weak acid and its salt solution (acidic buffer) or a weak base and its salt solution (basic buffer). Some examples of buffer solutions are:

Constituents	*pH Range*
Glycine + Glycine hydrochloride	1.0 — 3.7
Phthalic acid + Potassium hydrogen phthalate	2.2 — 3.8
Acetic acid + Sodium acetate	3.7 — 5.6
Disodium citrate + Trisodium citrate	5.0 — 6.3
Monosodium phosphate + Disodium phosphate	5.8 — 8.0
Boric acid + Borax	6.8 — 9.2
Borax + Sodium hydroxide	9.2 — 11.2
Disodium phosphate + Trisodium phosphate	11.0 — 12.0

Action of Acidic Buffer

Let us illustrate its action by considering a mixture sodium acetate and acetic acid in water.

$$CH_3COOH \rightleftharpoons CH_3COO^- + H^+ \text{ (feebly ionised)}$$
$$CH_3COONa \rightleftharpoons CH_3COO^- + Na^+ \text{ (highly ionised)}$$

As CH_3COONa is almost fully ionised, the acetate ions so produced suppress the ionisation of the acetic acid, so that the mixture contains unionised acetic acid molecules and a large number of acetate ions. When a few drops of an acid, say HCl, are added the H^+ ions from the acid added react with the CH_3COO^- ions to form unionised or feebly ionised acetic acid:

$$CH_3COO^- + H^+ \rightarrow CH_3COOH$$

On the other hand, when a few drops of an alkali, say, NaOH are added, the OH^- ions react with CH_3COOH to form unionised H_2O.

$$CH_3COOH + OH^- \rightarrow CH_3COO^- + H_2O$$

Thus, the addition of small amounts of the acid or alkali to a buffer solution does not alter the H^+ ions concentration of the buffer. In other words, the pH of the buffer remains unchanged.

Action of a Basic-buffer

Let us illustrate the action of a basic buffer by considering a mixture of NH_4OH and NH_4Cl in water. They exhibit the following equilibria

$$NH_4OH \rightleftharpoons NH_4^+ + OH^- \text{ (weakly ionised)}$$

$$NH_4Cl \rightleftharpoons NH_4^+ + Cl^- \text{ (completely ionised)}$$

The presence of the excess of NH_4^+ ions from completely ionised NH_4Cl suppresses the ionisation of NH_4OH, so that the mixture solution contains unionised NH_4OH molecules and a large number of NH_4^+ ions. When a few drops of alkali is added to this buffer solution, the OH^- ions produced from the alkali react with NH_4^+ present in solution to form unionised NH_4OH.

$$NH_4^+ + OH^- \rightarrow NH_4OH$$

On adding a few drops of the acid say HCl, the H^+ ions provided by the acid react with NH_4OH to give unionised H_2O.

$$NH_4OH + H^+ \rightarrow NH_4^+ + H_2O$$

Thus, the addition of small amounts of the alkali or acid to the above solution does not alter its H^+ ion concentration. In other words, the pH of buffer remains unchanged.

Mathematical Expression for the pH of an Acidic Buffer

Consider a solution having a weak acid HA and its salt BA, with a strong base. The ionisation of the weak acid is represented as:

$$HA \rightleftharpoons H^+ + A^-$$

or $$K_a = \frac{[H^+][A^-]}{[HA]} \quad \text{or} \quad [H^+] = K_a \frac{[HA]}{[A^-]} \qquad ...(1)$$

where K_a is the dissociation constant of the weak acid., The salt BA is almost completely ionised. Thus, high concentration of the A^- ions provided by almost complete ionisation of the salt BA suppresses the ionisation of the weak acid, so that it may be assumed that the free A^- ions are entirely due to the salt. Thus

$$[A^-] = [\text{Salt}] \text{ and } [HA] = [\text{Acid}] \qquad ...(2)$$

Substituting Eq. (2) in (1), we get

$$[H^+] = K_a \frac{[\text{Acid}]}{[\text{Salt}]}$$

Taking logarithm of both sides, we get

$$\log [H^+] = \text{loh } K_a + \log \frac{[\text{Acid}]}{[\text{Salt}]}$$

or $$-\log [H^+] = -\text{loh } K_a - \log \frac{[\text{Acid}]}{[\text{Salt}]}$$

or $$pH = pK_a - \log \frac{[\text{Acid}]}{[\text{Salt}]} \quad ...(3)$$

$$= PK_a + \log \frac{[\text{Salt}]}{[\text{Acid}]}$$

where $pK_a = -\log K_a$. Equation (3) is known as Henderson-Hasselbalch equation.

Mathematical Expression for the pH of a Basic Buffer

Consider a weak base BOH and its salt AB. The dissociation of the weak base may be written as

$$BOH \rightarrow B^+ + OH^-$$

or $$K_a = \frac{[B^+][OH^-]}{[BOH]} \text{ or } [OH^-] = \frac{K_b[BOH]}{[B^+]} \quad ...(1)$$

where K_b is the dissociation constant of the weak base.

The high concentration of B^+ ions produced by the complete ionisation of the salt, which thus suppresses the ionisation of the weak base so that it may be assumed that free B^+ ions in the solution are entirely due to the salt. Thus,

$$[B^+] = \text{Salt and } [BOH] = \text{Base}$$

Thus, equation (1) may be put as

$$[OH^-] = K_b \frac{[\text{Base}]}{[\text{Salt}]}$$

Taking the logarithm of the above equation, we obtain

$$\log [OH^-] = \log K_b + \log [\text{Base}] - \log [\text{Salt}]$$

or $$-\log [OH^-] = -\log K_b - \log [\text{Base}] + \log [\text{Salt}]$$

or $$pOH = pK_b + \log \frac{[\text{Salt}]}{[\text{Base}]} \quad ...(2)$$

But pH = 14 – pOH, Equation (2) becomes as

$$pH = 14 - \left(pK_a + \log \frac{[\text{Salt}]}{[\text{Base}]} \right) \quad ..(3)$$

Now $[H^+][OH^-] = k_w = 10^{-14}$ or $\log k_w = -\log 10 = -14$

or $$14 = -\log k_w \qquad ...(4)$$

From Eqs. (3) and (4), we get

$$pH = -\log k_w - \left(pK_b + \log \frac{[Salt]}{[Base]}\right)$$

$$= -\log k_w - \left(-\log K_b + \log \frac{[Salt]}{[Base]}\right)$$

$$= -\log \frac{k_w}{k_b} + \log \frac{[Base]}{[Salt]} \qquad ...(5)$$

HYDROLYSIS

Pure water is neutral and is feebly ionised into H^+ and OH^- ions as:

$$H_2O \rightarrow H^+ + OH^-$$

The concentration of H^+ and OH^- ions is same. Now when a salt is dissolved in water, it ionises into ions, *e.g.*, the salt AB will ionise as:

$$AB \rightarrow A^+ + B^-$$

The ions of the salt may react with the ions of water to give

$$A^+ + OH^- \rightarrow AOH$$

$$B^- + H^+ \rightarrow HB$$

Now, the solution will be acidic or alkaline depending upon the relative strengths of the acid HB and the base AOH. If the acid is strong and the base is weak, then the solution will be acidic in nature. If the base is strong and the acid is weak, the solution will be basic in nature. Thus, this phenomenon in which a salt interacts with water to produce either an acidic or alkaline solution is termed as hydrolysis. Actually, hydrolysis involves the interaction of the ions of a salt with the water to produce H^+ or OH^- ions in solution. Thus, it may be defined as:

"The interaction between the ions of salt and the ions of water to given free H^+ or OH^- ions in solution."

Example of Hydrolysis

Salt may be divided into four categories:

(a) Salts of Strong Acid and Strong Bases

These salt, when dissolved in water, produce a strong acid and a strong base, so that the relative proportions of H^+ and OH^- ions in solution remain unchanged and the solution remains neutral. IN simple words, it can be said that the salts of strong acids and bases do not undergo hydrolysis. Let us consider the dissolution of sodium chloride in water.

$$NaCl \rightarrow Na^+ + Cl^-$$

$$H_2O \rightarrow OH^- + H^+$$

$$Na^+ + OH^- \rightarrow NaOH \text{ (highly ionised)}$$

$$H^+ + Cl^- \rightarrow HCl \text{ (highly ionised)}$$

But the acid, HCl and the base, NaOH being equally strong are almost completely ionised. Thus, the concentration of H^+ and OH^- is not altered and the solution remains neutral. Examples of such salts are NaCl, KCl, Na_2SO_4, K_2SO_4, etc.

(b) Salts of Weak Acids and Strong Bases

These salts (like Na_2CO_3; CH_3COONa, KCN, etc.), when dissolved in water undergo hydrolysis in water to give an alkaline solution. For example,

$$CH_3COONa \rightarrow CH_3COO^- + Na^+$$

$$H_2O \rightarrow H^+ + OH^-$$

$$CH_3COO^- + H^+ \rightarrow CH_3COOH$$

$$Na^+ + OH^- \rightarrow NaOH$$

NaOH being a strong base is almost completely ionised and the concentration of OH^- ions in solutions is not altered. However, CH_3COOH being a weak acid is feebly ionised. Thus the concentration of H^+ ions in solution falls. Thus, the solution has more of OH^- ions than the H^+ ions. Hence, it is alkaline in nature.

(c) Salts of Strong Acids and Weak Bases

Examples are NH_4Cl, $CuSO_4$, etc. When such salts are dissolved in water, they produce acidic solutions. Consider the dissolving of ammonium chloride.

$$NH_4Cl \rightarrow NH_4^+ + Cl^-$$

$$H_2O \rightarrow H^+ + OH^-$$

$$NH_4^+ + OH^- \rightarrow NH_4OH$$

$$H^+ + Cl^- \rightarrow HCl$$

HCl being a strong acid is almost completely ionised and the concentration of the H^+ ions in solution is not affected. NH_4OH, however, is a weak base and feebly ionises to give only few OH^- ions in solution. Thus, the solution has a large concentration of the H^+ ions that the OH^- ions. Hence the solution is acidic in nature.

(d) Salts of Weak Acid and Weak Bases

Examples are CH_3COONH_4, $(NH_4)_2CO_3$, etc. Such salts when dissolved in water yield acidic or basic or even neutral solutions, depending upon the nature of the acid and the base produced. Consider the dissolving of ammonium acetate in water.

$$CH_3COONH_4 \rightarrow CH_3COO^- + NH_4^+$$

$$H_2O \rightarrow H^+ + OH^-$$

$$CH_3COO^- + H^+ \rightarrow CH_3COOH$$

$$NH_4^+ \; OH^- \rightarrow NH_4OH$$

Both CH_3COOH and NH_4OH are weak and are equally but feebly ionised. Hence the solution contains practically the same number of the H^+ and OH^- ions. Therefore, the solution is almost neutral. In case of $(NH_4)_2CO_3$, however, the solution is slightly basic because NH_4OH is relatively a stronger base than the H_2CO_3 acid.

MATHEMATICAL TREATMENT OF HYDROLYSIS

(a) Salt of Strong Acids and Weak Bases

(i) *Hydrolysis constant.* The hydrolysis of a salt of a strong acid and a weak base may be represented as:

$$A^+B^+ + H_2O \rightarrow HOH + B^- + H^+$$

or

$$A^+ + H_2O \rightarrow AOH + H^+$$

Applying the law of mass action to the above equilibrium,

$$K = \frac{[AOH][H^+]}{[A^+][H_2O]}$$

As water is taken in large excess, it means that its active mass remains practically constant. Therefore,

$$K[H_2O] = \frac{[AOH][H^+]}{A^+}$$

$$K_h = \frac{[AOH][H^+]}{A^+}, \text{ where } K_h = K[H_2O] \quad ...(1)$$

The constant K_h is called the hydrolysis constant of the salt.

(ii) *Relation between K_h, K_b and K_w.* The dissociation of the weak base produced in the hydrolysis of a strong acid and a weak base may be represented as:

$$AOH \rightarrow A^+ + OH^-$$

Applying the law of mass action,

$$K_b = \frac{[A^+][OH^-]}{[AOH]} \quad ...(2)$$

where K_b is the dissociation constant of the weak base.

Multiplying equations (1) and (2), we obtain

$$[H^+][OH^-] = K_h \times K_b$$

But $[H^+][OH^-] = K_w$, the ionic product of water

$$\because \quad K_w - K_h \times K_b$$

or

$$K_h = \frac{k_w}{K_b} \quad ...(3)$$

From equation (3), it is evident that the *weaker the base, the greater will be the hydrolysis constant of the salt.*

(iii) *Degree of hydrolysis:* Degree of hydrolysis of salt may be defined as:

"The fraction of the totai salt hydrolysed when the equilibrium has been established is called the degree of hydrolysis."

It is denoted by the letter '*h*'. In simple words, the extent to which the salt has been hydrolysed is known as the degree of hydrolysis. If *c* is the initial concentration of the salt in moles per litre and *h* is the degree of hydrolysis on the attainment of equilibrium, then the concentrations of various species will be represented as:

$$\underset{c(1-h)}{A^+} + \underset{\text{Excess}}{H_2O} \rightarrow \underset{ch}{AOH} + \underset{ch}{H^+} \quad ...(4)$$

Applying the law of mass action,

$$K_h = \frac{(ch)(ch)}{c(1-h)} \qquad ...(5)$$

$$= \frac{ch^2}{(1-h)} = \frac{ch^2}{(1-h)}$$

If the base is very weak, the value of h is very small as compared to unity. It means that 1 – h in the equation (5) may be replaced by 1, *i.e.*,

$$K_h = ch^2$$

or

$$h = \sqrt{\left(\frac{K_h}{c}\right)} \qquad ...(6)$$

From equation (3), we know

$$K_h = \frac{k_w}{K_c} \qquad ...(7)$$

Replacing the above value of K_h in equation (6), we obtain

$$h = \sqrt{\left(\frac{k_w}{cK_b}\right)} \qquad ...(8)$$

From the relation (8), it follows that the degree of hydrolysis of a salt of a weak base and a strong acid is:

(i) *directly proportional to the ionic product of water*

(ii) *inversely proportional to the value of K_b, and*

(iii) *inversely proportional to the value of c.*

From the above it follows that degree of hydrolysis increases if the base is weaker *i.e.*, if K_b is smaller, and also increases if the concentration decreases or the dilution increases.

As k_w increases much more than K_b, it means that the degree of hydrolysis at a given concentration increases with the increase in temperature.

(iv) Relation between pH, pK_b and c. From equation (4), it follows:

$$[H^+] = ch$$

$$= c\sqrt{\left(\frac{k_w}{cK_b}\right)} \qquad \text{[From equation (8)]}$$

$$= \sqrt{\left(\frac{ck_w}{K_b}\right)} \quad ...(9)$$

Taking logarithms and changing the signs throughout, Eq. (9) becomes as:

$$-\log [H^+] = -\frac{1}{2}\log k_w + \frac{1}{2}\log K_b - \frac{1}{2}\log c$$

$$pH = \frac{1}{2}pk_w - pK_b - \frac{1}{2}\log c \quad ...(10)$$

From equation (10), it follows that the pH of the solution must be less than $\frac{1}{2}$ pK_w and is less than 7.0 and so solutions of salts of the type under consideration will be acid solutions.

(b) Salts of Weak Acids and Strong Bases

(i) *Hydrolysis constant*: The hydrolysis of a salt of a weak acid and a strong base may be represented as:

$$(A^+ + B^-) + H_2O \rightarrow HB + A^+ + OH^-$$

$$B^- + H_2O \rightarrow HB + OH^-$$

Applying the law of mass action,

$$K = \frac{[OH^-][HB]}{[B^-][H_2O]} \quad ...(11)$$

As water is taken in large excess, its active mass remains practically constant. Therefore,

$$K[H_2O] = \frac{[OH^-][HB]}{[B^-]}$$

or

$$K_h = \frac{[OH^-][HB]}{[B]} \quad ...(12)$$

where K_h is known as the hydrolysis constant of the acid.

(ii) *Relation between K_h, K_a and k_w*: The dissociation of weak acid, HB, produced during hydrolysis may be represented as:

$$HB \rightleftharpoons H^+ + B^-$$

Applying the law of mass action,

$$K_a = \frac{[H^+][B^-]}{[HB^-]} \quad ...(13)$$

where K_a is known as the dissociation constant of the weak acid.,

Multiplying Eq. (12) by (13), we get

$$[H^+][OH^-] = K_a \times K_h$$

But $$[H^+][OH^-] = k_w$$

$\therefore$ $$k_w = K_a \times K_b$$

or $$K_b = \frac{k_w}{K_a} \quad ...(14)$$

From eq. (14), it follows that the hydrolysis constant, K_h, of the salt varies inversely as the dissociation constant, K_a, of the weak acid. Therefore, the weaker the acid, the greater is the hydrolysis constant of the salt.

(iii) *Degree of hydrolysis:* If c is the initial concentration fo the salt in moles per litre and h is the degree of hydrolysis on the attainment of equilibrium, the concentration of various species may be represented as:

$$\underset{c(1-h)}{B^-} + \underset{\text{excess}}{H_2O} \rightarrow \underset{ch}{HB} + \underset{ch}{OH^-} \quad ...(14A)$$

Applying the law of mass action,

$$K_h = \frac{(ch)(ch)}{c(1-h)} = \frac{ch^3}{1-h} \quad ...(15)$$

If the acid is very weak, h is negligible as compared to unity, *i.e.*, $1 - h \approx 1$. Thus, equation (15) becomes as:

$$K_h = ch^2$$

or $$h = \sqrt{\left(\frac{K_h}{c}\right)}$$

From equation (14), we have

$$K_h = \frac{k_w}{K_a}$$

$\therefore$ $$h = \sqrt{\left(\frac{k_w}{cK_a}\right)} \quad ...(16)$$

From equation (16), it follows that weaker the acid, greater is h, *i.e.*, the degree of hydrolysis. As value of k_w is also increased by a rise in temp., it means that the degree of hydrolysis 'h' will considerably increase with rise of temperature It is also seen that degree of hydrolysis 'h' also increases if the concentration decreases *i.e.*, dilution increases.

(iv) *Relation between pH, pK$_a$ and c:* From Eq. (14A), the concentration of $[OH^-]$ ion is given by

$$[OH^-] = ch$$

But $$[H^+][OH^-] = k_w$$

$$[H^+] = \frac{k_w}{ch} \quad \text{...(17)}$$

From equation (16), we have

$$h = \sqrt{\left(\frac{k_w}{cK_a}\right)}$$

Substituting the value of h in (17), we have

$$[H^+] = \frac{k_w}{c}\sqrt{\left(\frac{cK_a}{k_w}\right)} = \sqrt{\left(\frac{k_w K_a}{c}\right)} \quad \text{...(18)}$$

Taking logarithms and changing the signs throughout, equation (18) becomes as

$$-\log [H^+] = -\frac{1}{2}\log k_w - \frac{1}{2}\log K_a + \frac{1}{2}\log c$$

or $$pH = pk_w + \frac{1}{2}pK_e + \frac{1}{2}\log c \quad \text{...(19)}$$

From equation (19), it follows that pH or alkalinity of a solution of the salt of a weak acid and strong base increases with decreasing and strength, *i.e.*, increasing pK_a and increasing concentration.

(c) Salt of Weak Acids and Weak Bases

(i) *Hydrolysis constant*: The hydrolysis of the salt of a weak acid and a weak base may be represented as:

$$AB + H_2O \rightleftharpoons AOH + HB$$

Ionically, the above equilibria can be represented as:

$$A^+ + B^- + H_2O \rightleftharpoons \underset{\text{weak base}}{AOH} + \underset{\text{weak acid}}{HB} \qquad \text{...(19A)}$$

Applying the law of mass action

$$K = \frac{[AOH][HB]}{[A^+][B^-][H_2O]}$$

$$K[H_2O] = \frac{[AOH][HB]}{[A^+][B^-]}$$

$$K_h = \frac{[AOH][HB]}{[A^+][B^-]} \qquad \text{...(20)}$$

where K_h is the hydrolysis constant of salt.

(ii) Relation between K_h, k_w, K_a, and K_b: The dissociation of the weak acid and weak base may be represented as:

$$HB \rightarrow H^+ + B^- \qquad \text{...(21)}$$

and

$$AOH \rightarrow A^+ + OH^- \qquad \text{...(22)}$$

Applying the law of mass action to the above equilibria (21) and (22).

$$K_a = \frac{[H^+][B^-]}{[HB]} \qquad \text{...(23)}$$

$$K_b = \frac{[A^+][OH^-]}{[AOH]} \qquad \text{...(24)}$$

where K_a and K_b are the dissociation constants of the weak acid and base respectively. Multiplying Eqs. (20), (23) and (24).

$$[H^+][OH^-] = K_a \times K_b \times K_h$$

But

$$[H^+][OH^-] = k_w$$

$\therefore$

$$K_w = K_a \times K_b \times K_h$$

or

$$K_h = \frac{k_w}{K_a \times K_b} \qquad \text{...(25)}$$

(iii) *Degree of hydrolysis:* If c is the initial concentration of the salt in moles/litre and 'h' is the degree of hydrolysis, then the concentration of various species of equilibria (19A) may be represented as:

$$\underset{c(1-h)}{A^+} + \underset{c(1-h)}{B^-} + \underset{\text{excess}}{H_2O} \rightarrow \underset{ch}{AOH} + \underset{ch}{HB} \qquad \text{...(26)}$$

Applying the law of mass action,

$$K_h = \frac{[ch][ch]}{c(1-h)c(1-u)} = \frac{h^2}{(1-h)^2} \quad ...(27)$$

If h being very small, h may be neglected and 1 – h may be taken as unity. Thus, equation (27) becomes as:

$$K_h = h^2 \quad ...(28)$$

From Eq. (28) it follows that degree of hydrolysis of a weak acid and a weak base is independent of the dilution or concentration.

(iv) *Relation between pH, pK_w, pK_a and pK_b:* From equation (23), it follows that

$$[H^+] = K_a \frac{[HB]}{[B^-]} \quad ...(29)$$

From equation (26), we get

$$B^- = c(1-h) \text{ and } HB = ch \quad ...(30)$$

Substituting equation (30) in (29), we have

$$[H^+] = K_a \frac{ch}{c(1-h)} = K_a \frac{h}{1-h}$$

$$= K_a(K_h)^{1/2} \quad \text{[from equation (27)]}$$

$$= K_a\left(\frac{k_a}{K_a K_b}\right)^{1/2} \quad \text{[from equation (25)]}$$

$$= \left(\frac{k_w K_a}{K_b}\right)^{1/2}$$

$$= K_4(K_h)^{1/2} \quad \text{[from equation (27)]}$$

$$K_a = \left(\frac{K_w}{K_b}\right)^{1/2} \quad \text{[from equation (25)]}$$

$$= \left(\frac{k_w K_a}{K_a K_b}\right) \quad ...(31)$$

Taking logarithm of both sides obtain, we get

$$-\log [H^+] = -\frac{1}{2}\log k_w - \frac{1}{2}\log K_a + \frac{1}{2}\log K_b$$

$$pH = \frac{1}{2}pk_a + \frac{1}{2}pK_a - \frac{1}{2}pK_b \quad ...(32)$$

Now three different cases may arise:

(i) When $K_a = K_b$, it follows from equation (32), $pH = \frac{1}{2}pk_w = 7$, *i.e.*, the solution will be neutral.

(ii) When $K_a > K_b$; $pH < 7$, *i.e.*, the solution will be acidic.

(iii) When $K_a < K_b$; $pH > 7$, *i.e.*, the solution will be alkaline.

Determination of Degree of Hydrolysis

(a) From Colligative Properties

The colligative property of a solution number of the solute particles. Consider. 1 gm mole of the salt of weak acid and strong base like CH_3COONa which gets hydrolysed in the solution as:

$$\underset{1-x}{CH_3COO^-} + Na^+ + \underset{1-x}{H_2O} \rightarrow \underset{x}{CH_2COOH} + \underset{x}{Na^+} + \underset{x}{OH^-}$$

where x is the degree of hydrolysis. Then,

total number of particles before hydrolysis = 1 + 1 = 2

and the total number of particles after hydrolysis

$$= 1 - x + 1 - x + x + x + x$$

$$= 2 + x$$

$$\therefore \frac{\text{Observed freezing point depression}}{\text{Calculated freezing point depression}} = \frac{2+x}{2}$$

From the above relation, x can be calculated.

(b) From Conductance Measurements

The method is based upon the simple assumption that the equivalent conductance of an aqueous solution of a salt of a weak acid and strong base (or a weak base and strong acid) is due to the sum of equivalent conductances of strong base (or a weak base and strong acid) is due to the sum of equivalent conductances of

(a) the unhydrolysed salt and

(b) the free OH^- (or the H^+) ions produced by hydrolysis.

The weak acid or weak base formed during hydrolysis do not contribute anything towards the equivalent conductance of the salt solution at infinite dilution.

Procedure: The procedure involves the following steps:

(i) An aqueous solution fo salt of known dilution is prepared. Now determine the equivalent conductance at a certain dilution, say, V, *i.e.* λ_V.

(ii) Now mix the aqueous solution of the salt with excess of the weak acid or weak base as the case may be or the same which is produced during hydrolysis. This addition of weak acid or weak base checks the hydrolysis of the salt. Now determine the equivalent conductance of the mixture at the same dilution as before. Let it be λ'_V. Actually, this is the equivalent conductance of the unhydrolysed salt.

(iii) The equivalent conductance of the strong base or acid at infinite dilution is also determined by Kohlrausch's law. Let it be λ_∞.

Calculation: From the above values of the various equivalent conductances, the degree of hydrolysis may be calculated as follows:

Consider a salt AB of a strong acid and a weak base. Its hydrolysis may be represented as:

$$A^+B^- + H_2O \rightleftharpoons AOH + H^+ + B^-$$

or

$$\underset{1-h}{A^+} + H_2O \rightleftharpoons \underset{h}{AOH} + \underset{h}{A^+}$$

If h is the degree of hydrolysis, then

Concentration f unhydrolysed salt = 1 − h

Concentration of free H^+ ions = h

Equivalent conductance due to unhydrolysed salt = $\lambda'_V - (1 - h)$ and equivalent conductance due to H^+ ions = $\lambda_\infty \times h$

∴ Equivalent conductance of hydrolysed salt,

$$\lambda_V = \lambda_\infty \times h + \lambda'_V (1 - h)$$

or

$$h = \frac{\lambda_V - \lambda'_V}{\lambda_\infty - \lambda'_V}$$

This is known as Bredig's method.

(c) pH Determination Method

In case of the hydrolysis of a salt of a strong acid and a weak base.

$$A^+ + H_2O \rightleftharpoons AOH + H^+$$

and the value of $[H^+]$ is given as $[H^+] = ch$ ...(1)

where 'C' is the concentration of the salt in gm moles/litre and h is the degree of hydrolysis.

Taking logarithm of both sides of equation (1), we obtain

$$\log [H^+] = \log ch$$

or $$-\log [H^+] = -\log ch$$

or $$pH = -\log ch \quad ...(2)$$

In actual practice the aqueous solution of the salt is taken and its pH is determined with the help of pH meter. Thus, one can calculate the degree of hydrolysis.

(d) From the Dissociation Constants of the Weak Acids or Weak Bases

As we know in the case of

(i) a salt of a weak acid and a strong base

$$h = \sqrt{\left(\frac{k_w}{K_a \times C}\right)} \quad ...(1)$$

(ii) a salt of a weak base and a strong acid

$$h = \sqrt{\left(\frac{k_w}{K_b \times C}\right)} \quad ...(2)$$

(iii) a salt of a weak acid and a weak base

$$h = \sqrt{\left(\frac{k_w}{K_a K_b}\right)} \quad ...(3)$$

Knowing dissociation constant of acid or base (K_a or K_b) or both as the case may be, it is a simple matter to calculate the degree of hydrolysis of a given salt at a given concentration.

(e) Distribution Method

This method can also be used provided the free acid or base formed during hydrolysis is soluble in a immiscible solvent like benzene. Let us illustrate this principle by considering the hydrolysis of aniline hydrochloride.

When aniline hydrochloride is hydrolysed, the following equilibria are established in solution.

$$C_6H_5NH_3^+Cl^- + H_2O \rightarrow C_6H_5NH_3^+OH^- + HCl$$

$$\downarrow$$

$$C_6H_5NH_2 + H_2O$$

Applying the law of mass action, the hydrolysis constant K_b is given as:

$$K_h = \frac{[\text{Concentration of free base}]\,[\text{Concentration of free acid}]}{[\text{Concentration of unhydrolysed salt}]}$$

If c is the initial concentration of the salt in moles per litre, then the degree of hydrolysis is given by:

$$h = \left(\frac{K_h}{c}\right)^{1/2} \qquad ...(1)$$

The concentration of aniline formed by hydrolysis of aniline hydrochloride can be determined as follows:

(i) A known amount of aniline is distributed between benzene and water. The concentration of aniline in the benzene layer is measured by passing dry HCl gas through it and weighing the amount of aniline 'hydrochloride precipitated. Difference from the total weight of aniline taken gives the concentration of aniline in water layer. When distribution law is applied. distribution coefficient 'K' is given by

$$K = \frac{\text{Aniline in benzene}}{\text{Aniline in water}}$$

(ii) Now shake a known weight of aniline hydrochloride with a known weight of benzene and water. The free aniline formed distributes itself between benzene and water. Amount of the aniline in the benzene layer is determined by precipitating as aniline hydrochloride by passing dry HCl gas.

Then, calculate the amount of aniline in the water layer by using the distribution coefficient determined in the first step.

The sum of the weights of aniline in the benzene and water layers gives the total weights of aniline formed by hydrolysis. The concentration of free HCl (in moles per litre) will be the same as of aniline in moles/litre.

The concentration of unhydrolysed salt (c)

= [Initial concentration of the salt taken]

− [the loss of concentration due to hydrolysis]

= [Initial concentration of the salt]

− [Concentration of aniline formed by hyrolysis]

Substituting the various values in (1), the hydrolysis constant or the degree of hydrolysis of aniline hydrochloride can be determined.

(f) From E.M.F. Data

For details see the 'Applications of E.M.F. Measurements.'

STRENGTH OF AN ACID

According to the *ionic theory of ionization:*

(i) An acid is a substance which furnishes H^+ ions by itself or when H^+ dissolved in water.

(ii) The strength of an acid depends upon the concentration of free H^+ ions.

(iii) With the dilution, the degree of ionisation of an acid increases, thus increasing H^+ ions also.

(iv) An infinite dilution all acids are *completely ionised* and hence *they behave nearly equally strong at infinite dilution.*

(v) As degree of ionisation of an acid depends on its dissociation constant, so, the *dissociation constant of acid is the measure of its strength.*

(vi) Mineral acids like HCl, NHO_3, H_2SO_4, etc., which are largely ionised at all dilutions, are called *strong acids*, which oxalic, acetic, formic, benzoic acids, etc., which are *feebly ionised*, are called *weak acids*

Methods of Comparing Relative Strengths of Two Acids

1. Thomson's Thermal Method

This method is based on the fact that when a mixture of two acids is allowed to react with a base, the quantity of heat evolved will be the measure of their relative strengths. Suppose we want to compare the strength of HCl and H_2SO_4.

Let the heat of neutralization of 1 g equivalent of HCl by NaOH

$$= x \text{ cal}$$

and heat of neutralization of 1 g equivalent of H_2SO_4 by NaOH

$$= y \text{ cal}$$

Now treat mixture containing 1 g equivalent of each HCl and H_2SO_4, with 1 g equivalent of NaOH and let the heat produced

$$= z \text{ cal}$$

The two acids will react with NaOH in the ratio of their relative strengths, the stronger acid taking more part in neutralizing it. Suppose n g equivalent of HCl and (1 – n) g equivalent of H_2SO_4 have neutralized 1 g equivalent of NaOH; then

Heat evolved due to neutralization of n g equivalent of HCl by NaOH+ heat evolved due to neutralization of (1 – n) g equivalent of H_2SO_4 by NaOH

$$= z \text{ cal}$$

or $$nx + (1 - n)y = z$$

or $$n = \frac{z - y}{x - y}$$

Knowing n, the relative strengths of $HCl/H_2SO_4 = n/(1 - n)$ can be calculated.

2. Conductivity Method

The strength of an acid at any dilution depends upon its degree of dissociation. So, the relative strengths of two acids at same dilution is equal to the ratio of their degrees of ionisation. For a weak acid, the degree of ionisation is given by conductivity ratio. Thus

For acid I: $$\alpha_1 = \frac{\lambda_v}{\lambda_\infty}$$

For acid II: $\alpha_2 = \frac{\lambda'_v}{\lambda'_\infty}$

But at infinite dilution

$$\lambda_\infty = \lambda'_\infty$$

$$\therefore \quad \frac{\alpha_1}{\alpha_2} = \frac{\lambda_v}{\lambda'_v} = \frac{\text{Strength of acid I}}{\text{Strength of acid II}}$$

Thus, *the relative strengths of two weak acids is equal to the ratio of their equivalent conductivities at the same dilution.*

3. Dissociation Constants Method

(For weak acids only).

For acids HA_1 and HA_2 at some dilution (V), we have

$$K_1 = \frac{\alpha_1^2}{(1-\alpha_1)V}; \quad K_2 = \frac{\alpha_2^2}{(1-\alpha_2)V}$$

For weak acids $(1 - \alpha_1)$ and $(1 - \alpha_2)$ may be taken as unity.

$$\therefore \quad \frac{\alpha_1^2}{V} = K_1; \quad \frac{\alpha_2^2}{V} = K_2$$

or

$$V = \frac{K_1}{\alpha_1^2} = \frac{K_2}{\alpha_2^2}$$

or

$$\frac{\alpha_1}{\alpha_2} = \sqrt{\left(\frac{K_1}{K_2}\right)} = \frac{\text{Strength of acid I}}{\text{Strength of acid II}}$$

Thus, *relative strength of two weak acids is equal to the square root of the ratio of their dissociation constants.*

4. Ostwald's Volume Method

In this method change in volume is noted instead of noting the heat evolved. Let x cm^3 be the change in volume when 1 equivalent of each of NaOH and HA_1 react and y cm^3 be the change in volume when 1 equivalent each of NaOH and HA_2 react. Let z cm^3 be the change when 1 equivalent of NaOH and a solution containing 1 equivalent of each HA_1 and HA_2 are mixed. Calculating in the same manner.

$$n = \frac{z - y}{x - y}$$

$$\therefore \quad \frac{\text{Strength of } HA_1}{\text{Strength of } HA_2} = \frac{n}{(1-n)}$$

5. The Catalytic Method

It is well known that H^+ ions act as catalyst in many chemical reactions. The hydrolysis of esters and the inversion of cane sugar are catalysed by H^+ ions. To compare the strength of two acids HA_1 and HA_2, therefore, the velocity constants k_1 and k_2 of above first order reactions are determined in presence of these acids and the strengths are compared as

$$\frac{\text{Strength of } HA_1}{\text{Strength of } HA_2} = \frac{k_1}{k_2}$$

The strength of two bases can also be compared by the above mentioned methods. The conductivity method and comparison of dissociation constant method can be used as such. In catalytic process, we shall choose the base catalysed reactions such as condensation of acetone. In the Thomson's thermal and Ostwald's volume methods the two bases are neutralised by strong acids, say nitric acid.

DISSOCIATION CONSTANTS OF POLYBASIC ACIDS

Polybasic acids contain two or more ionisable hydrogen. They always dissociate in steps. The degrees of dissociation at different steps are different. Consider for example, the dissociation of phosphoric acid. It is a tribasic acid and ionises in three steps.

(i) $H_3PO_4 \rightleftharpoons H^+ + H_2PO_4^-$; $K_{a_1} = 7.5 \times 10^{-3}$ M

(ii) $H_2PO_4^- \rightleftharpoons H^+ + HPO_4^{2-}$; $K_{a_2} = 6.2 \times 10^{-7}$ M

(iii) $HPO_4^{2-} \rightleftharpoons H^+ + PO_4^{3-}$; $K_{a_3} = 4.8 \times 10^{-3}$ M.

The values of dissociation constants, K_{a_1}, K_{a_2} and K_{a_3} for the above three successive steps are taken from data book. These dissociation constants can be calculated in terms of the concentration, $[H^+]$, $[H_2PO_4^{1-}]$ and $[HPO_4^{2-}]$ and $[PO_4^{3-}]$ as follows:

$$K_{a_1} = \frac{[H^+][H_2PO_4^-]}{[H_3PO_4]}$$

$$K_{a_2} = \frac{[H^+][HPO_4^{2-}]}{[H_2PO_4^-]}$$

$$K_{a_3} = \frac{[H^+][HPO_4^{3-}]}{[HPO_4^{2-}]}$$

The ionisation constant in these three stages successively indicates that

$$K_{a_1} > K_{a_2} > K_{a_3}$$

The reason for the decrease in the dissociation constant values is that in the first dissociation the positively charged proton comes from a neutral molecule while in the second it is detached from a negatively charged molecule and in the third dissociation, it is detached from a doubly negatively charged molecule. The presence of negative charges makes the process difficult for the loss of proton.

COMMON ION EFFECT

Consider a simple weak electrolyte AB which ionises in solution as

$$AB \rightleftharpoons A^+ + B^- \qquad ...(1)$$

On applying the law of mass action, we get

$$K = \frac{[A^+][B^-]}{[AB]} \qquad ...(2)$$

Now, suppose to the above solution a strong electrolyte CB or AD *i.e.*, an electrolyte having A^+ or B^- as common ion with the electrolyte AB is added. These will ionise as

$$CB \rightleftharpoons C^+ + B^-$$

$$AD \rightleftharpoons A^+ + D^-$$

Suppose we have added AD to the equilibrium (1). Thus, the concentration of A in solution is increased. In order to keep K constant [equation (2)] at the given temperature, some of the A^+ added must combine with B^- to form unionised AB *i.e.*, the above equilibrium (i) is shifted backwards.

In other words, the degree of ionisation of AB is suppressed. *This decrease in ionisation of a weak electrolyte by the addition of a strong electrolyte having an ion common with the weak electrolyte is known as common ion effect.*

For example, addition of ammonium chloride suppresses the ionisation of ammonium hydroxide and hydrochloric acid suppresses the ionisation of hydrogen sulphide.

Applications of Common Ion Effect

(i) *In Analytical Chemistry*: In II group of qualitative analysis, the ionisation of weak acid H_2S

$$H_2S \rightleftharpoons 2H^+ + S^{2-}$$

is suppressed by the presence of strong acid HCl having a common ion H^+

$$HCl \rightleftharpoons H^+ + Cl^-$$

Due to the suppressed degree of dissociation of H_2S, the concentration of S^{2-} ions produced is low which causes the precipitation of sulphide of II group cations only.

Similarly, in III group, the ionization of weak base NH_4OH,

$$NH_4OH \rightleftharpoons NH_4^+ + OH^-$$

is suppressed by the prior addition of NH_4Cl which furnishes a high concentration of common ion NH_4^+.

$$NH_4Cl \rightleftharpoons NH_4^+ + Cl^-$$

Due to this low dissociation of NH_4OH, the OH^- concentration is quite low. This low OH^- ion concentration is sufficient only to precipitate hydroxides of Al^{3+}, Cl^{3+} and Fe^{3+}.

(ii) *Purification of Common Salt:* NaCl, as obtained from natural sources, is usually contaminated with small amounts of deliquescent impurities like $CaCl_2$ and $MgCl_2$. The purification of chloride solution is effected by passing HCl gas into a saturated solution of the impure sodium chloride solution is effected by passing HCl gas into a saturated solution of the impure sodium chloride.

The addition of HCl (containing Cl^-) causes the precipitation of only NaCl, but impurities remain in solution since the solution is not saturated so far as calcium and magnesium chlorides are concerned.

(iii) *Salting out of Soap:* In the soap manufacture, a solution of common salt is added to solution of soap (sod. stearate) to obtain solid soap. Addition of common salt (NaCl) increases the value of $[Na^+] \times [stearate^-]$ in solution until the solubility product of soap is exceeded, whereby soap gets precipitated.

(iv) *In Gravimetric Analysis:* When excess of precipitant (HCl or H_2SO_4) is added to solution containing Ag^+ or Ba^{2+} ions, complete precipitation of AgCl or $BaSO_4$ occurs due to the presence of excess of common ion Cl^- or SO_4^{2-}.

SOLUBILITY PRODUCT

Consider the *saturated* solution of a sparingly soluble binary electrolyte AB. There exist two equilibria simultaneously between undissolved solid, dissolved (but unionised) molecules and free ions, *i.e.*,

$$\underset{\text{(Undissolved solid)}}{AB} \overset{K_1}{\rightleftharpoons} \underset{\text{(Dissolved, but unionsed)}}{AB} \overset{K_2}{\rightleftharpoons} \underset{\text{(Ions in solution)}}{A^+ + B^-}$$

Applying the law of mass action, it follows that

$$\frac{[A^+][B^-]}{[(AB)_d]} = K_1; \qquad \frac{[(AB)_d]}{[(AB)_d]} = K_2$$

∴ On multiplying, we get

$$\frac{[A^+][B^-][(AB)_d]}{[(AB)_d][(AB)_s]} = \frac{[A^+][B^-]}{[(AB)_s]} = K_1 \times K_3$$

But the concentration of solid AB is constant, so that

$$[A^+][B^-] = \text{a constant} = K_s$$

where the constant K_s is known as the solubility product of AB.

In general, for an electrolyte, A_xB_y, which ionises as

$$A_xB_y \rightleftharpoons xA^{y+} + yB^{x-}$$

the solubility product is given by

$$K_s = [A^{y+}]^x\,[B^{x-}]^y$$

∴ where $[A^{y+}]$ and $[B^{x-}]$ are the ionic concentration in a *saturated solution*. Thus, solubility product of an electrolyte may be defined as the *maximum product of the concentrations of its constituent ions (expressed in g-ion per litre) in its solution, when each ionic concentration term being raised to the number of times the ion occurs in the equation representing the solution of 1 molecule of the electrolyte.* For example for

$$Ag_2CrO_4 \rightleftharpoons 2Ag^+ + CrO_4^{2-}$$

the solubility product, $K_s = [Ag^+]^2[CrO_4^{2-}]$.

Relation Between Solubility and Solubility Product

Consider the general form of electrolyte A_xB_y which dissociates as

$$\underset{S}{A_xB_y} \rightleftharpoons \underset{xS}{x[X^{y+}]} + \underset{yS}{y[X^{x-}]}$$

Let S gram mole/litre be the solubility of the electrolyte, then

$$[A^{y+}] = x \,.\, S \quad \text{and} \quad [B^{x-}] = y \,.\, S$$

$$\therefore \qquad K_s = [A^{y+}]^x[B^{x-}]^y$$

$$= [x \,.\, S]^x[y \,.\, S]^y = x^x \,.\, y^y \,.\, S^{x+y}$$

For example in case of $BaCl_2 \rightleftharpoons Ba^{2+} + 2Cl^-$; x = 1 and y = 2

$$Ks \rightleftharpoons 1^1 \,.\, 2^2\, S^{1+2} = 4S^3.$$

Analytical Applications of Solubility Product

From the relation:

$$K_s = [A^+]^x[B^-]^y$$

it is clear that:

(a) a solution in which $[A^+]^x[B^-]^y$ is *less than* K_s, the solution is unsaturated and more A_xB_y can be dissolved in it,

(b) a solution in which $[A^+]^x[B^-]^y$ is *equal* to K_s, *is just saturated*, and

(c) if anything is done to the solution which tends to make $[K^+]^x$ $[B^-]^y$ *greater than* K_s, then solid A_xB_y will be precipitated.

From the above discussion it is evident that if the product of concentration of ions in a saturated solution of an electrolyte becomes greater than the solubility product of the electrolyte, it will precipitate out. In qualitative analysis, separation and identification of metal ions such as Cu^{2+}, Zn^{2+}, etc., is possible only due to difference in solubility products of their sulphides. The above concepts of solubility product have become extremely useful in getting a clear picture of different processes in analytical chemistry. A few cases are illustrated below:

1. Qualitative Analysis of Cationic Mixtures

The identification of individual species of cations present in an unknown mixture of electrolytes involves a preliminary segregation of the cations into selected classifiable groups.

A simple system devised for this purpose is based on the selective precipitation of sparingly soluble compounds of the cations from aqueous solution of the electrolyte mixture.

In its most simplified form, for the more common cations only, these compounds fall into five groups. Within each group the anion is the same for the different cations and the solubility products are small and do not differ too greatly from one another:

Group I: The sparingly soluble chlorides derived from Pb^{2+}, Hg_4^{2+} and Ag^+ ions are precipitated by the addition of dilute HCl.

Chloride	K_s *at 25°C*
$PbCl_2$	1.6×10^{-5}
Hg_3Cl_2	1.3×10^{-15}
AgCl	1.8×10^{-16}

Group II: The sparingly soluble sulphides derived from Hg^{2+}, Pb^{2+}, Cu^{2+}, Cd^{2+} and Bi^{3+} ions are precipitated by the passage of H_2S through an acidic solution whose concentration of hydrogen ions is small.

Sulphide	K_s *at 25°C*
HgS	3×10^{-32}
PbS	8×10^{-28}
CuS	8×10^{-36}
CdS	7×10^{-27}
Bi_2S_3	1×10^{-96}

Group III: The sparingly soluble hydroxides derived from Fe^{2+}, Cr^{3+} and Al^{3+} ions are precipitated by the addition of ammonium hydroxide solution in the presence of excess ammonium chloride

Hydroxide	K_s *at 25°C*
$Fe(OH)_3$	6×10^{-28}
$Cr(OH)_3$	7×10^{-31}
$Al(OH)_3$	1.4×10^{-24}

Group IV: A further group of sparingly soluble sulphides, those derived from CO^{2+}, Ni^{2+}, Mn^{2+} and Zn^{2+} ions are precipitated by the passage of H_2S through a basic solution.

Sulphide	K_s *at 25°C*
CoS	8×10^{-23}
NiS	2×10^{-21}
MnS	1×10^{-11}
ZuS	8×10^{-25}

Group V: The sparingly soluble carbonates derived from Ca^{2+}, Sr^{2+} and Ba^{2+} ions are precipitated by the addition of ammonium carbonate solution in the presence of NH_4OH and NH_4Cl.

Carbonate	*Ks at 26°C*
$CaCO_3$	4.7×10^{-9}
$SrCO_3$	7×10^{-10}
$BaCO_3$	1.6×10^{-9}

The most common cations remaining in solution after the precipitation fo group V are Na^+, K^+ (all the simple salts of the alkali metals are water-soluble) and Mg^{2+} ions. The order of group precipitation is such that the precipitating reagents for later groups also precipitate in most cases cations in preceding groups, which must therefore the completely removed in the above prescribed order if they are not to interfere. For example,

(i) K_S of silver sulphide is 7×10^{-50} at 25°C. It means that silver sulphide will be precipitated in group II unless it has been completely removed as AgCl in group I.

(ii) Any of the cations normally precipitated as their sulphides in group II, will reappear in the group IV precipitate unless completely removed in group II.

(ii) K_S for $ZnCO_3$ at 25°C is 2.1×10^{-11}. Hence $ZnCO_3$ will be precipitated in group V unless it has been completely removed as ZnS in group IV.

2. *The Separation of the Sulphides into Groups II and IV*

The values of K_S for group II sulphide are smaller than those for group IV sulphides. Hence, it is possible to precipitate the sulphides of those cations which fall in group II selectively in the presence of those cations which fall in group IV if the $[S^{2-}]$ from the reagent H_2S is regulated such that it exceeds that required to reach the solubility products of the group II sulphides but is less than that required to reach the solubility produces of the group IV sulphides. This is done as follows:

H_2S is a weak diprotic acid for which:

(a) $$H_2S \rightleftharpoons H^+ + HS^-$$

$$\frac{[H^+][HS^-]}{[H_2S]} = K_{S_1} = 9.1 \times 10^{-3} \text{ at } 18°C$$

(b) $$HS^- \rightleftharpoons H^+ + S^{2-}$$

$$\frac{[H^+][S^{2-}]}{[HS^-]} = K_{S_2} = 1.2 \times 10^{-15} \text{ at } 18°C$$

Hence, on multiplying:

$$\frac{[H^+]^3[S^{2-}]}{[H_2S]} = 10^{-22}$$

$$\therefore \quad [S^{2-}] = \frac{[10^{-22}][H_2S]}{[H^+]^2}$$

If H_2S at 1 atmosphere is bubbled through water, for the saturated solution:

$$[H_2S] \sim 0.1 \text{ mole/litre}$$

$$\therefore \quad [S^2] = \frac{10^{-22} \times 10^{-1}}{[H^+]^2} = \frac{10^{-22}}{[H^+]^2}$$

i.e., the $[S^{2-}]$ is dependent on the $[H^+]$ and may be varied by varying the acidity of the solution.

If $\quad pH = 0(1 \text{ MHCl})$

$\therefore \quad [H^+] = 1$ mole/litre

or $\quad [S^{2-}] = 10^{-22}$ mole/litre

If $\quad pH = 12(0.01 \text{ M NaOH})$

$\therefore \quad [H^+] = 10^{-12}$ mole/litre

or $\quad [S^{2-}] = 10$ mole/litre

Hence if the acidity is too low, the corresponding $[S^{2-}]$ is large enough to exceed the requirements of the solubility products of both the group II and the group IV sulphides which are precipitated together in group II.

Conversely, if the acidity is too high, the $[S^{2-}]$ ion is suppressed to a degree when it is lower than that required to reach the solubility products of even the group II sulphides. These, therefore, remain in solution and are precipitated in group IV where the $[S^{2-}]$ in the basic solution is too high.

In practice, if the $[H^+]$ of the solution, after the precipitation of the group I chlorides is adjusted to 0.3 M prior to passing H_2S, the group II sulphides are precipitated selectively.

(a) Selective precipitation of the group III hydroxide: Three hydroxides only are precipitated in group III, those of Fe^{2+}, Cr^{3+} and Al^{3+} whose K_s values lie between 10^{-30} to 10^{-33}. However, the K_s values for the hydroxides of those cations which are normally precipitated in group IV as their sulphide are also relatively small, though of a higher order than the above there:

Hydroxide	K_s *at 25°C*
$Co(OH)_2$	2×10^{-16}
$Ni(OH)_2$	2×10^{-15}
$Mn(OH)_2$	1.6×10^{-13}
$Zn(OH)_2$	7×10^{-18}

Now, for a solution fo NH_3 in water the following equilibria exist:

$$NH_3 + H_2O \rightleftharpoons NH_4OH \rightleftharpoons NH_4^+ + OH^-$$

for which $\dfrac{[NH_4^+][OH^-]}{[NH_4OH]} = K_b = 1.8 \times 10^{-5}$ at 25°C

$\therefore \quad [NH_4^+][OH^-] = 1.8 \times 10^{-5}\,[NH_4OH]$

But $\quad [NH_4^+] = [OH^-]$

$\therefore \quad [OH^-]^2 = 1.8 \times 10^{-5}\,[NH_4OH]$

For 1 M NH_4OH :

$$[OH^-]^2 = 1.8 \times 10^{-5}$$

or $\quad [OH^-] = 4.24 \times 10^{-3}$ mole/litre

For 0.1 M NH_4OH:

$$[OH^-]^2 = 1.8 \times 10^{-5} \times 10^{-1} = 1.8 \times 10^{-6}$$

or $$[OH^-] = 1.34 \times 10^{-2} \text{ mole/litre}$$

Hence, if a solution of NH_4OH alone is used to precipitate the hydroxides in group III, the $[OH^-]$ is high enough to exceed the requirements of the solubility products for the hydroxides of the cations which fall in group IV. These are therefore precipitated together with the normal group III hydroxides.

It is therefore necessary to suppress the $[OH^-]$ concentration to such an extent where it exceeds the requirements of the solubility products for the group III hydroxides but is lower than is necessary to reach the solubility products for the hydroxides of the group IV cations.

This is done by the addition of an excess of NH_4Cl, when a buffer solution is produced consisting of the weak base, NH_4OH, and a salt of the weak base and a strong acid, NH_4Cl. NH_4Cl, being a salt, is a strong electrolyte:

$$NH_4Cl \rightarrow NH_4^+ + Cl^-$$

The excess NH_4^+ ions depress the ionisation of NH_4OH in order that K_b for NH_4OH may be maintained. Hence the $[OH^-]$ is suppressed to a point where only the group III hydroxides are precipitated.

(b) Quantitative Analysis: The solubility products of the silver halides are small:

At 25°C, K_s of AgCl $= 1.8 \times 10^{-10}$

K_s of AgBr $= 5.2 \times 10^{-13}$

K_s of AgI $= 8.3 \times 10^{-17}$

Hence, if a solution fo the soluble $AgNO_3$ is added to a solution containing a soluble halide, the halide ions are quantitatively removed from solution as the sparingly soluble silver salt. The concentration of halide ions remaining in solution when precipitation is just complete is very low and may be neglected.

For example, if $AgNO_3$ solution is added to NaCl solution, AgCl is precipitated. When precipitation is just complete, for the ions remaining in solution:

$$[Ag^+][Cl^-] = K_s = 1.8 \times 10^{-10}$$

But $$[Ag^+] = [Cl^-]$$

$\therefore \qquad [Cl^-] = (1.8 \times 10^{-10})^{1/2} = (1.34 \times 10^{-5})$ mole/litre

Thus precipitation of sparingly soluble silver halides is made the basis of quantitative determinations of the concentrations of solution of soluble halides. A solution of the halides of unknown concentration is titrated with a standard solution of $AgNO_3$. The end point of the reaction occurs when precipitation fo the sparingly soluble silver halide is just complete.

3. Using Excess of Reagent

The use of excess of reagent is essential to ensure complete precipitation in gravimetric analysis. The insoluble substances like AgCl, $BaSO_4$, etc., have very slight solubility but by adding excess of precipitant their solubility product is further exceeded. Thus, the ionic product $[Ba^{2+}][SO_4^{2-}]$ becomes still higher than the solubility product of $BaSO_4$, by using large excess of H_2SO_4.

$$BaSO_4 \rightleftharpoons Ba^{2+} + SO_4^{2-}$$

$$H_2SO_4 \rightleftharpoons 2H^+ + SO_4^{3-}$$

4. Use of K_2CrO_4 as Indicator in $AgNO_3$ Chloride Titrations

In silver nitrate-chloride titration, the $AgNO_3$ is added from the burette into the chloride solution to which some K_2CrO_4 solution is added. At first the added $AgNO_3$ precipitates AgCl, but not Ag_2CrO_4 because AgCl has a lower solubility (0.00145 g per litre) than Ag_2CrO_4 (0.0285 g per litre). When sufficient $AgNO_3$ has been added to precipitate all the AgCl, subsequent addition causes the formation of brick-red Ag_2CrO_4 precipitate, which serves to note the end-point.

IONIC PRODUCT OF WATER

Pure water is essentially a covalent compound. Nevertheless, it ionises very slightly and the following equilibrium is established:

$$H_2O \rightleftharpoons H^+ + OH^- \qquad ...(1)$$

where H^+ is a hydrogen ion and OH^- is a hydroxyl ion. The removal fo the one orbital electron from a hydrogen atom yields the positive hydrogen ion or hydrogen cation, which is in fact a bare proton. The bare proton has no separate existence in solution and owes its stability to solvation by a water molecule to give the hydronium ion, H_3O^+. The ionisation equilibrium of water is thus represented more accurately as:

$$2H_2O \rightleftharpoons H_3O^+ + OH^-$$

However, as many qualitative and quantitative considerations remain essentially the same for the unsolvated and the solvated proton, it is usual for convenience to refer to the unsolvated proton, H^+.

On applying the law of mass action to equation (1), we get

$$K = \frac{[H^+][OH^-]}{H_2O} \quad ...(2)$$

where, K is called the ionisation constant. As water is ionised to very slight extent (about 1 in 10^7 water molecules), the concentration of water molecules is very large as compared with H^+ and OH^- ions so that it can be regarded as practically constant.

Therefore, equation (2) may be written as

$$[H^+][OH^-] = K[H_2O] = k_w \quad ...(3)$$

where constant k_w is constant called the *ionic product of water.*

Conductivity measurements show that in pure water.

$$[H^+] = 10^{-7} \text{ mole/litre at } 25°C$$

Furthermore in pure water.

$$[H^+] = [OH^-]$$

Therefore, equation (3) may become as

$$[10^{-7}][10^{-7}] = k_w$$

or $$k_w = 10^{-14} \text{ at } 25°C \quad ...(4)$$

Significance of Equation (3) and (4)

For all aqueous solutions, the product of $[H^+][OH^-]$ may be taken to be equal to 10^{-14}. If extra H^+ ions are introduced, the ionisation of H_2O is suppressed to the point at which $[H^+][OH^-] = 10^{-14}$. Similarly, if extra OH^- ions are introduced, the ionisation of water is again suppressed to maintain the constancy of $[H^+][OH^-]$. From equation (3), it follows that the ionic product of water may be defined as *the product of concentration of H^+ and OH^- ions expressed in gram ions per litre.*

TRANSPORT NUMBER

The fraction of the total current carried by any one of the ionic species is known as transport number, transference number or Hittorf number of the species and may be denoted by sets of symbols like t_+ and t_-, t_a and t_c or n_c are n_a.

From the above definition, it follows that the transport number of an anion is

$$t_- = \frac{\text{Current carried by an anion}}{\text{Total current passes through the solution}} \quad ...(1)$$

and the transport number of cation is

$$t_+ = \frac{\text{Current carried by the cation}}{\text{Total current passed through the solution}} \quad ...(2)$$

But we know that the quantity of electricity carried by each ion is directly proportional to the speed of the concerned ions.

So. amount of electricity carried by anion $\propto$ speed of anion (u_a) amount of electricity carried by cation $\propto$ speed of cation (u_c).

$\because$ Total amount of electricity carried $\propto$ speed of anion + speed of cation

$$\propto u_a + u_c$$

Thus. the transport number of anion [From eq (1). t_- is]

$$t_- = \frac{u_a}{u_a + u_c} \quad ...(3)$$

and the transport number of cation [From eq. (2)]. t_+. is

$$t_+ = \frac{u_c}{u_a + u_c} \quad ...(4)$$

It follows from equation (3) and (4) that the transport number of an ion may be defined as the ratio of speed of the ion to the sum of the speeds of both cation and anion.

By adding Eqs. (3) and (4). we get

$$t_+ + t_- = \frac{u_a}{u_a + u_c} + \frac{u_c}{u_a + u_c} = \frac{u_a + u_c}{u_a + u_c} = 1 \quad ...(5)$$

$$t_+ + t_- = 1$$

From eq. (5). it follows that

(i) The sum of the transport numbers of all the species taking part in the transport of electricity is equal to unity. and

(ii) If the transport number of one of the ionic species is known. then that of the other can be easily calculated by using eqn. (5).

Experimental Determination of Transport Number

There are several methods for determining the transport numbers of ions, viz.

(i) Hittorf's Method

Principle : This method is based on the principle that "*During electrolysis, the fall of concentration around the electrode is proportional to the speed of ions moving away from that electrode.*"

Thus, the fall of concentration around the cathode ∝ speed of the anion

(u_a). ...(6)

and the fall of concentration around the anode ∝ speed of the cation

(u_c) ...(7)

Dividing eq. (6) by (7), we get

$$\frac{\text{Fall of concentration around the cathode}}{\text{Fall of concentration around the anode}} = \frac{u_a}{u_c}$$

Adding 1 to both sides, we get

$$1 + \frac{\text{Fall of concentration around the cathode}}{\text{Fall of concentration around the anode}} = 1 + \frac{u_a}{u_c}$$

or
$$\frac{\text{Fall of concentration around the anode + Fall of concentration around the cathode}}{\text{Fall of concentration around the anode}} = \frac{u_c + u_a}{u_c}$$

or
$$\frac{\text{Total fall of concentration}}{\text{Fall of concentration around the anode}} = \frac{u_c + u_a}{u_c}$$

Taking reciprocals of the abode equation.

$$\frac{\text{Fall of concentration around the anode}}{\text{Total fall of concentration}} = \frac{u_c}{u_c + u_a} \qquad ...(8)$$

From equation (4), we know $t_+ = \frac{u_c}{u_a + u_c}$

$$\frac{\text{Fall of concentration around the anode}}{\text{Total fall of concentration}} = t_+ \qquad ...(9)$$

Similarly,

$$\frac{\text{Fall of concentration around the cathode}}{\text{Total fall of concentration}} = t_- \qquad ...(10)$$

Working

(i) A dilute solution fo known concentration (N/10) of the salt containing the ions, of which the transport numbers are to be determined, is filled in the apparatus.

(ii) The apparatus is then connected to a battery, variable resistance and voltameter in series as shown in Fig. (2.15). A low current of the order of 1 – 20 milliamperes is passed for 2 to 3 hours.

If high current density is used, diffusion sets in and the solution sin the various compartments are mixed and thus leading to wrong results.

(iii) After the electrolysis the stop cocks connecting the U-tube with the side tubes are closed and the solutions are withdrawn from the various compartments separately. The concentration is determined by titration against KCNS solution. These should be no change in the concentration of the solution from the V-tube, *i.e.*, central compartment. Knowing the original concentration of the solution taken, the change in concentration is determined separately.

(iv) The weight of Ag(or Cu) deposited in the voltameter is also determined separately. This represents the total quantity of electricity passed during electrolysis.

Calculations: Calculation involves the following steps:

Case I : *When the electrodes are unattackable.* This condition can be obtained by using platinum electrodes. Suppose that,

(i) The weight of $AgNO_3$ in the a g of anodic solution before electrolysis = X gm.

(ii) The weight of $AgNO_3$ in the b g of anodic solution after electrolysis = Y gm.

(iii) The weight of $AgNO_3$ in the a g of anodic solution after electrolysis $= \frac{Y}{b} \times a = Z$

∴ Full in concentration around anode = (X – Z) gm.

Let the weight of Ag deposited in the voltameter = 2 gm

$$\frac{\text{Total weight of } AgNO_3 \text{ electrolysed}}{\text{Weight of Ag deposited in voltameter}}$$

$$= \frac{\text{Eq. wt. of } AgNO_3}{\text{Eq. wt. of Ag}}$$

$$\therefore \quad \frac{\text{Total weight of } AgNO_3 \text{ electrode}}{w} = \frac{170}{108}$$

or $$\text{Total weight of } AgNO_3 \text{ electrolysed} = \frac{w \times 170}{108} = W \text{ (say)}$$

Now the transport No. of Ag^+ ion is given by

$$t_{Ag^+} = \frac{\text{Loss in wt. of } AgNO_3 \text{ around anode}}{\text{Total weight of } AgNO_3 \text{ electrolysed}}$$

$$= \frac{X - Z}{W} \qquad ...(11)$$

and the transport No. of NO_3^- ions is given by

$$t_{NO_3^-} = 1 - t_{Ag^+} = 1 - \frac{X - Z}{W} \qquad ...(12)$$

Case II : *When the electrodes are attackable.* This condition can be obtained by using Ag anode in $AgNO_3$ solution.

In this case $Z \geq X$, *i.e.*, the concentration of $AgNO_3$ in the anodic compartment after electrolysis is more than that before electrolysis. This is due to the fact that NO_3^- ions after losing charge at the anode attack the silver anode to form $AgNO_3$. Thus, the concentration of $AgNO_3$ in the anodic compartment increases.

The fall in concentration of $AgNO_3$ in the anodic compartment due to the migration of the Ag^+ ions may be calculated as follows:

The amount of $AgNO_3$ formed by the discharge of NO_3^- ions at the anode = No. of the NO_3^- ions discharged.

= total $AgNO_3$ electrolysed W (say)

∴ The actual increase in conc. of anodic compartment = W.

But observed increase in conc. of anodic compartment = Z − X.

∴ Fall in concentration due to the migration of Ag^+ ions

= Actual increase − observed increase

= W − (Z − X)

Hence $$t_{Ag^+} = \frac{W - (Z - X)}{W} \qquad ...(13)$$

and $$t_{NO_3^-} = 1 - t_{Ag^+} = 1 - \frac{W - (Z - X)}{W} = \frac{Z - X}{W} \quad ...(14)$$

(2) Moving Boundary Method

Principle: It is based on measuring the rate of migration of one or both of the ionic species of the electrolyte, away from the similarly charged electrodes. This method was first of all suggested by Lodge and Whetham. It was modified by Masson (1899), Steel (1904) and Denison (1906). Finally it was the work of Macinnes (1923) whose work made the procedure more easy and simple.

In most of the methods, LiCl is not used as an indicator but instead of this cadmium is used as an anode which on electrolysis yields cadmium chloride which in turn itself acts as an indicator electrolyte.

(ii) The electrolytic cell consists of the shape as shown, the middle part of the left hand portion is vertical tube of uniform area of cross section.

(iii) The experimental solution NaCl is floated on top of a denser solution of $CdCl_2$.
At the bottom of the tube is fixed an anode of cadmium metal. Over the solution of NaCl a less dense solution of sodium acetate is floated.

(iv) The cathode used is of platinum. This is placed in a side tube so that the evolution of hydrogen gas at the cathode does not disturb the main liquid column.

Working

(i) A constant current is passed for 5 to 6 hours. The Na^+ ions move upwards towards the cathode closely followed by the Cd^{2+} ions. The sharp boundary moves gradually upwards. The movement of the boundary between NaCl and $CdCl_2$ can be easily followed on account of a difference in the refractive indices of the two solutions.

(ii) The rate of migration of Cl^- ions can also be observed as the acetate anions are slow moving than the chloride ions.

(iii) At the end of experiment, the distance through which the boundary has moved is noted. The time for which the current has passed is noted.

Calculations: Suppose a current of I amperes is passed for t seconds, the quality of electricity carried by Na^+ ions is given by

$$= t_+ \text{ It coulombs}$$

where t_+ is the transport number of the Na^+ cation

Suppose the concentration of NaCl solution = c gm eq. eq/c.c.

Suppose the area of cross section of the moving boundary = a cm^2.

Therefore, the amount of Na^+ ion that has migrated upwards

$$= \frac{t_+ It}{F} \text{ where F is Faraday.} \qquad ...(15)$$

Distance by which NaCl—$CdCl_2$ boundary moving during t seconds = x cm.

Volume of the solution cleared by the migration of Na^+ ions = ax

Number of gm equivalent of Na^+ ions transported = axc ...(16)

From Eqs. (15) and (16), we have $axc = \frac{t_+ It}{F}$

or
$$t_+ = \frac{F\,axc}{It} \qquad ...(17)$$

Similarly,
$$t_- = \frac{F \cdot ayc}{It} \qquad ...(18)$$

where y is the distance by which NaCl—CH_3COONa boundary moves during t send and t_- is the transport number of chloride anion.

Also
$$\frac{t_+}{t_-} = \frac{x}{y}$$

But
$$t_- = 1 - t_+$$

$\therefore$
$$\frac{t_+}{1 - t_+} = \frac{x}{y} \quad \text{or} \quad t_+ = \frac{x}{x+y} \qquad ...(19)$$

Similarly
$$t_- = \frac{x}{x+y} \qquad ...(20)$$

Thus, t_+ and t_- can be evaluated.

(3) From Ionic Mobility

The ionic mobility or conductance of an ion is directly proportional to the speed of the ion. Thus,

$$\lambda_+ \propto u_+$$

or $$\lambda_+ = Ku_+ \quad ...(21)$$

Similarly, $$\lambda_- \propto u_-$$

or $$\lambda_- = Ku_- \quad ...(22)$$

In Eqs. (21) and (22), u_+ and u_- are the speeds while λ_+ and λ_- are the ionic conductivities of the cation and anion respectively.

Now the transport number of cation is given by

$$t_+ = \frac{u_+}{u_+ + u_-}$$

$$= \frac{\frac{1}{K}\lambda_+}{\frac{1}{K}(\lambda_+ + \lambda_-)} \quad \text{[From Eqs. (21) and (22)]}$$

$$= \frac{\lambda}{\lambda_+ + \lambda_-} \quad ...(23)$$

But $$\lambda_+ + \lambda_- = \lambda_\infty \text{ (Kohlrausch's Law)}$$

$\therefore$ $$t_+ = \frac{\lambda_+}{\lambda_\infty} \quad ...(24)$$

Similarly $$t_- = \frac{\lambda_-}{\lambda_\infty} \quad ...(25)$$

From Eqs. (24) and (25), it follows that the transport number of cation and anion can be determined if one knows the values of λ_+, λ_- and λ_∞.

In this case, oxidation takes place at the zinc electrode and reduction at the hydrogen electrode, as shown below:

Zinc electrode $Zn\,(s) \rightleftharpoons Zn^{2+} + 2e^-$

Hydrogen electrode $2H^+ + 2e^- \rightleftharpoons H_2(g)$

$\therefore$ *The net reaction is* $Zn(s) + 2H^+ \rightleftharpoons Zn^{2+} + H_2(g)$ (1 atm)

Nernst's Equation for Single Electrode Potential

From the Nernst theory it follows that the potential difference of the double layer formed at one electrode is termed as single electrode potential.

Let us consider a metal of valency 'n' dipping in a solution

$$M \rightleftharpoons M^{+n} + ne$$

Let P_1 be the osmotic pressure of its ions, P_3, be the solution pressure of the metal and E be the electrode potential, *i.e.* the difference of potential between the metal and solution. Now pass the current through the electrode reversible until 1 gm. ion of the metal gets dissolved. It means that the electricity required for the dissolution of 1 gm. ion of the metal will be nF* coulombs. Hence,

the electrical work done = NF*E volt-coulombs

i.e. w_1 = nF*E volt-coulombs

where F* is one faraday whose value is equal to 96500 coulombs and w_1 is the electrical work done .Go on diluting the solution till the osmotic pressure gets reduced from P_1 to $P_1 - dP_1$. At the same time the difference of potential between the metal and the solution is now changed from E to E – dE. At the potential difference E – dE, the electrical work done (w_2) to cause the dissolution of 1 gm. ion of the metal is given by

$$w_2 = (E - dE)\ nF^* \text{ volt-coulombs}$$

The difference in electrical energy of work = $w_1 - w_2$

or
$$w_1 - w_2 = nEF^* - (E - dE)nF^*$$
$$= nF^*\ dE \text{ volt coulombs.}$$

The different in electrical energy is equal to the osmotic work done in transferring 1 gm. ion of the metal from P1 to P_1—dP_1 which is equal to the product of volume and change in pressure *i.e.*,

$$nF^*dE \rightleftharpoons VdP_1 \qquad ...(1)$$

where V is the volume of the solvent containing 1 gm. ion of the metal. For ideal solutions, osmotic pressures are directly proportional to the activities of ions, then V = RT/P_1. Thus, equation (2) reduces to

$$nF^*dE = \frac{RT\,dP_1}{P_1} \qquad ...(2)$$

$$nF^*\int dE = RT\int \frac{dP_1}{P_1}$$

$$nF^*E = 2.303\ RT \log P_1 + K \qquad ...(3)$$

where 'K' is the integration constant whose value will be evaluated as follows: when the solution pressure (P_2) is equal to osmotic pressure (P_1) of the ions, it means that no potential difference exists between the metal and the solution, *i.e.*, E = 0 and $P_1 = P_2$. Hence equation (3) becomes as

$$0 = 2.303 \text{ RT} \log P_2 + K$$

or

$$K = -2.303 \text{ RT} \log P_2 \quad ...(4)$$

Substituting the value of K from eq. (4) in (3), we get

$$nF^*E = 2.303 (\text{RT} \log P_1 - \text{RT} \log P_2)$$

$$E = 2.303 \frac{RT}{nF^*} \log \frac{P_1}{P_2} \quad ...(5)$$

Equation (5) is termed as the Nernst equation for electrode potential. Let us now consider tow similar electrodes dipping in two different solutions. In this case, the potential difference between the electrodes will be given by

$$E_1 - E_2 = \frac{2.303 \text{ RT}}{nF^*}\left(\log \frac{(P_1)_A}{P_2} - \log \frac{(P_1)_B}{P_2}\right)$$

$$= \frac{2.303 \text{ RT}}{nF^*} \log \frac{(P_1)_A}{(P_2)_B} \quad ...(6)$$

where $(P_1)_A$ and $(P_1)_B$ denote the osmotic pressure of the metal ions in two solutions A and B respectively. As we know that the osmotic pressure is proportional to concentration, it means, that equation (6) may be written as

$$E_1 - E_2 = \frac{2.303 \text{ RT}}{nF} \log \frac{(c_1)_A}{(c_2)_B}$$

STANDARD ELECTRODE POTENTIAL

When all the substances taking part in a reaction in a reversible cell are maintaining at unit activity, *i.e.*, in their standard states, the E.M.F. of any electrode in these conditions is known as standard electrode potential of the given cell (E°). If a reaction occurs by the passage of n Faradays in a cell in which the reactant and product have unit activities. the standard free energy change ΔG° is equal to – nE F*. Hence

$$-\Delta G^\circ = nE^\circ F^* \quad ...(1)$$

where F* is the Faraday of electricity. Let us now consider a reversible process (chemical reaction between A and B to form the products L and M)

$$aA + bB + ... \rightleftharpoons lL + mM + ...$$

When we apply the law of mass action to the above equilibrium, we get

$$K = \frac{a_L^l a_M^m}{a_A^a a_B^b \cdots} \quad ...(2)$$

As the increase in free energy ΔG^o at constant temperature is governed by the Vant Hoff's reaction isotherm, it, therefore, follows from this isotherm that

$$-\Delta G = RT \log_e K - RT \log_e \frac{a_L^l a_M^m}{a_A^a a_B^b \cdots} \quad ...(3)$$

If in the process the reactants and products are taken in their respective standard states, *i.e.*, at unity, then eq. (3) becomes as

$$-\Delta G^o = RT \log_e K \quad ...(4)$$

Eliminating ΔG^o between equations (1) and (4), we get

$$nE^o F^o = RT \log_e K \quad ...(5)$$

If the concentrations of reactants and products are taken at any arbitrary activities, then E.M.F. of the cell (E) is given by

$$\Delta G = -nEF^* \quad ...(6)$$

Substituting equations (6) and (5) in (3), we obtain

$$nEF^* = nE^oF^* - RT \log_e \frac{a_L^l a_M^m}{a_A^a a_B^b \cdots}$$

$$E = E^o - \frac{RT}{nF^*} \log_e \frac{a_L^l a_M^m}{a_A^a a_B^b \cdots} \quad ...(7)$$

Expression (7) is the general equation for the E.M.F. of any reversible cell in which the reactants and products are taken at any arbitrary activities. If a_A, a_B, a_C, ... and a_L, a_M, a_N, ... are unity, it means that equation (7) becomes as

$$E^o = E$$

So normal or standard electrode potential may be defined as:

"The potential of the electrode when the activities of the reactants and products are unity."

The earliest tabulation of standard electrode potential was mainly performed by Wilsmore and later revised by Abegg, Luther and Auerbach (1911).

However, Lewis and Randall recalculated the standard electrode potential from activity data. Let us now consider a metal of valency z_+, reversible with respect to M^{z+} ions, the electrode reaction is

$$M \rightarrow M^{z+} + z_+e^-$$

The equation for the electrode potential is taken as:

$$E_+ = E^o - \frac{RT}{z_+F*}\log_e \frac{a_M^+}{a_M} \qquad ...(9)$$

where a_M^+ is the activity of the cations in the solution and a_M is the activity of the solid metal. As the solution state of metal is generally taken as unity, it means that equation (9) takes the form

$$E_+ = E^o - \frac{RT}{z_+F*}\log_e a_M^+ \qquad ...(10)$$

Again consider an electrode A which is reversible to A^- ions. The electrode reaction is

$$M^{z-} \rightarrow A + z\_e$$

Proceeding in the same manner as we have done earlier, we can prove that

$$E_- = E^o + \frac{RT}{z_-F*}\log_e a_A^- \qquad ...(11)$$

From expressions (10) and (11), it follows that the general expression for any electrode reversible with respect to single ion of valency z± may be given as:

$$E_\pm = E^o \pm \frac{RT}{z \pm F*}\ln a_l \qquad ...(12)$$

where a_l is the activity of a particular species. In equation (12), the upper signs are taken for a positive ion while the lower ones for a negative ion. Putting R = 8.313 Joules, F* = 96500 coulombs and by converting Neperian to Briggsian logarithms, the factor 2.3026 is utilised, it means that equation (12) becomes as

$$E_{\pm} = E^{o} \pm 1.9835 \times 10^{-4} \frac{T}{z\pm} \log a_1$$

$$= E^{o} \pm \frac{0.05915}{z\pm} \log a_1 \text{ [when T = 298. 16°K]...(13)}$$

Effect of concentration and valence on electrode potential. We have already proved that

$$E_1 = E^{o} + \frac{0.0591}{n} \log a_1$$

From the above expression it is evident that a ten fold change in activity of an ion causes the electrode potential change by $0.0591/z\pm$ volts at 25°C. In a generalised manner, a change of activity by a factor of 10* changes the potential by $x \times 0.0591/z\pm$ volts at 25°. For monovalent ions, a ten fold change in concentration of solution will cause a change of 0.591 volts in monovalent and for bivalent ions, a change of 0.591/2 volts at 25°.